CHOLINERGIC MECHANISMS

Phylogenetic Aspects, Central and Peripheral
Synapses, and Clinical Significance

ADVANCES IN BEHAVIORAL BIOLOGY

Recent Volumes in this Series

CHOLINERGIC MECHANISMS

Phylogenetic Aspects, Central and Peripheral Synapses, and Clinical Significance

Edited by

Giancarlo Pepeu

Institute of Pharmacology
University of Florence
Florence, Italy

and

Herbert Ladinsky

Mario Negri Institute for Pharmacological Research
Milan, Italy

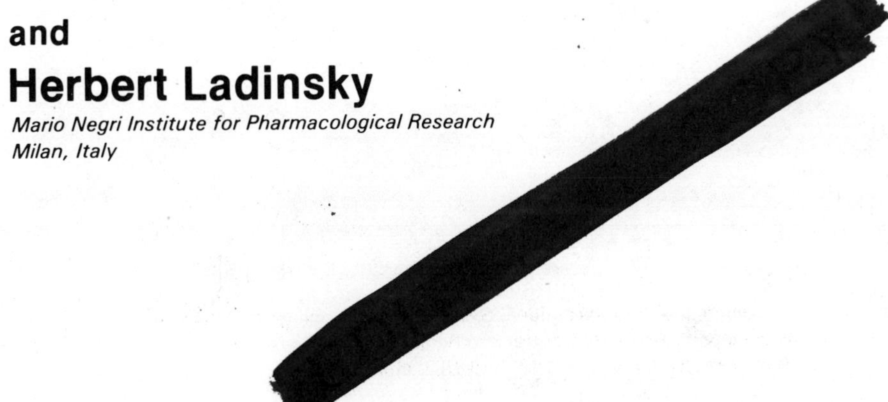

PLENUM PRESS • NEW YORK AND LONDON

Library of Congress Cataloging in Publication Data

Main entry under title:

Cholinergic Mechanisms: Phylogenetic Aspects, Central and Peripheral Synapses, and Clinical Significance.

Proceedings of the International Symposium on Cholinergic Mechanisms: Phylogenetic Aspects, Central and Peripheral Synapses, and Clinical Significance, held Mar. 11-15, 1980, in Florence, Italy.
Includes bibliographies and indexes.
1. Acetylcholine—Metabolism—Congresses. 2. Cholinergic receptors—Congresses. 3. Synapses—Congresses. 4. Parasympathomimetic agents—Testing—Congresses. I. Pepeu, Ginacarlo. II. Ladinsky, Herbert. III. International Symposium on Cholinergic Mechanisms: Phylogenetic Aspects, Central and Peripheral Synapses, and Clinical Significance (1980 : Florence, Italy) [DNLM: 1. Acetylcholine—Congresses. QV 122 161c 1980]
QP921.A25C48   612'.8042   81-11936
ISBN 0-306-40810-4   AACR2

Proceedings of an International Symposium on Cholinergic Mechanisms: Phylogenetic Aspects, Central and Peripheral Synapses, and Clinical Significance, held March 11-15, 1980, in Florence, Italy

© 1981 Plenum Press, New York
A Division of Plenum Publishing Corporation
233 Spring Street, New York, N.Y. 10013

SPONSORED BY

      Societa Italiana di Farmacologia

      International Union Pharmacology (IUPHAR)

      Consiglio Nazionale delle Ricerche

ACKNOWLEDGMENT

      The Symposium has been made possible by a
grant from the Consiglio Nazionale delle
Ricerche and by the generous support of
Fidia Research Laboratoreis, Abano Terme
(Padua) Italy.

ORGANIZING COMMITTEE

Lorenzo Beani
Department of Pharmacology
University of Ferrara
Ferrara, Italy

Edith Heilbronn
Unit of Neurochemistry and
    Neurotoxicology
University of Stockholm
Stockholm, Sweden

Donald J. Jenden
Department of Pharmacology
School of Medicine
University of California
Los Angeles, California

Alexander Karczmar
Department of Pharmacology
Stritch School of Medicine
Loyola University
Maywood, Illinois

Herbert Ladinsky
The Mario Negri Institute
for Pharmacological Research
Milan, Italy

Peter G. Waser
Institute of Pharmacology
University of Zurich
Zurich, Switzerland

Victor P. Whittaker
Max-Planck Institut for
Biomedical Research
Gottingen, FR Germany

Giancarlo Pepeu, Chairman
Department of Pharmacology
University of Florence
Florence, Italy

PREFACE

Every three years early Spring witnesses the convening of a
small group of neuroscientists from many parts of the world. They
are devoted to the study of acetylcholine, the oldest of the known
neurotransmitters.   To assess the level of knowledge, to take
stock of the still unsolved problems, to evaluate the practical
meaning of their findings, the cholinologists first met in 1969 in
the eerie atmosphere of a snow-covered Swedish forest, the second
in the Swiss Alps, the third facing the surf and foam of the Paci-
fic waves.   Finally in March 1980 Florence, her trees burgeoning
in the mild rain of March, confronted them with the temptation of
her monuments and museums.   But the cholinologists bravely dismis-
sed the temptations and went on with the presentations and discus-
sions of the many papers which form the present book.

The papers describe the most recent development of investigations on acetylcholine.   They open with an extensive coverage of
the developmental and phylogenetic aspects of the cholinergic
neurons.   The regulation of acetylcholine synthesis, release, postsynaptic, electrical and trophic effects in the peripheral cholinergic synapses including ganglia, neuromuscular junction, myen-
theric plexus and heart is then covered.   Special emphasis is given
to the cholinergic mechanisms in the eye.   But the largest part of
the papers are devoted to the central cholinergic synapse, to the
possibility of influencing its function through drugs and precursors
and to the clinical meaning of the most recent findings in this
realm.   The fast moving knowledge on the central cholinergic receptors, the elusive central cholinergic pathways, the complex interactions between acetylcholine and other neurotransmitter systems
also are the topic of several pages.

Many young scientists have joined in recent years the cholinergic field and to them was given the opportunity to present their
work together with that of many old hands in this business.   The
book is a blend of basic and clinically oriented papers and should
therefore be a useful source of ideas, techniques and references
for biochemists, physiologists, pharmacologists and neuropsychia-
trists.

We would like to express our gratitude to the members of the Scientific Committee and particularly to Dr. Whittaker, who gave valuable suggestions for the choice of the participants and the arrangement of the program, to the local committee to whom the participants owe the perfect organization of the meeting.

We are particularly grateful to Mrs. Flo Comes for her unique expertise and patience in editing the manuscripts and preparing them in the final form.

Last but not least we want to acknowledge the generous financial support of the Consiglio Nazionale delle Ricerche and of Fidia Research Laboratories which made the meeting possible.

Giancarlo Pepeu   Herbert Ladinsky
Florence   Milan

November 1980

CONTENTS

PERIPHERAL CHOLINERGIC SYNAPSES:
HEART, MYENTERIC PLEXUS AND NEUROMUSCULAR FUNCTION

CENTRAL CHOLINERGIC SYNAPSE:
<u>SYNAPTOSOMES AND BRAIN SLICES</u>

CENTRAL CHOLINERGIC SYNAPSE:
<u>CHOLINERGIC MECHANISMS IN THE EYE</u>

ACETYLCHOLINE RECEPTORS IN THE
CENTRAL NERVOUS SYSTEM

ACETYLCHOLINE BIOSYNTHESIS IN DEVELOPING CHOLINERGIC SYNAPSES

E. Giacobini and M. Marchi

Laboratory of Neuropsychopharmacology, Department of
Biobehavioral Sciences, University of Connecticut
Storrs, Connecticut 06268, USA

INTRODUCTION

In contrast to _in vitro_ conditions synapse formation _in situ_
represents a long term process (16) consisting of consecutive steps
of complex structural and biochemical modifications. On a chrono-
logical scale, synaptogenesis occurs not only during part of the
embryonic phase but also during the period following birth or
hatching. Recent studies have shown that in the avian peripheral
nervous system the process of synaptic growth may continue throughout adult life (30). In the present study chick ciliary ganglia
and their target organ (the iris) provides a useful model for investigating the process of synapse formation and growth (Fig. 1).
For a comparison, the development of synapses in sympathetic ganglia (lumbar) was examined in a parallel study (14). The possibility
of examining simultaneously cell bodies and terminals in the same
group of cells belonging to a small (3000 cells) and well defined
cholinergic population, represents a unique feature of this preparation (18). In this model, changes in biochemical parameters
related to acetylcholine (ACh) metabolism can be correlated step by
step to neurotransmission and morphology (15).

Our studies (14,15) suggest the presence of regulatory influences exerted by target organs as part of "trophic" interactions
with the innervating neurons. Such interactions seem to be both
modulatory and instructive for certain mechanisms of synapse

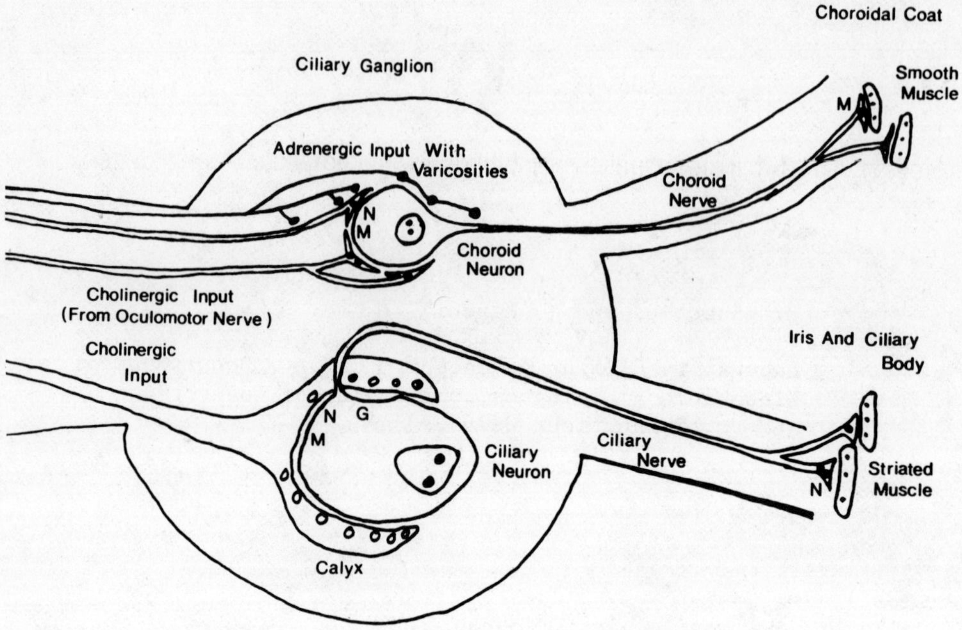

FIGURE 1: Diagram of chick ciliary ganglion and iris. N = nicotinic
receptors; M = muscarinic receptors; G = gap junctions.
Illustration not to scale. Modification after (14) and
(24).

formation. The present report describes our attempt to understand
how synaptic contacts are established between growing neurons and
their targets.

Differentiation of Autonomic Ganglia

Mesencephalic crest cells seem to possess some fundamental
markers of cholinergic metabolism from the onset of their migratory
phase (46). Acetylcholinesterase (AChE) activity has been detected in cells dissected from the neural tube and the neural crest
at 36, 48, 60 and 74 hr of incubation, prior to the appearance of
specific morphological characteristics of neuroblasts (35). This
may indicate an early differentiation of the cell population that
will later constitute the neurons of the ciliary ganglion (7). It

has, however, been suggested by several authors, mostly based on
results obtained in in vitro experiments, that stable cholinergic
differentiation might occur only at later stages, i.e. when the
cells have reached their final destination and have established
connections with their target organ. Two parasympathetic ganglia
deriving from neural crest ganglioblasts such as the ciliary ganglion and the ganglion of Remak express cholinergic differentiation
from a very early stage of development. The avian ciliary ganglion
shows choline acetyltransferase (CAT) and AChE activity as well as
ACh levels at 5 days of incubation (7,31). A high affinity uptake
mechanism for choline (Ch) is present in the fibers of the ciliary
ganglion innervating the primitive iris bud at 4-1/2 days of incubation (34). At 5 days of incubation Ch uptake is almost 80% of the
1 year value. In addition, quail ciliary ganglia have the ability
to convert [^{3}H]-Ch to [^{3}H]-ACh at 4 days of incubation (0.02 pmol/
ganglion)(46). At six days this synthetic capacity has increased
15-fold (29). Similarly, the Remak ganglion shows CAT activity and
is able to synthesize ACh at 4-1/2-5 days of incubation (29).

     Le Douarin et al. (29) demonstrated that the further differentia-
tion of presumptive parasympathetic ganglion cells (derived from a
"vagal" cholinergic level) which have stopped migrating, can be
modified if they are transplanted into a younger embryo at the trunk
crest "adrenomedullary" (adrenergic) level. The grafted ganglion
cells start migrating again and stop at the same site (adrenal) as
the host neural crest cells. Their further differentiation depends
on the site they have reached. When they reach the adrenal gland
they synthesize catecholamines. Thus, Le Douranin's experiments suggest that the differentiation of parasympathetic neurons may at least
partly depend on tissue unteractions even after the neural crest cells
have formed ganglionic structures.

     Our experiments on ciliary ganglia and irises at early stages
of ganglion formation (7,30-34) support such a hypothesis by demonstrating that interactions of the ganglion cells with the target
(5,7) can still substantially modify the neurotransmitter metabolism
(see section on recepto genesis). The experiments of Le Douarin
et al. (29) strongly suggest that the potential for developing into
cholinergic or adrenergic cells appears to be wide-spread in the
whole neural crest. The expression of either one or the other
phenotype is dependent on the tissue (mesenchymal) environment to
which the cells are subjected (8, 28).

     The factors that control the appearance and the choice of
neurotransmitter metabolism are still unknown. Neither is it known
whether the environment is instructive or merely permits the selective expression of certain prior decisions. Two recent hypotheses
about the developmental mechanisms involved in the chemical differentiation of sympathetic neurons have been suggested on the basis
of studies performed in culture of nerve cells in vitro (3,37,38).

Both hypotheses emphasize the influence of the environment in the cell's choice of neurotransmitter. A third hypothesis is based on our results derived by _in situ_ experiments (16). This hypothesis takes into account transynaptic regulation of neurotransmitter metabolism (14).

Innervation of the Target Organ

Normal Distribution and denervation changes of AChE and CAT: CAT is present in the normal ciliary ganglion, 60% in the presynaptic element and 40% in the ganglion cell body. AChE is 25% presynaptic and 75% postsynaptic (20). In the iris the distribution of CAT is at least 97% presynaptic while AChE is present almost entirely in the muscle (85%). The effect of preganglionic denervation is not confined to the degenerating terminals of the ganglion cells but is conveyed to the postsynaptic elements, i.e. the effect is transynaptic (20). The axoplasmic flow of CAT and AChE is related to the perikaryal concentration of these enzymes (20). It appears, therefore, that a transynaptic regulation of CAT and AChE synthesis in the ganglion and an intact innervation of the ganglion cells is required in order to maintain normal levels of these enzymes as well as of ACh (39) in the cell bodies. The activities of the non-specific enzymes LDH and coenzyme A (CoA) as well as total protein content, remain unchanged after denervation lending support to the theory that the transynaptic effect regulating the synthesis of CAT and AChE in the adult ganglion represents a specific mechanism (20).

Development of cholinergic enzymes: In a developmental study (7) we measured CAT and AChE activity in chick ciliary ganglia and irises from the fifth day of incubation until 1 week after hatching. The changes in enzyme activity were correlated in time with previous electrophysiological and morphological findings of synapse formation in the same structures (24-26).

Low levels of CAT activity were measured at stage (ST) 26 in the iris (Fig. 2b), which implies that as soon as the migration of the ganglion cell is completed (4 day of incubation, ST 25) the ganglion cells are biochemically differentiated and CAT is transported down their axons as they grow into the primitive iris (4-7 day of incubation). The specific activity of CAT and AChE rises during development and the increase is closely correlated to the onset and maturation of ganglionic (5-8 day of incubation) and iris synaptic transmission (9-10 day of incubation) (Figs. 2 and 3).

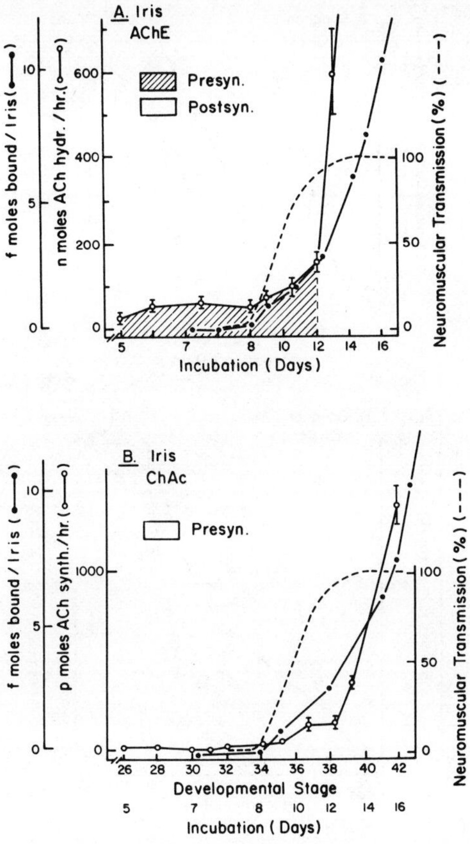

FIGURE 2: Relationship between CAT activity, AChE activity, α–Btx
binding and percent transmission in iris. Modification
after (18) and (7).

Two sets of data seem to be most significant: 1) the 200–fold
specific increase at CAT in iris nerve terminals which occurs at
ST 38 (Fig. 2b) probably reflects an increase in synthesis of the
enzyme in ganglionic cell bodies and suggests that the formation of
the iris neuromuscular junction triggers the enzyme induction. This
implies the fact that the cell responds to a signal retrogradely
ascending the axon from the terminals; and 2) the initial increase
of AChE specific activity in the ganglion occurs soon after transmission is established in all cells between ST 30 and 34 and is
probably mainly due to enzyme synthesis by the ganglion cells (Fig.
2a). This early increase of AChE specific activity is ganglia is chr-
onologically correlated to synapse formation. Since the time courses

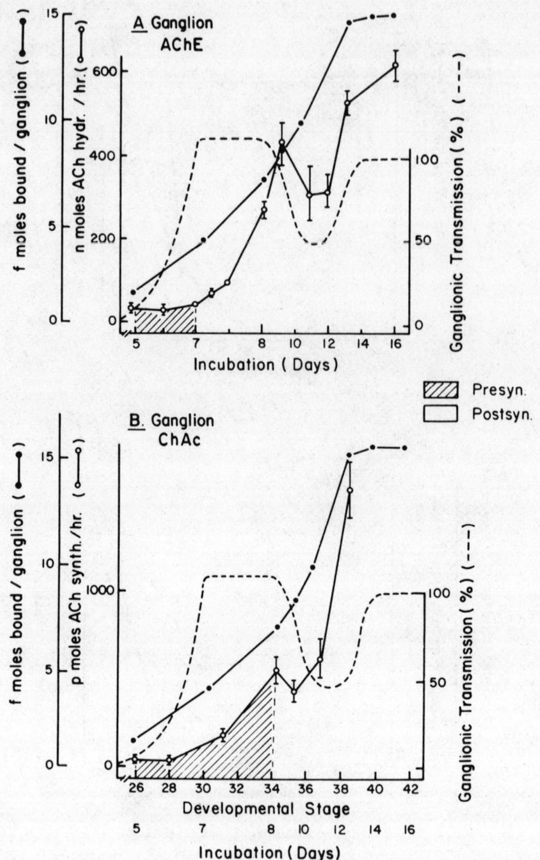

FIGURE 3: Relationship between CAT activity, AChE activity, α-Btx binding and percent transmission in ciliary ganglion. Modification after (18) and (7).

---

of specific induction for AChE and CAT in ganglia are different, it seems clear that the synthesis of these enzymes is regulated by separate mechanisms. In the iris there is a twofold increase in the specific activity of AChE after the formation of the neuromuscular junction, which probably reflects enzyme induction in the muscle subneural region (42) (Fig. 2a). This indicates that the specific induction of AChE in postjunctional cells is influenced by the prejunctional elements.

The growth of CAT activity is exponential in both ganglia and iris after ST 32, except for the period which coincides with the

cell death observed by Landmesser and Pilar (26)(Figs. 2 and 3). It
is also interesting to note that only very low levels of CAT are
present in both the ganglion and iris when 100% transmission is already present (Figs. 2 and 3). This suggests that the greatest induction of CAT occurs after functional connection has been established in the tissues. Since the initial increase in CAT seen in
the ganglion between ST 26 and 34 is not present in the iris (Fig.
2b) during these stages, it suggests that presynaptic elements are
responsible for this increase and that formation of functional synapses is closely followed in time by increased CAT activity in presynaptic terminals. Our results clearly show that during the normal
development of the ganglion, the formation of functional junctions
in the target organs result in a retrograde signal which reaches
the ganglion cell bodies and triggers a rapid (24-36 hr) and large
scale synthesis of enzymes needed to sustain the neurotransmitter
level at the axon terminal.

A similar event may take place in the presynaptic neurons of
the ganglion since CAT activity in the presynaptic terminal is very
low until functional innervation occurs. Thus, in both the ganglion and the iris, CAT activity seems to be induced in the presynaptic elements soon after functional synapses are formed.

This study (7,14) represents the first direct evidence linking
a physiological event, i.e. the establishment of functional neurotransmission, with the regulation of the neurotransmitter enzymes
in cell bodies and target organs (Fig. 1) within a specific period
of development.

Development of ACh and Ch Endogenous Levels

ACh and Ch levels were measured by means of a sensitive micromethod in developing irises, ciliary and sympathetic ganglia of the
chick, starting at 5-7 day incubation, and continuing up to 1 year
of age (31). In general, Ch levels parallel closely the levels of
ACh in each organ throughout development (Fig. 4).

Comparison of the ACh data obtained from the iris and ciliary
and sympathetic ganglia indicates that this neurotransmitter is
present in all three organs at relatively low levels (1-2 pmol,
ciliary and sympathetic ganglia; 13-43 pmol, iris) from 5 to 5 day
incubation. This is followed by a rapid ninefold increase (10-14
day incubation) in ACh levels in all three tissues (Figs. 4 and 6).
This increase in ACh levels during the first half of embryonic
development occurs at the time when neurotransmission reaches adult
levels in the iris (24)(Fig. 6). In the ciliary ganglion, however,
neurotransmission is fully developed by 7 day incubation, when ACh
levels are quite low. Thus, it appears that relatively low levels
of ACh may be sufficient for neurotransmission to occur under the

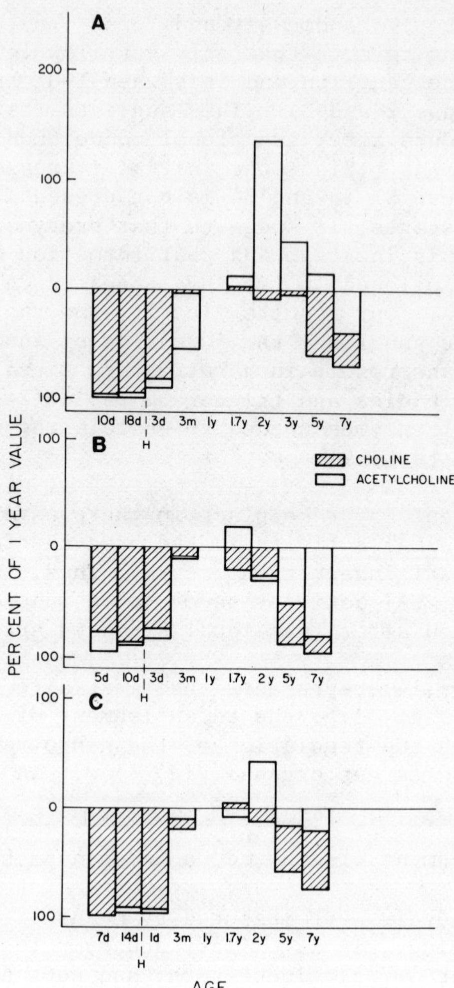

FIGURE 4: Variation in ACh and Ch total levels at various ages,
expressed as percent of 1 year values in (A) ciliary
ganglion, (B) iris, and (C) sympathetic ganglia. H =
hatching.

experimental conditions applied by the authors (24). It is, therefore, possible to conclude that minimal levels of ACh (and CAT
activity) are necessary to maintain functional ganglionic transmission during the initial phases of synaptogenesis, and at this
time the synaptic ultrastructural specialization is minimal (25,26).
At later stages there is a progressive increase of ACh and CAT
activities, although the synapses are still probably of low quantal

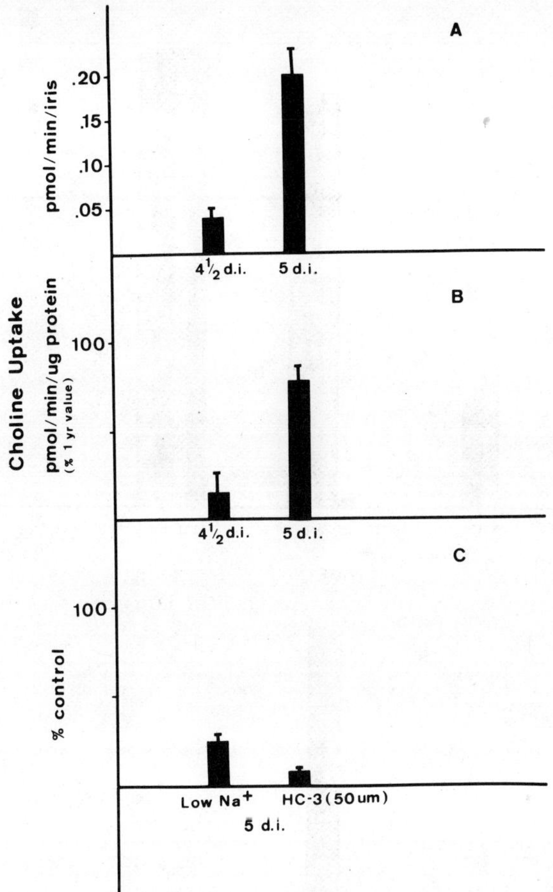

FIGURE 5: Characteristics of Ch uptake during early development.
A: uptake in 5 min; B: values in A, relative to 1 year
old value; C: inhibition by low Na+ and HC-3. Each point
is the mean ± S.E.M. of at least six determinations.
* p < 0.05; ** p < 0.001.

content with relatively few synaptic vesicles present in the
terminals (25,26).

In a subsequent study (30) we measured the endogenous levels
of ACh and Ch in sympathetic (lumbar) and parasympathetic (ciliary)
ganglia, and in the iris of the chick from 1 to 7 years of age
(Fig. 4).

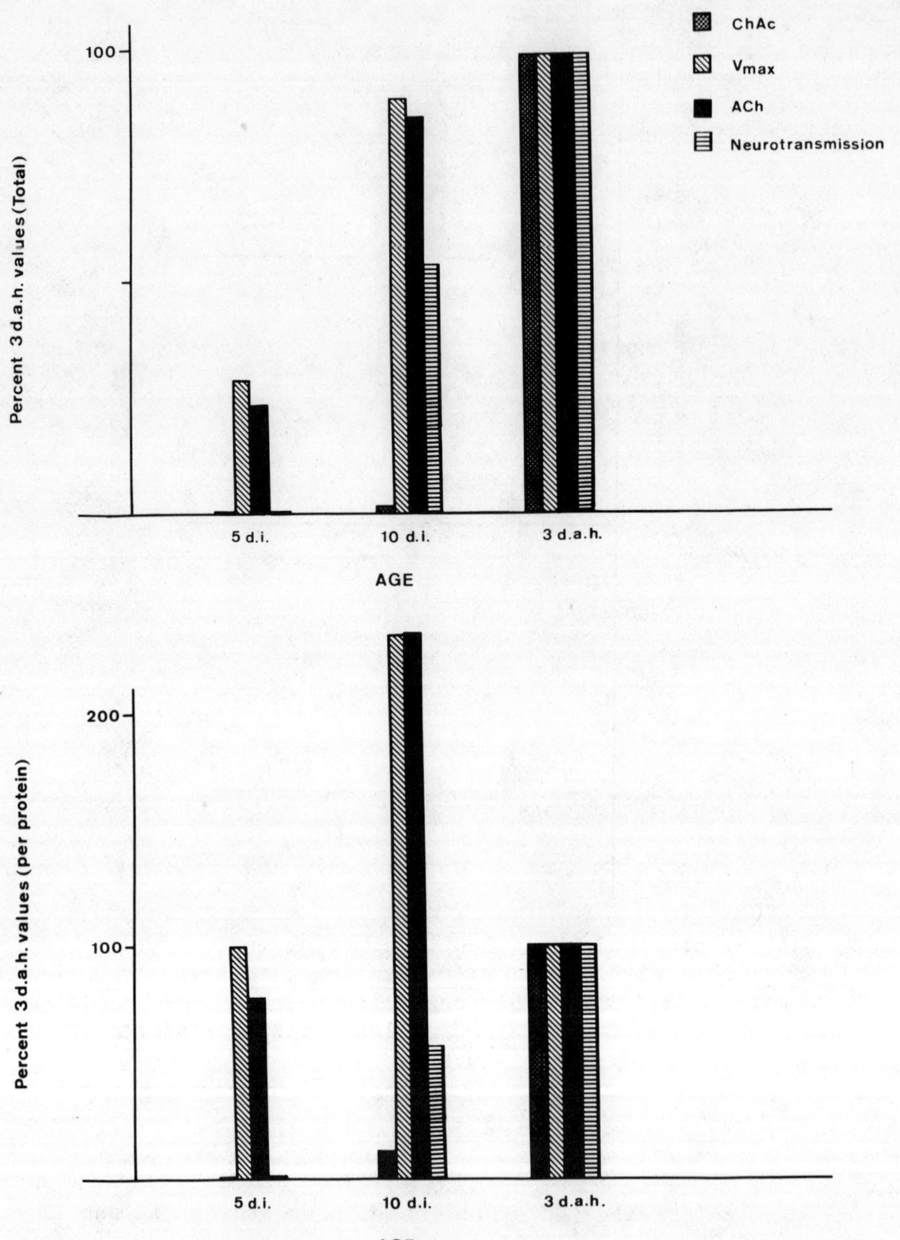

FIGURE 6: Levels of CAT activity and ACh, V_{max} of Ch uptake, and neurotransmission expressed in percent of days after hatching values. Each column represents the percent of at least six determinations. A = per total iris; B = per protein.

Levels of ACh and Ch seem to undergo a continuity of growth which is initiated in the first week of embryonic development and continues into adulthood up to 2 years of age (Fig. 4). The early increase in ACh and Ch levels are most marked in the ciliary ganglion. In the ciliary ganglion, morphological (12,36) and biochemical (14,15) observatiions show that this period corresponds to one of "continuous synaptic growth," In the second period, starting around 2 years, which can be defined as one of "synaptic regression," a phase of progressive and marked decrease in ACh and Ch levels is visible in all three organs examined (Fig. 4). In all three organs Ch levels parallel the ACh levels rather well, but the Ch/ACh ratio progressively decreases from 1 to 7 years from 3.5 to 1.5 (ciliary ganglia and iris) and from 2.5 to 1 (sympathetic ganglia). The low levels of ACh reached at 7 years in the three organs have clear implications with regard to the physiology of aging cholinergic synapses.

Receptogenesis

Development of Cholinergic Receptors
in Autonomic Ganglia and Iris in situ

The ontogeny of ACh receptors (AChR) in the ciliary ganglia and iris has recently been examined in our laboratory by studying the rate of appearance of α-bungarotoxin (α-Btx) binding sites in both organs (6). Specific, high affinity binding was found in both tissues (K_D: iris = 2.5 nM, K_D: ganglion = 2.7 nM). The binding is saturated above 10 nM toxin concentration and is inhibited by low concentrations (10^{-5} M) of the nicotinic antagonist tubocurarine. The binding appears first in ganglia and then in the iris, and is associated with a nicotinic cholinergic receptor in both tissues (Figs. 2 and 3). As shown in Fig. 5, the amount of binding in the iris begins to increase only after functional innervation of this organ is observed,i.e. at 8 day incubation, and continues to increase up to 4 months after hatching. In contrast, α-Btx binding in the ciliary ganglion increases fourfold between 7 and 11 day incubation, after which the amount of binding remains unchanged up to 4 months after hatching (Fig. 3).

In the ganglion, the highest rate of appearance of receptors occurs later than the onset of ganglionic transmission, but simultaneously with the rise in both AChE and CAT activity in the cell bodies (Fig. 3). In the iris, as in the ganglion, the development of the receptor and the increase in both enzyme activities is parallel and almost simultaneous (Fig. 2). This fact is interesting, since both α-Btx and AChE activity in the iris are localized to a

large extent in the muscle cells, while CAT activity is exclusively
presynaptic (20). If one considers the early period of transmission
and synaptogenesis in the ganglion and iris (i.e. between 5 and
14 days incubation, see Figs. 2 and 3), the following picture is
apparent; levels of receptor binding are clearly detectable in the
ganglion and iris before initiation of neurotransmission, while
only low CAT and AChE activities, probably located presynaptically,
are present (Figs. 2 and 3). Simultaneously with the maturation of
neurotransmission, receptor levels and AChE activity rapidly increase in the postsynaptic elements (terminals in the ganglia and
iris). When both the number of receptors and the percent neurotransmission have reached adult values in the ganglion (12 days
incubation), the CAT that has accumulated in ganglion cells begins
to be transported in large amounts down their axons to the terminals
(Fig. 2b). Increased levels of enzyme (CAT) can then begin to synthesize the amounts of ACh necessary to maintain neurotransmission
at the neuromuscular junction.

Effect of Cholinergic Receptor Blockade

The effect of _in ovo_ injections of nicotinic blockers on the
development of two neuronal systems in the chick embryo have been
examined (5). Chlorisondamine, a nicotinic ganglionic blocking
drug, caused a long term hypotrophy in both the ciliary ganglion
and the lumbar sympathetic ganglia. CAT activity was reduced in
both ganglia while AChE activity was reduced only in the ciliary
ganglion. Development of the iris, the end organ of the ciliary
ganglion, was not significantly altered by chlorisondamine. Repeated α-Btx injections _in ovo_ delayed development in the striated
iris muscle, while having little effect on the ontogeny of the
ciliary and sympathetic ganglia.

While normal synaptic transmission does not appear to be essential for the survival of developing neurons, blocking postsynaptic receptors significantly altered the developmental course in
the ganglia after ganglionic receptor blockade and in the iris after
neuromuscular junction receptor blockade. Thus, synaptic transmission may play a regulatory role in the process of synaptogenesis
(14,15).

Although the evidence obtained by using two specific receptor
binding agents confirms our previous view that functional innervation is correlated with the development of both the ciliary ganglia
and iris, the effect of peripheral receptor blockade at the target
organ (iris) is not comparable to the removal of the target organ
as performed by other authors (24,25,27).

In the presence of α-Btx binding, peripheral junctions can still develop;however, both general effects on weight and specific effects on enzymes are observed. These effects are both pre-and postsynaptic and are similar in the skeletal muscle and in the iris. The effect on ciliary cells is less pronounced, indicating that normal innervation of the iris is not the only signal allowing the survival of ganglion cells. Moreover, blockade of nicotinic receptors on ganglionic (ciliary and sympathetic) cells exerts a general hypotrophic effect on ganglionic development, involving both weight and enzymes.

In conclusion, normal synaptic transmission expressed by normal receptor density and function seems not to be essential for the survival of developing neurons. However, blockade of postsynaptic receptors significantly influences the development of both ganglia and target organs. Thus, we can conclude that synaptic transmission plays a regulatory role in synaptogenesis (14,15).

Receptogenesis Under <u>in vitro</u> Coditions

Ciliary ganglia were dissected from 8 day chick embros. Three to four ganglia were added to chick muscle culture differentiated in vitro. The muscle was grown on collagen coated 32 mm dishes (44). After 24-48 hr branched processes could be seen growing from the nerve explants. To measure the effect of the presence of ciliary ganglia on AChR levels, cultures were pulsed with ^{125}I-α-Btx (44). The amount of toxin binding in the co-cultures was significantly increased and almost doubled in 8 days in comparison to muscle culture alone. The rise of AChR levels seem to be specific since it was followed by a similar increase in phosphokinase and AChE activities. Similar results were obtained when the ciliary ganglia were enzymatically dissociated with trypsin prior to plating on muscle cultures (44). It was generally observed that in cocultures the rate of spontaneous contraction was higher than in the muscle alone. However, the level of AChR was not reduced as might have been expected from a higher rate of muscle activity.

In a series of parallel experiments of prolonged co-cultivation of mouse cord cells with rat muscle fibers under similar conditions, the number of cholinergic receptors was significantly decreased as compared to muscle cultures alone (44). This reduction in the amount of AChR was not a result of toxic effect of the nerve on the muscle, since other proteins, such as creatine phosphokinase, were almost unchanged. Autoradiography of co-cultures exposed to ^{125}I-α-Btx showed that in both the amount of clusters and uniformly distributed receptors were reduced by the prolonged co-cultivation (44).

The question, therefore, arises as to the reason for these changes in the amount of AChR. The mechanism that underlies the decrease in the level of AChR following prolonged co-cultivation is now known. Two possible explanations for this phenomenon are, the observed increase in muscle activity, or a putative neurotrophic influence of the nerve fiber on the muscle.

Development of a High Affinity Ch Uptake Mechanism
in the Chick Iris

At ST 25 (4-1/2 day incubation) nerve fibers, originating from the ciliary ganglion, are already present in the mesenchymal stroma of the embryonic eye, At ST 27 (6 day incubation) the ciliary nerves exhibit further growth. In the late part of development of the organ they constitute the only nerve supply to the iris sphincter muscle. .

In order to establish the earliest appearance of a Ch uptake mechanism in the iris, 200 µ wide sections were examined at 4-1/2 days incubation. At this age Ch uptake is already present (0.20 pmol/5 min/iris, Fig. 5a). This uptake represents only 15% of the 1 year value (Fig. 5b). Twelve hours later (5 day incubation) the uptake had already reached 1.00 pmol/5 min/iris, increasing almost 5 times from the 4-1/2 days incubation value (Fig. 5b). This uptake is strongly inhibited by low Na^+ and HC-3 (50 µM) and ouabain (Fig. 5c). Ch uptake at 5 days incubation is almost 80% of the 1 year value (Fig. 5c).

Our finding (34) demonstrate a close temporal relationship between the earliest detectable innervation of the target organ, the iris and the appearance of a selective high affinity uptake mechanism for Ch. It has been reported, in studies of the CNS, that even at early stages of postnatal development a low concentration of Ch is selectively accumulated by cholinergic terminals (47, 48). Based on these findings, this uptake was considered to be an early index of cholinergic synaptogenesis, i.e. formation and outgrowth of synaptic terminals.

In the adult nervous system, it is now well established that Ch uptake represents an essential feature of a functional cholinergic synapse (23). The process of maturation of such a mechanism is not well understood and is probably more difficult to study in CNS preparations than in the peripheral nervous system (15). By relating morphological to biochemical findings, we have demonstrated that a high affinity Ch uptake mechanism is present in the primitive nerve fibers starting from the earliest stages of innervation. The presence of Na^+ dependent high affinity uptake for the precursor Ch is detected as early as at 5 days incubation. At this stage, as demonstrated by our ultrastructural studies, the only neural components present in the iris bud are the numerous axons derived from the ciliary ganglion which are situated in the stroma. Therefore, this mechanism can be considered as an early expression of innervation rather than of synaptogenesis as previously suggested (47,48).

The parallel trend in the development of the V_{max} for high affinity Ch uptake and the ACh content strongly supports the role of Ch uptake as a regulatory mechanism which is rate limiting for the synthesis of ACh (Fig. 6). All the characteristics of high affinity Ch uptake are present from the beginning and the following developmental changes in uptake velocity correspond to an altered number of uptake sites (V_{max}) rather than to a change in the affinity of the Ch carrier (K_m). Ch uptake does not precede CAT activity as has been suggested for the CNS (47) but the two functions are present simultaneously. Neither does the ontogenetic development of high affinity Ch uptake parallel the development of CAT as suggested by studies in rat brain CNS (9,48). On the contrary, our data emphasize the different development of these two basic cholinergic attributes.

Development of the Relationship Among Uptake,
Neurotransmitter and Precursor Levels,
Enzyme Activities and Neurotransmission

In Table 1 and Fig. 6, Ch uptake, expressed as V_{max} is compared to several developmental parameters in the iris between 5 days incubation and 3 days after hatching. Our first observation is that high affinity Ch uptake, ACh and CAT and AChE activities are all present at a stage (5 days incubation) at which neurotransmission and α-Btx binding have not yet developed. At 5 days incubation CAT activity is only about 0.8 pmol/iris/5 min. The level of ACh present in the iris is only 13.3 pmol, which is less than 1/20 of the total Ch value. The question of a close relationship between uptake and CAT activity in the adult has been discussed (23). The data in Table 1 indicate close relationships during ontogeny between on the one hand, Ch uptake and ACh levels, and on the other hand, enzyme activities and neurotransmission. At early stages (5 days incubation) CAT activity is very low, which may explain, in part, the low level of ACh present. Absence of neurotransmission could be a reasonable explanation for the low level of CAT activity which follows more closely the trend of the development of physiological activity rather than that of ACh levels or uptake during the embryonic period (7). If ACh is already actively released at 5 days incubation, enough AChE activity is present to rapidly hydrolyze it (7).

TABLE 1: Developmental characteristics of cholinergic neurotransmission in the chick iris.

Age	Per Iris			Percent of 3 DAH		
	5 DI	10 DI	3 DAH	5 DI	10 DI	3 DAH
V_{max} (pmol/5 min/iris)	9.9	25.9	29.0	34	90	100
ACh (pmol/iris [31])	13.3	43.5	51.8	26	84	100
Ch (pmol/iris [31])	247.0	131.0	291.0	85	45	100
CAT (pmol/ACh synthesized or hydrolyzed 5/min/iris [7])	0.8	12.5	833.0	0.1	2	100
AChE (pmol/ACh synthesized or hydrolyzed 4 min/iris [7])	2.0	6.0	200.0	1	3	100
α-Btx binding (fmol/iris [6])	0	2.5	35.0	0	7	100
Neurotransmission (percent of 3 DAH [7])	0	50	100	0	50	100

DAH = days after hatching; DI = days incubation

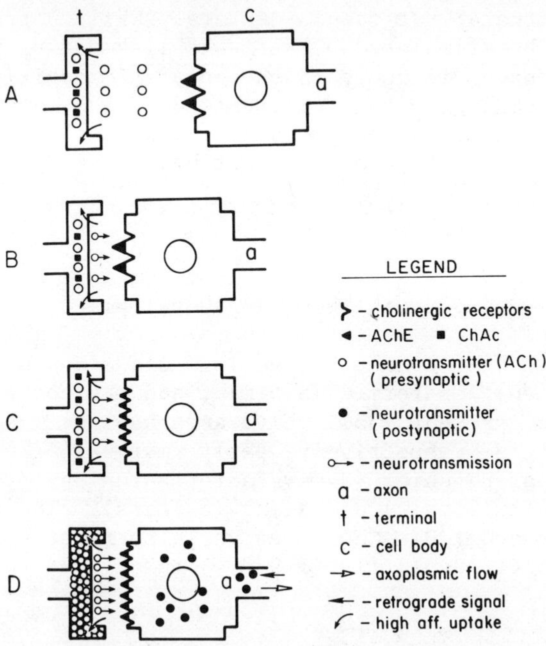

FIGURE 7: Developmental sequence of the basic components of a
          peripheral cholinergic synapse (see text). A = phase of
          innervation; B = phase of neurotransmission; C = phase of
          receptogenesis; D = phase of maturation.

At 10 days incubation, when neurotransmission has reached
about 50% of the 3 days after hatching value, CAT has increased 15-
fold representing, however, only 2% of the 3 days after hatching value
(Fig. 6a). However, both ACh and V_{max} have increased together by
only a factor of 3, to 250% of the 3 days after hatching value, on
a per protein basis (Fig. 6b). This can be related in time to
formation of neuromuscular junctions in the iris. At later stages
(3 days after hatching), the enzyme activity has increased 1000-fold,
while the capacity of Ch uptake has not significantly changed.

Our results do not support the hypothesis of a simultaneous
and related development of Ch uptake and CAT activity (9,43). These
parameters seem to develop independently, with CAT activity developing more slowly than uptake or ACh levels (Fig. 6). This may reflect the rate of synthesis or migration of CAT down axons during
this early stage of development, when less physiological demand is
present. The findings are in agreement with those of Shelton et al.

(45) in the postnatal rat fascia dentata. While both CAT and Ch
uptake seem to be stimulated by the establishment of neurotransmission, the tremendous increase in the enzymatic activity seen at
3 days after hatching, with no corresponding increase in Ch uptake,
favors correlation of the enzyme activity with the turnover of ACh
rather than with Ch uptake or ACh content.

Turnover of ACh During Development

The total $[^3H]$-C uptake in the iris was measured together with
the amount of $[^3H]$-ACh formed in 5 min, and the corresponding a-
mount of $[^3H]$-Ch present (33). Five ages have been selected representing the stages, respectively, early embryonic (10 days incubation), early post-hatching (3 days after hatching), young adult
(3 months), adult (1 year) and aging (5 years). The ratio between
$[^3H]$-Ch present and $[^3H]$-ACh formed was also reported. It was seen
that this ratio varies little (1.9–3.4) throughout the lifespan of
the animal.

The total amount of $[^3H]$-Ch taken up in the iris after 5 min
incubation with $[^3H]$-Ch is unvaried between 10 days incubation and
3 days after hatching, and then increases fivefold at 3 months (33).
A further significant (60%) increase is seen between 3 months and
1 year. At 5 years $[^3H]$-Ch uptake has decreased to levels close to
embryonic (10 days incubation) and early past-hatching (3 days after
hatching) stages. The amount of $[^3H]$-ACh formed is unvaried between
10 days incubation and 3 days after hatching, and increased eightfold at 3 months. A further 60% increase is seen between 3 months
and 1 year. A marked (sixfold) reduction is seen between this latter age and five years, which leads to a reduction in $[^3H]$-ACh
levels to embryonic or early post-hatching levels. The corresponding
amount of $[^3H]$-Ch present follows the same trend as $[^3H]$-ACh levels.

One of the most salient findings of this study is the remarkable stability of the ratio of $[^3H]$-Ch to $[^3H]$-ACh in the iris
terminals throughout the lifespan of the animal. This indicates
that the relationship between the $[^3H]$-Ch taken up in the iris and
the $[^3H[$-ACh formed per unit of time is relatively unvaried with
age.

These results stress the tendency of the cholinergic synapse
to maintain a constant ratio between the two parameters, $[^3H]$-Ch
and $[^3H]$-ACh, independently of the rate of synthesis from the available Ch and of the release and hydrolysis of ACh. These data complement our previous data on endogenous levels of ACh and Ch and
Ch uptake which are summarized in Table 1.

The ratio of endogenous ACh levels to [3H]-ACh formed does not appear to vary considerably up to 1 year, at 5 years this ratio is almost threefold higher. An indication of variations in the turnover rate of ACh with age can be found by applying the biosynthesis ratio method of Schuberth et al. (43). In this case an increase in turnover per iris up to 1 year is apparent, as well as a significant decrease between this age and 5 years. It should be emphasized that turnover changes of ACh reported per organ may relate to possible variations in the total number of terminals present at different ages.

It must be pointed out that our measurements of turnover may be influenced by age related modifications in compartmentation of ACh and Ch as well as variations in release mechanisms resulting from changes in number and structure of terminals or intraneuronal biochemical mechanisms.

CONCLUSIONS

Developmental Assembling of a Functional Synapse

Although no conclusive evidence is available (3,16,37,38), it seems that neural crest cells which will give rise to autonomic ganglia possess at least two basic markers of cholinergic metabolism such as CAT and AChE activity and the synthetic capacity for ACh. In addition, even before establishing contacts with the target organ, the iris, ciliary ganglion neurons exhibit a high affinity uptake mechanism for Ch. The early presence of cholinergic receptors on both ciliary ganglion cells and iris has also been demonstrated. How do these attributes and mechanisms fit into the developmental scheme? Is their appearance interrelated or do they represent separate manifestations of the developmental process? What is their relation to the establishment of synaptic transmission? Is the specialization of the primitive embryonic synapse comparable to the adult synapse?

Our findings clearly show that in both ciliary ganglion and iris, neurotransmission reaches adult values prior to the establishment of adult levels of ACh and cholinergic receptors as well as AChE and CAT activity. In the ciliary ganglion, the highest rate of appearance of cholinergic receptors is prior to the rise in cholinergic enzyme activity, while in the iris it is practically simultaneous. This implies that the postsynaptic system interacting with the neurotransmitter (i.e. AChR and AChE) starts developing with or even prior to the achievement of full capacity in neurotransmitter synthesis in the presynaptic terminals.

Based on these observations we can tentatively schematize the developmental sequence of a peripheral cholinergic synapse into four different phases, each one characterized by the appearance and consolidation of specific mechanisms (Fig. 7). Such a scheme suffers perhaps from an excessive rigidity and simplification, but it may prove useful as a testing hypothesis. It may possibly introduce an artificial separation between biochemical, morphological and functional aspects. However, it emphasizes the complexity of the picture and the fact that embryonic synapses might function under different conditions than adult ones. For example, it is possible that a relatively low rate of release of the neurotransmitter from a not yet fully developed presynaptic terminal might indeed be sufficient, not only to trigger neurotransmission "at a distance," but also to influence the subsequent phases of synaptogenesis. Biochemical, electrophysiological and morphological evidence for such a hypothesis seems to be very suggestive for the ciliary ganglion-iris preparation. However, further work is necessary in order to validate and extend it to other structures, including the CNS.

In earlier publications (14,15,17) we have proposed the following time course for peripheral synaptogenesis:

1.   Presence of a low number (low density) of cholinergic receptors and low levels of synthesis and release of the neurotransmitter (Phase of Innervation)(Fig. 7a). Absence of recognizable synaptic structure and postsynaptic specialization. At this stage the presynaptic terminals already display the capacity for accumulating the precursor (Ch) according to a high affinity, Na^+ and temperature dependent mechanism. This phase is characterized by an intensive building up of membrane components.

2.   Onset and maturation of neurotransmission, first in the ganglion and later at the target organ (Phase of Neurotransmission) (Fig. 7b). The uptake mechanism for the precursor in this phase reaches almost 100% capacity (V_{max}) of the adult synapse (Fig. 5b). This phase is characterized biochemically by an extremely active biosynthesis of the neurotransmitter (Figs. 2 and 3).

3.   Rapid and sustained increase in number and density of receptors and AChE activity in specialized areas of the ganglionic and target organ membrane area (Phase of Receptogenesis)(Fig. 7c). The functional activation of postsynaptic receptors might provide a feedback to trigger or sustain an intensive synthesis of specific proteins presynaptically, such as CAT.

4.   Structural and biochemical maturation and consolidation of the presynaptic and postsynaptic apparatus for synthesis and inactivation of the neurotransmitter (Phase of Maturation)(Fig. 7d). This phase can last for several weeks and extend to after hatching

(or after birth) periods. In the chick ciliary ganglion this phase
of "continuous growth" might persist up to the adult and mature
age of the animal (18-24 months).

5.   Phase of "Synaptic Regression." This phase, which is
characteristic of the aging animal, is analyzed in detail in a
parallel paper in this volume.

The developmental scheme suggested above is supported by results
obtained in structures other than autonomic ganglia such as muscle
endplate (4,19,21), adrenal gland (40), retina (49), myoblasts cul-
tivated in vitro (2,11,13), developing chick heart (41) and brain
(9,10,22). In chick embryo brain Enna et al. (10) found that both
CAT and AChE activity begin to develop rapidly at about the same
time as the muscarinic receptor. AChE activity in particular,
develops at a rate quite similar to the postsynaptic muscarinic
receptor. Baughman (1) found that the cholinergic system in the
chick retina develops in two stages. The first stage occurs relatively early in retina differentiation and is associated with increased ACh synthesis and storage and with a large rise in CAT
activity; the second stage occurs just before hatching, coinciding
with synaptogenesis and the appearance of visual function, and is
associated with further increases in ACh synthesis and storage and
with the development of high affinity Ch uptake. These two stages
are compatible with the developmental scheme proposed above (15,17).

ACKNOWLEDGEMENTS

These investigations were supported by USPHS grants NS-11496,
NS-11430, NS-15086 and NSF GB-41474 to E. Giacobini, and by grants
from the University of Connecticut Research Foundation. Dr. Marchi
was partially supported by a NATO (CNR) Senior Fellowship (No.217.9).

REFERENCES

1.   Baughman, R.W. (1979): IN Developmental Neurobiology of Vision
        (ed) R.D. Freeman, Plenum Press, New York.
2.   Betz, W. and Osborne, M. (1977): J. Physiol. 270:75-78.
3.   Bunge, R., Johnson, M. and Ross, C.D. (1978): Science 199:
        1409-1416.
4.   Burden, S. (1977): Develop. Biol. 57:317-329.
5.   Chiappinelli, V.A., Fairman, K. and Giacobini, E. (1978): Deve-
        lop. Neurosci. 1:191-202.

6. Chiappinelli, V. and Giacobini, E. (1978): Neurochem. Res.
    3:465-478.
7. Chiappinelli, V., Giacobini, E., Pilar, G. and Uchimura, H.
    (1976): J. Physiol. 257:749-766.
8. Cohen, A.M. (1972): J. Exp. Zool. 179:167-182.
9. Coyle, J.T. and Yamamura, H.I. (1976): Brain Res. 118:429-449.
10. Enna, S.J., Yamamura, H.J. and Snyder, S.H. (1976): Brain Res.
    101:177-183.
11. Fambrough, D. and Rash, J.E. (1971): Develop. Biol. 26:55-68.
12. Fiori, M.G. and Mugnaini, E. (1979): Amer. Soc. Neurosci. Abst.
    10:5.
13. Fischbach, G.D. and Cohen, S.A. (1973): Develop. Bio. 31:
    147-162.
14. Giacobini, E. (1978): IN Maturation of Neurotransmission (eds)
    E. Giacobini, G. Filogamo and A. Vernadakis, S. Kargar AG,
    Basel, pp. 41-64.
15. Giacobini, E. (1979): IN Neural Growth and Differentiation (eds)
    E. Meisami and M.A. Brazier, Raven Press, New York, pp.163-166.
16. Giacobini, E. (1980): IN Tissue Culture in Neurobiology (eds)
    E. Giacobini, A. Shahar and A. Vernadakis, Raven Press, New
    Your, pp. 187-204.
17. Giacobini, E. (1980): IN Study Week in Neurobiology Nerve Cells,
    Transmitters and Behavior, Pontifical Academy, Rome, 45:
    451-482.
18. Giacobini, E. and Chiappinelli, V. (1977): IN Synaptogenesis
    (ed) L. Tauc, Naturalia et Biologia Publications, Paris,
    pp. 89-116.
19. Giacobini, C., Filogamo, G., Weber, M., Boquet, P. and
    Changeux, J.P. (1973): Proc. Nat. Acad. Sci. USA 70:1708-1712.
20. Giacobini, E., Pilar, G., Suszkiw, J. and Uchimura, H. (1979):
    J. Physiol. 286:233-253.
21. Inestrosa, N.C. and Fernandes, H.L. (1977): Neurosci. Lett.
    5:91-93.
22. Kouvelas, E.D. and Greene, L.A. (1976): Brain Res. 113:111-126.
23. Kuhar, M.J. and Murrin, L.C. (1978): J. Neurochem. 30:15-21.
24. Landmesser, L. and Pilar, G. (1972): J. Physiol. 222:691-713.
25. Landmesser, L. and Pilar, G. (1974): J. Physiol. 241:715-736.
26. Landmesser, L. and Pilar, G. (1974): J. Physiol. 241:737-749.
27. Landmesser, L. and Pilar, G. (1976): J. Cell Biol. 68:357-374.
28. Le Douarin, N.M. (1977): IN Cell Interactions in Differentia-
    tion (ed) M. Karkinen-Jaaskelainen, Academic Press, London.
29. Le Douarin, N.M., Teillet, M.A., Ziller, C. and Smith, J.
    (1978): Proc. Nat. Acad. Sci. 75:2030-2034.
30. Marchi, M. and Giacobini, E. (1980): Develop. Neurosci. 3:39-48.
31. Marchi, M., Giacobini, E. and Hruschak, K. (1979): Develop.
    Neurosci. 2:201-212.
32. Marchi, M., Hoffman, D.W., Giacobini, E. and Fredrickson, T.
    (1980): Develop. Neurosci. 3:235-247.
33. Marchi, M., Hoffman, D.W., Giacobini, E. and Volle, R. (1980):
    Develop. Neurosci. (in press).

34. Marchi, M., Hoffman, D.W., Mussini, I. and Giacobini, E. (1980):
     Develop. Neurosci. 3:183-196.
35. Markow, R.M. (1979): Arkh. Anat. Histol. Embriol. (Russ.) 76:
     21-27.
36. Mugnaini, E. (1975): Amer. Soc. Neurosci. Abst. 1146:744.
37. Patterson, P.H. (1978): Ann. Rev. Neurosci. 1:1-17.
38. Patterson, P.H., Potter, D.D. and Furshpan, E.J. (1978): Sci.
     Amer. 239:50-59.
39. Pilar, G., Jenden, D.J. and Campbell, B. (1973): Brain Res.
     49:245-256.
40. Rotundo, R., Chiappinelli, V.A., Giacobini, E. and Denenberg,
     V. (1977): Int. Soc. Neurochem. Satellite Symp. Abst.
41. Sastre, A., Gray, D.B. and Lane, A.M. (1977): Develop. Biol.
     55:201-205.
42. Scarsella, G., Toschi, G., Chiappinelli, V.A. and Giacobini,
     E. (1978): Develop. Neurosci. 1:133-141.
43. Schuberth, J., Sparf, B. and Sundwall, A. (1969): J. Neurochem.
     16:605-700.
44. Shainberg, A., Shahar, A., Burstein, M. and Giacobini, E.
     (1980): IN Tissue Culture in Neurobiology (eds) E. Giacobini,
     A. Shahar and A. Vernadakis, Raven Press, New York, pp.25-34.
45. Shelton, D.L., Nadler, J.M. and Cotman, C.W. (1979): Brain
     Res. 163:263-275.
46. Smith, J., Fauquet, M., Ziller, C. and Le Douarin, N.M. (1979):
     Nature 282:853-855.
47. Sorimachi, M. and Kataoka, K. (1975): Brain Res. 94:325-336.
48. Sorimachi, M., Miyamoto, K. and Kataoka, K. (1974): Brain Res.
     79:343-346.
49. Vogel, Z. and Nirenberg, M. (1976); Proc. Nat. Acad. Sci. USA
     73:1806-1810.

AGING OF CHOLINERGIC SYNAPSES IN THE PERIPHERAL

AUTONOMIC NERVOUS SYSTEM

M. Marchi and E. Giacobini

Laboratory of Neuropsychopharmacology
Department of Biobehavioral Sciences
University of Connecticut
Storrs, Connecticut 06268 USA

INTRODUCTION

Aging of Cholinergic Synapses: Some Basic Questions

Only a limited amount of information is available about biochemistry and the pharmacology of aging synapses in the nervous
system. The majority of data reported in the literature refer to
the central nervous system (CNS) and not to the peripheral nervous
system. Although there seems to be a general tendency toward a
decrease in several neurochemical parameters such as enzyme activity
(13,19,43), receptor binding (31), or neurotransmitter levels (cf.
23,26), the results are sometimes contradictory.

By selecting the ciliary ganglion-iris preparation as a model
for aging, we can utilize the body of knowledge already accumulated
by studying the developmental and adult process (14,15,22,24,27-30,
36) and integrate it with our new notions of aging. Such a study
tends to envision aging of the nervous system as a part of a continuous and perhaps "normal" developmental process, beginning in the
embryo and slowly progressing until the end of the life of the individual. Before formulating some specific aims of our study, some
more general questions should be asked.

1.   Are there biochemical changes specifically related to aging, or can neurochemical decreases (e.g. neurotransmitter synthesis and receptor markers) be simply explained by a selective decrease in the number of cells or synapses? One of the most common interpretations of age related changes is a degeneration of nerve cells due to unknown factors acting on the aged neuron. However, in spite of the widely accepted belief in large scale neuronal loss during aging, a survey of the literature (9,17,18,35) does not support the view of a generalized phenomenon of neuronal death.

Ponzio et al. (35) showed that, in the rat, the concentration of catecholamines is not significantly decreased in the striatum and hypothalamus. However, the NE content in the brainstem, where the cell bodies of the neurons which project to different parts of the brain are located, does change. Therefore, at least in some areas, there is probably no loss of noradrenergic terminals due to aging. It may be that only selected populations of noradrenergic neurons projecting from the brainstem are lost or damaged as a consequence of aging. At present we do not know the mechanisms which could produce such a selective loss of a specific neuronal population. In the peripheral nervous system there is no evidence for neuronal loss due to aging. In the ciliary ganglion the number of cells present at 1 month old (approximately 2700) is the same as at 7 years (10). However, profound changes have been described in the target organ cholinergic innervation (22,23,26-29,31).

2.   Do age dependent modifications involve all parts of the neuron (cell body, axon, terminals), or are some parts more vulnerable or more affected than others? The information accumulated so far in our studies (23,26) on cholinergic mechanisms points to a selective vulnerability of the terminals during aging.

3.   Are some biochemical mechanisms related to neurotransmitters more affected than others? Our results indicate that uptake mechanisms may be more vulnerable in aging synapses.

4.   Are changes in neurotransmission due to a "cascade" phenomenon or to the sum of multiple and simultaneous defects taking place in the aging synapse? Our data (23,26) do not support the hypothesis of a "cascade" phenomenon but favor the hypothesis of a group of almostsimultaneous events such as decreased enzyme activity, altered uptake mechanisms and modified turnover of the neurotransmitter. Although all the neurochemical parameters examined so far show some evidence of decline, some are more severely impaired than others.

5.   Can biochemical damage explain a functional loss? We have seen that the levels of neurotransmitter synthesis in the iris during aging are as low as, or lower than that at embryonic ages. Could this phenomenon by itself lead to a great decrease in the function of the organ?

Variations in Enzyme Activities, Receptor Binding
and ACh levels during aging:
Growth·and Regression of the Adult Cholinergic Synapse

We have previously reported acetylcholine (ACh) and choline
(Ch) levels as well as the kinetics of acetylcholinesterase (AChE)
and Ch acetyltransferase (CAT), from early postnatal stages to advanced age (22,25). Three interesting observations can be made as
a consequence of our study. First, the K_m values for both CAT and
AChE seem to be stable throughout development and aging; second,
changes in V_{max} values correlate well with those for enzymatic
levels and third, the V_{max} values for CAT are consistently higher
than the V_{max} for Ch uptake supporting the role of the latter as a
rate limiting factor in ACh synthesis even at very early stages of
development (5 days of incubation). The levels of CAT, AChE and ACh
in the iris and ciliary ganglion are reported in Figs. 1 and 2.

Both CAT and ACh are concentrated at the neuromuscular junctions of the iris and represent reliable presynaptic markers of
cholinergic innervation (15,24). In the terminals of the iris,
total ACh levels and total CAT activity show two distinct periods
of change after hatching. The first, which can be defined as one
of "continuous growth," begins on the first days after hatching and
reaches its peak at 1 year of age. The second, which can be characterized as a period of "regression," starts around 1-2 years and
continues to 7 years (Fig. 1a-1c).

The interpretation of these variations is more difficult in
ciliary ganglia, since CAT and ACh are present both in pre- and
postganglionic components. Denervation experiments have shown that
ACh and CAT activity is distributed 60-70% preganglionic and 30-
40% postganglionic (15,34), respectively. These data, although
derived from adult birds, indicate that most ACh and Ch present in
the ciliary ganglia of the chick is related to presynaptic terminals.
We can, therefore, infer that the large increase up to 1 year in
CAT activity, and up to 2 years in ACh are most presynaptic. The
more rapid and pronounced decrease of CAT activity in the iris than
in the ganglion indicates either a more selective age related damage
to the neuromuscular junction, or a change in the axonal transport
of this enzyme to the periphery. It is interesting to note that
ACh levels continue to increase in the ganglion up to 2 years,
while they are already declining in the iris (Fig. 1c-2C). This
supports our hypothesis of a more pronounced vulnerability of ACh
biosynthesis at the periphery (23,26). This hypothesis is also
supported by the results of Tucek and Gutmann (42), showing that
neuromuscular junctions in the limb muscles of old rats possess a
reduced amount of enzyme for the synthesis of ACh and that the fall
in CAT activity is greater than the fall in number of muscle fibers.
A possible explanation for the continuous increase in CAT activity

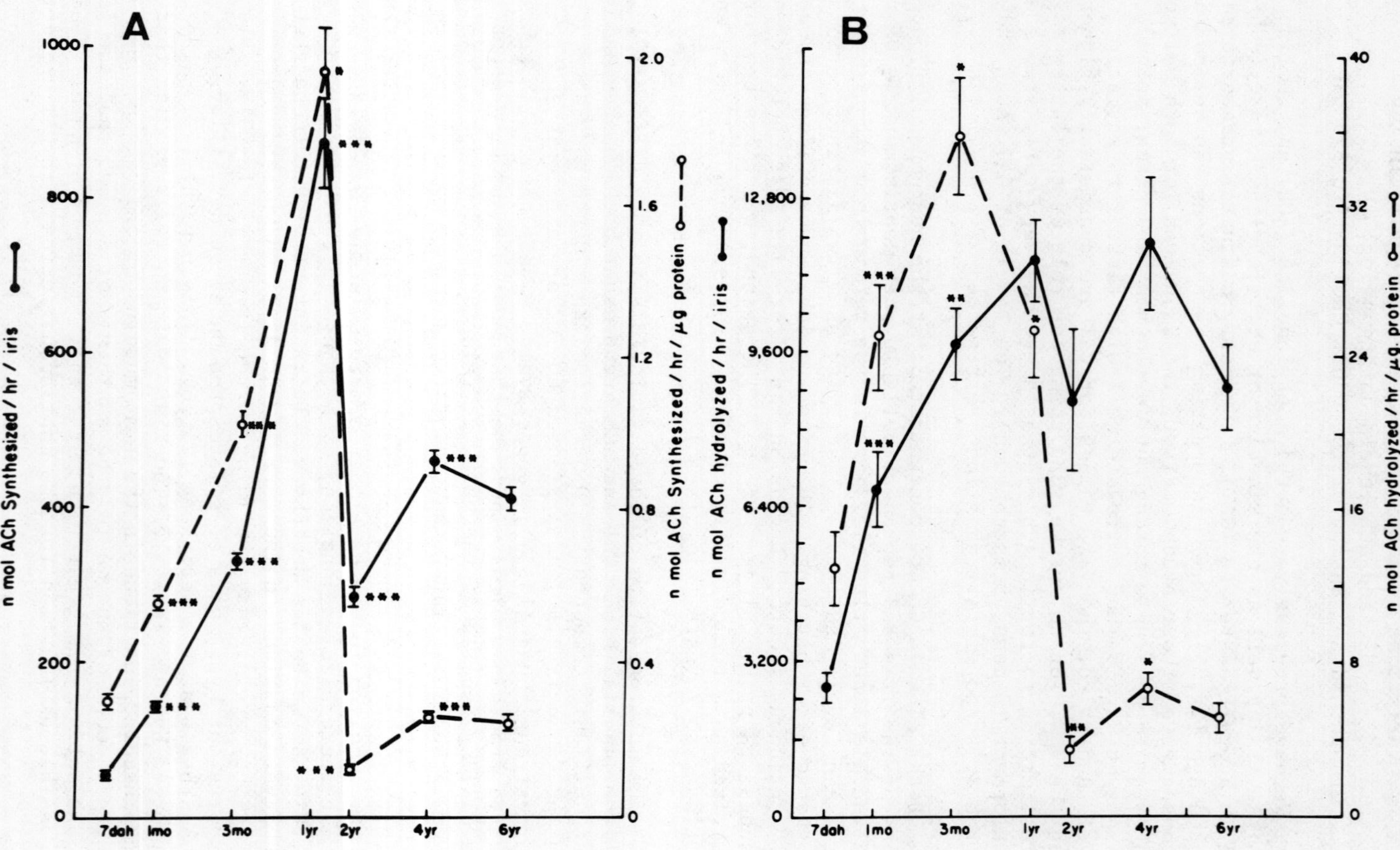
A
B
n mol ACh Synthesized / hr / iris
n mol ACh Synthesized / hr / µg protein
n mol ACh hydrolyzed / hr / iris
n mol ACh hydrolyzed / hr / µg protein
1000
800
600
400
200
0
2.0
1.6
1.2
0.8
0.4
0
12,800
9,600
6,400
3,200
0
40
32
24
16
8
0
7dah 1mo 3mo 1yr 2yr 4yr 6yr
7dah 1mo 3mo 1yr 2yr 4yr 6yr

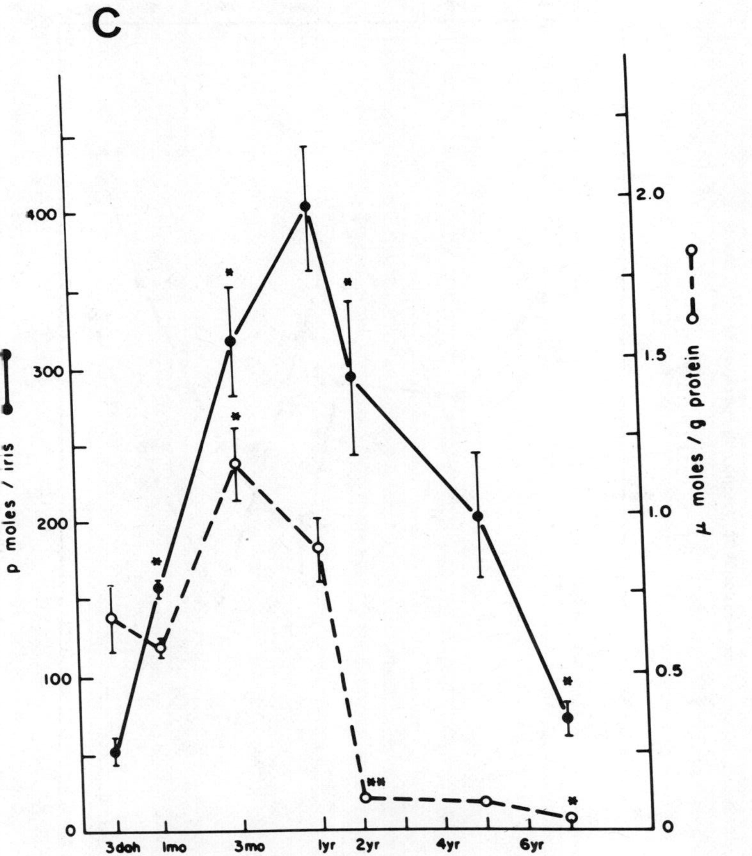

FIGURE 1: CAT (a) and AChE (b) activity in the chick iris. Each point represents the mean ± S.E.M. of 4-8 samples with duplicate determinations. ACh levels (c) are taken from Marchi et al. (24) and Marchi and Giacobini (22). Value significantly different from that of previous age.
* p < 0.05; ** p < 0.01; *** p < 0.001

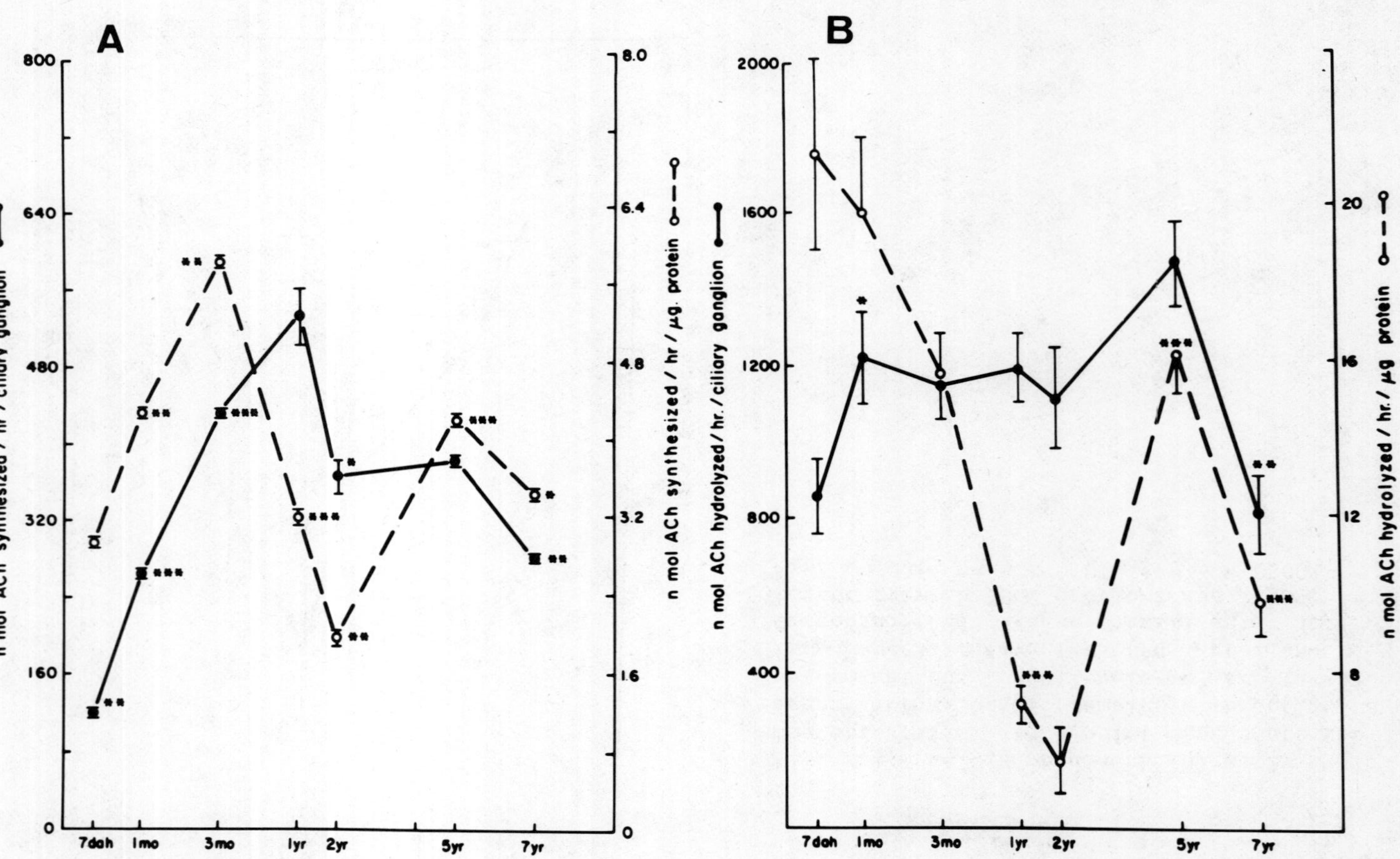
A
n mol ACh synthesized / hr / ciliary ganglion
n mol ACh synthesized / hr / μg. protein
800
640
480
320
160
0
8.0
6.4
4.8
3.2
1.6
0
7dah 1mo 3mo 1yr 2yr 5yr 7yr
B
n mol ACh hydrolyzed / hr. / ciliary ganglion
n mol ACh hydrolyzed / hr. / μg protein
2000
1600
1200
800
400
20
16
12
8
7dah 1mo 3mo 1yr 2yr 5yr 7yr

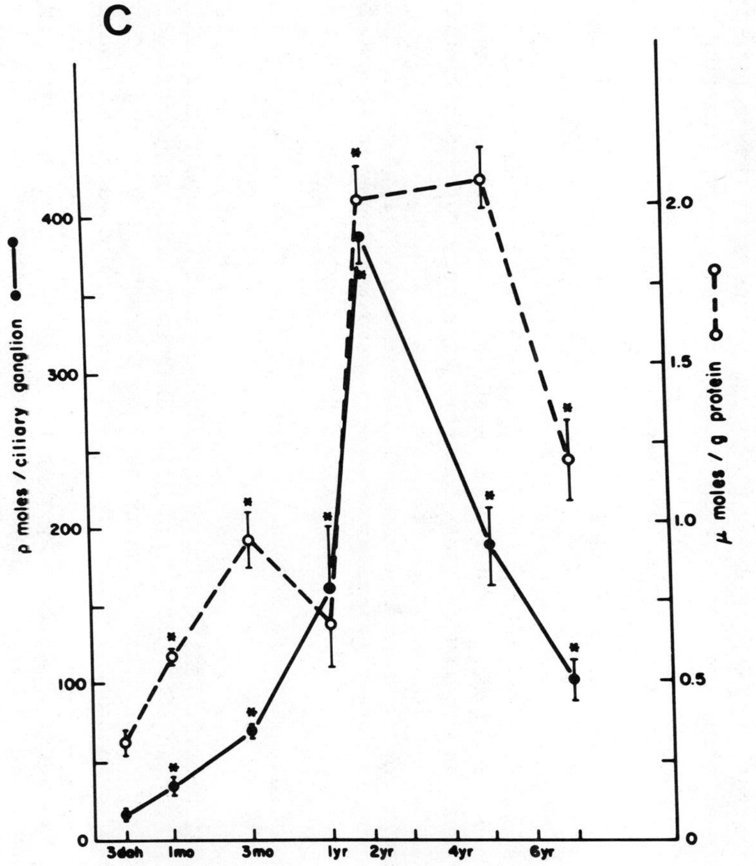

FIGURE 2: CAT (a) and AChE (b) activity in the chick iris. Each point represents the mean ± S.E.M. of 4-8 samples with duplicate determinations. ACh levels (c) are taken from Marchi et al. (24) and Marchi and Giacobini (22). Value significantly different from that of previous age.
* p < 0.05; ** p < 0.01; *** p < 0.001

and ACh in the adult ganglion is provided by the findings of
Mugnaini (33) and Fiori and Mugnaini (10) of a progressively increasing number of preganglionic fibers ending on ciliary cells.
While the number of cells in the chick ciliary ganglion do not vary
between 1 month and 7 years of age, the number of neurons displaying
invaginations from dendrites sprouting from the preganglionic fibers
increases with age, reaching its maximum at 2 years (at which time
almost one-third of ciliary neurons are invaginated). This sprouting
phenomenon decreases slowly but progressively, and in the 7 year old
chicken only 9-10% of the ciliary neurons show invaginations. These
morphological findings are consistent with our biochemical data of a
"continuous growth" up to 2 years. It remains to be established
whether the same phenomenon is true for other autonomic ganglia and
for central cholinergic synapses as well. Fisher and Easter (11)
reported that there were substantially more synapses (per cell or
per mm^2) in the inner plexiform layer in the brain of old (3-4 years)
goldfishes. Buell and Coleman (7) showed evidence for continued
dendritic growth in the parahippocampal gyrus of normal aged human
brain. Sprouting and degeneration of mammalian motor axons in
normal and deafferented skeletal muscles have been described by
Barker and Ip (3).

Changes in enzymatic specific activity in ganglia relate mainly to proteins of nervous origin while in the iris they relate to
the muscle component. Protein content in the ciliary ganglion and
iris increases after hatching up to 1 and 2 years, respectively,
and remains constant from these ages to 7 years (22,24).

The parallel increase in the iris of total specific CAT activity and protein from 7 days after hatching up to 1 year can be
related to a simultaneous growth of muscle and nerve tissue during
this period. The subsequent pronounced drop in specific activity
of the CAT between 1 and 2 years might be related to the continuous
increase in protein content. In this case both total and specific
activity are declining.

It is more difficult to assess possible changes in the postsynaptic pool of CAT in ciliary ganglia. However, the increase in
the iris from 7 days after hatching to 1 year, which parallels the
significant increase in the ganglion, indicates an accumulation of
the enzyme in the cell body and a continuous transport to the terminals. Since the number of cells is constant (10), an increase
in CAT activity might also depend on an increase in the size of
ciliary ganglion cells related to changes in the target tissue, the
iris. In fact, the iris continues to increase in both dry weight
and protein up to 2 years of age. However, both ganglionic weight
and proteins slow down by 1 year (8,22,24). The decrease in CAT
activity seen at later stages in the iris might reflect either a
lower synthesis in the ganglion or a slower transport.

Studies based on denervation (14,15) show that AChE is distributed 20% in the preganglionic terminals, 69% in ciliary ganglion cell somas and the remaining 11% in postganglionic axons. At the neuromuscular iris junctions 20% is found in the ciliary nerve terminals and 80% in the iris striated muscle. No results are available with regard to the distribution of AChE in avian sympathetic ganglia. However, in lumbar sympathetic ganglia of other vertebrates AChE is predominantly preganglionic (14). Two differences are evident in the post-hatching pattern of AChE activity with respect to CAT. First, AChE activity achieves it maximal activity more rapdily, i.e. at 1 month for the ciliary ganglion and at 1 year for this iris, second, total AChE activity, after having reached its peak, varies little during aging. A significant decrease is seen between 5 and 7 years, but only in the ciliary ganglion. Specific activity, however, starts to decrease very early (1-3 months) in both structures and reaches low levels, particularly in the iris. The marked decrease in specific AChE activity seen in mature and aging ciliary ganglia and irises can be compared to the effect of axotomy on the molecular forms of AChE (36), which are limited to the membranous fraction. In both ganglia and irises the same membranous form component IV (corresponding to the 17.5 S form), is markedly reduced by axotomy. Axotomy involves changes of ganglionic synapses and of neuronal membranes (6). The aging process could induce changes similar to those following denervation in specific AChE forms of the iris muscle. However, no specific studies of AChE molecular forms during aging are yet available to validate this hypothesis. Another factor which could account for peripheral changes in AChE activity is the effect of age on axoplasmic transport of enzymes. McMartin and O'Connor (32) reported that accumulation of AChE activity in rat sciatic nerves was strikingly reduced by old age. Although the general trend of both ACh and CAT during aging is similar and can be defined as that of a continuous decrease, we can see that the two events are not overlapping in time. For example, a CAT activity in the iris seems to achieve its lowest level very rapidly (at 2 years) while ACh still continues to decline between 2 and 7 years. This might be explained by the fact that CAT is present in the endings in amounts much greater than those needed to meet the demand for ACh synthesis during development. For example, in the iris CAT activity at 5-6 years is 245 pmol/µg protein/hr, while the rate of $[^3H]$-ACh formation following incubation with $[^3H]$-Ch at the same age can be estimated as 8 fmol/µg protein/hr (29). Because of this large excess, it seems improbable that CAT is the mechanism responsible for the effect of aging on ACh levels in the terminals. In conclusion, negative enzymatic changes in the autonomic ganglia and iris occur only when full maturity of the animals has been reached (1-2 years of age), and are concomitant with pronounced decreases in ACh levels. However, these conspicuous changes in enzymatic activity cannot solely account for the drop in ACh levels. Apparently the aging synapse is able to retain a

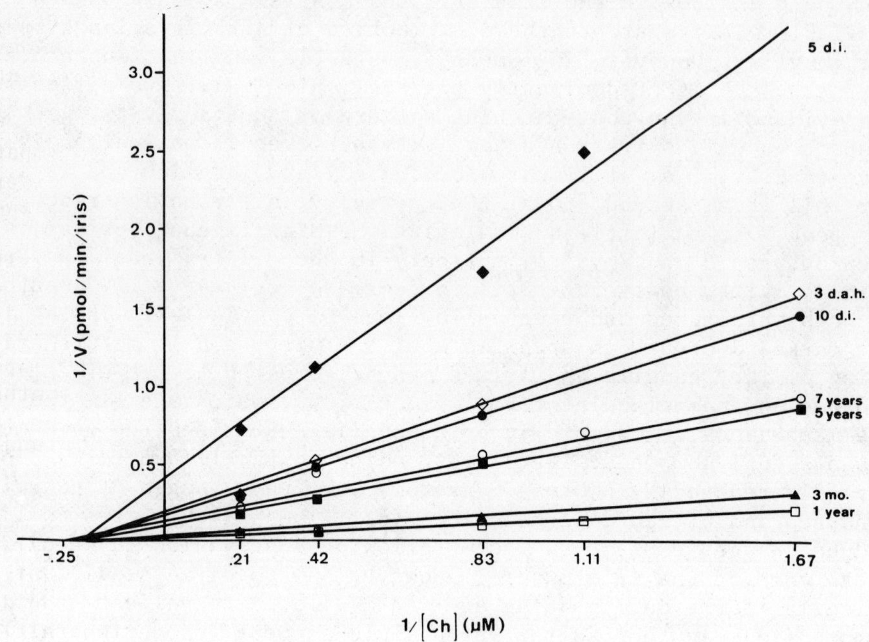

FIGURE 3: Kinetics of Ch uptake (Lineweaver-Burk plot). Each point
represents the mean of at least six determinations at
each of five concentrations (0.6-4.8 µM Ch). Intercepts
and slopes of lines calculated by linear regression.

considerable potential for synthesizing neurotransmitter up to the
last stages of life. This may indicate that the decline in high
affinity uptake of Ch is probably responsible for the low ACh levels
during aging. Recent results (31) show that the amount of α-bun-
garotoxin (α-Btx) binding which has increased continuously in the
iris from the period following hatching up to 7 months, decreases
continuously from 7 months to 5 years. In ciliary ganglia, on the
other hand, the amount of binding remains virtually unchanged up to
5 years. These results, compared on the one hand with the decline
in ACh, Ch content, Ch uptake and CAT activity and, on the other
hand, with the slight changes in AChE activity, support our view
that postsynaptic components are relatively less affected by aging
than presynaptic ones.

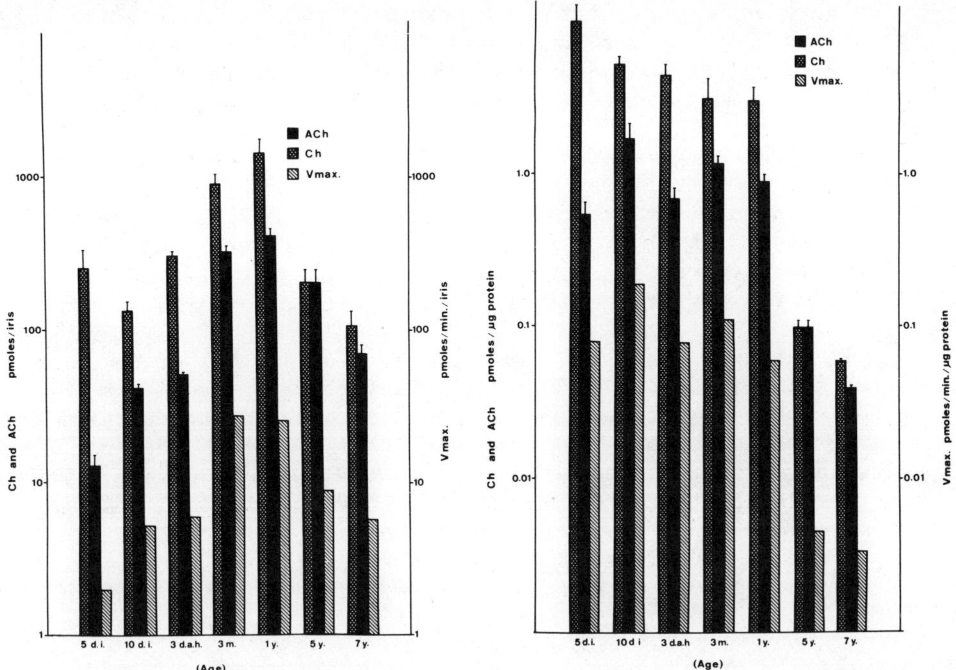

FIGURE 4: Levels of Ch and ACh, and V_{max} of Ch uptake. Each point
represents the mean ± S.E.M. of at least six determinations. Values significantly different than that of preceding age. * p < 0.05; ** p < 0.001. (A) calculated
per iris; (B) calculated per µg protein.

Ch Uptake and ACh Metabolism in the Aging Synape

The presence of two kinetically well defined Ch uptake mechanisms has been demonstrated in the chick iris (40). We have shown
(30) that the high affinity Ch uptake is present starting at the
earliest stage of development of the iris, i.e. at 5 days of
incubation.

Our results show that K_m values (per iris) remain stable
throughout development (Fig. 3) and do not change during aging. On
the contrary, V_{max} changes during both development and aging. The
Ch uptake can be correlated to Ch and ACh levels (27,30). In Fig. 4
the V_{max} values for Ch uptake are reported both per protein and per
iris, together with the endogenous levels of ACh and Ch throughout
the life span of the animal. Values expressed per iris represent
the total innervation of the organ at each stage while the data

calculated per protein reflect the density of the innervation relative to the size of the iris muscle tissue. All three parameters
decrease progressively and in a parallel fashion from 1 to 7 years.
However, the most significant declines, both in V_{max} and Ch levels,
are seen between 1 and 5 years of age. This might indicate the
period in which the most pronounced degenerative changes in the
peripheral cholinergic synapse take place. Morphological results
which could substantiate this view are still lacking. As shown in
Fig. 4a and 4b the V_{max} values decrease significantly at 5 years and
continue to decrease at 7 years. ACh and Ch levels follow a similar
trend, decreasing progressively from 1 year to 7 years.

To our knowledge, this dramatic decline in V_{max} of Ch uptake
during aging is unique. This change may represent modifications
in the number of uptake sites for Ch due to the aging process. It
is interesting to note that one characteristic which does not change,
either during development or aging, is the K_m of Ch uptake. The
constant and close correlation of ACh levels to V_{max} values, both
during development (27,30) and aging, points to a regulatory role
for Ch uptake in maintaining neurotransmitter levels. Obviously
the very low levels of ACh found at the lastest period studied (5-
7 years) may have functional implications for the target organ.

Changes in Ch levels during development and aging should be
interpreted in the light of the dynamic relation between Ch levels
and ACh biosynthesis. Of this measurable Ch, 90% is utilized in
the adult for the synthesis of ACh and 10% for the synthesis of
phosphorylcholine (2). We can, therefore, assume that a need for
the synthesis of phospholipids might be more pronounced in the period
up to 1-2 years, during which membranes may still be acitvely synthesized. Logically, therefore, the need for Ch in phospholipid
synthesis should be lower during aging. However, there are indications that synaptic structures are actively turning over during
aging in the iris (41).

ACh Metabolism in Aging Synapses

In a recent paper we described the metabolism of ACh in the
iris during development and aging (29). One of the salient findings
of this study was the remarkable stability of the ratio [3H]-ACh to
[3H]-Ch in the terminal of the iris throughout the life of the
animal (Table 1 and Fig. 5). This indicates that the relationship
between the [3H]-Ch present in the iris (as such, and as phosphorylcholine) and [3H]-ACh formed per unit of time, is relatively age
independent. The levels of [3H]-ACh formed are dependent, not only
on the rate of synthesis from the available [3H]-Ch, but also on
the release and hydrolysis of ACh.

TABLE 1: Relations among cholinergic parameters during development and aging of the iris of the chicken.

		AGE				
		10 DI	3 DAH	3 months	1 year	5 years
$\dfrac{[^3H]-Ch}{[^3H]-ACh}$	37°C	2.7	3.4	1.9	1.9	2.4
	27°C	–	–	2.8	–	2.8
	Ouabain (10^{-4} M)	–	–	1.9	–	2.5
$\dfrac{Ch}{ACh}$		19	6	3	3	1
$\dfrac{ACh}{[^3H]-ACh}$		54	83	65	50	142
$\dfrac{Ch}{[^3H]-Ch}$		60	13⸴	97	90	52
Ch uptake (V_{max})		5	6	27	26	9
ACh turnover		9	17	95	145	16

Ch uptake (V_{max}) = pmol/min/iris.   DI = days of incubation; DAH = days after hatching.

ACh turnover rate = pmol/min/iris (37) = $\dfrac{\dfrac{\text{percent dpm as } [^3H]-ACh}{\text{percent dpm as } [^3H]-Ch} \times Ch \text{ (nmol) g}}{t^o - t^i}$
$t^o - t^i$ = time of incubation

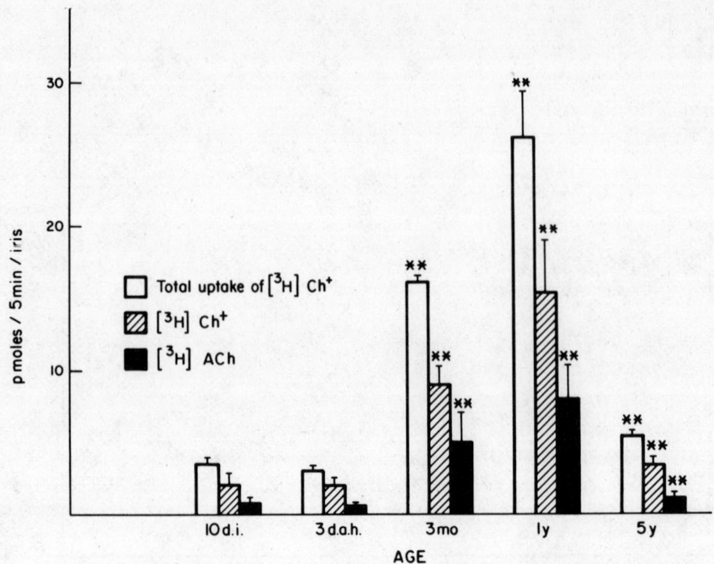

FIGURE 5: Uptake and metabolism of [^{3}H]-Ch in chick iris. Each bar
represents the mean $\pm$ S.E.M. of at least four determinations. Value significantly different than that of preceding age. * p < 0.05; ** p < 0.02.

Although we did not measure all of these parameters, our results emphasize the tendency of the cholinergic synapse to maintain
this ratio between the two parameters ([^{3}H]-ACh and [^{3}H]-Ch) (Fig. 5
and Table 1). In Fig. 5 the values of total [^{3}H]-Ch uptake in the
iris are reported, together with the amount of [^{3}H]-Ch present. The
five ages selected represent, respectively, early embryonic (10
days of incubation), early post-hatching (3 days after hatching),
young adults (3 months), adults (1 year), and aging (5 years)
stages. It can be seen that the ratio [^{3}H]-Ch/[^{3}H]-ACh varies
little (1.9-3.4) throughout the life span of the animal (Table 1).

The total amount of [^{3}H]-Ch taken up into the iris after incubation with [^{3}H]-Ch is unvaried between 10 days of incubation and
3 days after hatching and then increases fivefold at 5 months (Fig.
5). A further significant 60% increase is seen between 3 months
and 1 year, followed by a reduction at 5 years. [^{3}H]-ACh formed
is unvaried between 10 days of incubation and 3 days after hatching
and increases eightfold at 3 months; a further 30% reduction is
seen between this latter age and 5 years, which brings [^{3}H]-ACh
levels to embryonic or early post-hatching levels. The corresponding

amount of [3H]-Ch present follows the same trend as [3H]-ACh levels
(Fig. 5).

An indirect evaluation of the dynamics of this mechanism and
its relation to age can be provided by the analysis of previous
data obtained on endogenous levels of ACh and Ch, and Ch uptake
(22,24,27,30), which are reported in Table 1. The ratio of ACh
endogenous levels to [3H]-ACh formed does not appear to vary up to
1 year; however, at 5 years it is almost threefold higher (Table 1
and Fig. 5). The endogenous levels of ACh are decreased to 50% (22)
at the same age; therefore the changes in the ratio are a function
of the greater decrease (80%) in the formed [3H]-ACh (Fig. 5). This
is also indicated by the decrease of the ratio between Ch levels
and the [3H]-Ch present at 5 years (Table 1 and Fig. 5). In this
case the decrease in the [3H]-Ch is relatively less pronounced than
the decrease in endogenous Ch level. As a result of the apparently
slowed formation of [3H]-ACh we find, in the 5 year old terminals,
a relative increase in [3H]-Ch, as compared to the 1 year adult
(Fig. 5). An explanation for the apparent discrepancy between the
relatively constant [3H]-Ch/[3H]-ACh ratio and the continuous decrease of the Ch/ACh may be found in a decrease of the Ch uptake
not linked to ACh synthesis. This could imply a continuous decrease in the low affinity uptake for Ch, however, direct evidence
is presently not available to substantiate this view. As previously discussed (14,15), we have shown that the activity of CAT which
is presynaptically localized is reduced to 50%, while AChE activity
which is predominantly postsynaptic, is still 77% at 6 years of
age (25).

Given the large excess of CAT activity in the adult (25) in the
presence of an adequate supply of Ch and acetyl-coenzyme A (AcCoA)
these reduced levels of enzyme activity at 5 years may still be
sufficient to maintain physiological levels of ACh in the terminal.
It seems, therefore, less likely for CAT to be the site of effect
of aging on ACh levels in the terminals. However, a decrease in
CAT activity could be indicative of a reduced number of terminals
as well as of reduced synthesis or transport of the enzyme protein.
Age dependent changes in the affinity properties of the enzyme have
been ruled out (25) and variations involving properties of AcCoA
are not known. Many other studies (4,12,38,44) strongly indicate
that the high affinity Ch system present in the terminals is tightly coupled to ACh synthesis.

The percentage of Ch taken up by the high affinity system
which is converted to ACh in the adult iris seems to be relatively
unvaried (35-45%) under a variety of experimental conditions (40).
Our results show that the ratio between these two products does not
change significantly at the two ages, however, we can see that the
total [3H]-ACh found in the 5 year iris has decreased to 20% of
the adult value. Because this significant decrease relates to a

parallel 80% decrease in the total [^{3}H] accumulated by the tissue,
we may conclude that the net content of [^{3}H]-ACh or its rate of
synthesis is probably unvaried at 5 years. Other relevant results
should be considered. The threefold increase in the ratio ACh/
[^{3}H]-ACh (Table 1) would instead support the idea of a decreased
rate of ACh fomed at 5 years. Other indications of a possible decrease in the rate of synthesis of ACh might be derived indirectly
from our study on Ch uptake, as previously reported (27,30). We
have shown (27,30) that the V_{max} for Ch uptake and ACh levels follow
the same trend from 5 days of incubation to 7 years, which supports
the view of a regulatory role of Ch uptake for ACh levels, not only
in the adult (20), but also throughout development and aging. Ch
uptake (V_{max}) increases up to 3 months and decreases significantly
from 1 year to 7 years. These changes do not rule out possible age
dependent variations in the rate of Ch utilization for ACh synthesis.
Another indication of a reduced synthesis with aging may be the 50%
reduction in endogenous levels of ACh at 5 years as compared to 1
year. In this case the ratio ACh/Ch has also decreased from 3.5 to
1 (Table 1). The total pool of availabile Ch is therefore strongly
reduced. Although age dependency of Ch available for ACh synthesis
seems to be indicated by our results, we did not measure the extra-
neuronal Ch present in the membrane phospholipids which might also
vary with age.

An additional indication of variations in the turnover rate of
ACh with age can be found by applying the biosynthesis ratio method
of Schuberth et al. (37)(Table 1). In this case an increase in
turnover per iris up to 1 year is apparent, as well as a significant
decrease between this age and 5 years. It should be emphasized that
turnover changes of ACh reported per organ may relate to possible
variations in the total number of terminals present at different
ages.

In conclusion, we have seen significant decrease in activities
of AChE, CAT and Ch uptake and lower levels of ACh and Ch in the
aging iris. Further, our data indicate that the turnover rate of
ACh is slowed down in aging autonomic terminals. It must be pointed
out that our measurements of turnover may be influenced by age related modifications in compartmentation of ACh and Ch as well as
variations in release mechanisms resulting from changes in number
and structure of terminals or in intraneuronal biochemical mechanisms.

Changes in Characteristics of Ch Uptake During Aging

Some particular features of Ch uptake during development and
aging are reported in Fig. 6. Sensitivity to hemicholinium (HC-3;
0.5 x 10^{-5} M) decreases significantly between 5 days of incubation

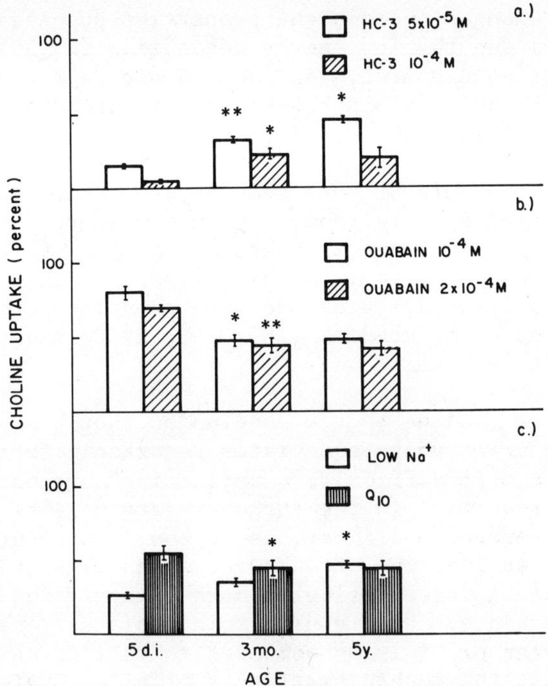

FIGURE 6: Changes in characteristics of Ch uptake with age.   Each point represents the mean $\pm$ S.E.M. of at least six determinations. Value significantly different than that of preceding age. * p < 0.05; ** p < 0.001. Inhibition by HC-3; (B) inhibition by ouabain; (C) uptake in low N$^+$ (25 mM) or at 27°C.

and 5 years (Fig. 6a). Ch uptake is inhibited by 85% at 5 days of incubation by 66% at 3 months and by only 56% at 5 years by HC-3 (Fig. 6a). At a higher concentration (10^{-4} M) no further inhibition is seen between 3 months and 5 years, although a significant difference is seen between 5 days of incubation and 3 months (Fig. 6a). Inhibition by ouabain (10^{-4} M) shows a reverse trend (Fig. 6b), increasing significantly from 20% of control at 5 days of incubation to 52% at 3 months. At 5 years, sensitivity to ouabain is the same as at 3 months. Ch uptake at 27°C decreases from 59% of control at 5 days of incubation to 45% at 3 months, and is not changed at 5 years (Fig. 6c). The lower sensitivity for ouabain and low temperature dependence seen at 5 days of incubation might relate to

developmental changes in membrane properties such as levels of
ATPase (1,5,45) and the low energy metabolism (21,39) found at the
early stages of development. Na$^+$ dependence decreases significantly
between 3 months and 5 years but no significant changes are seen
between 5 days of incubation and 3 months.

Finally, it should be observed that the characteristics of Ch
uptake do not develop simultaneously in cholinergic neurons but appear independently. Likewise, these attributes are not modified
simultaneously by the aging process. It seems, therefore, inap-
propriate to apply at all ages the same invariant criteria derived
from non-developmental studies (16,20) to discriminate between high
and low affinity uptake.

Our results lead us to the conclusion that biosynthesis of ACh
in aging peripheral synapses deviates remarkably form the adult
condition. The most difficult "quantitative" evaluation is the one
of a possible reduction in the turnover rate of ACh. Due to continuous age dependent modifications in pool distribution of both
ACh and Ch, determinations of turnover which might be performed in
the adult might be more difficultt both to apply and it interpret
in the aging synapse. Our results point to a decreased turnover of
ACh in the 5 year old iris as compared to earlier ages (25), but
fail to pinpoint the mechanism of such a fall. Obviously more work
is needed with alternative techniques to precisely identify the
cause of such a breakdown. A future study should include a more
detailed analysis of the reduced ACh metabolism, and the consequences
that this phenomenon might produce on the postsynaptic neuron (or
target organ). Biochemical studies on critical ages such as the
period 5-7 years in the chick, should be integrated with ultrastructural and physiological studies.

CONCLUSIONS

Toward a Biochemistry of Aging of Cholinergic Transmission

The main objectives of research on aging of the nervous system
can be condensed in the search for two kinds of information. First,
it is important to identify and characterize any mechanism responsible for gross malfunction in neurotransmission. Second, it is
necessary to pinpoint the period at which such degenerative changes
take place. The answers to these questions can provide the background which would be useful to the pharmacologist. The design of
drugs which might prevent or partially counteract the negative effect of aging on synaptic function will probably depend on the
characterization of the specific metabolic failure(s). From the

clinical point of view an early diagnosis of aging might help to preven, or at least reduce, any further damage.

Our study represents a first attempt to study systematically biochemical malfunctions of the peripheral cholinergic synapse by comparing three different periods of life of the nervous system. First we have analyzed changes occurring at the embryonic level and, second, we have followed these developmental changes into the adult life. Finally, we have studied the occurrence of definite changes taking place in the latter part of the adult life. ACh metabolism seems to gradually deteriorate in this period, leading to modifications which seem characteristic for the final period of life of the synapse.

ACKNOWLEDGEMENTS

These investigations were supported by USPHS grants NS-11496, NS-11430, NS-15086, NSF GB-41475, and by grants from the University of Connecticut Research Foundation to E. Giacobini. Dr. Marchi was partially supported by a NATO (CNR) Senior Fellowship (No.217.9).

REFEENCES

1. Alfei, L. and Venturini, B. (1972): Brain Res. 43:314-316.
2. Ansell, G.B. (1971): IN Chemistry and Brain Development (eds) R. Paoletti and A.N. Davison, Plenum Press, New York.
3. Barker, D. and Ip, M.C. (1966): Proc. Roy. Soc. B 163:538-554.
4. Barker, L.A. and Mittag, T.W. (1975): J. Pharmacol. Exp. Ther. 192:86-94.
5. Bignami, A., Palladini, G. and Venturini, G. (1966): Brain Res. 3:207-209.
6. Brenner, H.R. and Martin, A.R. (1976): J. Physiol. 260:159-175.
7. Buell, S.J. and Coleman, P.D. (1979): Science 206:854-856.
8. Chiappinelli, V. and Giacobini, E. (1978): Neurochem. Res. 3: 465-478.
9. Finch, C.E. (1973): Brain Res. 52:261-276.
10. Fiori, M.G. and Mugnaini, E. (1979): Soc. Neurosci. Abst. 10:5.
11. Fisher, L.J. and Easter, S.S. Jr. (1979): J. Comp. Neurol. 185:373-380.
12. Freeman, J.J. and Jenden, D.J. (1976): Life Sci. 19:949-962.
13. Frolkis, V.V., Bezrukof, V.V., Duplenko, Y.K. and Genis, E.D. (1972): Exp. Gerontol. 7:169-184.
14. Giacobini, E. (1978): IN Maturation of Neurotransmission (eds) E. Giacobini, G. Filogamo and A. Vernadakis, Karger, Basel, pp. 41-64.

15. Giacobini, E., Pilar, G., Suszkiw, J. and Uchimura, H. (1979):
     J. Physiol. 286:233-253.
16. Haga, T. and Noda, H. (1973): Biochem. Biophys. A 291:546-575.
17. Jacobson, M. (1978): IN Developmental Neurobiology, 2nd Ed.,
     Plenum Press, New York.
18. Kanowski, S. (1977): Prog. Neuropsychopharmacol. 1:249-256.
19. Kaur, G. and Kanungo, M.S. (1970): Indian J. Biochem. 7:122-125.
20. Kuhar, M.J. and Murrin, L.C. (1978): J. Neurochem. 30:15-21.
21. Larrabee, M.G. (1972); IN NGF and its Antiserum (eds) E.
     Zaimis and J. Night, Athlone, London, pp. 177-184.
22. Marchi, M. and Giacobini, E. (1980): Develop. Neurosci. 3:39-48.
23. Marchi, M., Giacobini, E. and Hoffman, D.W. (1980); IN Neural
     Regulatory Mechanisms During Aging, Alan R. Liss, Inc.,
     New York (in press).
24. Marchi, M., Giacobini, E. and Hruschak, K. (1979): Develop.
     Neurosci. 2:201-212.
25. Marchi, M., Hoffman, D.W. and Giacobini, E. (1980): Proc. Amer.
     Soc. Neurochem., p. 84.
26. Marchi, M., Hoffman, D.W. and Giacobini, E. (1980): IN Aging
     of the Brain and Dementia (ed) L. Amaducci, Raven Press, New
     York, pp. 159-166.
27. Marchi, M., Hoffman, D.W., Giacobini, E. and Fredrickson, T.
     (1980): Brain Res. 194:423-431.
28. Marchi, M., Hoffman, D.W., Giacobini, E. and Fredrickson, T.
     (1980): Develop. Neurosci. 3:235-247.
29. Marchi, M., Hoffman, D.W., Giacobini, E. and Volle, R. (1981):
     Develop. Neurosci. 3: (in press).
30. Marchi, M., Hoffman, D.W., Mussini, I. and Giacobini, E. (1980):
     Develop. Neurosci. 3:183-196.
31. Marchi, M., Yurkewicz, L., Giacobini, E. and Fredrickson, T.
     (1981): Develop. Neurosci. (in press).
32. McMartin, D.N. and O'Connor, J.A. Jr. (1979); Mech. Aging
     Develop. 10:241-248.
33. Mugnaini,E. (1975); Proc. Amer. Soc. Neurosci., p.744.
34. Pilar, G., Jenden, D.J. and Campbell, B. (1973): Brain Res.
     49:245-256.
35. Ponzio, F., Brunello, N. and Algeri, S. (1978): J. Neurochem.
     30:1617-1620.
36. Scarsella, G., Toschi, G., Chiappinelli, V.A. and Giacobini, E.
     (1978): Develop. Neurosci. 1:133-141.
37. Schuberth, J., Sparf, B. and Sundwall, A. (1969); J. Neurochem.
     16:605-700.
38. Schwartz, J.H. (1974): IN Synaptic Transmission and Neuronal
     Interaction, Raven Press, New York.
39. Seltzer, J.L. and McDougal, D.B. Jr. (1975): Develop. Biol.
     42:95-105.
40. Suszkiw, J.B. and Pilar, G. (1976): J. Neurochem. 26:1133-1138.
41. Townes-Anderson, E. and Raviola, G. (1978): J. Neurocytol. 7:
     583-600.

42. Tucek, S. and Gutmann, E. (1973): Exp. Neurol. 38:349-360.
43. Verkhratsky, N.S. (1970): Exp. Gerontol. 5:49-56.
44. Yamamura, H.I. and Snyder, S.H. (1973): J. Neurochem. 21:1355-
       1374.
45. Zaheer, N., Iqbal, Z. and Talwar, G.P. (1968): J. Neurochem.
       15:1217-1224.

THE CHEMICAL EMBRYOLOGY OF THE ELECTROMOTOR SYSTEM OF

TORPEDO MARMORATA

W.D. Krenz, T. Tashiro, K. Wachtler, V.P. Whittaker and
V. Witzemann

Abteilung Neurochemie, Max-Planck-Institut
fur Biophysikalische Chemie, Gottingen, FR Germany, and
Institut de Biologie Marine, Arcachon, France

INTRODUCTION

Although detailed morphological studies of the embryological
development of the electric organ of Torpedo were made in the
nineteenth century with the light microscope (1,12,18) relatively
little had been done with modern techniques with this important
model cholinergic system until recently, when electron microscopic
surveys were carried out (7,8,17). These have confirmed the conclusion of the earlier workers that electrocytes develop from
vertically orientated myotubes and have documented the differentation of the postsynaptic membrane and the onset of synaptogenesis.

In this study, started in 1976 and published in part in com-
munications, review articles and a thesis (9-11,14,24), we have followed, as a function of embryological development, five parameters:
choline acetyltransferase (ChAT), acetylcholinesterase (AChE),
nicotinic acetylcholine receptor protein (nAChR), electric organ
discharge in response to presynaptic stimulation (EOD) and myosin.
The first two of these enzymes responsible, respectively, for the
synthesis and breakdown of acetylcholine (ACh), have been followed
both in the electric lobe and differentiating electric organ. The
remaining three, indicative respectively of the biochemical differentiation of the postsynaptic membrane formation of a functional synapse and the transition from myotube to electrocyte, have been followed in the electric organ. Finally, changes in these parameters

have been correlated with morphological differentiation. This study
has been carried out as a necessary preliminary to detailed examination of the mechanisms of differentiation and synaptogenesis in this
system.

By way of introduction for those unfamiliar with reproduction
in this species, it should be mentioned that the _Torpedo_ is ovovi-
viparous, that is, the embryos, attached to fragile yolk-sacs, are
held in uterus-like expansions of the female genital tract until
birth, which occurs at a length of about 120-130 mm. At this stage
no external yolk-sac is visible. The _Torpedo_ grows throughout life,
though at a diminished rate after sexual maturation, and female
specimens of _Torpedo marmorata_ may attain 60 cm in length (16,17).

METHODS

Collection of Embryos

Pregnant female specimens of _Torpedo marmorata_ were collected
in successive summer seasons at the Institut Universitaire de
Biologie Marine, Arcachon, France, during the months of June to
September. The fish were obtained by diving and netting or snaring
in 2-5 m of water at low tide along the pilings of the channels
between the oyster beds in the Bassin d'Arcachon.

Removal and Preparation of Tissue

In the case of the smaller frozen embryos (20 mm or more), lobe
or presumptive electric organ tissue was punched out from thick
cryostat sections (19). With larger embryos, lobes and samples of
electric organ were removed by dissecting the frozen tissue. For
enzyme assays, tissue samples were homogenized in small glass tubes
with a pestle connected to a dentist's drill using enough 10 mM
EDTA in 0.5% Triton X-100 to give a 5-10% w/v tissue suspension;
of this, 10 µl was taken for each assay. For the separation of
myosin the homogenization medium was a buffer containing 2% sodium
dodecylsulphate (SDS), 0.1 M dithiothreitol, 0.08 M Tris HCl, pH 6.8,
10% v/v glycerol, and 0.001% bromphenol blue as tracking dye; the
tissue concentration was 70-75 mg.ml of medium^{-1} corresponding to
1.4 mg of protein$\cdot$ml^{-1} and the homogenate was heated at 100°C for
3 min to solubilize and denature protein. For nAChR determinations
by toxin binding assay, homogenization was carried out in a Virtis
homogenizer run at maximum speed for two 30 sec intervals, the tissue concentration was 50-250 mg$\cdot$ml^{-1} and the suspension medium was
10 mM phosphate buffer, pH 7.5.

Biochemical Studies

Enzyme assays: ChAT was measured (6) using [^{14}C]-acetyl-
coenzyme A (AcCoA) in a final concentration of 0.4 mM. Samples
(10 µl) of the incubation mixture and the homogenate were mixed
and incubated for 30 min at 30°C in small (4 x 10 mm) glass tubes
before extracting and counting the [^{14}C]-ACh formed, using acetonitrile as the organic phase. AChE was determined photometrically
(15), using acetylthiocholine iodide as the substrate. Samples
run in the absence of the specific butyrylcholinesterase inhibitor
tetraisopropylpyrophosphoramide showed only a small ($\sim$ 10%)
increase in activity with lobe tissue and none with electric organ;
thus (assuming a similar specificity for the inhibitor in Torpedo
as in the mammal) any contribution from butyrylcholinesterase could
be neglected. Protein was determined by Folin's reagent (15).

Myosin: Appropriate amounts of solubilized electric tissue
protein were submitted to slab gel electrophoresis in SDS and
stained with Coomassie brilliant blue dye. The amount of protein
applied to the gel was measured by the Amidoschwarz method (21).
The amount of myosin heavy chain in the sample was determined by
comparing the area under the peak corresponding to this protein in
a densitometric tracing of the gel with similar peaks obtained by
running known amounts of Torpedo muscle myosin heavy chain protein
eluted from a preparative SDS gel.

Receptor protein: nAChR was measured by ^{125}I-labelled α-
bungarotoxin binding (3,22). The tissue proteins were solubilized
by mixing the homogenate in 10 mM phosphate buffer with 1 vol of
phosphate buffer containing 0.2% Triton X-100; 25-100 µl samples
were incubated with excess toxin ($\simeq$ 0.1 µM) for 1-3 hr at room
temperature. As controls, samples of homogenate were used in which
the receptor sites had been blocked with a 5-10-fold excess of
"cold" α-bungarotoxin for 1 hr before addition of labelled toxin.
Samples (100 µl) of the toxin treated homogenate were applied to
Whatman DE 81 filter discs, washed (2) and counted in a Beckman
Biogamma gamma counter. A control samples of adult tissue was
included in each run.

Electrophysiological Studies

Measurements of EOD were made with 109 embryos from 39 gravid
females of 37-56 cm body length and 14 new-born Torpedo. Normally
experiments were done on the same day as the embryos were delivered.
When embryos that had been kept for a few days in running sea water
or uterine fluid were used, the results were discarded if the EODs
obtained were significantly below those of embryos from the same

uterus investigated immediately after delivery. Single or repetitive EODs were evoked in anaesthetized animals by surface stimulation of the electric lobe with single shocks or repetitive stimulation after craniectomy (25,26).

RESULTS

Biochemical Findings

The results are plotted in Figs. 1a (electric lobes) and in
1b (electric organ).

Lobe tissue: AChE and ChAT are both present in the lobe tissue
of adult fish, but whereas the ChAT concentration is of the same
order of magnitude as that in the electric organ, the AChE activity
is much lower. Expressed in another way, the AChE to ChAT ratio
(which each activity is expressed in common units, e.g. as nmol of
substrate transformed$\cdot$min$^{-1}\cdot$mg of protein^{-1}) is much lower in the
lobe (530) than in the electric organ (2600). This would be accounted for if in Torpedo, as in other systems, AChE while synthesized by and present in cholinergic neurons, is elaborated in
much greater amounts by the postsynaptic cholinoceptive cells. The
significance of the lobe AChE is examined further in the Discussion.

Perhaps the most interesting feature of the results with embryonic lobe is that cholinergic characteristics are not expressed
until 4-6 weeks after the lobe electromotor perikarya have differentiated and ceased dividing (at 20 mm; Ref. 20 and see Ref. 24).
ChAT is first detected at about 40 mm embryo length and rises
steadily to adult values at birth. This corresponds to the transition of the lobe neurons from a uni- to a multi-polar configuration
and to the appearance of polysomes in abundance in the cytoplasm.

Electric organ: Wheras in the electric lobe AChE and ChAT
rise together, with ChAT somewhat in the lead, AChE first appears
in the electric organ in significant amounts (> 1% of adult) at
about 55 mm and rises steadily to somewhat more than adult values
at birth, well ahead of ChAT; the latter appears to rise in a two-
step manner, but reaches only 50% of the adult value at birth.

nAChR is detectable in the organ as early as 28 mm (when it
is less than 5% of adult values); it remains at this low level until
about 55 mm and then begins to rise steeply to reach a maximum value
at birth. A 10% fall appears to occur thereafter as with AChE.

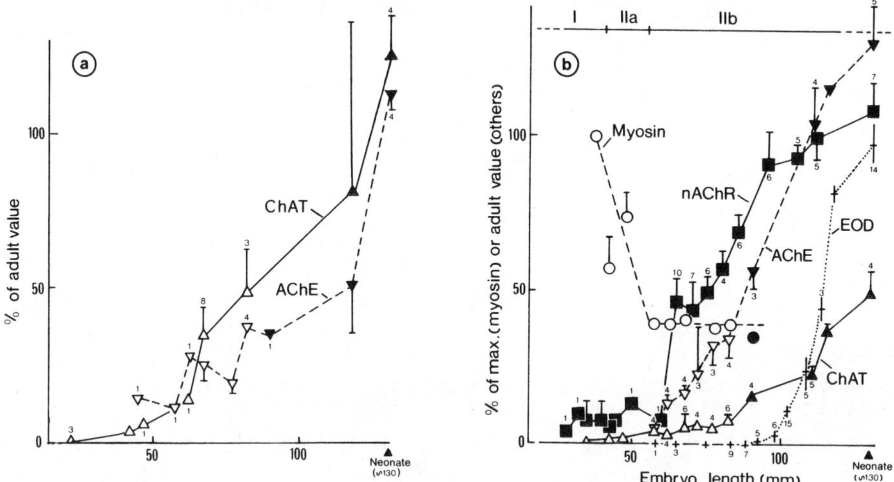

FIGURE 1: Concentrations of ChAT (triangles) and AChE (inverted triangles) in [a] lobe, [b] electric organ, and of myosin
heavy chain protein (circles), nAChR (squares) and EOD in
response to single stimuli (crosses) in [b], plotted as a
function of embryo length. Values are expressed as a percentage of maximum (myosin) or of adult values (others).
Points are observations on single embryos, or if more than
one, mean values; bars are the range (2 observations) or
SEM (more than 2) where these exceed the height of the
symbols; the small figures by the symbols indicate the
number of observations if these differ from 2. Open symbols, crosses indicate that values refer the material
pooled or results meaned from embryos varying ± 2 mm in
length from the value plotted. Maximum (or adult) values
(biochemical parameters are expressed per mg of protein)
were: myosin heavy chain, 40 µg; ChAT, 1.15 ± 0.17 (7)
(lobe) or 3.82 ± 0.39 (5)(electric organ); AChE, 612 ± 87
(7)(lobe) or 9930 ± 1280 (5)(electric organ) nmol substrate transformed.min^{-1}; nAChR, 85 ± 8 (12) pmol toxin
binding sites; EOD, 47.7 ± 1.2 (41) V. In [b] the length
ranges indicated by the Roman numerals are those during
which the following morphological take place: I, formation
of myotubes in the electric organ primordium ("myogenic
phase", Ref. 7) complete at 40 mm; II, differentation and
innervation of electrocytes ("electrogenic stage", Ref. 8)
comprising [a] horizontal flattening of vertically oriented myotubes with loss of myofibrils, complete at 55 mm,
[b] synaptogenesis, i.e. invasion of stacks of newly formed
electrocytes by ingrowing terminal axonal branches of electromotor nerves leading to the formation of mature synapses.

Myosin heavy chain component was investigated in the electric organ primordium beginning at 37 ± 2 mm when this is composed of myofibril-containing myotubes, and following through to 90 mm, when differentiation of the electrocyte is complete. There was a pronounced fall in the percentage of total tissue protein represented by heavy-chain protein between 37 ± 2 and 57 ± 2 mm, a period of growth during which the transformation of myotubes into primitive electrocyes (electroblasts, see Ref. 7) with loss of myofibrils occurs, to a new stable level. Since the protein content of the tissue did not change much during this time, the fall represents a true fall in the myosin heavy-chain protein content of the tissue.

Electrophysiological Findings

<u>Responses to a single stimuli</u>: When the electric lobes of an adult <u>Torpedo marmorata</u> are stimulated supramaximally a single EOD is evoked in both electric organs. This is a monophasic response representing the summed postsynaptic potentials generated in the individual, similarly orientated, electrocytes (13) in which the dorsal, uninnervated surface of each electrocyte, and therefore the dorsal surface of the entire organ, become positive with respect to the ventral. Its amplitude may reach 60 V, recorded in air (mean of 41 measurements, 48 ± 8 S.D.) and its duration is about 5 msec with a rising phase of 1.5-2.0 msec and a decay of 3-3.5 msec. EODs can also be evoked from late embryos which are similar to those recorded from adult fish but vary in amplitude and time characteristics with the size of the animals.

EODs in the microvolt range can be elicited in embryos of 60-70 mm length; these are not due to potentials in neighboring muscles as they are observed in isolated, innervated electric organs stimulated through one of their nerves. The single stimulus EOD increases in amplitude with embryo length until values comparable to those of adult are reached at 120-130 mm (Fig. 1b). Between 60 and 120 mm the increase of EOD with body length is exponential.

<u>Responses to repetitive stimuli</u>: It is known (26) that the single stimulus EOD is not a good indicator of the functional state of the electric organ. Therefore an attempt was made to study the fatiguability of the EOD in response to repetitive stimulation at different stages of development. This was felt to be of particular importance in view of the fact (evident from Fig. 1b) that in the younger embryos, ChAT concentrations rose more rapidly with bodylength than single stimulus EODs whereas in older embryos and newborn fish, such EODs had reached almost adult values at a point where ChAT had reached only 40% of adult values. One might, then, expect to find a greater fatiguability of the EOD in late embryos

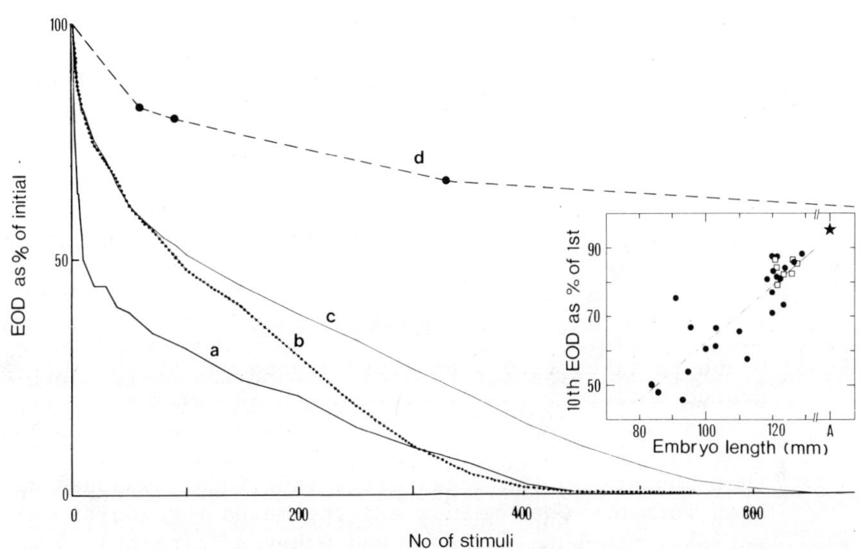

FIGURE 2: Typical decay curves for the EODs of embryos, new-born and
adult fish on repetitive stimulation at 5 Hz: embryos of
lengths a) 84, b) 122 mm; c) new-born fish of length
124 mm; d) adult. Initial EODs were a) 45 mV, b) 29,
c) 46, d) 32 V. Note the greater fatiguability of the
embryos and new-born compared to the adult. Inset: fatiguability (measured as the amplitude of the EOD recorded
at the 10th stimulus and expressed as a percent of the
initial EOD) as a function of body-length. Filled circles,
individual embryos; open squares, individual new-born;
asterisk, mean of 41 adults [A].

and new-born than in adults insofar as fatiguability is due to
depletion of transmitter stores rather than desensitization of
receptors (26).

The electric organs of 23 embryos (83 to 129 mm in length) and
7 new-born fish (121-127 mm body-length) were accordingly stimulated
via the electric lobe at 5 Hz frequency and the time course of the
decay of the repetitive EOD was followed. The results are presented in Fig. 2. It is clear that although single stimulus EODs
from late embryos and new-born fish may reach adult values, the

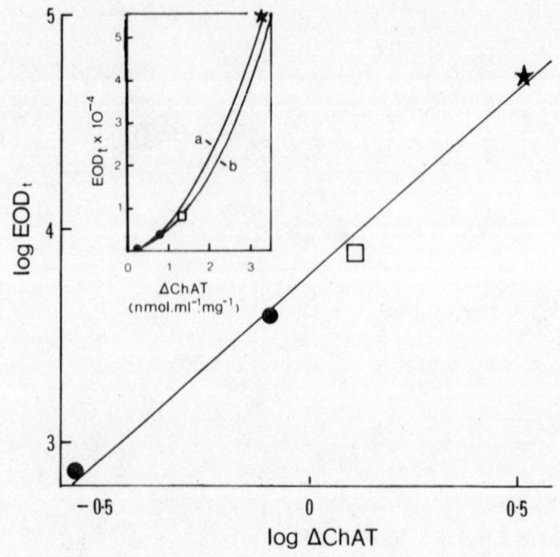

FIGURE 3: Graph to illustrate a possible dependence of the integrated output of the electric organ in response to repti-
tive stimuli (EOD_t) upon ChAT activity. Only "terminal"
ChAT (Δ ChAT) is considered, i.e. the additional ChAT
that appears in the organ after functional synapses have
been formed. ChAT values are the means for adults (asterisk), new-born (square) and embryos (circles) of the
same length for which measurements of EOD_t are available.
The inset diagram is a linear plot and the main diagram
a logarithmic plot of EOD_t vs Δ ChAT; the straight line of
best fit had r = 1.00, slope 1.73 and y intercept 3.8,
from which it may be deduced that EOD_t = 6300 $(\Delta \text{ ChAT})^{1.73}$
(insert, curve a). However, a quadratic relationship,
e.g. EOD_t = 5000 $(\Delta \text{ ChAT})^2$ (insert, curve b) fits the
results quite well.

fatiguability of the young organ is much greater than that of the
adult. This is shown, for example, by the smaller number of stimuli
required to reduce the EOD to a given percentage of the initial
value or by the greater fall in the EOD from the initial value after
a given number of stimuli. The latter measure of fatiguability
correlates particularly well with body-length (insert to Fig. 2).
However, a better measure of the functional state of the organ
which takes into account differences in the response to single
stimuli as well as differences in the fatiguability might be its
integrated output during a repetitive stimulus sufficient to reduce
output to (say) 1% of initial. In Fig. 3 this parameter is plotted
as a function of ChAT activity; the relationship, expressed

empirically, involves a power term; no doubt other factors are involved and may supervene in the smaller embryos.

DISCUSSION

Correlation of Chemical
and Morphological Events

We will now attempt to correlate the chemical events just described with the various stages of morphological development as recently defined in this laboratory (7,8,20).

<u>Myogenic and electrogenic phases</u>: The development of the electric organ has been divided into two main phases: in the first of these, referred to as "myogenic" (7), columns composed of myoblasts in the electric organ primordium fuse to form vertically orientated myotubes which contain prominent arrays of myofibrils; in the second, termed "electrogenic" (8) and beginning at about 40 mm embryo length, the myotubes lose their fibrillar structure and flatten to form electrocytes; the stacks of newly formed electrocytes are then invaded by axons terminating in growth cones, a process that eventually leads to the formation of mature synapses. The loss of myofibrillar structure and flattening of the myotubes is complete at about 55 mm. Using morphological criteria alone it is difficult to judge the conset of formation of mature synapses, but terminals having the morphological characteristics of those in the adult fish, i.e. a content of homogeneous clear spherical vesicles, some of which line the presynaptic membrane, glycogen granules and occasional mitochondria and apposed to junctional folds in the postsynaptic membrane, are seen with increasing frequency after about 70-80 mm embryo length.

<u>Myosin</u>: The fall in the content of myosin heavy-chain protein which we have observed in the electric organ primordium between 37 ± 2 and 57 ± 2 mm correlates well with the transformation of myotubes into electrocytes. Interestingly, myosin did not disappear from the organ entirely, but further work will be needed to decide whether it changes from a muscular to a non-muscular type and whether other typical muscle proteins are also affected.

<u>Receptor protein</u>: Our results show that synaptogenesis, seen morphologically as a continuous process of development, in fact takes place in two stages in which the formation of a receptor- (and esterase-) rich postsynaptic membrane largely precedes that of functional synapses. Thus nAChR, present up to 55 mm embryo length in

concentrations less than 5% of adult values, rises rapidly thereafter, reaching almost adult values at 90 mm, a stage of development in which EODs can be elicited only in the micro- or millivolt
range. Autoradiographic studies with radiolabelled α-bungarotoxin
(20) have indicated that toxin binding sites are, in fact, diffusely distributed in the plasma membranes of the myotubes and that
these become localized in the ventral membrane as soon as this
forms. Turnover studies and measurements of isoelectric points will
however be needed to decide to what extent localization of pre-
existing receptor occurs (as opposed to de novo synthesis) and
whether the receptor incorporated into the postsynaptic membrane
in the earlier stages of development differs from that of the later
embryo and adult in the same way as extra junctional, embryonic and
junctional receptor do in muscle.

ChAT: The ingrowth of axons derived from cell bodies which
have already begun to manifest their cholinergic character may account for the initial small rise of ChAT in the organ in the period
(60-90 mm) during which the postsynaptic membrane is undergoing
maximum biochemical differentiation. ChAT rises much more rapidly
in the electric lobe than in the organ. This is consistent with
the conventional notion that the site of synthesis of ChAT in cholinergic neurones is the cell body and that from there it is transported to the nerve terminals by axonal flow. This has not yet
been studied in embryonic electromotor fibers but is well established
to occur in the adult (4).

AChE: Since AChE rises more rapidly in the organ than it does
in the lobe it is clear that this enzyme is synthesized by the developing electrocyte. Histochemical studies (20) have enabled this
process to be followed. The extent to which the electromotor
neurones contribute to AChE at the synapse is unclear. The toxin
binding capacity of lobe tissue was undetectably low (< 1/10,000 of
that of the electric organ) so that the presence of cholinoceptive
cells within the lobe that might contribute AChE is unlikely;
furthermore, in the adult lobe, AChE is confined to the electromotor perikarya (23). However, axonal transport of this enzyme in
adult electromotor axons could not be demonstrated in vitro or in
vivo though transport was observed in control experiments with rat
sciatic nerve (4). A denervation study on the electric organ could
reveal the extent of the contribution of the presynaptic elements
to the total AChE of the organ but our attempts to do this have so
far been frustrated by the slow rate of nerve degeneration and the
attendant difficulty of keeping operated fish alive for the necessary
period of time.

EOD: The first electrophysiological responses correlate well,
as expected, with the appearance of mature synapses of adult type
in the organ and of appreciable amounts of ChAT. It is not clear

what significance is to be attached to the extremely small EODs
in the mV range obtained with embryos 60-80 mm in body length.
These may be due to the release of very small amounts of ACh from
growth cones which do contain vesicles of varying size (8) or to a
few mature synapses which have already formed at this stage. From
90-120 mm the single stimulus EOD continued to increase exponen-
tially reaching essentially adult values at the end of this phase,
corresponding to the proliferation of mature synapses to cover the
entire ventral surfaces of the electrocytes.

It is interesting that single stimulus EODs close to adult
values can be elicited at a stage of development when ChAT is less
than 50% of the adult value. For this reason, we examined the
response of the organ to repetitive stimuli, which gives a better
idea of its functional capacity. The results presented in Fig. 3
are consistent with the notion that the concentration of ChAT in
the nerve terminals is indeed a factor which limits output in immature or new-born fish. Presumably, in the shallow lagoons which
may serve as nurseries for the new-born, prey is available of a
size and strength which matches the discharge capacity of the
immature organ.

REFERENCES

1.   Babuchin, A. (1876): Arch. Anat. Physiol. Wiss. Med., 501-542.
2.   Blanchard, S.G., Quast, V., Reed, K., Lee, T., Schimerlik, M.I.,
     Vandlen, R., Claudio, T., Strader, C.D., Moore, H.-P.H, and
     Raftery, M.A. (1979): Biochemistry 18:1875-1883.
3.   Clark, D.G., Macmurchie, D.D., Elliott, E., Wolcott, R.G.,
     Landel, A.M. and Raftery, M.A. (1972): Biochemistry 11:
     1663-1668.
4.   Davies, L.P., Whittaker, V.P. and Zimmermann, H. (1977): Exp.
     Brain Res. 30:493-510.
5.   Ellman, G.L., Courtney, K.D., Andreas, V. and Featherstone,
     R.M. (1961): Biochem. Pharmacol. 7:88-95.
6.   Fonnum, F. (1974): J. Neurochem. 24:407-409.
7.   Fox, G.Q. and Richardson, G.P. (1978); J. Comp. Neurol. 179:
     677-697.
8.   Fox, G.Q. and Richardson, G.P. (1979): J. Comp. Neurol. 185:
     293-314.
9.   Fox, G.Q., Richardson, G.P., Tashiro, T. and Whittaker, V.P.
     (1977): Proc. Int. Union Physiol. Sci. 13:236.
10. Fox, G.Q., Richardson, G.P., Tashiro, T., Wachtler, K. and
     Whittaker, V.P. (1977): Proc. Int. Soc. Neurochem. 6:181.
11. Fox, G.Q., Tashiro, T., Wachtler, K. and Whittaker, V.P.
     (1977): Hoppe Seyler's Z. Physiol. Chem. 358:234.

12. Fritsch, G. (1890): Die Elektrischen Fische, Leipzig, von Veit.
13. Grundfest, H. (1957): Prog. Biophys. Biophys. Chem. 7:1-85.
14. Krenz, W.-D. (1978): Ph.D. dissertation, Gottingen, Federal
       Republic of Germany.
15. Lowry, O.H., Roseborough, N.J., Farr, A.L. and Randall, R.J.
       (1951): J. Biol. Chem. 193:265-275.
16. Mellinger, J. (1971): Bull. Biol. Fr. Belg. 105:165-218.
17. Mellinger, J., Belbenoit, P., Ravaille, M. and Szabo, T.
       (1978): Develop. Biol. 67:167-188.
18. Ogneff, J. (1897): Arch. Anat. Physiol. (Physiol. Abt.),
       270-304.
19. Palkovits, M. (1973): Brain Res. 59:449-450.
20. Richardson, G.P. (1979): Ph.D. Thesis, University of Sussex,
       United Kingdom.
21. Schaffner, W. and Weissman, C. (1973): Anal. Biochem. 56:
       502-514.
22. Schmidt, J. and Raftery, M.A. (1973): Anal. Biochem. 52:
       349-354.
23. Tsuji, S. (1977): Brain Res. 124:352-356.
24. Whittaker, V.P. (1977): Naturwiss. 64:606-611.
25. Zimmermann, H. and Whittaker, V.P. (1974): J. Neurochem. 22:
       435-450.
26. Zimmermann, H. and Whittaker, V.P. (1974): J. Neurochem. 22:
       1109-1114.

THE REGIONAL DISTRIBUTION OF ACETYLCHOLINE, CHOLINACETYLTRANSFERASE,

AND ACETYLCHOLINESTERASE IN VERTEBRATE BRAINS OF DIFFERENT

PHYLOGENETIC LEVELS

K. Wachtler

Institut fur Zoologie der Teirarztlichen Hochschule
3 Hannover 1, Germany

Comparative data on the occurrance and regional distribution
of acetylcholine (ACh) and its enzymes cholinacetyltransferase (ChAT)
and acetylcholinesterase (AChE) in the vertebrate brains have been
reported by several authors (for reviews see 3,4). Recent observations restricted to the determination of ChAT and AChE in a variety
of species can be found in several papers (6,10,13-16). In the
present paper an attempt to survey the regional distribution of
cholinergic components in vertebrates under phylogenetic aspects
is made. The data presented here were obtained from whole brains
and brain regions of 27 vertebrates, in most of which all three
components (ACh, Chat and AChE) were determined. In the selection
of species to be studied efforts were made to cover the main lines
of vertebrate evolution, which are indicated in Fig. 1.

MATERIAL AND METHODS

Twenty-seven species representing phylogenetic older and
younger stages of all eight vertebrate classes were studied either
for all three components (ACh, ChAT and AChE) or the enzymes only.
With the exception of those which are rare or difficult to obtain
(Chimaera, Polypterus, Lepidosiren, Caiman) at least five brains
were studied for each species.

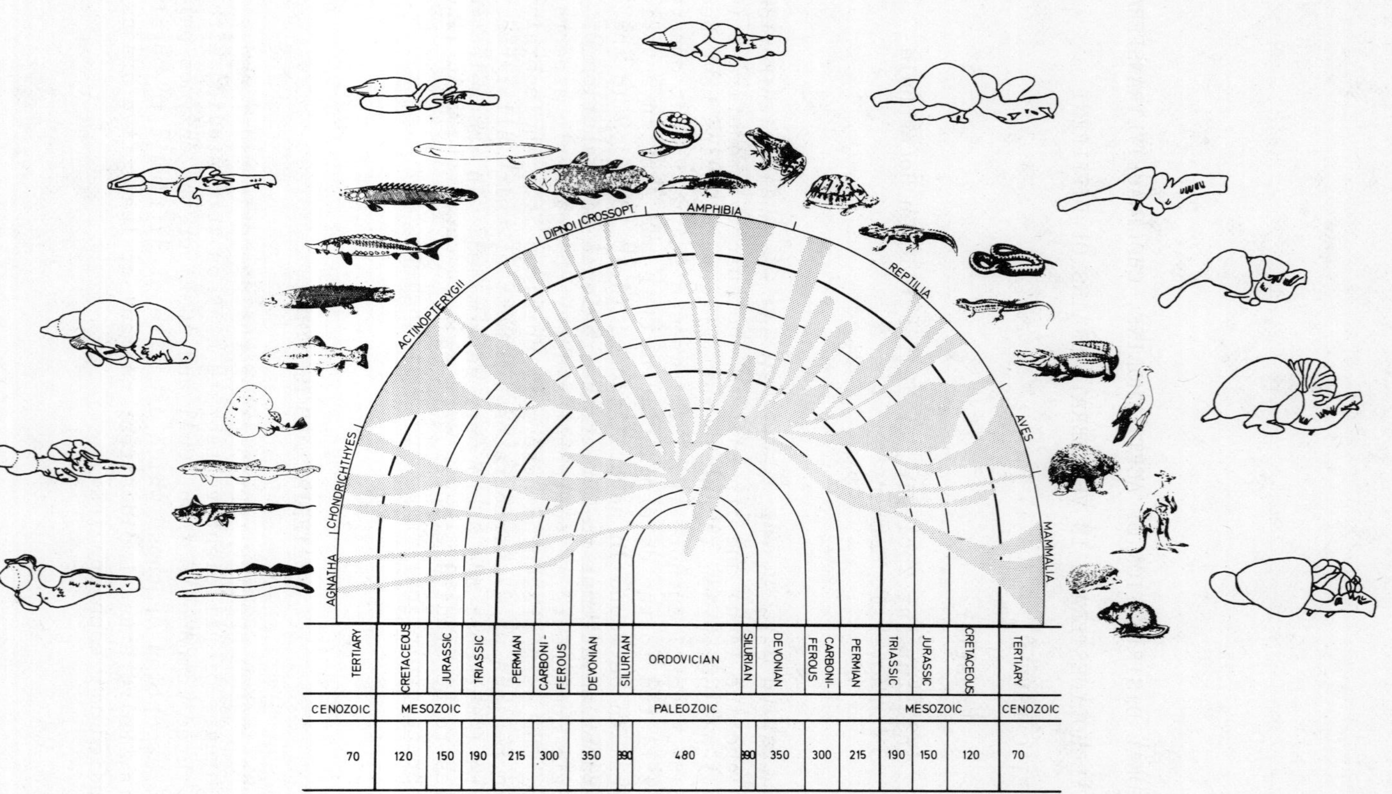

FIGURE 1: Survey of vertebrate phylogeny and side views of some representative brains (modified after several authors).

(1) Classis: AGNATHA
    Superordo: Petromyzontida
                 Petromyzon fluviatilis (lamprey)
    Superordo: Myxinoidea
                 Myxine glutinosa (hagfish)

(2) Classis: CHONDRICHTHYES
    Subclassis: Elasmobranchii
    Ordo: Selachii
                 Scyliorhinus canicula (dogfish)
    Ordo: Rajiformes
                 Torpedo maramorata (electric ray)
    Ordo: Chimaerae
                 Chimaera monstrosa

(3) Classis: ACTINIOPTERYGII
    Superordo: Chondrostei
                 Acipenser ruthenus x guldenstädti (sterlet)
    Superordo: Polypteriformes (bichirs)
                 Polypterus
                 Calamoichthys calabricus
    Superordo: Teleostei
                 Salmo gairdneri (trout)
                 Carassius gibelio (prussian carp)
                 Cyprinus (carp)
                 Anguilla (eel)

(4) Classis: DIPNOI (lungfish)
                 Lepidosiren paradoxa

(5) Classis: AMPHIBIA
    Ordo: Caudata (newts)
                 Cynops pyrrhogaster
    Ordo: Anura (frogs)
                 Rana temporaria
                 Xenopus laevis

(6) Classis: REPTILIA
    Ordo: Chelonia
                 Pseudemys scripta elegans (turtle)
    Ordo: Squamata
    Subordo: Lacertilia
                 Lacerta sicula (lizard)
    Subordo: Ophidia
                 Natrix natrix (snake)
    Ordo: Crocodilia
                 Caiman sclerops (caiman)

(7) Classis: AVES
    Ordo: Columbiformes
                 Columba livia (pigeon)
    Ordo: Psittaci
                 Melopsittacus undulatus (budgerigar)
    Ordo: Estrilidae
                 Taeniopygia guttata (Zebrafinch)

(8) Classis: MAMMALIA
    Ordo: Insectivora
                 Erinaceus europaeus (hedgehog)
                 Crocidura russula (shrew)
    Ordo: Rodentia
                 Rattus norvegicus f. domesticus (laboratory rat)
                 Mus musculus f. domesticus (laboratory mouse)

Dissection

All animals were killed by decapitation. Only one mammalian
species (Erinaceus) was previously anesthetized by an intramuscular
injection of 100 mg/kg ketanest (Ketalar: Parke-Davis) after this
injection had proved to be without detectable effect on the ChAT
activity in rats. This dissection was carried out in the cold room
at +5°C, the tissue was either directly transferred to chilled ACh
extractant (see below) or collected on pieces of alufoil kept on ice,
weighed and homogenized in cold 10 mM EDTA containing 0.5% Triton
X 100 (5). As immediate processing after homogenizing was not always possible all homogenates were stored in liquid nitrogen prior
to enzyme and protein determination for a period of up to several
weeks.

ACh determination

After quick removal the tissue was homogenized and extracted in
ice-cold acetone – 1 N formic acid 85:1 (10 x tissue vol) as re-
commended by Toru and Aprison (12). The homogenates were kept on

ice and centrifuged for 5 min ($\sim$ 10,000 g); either the whole supernatant or aliquots of it were dried in a vacuum centrifuge (1) and
stored in the cold room. The pellet was kept for protein determination. The extract was diluted in eserinized buffer before ACh determination on a leech muscle preparation (11) along with standards
of known ACh concentration. For control ACh samples were treated
with AChE.

ChAT was measured after the method of Fonnum (5) using ^{14}C-
acetyl-CoA in a final concentration of 0.4 mM. Ten µl of each incubation medium and homogenate were incubated for 30 min at 30°C in
7 ml scintillation vials (mini-vials). The reaction was stopped by
adding 2.5 ml iced 10 mM phosphate buffer (pH 7.4) followed by a
1 ml acetonitril containing 5 mg/ml Na-tetraphenyl-boron (Sigma) as
selective extractant of ^{14}C-ACh produced by the action of ChAT. After
the addition of 3.5 ml of Toluol-PPO-POPOP scintillator the radioactivity was counted in the organic phase using a Packard-Prias PLD
scintillation counter. ACh was determined photometrically according
to Ellman et al. (2). The protein content of the samples was determined after the method of Lowry et al (7).

RESULTS

Whole brain valus

As can be seen from Table 1 and Fig. 2, the three components
of the cholinergic system are detectable in all vertebrate brains
studied.

ACh: Ach was determined in amounts from 30 to 470 pmol/mg
protein. This is a range from highest (Myxine) to lowest (Melopsittacus) concentration of 15:1. In most species 100 to 300 pmol/mg
protein were observed. In cyclostomes, sharks, frogs and reptiles
concentrations of 200-300 pmol prevailed. An average of about 150
pmol was found in most fish and in mammals, whereas in bords only
30-80 pmol/mg protein were measured.

ChAT: The ChAT activity fluctuates more than the ACh content
and the AChE activity. Values between 11 pmol and 1.8 nmol/min/mg
protein were observed, which corresponds to a range of approximately
160:1 from highest to lowest activity (16). Fairly constant values
were found within closer related groups. In Torpedo, several modern
teleosts (in contrast to the ancient Actinopterygii) and Taeniopygia
extremely high ChAT activities were found (1 to 2 nmol/min/mg protein)

	ACH (nmol/mg Protein)			CAT (nmol/min/mg Protein)			ACHE (nmol/min/mg Protein)		
AGNATHA									
Petromyzon	(3)	0,242	± 0,028	(7)	0,054	± 0,01	(7)	14	± 2
Myxine	(4)	0,469	± 0,027	(4)	0,108	± 0,005	(4)	24	± 2
CHONDRICHTHYES									
Chimaera		–		(3)	0,383	± 0,017	(3)	22,3	± 2,3
Scyliorhinus	(3)	0,208	± 0,013	(4)	0,266	± 0,056	(4)	22,8	± 2
Torpedo	(1)	0,280		(4)	1,10		(4)	334	
ACTINOPTERYGII									
Acipenser	(3)	0,096	± 0,009	(5)	0,106	± 0,01	(5)	76,9	± 6
Calamoichthys	(3)	0,172	± 0,026	(3)	0,093	± 0,004	(3)	51,8	± 1
Polypterus	(1)	0192			–			–	
Salmo	(4)	0,115	± 0,006	(3)	1,81	± 0,03	(3)	135	± 7
Carassius	(2)	0,104	± 0,005	(3)	1,16	± 0,047	(3)	217	± 10
Cyprinus	(4)	0,178	± 0,007	(3)	1,63	± 0,081	(3)	286	± 8
Anguilla	(3)	0,103	± 0,007	(3)	0,714		(3)	124	
DIPNOI									
Lepidosiren	(1)	0,129		(1)	0,011		(1)	129	
AMPHIBIA									
Salamandra		0,162			0,03			73,3	
Triturus		–			0,013			142	
Cynops	(3)	0,117	± 0,016	(3)	0,032		(3)	204	
Rana		0,280		(4)	0,032		(4)	115	
Xenopus		0,236			0,055			133	
REPTILIA									
Pseudemys	(3)	0,325	± 0,017	(4)	0,286	± 0,036	(4)	62,8	± 5
Natrix		0,35		(1)	0,375		(1)	111	
Lacerta	(4)	0,239	± 0,017	(5)	0,42	± 0,05	(6)	240	± 50
Caiman		–		(4)	0,139	± 0,003	(4)	130	± 6
AVES									
Columba	(6)	0,081	± 0,007	(4)	0,36	± 0,03	(4)	166	± 10
Passer		0,078			–			–	
Taeniopygia	(3)	0,07	± 0,002	(3)	2,16	± 0,074	(3)	145	± 11
Melopsittacus	(3)	0,031	± 0,001	(3)	0,662	± 0,037	(3)	47	± 3
MAMMALIA									
Crocidura	(2)	0,159	± 0,015	(5)	0,216	± 0,04	(5)	74	± 10
Erinaceus		–		(3)	0,144	± 0,01	(3)	94	± 9
Mus		0,166		(4)	0,262	± 0,002	(4)	79	± 4
Rattus		0,115		(4)	0,24	± 0,011	(4)	76	± 3

TABLE 1: ACh, ChAT and AChE in whole brains of representative vertebrates (number of specimens in parenthesis, standard error to the right).

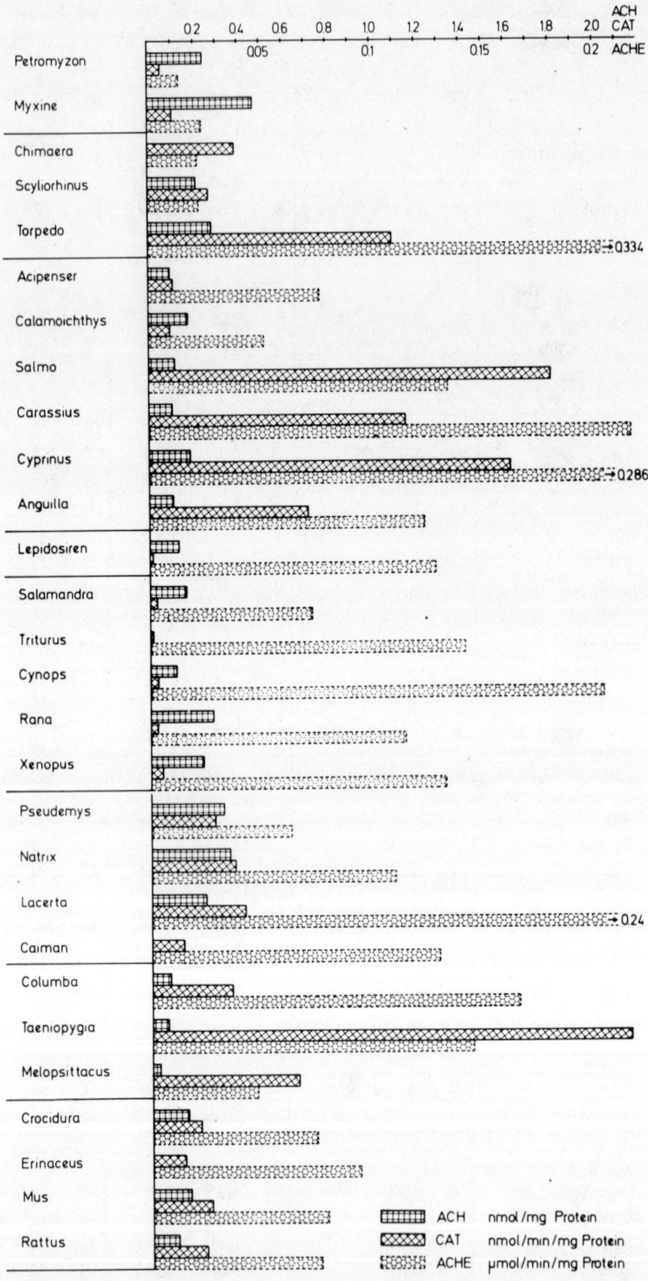

FIGURE 2: Same data as in Table 1, indicated by 3 columns for each species.

An average activity of 200-400 nmol/min/mg protein was found in
Scyliorhinus, Chimaera, the pigeon, reptiles and mammals. Extremely
low activities were observed in all amphibia (30-50 pmol and less),
in lamprey and lungfish.

AChE: The AChE activities determined varied relatively less.
Values of 14-334 nmol/min/mg protein were observed (that is approximately a ratio of 23:1 from highest to lowest values). Except from
the extreme value of Torpedo (334 nmol) values of around 150-250 nmol/
min/mg protein are characteristic for qhite a number of species. This
is the case for teleosts, amphibia, reptiles and most birds. Values
below 50-100 nmol were found in cyclostomes, chondrichthyes, ancient
actinotperygians and Melopsittacus, and were particularly consistent
in all four mammalian brains studied.

The analysis of whole brain values gives no indication of a
phylogenetic trend. There are, however, relatively consistent values
in closely related groups. The same ACh content of about 100-200
pmol/mg protein can go along with 1) both low ChAT and AChe activity as
in cyclostomes, 2) with low ChAT and average AChE activity as in
lungfish, amphibia, and some ancient actinopterygians, and 3) with
extremely high ChAT values and average or lower AChE activities as
in teleosts and some birds.

Regional analysis
(Tables 2,3; Figs. 3,4)

The diversity found in the whole brains makes a comparative
regional study difficult. For the ACh determination additional problems are posed by the loss of ACh during dissection, which is likely to show interspecific variation (due to varying AChE availability
and other possible factors).

As was pointed out by Hebb and Ratkovic (6), the regional values
are best compared on a relative basis. The regional amounts of the
components in percent of the whole brain amount divided by the regional protein percentage (ACh percent/protein percent; ChAT per-
cent/protein percent; AChE percent/protein percent) have proved to
be useful values (Table 3). From this presentation it can be seen
in which region the ratio is below or above an equal distribution
(the 1.0 ratio is indicated by a dotted line).

For some species the distribution pattern of ACh, ChAT and
AChE is illustrated by indicating the regional value in relation to
the whole brain average (which is taken as 100%) (Fig. 3). Regional
determinations in nmol/mg protein in twenty species are summarized
in Fig. 4.

		TEL	DIE	MES	MOB	CB
Petromyzon	ACH	0.109	0208	0248	0.322	-
	CAT	0.014	0025	0059	0.118	-
	ACHE	12	13	15	16	-
Myxine	ACH	0332	0.766	0251	0363	-
	CAT	0128	0.111	0122	0089	-
	ACHE	20	24	25	26	-
Chimaera	CAT	0026	0921	1.169	0425	0035
	ACHE	17	34	47	22	7
Scyliorhinus	ACH	0036	1307	0332	0331	0002
	CAT	0233	0372	023	0321	0174
	ACHE	20	39	22	15	18
Torpedo	ACH	0201	1903	0449	0194	0121
	CAT	0667	1192	119	134	0179
	ACHE	61	94	94	520	79
Acipenser	ACH	0024	014	0194	0133	0013
	CAT	0013	0067	0147	0266	0036
	ACHE	36	65	118	96	70
Calamoichthys	ACH	0038	0218	03	0269	0039
	CAT	0018	0054	0162	0208	0024
	ACHE	21	36	69	113	20
Polypterus	ACH	0023	014	0272	0465	0013
Salmo	ACH	0056	0164	0223	0072	0004
	CAT	0323	1882	2901	2624	014
	ACHE	58	199	168	111	101
Carassius	ACH	002	0088	0268	0131	0061
	CAT	0085	1308	1924	2001	0195
	ACHE	107	315	197	253	172
Cyprinus	ACH	0072	0345	0321	0211	0012
	CAT	0129	1904	2669	2374	0181
	ACHE	142	383	397	325	144
Anguilla	ACH	0018	0211	0173	0178	001
	CAT	0094	1052	2111	1101	0081
	ACHE	63	261	171	136	83
Lepidosiren	ACH	0026	0085	0262	0331	0194
	CAT	0004	0009	0016	0034	-
	ACHE	78	156	176	262	-

ACH nmol/mg Protein
CAT nmol/mg Protein · min
ACHE nmol/mg Protein · min

		TEL	DIE	MES	MOB	CB
Cynops	CAT	0015	0036	0048	0032	-
	ACHE	67	187	251	311	-
Rana	ACH	0128	0228	0563	025	-
	CAT	0016	0025	0046	0039	-
	ACHE	28	97	160	174	-
Xenopus	ACH	0085	0379	0575	0302	-
	CAT	0008	0065	0104	008	-
	ACHE	26	133	223	203	-
Pseudemys	CAT	0145	0176	041	0559	0139
	ACHE	61	48	97	81	27
Natrix	CAT	0245	0374	048	0404	-
	ACHE	87	106	126	125	-
Lacerta	ACH	0142	0292	0364	0302	0209
	CAT	0388	0426	0726	0654	0021
	ACHE	188	173	443	256	160
Cairnan	CAT	0078	0137	0235	0202	0042
	ACHE	84	85	137	115	228
Columba	ACH	0055	0155	0144	0172	0005
	CAT	027	04	089	085	0043
	ACHE	135	116	300	124	155
Taeniopygia	ACH	0059	0135	011	0167	0005
	CAT	1616	1642	5839	4319	0124
	ACHE	138	98	267	114	94
Melopsittacus	ACH	0026	0069	0053	0071	0003
	CAT	0596	0632	1504	1345	005
	ACHE	36	40	82	38	101
Crocidura	ACH	0169	0258	0244	0155	0014
	CAT	0251	0188	022	027	0149
	ACHE	63	52	70	79	106
Erinaceus	CAT	0178	0152	0159	0199	003
	ACHE	55	59	66	59	232
Mus	ACH	0139	0131	0084	0078	0019
	CAT	031	019	024	03	0046
	ACHE	86	68	95	89	32
Rattus	CAT	028	022	022	031	004
	ACHE	89	88	99	73	30

ACH nmol/mg Protein
CAT nmol/mg Protein · min
ACHE nmol/mg Protein · min

TABLE 2: ACh (nmol/mg protein), ChAT (nmol/min/mg protein) and AChE (nmol/min/mg protein) in their regional distribution.

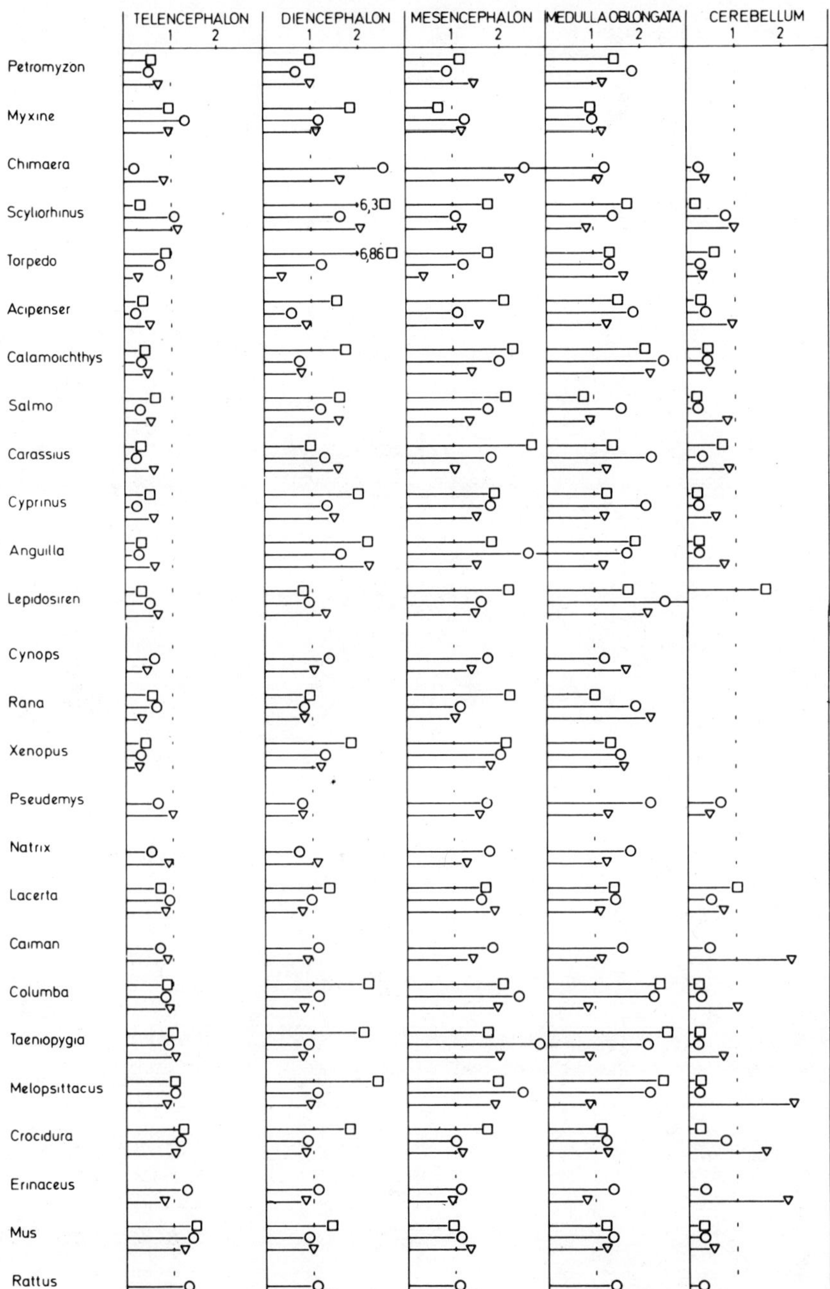

TABLE 3: Regional distribution of ACh (□), ChAT (O) and AChE (▽) given in ACh-percent/protein percent, ChAT-percent/protein percent, and AChE-percent/protein percent for each brain part.

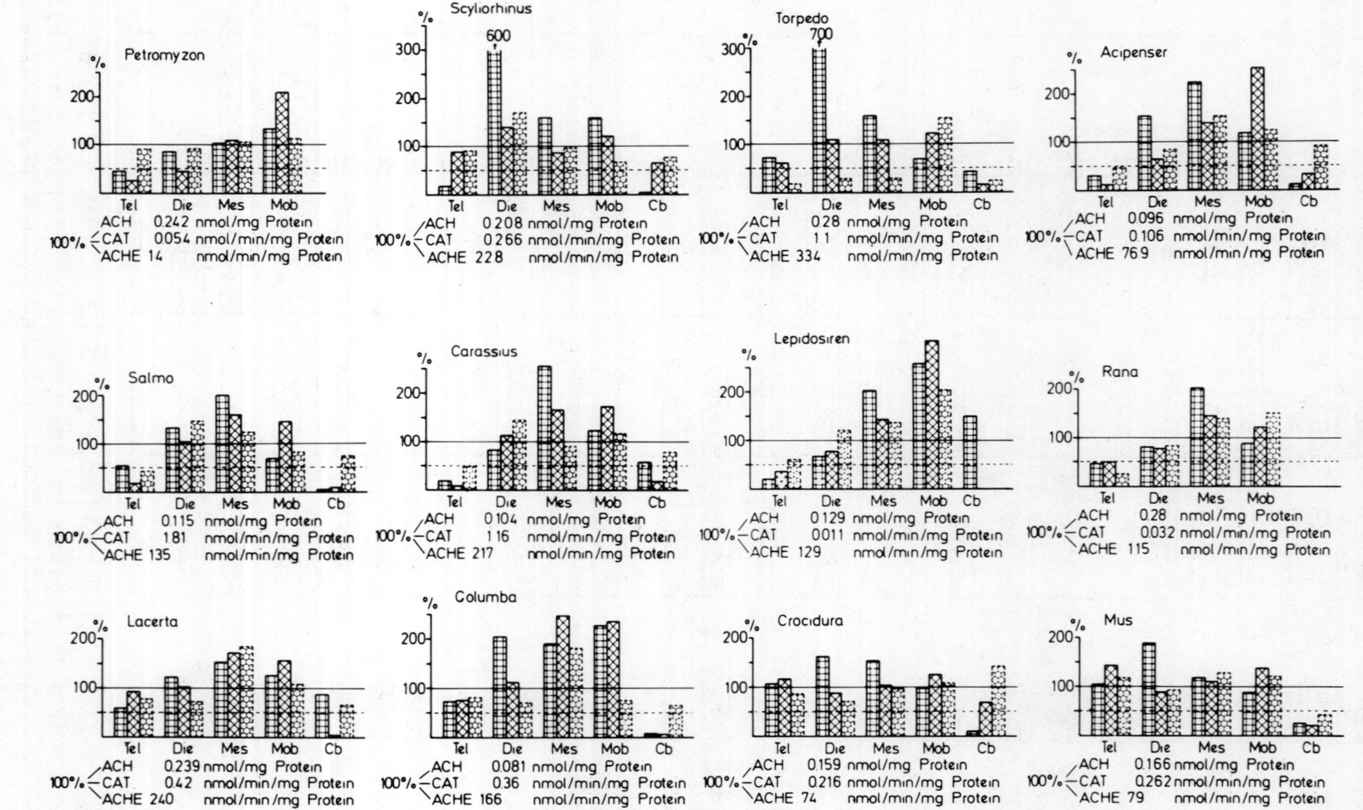

FIGURE 3: Regional activities in percent of the whole brain values (indicating the 100% line). The columns in each brain region indicate from left to right, ACh, ChAT and AChE.

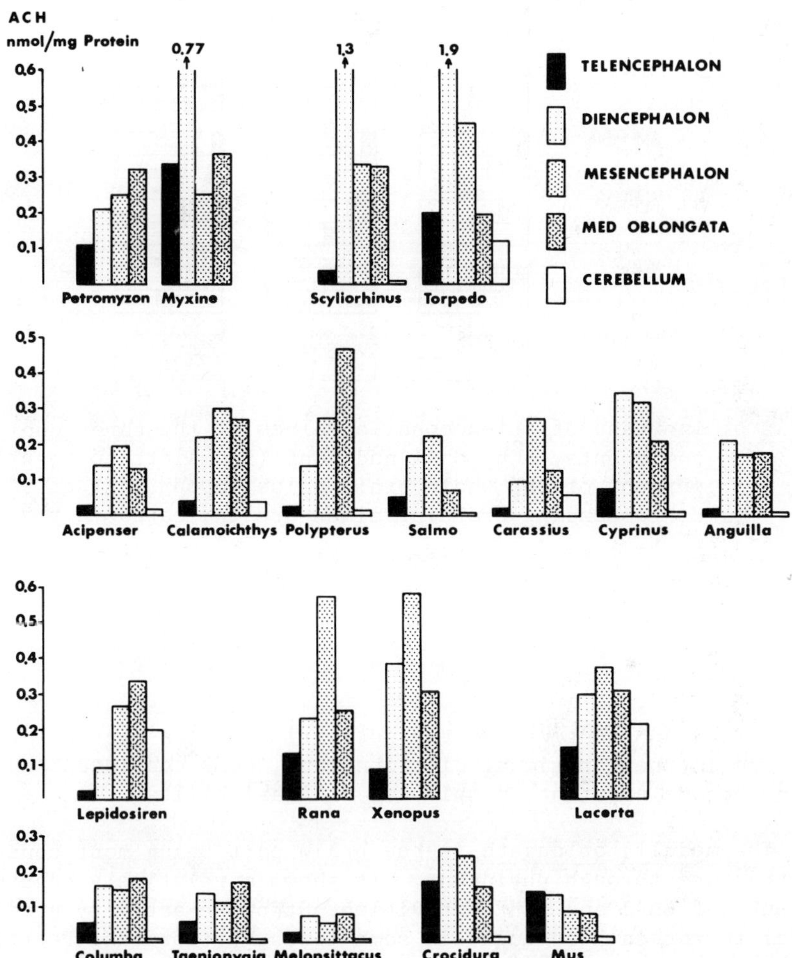

FIGURE 4: Regional ACh distribution in 20 vertebrates. There is good agreement within closer related groups.

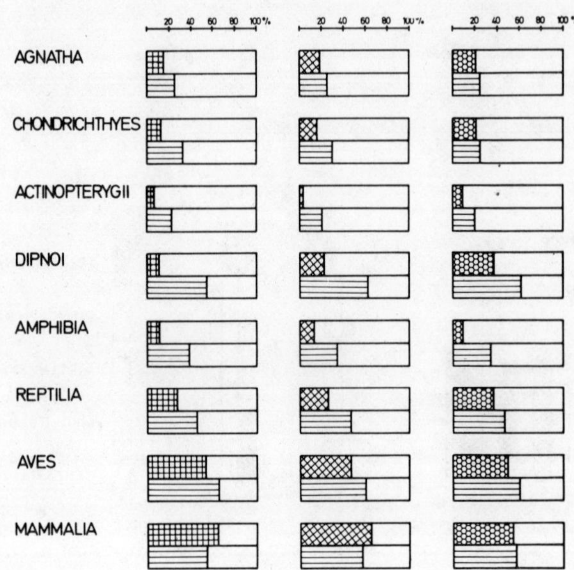

FIGURE 5: Increase of telencephalic values of the three cholinergic
          components, ACh, ChAT and AChE (from left to right) in
          percent of the whole brain value (upper part of each box).
          Corresponding protein values are indicated in the lower
          part of each box.

If the ACh content and the activities of its enzymes are
analyzed in their regional distributions, some consistent characteristics become apparent in spite of all variability.

The medulla oblongata is the brain part being most conservative
in structure through phylogeny. It shows a relatively constant ACh
content and ChAT and AChE activities. Stepwise activity gradient from
caudal to rostral was found in some lower vertebrates (Petromyzon,
Polypterus, Lepidosiren). The two dorsal differentiations on top of
the medulla oblongata, the cerebellum and the tectum are not only
specialized on different functions right from the beginning of vertebrate evolution, but apparently also rely consistently on different
transmitter systems. As far as the cholinergic system is concerned, in
most species only traces of it can be observed in the cerebellum. The
few exceptions, Torpedo, Lepidosiren and Lacerta, where cerebellar ACh
values of about 100 to 200 pmol were found need closer inspection.
Relatively high ChAT activity in the cerebellum was observed in some
sharks, and some teleosts. High cerebellar AChE activities were found

in some teleosts, the caiman and the budgerigar. In these cases no
corresponding ACh values were found.

The values for mesencephalon, the main optic and integration
center, particularly in lower verterates, are high in all classes and
often surpass those in the medulla oblongata (as in some fish; in frogs
and in lizards).

The two components of the prosencephalon, diencephalon and telencephalon, their function being primarily optic and olfactory, respectively, differ markedly in their use of ACh as transmitter. High
diencephalic activities were observed in all classes. In hagfish,
elasmobranchs and shrew, the diencephalon had the highest ACh content
in the whole brain. In the telencephalon, the region with the most
progressive evolution, low values prevail throughout the non-mammalian
vertebrates. The hagfish with its (in several respects) deviating forebrain structure, is the main exception to this rule. If the telencephalon values are compared with those of the remaining brain (on the basis
of their percentage of the whole brain value, corrected for the relative size) an increase of cholinergic activity can be observed with
increasing phylogenetic levels (Fig. 5). As far as can be concluded
from this admittedly rough approach it appears that every brain region
is characterized by a higher or lower proportion of neurons using ACh
as transmitter. As in comparative neuroanatomy a basic structural
design of the vertebrate brain can be recognized, this comparative
neurochemical approach reveals a basic ACh pattern which is kept
through all evolutionary changes. According to these observations
the main evolutionary changes in ACh distribution are found in the
telencephalon. If the ACh system plays an increasing role in the
telencephalon with increasing structural complexity, the question is
whether this increase can be ascribed to special system.s This is the
reason why we started to analyze the main subdivisions of the hemispheres in frogs, lizards and mice. These observations (Kortje and
Wachtler, in preparation) indicate that the relative contribution of
each subsystem to the total activity of the telencephalon is fairly
constant. The high cholinergic activity found in mammalian structures (basal ganglia, tuberculum olfactorium) can be demonstrated in
its amphibian and reptilian precursors. This is in good agreement
with Sakharov's concept (9) in which transmitter heterogeneity is
inherited from early homologous precursors.

The increase in cholinergic activity could be due to a structural elaboration of that fraction of neurons which already used ACh
at a lower level of differentiation. It would therefore be of special
interest to knwo whether in the telencephalic nuclei in question an
increased number of cholinergic receptor sites can be demonstrated
by binding studies. An increased dendritic arbcrization of telencephalic neurons from amphibia to reptiles and reptiles to mammals
has been shown in Golgi studies by Poliakov (8).

ACKNOWLEDGMENTS

I wish to thank Frau G. Jarosch for skillful technical assistance and for providing the figures and tables, Deutsche Forschungsgemeinschaft (Schwerpunktprogramm Biochemie des Nervensystems) for financial support.

REFERENCES

1. Albers, R.W. and Lowry, O.H. (1955): Anal. Biochem. 27:1829-1831.
2. Ellman, G.L., Courtney, K.D. and Andres, V.A. (1961): Biochem. Pharmacol. 7:88-95.
3. Fischer, H. (1971): Handbuch der vergleichenden Pharmakologie XXVI, Springer-Verlag, Heidelberg.
4. Florey, E. and Michelson, M.J. (1973): IN Comparative Pharmacology: International Encyclopedia of Pharmacology and Therapeutics (ed) M.J. Michelson, Pergamon Press, Oxford.
5. Fonnum, F. (1975): J. Neurochem. 24:407-409.
6. Hebb, C. and Raktovic, D. (1964): IN Comparative Neurochemistry (ed) D. Richter, Pergamon Press, Oxford.
7. Lowry, O.H., Rosebrough, N.J. and Farr, A.L. (1951): J. Biol. Chem. 193:205-275.
8. Poliakov, G.I. (1964): J. Hirnforsch. 7:253-273.
9. Sakharov, D.A. (1974): J. Neur. Transm. Suppl. XI, 43-49.
10. Sugden, P.H. and Newsholme, E.A. (1977): Comp. Biochem. Physiol. 56C:89-94.
11. Szerb, J.C. (1961): J. Physiol. 158:8-9.
12. Toru, M. and Aprison, M.H. (1966): J. Neurochem. 13:1533-1544.
13. Verschbinskaja, N.A. and Leibson, N.L. (1965): IN Functional Evolution of the Nervous System (Russian), Nauka, Leningrad.
14. Verschbinskaja, N.A. and Leibson, N.L. (1966): Roc. Akad, Sci. USSR Biol. Ser. 5 (Russian), pp.750-756.
15. Wachtler, K. (1973): Vergleichend histochemische Untersuchungen zur Acetylcholinesteraseverteilung im Telencephalon von Wirbeltieren, Thesis, Hannover.
16. Wachtler, K. (1980): Comp. Biochem. Physiol. 65C:1-16.

CHOLINE UPTAKE AND CHOLINE ACETYLTRANSFERASE IN BRAIN

OF DEVELOPING RATS MADE HYPOTHYROID WITH PROPYLTHIOURACIL

R.N. Kalaria, A.M. Kotas, A.K. Prince, R. Reynolds
and P.T-H. Wong

University of London, King's College
The Strand, London WC2R 2LS, England

INTRODUCTION

The development of the high affinity uptake of choline (Ch) in
frontal cortex of rat brain, as measured using synaptosome (P_2) fractions, was reported to be several days in advance of the development of choline acetyltransferase (ChAT, EC 2.3.1.6) activity, although adult values of both were evident by approximately 24 days
of age (17). Of the ^{3}H-Ch newly accumulated in small slices of
frontal cortex by sodium dependent uptake, a larger fraction was
acetylated in samples from adult rats than in samples from 7-14 day
olds (2). The greater ChAT activity in brain of adult rats, which
is accounted for largely by the basic molecular form of the enzyme,
suggested that ^{3}H-Ch is diluted into a pre-existing pool of Ch that
is smaller in mature than in immature cortex. This interpretation
was supported by an earlier observation of decreases during development in the concentration of Ch in rat brain, albeit of whole brain,
concurrent with increases in ChAT activity (12).

The development of ChAT activity in whole brain of rats was
retarded in propylthiouracil (PTU) induced thyroid deficiency, while
the Ch concentration was larger, up to 20 days, than in comparable
samples from untreated rats (12). Retarded development of ChAT
activity was also reported in the cerebral cortex of thyroid deficient rats. This was tentatively ascribed to smaller concentrations of ChAT in the synaptosomal fraction of cortex rather than to
smaller quantities of nerve terminals (18). However, morphological
as well as biochemical evidence suggested that neuronal differentiation in brain is retarded in neonatal thyroid deficiecy (4).

73

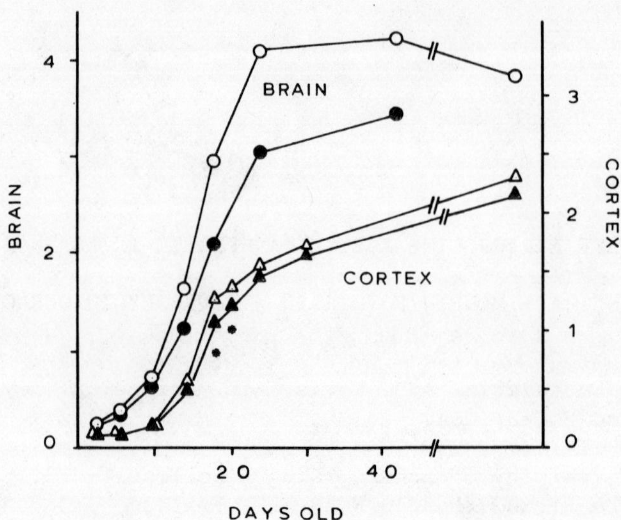

FIGURE 1: The effect of propylthiouracil on ChAT activity in brain
(whole brain minus cerebellum and brain stem) and cerebral
cortex of developing rats. Open and closed symbols represent brain (○ and ●) and cerebral cortex (△ and
▲) of untreated and PTU-treated rats, respectively.
Each point represents the mean value of independent assays
(μmol·hr^{-1}·g^{-1}) of ChAT from four animals (assays in quadruplicate), S.E.M. less than 2% of the mean. Brain ChAT
activity is significantly ($P < 0.05$) less in PTU-treated
rats than in controls from 14 days onwards, cortical
activity is significantly less only at days 17 and 20
($P < 0.05$).

Ch uptake has been taken as an indicator of the formation of
synapses by maturing cholinergic neurones (17). We therefore investigated its development in rat brain in neonatal thyroid deficiency.
The possibility was assessed of differential effects on ChAT activity
and Ch uptake in developing brain. The acetylation of ^{3}H-Ch newly
accumulated in brain slices taken from normally developing rats was
compared with that in samples taken from thyroid deficient neonates.

EXPERIMENTAL

Female Wistar rats with 3-4 day old litters (8-12 litter mates
of both sexes) and adult males (80-100 g) were fed for 50 days ex-

clusively with meal (41-B: Dixon & Sons, Herts, UK) containing PTU
(0.3% w/w). Controls were fed meal lacking PTU. Neonates were
selected randomly from three days of age, and adult males after
50 days treatment. Samples of cerebral cortex and corpus striatum
were immediately dissected, sliced (0.05 x 0.05 x approximately
1 mm) and suspended (10 or 20% w/v) in sodium free Tris-Krebs
solution at 0°C (19).

One sample of each suspension was treated with Triton X-100
(final concentration 0.5% v/v; 0°C, 2 hr) and then assayed for ChAT,
as described by Fonnum (9) (2 mM Ch; 110 μM ^{14}C-acetylCoA; 150 mM
NaCl; pH 7.0; 37°C). In some experiments (Fig. 1) dissected tissue
was homogenized (10% w/v in sodium-free medium) without prior slicing
then treated with Triton X-100 and assayed.

A second sample of the suspension of sliced tissue was assayed
for ^{3}H-Ch uptake. In all experiments, tissue from one animal was
dissected and assayed for uptake before a second animal was killed,
to minimize postmortem deterioration. In some experiments tissue
suspension (20-30 mg) was immediately incubated (37°C, 10 min) with
0.6 μM ^{3}H-Ch (12.5 μCi) plus 100 μM eserine salicylate in sodium
free medium and in Tris-Krebs medium containing 118 mM sodium (19).
Tissue was recovered and washed twice, with equal volumes of fresh
incubation medium, by centrifugation (1000 x g, 2 min) and then
extracted with formic acid-acetone mixture. Samples of the extract
were counted for ^{3}H and analyzed by high voltage electrophoresis
(6,19).

In other experiments the tissue was washed twice by centrifugation (1000 x g, 2 min) and by resuspension in sodium free medium
before being assayed. It was then immediately incubated with 0.5 μM
^{3}H-Ch (1 μCi) in the above media, recovered and washed by filtration
(pore size 1.2 μ: Millipore) solubilized with 2-ethoxyethanol (1 ml,
2 hr) and counted for ^{3}H in diotol scintillation fluid (10). Filter
paper blanks were measured under identical conditions. Prewashing
the tissue increased the measured total uptake of ^{3}H-Ch some 1.9-
fold. The filtration procedure improved the precision of the assays.
Tissue protein concentrations were assayed as described by Lowry et
al. (13).

Ch uptake was expressed in the following ways: TU, the total
uptake in the medium containing sodium; SFU, the uptake in the
sodium free medium; SDU, sodium dependent uptake calculated by correcting total uptake for the corresponding uptake in sodium free
medium (i.e. TU - SFU). Ch uptake and ChAT activities are expressed
throughout as values per gram tissue.

TABLE 1:   Example of the effects of propylthiouracil on development
in rat cerebral cortex and corpus striatum

| | Cortex | | Striatum | |
age	ChAT (μmol.h^{-1}.g^{-1})	3H-choline uptake* (nmol.h^{-1}.g^{-1})	ChAT (μmol.h^{-1}.g^{-1})	3H-choline uptake* (nmol.h^{-1}.g^{-1})
a: run 4		7.14 (0.45)		12.51 (0.69)
8 - control	1.07 (0.06)	2.16 (0.07)	6.71 (0.27)	2.33 (0.10)
		4.98 (0.48)		10.19 (0.62)
24 days				
		7.29 (0.35)		10.07 (0.47)
8 - PTU	1.04 (0.07)	2.28 (0.04)	3.80 (0.23)	2.19 (0.08)
		5.01 (0.35)		7.88 (0.46)
% deficit	-	-	43% (P < 0.001)	20% (P < 0.01)
				23% (P < 0.01)
		8.99 (0.46)		18.81 (1.58)
8 - control	2.05 (0.12)	2.45 (0.09)	10.83 (0.66)	2.39 (0.14)
		6.55 (0.39)		16.42 (1.53)
46 days				
		8.46 (0.59)		13.22 (0.75)
8 - PTU	1.79 (0.10)	2.14 (0.09)	6.07 (0.23)	2.33 (0.10)
		6.32 (0.56)		10.90 (0.76)
% deficit	-	-	44% (P < 0.001)	30% (P < 0.02)
				34% (P < 0.02)
b: run 6		5.64 (0.22)		14.50 (1.13)
6 - control	1.00 (0.03)	2.05 (0.06)	7.03 (0.31)	2.45 (0.17)
		3.59 (0.20)		12.05 (1.15)
24 days				
		6.08 (0.34)		10.85 (0.83)
6 - PTU	0.85 (0.03)	2.11 (0.13)	3.88 (0.09)	2.77 (0.17)
		3.97 (0.27)		8.08 (0.80)
% deficit	15% (P < 0.01)	-	45% (P < 0.001)	25% (P < 0.02)
				31% (P < 0.02)
		7.68 (0.24)		21.45 (1.01)
7 - control	2.46 (0.07)	2.24 (0.06)	13.90 (0.41)	2.17 (0.09)
		5.43 (0.26)		19.28 (1.05)
42 days				
		9.01 (1.75)		14.83 (1.65)
3 - PTU	2.33 (0.12)	1.89 (0.20)	7.20 (0.45)	2.32 (0.21)
		7.12 (1.55)		12.51 (1.50)
% deficit	-	-	48% (P < 0.001)	31% (P < 0.01)
				35% (P < 0.01)

Prewashed slices were incubated (20 mg tissue) with 0.5 μM ^{3}H-Ch,
then isolated and washed by filtration (see Experimental section).
Uptake and ChAT activities are mean values (S.E.M. in parentheses).
* Uptake is represented as the total in medium containing sodium
(upper figure), as sodium dependent uptake (lower figure), and as uptake in sodium free medium (middle figure). Where statistically significant, deficits caused by PTU are represented as percent of control
values. Deficits in uptake are given for total uptake (upper figure)
and for sodium dependent uptake (lower figure).

TABLE 2: Effect of PTU on ChAT activity and the metabolism of ^{3}H-Ch in corpus striatum of developing rats.

age (days)		ChAT (μmol.h^{-1}.g^{-1})	3H-ACh[*] (%)	3H-PCh[†] (%)
24	control (n = 4)	12.20 (0.80)	63.0 (3.6)	4.9 (0.6)
			85.6 (2.6)	-
	PTU (n = 4)	7.64 (0.74)[‡]	65.0 (3.9)	4.9 (0.1)
			83.2 (5.9)	-
50	control (n = 4)	9.56 (0.95)	74.9 (3.1)	0.9 (0.1)
			91.3 (3.3)	-
	PTU (n = 5)	5.21 (0.18)[§]	68.4 (3.0)	2.1 (0.3)
			90.5 (3.2)	-

Striatal slices were incubated (20 mg tissue) with 0.6 μM ^{3}H-Ch (12.5 μCi), then isolated and washed by centrifugation. Radiolabel accumulated by the tissue was analyzed by high voltage electrophoresis (see Experimental section). * Labelled ACh is represented as percent (S.E.M. in parentheses) total uptake (upper figure) and as percent sodium dependent uptake (lower figure).† Labelled phosphorylcholine (^{3}H-PCh) is represented only as percent total uptake. In this series of experiments there was no evidence that the PTU diet affected either total uptake, or sodium dependent uptake in striatum (but see Results section, and Table 1) or the fraction of ^{3}H-Ch taken up that was acetylated. ChAT activities were represented as the mean values (S.E.M. in parentheses), 4 to 5 animals. There was clear evidence of less ChAT activity (46% less) in striatum of PTU treated neonates than in samples from controls of comparable age. ‡ P < 0.02; § P< 0.01, Student's two tailed t-tests.

RESULTS

In preliminary experiments PTU treatment of adult male rats resulted in smaller concentrations of serum thyroxine (< 13 compared with 43.5 nmol$\cdot$1^{-1}, S.E.M., 3.5; thyroid wt. five times control) than in untreated controls. Gain in body weight was also retarded, although brain weight was not affected. PTU treatments starting with litters of 3 to 4 day olds also resulted in markedly smaller body weights throughout development (40 to 60% less than control weights). Effects on the cerebral cortex, however, were only apparent from 15 days of age, when weights were 10-20% less than control values. At this age, the weights of thyroid gland were 39.0 mg (S.E.M., 2.3) compared with weights of 17.3 mg (S.E.M., 1.1) in untreated controls of the same age. ChAT activity of brain reached maximal values in normally developing 24-30 day olds, and

PTU retarded this development. Activities were significantly less
than in samples from untreated neonates of the same age, from 14 days
onwards (Fig. 1) and remained approximately 20% less in 40 day olds.
In contrast, ChAT in the cerebral cortex was minimally affected by
PTU.

In further experiments, ^{3}H–Ch uptake and acetylation, as well
as ChAT activity, were assayed in cortex at 24 to 25 days of age,
when acetylation in normally developing neonates was close to maximal and at 40 to 50 days, when untreated rats exhibited the characteristics of cortical maturity (Table 3). The effect of the PTU
diet on the development of the corpus striatum was also assessed
using samples from 24 to 50 day olds. This tissue was not sufficiently well defined in 14 day olds for confident dissection.

In six independent series of PTU treatments ChAT activity in
the cortex was significantly less (10–20%) than in control samples
on five of the seventeen occasions on which is was assayed (including
the two preliminary experiments, see Fig. 1) while ^{3}H–Ch uptake was
significantly less (approximately 15%) than control values only once
in twelve occasions. In contrast, there was clear evidence that ChAT
activity in the striatum in PTU treatments was significantly less
(35–50%) than in samples from untreated controls on all nine occasions it was assayed. ^{3}H–Ch uptake in striatum, however, was
significantly less (20–40%) than control values on six of the ten
occasions it was assayed. When the more precise assay using filtration on Millipore filters was used (see Experimental section) this
latter evidence was more certain (Table 1). Although the development of striatal ChAT activity was clearly retarded by PTU, there
was no evidence that the fraction acetylated, of the ^{3}H–Ch taken up
into striatal slices, was different from the fraction acetylated in
samples from untreated controls of the same age (Table 2). Similarly
unchanged acetylations were apparent in cortical slices (Table 3).

Protein concentrations in cortex and corpus striatum were minimally affected by the PTU diet. Typical values at 24 days were
respectively (mg/g tissue)(S.E.M. in parentheses): 92.8 (1.6) and
114.7 (2.9) for the controls and 87.1 (2.2) and 114.4 (3.1) for
the PTU treated animals. At 42 days these values were 90.0 (1.4)
and 124.3 (1.8); 92.6 (2.9) and 118.4 (2.5).

In control experiments the effect of the PTU diet on cortex and
striatum of adult rats was investigated. Although these animals
displayed the characteristics of thyroid deficiency (ibid) there was
no evidence of differences between test and control values of ChAT
activity or ^{3}H–Ch uptake in cortex or striatum. The fraction of ^{3}H–
Ch taken up into cortex, that was acetylated was also unaffected by
PTU.

DISCUSSION

The development of ChAT activity in whole brain was retarded by the PTU diet (Fig. 1) as described previously (12). However, the effects on cortical ChAT activity were less clear (Fig. 1) than previously reported (18). This difference in results for cortex could well arise from the different procedures used to induce thyroid deficiency. In the previous work, rats born to females fed a low iodine diet during gestation were treated with 131I at day 1 of age. This procedure would seem likely to cause thyroid deficiency at an earlier stage of development than dietary PTU administered, as here, to lactating females when the litters are 3-4 days old (see Experimental section). The likely importance of the the time of onset of thyroid deficiency during development is thus immediately emphasized.

That the development of striatal ChAT was persistently retarded (Table 1) by PTU (although effects in the cortex were uncertain) was consistent with the observations that the development of whole brain ChAT was retarded in PTU treatments (12; and Fig. 1); striatum is some sixfold richer in ChAT than is cortex (Tables 1-3). As has been discussed previously (5) undernutrition may contribute to biochemical and neurochemical responses evoked by experimentally induced thyroid deficiency. Indeed, the development of ChAT and glutamate decarboxylase in all regions of neonatal rat brain was retarded by undernutrition initiated on the sixth day of pregnancy (15). In brain of 10 day old neonates, ChAT activities were 30-60% less than control values. However, even though nutritional deprivation was continued, the enzyme activities recovered to control values by 21 days of age. This contrasts with our present results, which always showed deficits in ChAT activity at 24 and 40-50 days, with no evidence of recovery (Tables 1 and 2). It seems clear, therefore, that the present retardation in the development of ChAT, and by implication the retarded development of ^{3}H-Ch uptake, caused by the PTU diet, is the result of thyroid deficiency.

In the cerebrum, postnatal neurogenesis is largely restricted to microneurones (1) with some possible exceptions. For example, in the caudate nucleus and nucleus accumbens septi, appreciable numbers of large neurones may be formed up to 2 to 6 days after birth (7). Furthermore, it is thought that thyroid deficiency induced soon after birth retards neuronal differentiation in whole cerebrum. without greatly affecting cell proliferation (3). If this is so, neurones still dividing after birth and reaching the stage of differentiation relatively late will be likely to be affected by thyroid deficiency induced by PTU after four days of age (see Experimental section). Neurones that have finished dividing and are closer to the stage of differentiation at birth, however, seem less likely to be at risk.

TABLE 3: ChAT activity and the metabolism of ^{3}H-Ch in cerebral cortex of developing rats.

age (days)		ChAT (μmol.h^{-1}.g^{-1})	3H-ACh[*] (%)[‡]	3H-PCh[†] (%)[‡]
14	control (n = 5)	0.91 (0.03)	31.8 (1.5)	12.5 (0.5)
			64.2 (4.8)	-
	PTU (n = 5)	0.93 (0.01)	35.0 (2.3)	11.5 (0.5)
			63.5 (6.0)	-
24	control (n = 5)	1.91 (0.11)[**]	54.4 (2.9)	1.9 (0.1)
			84.1 (4.8)	-
	PTU (n = 7)	1.71 (0.08)[**]	51.1 (2.4)	2.5 (0.1)
			83.0 (4.8)	-
50	control (n = 6)	1.50 (0.10)[**]	57.1 (1.6)	3.0 (0.2)
			81.9 (5.6)	-
	PTU (n = 6)	1.45 (0.06)[§]	56.8 (1.7)	3.4 (0.3)
			77.9 (3.1)	-

Cortical slices were incubated (14 day olds, 30 mg tissue; 24 and 50 day olds, 20 mg); with 0.6 μM ^{3}H-Ch (12.5 μCi). ChAT activity and the metabolism of ^{3}H-Ch were assayed and represented here * ‡ as described in Table 3. There was no evidence that the PTU diet affected either total or sodium dependent uptake, or the fraction of ^{3}H-Ch taken up that was acetylated. In this series of experiments there was also no evidence that the PTU diet affected cortical ChAT activity (but see Table 1b). ‡ Percentage conversions for 24 and 50 day olds differed significantly ($P < 0.05$) from values for 14 day olds. ChAT activities at 24 and 50 days were significantly greater than at 14 days: ** $P < 0.01$; § $P < 0.05$. Student's two tailed t-test used throughout.

The present observations that the PTU diet caused marked retardation of the development of ChAT and ^{3}H-Ch uptake in the striatum, while effects in the cortex are uncertain, support the above interpretation. Cholinergic neurones in the striatum are predominantly interneurones (11,14) and, significantly, the development of Ch uptake is more protracted in striatum than in, for example, cortex or hippocampus (17). Cholinergic neurones in neonatal striatum, therefore, would seem likely to be at risk from the relatively late onset of thyroid deficiency caused by feeding PTU to lactating females. In contrast, there may be few cholinergic interneurones in cortex, so that, although their development may be retarded, the overall effect on ChAT activity and particularly on ^{3}H-Ch uptake would be difficult to detect.

One explanation of the impaired development of ^{3}H–Ch uptake in striatum may be that the formation of cholinergic nerve terminals is impaired. That thyroidectomy at 1 day old, with ^{131}I, retarded the development of glucose metabolism as well as protein synthesis, in the cerebrum of neonates, was interpreted as an effect on the differentiation of neurones (3). However, while ChAT activity was impaired in cortex, synaptosomal protein concentrations at control values suggested the number of nerve terminals was unaffected (18). The present results showed protein concentrations in the striatum changed little between 24 and 42 days of age, and were also unaffected by thyroid deficiency (see Results section), even though ^{3}H–Ch uptake, as well as ChAT activity, was impaired. This raises the possibility that, although the production of terminals was unaffected by the late onset of thyroid deficiency, increases in Ch uptake and in ChAT activity associated with the final maturing of cholinergic neurones were impaired.

During the development of cortex in untreated controls there occurs an increase, from 14 days, in the fraction of the ^{3}H–Ch taken up that is acetylated (Table 3). This increase is concurrent with increases in ChAT activity. However, ^{3}H–Ch uptake is approaching maximal values by 14 days (17), suggesting the production of nerve terminals is close to completion. Thus increasing concentrations of ChAT appear to control the acetylation of newly transported ^{3}H–Ch. However, increasing intraneuronal capacity for glycolysis during development (4,8) may be an important factor.

Thyroidectomy at 1 day old retarded the development of glycolysis in rat cerebrum (3) and the normal development of cytoplasmic pacemaker enzymes in glycolysis was impaired by 35–65% (16). Furthermore, PTU induced thyroid deficiency retarded the development of ChAT in the striatum by as much as 48% (Tables 1–2). In spite of these limitations placed on development, there was no evidence that the acetylation of ^{3}H–Ch was affected (Table 2). Thus impaired ChAT activity and ^{3}H–Ch uptake may both reflect the presence of fewer cholinergic nerve terminals than in normally developing striatum at the same age. If so, these fewer terminals would seem likely to be unaffected in their capacity for uptake and acetylation of ^{3}H–Ch. However, the present results do not exclude the possibility that impaired ChAT activities and ^{3}H–Ch uptake reflect, at least in part, intraterminal deficiencies. Deficiencies in ChAT activity of the magnitude observed may not be sufficient to affect significantly the synthesis of ^{3}H–ACh by kinetically coupled uptake and acetylation.

SUMMARY

Thyroid deficiency induced in neonatal rats between 4 and 15 days of age caused markedly impaired development of ChAT activity and Ch uptake in the corpus striatum. In contrast, there were only a few instances of similar effects in the cortex, where uptake in particular was insensitive to the thyroid deficiency. Impaired development of this kind was evident at 24 and 40 days of age. This contrasted with the recently reported, transient effects of undernutrition on ChAT activities in all regions of rat brain, during the first 21 days of age. The differential effect on brain of the late onset of postnatal thyroid deficiency, seemed likely to reflect the timing of the differentiation of cholinergic interneurones in the striatum. Although the development of striatal ChAT activity was retarded by as much as 49%, the fraction of ^{3}H-Ch taken up that was acetylated was unaffected. The terminals, therefore, that were able to develop in the striatum seemed likely to be normal in capacity for uptake and acetylation of Ch. The alternative is of intraterminal deficiencies in ChAT activity.

REFERENCES

1. Altman, J. (1969): IN Handbook of Neurochemistry Vol. 2 (ed) A. Lajtha, Plenum Press, New York, pp. 137-182.
2. Atterwill, C.K. and Prince, A.K. (1978): J. Neurochem. 31: 719-725.
3. Balázs, R. (1971): IN Influence of Hormones on the Nervous System (ed) D.H. Ford, Karger, Basel, pp. 150-164.
4. Balázs, R. (1976): IN Perspectives in Brain Research: Progress in Brain Research (eds) M.A. Kormer and D.F. Swaab, Elsevier, Amsterdam, pp. 139-159.
5. Balázs, R., Patel, A.J. and Lewis, P.D. (1977): IN Biochemical Correlates of Brain Function and Structure (ed) A.N. Davison, Academic Press, New York, pp. 43-83.
6. Burgess, E.J., Atterwill, C.K. and Prince, A.K. (1978): J. Neurochem. 31:1027-1033.
7. Das, G.D. and Altman, J. (1970): Brain Res. 21:122-127.
8. Diamond, I. and Fishman, R.W. (1973): J. Neurochem. 21:1043-1050.
9. Fonnum. F. (1975): J. Neurochem. 24:407-409.
10. Herberg, R.J. (1960): Anal. Chem. 32:42-46.
11. Kataoka, K., Bak, I.J., Hassler, R., Kim, J.S. and Wagner, A. (1974): Exp. Brain Res. 19:217-227.
12. Ladinsky, H., Consolo, G.P. and Garattini, S. (1972): J. Neurochem. 19:1947-1952.
13. Lowry, O.H., Rosenbrough, N.J., Farr, A. and Randall, R.J. (1951): J. Biol. Chem. 193:265-275.

14. McGeer, P.L., McGeer, E.G., Fibiger, H.C. and Wickson, V. (1971): Brain Res. 35:308-314.
15. Patel, A.M. del Vecchio, M. and Atkinson, D.J. (1978): Dev. Neurosci. 1:41-53.
16. Schwark, W.S., Singhal, R.L. and Ling, G.M. (1972): J. Neurochem. 19:1171-1182.
17. Sorimachi, M. and Kataoka, K. (1975): Brain Res. 94:325-336.
18. Valcana, T. (1971): IN Influence of Hormones on the Nervous System (ed) D.H. Ford, Karger, Basel, pp. 174-184.
19. Wong, P.T-H. and Prince, A.K. (1979): Neuropharmacology 18: 511-513.

MORPHOLOGICAL ANALYSIS OF CHICK EMBRYO HEART DEVELOPMENT

L. Sisto-Daneo and G. Filogamo

Istituto di Anatomia Umana Normale
Universita di Torino, Corso Massimo d'Azeglio 52
10126 Torino, Italy

INTRODUCTION

It is no longer open to serious doubt that motor root fibers
emerging from the neural tube come directly into contact with the
somitic mesodermic masses, and in the presence of neural crest
cells, exert an inductive influence on the determination of myoblasts (2). A determinant neural influence is also accepted with
regard to other muscular structures of non-somitic origin, such as
the intrinsic eye muscles. This, indeed, may be felt even earlier
than the present evidence shows.

Myocardium, on the other hand, poses a different problem;
evidence against neural determination can be cited. On the one
hand, there is the morphological and functional precocity of the
heart bud muscle cells: in the chick embryo they are actively contractile at the 8-somite stage (5). On the other hand, there is
the apparent lateness of the morphological relations between such
myocytes and the nerve fibers: these would seem to appear only
after the 96th hour (1).

Our preliminary investigations have been directed to a reex-
amination in the chick embryos of the relations between the mesodermic structures responsible for the development of the myocardium,
the nerve cells and the neural crest cells. Light and electron
microscopy were used with the aid of Karnovsky's technique for AChE.

RESULTS

At the 24th hour, Stage 6 HH (3), the mesodermic cells of the
presumable heart area, which extends forwards and backwards astride
the primitive node to the side of the median line, lying nearest to
the median line (7), are in contact with the cells of the neural
groove, which is already in the form of a tight U (Fig. 1). This
relation takes the form of distinct filopodia running across the
groove as far as the mesodermic cells (Fig. 2). There is no other
morphological differences between the contacted and non-contacted
mesodermic cells.

At the 27th hour, stage 8 HH, 7-8-somite, the heart bud consists of two contiguous endothelial tubes running caudally to the
omphalomesenteric veins. The buds lie cranially to the first somite
and ventrally with respect to the endoderm of the foregut. The
tubes are covered by the splanchnopleure, which form dorsally distinct angle with the somatopleure. As shown by Manasek (4) and
Viragh and Challice (8), the cells of the splanchnopleure, a long
section of which is thickened, have the appearance of what may be
presumed to be myoblasts, particularly at the level of the heart
tube's caudal segment.

At the 42nd hour, in 19-somite embryos (stage 13 HH) and at the
following stages, the splanchnopleure cells continue a stichotropic
ventral migration to cloak the endothelial tube; a positive Karnovsky reaction can be demonstrated in these cells (Figs. 3 and 4).

Nerve fibers were observed at the dorsal part of the dorsal
aortas in 52-hour embryos. In the same embryos, small groups of
globose cells were noted between the side wall of the aorta, the
anterior end of the somite, and the beginning of the splanchnopleure.

From the 9-10-somite stage onwards, the narrow corridor between
the foregut endoderm and the thickened splanchnopleure displays
individual or groups of elongated cells, with their major axis lying
lengthwise dorsoventrally. In addition, these cells, tested at the
42nd hour, display a positive Karnovsky reaction. Their feature
is similar to that of the neural crest cells found in the space
between the neural tube and the somite medial wall, emerging from
the crests themselves.

CONCLUSIONS

Since the first heart myoblasts contract at the 8-somite stage,
their determination is earlier than that of the somite musculature.

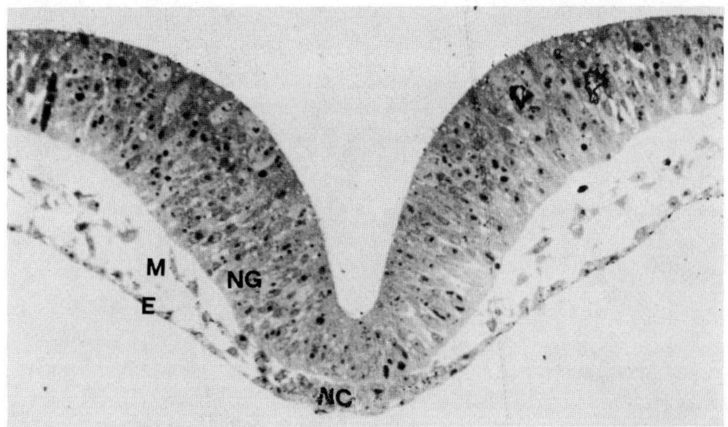

FIGURE 1: Cross-section through the presumable heart area in a
            24 hr chick embryo (stage 6 HH). Magnification x 1200.
            NG = neural groove; E = endoderm; M = mesoderm;
            NC = notochord

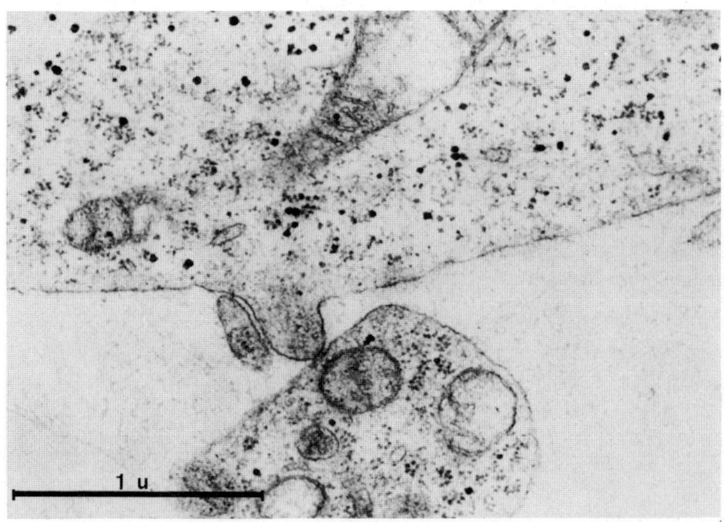

FIGURE 2: At last; Filopodia from the neural groove cells in con-
           tact with mesodermic cells x 48000.

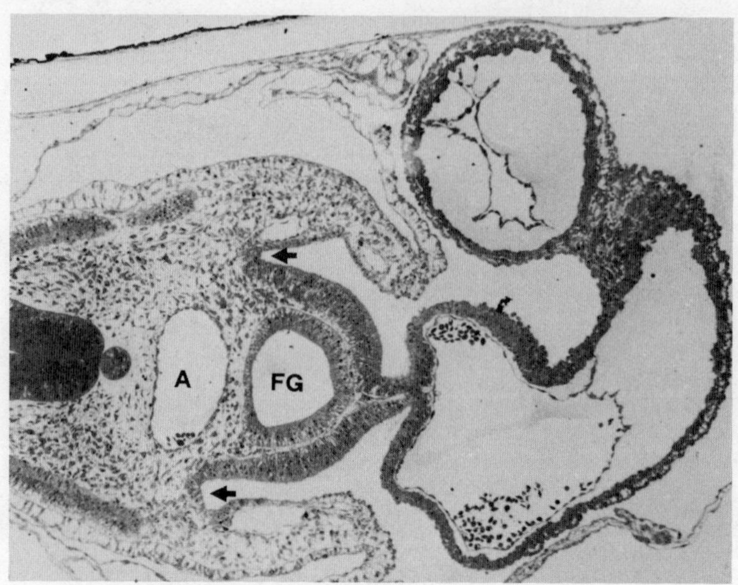

FIGURE 3: Cross-section through lowest 1/3 of heart bud in 48 hr
chick embryo (stage 16 HH). Magnification x 400. Arrows
point to angle between somatopleure (laterally) and
splanchnopleure (medially). FG = foregut; A = aorta

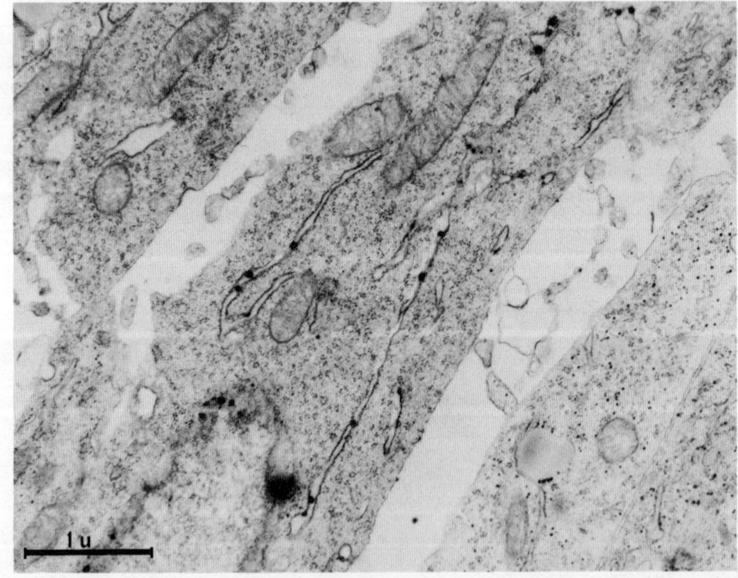

FIGURE 4: Section of 42 hr chick embryo (stage 13 HH) splanchnopleure through atrial portion of heart bud. Positive
intracellular Karnovsky reaction x 24000.

Our data do not show which factor determines the myogenic line, nor even the moment when such determination commences. Two facts, however, would seem to have been established.

In the first place, very early contacts are formed between the filopodia protruding from the neural groove cells and the mesodermic cells of the presumed heart area, which are destined to constitute the splanchnopleure covering the cardiac tube. Only a few hours later, the filopodia emerging from the neural tube determine the somitic musculature. In the second place, there are close territorial relations between the migrating neural crest cells and the thickened, already muscular, segment of the splanchnopleure. These crest cells, which may give rise to the conduction cells, will give rise certainly to part at least of the cardiac intraparietal nervous cells scattered in most of the heart wall by the third-fourth day (1,6). In our view they may be regarded, during their migration from the neural tube, as possible myogenic starters of the splanchnopleure.   These two facts are not mutually exclusive: a first nervous influence, followed by a second on the part of the crest cells, is an acceptable hypothesis. Further work is in progress to test it.

ACKNOWLEDGEMENT

This work was supported by the CNR.

REFERENCES

1.   Abel, W. (1912): J. Anat. 47:37-72.
2.   Filogamo, G., Peirone, S. and Sisto-Daneo, L. (1978): In
        Maturation of Neurotransmission, Karger, Basel, pp.1-9.
3.   Hamburger, V. and Hamilton, H.L. (1951): J. Morph. 88:49-92.
4.   Manasek, F.J. (1968): J. Morph. 125:329-366.
5.   Olivo, O.M. (1925): Arch. Exp. Zellforsch. Geweb. 1:427-500.
6.   Ramon, S. y Cajal (1929): In Etudes sur la neurogenese de
        quelques vertebres, Madrid, pp.19-69.
7.   Rudnik, D. (1937): Anat. Rec. 70:351-368.
8.   Viragh, S. and Challice, C.E. (1973): J. Ultrastruct. Res.
        42:1-24.

PLACENTAL CHOLINE ACETYLTRANSFERASE PURIFICATION AND PROPERTIES

C. Froissart and R. Massarelli

Institute of Biological Chemistry, Faculty of Medicine
and Neurochemistry Center of CNRS
11, rue Humann
67085 Strasbourg Cedex, France

INTRODUCTION

Several attempts have been made in the past to purify choline-
acetyltransferase (CAT) (EC. 2.3.1.6). Some authors (4,6,9,10)
have obtained a purified enzyme with a very high specific activity.
However, their preparations showed several bands in SDS polyacrylamide gel electrophoresis (PAGE), and they concluded that they were
heterogeneous. In addition, it was observed (6,9) that the enzyme
was not highly antigenic even though their immunserum, while not
monospecific, was able to inhibit up to 98% of the enzyme activity.
In contrast, other authors (1,12) have obtained an enzyme with a
very low specific activity compared to the previous group and have
concluded that their preparations were pure even though several
bands were observed on SDS PAGE (1). They found CAT to be highly
antigenic and claimed that their antisera were monospecific. However, only 50% of the enzyme activity was inactivated by the antisera. Some discrepancies wre also observed regarding the physicochemical properties of CAT (for review, see Ref. 7).

We tried to purify CAT from human placenta using a variety of
procedures. However, regardless of the final specific activity
(ranging from 0.174 to 6.6 µmol of acetylcholine (ACh) synthesized
per mn and per mg of proteins), SDS PAGE showed one to three major
bands and several minor ones (2). From these results we concluded
that either we were co-purifying a contaminant protein(s) which
could stick to CAT or that different forms of CAT exist _in vivo_ or
appear during purification.

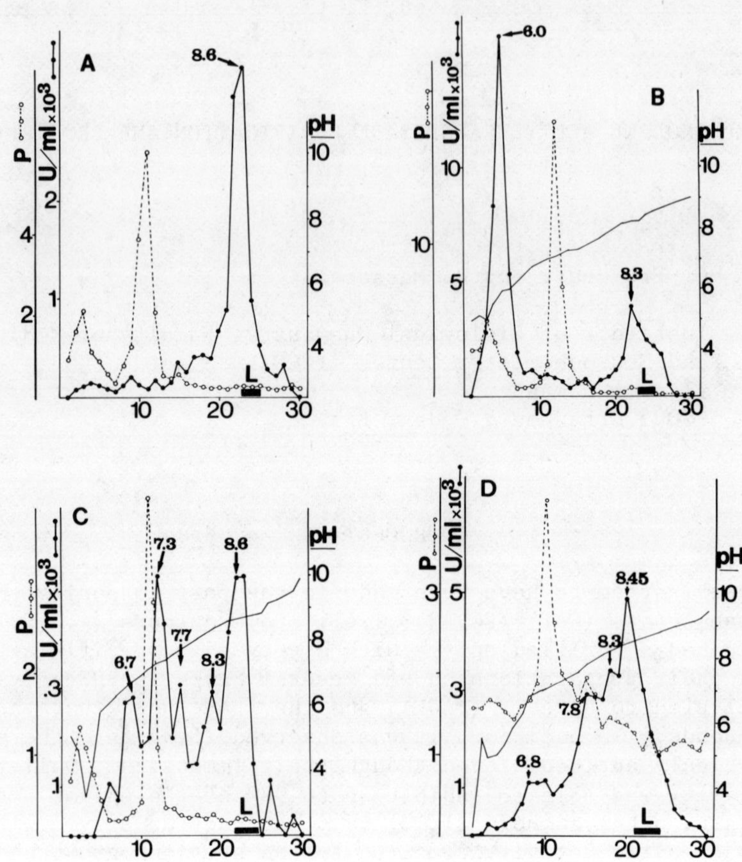

FIGURE 1: Isoelectric focussing of human placenta CAT. A) HSS (11 mg
           protein/ml).   B) Concentrated HSS (85 mg protein/ml).
           C) HSS (11 mg protein/ml) stored at 4°C for 24 hr.
           D) Partially purified CAT (homogenization, acidification
           to pH 5.1, ammonium sulfate precipitation 30-60%, dialysis,
           DEAE-cellulose) (3 mg protein/ml)· L) L = load region;
           P = protein expressed as arbitrary units; U/ml = μmol of
           ACh synthesized/mn/ml; arrows indicate significant pH.

TABLE 1: Physicochemical properties of CAT forms.

	Basic Form		Mixture of of Basic and Acidic Forms	
	+NaCl	−NaCl	+NaCl	−NaCl
Optimum T (°C)	37	37	45	45
Activation Energy (KCal/mol)	13.74	13.46	17.63	27.02
Optimum pH	8.0	8.3	8.6	8.5
Heat Inactivation	3.1 mn*	2.5 mn*	1.5 mn**	0.94 mn**
Inhibition by NVP 10^{-4} M	90%	--	98.5%	--
K_m AcCoA (µM)	55.5	2.3	19; 15***	1.1
K_m Ch (µM)	800	133	690; 440***	166
Reaction Mechanism	Sequential		Ping-Pong***	

The experiments were performed after isoelectric focussing (IEF).
Due to the reversibility of the phenomenon, after collection of
the acidic form from the IEF gel, a mixture of acidic and basic
form is obtained (3).
* 50% of enzymatic inactivation at 45°C
** 50% of enzymatic inactivation at 50°C
*** Values obtained after concentration of the mixture, i.e. they
     represent the kinetics of only the acidic form

It was thus decided to study the physicochemical characteristics of CAT from human placenta. The results have shown that some
properties may, at least partially, explain the discrepancies observed in the literature and the difficulties encountered in puri-
fying the enzyme.

MATERIALS AND METHODS

Human placenta at term, obtained from the Service de Gyne-
cologie et d'Obstetrique II (Hospices Civils de Strasbourg), were
kept at 4°C until homogenization (1-2 hr, maximum interval). Pieces
of placenta were homogenized in 5 vol of 10 mM Tris-HCl buffer,
containing 10 mM EDTA and 0.2 mM DTT pH 8.2 (buffer A), centrifuged
(105,000 g, 60 mn) and the supernatant (HSS = high speed supernatant)
collected. Aliquots of the HSS were used for isoelectric focussing
or they were previously concentrated by ultrafiltration in a dialysis
tube under vacuum or by ammonium sulfate precipitation and subsequently dialysed against buffer A. Preparative isoelectric

focussing and the enzyme activity measurement were performed as
prviously described (3).

RESULTS

The isoelectric focussing pattern of the HSS of human placenta
(11 mg protein/ml) showed only one molecular form of CAT with a
pI of 8.6 (Fig. 1A). When the same HSS was concentrated (85 mg
protein/ml) by various means (vacuum dialysis or ammonium sulfate
precipitation) without any loss of activity, a major form was observed at pH 6 (1B). This transformation of the enzyme from a basic
to a more acidic form was found to be exclusively dependent upon
the protein concentration (3), a reversible phenomenon (3), not
confined to placenta CAT since it was also observed for the rat
brain CAT (3) and was also found to occur with a partially purified
preparation (3). If the HSS (11 mg protein/ml) was stored at 4°C
for 24 hr, the isoelectric focussing pattern changed dramatically
(Fig. 1C), even though no significant loss (less than 10%) of enzyme
activity was observed. This phenomenon was also observed during
purification of the enzyme (Fig. 1D) and with rat brain CAT (2).
Moreover, these "aged" forms were irreversible.

What are these forms? Are they due to the action of proteases?
This possibility however, can be excluded for the acidic form
(pI 6.0), which represents a reversible phenomenon. Do they represent aggregates (homo- or heteroaggregates)? Do they represent
conformational changes of the enzyme? While the "aged" forms can be
considered as artefacts of the purification procedure, it is more
difficult to make the same conclusion for the form produced by concentration (pI 6.0) since the protein concentration needed to observe this form was equivalent to that _in vivo_.

This led us to study the physicochemical characteristics of
the acidic (pI 6.0) and the basic (pI 8.6) forms. The results are
summarized in Table 1. The most interesting differences concern
the kinetic properties. The acidic form showed a higher affinity
for the substrates than the basic one and had a ping-pong mechanism,
whereas the basic one had a sequential mechanism. Moreover, the
presence of salts (0.3 M NaCl) increased the V_{max} for both forms,
but decreased the affinity of the enzyme for the substrates.

DISCUSSION

The properties observed for the acidic form and the mixture of
both acidic and basic forms were similar to those reported by
Schuberth (11) while the basic form showed the same properties

except for the heat inactivation) reported by Hersh and Peet (5).

Our results show that the appearance of various forms of CAT and the differences observed among them may account for some of the discrepancies reported in the literature and might explain the difficulty encountered during purification, especially if the forms represent heteroaggregates.

Regarding the _in vivo_ situation it is difficult to state which form is present in the organ - the basic diluted form or the acidic one which approaches the protein concentration _in vivo_. Considering the differences in the kinetic characteristics of the two forms this might have little more than a philosophical interest. However, these differences may not have a great importance in the _in vivo_ regulation of ACh metabolism. Consequently, taking into consideration all relevant data, we may tentatively conclude that the metabolism of ACh may be regulated by mass action of the precursors and products, a suggestion which is now ten years old (8).

REFERENCES

1.   Chao, L.P. and Wolfgram, F. (1973): J. Neurochem. 20:1075-1081.
2.   Froissart, C. (1979): Doctorate thesis at University Louis
         Pasteur, Strasbourg, France.
3.   Froissart, C., Basset, P., Mandel, P. and Massarelli, R. (1979):
         FEBS Lett. 100: 276-280.
4.   Hersh, L.B., Coe, B. and Casey, L. (1978): J. Neurochem. 30:
         J. Neurochem. 30:1077-1085.
5.   Hersh, L.B. and Peet, M. (1978): J. Neurochem. 30:1087-1093.
6.   Malthe-Sørenssen, D., Lea, T., Fonnum, F. and Eskeland, T.
         (1978): J. Neurochem. 30:35-46.
7.   Mautner, H.G. (1977): CRC Crit. Rev. Biochem. Nov.:341-370.
8.   Potter, L.T., Glover, V.A.S. and Saelens, J.K. (1968): J. Biol.
         Chem. 243:3864-3870.
9.   Rossier, J. (1976): J. Neurochem. 26:543-548.
10.   Ryan, R.L. and McClure, W.O. (1979): Biochemistry 18:5357-5365.
11.   Schuberth, J. (1966): Biochim. Biophys. Acta 122:470-481.
12.   Singh, V.K. and McGeer, P.L. (1974): Life Sci. 15:901-913.

THE RELATIONSHIP BETWEEN CHOLINE UPTAKE AND ACETYLCHOLINE SYNTHESIS

IN A SYMPATHETIC GANGLION

B. Collier and S. O'Regan

Department of Pharmacology, McGill University
McIntyre Building, 3655 Drummond
Montreal H3G 1Y6, Canada

INTRODUCTION

Acetylcholine (ACh) synthesis in cholinergic nerve terminals
is regulated so that the rate of synthesis is slow when the nerve
terminal is at rest and is much accelerated when the nerve terminal
is actively releasing neurotransmitter. However, the mechanism by
which this control of synthesis rate is exerted is not known with
any certainty (see reviews, 23,36). The present work concerns the
role played by choline (Ch) delivery into the nerve terminal in
this process of ACh synthesis regulation; it is a direct extension
of the work reported in the proceedings of the previous meeting (10).

Ch delivery from an extracellular site to Ch acetyltransferase
within the preganglionic nerve terminals of the sympathetic ganglion
was clearly shown to be essential for ACh synthesis in 1961 (7). It
is now established that this process of Ch delivery is one of active
transport and it is usually considered that Ch uptake for ACh synthesis in cholinergic nerve terminals is a high affinity, sodium
dependent transport system (see reviews, 2,17,20,27,35). The importance of this Ch uptake mechanism to ACh synthesis cannot be denied
(despite, 18), but whether Ch transport activity, alone, regulates
ACh synthesis, as has been suggested (e.g. 3,15,19,33), is less
certain.

Our approach to this problem of the relationship of Ch delivery
to ACh synthesis is to compare changes in Ch transport activity to
changes in the rate at which ACh is synthesized in the cat's perfused
superior cervical ganglion. Ch analogs are used to assess changes

in Ch transport activity because they can be chosen such that, under the conditions of the experiment, they serve as substrates for the Ch uptake mechanism but not for Ch acetyltransferase. Thus, their use distinguishes changes in uptake from changes in metabolism (see 9-11), a situation not easily achieved using Ch itself. With the analogs, conditions can be identified that alter Ch transport activity; we can then present the tissue with Ch instead of analog, and test whether altered Ch transport activity is associated with altered ACh synthesis in the direction predicted if Ch uptake regulates transmitter synthesis.

METHODS

The experiments were done on cats anesthetized with chloralose (80 mg/kg i.v.) after induction with N_2O and halothane in O_2. Superior cervical ganglia were perfused (see 9 for method and references) with Krebs solution; the preganglionic nerve was cut low in the neck and stimulated when necessary at 20 Hz. When ACh release was measured, the perfusion fluid contained eserine (10^{-5} M), and in other experiments no anticholinestearse agent was used. Ch was added to the perfusion medium where indicated, its concentration was 10^{-5} M, unless indicated otherwise. When the accumulation of Ch analogs was measured, $[^3H]$-homocholine (2 Ci/mmol; 10^{-6} M) or $[^{14}C]$-triethylcholine (8.6 mCi/mmol; 10^{-5} M) was added to the perfusion medium. Ionic manipulations of the Krebs solution are indicated in the text; when Ba^{++} replaced Ca^{++}, $MgCl_2$ was used instead of $MgSO_4$.

ACh or Ch was measured in tissue extracts or in the ganglion's effluent (collected from the transverse prevertebral vein) by a radioenzymic method using Ch phosphokinase (14). Ch analogs accumulated by ganglia was corrected for metabolism: phosphates were measured as the proportion of radioactivity in an aliquot of tissue extract not extracted by tetraphenylboronate in heptanone; acetyl esters were tested for by measuring the proportion of tetraphenyl-boronate-extractable radioactivity that was not phosphorylated by Ch phosphokinase (see 9 for details).

RESULTS

A.   The relationship between Ch delivery and ACh synthesis
      during stimulation: The initial experiment (9) was to test whether preganglionic nerve stimulation accelerated Ch transport activity, as measured by Ch analog accumulation, as it should if Ch uptake regulates ACh synthesis and if stimulation enhances

TABLE 1:   Accumulation of Ch analogs by cat superior cervical
ganglia during preganglionic nerve stimulation.

Condition Stimulation	Percent Increased Accumulation During Stimulation		
	Homocholine	Triethylcholine	HBTM
10 Hz	104 ± 8	50 ± 5	
20 Hz	221 ± 10	111 ± 10	10 ± 5
20 Hz + HC-3	9 ± 5	-2 ± 5	
20 Hz + Tuboc	209 ± 9	108 ± 10	
20 Hz + Atropine		117 ± 8	
20 Hz + Na$^+$	3 ± 5		

HBTM = hydroxybutyltrimethylammonium
HC-3 = hemicholinium-3
Tuboc = d-tubocurarine

ACh synthesis. The test was clearly positive (Table 1): during
20 min stimulation at 20 Hz, analog accumulation is increased 2-3
fold. This increase is frequency dependent, and has the characteristics expected if it represents increased Ch transport into preganglionic nerve terminals. Thus (see Table 1), the phenomenon is
sensitive to hemicholinium, but not other drugs. It is not seen
with the Ch analog, 4-hydroxybutyltrimethylammonium, which is not
a substrate for the Ch uptake mechanism; and it is Na$^+$-dependent.

Although the results summarized in the paragraph above suggest
a physiological importance for enhanced Ch transport activity during
synaptic activity, increased Ch uptake does not always lead to increased ACh synthesis. This was first shown (Fig. 1) in the presence
of raised (18 mM) MgSO$_4$, a condition that allows stimulation induced
activation of Ch transport activity (part a), reduces ACh release
(part b), but does not produce an increase in tissue ACh level when
Ch is provided to the ganglion (part c). Thus, increased Ch uptake
does not necessarily lead to increased ACh synthesis, and a similar
situation is apparent (Fig. 2) when Ba^{++} replaces Ca^{++} in the medium
perfusing the ganglion. Barium supports Ch analog uptake (part a),
does not fully support ACh release (part b; see also 24,32), but,
again, increased Ch uptake does not result in increased ACh content
(part c).

The most obvious conclusion from the above results is that some
factor other than Ch delivery to Ch acetyltransferase plays a dom-
inant role in regulating ACh synthesis in ganglia stimulated under
the conditions described above. Thus, at least in ganglia, the

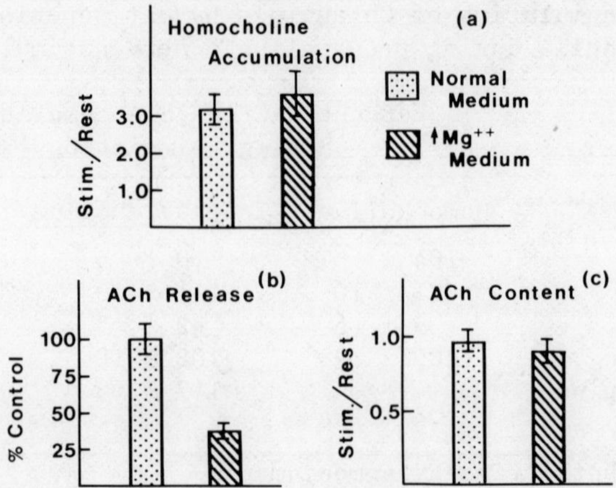

FIGURE 1: Homocholine accumulation and ACh turnover in ganglia perfused with high Mg++. a) The increased homocholine uptake
during stimulation; b) the release of ACh during stimulation; c) the ACh content of ganglia following stimulation
in the presence of Ch.

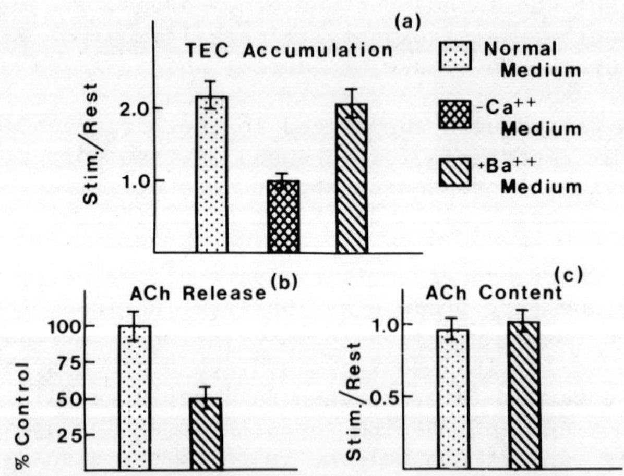

FIGURE 2: Triethylcholine accumulation and ACh turnover in ganglia
perfused with medium in which Ba++ replaced Ca++. a) The
increased triethylcholine uptake during stimulation; b) the
release of ACh during stimulation; c) the ACh content of
ganglia following stimulation in the presence of Ch.

precursor-product relationship between Ch and ACh levels suggested by Wurtman (see 37,41,42) appears not to hold. Wurtman's suggestion that Ch loading can alter tissue ACh levels is based mainly upon findings that Ch administered to rats results in increased ACh levels in certain areas of brain (see also 16,21), although this is not a consistent finding in normal animals (12,22,26,39). To test the effects of exposing ganglia to raised Ch levels, we have measured tissue ACh following perfusion with Krebs solution containing normal Ch (10^{-5} M), or raised Ch (10^{-4} or 10^{-3} M). Although increased Ch in the perfusion medium increased tissue Ch levels, it did not increase ganglionic ACh content in resting or stimulated ganglia (Table 2). Nor did increased Ch supply increase ACh release; this was measured in the presence of eserine, and release during perfusion with 10^{-4} M Ch was 95 $\pm$ 10% of that during exposure to normal levels of Ch.

Thus the experiments described in this first section consistently show that increased Ch delivery to preganglionic terminals does not increase ACh synthesis during rest or during 20 Hz stimulation.

B.   The relationship between Ch delivery and ACh synthesis following stimulation: These experiments were initiated for two reasons. First, there is considerable evidence that, in synaptosomes and certain other tissue (e.g. 1,25,30,38,40), depolarization is followed by a period of enhanced Ch uptake that in these preparations likely relates to transmitter replenishment. We wanted to test whether a post-stimulation increase of Ch transport activity was apparent in the ganglion of the cat. This tissue contains normal transmitter stores at the end of a period of stimulation (7) and does not need to replenish transmitter. However, and second, there is now appreciable supporting evidence for the early suggestion (29) that ganglia show a rebound increase in ACh levels following a period of conditioning stimulation (see 6,8,13), and we wanted to test whether this rebound ACh synthesis is dependent upon increased Ch transport activity; as in the earlier experiments, this was assessed using Ch analog accumulation.

For these experiments, both left and right superior cervical ganglia were perfused with Krebs solution containing Ch; one ganglion (the conditioned) was stimulated via its preganglionic nerves for 1 hr at 20 Hz, and the other ganglion (the control) was at rest. Following this, both ganglia were perfused, at rest, with medium containing [^{3}H]-homocholine. Ganglia that had been conditioned accumulated more Ch analog than did their contralateral controls (Table 3a), showing that conditioning leads to a post-stimulus increase of Ch transport activity. It should be noted that this post-stimulation effect on analog accumulation is smaller in magnitude

TABLE 2: Effect on tissue ACh and Ch levels of perfusing ganglia
          with Krebs medium containing increased Ch.

Ch Concentration	Content (test/control)	
	Ch	ACh
10^{-4} M	3.7	0.97
10^{-3} M	8.6	1.0

Control ganglion perfused with medium containing Ch (10^{-5} M).

TABLE 3: Post-stimulation increase of homocholine accumulation and
          of ACh content of ganglia.

Conditioning	Uptake	Percent Increase Over Control	
		Homocholine Uptake*	ACh Content**
a)Normal	Normal	70 ± 12	48 ± 10
b)Normal	-Na	0 ± 2	10 ± 8
Normal	-Ca**	75 ± 10	
Isethionate	Isethionate	79 ± 9	
c)-Ca^{++}	Normal	12 ± 9	-6 ± 4
Tubocurarine	Normal	74 ± 12	34 ± 6

Conditioning column describes composition of perfusion medium
during conditioning; uptake column describes composition of perfusion medium in the post-conditioning period.
* = Measured as accumulation during the post-conditioning period.
** = Measured following exposure of ganglion to Ch (10^{-5} M) during
      post-conditioning period.

than is that measured during stimulation; but it is not simply an
overshoot from the change that occurs during stimulation, because
the post-stimulus increase in homocholine accumulation is less when
measured during the first 6 min following the end of the period of
stimulation than when it is measured during 15 min following stimulation. Thus, the phenomenon appears to be one of rebound.

Some properties of the rebound increase in homocholine uptake
that follows conditioning are shown by the results summarized in
Table 3b. To test the Na$^+$-dependence of the effect, ganglia were
conditioned as described in the previous paragraph and homocholine
accumulation during the post-conditioning period was tested in a

medium containing LiCl instead of NaCl: enhanced accumulation was Na^+-dependent. To test the Ca^{++} dependence of the phenomenon, similar experiments were done except that homocholine accumulation was tested in a medium containing no added $CaCl_2$, and to which had been added EGTA: accumulation was Ca^{++}-independent. To test whether post-conditioning enhanced Ch transport activity was the result of stimulation induced after-hyperpolarization in nerve terminals, the effect of substituting isethionate for chloride was tested; this substitution increases the after-hyperpolarization, at least in unmyelinated nerve fibers (28,34), but it did not increase the post-stimulation increase of homocholine accumulation. Thus, the phenomenon, as measured, is Na^+-dependent, and Ca^{++}- and Cl^--independent.

The experiments whose results are summarized by Table 3c were designed to test whether the rebound increase in homocholine accumulation required transmitter to be released during the conditioning stimulation period. For this test, ganglia were conditioned during perfusion with a medium that was Ca^{++}-free; post-stimulation uptake of homocholine was measured in normal medium. Ganglia conditioned in the absence of Ca^{++} did not show the rebound increase of Ch transport activity; they accumulated no more homocholine than did their contralateral controls which were treated in exactly the same way as the test ganglia except they were not stimulated. The absence of the post-conditioning increase of homocholine accumulation in these experiments was more likely the result of the failure of transmitter release during conditioning than the result of the absence of an action of released ACh, because tubocurarine (3 x 10^{-5} M), which blocked synaptic transmission, did not alter the post-stimulation increase of analog uptake. Thus, it appears that transmitter release is necessary for the phenomenon of rebound increase in Ch transport activity to manifest itself, although we can not yet eliminate the possibility that it is Ca^{++} entry _per se_ rather than transmitter release during conditioning that is the essential factor.

The most important characteristic of the post-conditioning increase in Ch transport activity described above is that it appears to be associated with increased ACh synthesis. Thus, when conditioning was done in the way described above, and Ch was available to the ganglion during the rest period following conditioning, tissue ACh content increased (Table 3a), a result which is consistent with reports by others (6,8,13). Furthermore, the increased ACh synthesis following conditioning used extracellular Ch, because when the Ch available during this time was [3H]-labelled, much more [3H]-ACh was formed in the conditioned ganglia than in control ganglia (Table 4).

So far, under all experimental conditions tested, increased Ch uptake activity, revealed by the homocholine accumulation measures, has resulted in increased ACh synthesis when Ch is available (Table 3).

TABLE 4: Synthesis of [^{3}H]-ACh from [^{3}H]-Ch in ganglia following
          conditioning.

	Conditioned	Control	Difference
[^{3}H]-ACh (dpm)	36,480	1,280	35,200
Total ACh (pmol)	1,817	1,252	565
S.A. ACh (dpm/pmol)	20.1	1.0	62.3

Thus in the absence of Na$^+$ in the perfusate during the period
following conditioning, which abolished rebound homocholine uptake
(see above), no rebound ACh formed. Similarly, when transmitter
release was prevented during conditioning, which abolished rebound
homocholine uptake, no rebound ACh formed. Conversely, when transmitter action was prevented during conditioning, which did not reduce rebound homocholine uptake, rebound ACh did form. This last
result differs from one previous report (8), for reasons that are
not yet clear, although the present experiments used a somewhat
different experimental design.

DISCUSSION

The results described in section B above show a good relationship between increased Ch transport activity and increased ACh synthesis following a period of conditioning stimulation; in this
situation Ch uptake appears to regulate ACh synthesis. This contrasts with the results described in section A above, which showed
that, during stimulation, increased Ch transport activity is not
necessarily associated with increased ACh synthesis; in this situation Ch uptake appears not to regulate ACh synthesis. Thus, the
relationship between Ch uptake and ACh synthesis appears to change,
and to depend upon the functional state, or the past history, of
the synapse. The mechanism by which this change occurs can only
be speculated upon. It could involve an altered ionic balance within the nerve terminal, because continued stimulation is likely to
induce such and Ch acetyltransferase activity has been suggested to
be regulated by ionic changes (31). The present results would favor
intracellular Ca^{++} if the phenomenon is to have an ionic basis. Alternatively, the change could involve an altered intraterminal ACh
distribution; although, as emphasized above, tissue ACh levels in
ganglia are not altered at the end of the conditioning stimulation,
the intracellular compartmentation of ACh may be (see 5), and such
a change could alter factors that regulate transmitter synthesis.

Finally, the apparent altered relationship between Ch uptake and
ACh synthesis could involve the depletion of an endogenous regulator
of ACh synthesis, or the accumulation of an endogenous activator of
ACh synthesis, such as was postulated by Behrens et al. (4) to be
transported by axonal transport mechanisms. This suggestion was
made to account for their finding on skeletal muscle that "Rosenblueth's Phenomenon," which may be similar to rebound ACh formation
in ganglia, is dependent upon the length of the cut nerve.

The change in the relationship between Ch delivery and ACh
synthesis with altered functional state might indicate that the effect of raised plasma Ch on ACh turnover can similarly change with
functional state, and there is modest evidence supporting this
(see also chapter by Wecker, This Volume). First, London and Coyle
(22) have clearly shown a greater effect of administered Ch on
striatal ACh levels in rats with a partial striatal lesion induced
by kainic acid than on intact striatum. Second, we have preliminary
results showing that exposing conditioned ganglia to an increased
Ch concentration (10^{-4} M) produces a greater rebound ACh increase
than occurs in ganglia exposed to normal (10^{-5} M) Ch levels.
Whether therapeutic advantage can be taken of this is not yet clear,
but it is entirely possible that a disturbance of central cholinergic function might result in altered neuronal activity of
residual cholinergic neurones, and this might render them more
responsive to manipulation by precursor.

ACKNOWLEDGEMENT

This research was supported by the Medical Research Council
of Canada.

REFERENCES

1.   Barker. L.A. (1976): Life Sci. 18:725-732.
2.   Barker. L.A. (1979): IN Brain Acetylcholine and Neuropsychiatric
        Disease (eds) K.L. Davis and P.A. Berger, Plenum Press, New
        York, pp. 515-531.
3.   Barker, L.A. and Mittag, T.W. (1975): J. Pharmacol. Exp. Ther.
        192:86-94.
4.   Behrens, M.I., Lorenzo, D., Fernandez, O. and Luco, J.V.
        (1979): Brain Res. 179:37-47.
5.   Birks, R.I. (1974): J. Neurocytol. 3:133-160.
6.   Birks, R.I. and Fitch, S.J.G. (1974): J. Physiol. 240:125-134.
7.   Birks, R.I. and MacIntosh, F.C. (1961): Canad. J. Biochem.
        Physiol. 39:787-827.

8. Bourdois, P.S., McCandless, D.L. and MacIntosh, F.C. (1975):
     Canad. J. Physiol. Pharmacol. 53:155-165.
9. Collier, B. and Ilson, D. (1977): J. Physiol. 264:489-509.
10. Collier, B., Ilson, D. and Lovat, S. (1977): IN Cholinergic
     Mechanisms and Psychopharmacology (ed) D.J. Jenden, Plenum
     Press, New York, pp. 457-464.
11. Collier, B. and O'Regan, S. (1980): IN Ontogenesis and Func-
     tional Mechanisms of Peripheral Synapses (ed) J. Taxi,
     Elsevier-North Holland, Amsterdam, pp.65-76.
12. Flentge, F. and Van den Berg, C.J. (1979): J. Neurochem. 32:
     1331-1333.
13. Friesen, A.J.D. and Khatter, J.C. (1971): Canad. J. Physiol.
     Pharmacol. 49:375-381.
14. Goldberg, A.M. and McCaman, R.E. (1973): J. Neurochem. 20:1-8.
15. Guyenet, P.G., Lefresne, P., Beaujouan, J.C. and Glowinski, J.
     (1975): IN Cholinergic Mechanisms (ed) P.G. Waser, Raven
     Press, New York, pp. 137-144.
16. Haubrich, D.R., Wang, P.F.L., Clody, D.E. and Wedeking, P.W.
     (1975): Life Sci. 17:975-980.
17. Jope, R.S. (1979): Brain Res. Rev. 1:313-344.
18. Kessler, P.D. and Marchbanks, R.M. (1979): Nature 279:542-544.
19. Kuhar, M.J. (1979): IN The Cholinergic Synapse (ed) S. Tucek,
     Elsevier, Amsterdam, pp. 71-76.
20. Kuhar, M.J. and Murrin, L.C. (1978): J. Neurochem. 30:15-21.
21. Kuntscherova, J. (1972): Physiol. Bohem. 21:655-660.
22. London, E.D. and Coyle, J.T. (1978): Biochem. Pharmacol. 27:
     2962-2965.
23. MacIntosh, F.C. and Collier, B. (1976): IN Neuromuscular
     Junction (ed) E. Zaimis, Springer-Verlag, Berlin, pp. 99-228.
24. McLachlan, E.M. (1977): J. Physiol. 267:497-518.
25. Murin, L.C. and Kuhar, M.J. (1976): Molec. Pharmacol. 12:
     1082-1090.
26. Pedata, F., Wieraszko, A. and Pepeu, G. (1977): Pharmacol.
     Res. Commun. 9:755-761.
27. Pilar, G. and Vaca, K. (1979): IN The Cholinergic Synapse
     (ed) S. Tucek, Elsevier, Amsterdam, pp. 97-106.
28. Rang, H.P. and Ritchie, J.M. (1968): J. Physiol. 280:559-572.
29. Rosenblueth, A., Lissak, K. and Lanari, A. (1939): Amer. J.
     Physiol. 128:31-44.
30. Roskoski, R. (1978): J. Neurochem. 30:1357-1361.
31. Rossier, J., Spantidakis, Y. and Benda, P. (1977): J. Neuro-
     chem. 29:1007-1012.
32. Silinsky, E.M. (1977): Brit. J. Pharmacol. 59:215-217.
33. Simon, J.R., Atweh, S. and Kuhar, M.J. (1976): J. Neurochem.
     26:909-922.
34. Smith, I.C.H. (1979): J. Physiol. 294:135-144.
35. Speth, R.C. and Yamamura, H.I. (1979): IN Choline and Lecithin
     in Brain Disorders (eds) A. Barbeau, J.H. Growdon and R.J.
     Wurtman, Raven Press, New York, pp. 129-139.

36. Tucek, S. (1978): Acetylcholine Synthesis in Neurons, Chapman
        and Hall, London, pp. 1-259.
37. Ulus, I.H., Wurtman, R.J., Scally, M.C. and Hirsch, M.J.
        (1977): IN Cholinergic Mechanisms and Psychopharmacology
        (ed) D.J. Jenden, Plenum Press, New York, pp. 525-538.
38. Vaca, K. and Pilar, G. (1979): J. Gen. Physiol. 73:605-628.
39. Wecker, L. and Dettbarn, W.D. (1979): J. Neurochem. 32:961-967.
40. Weiler, M.H., Jope, R.S. and Jenden, D.J. (1978): J. Neurochem.
        31:789-796.
41. Wurtman, R.J. (1979): IN Choline and Lecithin in Brain Dis-
        orders (eds) A. Barbeau, J.H. Growdon and R.J. Wurtman,
        Raven Press, New York, pp. 1-12.
42. Wurtman, R.J. and Growdon, J.H. (1979): IN Brain Acetylcholine
        and Neuropsychiatric Disease (eds) K.L. Davis and P.A. Berger,
        Plenum Press, New York, pp. 461-481.

MULTIPLE MECHANISMS IN GANGLIONIC TRANSMISSION

N.J. Dun and A.G. Karczmar

Department of Pharmacology, Loyola University
Stritch School of Medicine
Maywood, Illinois   60153 USA

INTRODUCTION

Transmission in sympathetic ganglia is complex; in addition to
the fast excitatory postsynaptic potential (f-epsp) which represents
a primary synaptic pathway of the ganglia, there are several slow
synaptic potentials, both excitatory and inhibitory that may serve
to modulate the primary transmission; these include slow excitatory
and late slow excitatory postsynaptic potentials (s-epsp and ls-epsp,
respectively) and slow inhibitory postsynaptic potential (s-ipsp)
(3,14,18,19,23). Ganglionic transmission is also a multitransmitter
phenomenon. There is now evidence that in addition to the classical
transmitter, acetylcholine (ACh), a number of other substances may
serve as transmitters or modulators in the ganglia. In recent years,
immunohistofluorescent techniques have demonstrated the presence of
various peptides, such as substance P or a related peptide, and
enkephalins or enkephalin-like substances, in peripheral media
(cf. 28). We present here some of our findings indicating that
substance P and enkephalins may indeed participate in ganglionic
transmission.

SUBSTANCE P

In a number of sympathetic ganglia, such as the bullfrog sympathetic ganglia and inferior mesenteric ganglia of the guinea pig,
repetitive preganglionic stimulation elicits a slow depolarization,
ls-epsp, which lasts for seconds to minutes. Contrary to the other
ganglionic potentials, f-epsp, s-epsp and s-ipsp, the ls-epsp is

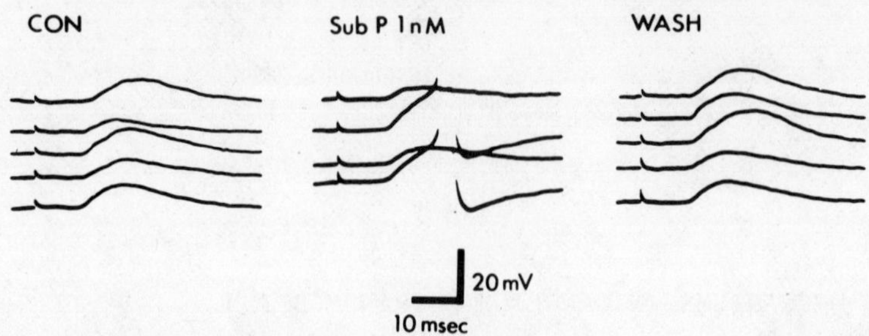

FIGURE 1: Enhancing effect of substance P on the amplitude of f-epsp
of an inferior mesenteric ganglion cell of guinea pig.
Consecutive subthreshold f-epsp's were elicited by preganglionic stimulation at 1 sec intervals. Substance P
(1 nM) was applied to the ganglion cell by superfusion
for 1 min which caused little change in resting membrane
potential but enhanced the amplitude of f-epsp to suprathreshold level in every other response. The f-epsp returned to subthreshold level after washing with Krebs
solution.

not abolished by cholinergic nicotinic and/or muscarinic antagonists;
accordingly, this ls-epsp is referred to as non-cholinergic potential (6,22,24). We have been interested in identifying the possible
transmitter(s) mediating the non-cholinergic potential, and in the
last few years we excluded a number of transmitter substances, such
as serotonin, histamine and gamma-aminobutyric acid (GABA) as
potential candidates. Our investigation of the possibility that
substance P may be the transmitter for non-cholinergic potential in
guinea pig inferior mesenteric ganglia was prompted by the finding
that substance P or a closely related peptide is present in high
density in nerve endings of these ganglia (11).

The membrane effects of substance P on single inferior mesenteric ganglion cells was studied using intracellular recording
techniques. Substance P (10-100 nM) when applied to the ganglion
cells by superfusing caused a slow membrane depolarization, which
was often accompanied by intense neuronal discharge (6). The depolarization was associated with either a decrease or an increase
in membrane resistance, which suggests that substance P can cause
membrane depolarization by two different ionic mechanisms. It is
of interest that non-cholinergic potentials also is associated with
two types of membrane resistance change (6,22). Thus, substance P
induced depolarization closely mimics the non-cholinergic ls-epsp.

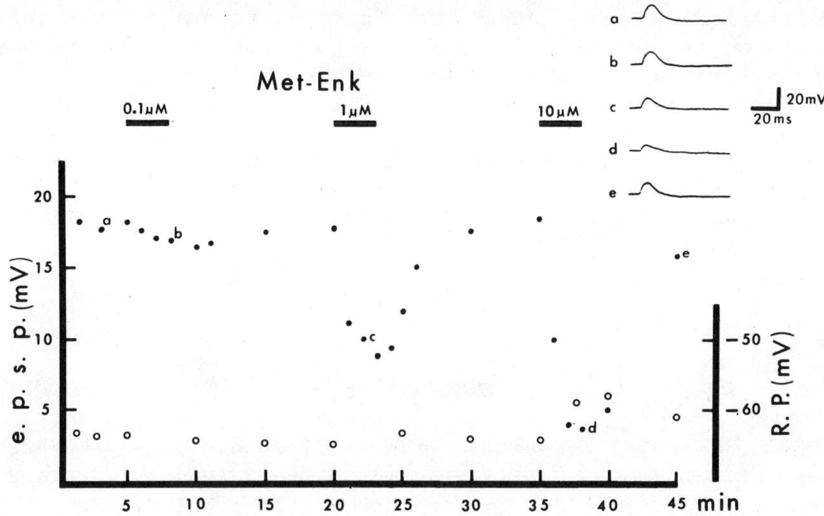

FIGURE 2: Effects of three different concentrations of met-enkephalin
on the amplitude of f-epsp and resting membrane potential
of a rabbit superior cervical ganglion cell. Met-enkephalin was applied to the ganglion cell for a period indicated by the solid bars. The solid and open circles
represent the amplitude of f-epsp and resting membrane
potential, respectively. Representative tracings taken
from the graph can be seen in the upper right hand column.

In addition, we found that desensitization of the ganglia with exogenously applied substance P prevented the generation of the non-
cholinergic potential. This indicates that the postsynaptic receptors mediating substance P depolarization and the ls-epsp are
pharmacologically similar (6).

In the course of our study, we found that substance P may have
an additional action on synaptic transmission, as it can potentiate
the f-epsp. Substance P in concentrations of 1-10 nM which caused
little or no change in membrane potential or membrane resistance
increased the amplitude of f-epsp, and often converted the subthreshold f-epsp into a suprathreshold action potential; one of
the typical experiments is shown in Fig. 1.

In summary, substance P may fulfill the following physiological
roles in sympathetic ganglia. First, when it is released in sufficiently large quantities, as in the case of repetitive stimulation

of preganglionic fibers, substance P may initiate a non-cholinergic
transmission as its application to the ganglion can cause neuronal
discharge. Second, when it is released in small quantities, substance P may modulate and enhance the response of subthreshold f-
epsp. It should be added that in other ganglia or species, ls-epsp
may be mediated by peptides other than substance P. For example,
luteinizing hormone-releasing hormone may be the transmitter mediating the non-cholinergic transmission in bullfrog sympathetic
ganglia (13).

ENKEPHALINS

The presence of enkephalin-like peptides has been demonstrated
by means of the immunohistofluorescent methods in nerve endings of
sympathetic ganglia (28). In this study, the rabbit and rat superior cervical ganglia were used.

The most conspicuous effect of met- or leu-enkephalins when
applied to the sympathetic neurons is a depression of the amplitude
of f-epsp. At concentrations of 0.1 to 10 µM met-enkephalin caused
from slight to very marked depression of f-epsp; only at the concentration of 10 µM, met-enkephalin was capable of inducing a
slight membrane depolarization (Fig. 2). The effects of met-enkephalin were fully reversible upon wash with drug-free Krebs solution. Met- and leu-enkephalins were about equally potent in de-
pressing the amplitude of f-epsp, and species difference between
rat and rabbit superior cervical ganglia was not detected. However, the depressant effects of enkephalins were observed in only
about 60% of the 30 neurons tested. Hence, only a limited number
of neurons in superior cervical ganglia are sensitive to the action
of enkephalins.

Synaptic depression caused by enkephalins could be due to
either a reduction of ACh output from nerve fibers or a decrease
of postsynaptic sensitivity to ACh. Enkephalins in concentrations
which reduced significantly the amplitude of f-epsp did not alter
the membrane conductance indicating that the attenuation of the
f-epsp is not due to a shunting effect of enkephalins on the postsynaptic membrane. In keeping with this finding it was noticed
that threshold level for initiation of ganglionic action potential
induced by intracellular stimulation was not changed in the presence
of enkephalins. An additional study for differentiating between
pre- and postsynaptic site of action of enkephalins was carried out
by recording simultaneously the f-epsp and ACh potentials elicited
by iontophoresis of ACh onto the surface of ganglion cells. In
seven neurons, enkephalins depressed f-epsps without affecting

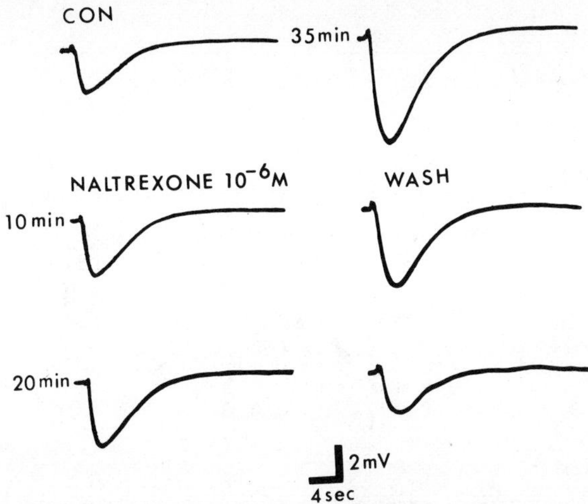

FIGURE 3: The effect of naltrexone (1 μM) on the amplitude of slow
surface potentials of a rabbit superior cervical ganglion
recorded by means of the sucrose gap method. The ganglion
was continuously superfused with a Krebs solution containing d-tubocurarine (0.1 mM) which suppressed the
nicotinic compound action potential. The slow potentials
were elicited every 5 min by repetitive preganglionic
stimulation (30 Hz, 1 sec). Naltrexone applied to the
ganglion caused a progressive increase of the amplitude
of P and LN potentials; the effect was reversible upon
wash.

responses to ACh or the resting membrane potential. Also, it should
be pointed out that the depressant effect of enkephalins was fully
prevented or reversed by opiate antagonists, naloxone or naltrexone.
It appears then that exogenous enkephalins reduce ACh release by
acting on specific opiate receptors located on preganglionic nerves.

The question whether endogenous enkephalins play a role in
ganglionic transmission was investigated by studying the effects of
opiate antagonists on synaptic transmission on the assumption that
if endogenously released enkephalins affect transmission, this effect would be unmasked by opiate antagonist. Naloxone (1 μM) or
naltrexone (1 μM) has little effectzon the f-epsp elicited by single
preganglionic stimulation. However, these opiate antagonists
markedly potentiated slow potentials in the majority of the

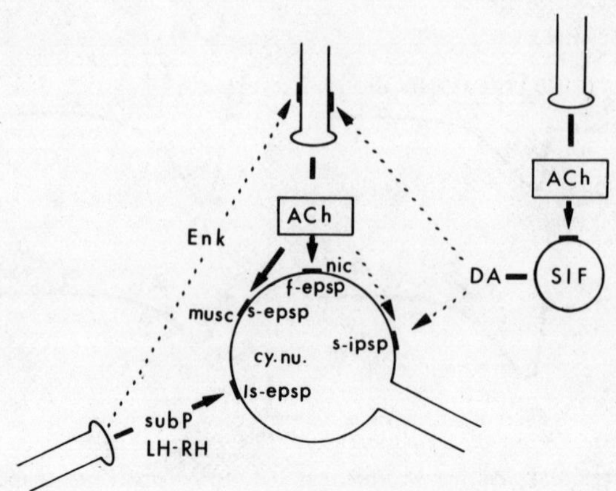

FIGURE 4: Diagram of the proposed synaptic functions of various
transmitters in a sympathetic ganglion. The solid and
broken line represent the excitatory and inhibitory action
of a putative transmitter, respectively. ACh = acetylcholine; DA = dopamine; nic = nicotinic; musc = muscarinic;
SIF = small intensely fluorescent cell; LH-RH = luteinizing
hormone-releasing hormone; ENK = enkephalins; SubP = substance P; cy.nu. = cyclic nucleotides; f-epsp = fast
excitatory postsynaptic potential; s-epsp = slow epsp;
ls-epsp = late slow spsp; s-ipsp = slow inhibitory postsynaptic potential.

preparations, naltrexone being more potent in this respect than
naloxone. The enhancing effect of naltrexone on the slow potentials
recorded by means of the sucrose-gap method is shown in Fig. 3. In
about 40 min naltrexone caused about twofold increase in the amplitude of P potential (extracellular counterpart of the s-ipsp) and
nearly doubled the response of LN (extracellular counterpart of the
s-epsp). The effect of naltrexone was reversible after washing the
ganglion with Krebs solution. It appears then that endogenous
enkephalins exert an inhibitory effect on the terminal, and that
opiate antagonists induce disinhibition, thus increasing ACh output.
It may have been expected that this effect will be demonstrated
with respect to the slow potentials rather than the f-epsp, as the
repetitive stimuli that must be used to elicit the slow potentials
should generate considerable release of endogenous enkephalins.

The magnitude of increase in the slow potentials induced by opiate antagonists varied considerably in individual ganglia; increase was not detected in several ganglia. In this context, it is noteworthy that superior cervical ganglia of the rat differed markedly in the intensity of their enkephalin immunoreactivity (28).

In summary, our results indicate that endogenous enkephalins or enkephalin-like peptides may have a regulatory effect on ACh release. Especially when preganglionic fibers are stimulated repetitively enkephalins may exert, via opiate receptors located on preganglionic terminals, a negative feedback inhibition of excessive release of ACh. On the other hand, in the case of low frequency of stimulation, the modulatory effect of enkephalins on ACh release may not be readily detected. It is of interest that this role of enkephalins may be generalized from the superior cervical ganglia to parasympathetic ganglia, as enkephalins may also exert a presynaptic inhibitory action on parasympathetic nerve terminals in the myenteric plexus of the guinea pig (27).

OTHER ENDOGENOUS SUBSTANCES

A number of putative transmitter substances, histamine, serotonin, GABA, prostaglandins and catecholamines have been shown to be either release or present in various sympathetic ganglia (for ref. see 28). The effects of GABA and catecholamines on ganglionic transmission have been studied more extensively, as Adams and Brown (1) reported that GABA depolarizes the postsynaptic membrane by a Cl-dependent mechanism while Koketsu and colleagues (15) proposed that the major effect of GABA on ganglionic transmission is to depress the release of ACh from preganglionic fibers (15).

The action of acetcholamines in sympathetic ganglia appears to be complex. Libet and his colleagues (19) suggested that catecholamines are responsible for the generation of the s-ipsp which may modulate neuronal excitability by increasing the resting membrane potential (19). In the case of the rabbit superior cervical ganglia the catecholamine in question may be dopamine and it was proposed that it is release from an interneuron, the small intensely fluorescent (SIF) cell, which is activated muscarinically by ACh released from preganglionic fibers (21). In addition, Kobayashi et al. (16) proposed recently that dopamine increases neuronal excitability by causing long lasting facilitation of the s-epsp. Still another proposal is that of Weight and Padjen (29) who suggested that in amphibian sympathetic ganglia the s-ipsp is generated by a direct muscarinic action of ACh. The issue whether the s-ipsp is generated disynaptically by a catecholamine or monosynaptically via direct

muscarinic action of ACh has not been resolved satisfactorily (3).
Furthermore, we have obtained evidence indicating that the primary
effect of catecholamines when released within the ganglia is not
exerted postsynaptically; instead, it concerns α-adrenoceptive sites
located on pre-terminal regions and leads to depression of the
release of ACh (4,7).

Little is known about the role of serotonin, histamine and
prostaglandins on ganglionic transmission (25). Similarly, the
function of several peptides such as neurotensin, somatostatin and
vasoactive intestinal peptide which are present in the ganglia (28)
remains to be explored in future studies.

Finally, the interesting finding that preganglionic stimulation
increased the level of ganglionic cyclic nucleotides, cyclic AMP
and cyclic GMP, raised the possibility that the synaptic action of
transmitters may depend on the phosphorylation of specific membrane
proteins by cuclic nucleotides, and that cyclic AMP and cyclic GMP
are the intracellular second messengers for the action of dopamine
and ACh in the generation of ganglionic s-ipsp and s-epsp, respectively (10). This latter hypothesis received little support from
recent investigations including those from our laboratory (5,8,9).
The role of cyclic nucleotides in ganglionic transmission remains
to be established.

CONCLUSIONS

What emerges from this description is a complexity of ganglionic sites and mechanisms that provide for and modulate transmission
(cf. Fig. 4). Full understanding of these mechanisms requires conclusive identification of transmitters and modulators involved and
of the pertinent circuitry. Yet even in the case of the much studied
P potential or the s-ipsp it cannot be decided today whether it is
generated di- or mono-synaptically (3) or, for that matter, whether
the inhibitory modulations induced by ganglionic catecholamines are
generated primarily post- or presynaptically ((20; cf. Fig. 4).
Even less well known, is the topography of the peptidergic and
enkephalinergic responses. Indeed, it cannot be stated at this
time whether the peptidergic responses originate from preganglionic
fibers which are separate from the cholinergic terminals (as shown
in Fig. 4) or are generated from the latter jointly with ACh. This
possibility of the release of a plurality of neurotransmitters from
a single neuron that may have been startling a few years back appears to be very plausible today For example, the coexistance of
serotonin and substance P in certain neurons of the rat medullary
raphe and interfascicularis hypoglossi cell groups has been

demonstrated recently (2). It is also possible that substance P is released from non-autonomic afferent fibers (12).

It is evident on several grounds that the ganglion may constitute a useful model for brain studies. First, just as in the case of the brain (17) ganglionic transmission appears to be modulated by slow potentials and by long lasting changes in neuronal excitability. Indeed, it bears stressing that some of the slow potentials discussed here have a latency of several seconds and duration of even minutes. This, of course, lends itself to the study in the ganglion of phenomena heretofore considered characteristic for brain function alone. Second, the ganglion may serve readily for the analysis of complex interactions that depend on several transmitters, on complex circuitry, and on multiple pre- and post-synaptic sites. Lastly, the ganglion appears to be particularly pertinent for the study of the action of substances that come to fore only recently, namely enkephalins, other peptides and hypophyseal-pituitary factors, and prostaglandins. It is important in this context that phenomena of opiate and enkephalin-induced tolerance and withdrawal can be duplicated in the case of parasympathetic ganglia and the myenteric plexus (26). The exploitation of the ganglia along these several lines should be particularly rewarding as the phenomena in question, including, e.g. their ionic mechanisms and membrane characteristics, cannot be readily investigated in the CNS.

ACKNOWLEDGEMENTS

Published and unpublished investigations carried out in these laboratories were supported in part by NIH grants NS-06455 and NS-15848 and BRSG RR 05-368.

REFERENCES

1.   Adams, P.R. and Brown, D.A. (1975): J. Physiol. (Lond) 250: 85-120.
2.   Chan-Palay, V., Jonsson, G. and Palay, S.L. (1978): Proc. Nat. Acad. Sci. 75:1582-1586.
3.   Dun, N.J. (1980): Fed. Proc. (in press).
4.   Dun, N. and Karczmar, A.G. (1977): J. Pharmacol. Exp. Ther. 200:328-335.
5.   Dun, N.J. and Karczmar, A.G. (1977): J. Pharmacol. Exp. Ther. 202:89-96.

6. Dun, N.J. and Karczmar, A.G. (1979): Neuropharmacology 18:
    215-218.
7. Dun, N. and Nishi, S. (1974): J. Physiol. (Lond) 239:155-164.
8. Dun, N.J., Kaibara, K. and Karczmar, A.G. (1977): Science
    197:778-780.
9. Dun, N.J., Kaibara, K. and Karczmar, A.G. (1978): Brain Res.
    150:658-674.
10. Greengard, P. (1976): Nature 260:101-102.
11. Hokfelt, T., Elfvin, L.G., Schultzberg, M., Goldstein, M. and
    Nilsson, G. (1977): Brain Res. 132:29-41.
12. Hokfelt, T., Kellerth, J.O., Nilsson, G. and Pernow, B. (1975):
    Science 190:889.
13. Jan, Y.N., Jan, L.Y. and Kuffler, S.W. (1979): Proc. Nat. Acad.
    Sci. 76:1501-1504.
14. Karczmar, A.G. and Dun, N.J. (1978): In Psychopharmacology:
    A Generation of Progress (eds) M.A. Lipton, A. DiMascio and
    K.F. Killam, Raven Press, New York, pp. 293-305.
15. Kato, E., Kuba, Y. and Keketsu, K. (1978): Brain Res. 153:
    398-402.
16. Kobayashi, H., Hashiguchi, T. and Ushiyama, N.S. (1978):
    Nature 271:268-269.
17. Krnjevic, K. (1974): Physiol. Rev. 54:418-540.
18. Kuba, K. and Koketsu, K. (1978): In Progress in Neurobiology 11,
    Pergamon Press, Oxford.
19. Libet, B. (1970): Fed. Proc. 29:1945.
20. Libet, B. (1979): In Integrative Functions of the Autonomic
    Nervous System (eds) C. McC. Brooks, K. Koizumi and A. Sato,
    University of Tokyo Press, Tokyo.
21. Libet, B. and Owman, C. (1974): J. Physiol. (Lond) 237:635-646.
22. Heild, T.O. (1978): Brain Res. 140:231-239.
23. Nishi, S. (1974): In The Peripheral Nervous System (ed) J.I.
    Hubbard, Plenum Press, New York, pp. 225-255.
24. Nishi, S. and Koketsu, K. (1968): J. Neurophysiol. 31:109-121
25. Nishi, S., Karczmar, A.G. and Dun, N.J. (1978): In Advances
    in Pharmacology and Therapeutics, Vol. 2 (ed) P. Simon,
    Pergamon Press, New York.
26. North, R.A. and Karras, P.J. (1978): Nature 272:73-74.
27. North, R.A., Katayama, Y. and Williams, J.T. (1979): Brain
    Res. 165:67-77.
28. Schultzberg, M., Hokfelt, T., Lundberg, J.M., Terenius, L.,
    Brandt, J., Elde, R.P. and Goldstein, M. (1979): Neuroscience
    4:249-256.
29. Weight, F.F. and Padjen, A. (1973); Brain Res. 55:225-234.

PGE$_1$-INDUCED cAMP BIOSYNTHESIS IN THE SUPERIOR CERVICAL GANGLION

OF DIFFERENT ANIMAL SPECIES

V. Perri, O. Belluzzi, C. Biondi, P.G. Borasio, A.
Capuzzo, M.E. Ferretti and A. Trevisani

Institutes of Human Physiology and General Physiology
University of Ferrara
44100 Ferrara, Italy

INTRODUCTION

Some years ago Greengard and coworkers (5-7) put forward a
comprehensive hypothesis to explain the genesis of the slow synaptic potentials observed in curarized mammalian sympathetic ganglia
after tetanization of preganglionic fibers. Following this hypothesis the slow synaptic potentials were due to an increase of
cyclic nucleotides levels brought about by the interaction of
either acetylcholine (ACh)(slow EPSP) or dopamine (DA)(slow IPSP)
with specific receptors located on ganglion neuron membranes. In
particular, according to Greengard's hypothesis, prostaglandin E$_1$
(PGE$_1$) would act as a DA antagonist reducing the amplitude of the
slow IPSP extracellularly recorded from isolated ganglion preparations. Since DA was supposed to increase cAMP levels via activation of adenylate cyclase, PGE$_1$ would exert an inhibitory action
on this enzyme.

In an attempt to test this hypothesis we found that PGE$_1$ was
as effective as DA in stimulating cAMP synthesis in bovine superior
cervical ganglion slices (13,14). Moreover, when ganglion slices
were treated with appropriate amounts of DA and PGE$_1$ in combination,
the increase in cAMP levels was much more than additive; the synergistic effect of PGE$_1$ plus DA reached about 400% of the control
value.

In the present study we have extended our observations to determine the role of PGE$_1$ in the processes of ganglionic impulse

transmission and to investigate whether both DA and PGE$_1$ exhibit
effects similar to those seen in sympathetic ganglia of other
animal species.

MATERIALS AND METHODS

Superior cervical ganglia from rat, rabbit, guinea pig and
calf were used. Preparation of synaptosomes from calf ganglia,
adenylate cyclase activity determination, incubation of ganglia
in vitro and assay of cAMP were performed as previously described
(13). Electric stimulation of ganglia was as described by Perri
et al. (11); calf ganglionic slices were electrically stimulated
according to the method described by Beani et al. (2). PGs of the
E type, released from the same ganglion in resting condition and
after stimulation, were measured in the perfusion medium by radioimmunoassay as described by Bartolini et al. (1). PGEs were extracted and purified according to the method of Salmon and Karim
(12).

RESULTS

To determine on which structures the adenylate cyclase syn-
ergistically stimulated by combinations of DA plus PGE$_1$ is located,
synaptosome fractions were isolated from bovine superior cervical
ganglia and adenylate cyclase activity was tested. Results shown
in Fig. 1 demonstrate that enzyme activity is significantly stimulated by PGE$_1$, whereas DA appears to be quite ineffective. Accordingly, the combination of DA plus PGE$_1$ failed to reproduce the
synergistic effect observed in intact ganglia preparations. Since
the synaptosomal fraction, when observed at the electron microscope,
appeared to be almost entirely composed by presynaptic elements,
our results, taken altogether, suggest the presence in the ganglia
of two different adenylate cyclases: one DA sensitive and PGE$_1$
modulated located on the membrane of principal neurons, and the
second, sensitive to PGE$_1$ alone, on cholinergic terminals.

In Fig. 2 the effects of catecholamines and PGE$_1$ on cAMP levels
in superior cervical ganglia of four different animal species are
shown. The results indicate that DA is highly effective only on
bovine ganglia and noradrenaline only on rat ganglia, whereas their
activity is either slight or even absent in preparations of ganglia
from other animal species. On the contrary, PGE$_1$ was highly effective on the ganglia of all the animal species tested. However,
the synergistic effect produced by combinations of DA plus PGE$_1$ on
bovine ganglion slices was never observed in either rat, rabbit or
guinea pig ganglion preparations.

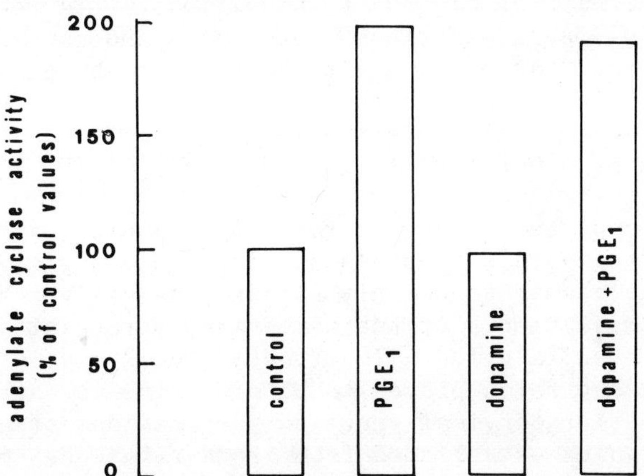

FIGURE 1: The effects of PGE₁ and DA on adenylate cyclase activity of synaptosomes of bovine ganglia. PGE₁ was present at 4 µg/ml; DA was 0.1 mM.

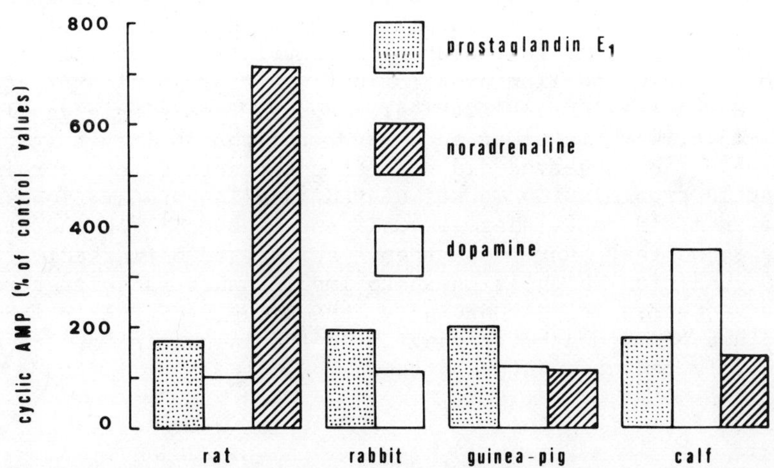

FIGURE 2: The effects of PGE₁, noradrenaline and DA on cAMP levels in isolated preparations of superior cervical ganglia of different animal species. PGE₁ was used at 4 µg/ml. Noradrenaline and DA were 0.1 mM.

Electrical stimulation of preganglionic fibers has already
been shown by Kalix et al. (9) to significantly increase cAMP levels
in rabbit ganglion preparations. Greengard and coworkers interpreted this result as being due to DA released by ganglion interneurons and acting on a DA-sensitive adenylate cyclase. By analogy
Greengard suggested this interpretation to be valid also for autonomic ganglia of other mammals.

Results reported in Fig. 2 raise some doubt on the general
validity of Greengard's hypothesis, since guinea pig, rat and even
rabbit ganglia appear to be only slightly sensitive or quite insensitive to DA action. At the same time, our results may offer
an alternative explanation. In effect, Fig. 2 suggests that PGE_1
is the only agent which proved to be effective in increasing cAMP
levels in superior cervical ganglion preparations of all the animals
tested. It can be established from cross reference to Table 1 and
Fig. 2 that PGE_1 is the only tested substance able to qualitatively
and quantitatively mimic the effects of electrical stimulation of
preganglionic fibers on cAMP levels in rat, rabbit, guinea pig and
bovine ganglia. In guinea pig ganglion preparations, these observations could be further substantiated by quantitative determination of PGE liberated in the incubating medium during a 10 min
period of high frequency stimulation of the cervical sympathetic
trunk. Table 1 shows that the amounts of PGE-like material released were remarkably increased by electrical stimulation.

The involvement of cyclic nucleotides in the processes of
synaptic transmission in autonomic ganglia has been recently questioned since the administration of appropriate amounts of cAMP or
cGMP to isolated ganglion preparations, very often did not give
rise to any appreciable electrical response from single ganglion
neurons (3,4,10). In order to exclude the possibility that the
increase in cAMP observed did not reflect events directly related
to synaptic transmission we stimulated ganglion preparations incubated in media containing low Ca^{++} and high Mg^{++} concentration.
Table 2 shows that cAMP biosynthesis appears to be markedly reduced
by this condition which is known to impair neurotransmitter release
from nerve terminals. Accordingly, the amounts of PGE_1 released
from guinea pig sympathetic ganglia in the incubating medium are
also sharply reduced (data not shown).

DISCUSSION

The data presented in this paper clearly indicate that PGs of
the E series are able to increase cAMP levels in the rat, rabbit
and guinea pig superior cervical ganglia. Our results also confirm
and extend previous observations (13,14) concerning the effect of
PGE_1 and DA in the bovine superior cervical ganglion, revealing the

TABLE 1: Effect of preganglionic stimulation on cAMP and PGE synthesis in superior cervical ganglia of different animal species.

ANIMAL	PERCENT OF CONTROL VALUES	
	cAMP	PGE
Rat	190	---
Rabbit	204*	---
Guinea pig	197	199
Calf	185**	---

*   From Kalix et al. (9).
**   Experiments carried out on ganglion slices.

TABLE 2: Effect of electric stimulation (20 Hz) on cyclic AMP levels in guinea pig and rat superior cervical ganglion in the presence of a low Ca^{++}, high Mg^{++} solution.

EXPERIMENTAL CONDITIONS	Guinea Pig	Rat
	Cyclic AMP (% of control values)	
	Control Ganglia	
Ca^{++} 2.5 mM; Mg^{++} 1.15 mM	100	100
	10.6 ± 0.3*	3.0 ± 0.3*
	Stimulated Ganglia	
	197 ± 15	263 ± 35
	Control Ganglia	
Ca^{++} 0.5 mM; Mg^{++} 12.0 mM	100	100
	7.3 ± 0.3*	4.2 ± 0.7*
	Stimulated Ganglia	
	133 ± 17	157 ± 39

* Data are expressed as pmol cyclic AMP/ganglion ± S.E.M.

presence of an adenylate cyclase sensitive to PGE_1 in the synaptosome fraction of the same tissue.

As has already been pointed out, cAMP levels have been suggested as a modulating factor for nervous transmission in peripheral ganglia. DA released from an interneuron binds to a receptor coupled to an adenylate cyclase causing cAMP levels to increase. In turn the cyclic nucleotide would be responsible for a hyperpolarizing response (5-7).

Our data indicate that this mechanism cannot be applied to sympathetic ganglia of all animal species. We observed a highly variable degree of sensitivity to DA and noradrenaline of the cAMP generating system in the ganglia of rabbit, rat, guinea pig and calf. At variance with the different responses to catecholamines, the ganglia of all the animal species tested were very responsive to PGE_1. We consistently observed a significant rise in cAMP levels when the ganglia were incubated in the presence of PGE_1. These data are in agreement with those obtained by electrical stimulation of the ganglia. Following preganglionic stimulation of rat and guinea pig superior cervical ganglia we observed an increase of cAMP concentrations of the same order of magnitude as that obtained after PGE_1 treatment. In addition, intramural stimulation of bovine ganglion slices gave rise to a similar increase.

We can presume the existence of a functional correlation linking PGs of the E type and cAMP levels to the processes of nervous transmission occurring in mammalian sympathetic ganglia. Preliminary experiments referred to in this paper and which are being reported in full elsewhere are in agreement with such a hypothesis since we could detect a marked increase of PGE production and release following electrical stimulation of the guinea pig superior cervical ganglion (15).

PGEs are known to play a role in autonomic nervous function by regulating neurotransmitter release (8). Similarly, in mammalian sympathetic ganglia PGEs might be involved as neuromodulator of nerve impulse transmission reducing the output of ACh from presynaptic terminals.

ACKNOWLEDGEMENTS

This work was supported in part by Grant No. CT. 79-01868 to Professor V. Perri from the Consiglio Nazionale delle Ricerche.

REFERENCES

1.   Bartolini, G., Meringolo, C., Orlandi, M. and Tomasi, V. (1978): Biochim. Biophys. Acta 530:325-332.
2.   Beani, L., Bianchi, C., Giacomelli, A. and Tamberi,F. (1978): Eur. J. Pharmacol. 48:179-193.
3.   Dun, N.J., Kaibara, K. and Karczmar, A.G. (1977): Science 197:778-779.
4.   Gallagher, J.P. and Shinnick-Gallagher, P. (1977): Science 198:851-852.

5. Greengard, P. (1976): Nature 260:101–108.
6. Greengard, P. and Kebabian, J.W. (1974): Fed. Proc. 33: 1059–1067.
7. Greengard, P., McAfee, D.A. and Kebabian, J.W. (1972): Adv. Cyclic Nucleotide Res. 1:337–355.
8. Hedqvist, P. (1977): Ann. Rev. Pharmacol. 17:259–279.
9. Kalix, P., McAfee, D.A., Schorderet, M. and Greengard, P. (1974): J. Pharmacol. Exp. Ther. 188:676–687.
10. Libet, B. (1979): Life Sci. 24:1043–1058.
11. Perri, V., Sacchi, O. and Casella, C. (1970): Quart. J. Exp. Physiol. 55:25–35.
12. Salmon, J.A. and Karim, S.M.M. (1975): In Advances in Prostaglandin Research, Vol. 1, (ed) S.M.M. Karim, MTP Press, Lancaster, pp. 25–85.
13. Tomasi, V., Biondi, C., Trevisani, A., Martini, M. and Perri, V. (1977): J. Neurochem. 28:1289–1297.
14. Tomasi, V., Trevisani, A., Biondi, C., Capuzzo, A. and Perri, V. (1977): Biochem. Soc. Trans. 5:520–523.
15. Trevisani, A., Biondi, C., Borasio, P.G., Capuzzo, A., Ferretti, M.E., Belluzzi, O. and Perri, V. (1980) (in press).

QUESTIONS RAISED BY THE ELECTRON MICROSCOPIC LOCALIZATION OF

ACETYLCHOLINESTERASE AND BUTYRYLCHOLINESTERASE IN NORMAL AND

DENERVATED SUPERIOR CERVICAL GANGLIA IN THE CAT

G.B. Koelle, R. Davis, W.A. Koelle, G.A. Ruch,
K.K. Rickard and U.J. Sanville

Department of Pharmacology, University of Pennsylvania
Medical School, Philadelphia, Pennsylvania   19104 USA

With the development of the copper thiocholine (CuThCh) histochemical method for the light microscopic (LM) localization of
cholinesterases over 30 years ago (15) and the application of its
modifications (9-11,16) to the normal and preganglionically denervated superior cervical ganglion (SCG) of the cat, the distributions of acetylcholinesterase (AChE) and butyrylcholinesterase (Bu-
ChE) in this tissue appeared to have been established. In the normal ganglion, AChE was present in high concentrations in the perikarya of occasional (< 1%) ganglion cells (6) and in traces in the remainder, and in considerable concentration throughout the neuropil
(Fig. 1). Following preganglionic denervation, staining for AChE
disappeared completely from the neuropil but remained unchanged in
the perikaya of the ganglion cells. It was therefore concluded
that staining of the neuropil represented enzyme confined solely to
the preganglionic fibers and their terminals. Staining for BuChE
was intense throughout the neuropil and completely absent from the
perikaya of the ganglion cells; following denervation, there was
only a slight decrease in the intensity of staining. From these
and related findings (1,2) it appeared that BuChE was present only
in capsular glial cells. These findings were consistent with the
results of quantitative determinations in normal and preganglionically denervated ganglia (14,25). Pharmacological studies provided
a hypothetical explanation for the function of the presynaptically
located AChE (12,30).

More recently, the bis-(thioacetoxy) aurate (I), or $Au(TA)_2$,
method with appropriate controls has permitted direct visualization

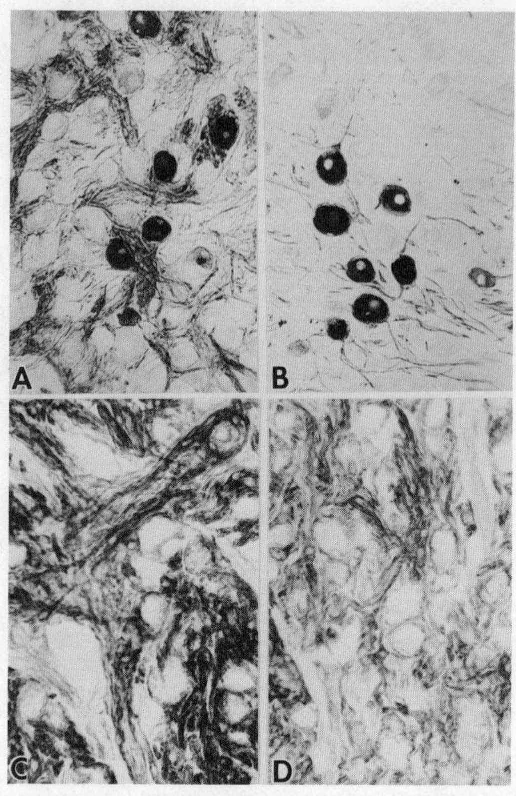

FIGURE 1: Light micrographs of cat ganglia stained by the CuThCh
method. Magnification X 270. A = stellate ganglion, AChE,
normal. The only difference between this tissue and the
SCG is the considerably higher percentage (18%) of ganglion cells that show heavy staining for AChE in the
former (6). B = stellate ganglion, AChE, denervated.
C = SCG, BuChE, normal. D = SCG, BuChE, denervated.

by electron microscopy (EM) of sites of AChE and BuChE activity in
normal and denervated cat SCG (4). AChE is present not only throughout the axolemmas and terminal membranes of preganglionic fibers,
but at the plasmalemmas of the dendrites and perikarya of the
ganglion cells as well; following preganglionic denervation, it
disappears completely from all these sites (Fig. 2). BuChE is present at the identical postsynaptic locations on the perikarya and
dendrites of the ganglion cells and shows only a slight decrease at
these sites following preganglionic denervation. The presence of
other $Au(TA)_2$-hydrolyzing enzymes in the glial cells and endoplasmic
reticulum (ER) of the ganglion cells precludes the positive

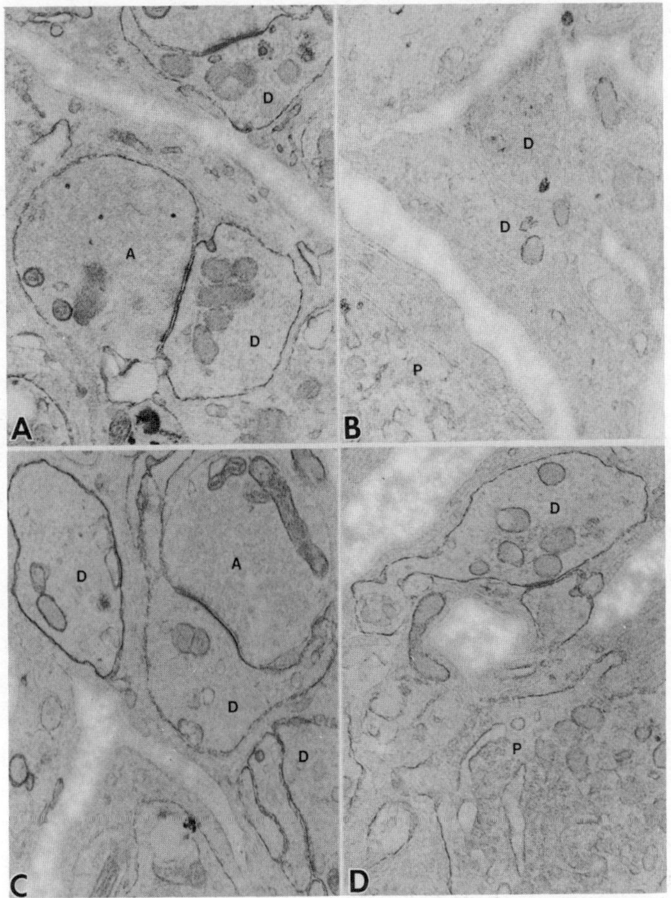

FIGURE 2: Electron micrographs of cat SCG stained by the Au(TA)$_2$
method. Magnification X 31,000. A = AChE, normal. Note
staining of plasmalemmas of axonal terminals (A), dendrites (D), and perikarya (P). B = AChE, denervated.
Note complete disappearance of plasmalemmal staining.
C = BuChE, normal. Staining of plasmalemmas of dendrites
only. D = BuChE, denervated. Only change detectable is
slight decrease in intensity of staining of dendritic
membranes.

identification of either enzyme at these sites. However, findings
are consistent with earlier ones with the CuThCh method, that low
concentrations of AChE but not of BuChE are present in the ER of
ganglion cells, and that the glial cells may contain low concentrations of BuChE.

These findings have raised several questions which cannot be answered at present:

By what mechanism is AChE maintained normally at postsynaptic sites, from which is disappears within a few days following preganglionic denervation?

One possibility is that the preganglionic fibers release a neurotrophic factor which is necessary for the maintenance of the enzyme at postsynaptic sites, as has been demonstrated at the motor endplate in cultured preparations of skeletal muscle (3,7,22,24). Alternatively, AChE may be released continually by the presynaptic terminals and diffuse through intercellular spaces to become localized at dendritic and perikaryonal membranes (27,28). The latter explanation seems less likely because of the extensive postsynaptic distribution of AChE at sites remote from synaptic contact, and the apparent absence of AChE from intercellular spaces (4).

What is the site of synthesis of the BuChE at the ganglion cell dendritic and perikaryonal membranes, in view of its apparent absence from the ER?

Three possibilities can be offered: 1) BuChE is taken up from the plasma, as has been proposed for the cuticular proteins of arthropods (23); 2) the enzyme is synthesized in capular glial cells and transposed to the plasmalemmas of the ganglion cells; and 3) BuChE is present in the ER of the ganglion cells at subliminal concentrations as detectable by the CuThCh method. At present, there is no conclusive evidence to support or exclude any of these possibilities.

Is BuChE a precursor of AChE?

The physiological function of BuChE has not been established, inasmuch as its selective inactivation produces no detectable pharmacological or pathological effects (13,26). The finding of its identical localization with that of AChE at the dendritic and perikaryonal membranes of the ganglion cells suggested that it might function as a post-transcriptional precursor of AChE (17). From structure-activity relationship studies of a large series of organophosphorous compounds, it was concluded independently by two groups of investigators (8,21) that the functional groups of the anionic and esteratic sites of AChE and BuChE are probably identical, and that the differences between the two enzymes are due to the locations and structures of the hydrophobic areas adjacent to either site. By analogy with the present proposal, one of the changes involved in the activation of chymotrypsinogen-A to α-chymotrypsin is the formation or unmasking of a hydrophobic pocket to which the side chains of specific substrates bind (5,29).

In support of this proposal it was shown that when BuChE was selectively alkylphosphorylated by the repeated administration of iso-OMPA, the rate of regeneration of AChE following sarin was reduced significantly in several autonomic ganglia of the cat (18). This finding could not be repeated with whole homogenates of the SCG of the rat in which the pseudocholinesterase is a propionylcholinesterase (PrChE)(20). However, when the aqueous soluble fractions of rat SCG were examined, significant suppression of AChE regeneration was detected at 6 hr (Koelle, Koelle and Rickard, in preparation). Measurement of the rates of fall in the AChE activities of homogenates of the SCG of both cats and rats incubated in vitro showed that this process was accelerated in both species by the prior inactivation of BuChE or PrChE by iso-OMPA administered in vivo (19). Thus, most of these findings are consistent with the present working hypothesis. Studies are now in progress to attempt to answer these several questions raised by the recent EM findings.

ACKNOWLEDGEMENTS

The research presented in this paper was supported by USPHS, NIH research grants NS-00282-01-27 from the National Institue of Neurological and Communicative Disorders and Stroke.

REFERENCES

1.   Bulbring, E., Philpot, F.J. and Bosanquet, F.D. (1963): Lancet 1:865-866.
2.   Cavanagh, J.B., Thompson, R.H. and Webster, G.R. (1954): Quart. J. Exp. Physiol. 39:185-197.
3.   Davey, B., Younkin, L.H. and Younkin, S.G. (1979): J. Physiol. 289:501-515.
4.   Davis, R. and Koelle, G.B. (1978): J. Cell Biol. 78:785-809.
5.   Freer, S.T., Kraut, J., Robertus, J.D., Wright, H.T. and Xuong, Ng.H. (1970): Biochemistry 9:1997-2009.
6.   Holmstedt, B. and Sjoqvist, F. (1959): Acta Physiol. Scand. 47:284-296.
7.   Jessell, T.M., Siegel, R.E. and Fischbach, G.D. (1979): Proc. Nat. Acad. Sci. 76:5397-5401.
8.   Kabachnik, M.I., Brestkin, A.P., Godovikov, N.N., Michelson, M.J., Rozengart, E.V. and Rozengart, V.I. (1970): Pharmacol. Rev. 22:355-388.
9.   Koelle, G.B. (1950): J. Pharmacol. Exp. Ther. 100:158-179.
10. Koelle, G.B. (1951): J. Pharmacol. Exp. Ther. 103:153-171.

11. Koelle, G.B. (1955): J. Pharmacol. Exp. Ther. 114:167–184.
12. Koelle, G.B. (1962): J. Pharm. Pharmacol. 14:65–90.
13. Koelle, G.B. (1963): IN Cholinesterases and Anticholinesterase Agents (ed) G.B. Koelle, Springer-Verlag, New York.
14. Koelle, G.B., Davis, R. and Koelle, W.A. (1974): J. Histochem. Cytochem. 22:244–251.
15. Koelle, G.B. and Friedenwald, J.S. (1949): Proc. Soc. Exp. Biol. 70:617–622.
16. Koelle, W.A. and Koelle, G.B. (1959): J. Pharmacol. Exp. Ther. 126:1–8.
17. Koelle, W.A., Koelle, G.B. and Smyrl, E.G. (1976): Proc. Nat. Acad. Sci. 73:2936–2938.
18. Koelle, G.B., Koelle, W.A. and Smyrl, E.G. (1977): J. Neurochem. 28:313–319.
19. Koelle, G.B., Rickard, K.K. and Ruch, G.A. (1979): Proc. Nat. Acad. Sci. 76:6012–6016.
20. Koelle, G.B., Rickard, K.K. and Smyrl, E.G. (1979): J. Neurochem. 33:1159–1164.
21. Langel, U. and Jarv, J. (1978): Biochim. Biophys. Acta 525:122–133.
22. Lentz, T.L. (1974): Exp. Neurol. 45:520–526.
23. Neville, A.C. (1975): Biology of the Arthroped Cuticle, Vol.4/5, Springer-Verlag, New York.
24. Oh, T.H. and Markelonis, G.J. (1978): Science 200:338–339.
25. Sawyer, C.H. and Hollinshead, W.H. (1945): J. Neurophysiol. 8:137–153.
26. Silver, A. (1974): The Biology of Cholinesterases, American Elsevier Publ. Co., New York.
27. Skau, K.A. and Brimijoin, S. (1978): Nature 275:224–226.
28. Somogyi, P., Chubb, I.W. and Smith, A.D. (1975): Proc. Roy. Soc. Lond. B 191:271–283.
29. Steitz, T.A., Henderson, R. and Blow, D.M. (1969): J. Molec. Biol. 46:337–348.
30. Volle, R.L. and Koelle, G.B. (1961): J. Pharmacol. Exp. Ther. 133:223–240.

ACETYLCHOLINE METABOLISM IN PC12, A CLONAL CELL LINE

ON SECRETORY CELLS

B.D. Howard and W.P. Melega*

Departments of Biological Chemistry and Pharmacology*,
School of Medicine, University of California, Los
Angeles, California 90024 USA

PC12 is a clonal line of rat pheochromocytoma (7). PC12
cells synthesize catecholamines (primarily dopamine) and acetylcholine (ACh), store each in different granules and exhibit a Ca^{++}-
sensitive, depolarization-induced secretion of each (8,9,15,19).
These properties make it possible to compare the metabolism and
granular storage of two different transmitters produced by the
same secretory cell. In this paper we summarize our use of PC12
cells for such studies, with the emphasis being on ACh.

Culture and General Properties of PC12 Cells

We generally culture the PC12 cells in plastic petri dishes
or flasks, to which they readily attach.   The cells are grown as
described (15) at 37°C in Dulbecco's modified Eagle's medium.
For experiments involving incubation in defined buffers, the cells
were incubated at 37°C while still attached to plastic dishes or
flasks in a cell incubation buffer consisting of 60 mM sucrose,
10 mM glucose, 130 mM NaCl, 4.8 mM KCl, 1.3 mM $CaCl_2$, 1.2 mM
$MgSO_4$, 1.2 mM KH_2PO_4, 25 mM 4-(2-hydroxyethyl)-1-piperazine-
ethanesulfonic acid (Hepes), pH. 7.3. Where indicated, the KCl
concentration was raised to 55 mM to provide a "high K^+" buffer
to induce secretion from the cells. In this case the NaCl concentration was lowered correspondingly to maintain isoosmolarity.
When the experiments involved measurement of secreted products,
all incubation buffers contained 40 µM eserine.

PC12 cells are spherical in shape. Electron microscopy shows that the cells contain dense core granules that are similar in appearance to chromaffin granules of normal adrenal medullary cells but smaller in size and fewer in number (15). We did not observe translucent granules similar to those found in cholinergic nerve terminals. In PC12 cells ACh appears to be contained in dense core granules; Schubert and Klier (19) examined a homogenate of reserpine-treated PC12 cells and found that a sucrose gradient fraction enriched in ACh storing granules contained dense core granules and few other membrane structures.

In our hands PC12 cells contain 20 times as much dopamine as norepinephrine; thus our examination of catecholamine metabolism in these cells has been limited to dopamine. Table 1 shows the levels of dopamine and ACh in PC12 cells and in various fractions of cell homogenates. The recovery of these compounds in these fractions was 88-92%. The ACh and dopamine in the S_2 fraction were nonparticulate in that these compounds failed to sediment when the S_2 fraction was centrifuged at 62,000 g for 40 min, a procedure sufficient to sediment synaptic vesicles isolated from rat brain (22).

Approximately 60% of the cells' dopamine and 35% of the ACh are particulate and sediment in the mitochondrial fraction (P_2). Most of the dopamine and ACh in the P_2 fraction was found to band in a sucrose density gradient at regions determined by Schubert and Klier (19) to be enriched in granules and transmitter. Therefore, we have assumed that the bound ACh and dopamine in the P_2 fraction are in storage granules and the P_2 fraction was used without further purification for most of our studies on isolated granules.

Choline Uptake

We have found that PC12 cells exhibit a carrier mediated uptake of choline (Ch).   Under our standard conditions of incubation (see Table 2 legend) the amount of [^{3}H]Ch accumulated by the cells was linear with time of incubation for approximately 8 min.   The K_m for Ch transport was 12 μM as determined from a Lineweaver-Burk plot.

Table 2 gives the effects of various conditions of incubation on the uptake of Ch by PC12 cells. Ch uptake was temperature dependent, Na^+ independent and only slightly inhibited by 0.1 mM ouabain. Hemicholinium-3 (HC-3), a very potent inhibitor of Ch uptake into mammalian brain synaptosomes, was not an effective inhibitor of Ch uptake by PC12 cells. The IC_{50} of HC-3 was approximately 50 μM.

Table 2 also shows that Ch uptake into PC12 cells was not blocked by iodoacetate, oligomycin or sodium azide, three inhibitors

TABLE 1: Content of dopamine and acetylcholine in whole cells and
         in subcellular fractions of PC12.

Fraction	Dopamine	Acetylcholine
Whole cells	100 ± 12	100 ± 11
P_1	12 ± 1	3 ± 0.6
P_2	57 ± 2	32 ± 1
S_2	23 ± 3	53 ± 1

The P_1 fraction was the pellet fraction obtained after centrifuging
a cell homogenate at 1000 x g for 10 min. The resulting supernatant
suspension was centrifuged at 20,000 x g for 30 min to give the P_2
(pellet) and S_2 (supernatant) fractions. The results are expressed
as percentage of dopamine of ACh found in whole cells, which contained 10.6 ± 1.2 nmol of dopamine per mg of protein and 3.8 ± 0.4
nmol of ACh per mg of protein. The values are means $\pm$ S.D. for
three samples.

TABLE 2:   Effect of various agents on choline accumulation by
            PC12 cells.

Condition of Incubation	Choline Accumulation (percent of control)
Control	100 ± 6 (4)
$0°C$	1 ± 1 (4)
Omit Na^+	145 ± 14 (4)
Ouabain (0.1 mM)	84 ± 3 (2)
Hemicholinium-3 (5 µM)	103 ± 2 (2)
Hemicholinium-3 (50 µM)	52 ± 4 (4)
Iodoacetate (1 mM)	109 ± 3 (4)
Oligomycin (4 µg/ml)	123 ± 1 (2)
Azide (3 mM)	105 ± 12 (2)
Iodoacetate (1 mM) + Azide (3 mM)	266 ± 30 (6)
Iodoacetate (1 mM) + Azide (3 mM, 4°C)	5 ± 2 (2)
Iodoacetate (1 mM) + Oligomycin (4 µg/ml)	134 ± 8 (2)

The cells were preincubated for 10 min in incubation buffer modified as indicated and subsequently incubated for 4 min with 5 µM
$[^3H]$Ch (0.25 µCi/ml). The cells were washed and harvested, and the
accumulated radioactivity was counted by scintillation spectrometry.
When sodium was omitted from the buffer, the sucrose concentration
was correspondingly increased to maintain the osmolarity. Ch accumulation values, which are expressed as the percent of Ch accumulated by control cells, are the means $\pm$ S.E. for the number of incubations given in parentheses. Control cells accumulated $107 \pm$
6 pmol of Ch per min per mg of protein.

of energy metabolism. One of these inhibitors, oligomycin, caused a slight increase in Ch uptake. Under the conditions employed for Table 2, only iodoacetate, which inhibits glycolysis, decreased cellular pools of ATP (20% of control levels). Oligomycin and sodium azide, which inhibit oxidative phosphorylation, did not alter ATP stores in PC12 because the cells derive a major portion of their ATP from glycolysis rather than oxidative phosphorylation (E. Reynolds, unpublished observation). The data of Table 2 suggest that Ch is transported into PC12 cells by a facilitated diffusion mechanism. In this respect PC12 cells differ from cholinergic axon terminals of mammalian brain where Ch transport occurs by a Na^+-dependent, energy dependent process (13), suggesting an active transport mechanism.

An unexpected effect was obtained by treatment of the cells with iodoacetate and sodium azide simultaneously. Under this condition Ch uptake increased approximately 2.5-fold (Table 2). This marked increase in Ch uptake did not occur when the cells were treated with these agents at 4°C or with a combination of iodoacetate and oligomycin. At present we cannot explain this effect.

It has been found that Ch uptake into nerve terminals of mammalian brain is increased after the terminals have been depolarized prior to the period of Ch uptake (13). A similar phenomenon occurs in PC12 cells as is shown in Table 3. Cells that were exposed for 10 min to a depolarizing level of K^+ (55 mM) and then returned to the normal low K^+ buffer prior to the addition of [³H]Ch took up significantly more label than did control cells that were never exposed to the high K^+ buffer. However, the presence of 55 mM K^+ during the incubation with labelled Ch inhibited the uptake of Ch.

Acetylcholine Synthesis

Under the conditions of the incubations used for our studies, the [³H]Ch taken up by the PC12 cells is rapidly acetylated (15). The cellular level of [³H]ACh reaches a plateau by 10 to 15 min after the addition of [³H]Ch. However, the loading of the newly synthesized [³H]ACh into storage granules in the cells occurs comparatively slowly. The amount of [³H]ACh in granules increases linearly with time for at least 120 min; at that time [³H]ACh in granules is about 30% of total cell [³H]ACh (15).

We have also used [²H₄]Ch to label PC12 stores of ACh; under such conditions ACh levels were measured by gas chromatography mass spectrometry (GCMS) with the techniques of Freeman et al. (6). This procedure allows the simultaneous determination of newly synthesized ACh ([²H₄]ACh) and unlabelled ACh, which, at least in part, reflects older ACh stores. The fraction of ACh labelled during a

TABLE 3:   Choline accumulation by PC12 cells as a function of K^+ concentration in the incubation buffer.

K^+ Concentration (mM)		Choline Accumulation
Preincubation	Incubation	
6	6	118 ± 8
55	6	142 ± 3
55	55	92 ± 2

The cells were preincubated for 10 min in incubation buffer and then for 5 min in buffer containing 5 µM [^{3}H]Ch (0.25 µCi/ml). The K^+ concentrations of the buffers are listed. After the 5 min incubation the cells were washed and harvested and the radioactivity was counted. The values for Ch accumulation (pmol/min/mg of protein) are means ± S.E. for 4-6 incubations.

TABLE 4:   Effect of various agents on the levels of cell [^{3}H]acetylcholine, granule [^{3}H]acetylcholine and cell ATP.

Treatment	[^{3}H]Acetylcholine		Cell ATP
	Cell	Granule	
Control	100 ± 2 (4)	100 ± 9 (4)	100 ± 5 (3)
Nigericin (0.2 µg/ml)	106 ± 8 (2)	23 ± 1 (2)	84 ± 9 (3)
FCCP (4 µM)	96 ± 15 (4)	23 ± 8 (4)	95 ± 10 (3)
FCCP (1 µM)	108 ± 21 (4)	49 ± 1 (4)	98 ± 9 (3)
Valinomycin (3 µg/ml)	96 ± 14 (6)	111 ± 15 (6)	96 ± 8 (3)
Valinomycin (3 µg/ml) + FCCP (1 µM)	78 ± 16 (6)	18 ± 5 (6)	99 ± 9 (3)
DCCD (40 µM)	88 ± 14 (6)	43 ± 8 (6)	112 ± 3 (3)
Oligomycin (4 µg/ml)	123 ± 23 (6)	100 ± 18 (6)	104 ± 9 (3)
Iodoacetate (0.5 mM)	157 ± 21 (6)	61 ± 9 (6)	30 ± 11 (3)

The cells were incubated with [^{3}H]Ch and the agents listed for 30 min at 37°C in cell incubation buffer. After the incubation [^{3}H]ACh was measured as described (22) in samples of whole cells and a granule fraction obtained from a cell homogenate. ATP was measured in extracts of unfractionated cells after a similar incubation but without [^{3}H]Ch. The results, which are expressed as percentage of control values, are the means ± S.D. for the number of incubations given in parentheses. Contral ACh values were 30,200 ± 630 and 16,000 ± 1,500 dpm per mg of protein for whole cells and granules respectively. Control cellular ATP was 24.8 ± 1.1 nmol per mg of protein.

30 min incubation with $[^2H_4]$Ch varied with the concentration of
$[^2H_4]$Ch in the incubation medium. When extracellular $[^2H_4]$Ch was
10 μM, approximately 25% of the ACh was labelled.

The time courses of labelling of total cell ACh and ACh in
the granule fraction of cell homogenates, when the cells were incubated with $[^2H_4]$Ch, were similar to those described above when
ACh was labelled with $[^3H]$. The level of total cell $[^2H_4]$ACh and
$[^2H_4]$ACh in granules was affected by cholinesterase activity. The
level of each increased approximately twofold when the incubation
buffer also contained the cholinesterase inhibitor, eserine, at
40 μM.

Acetylcholine Turnover and Release

We have measured the turnover of ACh in PC12 cells that had
been incubated with $[^2H_4]$Ch for 30 min, washed, and further incubated in buffer that contained unlabelled Ch. The cellular levels
of $[^2H_4]$ACh decreased rapidly after $[^2H_4]$Ch was washed from the
cells, and by 10 min it was 30% of the value determined at the
beginning of the post-washout incubation period. The level of
total ACh did not change much during this time. $[^2H_4]$ACh in the
granules turned over much more slowly than did total cell $[^2H_4]$-
ACh. The decrease in cellular $[^2H_4]$Ach was likely due primarily to
hydrolysis by cellular cholinesterase because when 40 μM eserine was
present, the cellular level of $[^2H_4]$ACh declined only slightly.

Exposure of the cells to 55 mM K^+ induces the release of ACh
from the cells in a Ca^{++} dependent fashion (9,15,19). We have
determined that approximately 9% of the $[^2H_4]$ACh formed after a
30 min incubation with $[^2H_4]$Ch is released by the cells upon
exposure of the cells to 55 mM K^+ for 5 min. Under these conditions the specific activities of the total cell ACh, granule ACh,
and released ACh were 0.24, 0.1 and 0.14, respectively. The
specific activity is the mole fraction:

$$\frac{[^2H_4]ACh}{[^2H_4]ACh + [^2H_0]ACh}$$

The specific activities given above indicate that newly synthesized
ACh was not preferentially released as it appears to be under other
conditions in some other cholinergic systems (4,16,17).

the ability of FCCP to decrease the storage of catecholamines in
isolated synaptic vesicles, chromaffin granules and PC12 granules
(15,21,22) and to dissipate the proton gradient across chromaffin
granule membranes (11). In these latter cases the effect can be
attributed to a valinomycin-induced influx of K^+, which facilitates
the FCCP induced efflux of protons from the catecholamine storing
vesicles or granules (11). The net loading of [^{3}H]ACh into PC12
granules was also inhibited by DCCD, but it was not affected by
oligomycin. Neither of these ATPase inhibitors significantly
altered total cell levels of [^{3}H]ACh.

All of the ionophores and ATPase inhibitors listed in Table 4
can inhibit mitochondrial ATP synthesis. However, under the conditions used for the experiments discussed above, none of these
agents, with the exception of nigericin, caused any depletion of
cellular stores of ATP (Table 4). Nigericin caused only a slight
decrease in cellular ATP. Thus, it is unlikely that any of the
ionophores or ATPase inhibitors exerted their effects on granular
storage of ACh by altering cellular pools of ATP.

Iodoacetate, an inhibitor of glycolysis, did markedly reduce
cellular ATP (Table 4). Unexpectedly, the iodoacetate treatment
substantially increased the level of [^{3}H]ACh in the PC12 cells;
we have not investigated the mechanism of this effect. In spite
of the increase in total cell [^{3}H]ACh, the level of [^{3}H]ACh in
granules was decreased by the treatment with iodoacetate.

Our results indicate that the ionophores and ATPase inhibitors
we examined do not affect ACh storage in PC12 granules by reducing
cellular stores of ATP. Rather it is more likely that these agents
exert their effects on ACh storage by acting directly on the
storage granules as they do when they alter catecholamine storage
in PC12 granules (15), chromaffin granules (1,3,5,10) and catecholamine storing synaptic vesicles from brain (22).

The ability of the proton ionophores FCCP and nigericin to
inhibit ACh accumulation by PC12 granules indicates that a protonmotive force (14) across the granule membrane functions in the
transport of ACh into the granules. The protonmotive force has
two components; a transmembrane proton concentration gradient and
a transmembrane potential, either or both of which could be the
driving force for ACh transport.

Iodoacetate induced a depletion of cellular stores of ATP and
caused a significant decrease in the accumulation of ACh by PC12
granules. These results are consistent with ACh transport into
PC12 granules being an ATP dependent process. The involvement of
and ATPase in this process is suggested by the ability of DCCD to
inhibit ACh loading into the granules. DCCD is known to inhibit

Acetylcholine Transport Into Storage Granules

Studies with cell free preparations of chromaffin granules and synaptic vesicles indicate that the transport of catecholamines into these particles is driven by a transmembrane pH gradient (inside acidic) and/or membrane potential, which are established by the activity of a proton-translocating Mg^{++}-ATPase present in the membrane of the catecholamine containing storage granules or vesicles (1,3,5,10,12,22). This ATPase activity and the ATP stimulated uptake of catecholamines into isolated synaptic vesicles and chromaffin granules are inhibited by N,N'-dicyclohexylcarbodiimide (DCCD) and 4-chloro-7-nitrobenzofurazan (NBD-Cl), but not by oligomycin or efrapeptin, two inhibitors of mitochondrial ATPase (1,5, 12). Catecholamine transport into the vesicles and granules is also blocked by ionophores that make membranes permeable to protons (1,3,12). We have found that this mechanism of granular transport of catecholamines also operates for isolated PC12 granules and for the granules in intact cells (15).

Little is known about the mechanism by which ACh is transported into synaptic vesicles of cholinergic nerve terminals. However, recent studies have demonstrated that cholinergic synaptic vesicles from the elasmobranch <u>Torpedo</u> contain a Ca^{++}/Mg^{++}-ATPase with kinetic properties intriguingly similar to those of the ATPase that functions in the transport of catecholamines into catecholaminergic vesicles (2,18). The ATPase of <u>Torpedo</u> cholinergic vesicles is inhibited by DCCD and NBD-Cl, but not by oligomycin or efrapeptin (18). Unfortunately, since specific loading of ACh into isolated synaptic vesicles has not been demonstrated, it is not possible at present to examine the effects of the ATPase inhibitors and ionophores on ACh loading using preparations of isolated vesicles. As an alternative procedure, Toll and Howard (23) examined the effect of various inhibitors of energy metabolism on ACh synthesis and uptake into storage granules in intact PC12 cells. They found that the net uptake of ACh into the PC12 granules is blocked by agents known either to inhibit the ATPase activity of <u>Torpedo</u> synaptic vesicles or dissipate proton electrochemical gradients. The effects do not seem to be attributable to a decrease by the agents of cellular stores of ATP.

Table 4 describes the effect of these agents on cell and granule levels of newly synthesized [^{3}H]ACh 30 min after addition of [^{3}H]Ch. Several of the agents caused a substantial decrease in the granule store of [^{3}H]ACh without affecting total cellular [^{3}H]-ACh. Two of them were the ionophore nigericin, which causes a H^+/K^+ exchange and the proton ionophore carbonyl cyanide p-trifluoromethoxyphenylhydrazone (FCCP). The K^+ ionophore valinomycin by itself had no effect on cell or granule levels of [^{3}H]ACh; however, it potentiated the effect of FCCP. Valinomycin also potentiates several proton-translocating ATPases including the ATPase of chromaffin granules (1). Cholinergic synaptic vesicles from Torpedo contain an ATPase that is sensitive to DCCD but the PC12 cholinergic granules have not been sufficiently purified to determine whether they also contain a DCCD sensitive ATPase (15).

Upon treatment with nerve growth factor PC12 cells differentiate into neurites that also store and secrete both dopamine and ACh (7-9,20). The studies described here for undifferentiated PC12 should be repeated with the differentiated PC12 neurons, which might have different patterns of metabolism of dopamine and ACh.

REFERENCES

1. Bashford, C.L., Casey, R.P., Radda, G.K. and Ritchie, G.A. (1976): Neuroscience 1:399-412.
2. Breer, H., Morris, S.J. and Whittaker, V.P. (1977): Eur. J. Biochem. 80:313-318.
3. Casey, R.P., Njus, D., Radda, G.K. and Sehr, P.A. (1977): Biochemistry 16:972-977.
4. Collier, B. (1969): J. Physiol. 205:341-352.
5. Flatmark. T. and Ingebretsen, O.C. (1977): FEBS Lett. 78:53-56.
6. Freeman, J.J., Choi, R.L. and Jenden, D.J. (1975): J. Neurochem. 24:729-734.
7. Greene, L.A. and Tischler, A.S. (1976): Proc. Natl. Acad. Sci. 73:2424-2428.
8. Greene, L.A. and Rein, G. (1977): Brain Res. 129:247-263.
9. Greene, L.A. and Rein, G. (1977): Nature 268:349-351.
10. Johnson, R.G. and Scarpa, A. (1976): J. Biol. Chem. 251: 2189-2191.
11. Johnson, R.G. and Scarpa, A. (1976): J. Gen. Physiol. 68: 601-631.
12. Johnson, R.G. and Scarpa, A. (1979): J. Biol. Chem. 254: 3750-3760.
13. Kuhar, M.J. and Murrin, L.C. (1978): J. Neurochem. 30:15-21.
14. Mitchell, P. (1976): Biochem. Soc. Trans. 4:399-430.
15. Rebois, R.V., Reynolds, E.E., Toll, L. and Howard, B.D. (1980): Biochemistry 19:1240-1248.
16. Richter, J.A. and Marchbanks, R.M. (1971): J. Neurochem. 18: 691-703.
17. Richter, J.A. and Marchbanks, R.M. (1971): J. Neurochem. 18: 705-712.
18. Rothlein, J.E. and Parsons, S.M. (1979): Biochem. Biophys. Res. Commun. 88:1069-1076.
19. Schubert, D. and Klier, F.G. (1977): Proc. Natl. Acad. Sci. 74:5184-5188.

20.   Schubert, D., Heinemann, S. and Kidokoro, Y. (1977): Proc.
         Natl. Acad. Sci. 74:2579-2583.
21.   Schuldiner, S., Fishkes, H. and Kanner, B.I. (1978); Proc.
         Natl. Acad. Sci. 75:3713-3716.
22.   Toll, L. and Howard, B.D. (1978): Biochemistry 17:2517-2523.
23.   Toll, L. and Howard, B.D. (1980): J. Biol. Chem. 255:
         1787-1789.

REGULATION OF CHOLINERGIC TRANSMISSION IN ADRENAL MEDULLA

E. Costa, A. Guidotti, I. Hanbauer*, T. Hexum**,
L. Saiani and H.-Y.T. Yang

Laboratory of Preclinical Pharmacology,
National Institute of Mental Health
St. Elizabeths Hospital, Washington, D.C. USA, and
*Hypertension-Endocrine Branch, NHLB Institute
National Institute of Health, Bethesda, Maryland, and
**On leave from: Department of Pharmacology,
University of Nebraska, Omaha, Nebraska

INTRODUCTION

Enkephalin-like peptides co-exist in splanchnic axons with
acetylcholine (ACh) (7); therefore, when nerve impulses release
these peptides together with ACh, the enkephalin-like peptides could
interact with regulatory sites of ACh receptors and modify the
responses that are elicited by this amine. Since we are testing
this hypothesis in this presentation we have termed the ACh receptor
of chromaffin cells "the receptor for the primary transmitter" and
the peptide receptor will be referred to as the "co-transmitter
receptor." It appears that in various cholinergic axons a variety
of neuropeptides may be stored with ACh (3). The co-existence of
polypeptides and amine transmitters has generated great interest
and much speculation. Among others, the following trends of
thinking have emerged: 1) at the postsynaptic receptors, the sensitivity of recognition sites for primary transmitters is continuously
modified by a number of factors, including the amount of co-transmitter that is released from the nerve terminals together with the
primary transmitter (5,6) and 2) it appears that the co-existence
of a primary transmitter with a given co-transmitter is not obligatory. For instance, ACh co-exists with enkephalin-like peptides in
splanchnic axon but in other axons it is associated with VIP (3).
Moreover, in chromaffin cells, enkephalin-like peptides co-exist

with catecholamines (2,5). Although we still wonder whether these peptides are released extraneurally together with the primary transmitter, and since chromaffin cell membranes contain high affinity recognition sites for opiates (1), it is plausible that enkephalin-like peptides stored in splanchnic axons act on the membrane of the cell they innervate. The study of such receptors may help to answer questions concerning the mode of interaction of receptors for primary transmitters and co-transmitters. Among a number of possibilities that could be investigated, we have selected to study whether the activation of opiate receptors modifies the amount of catecholamines released by the activation of nicotinic cholinergic receptors. The results of our studies support the view that opiate receptor activation inhibits the secretion of catecholamines elicited by nicotine (6).

Characterization of Opiate Peptides
in the Adrenal

The molecular heterogeneity of opiate peptides stored in the adrenal gland is well established (9); however, the physiological significance of such heterogeneity is not completely understood. Various purification procedures have detected a number of enkephalin-like opiate peptides in adrenal tissue. These peptides had different molecular weights, some of them weigh 10,000 or more, others are smaller (Table 1).

The data were obtained with met[5]-enkephalin antibody often associated with trypsinization. Leu-enkephalin peptides are not detected with met[5]-enkephalin antibodies. These peptides are present but in a much smaller quantity and with a lower degree of heterogeneity. They are not included in Table 1. The data of Table 1 show that when different tissue preparations containing chromaffin cells are studied for their peptide content, the greatest variety of molecular forms of immunoreactive enkephalin-like peptides can be detected in extracts of total adrenal tissues. To obtain a complete spectrum of the multiplicity of the molecular forms of these enkephalin-like peptides, one must compare the results obtained with a number of procedures, including tryptic digestion and antisera specifically directed against (leu[5])- or (met[5])-enkephalins. The omission of tryptic digestion and the use of radioreceptor assay combined with antibodies directed toward (leu[5])-enkephalin reveals a profile of the various molecular forms of enkephalin-like peptide which has a low degree of heterogeneity similar to that reported by Viveros et al. (8). The validity of such a comment concerning the methodology can be appreciated when one compares the complexity of the pattern of immunoreactivity detected with (met[5])-enkephalin antibodies (Table 1) and the simple profile detected in the same

TABLE 1: Met[5]-enkephalin-like immunoreactive peptides in adrenal tissue preparations and in the blood of effluent from adrenal.

Tissue	MW 2000		MW 2000–1000		MW 1000	
	No. Peptides Detected	Relative Abundance	No. Peptides Detected	Relative Abundance	No. Peptides Detected	Relative Abundance
Adrenal gland	3	+++++	2	+++	3	++
Chromaffin cells	2 or more	+++++	?	–	1	+
Chromaffin granules	3	+++++	–	Unknown	–	+
Splanchnic nerves	–	+	–	Unknown	–	+
Blood plasma	2 or more	+	–	Unknown	–	+

2 or more = Two main fractions and other minor fractions

? = possibly in a very small quantity

– = undetermined

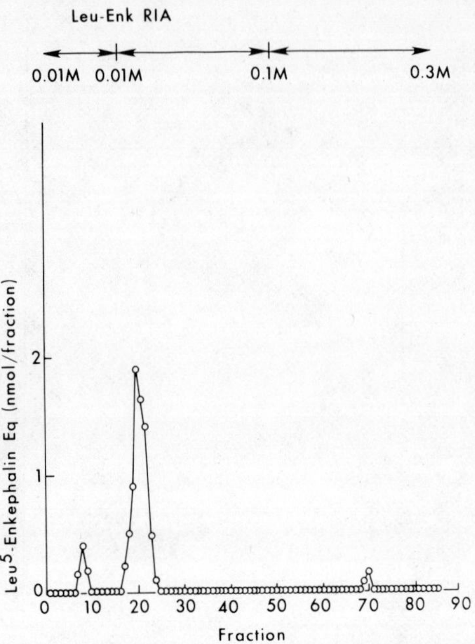

FIGURE 1: Twenty grams of bovine adrenal medulla were homogenized
in 5 vol of 1 N CH$_3$COOH and the homogenate was then boiled
and centrifuged. The extraction was repeated once on the
precipitate. The combined supernatant was lyophilized and
extracted with acetone: 1 N CH$_3$COOH (70:30) mixture. The
soluble fraction, after removal of the acetone by petrole-
um ether, was lyophilized and subjected to a SP, Sephadex
C-25 column chromatography. The column, 1.5 x 30 cm pre-
equilibrated with 0.01 M pyridine acetate (pH 5.6) was
eluted with 80 ml of the same buffer. The column was then
eluted with a linear gradient formed by 75 ml each of the
0.01 M and the 0.01 M pyridine acetate buffer (pH 5.6) and
another gradient formed by 75 ml each of 0.1 M and 0.3 M
of the same buffer. The eluates were lyophilized, tryp-
sinized and then radioimmunoassayed with leu^5-enkephalin
antiserum.

preparation using (leu^5)-enkephalin antibodies (Fig. 1), the profile
for (leu^5)-enkephalin remained unchanged. This finding is very much
in line with the findings discussed by Viveros et al. (8) in a preceding report. Because of the simplicity of the molecular heterogeneity of enkephalin-like polypeptides, these authors suggested
that the adrenal gland contains only enkephalin(s) that are mainly

stored in the adrenal chromaffin cells. Undoubtedly, as shown in
Table 1, extracts of primary cultures of chromaffin cells show a
less complicated pattern of polypeptide heterogeneity than extracts
of total adrenal gland. As shown in Table 1, the primary cultures of
chromaffin cells appear to lack a class of peptides with a molecular
weight between 1000 and 2000 dalton; in addition, these cells contain a very insignificant amount of enkephalins (molecular weight
below 1000 dalton). This finding could be interpreted to indicate
that the peptides with molecular weights of 1000 to 2000 dalton may
be formed during secretion, and, in fact, the culture medium does
contain more low molecular weight enkephalin-like peptides than the
chromaffin cells. When the ACh secreted by the axon terminals of
the splanchnic axons is added to the medium, it stimulates the
secretion of enkephalin-like peptides which include high and low
molecular weight forms (Stine, unpublished observations, this laboratory).

These results stimulated further work on the secretion mechanism. This work was carried out in the free moving dog with a
permanent cannula in the adrenal vein. In this preparation, the
i.v. injection of methacholine stimulates the secretion of enkephalin-like peptides from the adrenal gland. This is revealed by
the presence of opiate peptides in the blood collected from the
adrenal vein. Normally, this blood contains more (met^5)-enkephalin-
like material than the arterial blood or the blood from the femoral
vein. In the dog there is more than one receptor that can be activated by neurally released ACh and at least theoretically, several
receptors could participate in the regulation of the secretion of
enkephalin-like peptides from medulla. Indirect evidence (8) indicates that muscarinic and nicotinic receptors stimulate enkephalin
secretion. We do not know whether both of these receptors are
located in the chromaffin cell membrane. In fact, some receptors
could be located on the membrane of splanchnic axon terminals. In
the dog, secretion of peptides into the blood of the adrenal vein
can be evoked either by electrical stimulation of the splanchnic
nerves (20 volts, 3 sec, 0.1 msec) or by the injection of morphine.
The increase in the secretion of adrenal peptides elicited by electrical stimulation of the splanchnic cannot be blocked by naloxone
pretreatment, whereas the increased peptide secretion elicited by
morphine is blocked by naloxone. Furthermore, morphine fails to
elicit a peptide secretion from denervated adrenals. It can be
inferred that morphine causes the secretion of opiate peptides from
adrenals of dog by acting on a neurally mediated process. Clearly,
the opiate receptors present in the chromaffin cells are not operative in causing secretion of enkephalin-like peptides from adrenal
medulla. In fact, even in primary cultures of bovine chromaffin
cells the addition of morphine alone does not appear to be a secretory stimulus for enkephalin-like peptides.

The necessity of innervation brings up the question of whether splanchnic nerves contain enkephalin-like peptides and opiate peptide receptors. As shown in Table 1, the splanchnic nerves in dog contain high and low molecular weight peptides, whose characteristics will now be studied in detail. Finally, in the dog with cannulated adrenal vein, electrical stimulation of the splanchnic nerve causes a conspicuous increase in the secretion of low molecular weight opiate peptides. High molecular weight enkephalin-like material also is present in the blood effluent from the adrenal, but during stimulation, these enkephalin-like peptides do not appear to increase as much as the low molecular weight enkephalin-like peptides. Since the blood from the adrenal vein contains more enkephalin-like peptides than that from the femoral artery, or femoral vein, it can be concluded that this material comes from the adrenal. The stability of the high molecular weight enkephalin-like peptides in blood is greater than that of the low molecular weight enkephalin-like peptides (Harsing, unpublished observation, this laboratory), suggesting that in blood the steady-state of the latter is maintained either by continuous formation from the high molecular weight enkephalin-like peptides or that blood contains a specific enkephalin carrier. We are currently investigating whether the high molecular weight enkephalin-like peptides that are secreted from the adrenal protect the low molecular weight peptides from enzymatic degradation.

Action of Opiates on the Nicotine Mediated
Secretion of Catecholamines

The stimulation of nicotinic receptors releases catecholamines stored in bovine chromaffin cells maintained in culture. This secretion is Ca^{++}-dependent (4). D-Tubocurarine and hexamethonium can abolish the release of catecholamines elicited by nicotine. One million cells contain about 20 nmol of catechalamines. This catecholamine storage is composed of 75% epinephrine, 17% norepinephrine and 7% dopamine. The composition of the catecholamines released by nicotine includes 75% epinephrine, 17% norepinephrine and 7% dopamine. In total, a maximally active dose of nicotine releases 3 nmol of catecholamines from 1 million chromaffin cells in 3 min. In the same cell population, this release can be repeated at least three consecutive times. The number of ACh receptors can be measured with the Bmax of α-bungarotoxin (α-Btx) binding, which binds to the cell almost irreversibly. One million chromaffin cells can bind about 5 fmol of ^{125}I-α-Btx. Moreover, α-Btx can cause a modest release of catecholamines from chromaffin cells. This release has about one-fifth of the release rate elicited by nicotine and about double the rate of the spontaneous release. When the α-Btx (10^{-6} M) is incubated for several hours with chromaffin cells, it completely blocks the catechalamine releasing action of nicotine (10^{-5} M). This

TABLE 2: K_D of various opiates for receptors of bovine adrenal
medulla and frontal cortex.

^{3}H-Ligand	K_D (nM)	
	Adrenal Medulla	Frontal Cortex
Etorphine	1.2	1.4
Diprenorphine	0.9	1.3
Naloxone	3.5	4.5
D-Ala^2-leu-enkephalin	0.8	0.5
Dihydromorphine	2.2	2.5
Ethylketazocine	1.4	2.1

TABLE 3:   Characterization of opiate receptors in membranes of
adrenal medulla and frontal cortex.

Radioactive Ligand	Displacer	K_i (nM) Adrenal Medulla	K_i (nM) Frontal Cortex
^{3}H-Etorphine	Etorphine	1.4	1.2
	Morphine	300.0	150.0
	D-ala^2-leu-enkephalin	400.0	150.0
^{3}H-Dihydro-morphine	Etorphine	1.0	1.0
	Morphine	2.5	2.5
	D-ala^2-leu-enkephalin	100.0	13.0
^{3}H-D-Ala^2-leu-enkephalin	Etorphine	0.8	0.4
	Morpine	50.0	30.0
	D-Ala^2-leu-enkephalin	0.5	0.4

finding prompted us to investigate whether the stimulation of the
opiate receptors present in the membrane of the chromaffin cells
(whose activation fails to elicit a secretion of either catecholamines or opiate peptides stored in chromaffin cells) could modulate
the secretory activity of nicotine. It was found that a number of
receptor agonists antagonized the stimulant action of nicotine when
present in appropriate concentrations: the ED_{50} of etorphine was
greater than 10^{-7} M, that of β-endorphin was 10^{-6} M, whereas, D-ala^2-
met^5-enkephalin amide or morphine were less active, the ED_{50} ranging

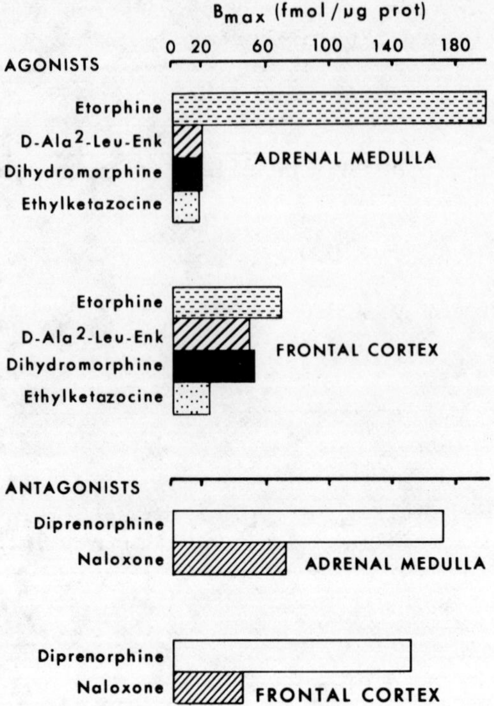

FIGURE 2: Bmax for the binding of various agonists and antagonists
to crude synaptic membrane prepared from bovine adrenal
medulla and frontal cortex.

around 10^{-4} to 10^{-3} M.   The ranking order of the potency of the
various opiate agonists tested as inhibitors of the secretion of
catecholamines elicited by nicotine did not appear to follow any of
the ranking orders described for μ, δ, ξ, β-receptors.   To clarify
this discrepancy we characterized the opiate receptors located in
the membranes of chromaffin cells by measuring the Bmax and K_D of
these recognition sites for the various ligands.

The data reported in Table 2 show that the K_D of the receptors
present in the membranes of bovine adrenal medulla are very similar
to those present in the membranes prepared from bovine cerebral
cortex.   However, when the binding capacity of both receptor families is tested with agonists and antagonists (Fig. 2), important
differences emerge.   The binding capacity of the receptors present
in membranes prepared from bovine adrenal medulla is three times

that of those from bovine frontal cortex when ^{3}H-etorphine is used
as the ligand. In contrast, the capacity to bind D-ala^2-leu-enkep-
haline and dihydromorphine, and ethylketozocine is three times
greater in the membranes of frontal cortex than in those of adrenal
medulla (Fig. 2). There were no important differences observed
with diprenorphine but the membranes of cerebral cortex have half
the binding capacity for naloxone when compared with those of adrenal medulla (Fig. 2). These differences are reflected in the
measurements of K_i for etorphine, morphine and D-ala^2-leu-enkephalin
using ^{3}H-etorphine, ^{3}H-dihydromorphine and ^{3}H-D-ala^2-leu-enkephalin
as the ligands (Table 3). The ratio of the K_i of morphine and D-
ala^2-leu-enkephalin to displace ^{3}H-etorphine is 2 to 3 times greater
in the medulla than in the frontal cortex, indicating that the receptor in the medulla has peculiar properties toward etorphine
(Table 3). These properties are different from what one would expect if the receptor were of the δ type. This difference is confirmed by the data reported in Table 3 using ^{3}H-dihydromorphine and
^{3}H-D-ala^2-leu-enkephalin as ligands.

Molecular Mechanism for the Interactions
Between Opiate and Cholinergic Receptors

When ^{125}I-α-Btx was used to measure the density of cholinergic
receptors, it was found that the number of cholinergic receptors
was smaller if the measurement was performed in the presence of
opiate receptor agonists. This decrease appears to be due to an
allosteric mechanism because a reciprocal plot of the concentrations
of a given opiate agonist vs the Bmax of cholinergic receptor binding
fails to document a competitive interaction. Moreover, it was found
that the ranking order of the dose of opiates that prevent the
binding of 50% of the ^{125}I-αBtx was equal to the dose that inhibits
the release of catecholamines elicited by nicotine by 50%. This
relationship indicates that these two phenomena which appear to be
allosteric in nature are associated. The molecular nature of this
interaction is being investigated.

CONCLUSIONS

Opiate peptides that are immunoreactive to (met^5)- or (leu^{5}0-
enkephalin antibodies represent a heterogeneous class of compounds
which are stored in splanchnic (Table 1) and in chromaffin cells
(Table 1). They can be divided into high, intermediate and low
molecular weight forms. Though these various classes of compounds
have not been characterized biologically it is known (9) that the
high molecular weight forms are devoid of opiate receptor affinity,

whereas the low molecular weight forms bind to the opiate receptors. Authentic (met[5])- and (leu[5])-enkephalins have been identified in the adrenal tissues and in the blood effluent from the cannulated adrenal vein of the dog. Because when either of the two pentapeptides is added to blood it is rapidly destroyed, one wonders whether these pentapeptides are secreted bound to a protective protein. Alternatively, since the pentapeptides and high molecular weight enkephalin-like compounds are present in blood, it should be investigated to see whether a steady-state of enkephalins is maintained by metabolizing the high molecular weight enkephal in-like compounds into enkephalins

Finally, one might even think that either the high molecular weight enkephalin-like compounds are carried to specialized sites in the periphery where they can perform the physiological action or that they are competitive inhibitors of the enzyme that rapidly destroys enkephalin in the blood. More conservatively, the multiple forms of enkephalin-like peptides detected in adrenal medulla could be a reflection of a multiplicity of actions performed by these opiat peptides. In support of this view is the finding that intermediate molecular weight forms of enkephalin-like peptides cannot be metabolized into enkephalins; perhaps these compounds have a specific activity profile. We have found that the secretion of opiate peptides from medulla is stimulated by ACh, morphine and electrical stimulation of the splanchnic nerve. The secretion of peptides elicited by electrical stimulation cannot be blocked by naloxone or atropine, suggesting that nicotinic receptors may be operative. Morphine acts indirectly because it fails to cause secretion when injected into dogs with denervated adrenal or when added to the primary cultures of bovine chromaffin cells. In these cell cultures the secretion of opiate peptides appears to be mediated only by nicotinic receptors. In contrast, the secretion from dog adrenal can be elicited by methacholine. Preliminary data seem to support the view that methacholine action as well, requires the presence of innervation. Splanchnic nerves contain ACh and enkephalin-like peptides of high and low molecular weight. Since chromaffin cells contain opiate and nicotinic receptors we have studied whether opiate peptides modify the action of ACh on chromaffin cells. If the splanchnic peptides act on the opiate receptors of chromaffin cells, it is possible that these peptides mimic the action of exogenous opiates. Since the latter reduces the number of nicotinic receptors available and curtails the ability of a given dose of nicotine to release catecholamines, it is proposed that endogenous peptides exert a similar action when released from the splanchnic by nerve impulses. This modulation may be an example of a more generalized action that the opiate peptides secreted from the adrenal medulla could exert on cholinergic receptors located in periphery. It is suggested as a working hypothesis that blood borne enkephalin-like peptides secreted from the adrenal medulla could reach peripheral ACh receptors and reduce their number and/or excitability. On more general grounds

it is proposed that the co-existence of neuropeptides and primary
transmitters in splanchnic axons may be an index of a functional
interaction between primary transmitter and co-transmitter neuropeptide at postsynaptic receptors on chromaffin cells. Perhaps the
primary transmitter is released by every nerve impulse that reaches
the terminal, whereas the neuropeptide co-transmitter is released
only under particular circumstances. The latter, when released,
modulates the number of the receptors available for the primary
transmitter. Similarly, in addition, the opiate peptides secreted
from the adrenal medulla may be viewed as modulators for the number
of peripheral receptors to ACh and, perhaps, other transmitters.
This hypothesis is currently being tested.

REFERENCES

1.   Chavkin, C., Cox, B.M. and Goldstein, A. (1978): Molec.
        Pharmacol. 15:751-753.
2.   Costa, E., Di Giulio, A.M., Fratta, W., Hong, J.S. and Yang,
        H.Y.T. (1979): IN Catecholamines: Basic and Clinical Frontiers
        (eds) E. Usdin, I.J. Kopin and J.D. Barchas, Pergamon Press,
        Oxford, pp. 1020-1025.
3.   Hokfelt, T., Lundberg, J.M., Schultzberg, M., Johansson, O.,
        Ljungdahl, A. and Rehfeld, J. (1980): IN Advances in Bio-
        chemical Psychopharmacology Vol 22 (eds) E. Costa and M.
        Trabucchi, Raven Press, New York, pp.1-23.
4.   Kumakura, K., Guidotti, A. and Costa, E. (1979): Molec.
        Pharmacol. 16:865-876.
5.   Kumakura, K., Guidotti, A., Yang, H.Y.T., Laiani, L. and
        Costa, E. (1980): IN Advances in Biochemical Psychopharma-
        cology Vol. 22 (eds)E. Costa and M. Trabucchi, Raven Press,
        New York, pp.571-580.
6.   Kumakura, K., Karoum, F., Guidotti, A. and Costa, E. (1980):
        Nature 283:489-492.
7.   Schultzberg, M., Hokfelt, T., Terenius, L., Elfvin, L-G.,
        Lundberg, J.M., Brandt, J., Elde, R.P. and Goldstein, M.
        (1979): Neuroscience 4:249-270.
8.   Viveros, O.H., Diliberto, E.J.Jr., Hazum, E. and Chang, K-J.
        (1979): Molec. Pharmacol. 16:1101-1108.
9.   Yang, H.Y.T., Di Giulio, A.M., Fratta, W., Hing, J.S., Majane,
        E.A. and Costa, E. (1980): Neuropharmacology 19:209-215.

EFFECT OF LATRODECTUS MACTANS TREDECIMGUTTATUS VENOM ON SYNAPTIC

TRANSMISSION OF CHICKEN CILIARY GANGLION IN VITRO

L. Bolzoni, A. Meneguz* and P. Paggi

Istituto di Fisiologia Generale, Universita di Roma, and
*Istituto Superiore di Sanita, Roma, Italy

INTRODUCTION

The venom of the black widow spider (BWSV), Latrodectus mactans
tredecimguttatus, contains a neurotoxin causing a massive transmitter release from various vertebrate terminals (3,6,9,13). The
pharmacological and physiological effects of this venom, in vivo
and in vitro, are well known (1,7,15). On the other hand, its
mechanism of action is still not clear, despite the fact that much
evidence on the mode of action of this venom at vertebrate neuromuscular junctions has been collected in recent years (5). Some
insight into this problem was furnished by previous investigations
on the rat sympathetic ganglion (13,14).

The avian ciliary ganglion is particularly interesting in the
study of this problem because of its morphological and physiological
characteristics. This parasympathetic ganglion contains two cholinergic neuronal populations, i.e. ciliary and choroid neurones
(8). For both the ciliary and choroid cells the transmission is
cholinergic, with the exception being that the electrical coupling
is restricted to the ciliary cells (10). In the present paper we
describe the effect of BWSV on synaptic transmission and acetylcholine (ACh) release in this parasympathetic ganglion in vitro.

MATERIALS AND METHODS

Ciliary ganglia from 30-40 day old chickens were excised and
incubated at 23°C as previously reported (1). The response to

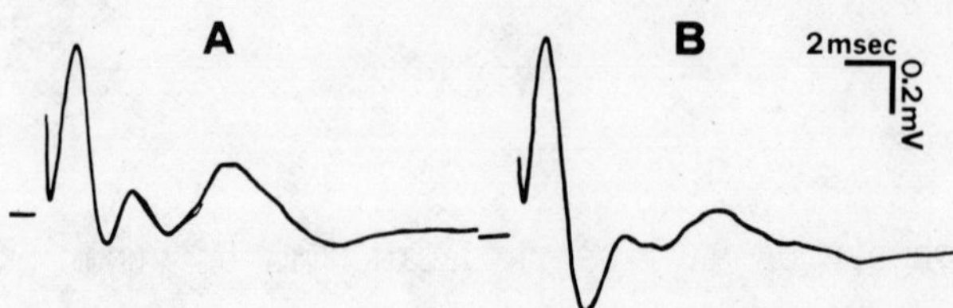

FIGURE 1: Effect of black widow spider venom on synaptic trans-
         mission in the chicken ciliary ganglion. The responses
         were elicited by supramaximal stimulation of the pre-
         ganglionic trunk (oculomotor nerve) and recorded from
         the postganglionic trunks (ciliary and choroid nerves
         together). ˌA = record obtained before the addition of
         black widow spider venom; B = record obtained after
         45 min incubation in the presence of 20 µl black widow
         spider venom.

preganglionic stimulation, recorded from the ciliary and choroid
nerves together, was a 3-peaked compound action potential reflecting both the electrical coupling of the ciliary cells (first peak)
and the cholinergic transmission through the ciliary and choroid
cells (second and third peaks, respectively)(11)(Fig. 1A). The
preganglionic trunk was stimulated supramaximally (at 5 Hz) and
the amplitude of the compound action potentials was recorded. The
observed depression consequent to the addition of BWSV was expressed as percent difference between the amplitude of the action potentials recorded before and after BWSV addition. The ACh content
of the tissue (ganglion body + about 1 mm of pre- and 1 mm of postganglionic nerves) and the incubation medium was assayed on the
guinea pig ileum according to the method of Mayer and Michalek (12).
In order to determine the ACh released into the incubation medium
and that contained in the ganglion, the resting ganglion was eserinized (5 µg/ml prostigmine; Roche, Milan) and incubated for 1 hr
in the presence of BWSV. The BWSV was prepared from adult female
Latrodectus glands (kindly provided by Dr. N. Frontali, Istituto
Superiore di Sanita, Rome), homogenized in 0.04 M Tris-HCl, pH 7.5,
containing 0.15 M NaCl, 0.1 mM dithioerythritol and centrifuged
at 12,000 x g for 20 min at 4°C. Only the supernatant was used
(10 µl of the supernatant corresponded to 0.65 glands).

RESULTS

When BWSV was added to the incubation medium in a final concentration of 20 μl/ml, the height of the chemically transmitted action potential was depressed by 56 $\pm$ 4%, the onset of the effect being observed within 12 $\pm$ 2 min and the maximum in 43 $\pm$ 4 min (mean $\pm$ S.E.M. of 13 experiments)(Fig. 1). The first peak (due to electrical coupling) was not impaired during 45 min of incubation in the presence of BWSV. This peak often was decreased after 1 hr incubation with the venom.

Under similar experimental conditions the ACh content in the ganglion decreased from 77 $\pm$ 2 ng ACh/ganglion to 59 $\pm$ 1 (-23%) and the ACh released into the medium increased from 157 $\pm$ 4 ng ACh/ml incubation medium to 282 $\pm$ 7 (+80%). The reported data are the average ($\pm$ S.E.M.) of 5 experiments. When a higher venom concentration (30 μl/ml) was employed, the decrease in ganglion ACh content reached 57% and the ACh released into the medium was increased by 110%.

DISCUSSION

The absence of any impairment of electrical coupling for at least 45 min of incubation with BWSV, excludes the possibility that the venom exerts its immediate action on chemical transmission as a depolarizing agent. On the other hand, the decrease in amplitude often observed after 1 hr of incubation in the presence of the venom could be due to an alteration of the synaptic structure. Indeed, this period of incubation with BWSV has been previously reported to cause swelling of nerve terminals in the frog neuromuscular junction (2).

It has been seen that a 20 μl/ml concentration of BWSV (equivalant to 1.3 glands/ml) is able to stimulate ACh release but probably is not sufficient to deplete the nerve terminals. In fact, a higher concentration of 30 μl/ml of venom produced a greater depletion of ganglion ACh. This concentration of BWSV necessary to affect the ciliary ganglion is very high compared to that affecting mouse neuromuscular junction (0.08 glands/ml) (4), but is in the same range as that effective on rat sympathetic ganglion (0.5-1 glands/ml)(13).

ACKNOWLEDGEMENTS

This research was partially supported by C.N.R. ct.n. 79. 01990.04.

REFERENCES

1. Conti-Tronconi, B., Gotti, C., Paggi, P. and Rossi, A. (1979): Brit. J. Pharmacol. 66:33-38.

2. Clark, A.W., Hurlbut, W.P. and Mauro, A. (1972): J. Cell Biol. 52:1-14.

3. Frontali, N. (1972): Brain Res. 37:146-148.

4. Gorio, A., Hurlbut, W.P. and Ceccarelli, B. (1978): J. Cell Biol. 78:716-733.

5. Gorio, A., Rubin, L.L. and Mauro, A. (1978): J. Neurocytol. 7:193-205.

6. Granata, F., Paggi, P. and Frontali, N. (1972): Toxicon 10:551-555.

7. Grasso, A., Alema, S., Rufini, S. and Senni, M.I. (1980): Nature 283:774-776.

8. Hess, A. (1965): J. Cell Biol. 25:1-19.

9. Longenecker, H.E. Jr., Hurlbut, W.P., Mauro, A. and Clark, A.W. (1970): Nature 225:701-703.

10. Martin, A.R. and Pilar, G. (1963): J. Physiol. 168:464-475.

11. Marwitt, R., Pilar, G. and Weakly, J.N. (1971): Brain Res. 25:317-334.

12. Mayer, O. and Michalek, H. (1971): Biochem. Pharmacol. 20:3029-3037.

13. Paggi, P. and Rossi, A. (1971): Toxicon 9:265-269.

14. Paggi, P. and Toschi, G. (1972): Life Sci. 11:413-417.

15. Sampayo, R.R.L. (1944): J. Pharmacol. Exp. Ther. 80:309-322.

REGULATION OF ACETYLCHOLINE SYNTHESIS AND RELEASE

IN THE ISOLATED HEART

K. Loffelholz, R. Lindmar and W. Weide

Department of Pharmacology, University of Mainz
Obere Zahlbacher Strasse 67
D-6500 Mainz, Federal Republic of Germany

INTRODUCTION

In the two decades following Loewi's work (8,9) on the frog
heart several research groups tried to use the isolated heart preparation from higher vertebrates for further investigations on
synthesis, release and inactivation of acetylcholine (ACh). However
the results were discouraging because the overflow of ACh during
vagal stimulation was near, or below, the limit of the assay.

Some years ago, while studying adrenergic mechanisms of the
chicken heart, we found not only an overflow of noradrenaline and
adrenaline (1), but also a large output of ACh into the perfusate
even in the absence of physostigmine (2,6). During the last years,
the perfused chicken heart proved to be a useful synaptic model for
studying effects of substances such as choline (Ch)(3) and pentobarbital (7) on the release of ACh. In the present study we want
to present recent data on the regulation of synthesis and release
of ACh in this preparation.

METHODS

ACh in the perfusate was determined by bioassay on the guinea
pig ileum. The identity with ACh was confirmed in 36 samples by
additional determination using the radioenzymatic method described
by Goldberg and McCaman (5). This method was also used to measure
Ch in the perfusate.

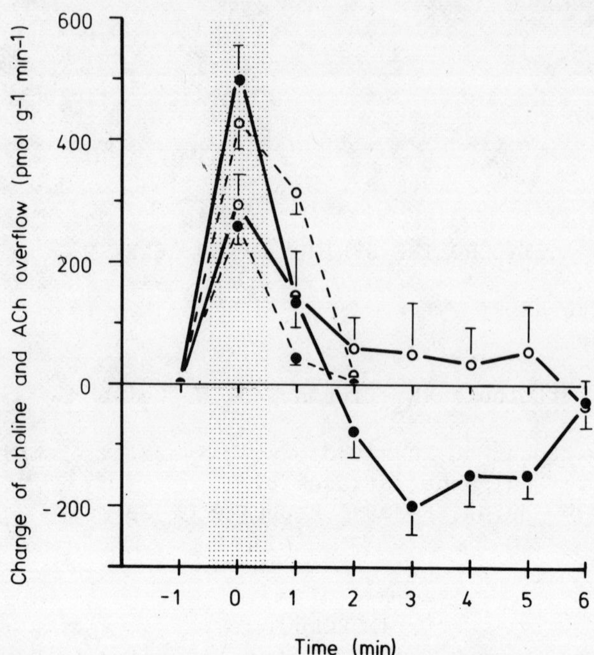

Time (min)

FIGURE 1: Effects of vagal stimulation and exogenous ACh on Ch and
ACh overflow into the perfusate of the chicken heart. The
nerves were stimulated at 20 Hz for 1 min (closed circles)
or 2×10^{-7} M ACh was infused for 1 min (open circles).
Ordinate, change in the overflow of Ch (solid lines) and
ACh (broken lines) evoked by nerve stimulation or exogenous ACh (shaded area). Given are means $\pm$ S.E.M. of 20
(nerve stimulation) or 4 experiments (ACh infusion).

RESULTS

The heart was perfused with Tyrode's solution (20 ml/min) at
36°C. At rest, the efflux of Ch ranged between 1.3 and 4.5 nmol
$g^{-1}min^{-1}$ (n = 66), whereas the spontaneous overflow of ACh was only
22 ± 4 (n = 18), or 67 ± 12 pmol $g^{-1}min^{-1}$ (n = 6) in the absence or
presence of physostigmine, respectively.

Figure 1 shows the changes in Ch and ACh overflow induced by
stimulation of both cervical vagus nerves for 1 min at 20 Hz (1 ms,
40 V). The experiments were carried out in the absence of cholinesterase inhibition. Vagal stimulation increased the overflow of
ACh by 239 pmol/g and of Ch by 499 pmol/g (n = 20) during stimulation, which means that about two-thirds of the ACh released was

hydrolyzed. The rate of washout was rapid (half-time < 1 min) for
both ACh and Ch. After the initial increase, the Ch efflux fell to
below the prestimulation level in the second minute after stimulation, reached a "negative maximum" in the third minute and then
gradually returned to the prestimulation level. The area under the
"negative Ch curve" represents Ch which was removed from the extracellular space upon vagal stimulation. We assumed that this removal
was caused by neuronal uptake to support ACh synthesis and a replacement of the ACh released. The following experiments were
designed to prove this hypothesis.

In order to prove whether the removal of Ch was mediated by
cholinoceptors, ACh (2×10^{-7} M) was infused for 1 min instead of
nerve stimulation (Fig. 1). Most of the infused ACh occurred unmetabolized in the perfusate; 41% was hydrolyzed to Ch. Both treatments, ACh infusion and vagal stimulation, increased initially ACh
overflow and Ch efflux and caused cardioinhibition. However, the
Ch efflux fell to below the prestimulation level only after stimulation and not after infusion of ACh. It is concluded that the removal of Ch observed after nerve stimulation was not a postsynaptic
phenomenon.

Hemicholinium-3 (HC-3) (2×10^{-5} M) when present throughout the
experiment decreased the area under the "negative Ch curve" by 65%
but did not change "ACh release" (Fig. 2). ACh release is the sum
of the increases in ACh and Ch overflow evoked by stimulation which
is expected to be close to the amount of ACh released into the
extracellular space. 10^{-6} M Physostigmine reduced the ACh release
by about 40%, an effect which is probably due to negative feedback
inhibition. After inhibition of the cholinesterase activity, the
overflow of unmetabolized ACh increased from 239 $\pm$ 28 (n = 20) to
654 $\pm$ 73 pmol/g (n = 13), whereas the efflux of Ch was not increased.
The area under the negative Ch curve was unchanged by physostigmine
(Fig. 2). To sum up, the observed removal of extracellular Ch was
sensitive to HC-3, was not a consequence of the initial increase of
the Ch efflux and had a size that was equimolar to the preceding
release of ACh (in the presence of physostigmine).

Under control conditions (first pair of columns in Fig. 2),
the removal of Ch is probably larger than the area under the negative Ch curve, since the formation of Ch from released ACh would
reduce or even reverse the negative change in Ch efflux, if both
events occurred in parallel for some time. Thus, in order to study
the factual timecourse of the Ch removal it was necessary to inhibit
the cholinesterase activity. Figure 3 (lower diagram), shows the
negative Ch curve in the presence of physostigmine. The removal of
Ch was small and insignificant during the 1 min period of stimulation, reached the maximum in the second minute after stimulation
and returned to prestimulation level in the sixth minute. If the

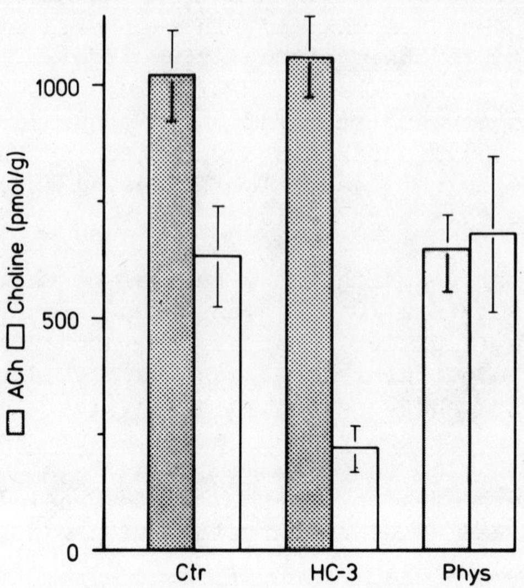

FIGURE 2: Effect of hemicholinium-3 and physostigmine on "ACh release"and "area under the negative Ch curve." ACh release
(shaded column) is the sum of the increases in ACh and Ch
overflow evoked by nerve stimulation (see Fig. 1). "Negative Ch curves" (open column) are shown in Figs. 1 and 3.
Hemicholinium-3 (2×10^{-5} M) and physostigmine (10^{-6} M)
were present at least 10 min prior to nerve stimulation.
The data are expressed as means $\pm$ S.E.M. of 18 (control =
Ctr), 7 (hemicholinium-3 = HC-3) and 13 (physostigmine =
Phys) experiments.

negative Ch curve reflects neuronal net uptake caused by nerve
stimulation, ^{14}C labelling of newly synthesized ACh should be most
effective when ^{14}C-Ch (1.6×10^{-7} M; 1 min) was infused shortly
after stimulation and should be the least effective when applied
1 or 2 min before or 5-10 min after stimulation. The efficacy of
the labelling procedure was determined by measuring the ^{14}C-ACh
release evoked by a second period of stimulation 20 min after the
first, the activating one.

^{14}C-Ch was infused at seven different times before, during and
after the first stimulation, as indicated in Fig. 3 (upper diagram),
by the arrows a to g. The respective release of ^{14}C-ACh evoked by
the second stimulation was smallest under conditions a and b (infusion before stimulation) and reached a maximum at d, when ^{14}C-Ch
had been infused in the first minute after stimulation. The

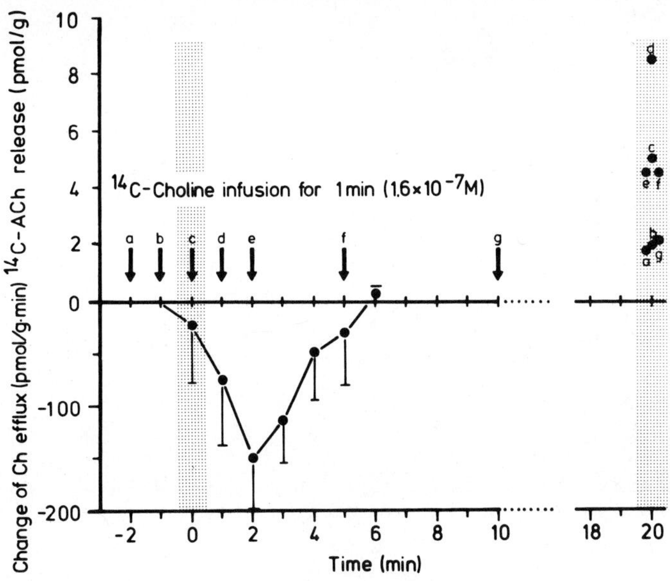

FIGURE 3: LOWER DIAGRAM: The "negative Ch curve" in the presence of
          physostigmine. The nerves were stimulated at 20 Hz for
          1 min (zero time, shaded area). Ordinate, decrease in Ch
          efflux. The data are expressed as means $\pm$ S.E.M. of
          13 experiments.
          UPPER DIAGRAM: Variation of the ^{14}C-ACh release evoked
          by nerve stimulation with the time of ^{14}C-Ch infusion
          before, during and after an "activating" stimulation. The
          nerves were stimulated for 1 min at 10 Hz at zero time
          and after 20 min (shaded areas). ^{14}C-Ch was infused for
          1 min as indicated by the arrows a to g. Ordinate, release
          of ^{14}C-ACh in response to the second stimulation (closed
          circles; the letters indicate the time of ^{14}C-Ch infusion;
          see arrows). The data are expressed as mean of 4-20
          experiments. For further details see Fig. 4.

prestimulation level was observed at g (infusion 10 min after stimulation). If these values were plotted against the time of their
respective ^{14}C-Ch infusion, the resulting curve looks close to a
mirror image of the negative Ch curve (Fig. 3, lower diagram).
However, the maxima of these curves differ by 1 min. This difference
has presumably methodical reasons. Changes in the Ch concentration
of the extracellular space appear in the perfusate, which is our
assay material, after a delay of approximately half a minute due to
the rate of extracellular washout of Ch (half-time < 1 min, see
above). Thus, the negative Ch curve should be shifted to the left

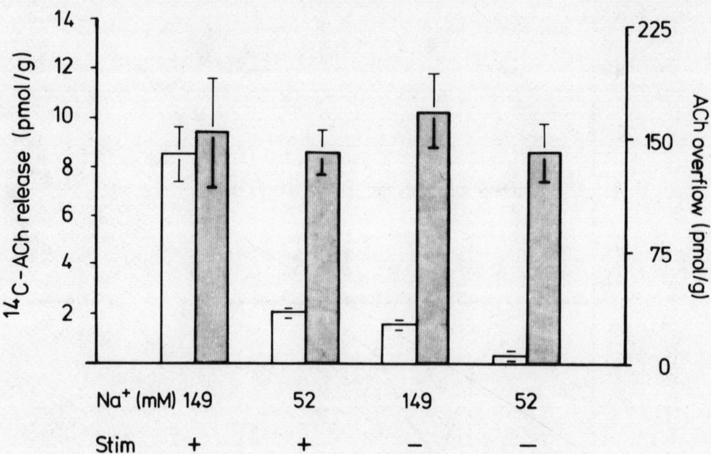

FIGURE 4: Activation by electrical nerve stimulation and Na$^+$-
dependence of the ^{14}C-Ch uptake leading to the release
of labelled ACh. The design of the experiments is shown
in the upper diagram of Fig. 3. ^{14}C-Ch was infused for
1 min beginning 19 min before the second period of stimulation. In some experiments (as indicated) the first
stimulation (Stim) was deleted and the Na$^+$ concentration
was reduced during and for 7 min after ^{14}C-Ch infusion.
Shaded columns, overflow of nonlabelled ACh. Open columns,
release of ^{14}C-ACh. The data are expressed as means $\pm$
S.E.M. of 4 experiments each.

by less than a minute when this factor is taken into account.

Finally we investigated the release of labelled ACh when the
first stimulation was deleted. Under this condition, the maximum
release of ^{14}C-ACh dropped from 8.6 (n = 20) to 1.8 pmol/g (n = 6),
which is regarded as basic release. All values above this basic
release are considered to be increased by the activating stimulation (experiments c-f in Fig. 3).

The uptake of ^{14}C-Ch leading to the basic and the activated
release of ^{14}C-ACh was found to be dependent on extracellular Na$^+$
(Fig. 4). In all experiments ^{14}C-Ch was infused 19 min before the
second stimulation and the Na$^+$ concentration was reduced from 149
to 52 mM during infusion of ^{14}C-Ch and in the following 7 min.
Sucrose was used to maintain isotonicity. Low extracellular Na$^+$
caused a reduction of the basic and the activated release of ^{14}C-

ACh by 70-80%. The Na^+ concentration was normal during all periods
of nerve stimulation. Therefore the overflow of unlabelled ACh was
equal at all conditions tested. It is concluded that both experimental conditions, reduction in the Na^+ concentration and activating
nerve stimulation, did not affect the release process itself, but
changed the uptake of ^{14}C-Ch.

DISCUSSION

The spontaneous efflux of Ch ranging between 1.3 and 4.5 nmol
$g^{-1}min^{-1}$ from the isolated chicken heart was about 30 times higher
than the spontaneous output of ACh (67 pmol $g^{-1}min^{-1}$; physostigmine
present). Therefore, the major source of the spontaneous Ch efflux
is not ACh but presumably Ch-containing phospholipids of the myocardium.

It has been shown previously that ACh synthesis in autonomic
nerves is dependent on uptake of extracellular Ch (for review see
Ref. 10). Since the neuronal net uptake of Ch is increased after
the release of ACh, vagal stimulation had been expected to cause a
transient decrease in the spontaneous Ch efflux from the heart into
the perfusate. It was found that vagal stimulation for 1 min caused
a biphasic change in the Ch efflux when the cholinesterase activity
was not inhibited. The initial increase due to hydrolysis of released ACh was followed by a decrease that seemed to reflect the
timecourse of the stimulation-induced neuronal net uptake of Ch.
The decrease was not a postsynaptic phenomenon, was sensitive to
HC-3 and its size (area under the negative Ch curve) was equimolar
to the preceding release of ACh.

The possibility of measuring neuronal net uptake of Ch indirectly by determining removal of extracellular Ch is advantageous in the
light of the following aspects. First, a direct determination of
the Ch taken up by the cardiac nerves appears to be difficult in the
heart because of the preponderance of nonneuronal cells. Second,
the timecourse of the neuronal net uptake activated by a 1 min period
of nerve stimulation could be measured in a single heart preparation
which would not be possible by "direct" estimation of the amount of
Ch taken up. Third, the possibility of measuring the neuronal net
uptake of Ch in the perfused chicken heart is appreciated in particular because this preparation proved to be a useful tool for studying
release (see Introduction) and inactivation of released ACh. It was
possible to determine the fraction of ACh that was hydrolyzed to Ch
after its release and, for example, to show a reduction by physostigmine of the evoked release of ACh without using a radioactive
method (11). Hence the present study describes the temporal and

quantitative relationship between the release of ACh and the consecutive net uptake of Ch caused by a 1 min period of nerve stimulation. The major disadvantage of the removal technique is given by the fact that the maximum rate V_{max} and the half-saturation constant K_m cannot be determined.

After the removal technique has been discussed, the major results obtained by this technique will be summarized briefly. The main finding was the time course of the neuronal net uptake of Ch activated by vagal stimulation. Activation and inactivation of the Ch uptake were slow processes and the maximum rate of uptake occurred 2 min after nerve stimulation. This time course was corroborated by the result showing a fivefold increase in the specific activity of ^{14}C-ACh released by nerve stimulation when ^{14}C-Ch had been infused 1-5 min after the activating period of nerve stimulation; the maximum occurred in the first minute after stimulation. The uptake of ^{14}C-Ch leading to both basic (without preceding stimulation) and activated release of ^{14}C-ACh was highly dependent on extracellular Na^+ (Fig. 4) and was essentially blocked by 10^{-6} M HC-3 (Lindmar, Loffelholz and Weide, unpublished observation). In the previous literature, a delay between depolarization evoked release of ACh and consecutive Ch uptake has not been shown after repetitive electrical nerve stimulation because of methodical reasons but was found after a sustained depolarization-repolarization cycle, for example, in the chick ciliary nerve-iris preparation using high K^+ solution or veratridine (12) although a time course of the uptake activation and inactivation has not as yet been described. The delayed occurrence of the Ch uptake after sustained depolarization does not necessarily imply that a similar delay will occur after repetitive electrical stimulation, because sustained depolarization reduces Ch uptake and thereby may postpone the Ch uptake to the repolarization phase (12). The present results clearly show, however, such a delay also after electrical stimulation at 10 Hz.

The present results obtained in the chicken heart are compatible with the view that the resting and activated neuronal transport of Ch is due to the Na^+ dependent, high affinity uptake of Ch (4,13). Moreover, the transport may be activated by a reduced concentration of cytoplasmic Ch or, in other words, by a reduced Ch gradient. It is highly unlikely that after-hyperpolarization or an altered Na^+ concentration following the 1 min period of nerve stimulation were directly involved in the activation of the net uptake under the present conditions because of the long delay of up to 5 min between nerve stimulation and uptake activation. The finding that the uptake of Ch was equimolar to the preceding release of ACh is consistent with the view that the synthesis of ACh is controlled by the law of mass action and is totally dependent on the supply with extracellular Ch.

ACKNOWLEDGEMENTS

This work was supported by a grant from the Deutsche Forschungsgemeinschaft. We wish to thank Miss U. Wolf for excellent technical assistance.

REFERENCES

1.   DeSantis, V.P., Langsfeld, W., Lindmar, R. and Loffelholz, K.
       (1975): Brit. J. Pharmacol. 55:343-350.
2.   Dieterich, H.A., Kaffei, H., Kilbinger, H. and Loffelholz, K.
       (1976): J. Pharmacol. Exp. Ther. 199:236-246.
3.   Dieterich, H.A., Lindmar, R. and Loffelholz, K. (1978): N.-S.
       Arch. Pharmacol. 301:207-215.
4.   Haga, T. and Noda, H. (1973): Biochim. Biophys. Acta 291:
       564-575.
5.   Goldberg, A.M. and McCaman, R.E. (1974): IN Choline and Acetyl-
       choline. Handbook of Chemical Assay Methods. (ed) I. Hanin,
       Raven Press, New York, pp. 47-61.
6.   Kilbinger, H. and Loffelholz, K. (1976): J. Neur. Trans. 38:9-14.
7.   Lindmar, R., Loffelholz, K. and Weide, W. (1979): J. Pharmacol.
       Exp. Ther. 210:166-173.
8.   Loewi, O. (1921): Pflug. Arch. Ges. Physiol. 189:239-242.
9.   Loewi, O. and Navratil, E. (1926): Pflug. Arch. Ges. Physiol.
       208:689-696.
10. MacIntosh, F.C. and Collier, B. (1976): IN Neuromuscular Junc-
       tion. Handbook of Experimental Pharmacology. (ed) E. Zaimis,
       Springer-Verlag, New York, pp. 99-228.
11. Szerb,J.C. and Somogyi, G.T. (1973): Nature New Biology 241:
       121-122.
12. Vaca, K. and Pilar, G. (1979): J. Gen. Physiol. 73:605-628.
13. Yamamura, H.I. and Snyder, S.H. (1973): J. Neurochem. 21:
       1355-1374.

MODULATION BY SCOPOLAMINE, ACETYLCHOLINE AND CHOLINE OF THE EVOKED

RELEASE OF ACETYLCHOLINE FROM THE GUINEA PIG MYENTERIC PLEXUS:

EVIDENCE FOR A MUSCARINIC FEEDBACK INHIBITION OF ACETYLCHOLINE

SECRETION

H. Kilbinger, R. Kruel and I. Wessler

Pharmakologisches Institut der Universitat Mainz
D-6500 Mainz, West Germany

INTRODUCTION

There is evidence that the release of acetylcholine (ACh) from
the guinea pig myenteric plexus is controlled via presynaptic
muscarine receptors. Muscarinic antagonists such as atropine enhance the release evoked by either electrical field stimulation, by
nicotinic drugs or by high K^+ concentrations (4,7,10). On the other
hand, the muscarinic agonist oxotremorine inhibits the evoked
release of ACh (7). A comparable feedback inhibition has been
described for the release of ACh from central cholinergic nerves
(for review, see Ref. 12). However, it has so far not been shown
whether the physiological transmitter itself is able to depress the
release of neuronal ACh. We have, therefore, studied the effects
of extracellular ACh on the release of labelled neuronal ACh from
the longitudinal muscle-myenteric plexus preparation of the guinea
pig. In addition, the effects of scopolamine and choline (Ch) on
the evoked outflow of labelled ACh in the absence of a cholinesterase inhibitor were investigated. The measurement of ACh release
from the myenteric plexus without cholinesterase inhibition was
based on the method described by Szerb (11); incubation of the
myenteric plexus preparation with $[^3H]$-Ch leads to the formation of
$[^3H]$-ACh in the tissue. Electrical field stimulation causes an outflow of $[^3H]$-Ch which originates from the release of $[^3H]$-ACh that
is hydrolyzed by cholinesterase.

METHODS

Longitudinal muscle strips with the myenteric plexus attached were prepared from the guinea pig ileum. Two strips were suspended in a 2 ml organ bath under a tension of 1 g and superfused with Tyrode's solution (1 ml/min) that contained 1 µM Ch. After a 30 min equilibration period superfusion was stopped and the strips were incubated with [methyl-^{3}H]-Ch (5 µCi/ml). The strips were stimulated with square wave pulses (1 msec; 0.2 Hz) during this labelling period in order to increase the turnover rate of neuronal ACh. Two platinum electrodes were used to stimulate the strips. At the end of the incubation with [^{3}H]-Ch the strips were superfused with Tyrode's solution that contained, in most of the experiments, 10 µM hemicholinium-3 (HC-3). After a 60 min washout period the strips were stimulated three times (S_1 - S_3) at 1 Hz (3 min). The superfusate was collected in 2 min samples. 0.5 ml of the superfusate was added to 7 ml of a Triton-toluene scintillator and the tritium was measured by liquid scintilation spectrometry. The stimulation evoked outflow of tritium was obtained from the difference of the total tritium outflow during stimulation and the following 11 min and the calculated spontaneous outflow. Stimulation periods started 6 min (S_1), 26 min (S_2) and 67 min (S_3) after the 60 min washout period. Drugs were added to the superfusate 21 min before S_3. Their effect on the evoked outflow of tritium is expressed as the ratio S_3/S_2. It has been shown (9) that the outflow of tritium induced by electrical field stimulation of the myenteric plexus originates from the release of [^{3}H]-ACh. Therefore, the stimulation-evoked outflow of tritium (difference between total and spontaneous outflow) is designated as release of [^{3}H]-ACh.

RESULTS AND DISCUSSION

Increase by Scopolamine of the Evoked Release of [^{3}H]-Acetylcholine

Scopolamine was added to the superfusate 21 min before S_3. Figure 1 shows that scopolamine increased the stimulation evoked release of [^{3}H]-ACh in a concentration dependent manner. The facilitatory effect of a maximal effective concentration of scopolamine was more marked at a frequency of 3 Hz (increase by 92%; $p < 0.001$ vs control) than at 1 Hz (increase by 37%; $p < 0.001$). The spontaneous outflow of tritium was not affect by scopolamine.

In some experiments the effect of scopolamine was studied in the presence of the cholinesterase inhibitor physostigmine (10^{-7} M). Physostigmine was added to the superfusate at the end of the 60 min

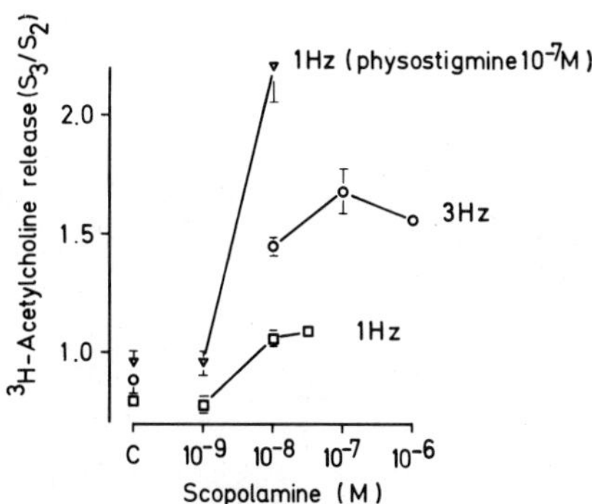

FIGURE 1: Increase by scopolamine of the evoked release of $[^3H]$-ACh.
The strips were stimulated at 1 or 3 Hz with 180 pulses
during each stimulation period(S_1, S_2, S_3). Ordinate, ratio
of $[^3H]$-ACh release evoked by S_3 and S_2. S_3, outflow of $[^3H]$-
ACh in the presence of scopolamine. C = control experi-
ments. Means ± S.E.M. of 3-7 experiments.

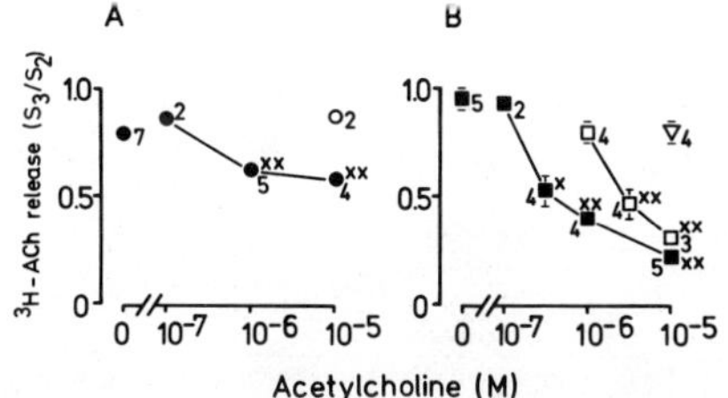

FIGURE 2: Inhibition by exogenous ACh of the evoked release of $[^3H]$-
ACh. The strips were stimulated three times at 1 Hz
(180 pulses). Ordinates, ratio of $[^3H]$-ACh release evoked
by S_3 and S_2. S_3, outflow in the presence of ACh (● ,
■), 10^{-9} M scopolamine plus ACh (□) or 10^{-8} M
scopolamine plus ACh (○ , ▽). Left panel, experiments without cholinesterase inhibition. Right panel,
experiments in which 10^{-7} M physostigmine was present in
the superfusate. Means ± S.E.M. of the number of experiments indicated. Significance of difference from controls: * p < 0.01; ** p < 0.001.

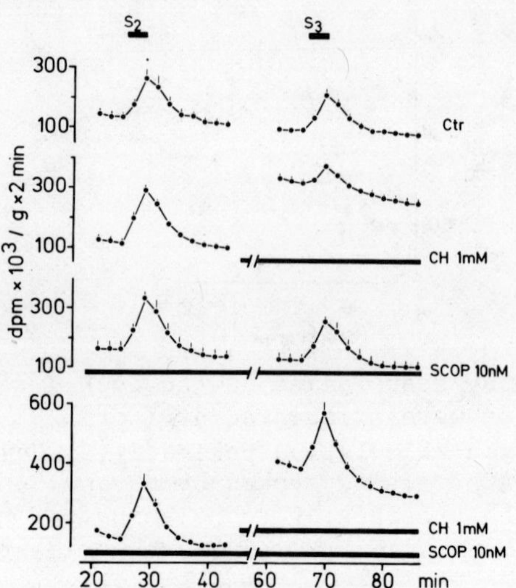

FIGURE 3: Effects of Ch alone or in combination with scopolamine on
          spontaneous and evoked outflow of tritium in the presence
          of HC-3. Strips were incubated with 1 µM [^{3}H]-Ch and sub-
          sequently superfused with Tyrode's solution containing
          10 µM HC-3. Stimulation periods S_2 and S_3 (1 Hz, 3 min)
          are indicated at the top of the Fig. 3. Horizontal bars
          indicate superfusion with 1 mM Ch, 10 nM scopolamine
          (SCOP) and Ch plus scopolamine. CTR indicates control
          experiments. Given are the means ± S.E.M. of 3-6
          experiments.

washout period and remained in the medium throughout the experiment.
Figure 1 shows that the facilitatory effect of scopolamine was much
more pronounced in the presence than in the absence of physostigmine.
This more marked increase was probably due to the fact that physostigmine inhibited the evoked release of [^{3}H]-ACh already during
the control stimulation period S_2. Scopolamine, by overcoming this
inhibition, largely enhanced the evoked release. Previous experiments have already shown that antimuscarinic drugs increase the
evoked release of endogenous ACh determined in the presence of a
cholinesterase inhibitor (7,8). The present experiments show that
scopolamine facilitates the stimulation evoked outflow of ACh also
under more physiological conditions, i.e. in the absence of a
cholinesterase inhibitor.

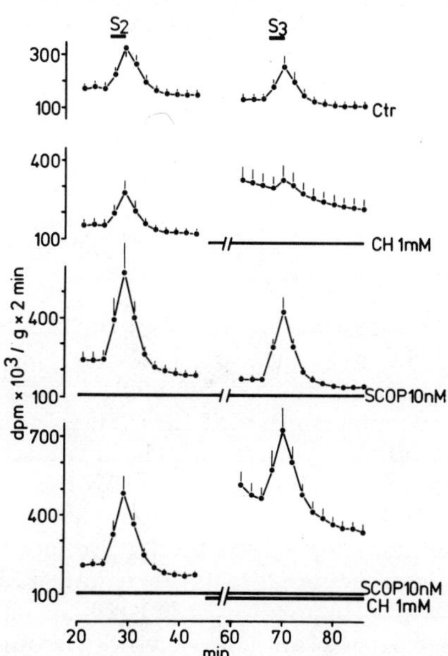

FIGURE 4: Effects of Ch alone or in combination with scopolamine on
         spontaneous and evoked outflow of tritium in the absence
         of HC-3. Strips were incubated with 1 μM [^{3}H]–Ch and sub-
         sequently superfused with Tyrode's solution. For details
         see legend to Fig. 3. Given are means ± S.E.M. of 3–7
         experiments.

Inhibition by acetylcholine and choline
of the evoked release of [^{3}H]–acetylcholine

<u>Effects of acetylcholine</u>

    Figure 2 (left panel) shows that 10^{-6} and 10^{-5} M exogenous ACh
caused slight but significant reductions in the evoked release of
[^{3}H]–ACh. Scopolamine (10^{-8} M) completely antagonized the inhibitory effect of 10^{-5} M ACh. In another series of experiments the
cholinesterase inhibitor physostigmine (10^{-7} M) was added to the
superfusate at the end of the 60 min washout period and remained
in the medium throughout the experiment. In the presence of physostigmine exogenous ACh caused a marked inhibition of the evoked
release of [^{3}H]–ACh (Fig. 1, right panel). The maximal effect
obtained with 10^{-5} M ACh (76% reduction) was completely reversed

by 10^{-8} M scopolamine. In the presence of 10^{-9} M scopolamine the concentration-response curve for the inhibitory effect of exogenous ACh was shifted to the right. From the interaction experiments with scopolamine it is concluded that exogenous ACh specifically reduces the evoked release of labelled ACh by an activation of muscarine receptors.

Effects of choline

Figure 3 illustrates the effects of Ch on both spontaneous and evoked outflow. In the presence of 1 mM Ch the stimulation evoked release of $[^3H]$-ACh was depressed by 65%. Figure 3 shows that Ch increased the spontaneous efflux of tritium. This increase may reflect displacement of $[^3H]$-Ch from the tissue by Ch. It might therefore be supposed that the decrease by Ch of the evoked outflow of $[^3H]$-ACh is due to a dilution of the specific radioactivity of the releasable neuronal pool of ACh. To check this point, interaction experiments with scopolamine were carried out. Scopolamine (10 nM) was added to the superfusion fluid at the end of the 60 min washout period and remained in the medium throughout superfusion. In the presence of scopolamine alone (Fig. 3) the ratio S_3/S_2 was not significantly different from that of control experiments. If, in addition, 1 mM Ch was added to the medium, the spontaneous outflow was accelerated to the same extent as in the absence of the antagonist; yet Ch failed to inhibit the evoked release (Fig. 3). Thus, these experiments suggest that exogenous Ch inhibits the stimulation evoked release of $[^3H]$-ACh via stimulation of muscarine receptors.

In the experiments described so far, HC-3 had been added to the superfusate in order to prevent the reuptake of Ch. Since this drug also exhibits antimuscarinic effects (1,2) we have investigated whether the inhibitory action of Ch is also seen if HC-3 is omitted from the superfusion fluid. In control experiments the evoked release of $[^3H]$-ACh during S_2 was similar, irrespective of whether HC-3 was present in the superfusate or not (cf. Figs. 3 and 4). Likewise, the decline in the evoked outflow of $[^3H]$-ACh (S_3/S_2) was similar under both experimental conditions. In the absence of HC-3 Ch (1 mM) depressed the stimulation evoked release of $[^3H]$-ACh to the same degree as in experiments in which HC-3 was present (Fig.4). The inhibitory effect was fully antagonized by 10 nM scopolamine. Comparison of Figs. 3 and 4 shows that the acceleration of spontaneous outflow of tritium in the presence of 1 mM Ch was similar in the absence and presence of HC-3.

CONCLUSIONS

Extracellular ACh inhibits the stimulation evoked release of labelled neuronal ACh. This inhibition is mediated via muscarine receptors since it is antagonized by low concentrations of scopolamine. Scopolamine causes a significant increase of the evoked release of labelled ACh even in the absence of a cholinesterase inhibitor. This suggests that the negative feedback mechanism of ACh release occurs under physiological conditions.

The evoked release of labelled ACh is also depressed by 1 mM Ch. It has been shown previously that the injection of high doses of Ch to rats and guinea pigs leads to increased levels of ACh in the central nervous system and in peripheral organs (3,5,6). Our results suggest that this increase in ACh content might not only result from the enhanced availability of the precursor for ACh synthesis but rather from the inhibition of ACh release which, in turn, leads to an accumulation of the transmitter.

ACKNOWLEDGEMENTS

The skilled technical assistance of Miss A. Mergler and Miss B. Riffel is gratefully acknowledged. This work was supported by a grant from the Deutsche Forschungsgemeinschaft.

REFERENCES

1. Bertolini, A., Greggia, A. and Ferrari, W. (1967): Life Sci. 6:537-543.
2. Bieger, D., Lullmann, H. and Wassermann, O. (1968): Naun.-Schmied.Arch. Pharmak. Exp. Path. 259:386-393.
3. Cohen, E.L. and Wurtman, R.J. (1975): Life Sci. 16:1095-1102.
4. Dzieniszewski, P. and Kilbinger, H. (1978): Eur. J. Pharmacol. 50:385-391.
5. Haubrich, D.R., Wang, P.F.L., Clody, D.E. and Wedeking, P.W. (1975): Life Sci. 17:975-980.
6. Haubrich, D.R., Wedeking, P.W. and Wang, P.L.F. (1974): Life Sci. 14:921-927.
7. Kilbinger, H. (1977): Naun.-Schmied.Arch. Pharmacol. 300:145-151.
8. Kilbinger, H. and Wagner, B. (1979): In Presynaptic Receptors (eds) S.Z. Langer, K. Starke, M.L. Dubocovich, Pergamon Press, Oxford, pp. 347-351.
9. Kilbinger, H. and Wessler, I. (1980): Neuroscience 5:1331-1340.

10. Sawynok, J. and Jhamandas, K. (1977): Canad. J. Physiol.
       Pharmacol. 55:909-916.
11. Szerb, J.C. (1976): Canad. J. Physiol. Pharmacol. 54:12-22.
12. Szerb, J.C. (1979): In Presynaptic Receptors (eds) S.Z.
       Langer, K. Starke, M.L. Dubocovich, Pergamon Press, Oxford,
       pp. 293-298.

KINETIC STUDIES ON THE RELEASE OF [^{3}H]-ACETYLCHOLINE FROM GUINEA

PIG MYENTERIC PLEXUS: DIFFERENCE IN THE EFFECTS OF MORPHINE AND

REDUCED CALCIUM INFLUX

J.C. Szerb

Department of Physiology and Biophysics
Dalhousie University
Halifax, Nova Scotia, B3H 4H7 Canada

The guinea pig ileum and later the myenteric plexus-longitudinal muscle preparation derived from it, have been used extensively as convenient in vitro models for demonstrating the specific
effects of opiates on neural function (8). Opiates reduce the contractions produced by low frequency stimulation, decrease the
resting release of acetylcholine (ACh) and the release produced by
low frequency stimulation but not by high frequency stimulation (11).
The mechanism whereby opiates depress the release of ACh from the
myenteric plexus is still not understood. One of the problems in
investigating the action of opiates in this preparation is the
impossibility of simultaneously measuring the effect of opiates on
ACh release and on mechanical activity; cholinesterise inhibitors,
which have to be used for measuring ACh release, cause a steady contraction of the preparation which precludes the detection of mechanical activity induced by stimulation. To overcome this problem,
the method developed for measuring labelled ACh release from brain
slices in the absence of cholinesterase inhibitors (13) was adapted
for the myenteric plexus-longitudinal muscle preparation. After
labelling part of the ACh stores with [^{3}H]-choline (Ch), the spontaneous release of radioactivity measured in the presence of physostigmine consists of a mixture of labelled ACh, Ch and phosphorylcholine. However, the increase in efflux of radioactivity evoked
by electrical stimulation is due exclusively to an increase in the
effux of [^{3}H]-ACh (18,20). Therefore it can be assumed that even
in the absence of a cholinesterase inhibitor the increase in the

efflux of total radioactivity resulting from stimulation reflects
quantitatively the release of labelled ACh, provided that the uptake
of [^{3}H]-Ch derived from [^{3}H]-ACh is prevented by hemicholinium-3
(HC-3).

METHODS

Guinea pig ileum myenteric plexus-longitudinal muscle preparations were suspended in a narrow 0.5 ml bath at 37°C and were superfused at a rate of 0.4 ml/min. The mechanical activity of the preparation was recorded isotonically. During supraxamimal 0.1 Hz
stimulation the preparation was superfused with 1 μM [^{3}H]-Ch for
15 min. Stimulation was then stopped and superfusion with Krebs
solution containing 1 μM unlabelled Ch and 10 μM HC-3 was begun.
Sample collection was started 55 min after the end of labelling.
After the collection of 5 samples without stimulation, releasing
stimulation was started using either supramaximal (1.3 x maximal)
or submaximal (0.67 x maximal) stimulation and this was continued
for 64 or 80 min during which 20 or 22 samples were collected. After
the stimulation, 10 further samples were collected without stimulation. The evoked release of the label was obtained from the differences between the total radioactivity and the calculated spontaneous efflux in samples collected during stimulation. Spontaneous
efflux was calculated by interpolating between the efflux obtained
before and after stimulation, assuming an exponential decline in
spontaneous efflux. The efflux was expressed as pmol.g^{-1}.min^{-1} based
on the specific activity of [^{3}H]-Ch in the perfusion medium. Kinetic
analysis of the evoked release was based on the assumption that the
efflux of [^{3}H]-ACh from a limited store follows first order kinetics,
i.e. the amount of [^{3}H]-ACh released over a short period of time is
proportional with the amount of [^{3}H]-ACh in the store. Therefore
plotting the log of efflux against time gives a straight line, the
slope of which is the rate constant(k) and the zero intercept the
log of the product of the rate constant and the initial pool. Thus,
dividing the antilog of the zero intercept by k gives the size of
the pool from which release has occurred. In instances where release
proceeded with not one but two rate constants, the two components
were separated by the "peeling" method. Details of the kinetic analysis have been described (6).

RESULTS

Supramaximal stimulation caused an almost fourfold increase in
the effux of the label which then rapidly declined during continuing

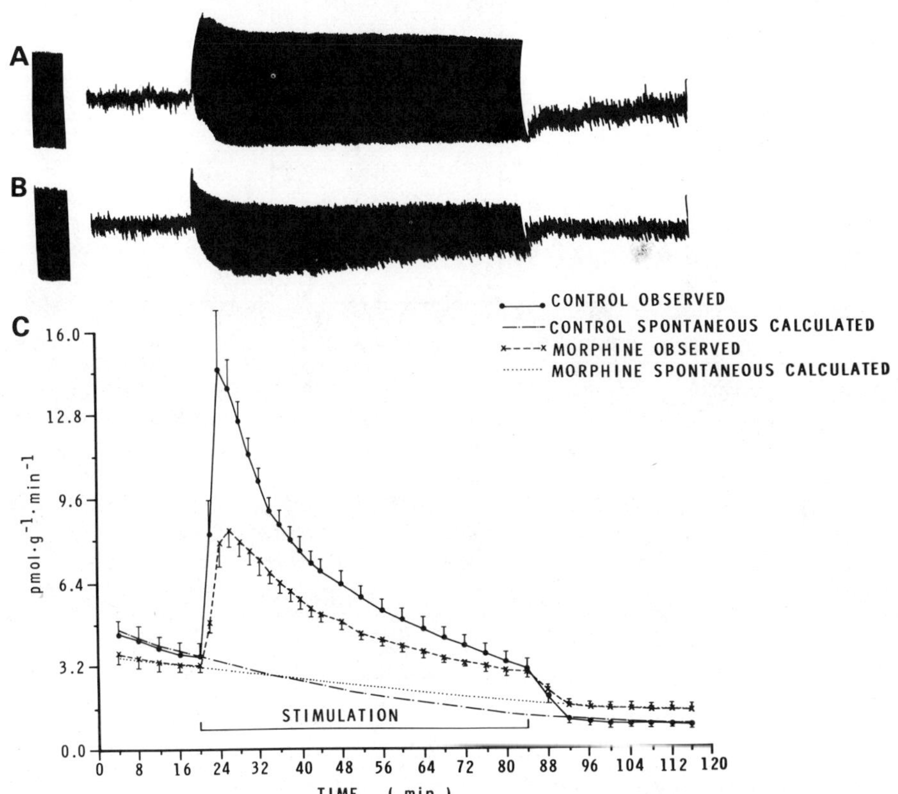

FIGURE 1: The effect of supramaximal stimulation on contractions
and [^{3}H]-ACh release in the absence or presence of morphine (1 μM). A and B are examples of isotonic contractions recorded while measuring [^{3}H]-ACh release. Contractions shown at the beginning of the tracings are
those caused by supramaximal stimulation during labelling.
A, contractions in the absence of morphine, stimulation
with 45 V during labelling and release; B, contractions
in the presence of morphine (1 μM), stimulation with 47 V
during labelling and release. Superfusion with morphine
started after labelling; C, release of the label ●—●
in the absence, X---X in the presence of morphine. Average of 6 experiments, bars denote S.E.M. – · – · calculated spontaneous release in the absence and ····· in the
presence of morphine.

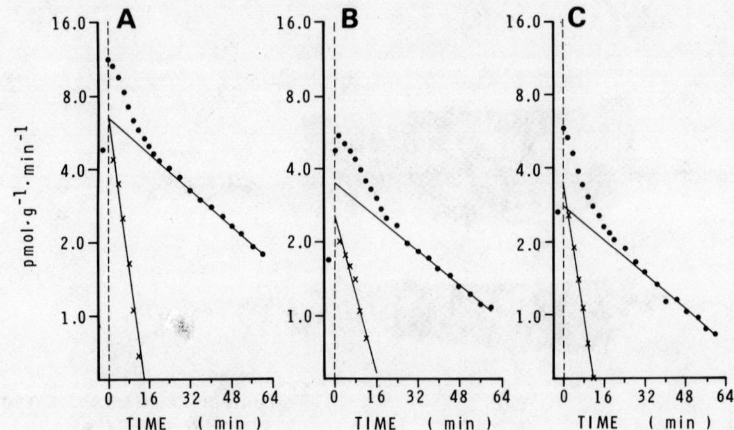

FIGURE 3: Semilogarithmic plots of the average evoked release against time. A and B supramaximal releasing stimulation, A in the absence, B in the presence of 1 μM morphine; C, submaximal releasing stimulation in the absence of morphine. Dots indicate total release, crosses calculated fast release. Lines show the calculated regression lines for slow and fast release.

absence of morphine (C). The release by submaximal stimulation in the presence of morphine was too small and erratic to be analyzed kinetically. It is apparent that the evoked release had an initial fast and a later slow phase. The two components were separated in each experiment by the peeling method as shown in Fig. 3. Results of the kinetic analyses are summarized in Table 1. The rate of fast release was significantly slowed down by morphine but not by reducing the intensity of stimulation to submaximal. The size of the pool from which the fast release occurred was decreased slightly both by morphine or by reduced intensity of stimulation. While morphine decreased the rate of the fast efflux it did not change the rate of the slow efflux but only the size of the pool from which the slow release occurred and this was similar to the effect of reduced intensity of stimulation.

The similarity of the effects of morphine and of submaximal stimulation on the kinetics of the slow efflux suggested that morphine may decrease the release of ACh from the slowly releasing pool by decreasing the number of cholinergic fibers excited by stimulation and not by interfering with the release of ACh itself. The next series of experiments were designed to test this conclusion by

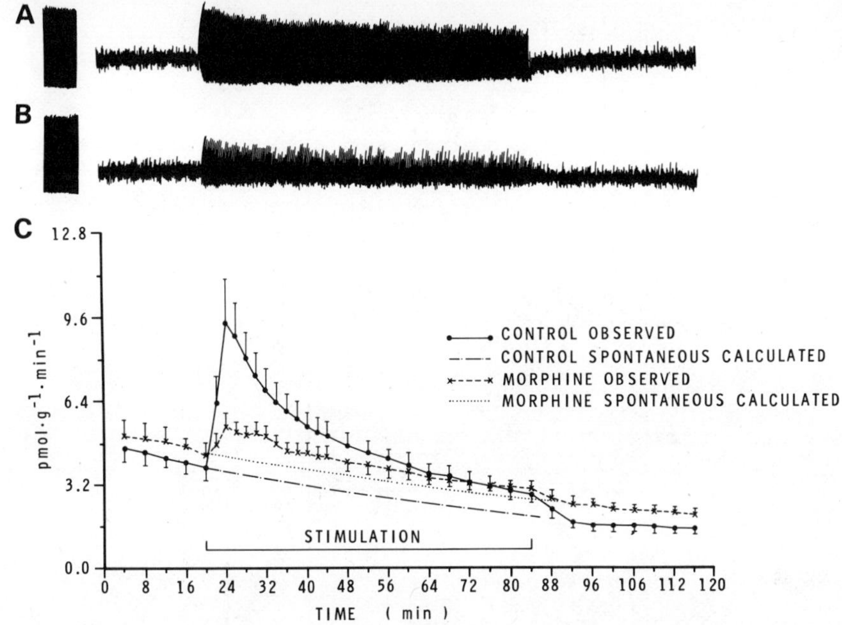

FIGURE 2: The effect of submaximal stimulation on contractions and
[³H]-ACh release in the absence or presence of morphine
(1 µM). Symbols same as in Fig. 1. A and B examples of
isotonic contractions recorded while measuring [³H]-ACh
release. A, in the absence of morphine, stimulation with
45 V during labelling and 22 V during release; B, in the
presence of morphine after labelling, labelling with 47 V,
releasing with 24 V; C, release of label in the presence
or absence of morphine. Averages of 6 experiments.

stimulation (Fig. 1). Contractions recorded simultaneously also
declined due to the presence of HC-3. In the presence of morphine
the initial evoked release was only half as much as in controls and
twitches recorded were also depressed (Fig. 1). Submaximal releasing
stimulation evoked less release and produced smaller twitches than
supramaximal stimulation (Fig. 2). Morphine reduced twitches and
release produced by submaximal stimulation more than that induced
by supramaximal stimulation (Fig. 2). Naloxone (0.5 µM) applied
during stimulation in the presence of morphine caused a large increase in efflux and an increase in contractions (not shown).

Figure 3 illustrates semilogarithmic plots of the average evoked
release due to supramaximal stimulation in the absence (A), in the
presence of morphine (B) and due to submaximal stimulation in the

TABLE 1: Kinetic parameters of the evoked release of $[^3H]$-ACh.

Releasing Stimulation	Drug	Fast Release		Slow Release	
		Initial Pool $(pmol\ g^{-1})$	Rate Constant (min^{-1})	Initial Pool $(pmol\ g^{-1})$	Rate Constant (min^{-1})
Supramaximal	–	55.7 ±5.3	0.1811 ±0.0167	322 ±31	0.0215 ±0.0013
Supramaximal	Morphine (1 μM)	34.6 ±7.2*	0.0862 ±0.0072***/+	182 ±18**	0.0201 ±0.0011
Submaximal	–	34.9 ±6.3*	0.1460 ±0.0216	142 ±20***	0.0197 ±0.0010

Means ± S.E. of 6 observations. Significance of differences (P) from supramaximal control:
* P < 0.05;   ** P < 0.01;   *** P < 0.001.
+ Significantly different (P < 0.05 from submaximal control

comparing the kinetics of $[^3H]$-ACh release in the presence of morphine with conditions known to affect release of ACh directly by lowering the Ca^{++} content of the perfusion medium (from 2.52 to 1.26 V_{max}) and by using oxotremorine. Oxotremorine has been shown to depress ACh release from the myenteric plexus by acting on presynaptic muscarinic receptors (7) while the rate of the Ca^{++} dependent release of ACh is expected to be influenced by external Ca^{++} concentration. However, when the low Ca^{++} medium, and especially oxotremorine (50 nM), were superfused, the preparation went into a prolonged contraction which resulted in a decreased excitability to field stimulation of neurons in the myenteric plexus (19). To overcome this prolonged contracture, 0.5 mM $MnCl_2$, which is known to inhibit the tonic phase of contraction induced by muscarinic agonists in this preparation (14), was also added during releasing stimulation.

Semilogarithmic plots of the evoked release in the presence of only Mn^{++} and of Mn^{++} with low Ca^{++}, oxotremorine, or morphine are shown in Fig. 4. It is evident that with Mn^{++} present release occurred at a single rate similar to the slow rate observed in the absence of Mn^{++}. The single rate was reduced by lowering Ca^{++} or by using oxotremorine without change in the releasable pool size (Fig. 4; Table 2). On the other hand, morphine in the presence of Mn^{++} reduced the releasable pool size but not the rate of release. These effects were prevented by the appropriate antagonists (Table 2).

DISCUSSION

The evoked release of $[^3H]$-ACh in the absence of Mn^{++} proceeded with an initial rapid and a late slow phase and these two components were affected differently by morphine and Mn^{++}. Morphine reduced the rate constant of the fast efflux but not that of the slow efflux while Mn^{++} caused the fast efflux to disappear without affecting the slow efflux. There are several lines of evidence to suggest that the two phases of evoked release originate from two populations of cholinergic neurons in the myenteric plexus. The likely source of the fast release is the spontaneously firing neuronal elements observed with extracellular (5,16,21) or intracellular (22) recording. The spontaneous firing of these neurons is suppressed by tetrodotoxin and by opiates as is the spontaneous release of ACh (12). The spontaneous release of ACh is depressed by small concentrations of Mn^{++} (3) as is the fast evoked release, without affecting either the twitch response (3) or the slow evoked release. Assuming that each action potential releases a constant fraction of ACh present in the terminals of repetitively firing neurons, a reduction in the rate of firing of these neurons by morphine would be expected to decrease the rate of $[^3H]$-ACh released per time, as observed here.

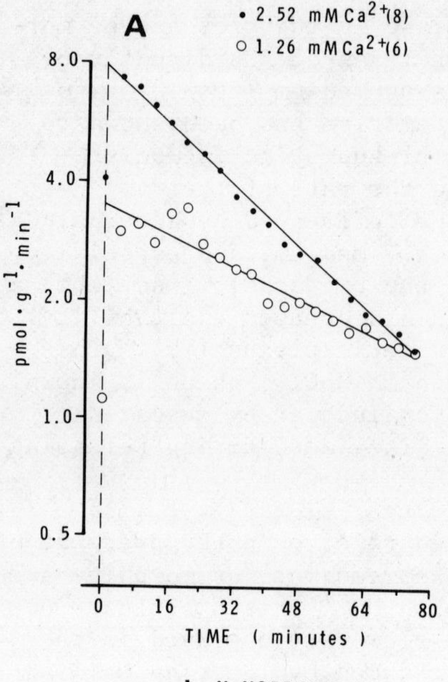

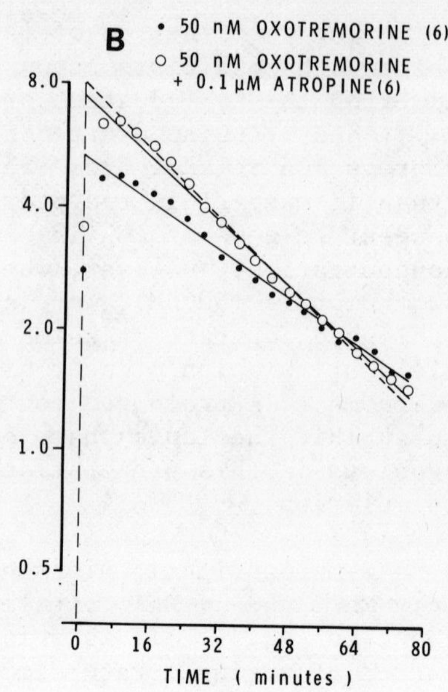

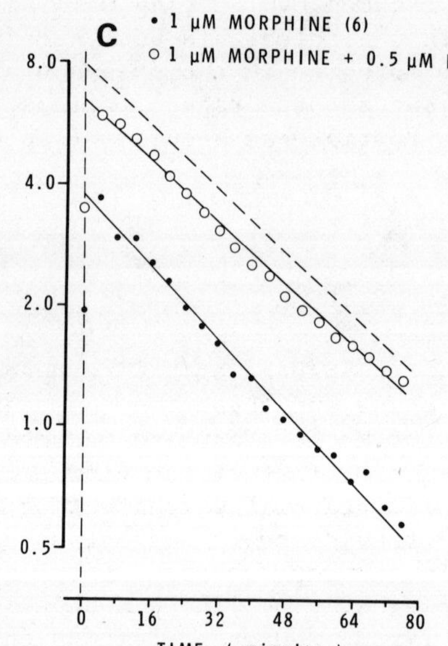

FIGURE 4: Semilogarithmic plots of the average evoked release in the presence of 0.5 mM $MnCl_2$.

A - Normal Krebs •, 1.26 mM Ca ⧺ O;
B - 50 nM oxotremorine •, 50 nM exotremorine + 0.1 µM atropine O;
C - 1 µM morphine •, 1 µM morphine + 0.5 µM naloxone O.
Broken lines in B and C are the same control effluxes as shown in A. (From Szerb (19) with permission of Publisher).

The slow release, whose rate if not affect by morphine, is likely to originate from a different population of myenteric neurons, those that do not fire spontaneously, produce only one or two action potentials upon stimulation of their processes and their action potential is followed by prolonged hyperpolarization (9,10,23).

TABLE 2:   Kinetic parameters of the evoked release of $[^3H]$–ACh in the presence of 0.5 mM $MnCl_2$.

Perfusion Fluid	n	Pool (pmol. g^{-1})	Rate Constant (min^{-1})
Normal Krebs	8	364 ±33	0.0228 ±0.0016
Ca^{++} (1.26 mM)	6	317 ±59	0.0124 ±0.0014**
Oxotremorine (50 nM)	6	323 ±39	0.0161 ±0.0004*
Ca^{++} (1.26 mM) + Atropine (0.1 μM)	6	317 ±51	0.0140 ±0.0012**
Oxotremorine (50 nM) + Atropine (0.1 μM)	6	352 ±50	0.0220 ±0.0015
Morphine (1 μM)	6	148 ±23**	0.0236 ±0.0023
Morphine (1 μM) + Naloxone (0.5 μM)	6	305 ±48	0.0218 ±0.0009

Significantly different from control: * P < 0.01;   ** P < 0.001

These units are hyperpolarized and their membrane resistance reduced by morphine and as a result a number of them fail to respond in the presence of morphine to previously effective focal stimulation (10). Some of these neurons which fail to respond to stimulation in the presence of morphine will not contribute to the evoked release of $[^3H]$–ACh and the size of the releasable pool will decrease compared to the controls, as observed in these experiments. An increased effectiveness of morphine in reducing twitches resulting from sub-maximal stimulation observed here and previously by others (3,4,17) also indicates that morphine increases the threshold of excitation of neurons in the myenteric plexus. The unchanged rate of slow re-lease from a reduced pool in the presence of morphine suggests that morphine does not affect the release of ACh from neurons which on-tinue to fire in the presence of morphine. The effect of morphine on the kinetic of the slow release can be clearly distinguished from that of low Ca^{++} or of oxotremorine which by interfering with the release process, decreased the rate of slow release.

In summary, these observations suggest that morphine reduces the release of ACh in the myenteric plexus not by interfering with the release process itself but by decreasing the excitability of cholinergic neurons, possibly by an action on the neural membrane. Opiate receptors have been localized along the medullary fibers of the vagus by autoradiography (1), supporting the conclusion that opiates may act by interfering with action potential conduction in unmyelinated nerves rather than with transmitter release. In addition, the excitability to antidromic stimulation of central nociceptive terminals in the cord is decreased by opiates (2,15), suggesting that the mechanism of action of opiates in the central nervous system and in the myenteric plexus may be similar.

ACKNOWLEDGMENT

This work was supported by the Medical Research Council of Canada.

REFERENCES

1. Atweh, S.F. and Kuhar, M.J. (1977): Brain Res. 124:53-67.
2. Carstens, E., Tulloch, I., Zieglgansberger, W. and Zimmermann. M. (1979): Pflug. Arch. 379:143-147.
3. Cowie, A.L., Kosterlitz, H.W. and Waterfield, A.A. (1978): Brit. J. Pharmacol. 64:565-580.
4. Cox, B.M. and Weinstock, M. (1966):Brit. J. Pharmacol. 27;81-92.
5. Dingledine, R. and Goldstein, A. (1976): J. Pharmacol. Exp Ther. 196:97-106.
6. Down, J.A. and Szerb, J.C. (1980): Brit. J. Pharmacol. 68:47-55.
7. Kilbinger, H. (1977): N.-S. Arch. Pharmacol. 300:245-151.
8. Kosterlitz, H.W. and Waterfield, A.A. (1975): Ann. Rev. Pharmacol. 15:29-48.
9. Mishi, S. and North, R.A.(1973): J. Physiol.(Lond.) 231:471-491.
10. North, R.A. and Tonini, M. (1977): Brit. J. Pharmacol. 61: 541-549.
11. Paton, W.D.M. (1957): Brit. J. Pharmacol. 11:119-127.
12. Paton, W.D.M., Vizi, E.S. and Zar, M.A. (1971): J. Physiol. (Lond.) 194:13-33.
13. Richardson, I.W. and Szerb, J.C. (1974): Brit. J. Pharmacol. 52:499-507.
14. Rosenberger, L.B., Ticku, M.K. and Triggle, D.J. (1979): Canad. J. Physiol. Pharmacol. 57:333-347.
15. Sastry, B.R. (1979): Neuropharmacology 18:367-375.
16. Sato, T., Takayanagi, I. and Takagi, K. (1973): Jap. J. Pharmacol. 23:665-671.
17. Sawynok, J. and Jhamandas, K. (1976): J. Pharmacol. Exp. Ther. 197:379-390.
18. Szerb, J.C. (1976): Canad. J. Physiol. Pharmacol. 54:12-22.
19. Szerb, J.C. (1980):N.-S. Arch Pharmacol. 311:119-127.
20. Wikberg, J. (1977): Acta Physiol. Scand. 101:302-317.
21. Williams, J.T. and North, R.A. (1979): Brain Res. 165:57-65.
22. Wood, J.D. and Mayer, C.J. (1978): Pflug. Arch. 374:265-275.
23. Wood, J.D. and Mayer, C.J. (1979): J. Neurophysiol. 42:569-581.

Na^+-K^+-ACTIVATED ATPase AND NON-QUANTAL/CYTOPLASMIC

RELEASE OF ACETYLCHOLINE

E.S. Vizi

Department of Pharmacology
Semmelweis University of Medicine
Budapest, Hungary

INTRODUCTION

Essentially there are two hypotheses which attempt to explain
how transmitters are released from axon terminals (cf. 14,16,20).
Exocytosis assumes that transmitters are packed in vesicles which,
after having fused with the membrane of the axon terminal, discharge
their contents into the external cleft. The first observation of
the location of a transmitter in vesicles at nerve endings dates
back to 1952; Blaschko and Welch for noradrenaline (7) and Fatt and
Katz for acetylcholine (ACh) (16). Another hypothesis is that the
transmitter is released from the cytoplasm (7,13,27,28,38,39). However, none of them can explain how transmitters cross the membrane
and what kind of membrane mechanism operates to make the membrane
permeable to the transmitters.

There is accumulating evidence that in cholinergic neurones
ACh appears both in synaptic vesicles and in free form in cytoplasm.
While the association of ACh and NA with vesicles is beyond doubt,
convincing evidence is still lacking regarding the relative importance of such structures for transmitter release phenomena. Although Sir Bernard Katz stated "...that transmitter substances are
discharged from many, and possibly from most, chemical nerve terminal in the form of discrete multi-molecular packets is probably
no longer questioned," there is some evidence that ACh is released
in non-quantal form (18,47,48). In addition, Katz and Miledi (23,
24) and Miledi et al. (29) showed a steady leakage of ACh from
nerve terminals also.

TABLE 1:   Ouabain induced release of ACh from cerebral cortex slices
           in calcium free Krebs solution.

Ouabain Concentration	Non-Treated	Ouabain Treated	P
2×10^{-6} M	21.2 ± 2.0	43.8 ± 3.6	< 0.05
2×10^{-5} M	17.2 ± 4.2	86.3 ± 12.9	< 0.01
2×10^{-4} M	24.5 ± 1.3	213.0 ± 24.7	< 0.001
2×10^{-3} M	27.8 ± 3.0	213.6 ± 37.7	< 0.01

The calcium free Krebs solution contained 1 mM EGTA.
Figures are means $\pm$ S.E.M. (8).

The main problem is, therefore, where does the transmitter
release come from. Is it released from the vesicle or from the
cytoplasm, or from both? What mechanism allows the membrane to be
let through the transmitter? Na^+-K^+ activated ATPase of the basal
membrane has been shown to play a crucial role in this mechanism
(cf. 40,44,45).

Membrane ATPase and Transmitter Release

Different drugs (ouabain, digoxin, N-ethylmaleimide [NEM], p-
chloromercuribenzoate [PCMB], quinidine, chlorpromazine, ethanol)
or conditions (Na-deprivation, K-deprivation) and some particular
cations (Ca^{++}, Ba^{++}, Cu^{++}) which are known to inhibit the activity
of Na^+-K^+ activated ATPase (cf. 35,36), increase the release of
various transmitters (cf. 40,44,45). Ouabain releases ACh from
different nerves and synaptosomes (19,32,40,43,44). Since physiological release of neurotransmitters is dependent on calcium entry
into the neuron, the question arises as to whether calcium entering
the neurone during depolarization may inhibit ATPase so that inhibition of ATPase is the underlying mechanism in the physiological
release of neurotransmitters.

The Role of Ca^{++} as Inhibitor of Membrane Adenosine 5-Triphosphatase in Triggering Transmitter Release

The crucial role or the actual activity of Na^+-K^+ activated
ATPase in transmitter release, is favored by those experimental
observations in which it was shown that drugs and conditions able

DEPOLARIZATION OF NERVE TERMINALS

↓

Na^+-INFLUX

↓

Ca^{++}-INFLUX

↓

INHIBITION OF (Na^+-K^+)-ACTIVATED ATPase←OUABAIN, PHMB, NEM, Ba^{++},
K-DEPRIVATION, ETC.

↓

DIFFUSION OF TRANSMITTER THROUGH THE MEMBRANE

↓

STIMULATION OF (Na^+-K^+)-ACTIVATED ATPase
(BY ACCUMULATED Na^+ OR BY Mg^{++})

↓

TERMINATION OF TRANSMITTER RELEASE

FIGURE 1: Model of the proposed mechanism by which (Na^+-K^+)-activated
ATPase acts as a trigger of ACh release. It is proposed
that Ca entry into the membrane as a result of the change
in Na concentration during the action potential (1-3) and
the subsequent transient inhibition of (Na^+-K^+)-activated
ATPase followed by stimulation of the enzyme is the underlying mechanism in the physiological release of transmitters.

---

to reduce ATPase activity release transmitter even in the absence
of external calcium ions (44), e.g. sodium or potassium deprivation
or ouabain administration (31,40-42) and ethanol (33) which is also
an inhibitor of membrane ATPase (21,37) produce ACh release even in
absence of (Ca^{++}) and presence of EGTA. Calcium independent release of nonadrenaline in response to sodium deprivation (17,25)
and to ouabain (rat vas deferens: 41) was also shown. A similar
observation was made (22) on guinea pig vas deferens. In addition,
it was observed that ouabain releases cytoplasmic Na from rabbit
heart (26), which has been pretreated by MAO inhibitor and reserpine. Nakazato et al. (30) however, failed to show that ouabain
is able to release NA from guinea pig vas deferens when Ca^{++} was
not present in the medium. An accelerated release of ACh was indicated by an increase of frequency of mepp (4-6,15,47) occurring in

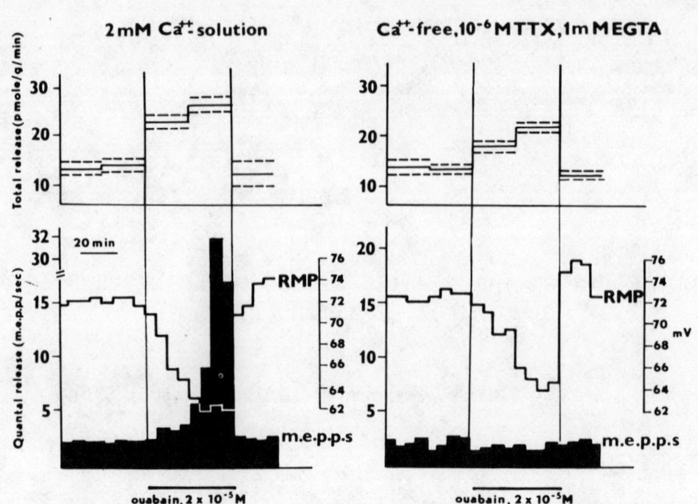

FIGURE 2: Effect of ouabain on quantal and total release of ACh and
on resting membrane potential (RMP, mV) of mouse diaphragm
in medium with 2 mM Ca and in Ca-free (1 mM EGTA) solution.
TTX = tetrodotoxin. Temperature, 37°C. When ACh release
was measured, 2 µg/ml eserine sulfate was used. Figure
redrawn from Vizi and Vyskocil (47).

the absence of external calcium and when the membrane ATPase is
inhibited. Carmody (10) observed that NEM ehhanced the frequency
of mepp and the amplitude of endplate potentials even in the absence
of Ca^{++}. It is suggested, therefore, that under physoilogical conditions, calcium ions, which cross the membrane as a result of the
change in Na^+ concentration during the action potential (1-3) may
act by inhibiting the membrane ATPase. The transient inhibition of
enzyme activity might result in a change in the permeability of the
membrane; it becomes permeable to transmitters. Calcium has in fact
been shown to be a strong inhibitor of Na^+-K^+-activated ATPASE (36).

The fact that Ba^{++} can replace Ca^{++} both in resting transmitter
release (12,17,31,40,42) and in stimulation evoked release (31,42)
and that Ba^{++} is more effective in ACh release (31,42) and a much
more potent inhibitor of membrane ATPase (34) than Ca^{++}, strongly
support this hypothesis.

Table 1 shows the effect of ouabain on cerebral cortical slices
of the rat when no Ca ions were present. Ouabain, in the concentration range of 2 x 10^{-6}-2 x 10^{-3} M significantly enhanced the

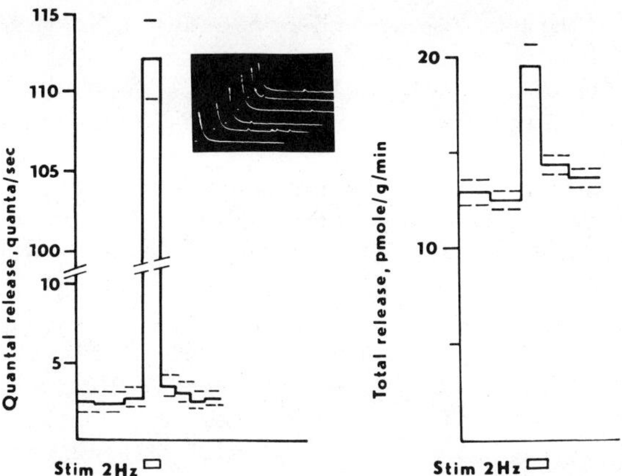

FIGURE 3: Quantal and total release of ACh from mouse diaphragm.
Stimulation frequency: 2 Hz (10 min). Oscillograph records
(see insert) of endplate and mepps from the cut disphragm,
from which quantum content was calculated (ratio of epp
and mepp amplitude). Total release was measured in
eserinized Krebs solution. Figure redrawn from Vizi and
Vyskocil (47).

---

release in the absence of Ca and presence of 1 mM EGTA. This fact
indicates that when Na^+-K^+-activated ATPase in inhibited, Ca^{++} is
not needed for release (Fig. 1).

Membrane ATPase and Quantal Release

It has been shown that cardiac glycosides cause a slow but
progressive increase in quantal release from a variety of tissues.
This has been shown for rat, mouse and frog motor nerve terminals
(4,6,8,15,47). In our experiments (47) ouabain at a concentration
of 2×10^{-5} M enhanced total release of ACh almost immediately from
14.0 ± 1.3 to 22.9 ± 1.9 pmol/g min measured in the first 20 min
and 26.5 pmol/g min estimated 20-40 min after administration. Quantal release changed more slowly (Fig. 2). Branisteanu et al. (18)
also showed that ouabain increased both spontaneous and evoked
transmitter release in frog neuromuscular junction. In a concentration of 10^{-5} M, it enhanced mepp frequency by a factor of almost
200.

Membrane ATPase and Non-Quantal Release

The question arose as to whether the increase in both total
(Quantal + non-quantal) and quantal release depends on Ca influx.
Fig. 2 shows that ouabain enhanced the total but not the quantal
release under condition when Ca was deprived. This fact indicates
that the non-quantal release is enhanced by membrane ATPase inhibition even when Ca^{++} is deprived. Vyskocil and Illes (48) provide
electrophysiological evidence that the non-quantal release of ACh
can be enhanced by ouabain. The non-quantal release produced a
minor depolarization of the postjunctional membrane of the mouse
diaphragm fiber, which can be revealed by local curatization of the
endplate zone of anticholinesterase treated muscles. Under this
condition, the hyperpolarization of 40 μV was ehhanced to 1.43 ±
0.3 mV in response to ouabain. In addition, we succeeded in showing
(43,48) that the stimulation of membrane ATPase leads to an inhibition of ACh release and a reduction of non-quantal release.

The question now arises as to what is the proportion of quantal
and non-quantal release of ACh. For the electrophysiological estimation of quantal release, a sectioned preparation was used to obtain
epps and mepps during nerve stimulation with the presence of any
blocking drug. The quantal content of epps was estimated during a
stimulation at a frequency of 2 Hz and the mean increase in number
of quanta was compared with the total amount of ACh released from
another muscle during indirect stimulation (for details see 47).
It was found that the total output of ACh is increased by only 30%
during nerve stimulation (Fig. 3). This fact indicates that 70%
is still non-quantal. Quantal release, however, increased from
2.5 ± 0.8 mepp/sec before stimulation to 112.3 ± 2.7 quanta/sec,
i.e. a 45-fold increase. This difference in proportional increase
of total and quantal release provided an opportunity to estimate
the quantal contriubtion to the total ACh output at rest (47). The
non-quantal fraction at rest proved to be 99% of the total release,
indicating that the quantal release at rest is only 1% of the total
release.

Non-Synaptic Interaction of Axon Terminal and Target Cell

In contrast to the situation in the neuromuscular junction,
where the distance between the axon terminal and affector cell is
about 5 nm, there is a remote (100-200 nm) innervation of target
cells in the neurovegatative system and in some part of the central
nervous system (e.g. noradrenergic innervation of the cerebral
cortex or nigro-striatal dopaminergic input). The existance of

such a long distance between varicosities and target cells raises
important problems concerning the role of quantal and non-quantal
release. It is suggested that besides quantal release, the non-
quantal release does make sense as romote control of a target cell.
In this case transmitter/modulator reaches its target cell by diffusion, an event that takes a longer time. In addition, during the
diffusion the molecules released either in quantal form or from the
cytoplasm certainly become mixed up.

In longitudinal muscle-Auerbach plexus preparation, where the
distance between axon terminals and effector smooth muscle cells is
rather large, ouabain in a concentration of 10^{-6} M releases ACh (31,
44) and in the absence of cholinesterase inhibition, is able to
produce an atropine sensitive contraction. This fact indicates that
the concentration of ACh released by ouabain is high enough to produce contraction of smooth muscle cells.

DISCUSSION

From these data it seems very likely that Na^+-K^+-activated
ATPase plays a trigger role in transmitter release. Its actual
activity makes the membrane ready to let through transmitter or
terminate the release. Under physiological conditions Ca^{++} influx
results in a transient inhibition of membrane ATPase and transmitter
release while stimulation of the enzyme, possibly caused by the
termination of Ca^{++} induced inhibition (either by the accumulated
Na^+ or by Mg^{++}) makes the membrane again impermeable to the transmitter. This quantization mechanism is, therefore, independent of
whether the transmitter is released from vesicles or from the soluble
cytoplasm (43,44). Cooper and Meyer (11, and this volume) provided
evidence that during electrical field stimulation, or veratridine
administration, the inhibition of membrane ATPase and the release
of ACh from synaptosomes are correlated. This mechanism seems to be
crucial both in cytoplasmic and in quantal release (44,49). It is
tempting to speculate that cytoplasmic release is an ancient form
of the release. Where the target cells are far from the nerve terminals, the advantage of the quantal release, e.g. to produce a
package or a very high concentration of transmitter in the vicinity
of receptors, is not effective anymore since during diffusion the
transmitter molecules become diluted. In this case the cytoplasmic
release should play a rather critical role in maintaining the ef-
fecient concentration of transmitter. In this context, the vesicular release of a transmitter would be a more sophisticated form
of transmission. Its role would be critical in those cases where
an extremely high concentration of transmitter is needed (e.g. in
the neuromuscular junction where inactivating cholinesterase activity is high. The transmitter is released in packets and the basic

features of the chemically transmitting synapse are the presence of
a cleft (5-60 nm) and specialized pre- and postsynaptic membranes.

The question arises as to whether two alternative mechanisms
of transmitter release can be considered at different synaptic types,
or at the same synapse? Since even at the neuromuscular junction
there is a non-quantal form of ACh release during rest (only 1% is
quantal) and at stimulation 30% of total release is quantal, it
is suggested that cytoplasmic release might be of physiological
importance in transmission.

Because this argument could hold for any transmission site
where the interaction is non-synaptic (e.g. noradrenergic input in
cerebral cortex) (46), it is suggested that cytoplasmic release of
a transmitter has to be taken into account as an important mechanism in relaying information.

REFERENCES

1. Baker, P.F. (1972): Prog. Biophys. Molec. Biol. 24:177-223.
2. Baker, P.F. and Blaustein, M.P. (1968): Biochim. Biophys.
     Acta, 150:167-170.
3. Baker, P.F., Blaustein, M.P., Hodgkin, A.L. and Steinhardt, R.
     A. (1969): J. Physiol. (Lond.) 200:431-458.
4. Baker, P.F. and Crawford, A.C. (1975): J. Physiol. (Lond.)
     247:209-226.
5. Birks, R.I. (1974): J. Neurocytol. 3:133-160.
6. Birks, R.I. and Cohen, M.W. (1968): Proc. Roy. Soc. Lond. B
     170:381-399.
7. Blaschko, H. and Welch, A.D. (1953): N-S. Arch. 219:17-24.
8. Branisteanu, D.D., Proca, B. and Haulica, I.D. (1979): J.
     Pharmacol. Exp. Ther. 209:31-37.
9. Burnstock, G. (1970): IN Smooth Muscle (eds) L. Bulbing and
     M.F. Shuba, Raven Press, New York, pp. 357-367.
10. Camody, J.I. (1978): Eur. J. Pharmacol. 47:457-462.
11. Cooper, J. and Meyer, E.M. (this volume).
12. Douglas, W.W., Lywood, D.W. and Straub, R.W. (1961): J.
     Physiol. (Lond.) 156:515-522.
13. Dunant, Y., Gautron, J., Israel, M., Lesbats, B. and Manar-
     anche, R. (1972): J. Neurochem. 19:1987-2002.
14. Dunant, Y. and Israel, M. (1979): Trends Neurosci. 2:30-32.
15. Elmquist, D. and Feldman, D.S. (1965): J. Physiol. (Lond.)
     181:487-497.
16. Fatt, P. and Katz, B. (1952): J. Physiol. (Lond.) 117:109-128.
17. Garcia, A.G. and Kirpekar, S.M. (1973): Brit. J. Pharmacol.
     Chemther. 47:729-747.

18. Gorio, A., Hurnblut, W.P. and Ceccarelli, B. (1978): J. Cell.
     Biol. 78:716-733.
19. Grewaal, D.S. and Quastal, J.H. (1973): Biochem. J. 132:1-14.
20. Israel, M., Dunant, Y. and Manaranche, R. (1979): Prog. Neuro-
     Biol. 13:237-275.
21. Israel, Y., Kalant, H. and Laufer, K. (1965): Biochem. Phar-
     macol. 14:1803-1814.
22. Katsurgai, T., Fukuski, Y. and Suzuki, T. (1978): Eur. J.
     Pharmacol. 47:407.
23. Katz, B. and Miledi, R. (1965): The Release of Neural Trans-
     mitter Substances, University Press, Liverpool.
24. Katz, B. and Miledi, R. (1971): Proc. Roy. Soc. B 196:59-72.
25. Lastowecka, A. and Trifaro, J. (1974): J. Physiol. (Lond.)
     236:681-705.
26. Lindmar, R. and Loffelholz, K. (1974): N-S. Arch. Pharmacol.
     284:93-100.
27. Marchbanks, R.M. (1978): Trends Neurosci. 1:83-84.
28. Marchbanks, R.M. and Israel, M. (1971): J. Neurochem. 18:
     439-448.
29. Miledi, R., Molenaar, P.C. and Polak, R.L. (1977): Proc. Roy.
     Soc. Lond. B 197:285-297.
30. Nakazato, Y., Ohga, A. and Onoda, Y. (1978): J. Physiol. (Lond.)
     278:45-47.
31. Paton, W.D.M., Vizi, E.S. and Zar, M.A. (1971): J. Physiol.
     (Lond.) 215:819-848.
32. Poulsen, J.H. (1974): Pflug. Arch. 350:281-286.
33. Quastel, D.M.J., Hackett, J.T. and Cooke, J.D. (1971): Science
     172:1034-1036.
34. Rendi, R. (1966): Biochim. Biophys. Acta 119:621-630.
35. Schwartz, A., Lindenmayer, G.E. and Allen, J.C. (1975):
     Pharmacol. Rev. 27:3-105.
36. Skou, J.C. (1960): Biochim. Biophys. Acta 42:6-23.
37. Sun, A.Y. and Samorajski, T. (1970): J. Neurochem. 24:161-164.
38. Tauc, L. (1979): Biochem. Pharmacol. 27:3493-3498.
39. Tauc, L., Hoffmann, A., Tsuji, S., Hinzen, D.H. and Faille,
     L. (1974): Nature 250:496-498.
40. Vizi, E.S., 1972): J. Physiol. (Lond.) 226:95-117.
41. Vizi, E.S. (1975): IN Subcortical Mechanism and Sensorimotor
     Activities (ed) T.L. Frigyesi, Hans Huber, Bern, pp. 63-89.
42. Vizi, E.S. (1975): IN Cholinergic Mechanisms (ed) P.G. Waser,
     Raven Press, New York, pp. 199-211.
43. Vizi, E.S. (1977): J. Physiol. (Lond.) 267:261-280.
44. Vizi, E.S. (1978): Neuroscience 3:367-384.
45. Vizi, E.S. (1979): Prog. Neurobiol. 12:181-290.
46. Vizi, E.S. (1980) Neuroscience (in press).
47. Vizi, E.S. and Vyskocil, F. (1979): J. Physiol. 286:1-14.
48. Vyskocil, F. and Illes, P. (1977): Pflug. Arch. 370:295-297.
49. Zimmermann, H. (1979): Neuroscience 4:1773-1804.

HYDROLYSIS OF ACETYLCHOLINE BY FROG SKELETAL MUSCLE

R. Miledi, P.C. Molenaar* and R.L. Polak**

Department of Biophysics, University College London
Gower Street, WC1E 6BT, United Kingdom, and
*Department of Pharmacology, Sylvius Laboratories
Leiden University Medical Center, 72 Wassenaarseweg
2333 AL Leiden, The Netherlands, and
**Medical Biological Laboratory TNO, 139 Lange Kleiweg
2288 GJ Rijswijk, The Netherlands

There are several reports showing that intact muscles hydrolyze acetylcholine (ACh) at a much slower rate than their homogenates (see for instance Refs. 4,9,10,16). Marnay and Nachmansohn
(10) suggested that in intact muscle the relative slowness of the
hydrolysis of added ACh by cholinesterase (ChE) is caused by slow
diffusion of ACh through the muscle tissue. In mammalian muscle
the difference in ChE activity between intact and homogenized preparations has been attributed, in part or fully, to the liberation
of intracellular ChE by the homogenization. This and other considerations led to the suggestion that the data obtained with intact
preparations are more relevant to the physiological function of the
enzyme at the endplates than those obtained with homogenates (9,16).
On the other hand, it could be that the _in situ_ synaptic activity
of ChE, the "functional activity," of the enzyme, is severely dis-
torted by hindred diffusion, when it is measured with exogenous substrates on intact tissue.

We tried to estimate the functional ChE activity in the sartorius muscle of the frog by studying ACh hydrolysis in both intact
and homogenized muscles and found that hydrolysis measured in the
homogenates is a better index for functional ChE than that measured
in intact tissue, especially for evaluating the effect of ChE inhibitors. However, measuring activity in the homogenate still

provides a crude index because homogenates contain junctional as well as extrajunctional and intracellular enzyme, the junctional ChE being overestimated by about 40%.

The experiments were made on sartorius muscles of male and female frogs (Rana temporaria). The muscles were dissected at room temperature and the nerves were cut close to their entry into the muscle. The sartorius is about 35 mm long and consists of parallel fibers, most of which are innervated at two points about 15 mm apart. Only a small area of the muscle, a pelvic segment of about 10 mm long is completely free of endplates (6,11,13,18).

Muscles and muscle segments were incubated in Ringer (in mM): NaCl, 115.6; KCl, 2.0; CaCl$_2$, 1.8; Na-phosphate, 2.0 (pH 7.0). Homogenization of the tissue was done in Ringer, in a glass-in-glass homogenizer, after cutting the tissue into fine fragments with scissors.

We assayed ChE radiochemically using [^{3}H-acetyl]choline as a substrate. After acidification of the reaction mixture with acetic acid the [^{3}H]-acetate formed by the reaction was separated from [^{3}H]-ACh by extraction into a toluene scintillator layer containing n-amylalcohol, as described by Potter (20) and Buckley and Heaton (2). For the assay of ChE in intact muscles and muscle segments, these were incubated at 20°C, pinned onto a petri dish with Sylgard, in 1 ml Ringer containing [^{3}H]-ACh. For the assay of ChE in homogenates, dilutions of these in Ringer were tested and the reaction was stopped by the addition of 0.4 ml 10 M acetic acid. For each assay we prepared blanks by incubating [^{3}H]-ACh containing medium in the absence of tissue; 100% hydrolysis values were obtained similarly but in the presence of excess electric eel AChE (Boehringer). In order to lower the blank, the stock [^{3}H]-ACh was purified every week by precipitation in the presence of KI$_3$ as described earlier (19), but in the absence of any carrier.

There were considerable variations in the rate of hydrolysis of ACh by muscles from different frogs. However, the activities in sartorii from the same animal were usually similar. An example of this is given in Table 1, which shows the hydrolysis of [^{3}H]-ACh by two intact muscles. The diffusion of [^{3}H]-ACh into and the subsequent release of [^{3}H]-acetate from the muscle is fairly rapid, which is illustrated by the fact that the amount of [^{3}H]-acetate collected in the medium during the first 15 min was already close to the level of collection during the following periods. Subsequently, when the [^{3}H]-ACh was replaced by Ringer, there was some residual release of [^{3}H]-acetate, of which 70% was already released in 15 min. Table 1 further shows that the rate of spontaneous hydrolysis of [^{3}H]-ACh (224-150 = 74 cpm in 15 min) was about 100 times less than that caused by the muscle.

TABLE 1: Hydrolysis of $[^3H]$-ACh by intact sartorius muscle of the frog.

| Medium | Time (min) | Formation of $[^3H]$-acetate (cpm) | | |
		Left Muscle	Right Muscle	No Tissue
$[^3H]$-ACh + Ringer	0			150 ± 8 (4)
$[^3H]$-ACh + Ringer	0-15	4990	5118	226
$[^3H]$-ACh + Ringer	15-30	6050	7133	221
$[^3H]$-ACh + Ringer	30-45	5545	6492	226
$[^3H]$-ACh + Ringer	45-60	5873	6240	223
Ringer	60-75	1590	1494	
Ringer	75-90	418	421	
Ringer	90-105	220	231	

Muscles were incubated at 20°C in medium containing 1 µM (81,000 cpm) $[^3H]$-ACh during the first four periods of 15 min. Thereafter the muscles were incubated in Ringer in the absence of $[^3H]$-ACh. One experiment on two muscles from one frog.

TABLE 2: The difference in apparent ChE activity between intact and homogenized muscle.

| $[^3H]$-ACh (M) | Formation of $[^3H]$-acetate (pmol/min) | | $\frac{b}{a}$ |
	Intact Muscle (a)	Homogenized Muscle (b)	
10^{-8}	0.053± 0.04	1.2 ± 0.2	23
10^{-7}	0.51 ± 0.08	14.2 ± 2.7	28
10^{-6}	4.9 ± 0.6	110 ± 6	22
10^{-5}	50 ± 4	910 ± 90	18
10^{-4}	470 ± 35	6400 ± 550	14
10^{-3}	3400 ± 380	17000 ± 1450	5

Hydrolysis of $[^3H]$-ACh at different concentrations by a homogenate (from left muscle) and intact (right) muscle. Mean ± S.E.M. of two experiments. The activities were determined in triplicate.

As expected, homogenates of sartorius muscles hydrolysed $[^3\mathrm{H}]$-ACh at a greater rate than intact preparations (Table 2). There was approximately a twentyfold difference in activities, assayed in the presence of 10^{-8} to 10^{-5} M ACh. In this range, the activity increased linearly as a function of ACh concentration. At concentrations higher than 10^{-5} M the activity in the homogenates levelled off, as did the activity in intact muscle above 10^{-4} M ACh. Consequently, the difference in activities between homogenized and intact muscles gradually decreased from 20 to about 5 times at 10^{-3} M ACh.

Is the effect of homogenization on ACh hydrolysis due to an effect on the V_{max} (e.g. by the liberation of intracellular ChE) or due to an effect on the K_m? To answer this question we analysed the activities at concentrations of ACh up to 30 mM, in order to obtain K_m and V_{max} values from Lineweaver-Burk plots. Since it is well known that ChE is highly concentrated at the endplates (see for instance Ref. 3) and since there could be a difference in the properties of junctional and extrajunctional ChE (cf. Refs. 8,17), we performed this analysis on endplate-containing segments (e.p.) and endplate-free segments (non-e.p.) containing the muscle tendon junction.

The V_{max} for ChE in the e.p. segments was 10 to 20 times higher than in the non-e.p. pieces (Table 3). In great measure this difference was due to the differences in the length of the segments (29 and 6 mm), but after correction for length (last column Table 3) there was still nearly a fourfold difference which was not changed appreciably after homogenization. This difference agrees reasonably well with the finding by Marnay and Nachmansohn (10) that the ChE activity in frog sartorius homogenates containing 5 mM ACh was 5-6 times higher in e.p. than in non-e.p. homogenates.

Homogenization caused an increase in the V_{max} in e.p. segments from 19 to 27.6 pmol/cm fiber/min and in the non-e.p. segments from 4.7 to 9.4 pmol/cm fiber/min. This is probably due to the release of intracellular ChE (Cf. Ref. 12) brought about by the homogenization (see below).

As shown in Table 3, the apparent K_m for ACh in non-homogenized e.p. segments was as high as 7.6 mM, whereas that in non-e.p. segments was only 1.4 mM. Homogenization reduced the K_m's for ACh to 0.23 mM and 0.34 mM in e.p. and non-e.p. homogenates, respectively. These two values are not significantly different from each other. The large reduction in the K_m's brought about by homogenization suggests that the access of ACh from the incubation medium to the ChE in the intact tissue was impeded by diffusion barriers. The finding that the K_m of ChE in intact e.p. segments (the main part of which is junctional ChE) was much higher than that in intact

TABLE 3:   Michaelis–Menten constants of ChE derived from different
           compartments in the muscle.

	n	Length (mm)	K_m (mM)	V_{max} (nmol/min)	V_{max} (pmol/cm fiber/min)
Homogenized Tissue					
e.p.	4	29	0.23 ± 0.03	48 ± 2.4	27.6
non-e.p.	2	6	0.34 ± 0.13	3.4 ± 0.23	9.4
Pieces of Tissue					
e.p.	4	29	7.6 ± 0.9	33 ± 3.6	19.0
non-e.p.	2	6	1.4 ± 0.1	1.7 ± 0.16	4.7

Sartorius muscles were divided into endplate-containing (e.p.) and
endplate-free (non-e.p.) segments (see Refs. 6,18) as described by
Miledi et al. (13).   The hydrolysis was estimated at different
[^{3}H]-ACh concentrations in the medium.   Thereafter the ACh was
washed from the tissue, the segments were homogenized and the
hydrolysis rate was measured again. K_m and V_{max} were derived from
Lineweaver–Burk plots.   For the calculations in the table (last
column) it was assumed that the sartorius contains 600 fibers. All
muscles were stretched to a length of 35 mm.   Presented are the
means of n experiments $\pm$ S.E.M.

TABLE 4:   Solubilization of extracellular ChE by collagenase.

	Control (nmol/min/g) a	Collagenase (nmol/min/g) b	$\frac{b}{a}$	P_2*
Total in muscle	$2.3 \pm 0.25(7)$	$0.5 \pm 0.12(6)$	0.22	0.0001
Total released into medium	$0.07 \pm 0.03(4)$	$1.2 \pm 0.18(6)$	17	0.001
In e.p. segments	$2.6 \pm 0.42(5)$	$0.5 \pm 0.07(5)$	0.19	0.003
In non-e.p. segments	$0.7 \pm 0.15(5)$	$0.3 \pm 0.11(5)$	0.43	0.09

The muscles were treated with 3 mg collagenase (Sigma) per ml
Ringer for 90 min at 20°C. Subsequently the collagenase was removed,
replaced by Ringer and the muscles kept overnight at 5°C. The
muscles, or e.p. and non-e.p. segments (see Legend Table 3), were
then homogenized and ChE activity was assayed at 20°C in the presence of 1 μM [^{3}H]-ACh. Values are means $\pm$ S.E.M. with the number
of muscles in parentheses.   * Students t-test.

non-e.p. segments, strongly suggests that the access of ACh to the
junctional ChE is hindered further by an additional diffusion barrier, viz. the narrow synaptic gap.

The question arises as to why the diffusion barrier so pro-
foundly affected the K_m for ACh in intact muscle segments. Possibly there is a gradient of ACh concentration between the ChE on
the muscle membrane and the medium, caused by the fact that ACh is
hydrolysed more rapidly by the ChE than it is supplied by diffusion
(10). Consequently a lower ACh concentration will be maintained
at the site of the enzyme than in the medium. This effect will
produce an even lower ACh concentration near the endplates where
the junctional ChE is concentrated. Here the synaptic gap constitutes an additional diffusion barrier. The junctional ChE near the
outside of the synaptic gap probably hydrolyses most of the ACh
which arrives from the outside, and the more internally located
ChE probably does not come into play unless the ACh concentration
in the medium is extremely high. Only under this condition is the
full capacity (V_{max}) of the junctional ChE disclosed. This would
explain why the apparent K_m is so high and why it is reduced some
30 times by homogenization (see Table 3).

An alternative explanation of the difference in apparent ChE
activity between intact and homogenized muscle is that ACh at high
concentrations migrates through the cell membranes, and then is
hydrolysed by cytoplasmic ChE. This possibility has been proposed
for the iris of the rat by Lund-Karlsen and Fonnum (9) on similar
evidence, namely, that homogenization of the iris of the rat causes
a very large increase in ACh hydrolysis and that this effect virtually disappears at high concentrations of ACh. In the sartorius
there is virtually no uptake of ACh by the fibers up to a concentration of 0.1 mM (15), but it cannot be excluded that a high concentrations of ACh, say 10 mM, there is a substantial uptake. Even
if such uptake really were to take place, it could only lead to an
important increase of the activity of intact preparations, if the
greater part of the ChE is intracellularly located. However, this
is not the case, as shown by the following experiments, in which
muscles were treated with collagenase. By this procedure most if
not all of the extracellular ChE is dissolved (1,5). Table 4 shows
that after treatment with collagenase about 20% of the total ChE
activity was left in the muscles, whereas 50 to 60% was recovered
in the medium. Apparently some activity was lost, possibly through
some denaturation of ChE by collagenase. Table 4 shows that in the
e.p. part of the muscle 20% of the activity was left after treatment with collagenase, and in the non-e.p. part a somewhat higher
proportion, but this difference was not statistically significant.
These results indicate that intracellular ChE is perhaps no more
than about 20% of the total. The remaining extracellular activity,
80%, is roughly 60% junctional and 20% extrajunctional (including

tendon). Since, as shown in Table 3, the V_{max} of the ChE in intact
e.p. segments is about 70% of that in e.p. homogenates (in which
the intracellular enzyme contributes to the total activity), and
since the surface enzyme represents at least a similar proportion
(80%) of the total activity according to the results obtained with
collagenase, we believe that the V_{max} for ChE in intact e.p. segments represents essentially that of the extracellular enzyme (in
which the junctional ChE is the main part). That is, in intact
sartorius muscle segments the intracellular enzyme does not contribute significantly to the observed V_{max}. This leaves the diffusion barriers (especially the synaptic cleft) combined with the
fast action of the ChE as responsible for the fact that ACh is
hydrolysed much more slowly by intact muscle segments than by their
homogenates. If this is true, ChE inhibitors, by decreasing ChE
activity, should decrease the concentration gradient between ACh in
the medium and ChE on the synaptic membrane and thus reduce the
difference between the rates of ACh hydrolysis in intact and
homogenized preparations. Indeed, at 98-99% inhibition of ChE in
the homogenates, i.e. when the residual activity was just measurable
above the blank, we found that the hydrolysis of ACh in intact
tissue reached a value of about 0.6 times that in the homogenized
tissue, a ratio which approaches that between the V_{max} values
(19.0/27.6 = 0.69; see Table 3). In other words, ChE inhibition
allows the ACh to penetrate deeper into the synaptic gap, causing
a higher proportion of the remaining active enzyme to exercise
its action.

It is well known that ACh applied either in the bath or ionophoretically induces a much greater electric response when the ChE
is inhibited. The relative weakness of the ACh effect in preparations in which ChE is not inhibited may have an explanation similar
to that of the slowness of the hydrolysis of applied ACh in intact
preparations, since both ACh receptors and ChE are confined in the
same synaptic space.

Our results imply that measuring ACh hydrolysis in intact
tissue is not a good index of the activity of junctional ChE. Only
about 1/20 of the junctional activity is measured in this way, the
rest being mainly the activity of extrajunctional ChE which is more
easily accessible to the ACh, because of smaller diffusion barriers.
This probably also applies to neuromuscular junctions in other
species, since they generally contain a high density of ChE in a
confined space. Particular care should be exerted with the interpretation of the effects of drugs, or the effect of surgical procedures, such as nerve section, on the "ChE activity" in intact muscle
since the activity is not proportional to the number of ChE molecules at the junction. An example of this is given by the fact that
0.15 µM neostigmine, which causes about 30% inhibition of ChE when
assayed in a sartorius homogenate, caused a marked increase in the

amplitude and half decay time of the miniature endplate currents, whereas the inhibition of the hydrolysis by intact muscles by this concentration of the drug is at most a few percent (Miledi and Molenaar, unpublished observations).

ACKNOWLEDGEMENTS

We thank Mrs. P. Braggaar-Schaap and Miss J.W.M. Tas for their excellent technical assistance. Financial support by FUNGO/ZWO, the MRC and the Wellcome Trust is gratefully acknowledged.

REFERENCES

1. Betz, W. and Sakmann, B. (1973): J. Physiol. 230:673-688.
2. Buckley, G.A. and Heaton, J. (1968): J. Physiol. 199:743-749.
3. Couteaux, R. (1963): Proc. R. Soc. Lond. B. 158:457-480.
4. Feng, T.P. and Ting, Y.C. (1938): Chin. J. Physiol. 13:141-144.
5. Hall, Z.W. and Kelly, R.B. (1971): Nature New Biol. 232:62-63.
6. Katz, B. and Kuffler, S.W. (1941): J. Neurophysiol. 6:99-110.
7. Katz, B. and Miledi, R. (1977): Proc. R. Soc. Lond. B. 196:59-72.
8. Liu, A.Y.C. and Mittag, T. (1974): Molec. Pharmacol. 10:283-292.
9. Lund Karlsen, R. and Fonnum, F. (1977): J. Neurochem. 29:151-156.
10. Marnay, A. and Nachmansohn, D. (1938): J. Physiol. 92:37-47.
11. Miledi, R. (1960): J. Physiol. 151:1-23.
12. Miledi, R. (1964): Nature (Lond) 204:293-295.
13. Miledi, R., Molenaar, P.C. and Polak, R.L. (1977): Proc. R. Soc. Lond. B. 197:285-297.
14. Miledi, R., Molenaar, P.C. and Polak, R.L. (1977): In Cholinergic Mechanisms and Psychopharmacology, (ed), D.J. Jenden, Plenum Press, New York, pp. 377-386.
15. Miledi, R., Molenaar, P.C. and Polak, R.L. (1980): J. Physiol. 309:199-214.
16. Mittag, T.W., Ehrenpreis, S. and Hehir, R.M. (1971): Biochem. Pharmacol. 20:2263-2273.
17. Namba, T. and Grob, D. (1970): J. Clin. Invest. 49:936-942.
18. Pezard, A. and May, R.M. (1937): C.R. Soc. Biol. Paris 124: 1081-1083.
19. Polak, R.L. and Molenaar, P.C. (1974): J. Neurochem. 23:1295-1297
20. Potter, L.T. (1967): J. Pharmacol. Exp. Ther. 156:500-506.

EARLY EFFECTS OF DENERVATION ON ACETYLCHOLINE AND

CHOLINE ACETYLTRANSFERASE IN SKELETAL MUSCLE

R. Miledi, P.C. Molenaar* and R.L. Polak**

Department of Biophysics, University College London
Gower Street, London WC1E 6BT, England, and
*Department of Pharmacology, Sylvius Laboratories
Leiden University Medical Center, 72 Wassenaarseweg
2333 AL Leiden, The Netherlands, and
**Medical Biological Laboratory TNO, 139 Lange Kleiweg
2288 GJ Rijswijk, The Netherlands

INTRODUCTION

The effects of motor nerve section on the physiology and ultra-
structure of nerve and muscle have been studied in great detail (see
for instance 2,18,27). After a delay which is dependent on the
length of the nerve stump, the transmission of impulses fails, due
to disintegration of motor nerve terminals. At this time motor
axons still conduct impulses but they also degenerate later after
denervation. In the frog the muscle membrane at the junctions is
not left naked because Schwann cells take the place previously occupied by the motor terminals. In the rat this occurs only
transiently.

It has been previously shown that the amounts of acetylcholine
(ACh) and choline acetyltransferase (CAT) are greatly reduced in
denervated mammalian muscle (1,8-10,26). These studies (except 10)
were made long after the disappearance of axons and terminals, when
secondary, atrophic, changes may have taken place in the muscle.

We tried to examine the localization of ACh and CAT in skeletal
muscle of both the frog and the rat, by studying the early changes
of these substances after denervation. As expected, our studies
show that ACh and CAT are located predominantly in the motor nerve
terminals. Furthermore, some CAT, but practically no ACh, is
present in the motor axons. No CAT was detected in the muscle fibers

205

of the frog, but a little is present in muscle fibers of the rat; in both species the muscle fibers contain a substantial amount of ACh.

METHODS

The experiments were made on sartorius muscles of male and female frogs (Rana temporaria) and hemidiaphragms of young male Wistar rats (70-100 g). In the frogs the sartorius of one leg was denervated under ether anesthesia by cutting the sciatic nerve in the pelvis (14). The muscle of the opposite leg served as a control. The sartorius muscles were dissected at room temperature and the nerve was cut close to its entry into the muscle. Before homogenization most muscles were divided into an endplate-free (11,22) pelvic segment (non-e.p.), approximately 4 mm long, and a longer segment containing all the endplates (e.p.)(15). In some experiments the muscles were incubated in Ringer of the following composition (mM): NaCl 115.6; KCl 2.0; $CaCl_2$ 1.8; Na-phosphate 2.0 (pH 7.2). In the rats the left hemidiaphragm was denervated under ether anesthesia by transection of the phrenic nerve in the thorax about 5 mm from its entrance into the muscle (cf. 18). The hemidiaphragms were dissected free from the ribs and placed in a 95% O_2/5% CO_2-bubbled Ringer of the following composition (mM): NaCl 114; KCl 4.5; $NaHCO_3$ 25; NaH_2PO_4 1; $CaCl_2$ 2; $MgSO_4$ 1; glucose 11. Before homogenization most muscles were divided into e.p. (70 mg) and non-e.p. (30 mg) strips. The e.p. strips were made much broader than in the experiments by Hebb et al. (9), because the endplates were not so sharply located in the muscles of our rats.

ACh was measured by pyrolysis mass fragmentography with deuterated ACh as an internal standard, as described earlier (15,24).

For the measurement of CAT the tissue was homogenized in a solution containing 0.1% (w/v) Triton X100 and 100 mM K-phosphate (pH 7.9). ACh synthesis was determined according to Braggaar-Schaap (3). The composition of the incubation medium was (final concentration in 70 µl): [^{14}C]-acetyl CoA (Amersham) 50,000 dpm diluted with unlabelled acetyl CoA to 25 µM (frog sartorius) or to 80 µM (rat diaphragm): NaCl 50 mM; K-phosphate 50 mM; physostigmine salicylate 0.2 mM: Triton X100 0.06% (w/v). The temperature was 20°C for frog sartorius and 37°C for rat diaphragm. For blanks the tissue was incubated in the presence of 1 µg electric eel acetylcholinesterase (Boehringer) but in the absence of physostigmine.

RESULTS AND DISCUSSION

Acetylcholine in Frog Sartorius Muscle

We have reported earlier (15) that freshly dissected sartorius contains about 45 pmol ACh. The bulk of this ACh is localized in the e.p. region of the muscle, suggesting that it is contained in nervous elements, but some ACh is present in the non-e.p. part. In the period between 2 and 8 days after denervation the ACh in the e.p. region of the muscle decreases by about 30 pmol and the ACh concentration in the e.p. region becomes equal to that in the non-e.p. segment. The ACh in the non-e.p. region itself does not change after denervation (15). From these observations it was concluded that the 30 pmol of ACh which disappears from the muscle within the first 8 days after denervation, is localized in the motor nerve terminals since these are known to degenerate in the same period (2). Although the intramuscular myelinated nerves do not degenerate until a few weeks have elapsed (6,28), the bulk of the ACh persisting after denervation is not contained in them. Otherwise after denervation the ACh concentration in the e.p. region would remain higher than in the non-e.p. region. For a similar reason only a small part of the persisting ACh appears to be located in the Schwann cells which replace the degenerated nerve terminals at the endplates and from which, after an electrical silence of a few days, ACh quanta are released at a low frequency (2). On different grounds the amounts of ACh in the intramuscular branches of the nerve has been estimated to be less than 5 pmol and that in the Schwann cells 0.5 pmol (15).

Acetylcholine Release from Satorius

Muscles were pretreated with diethyl dimethylpyrophosphonate (DEPP), a very effective, irreversible inhibitor of frog muscle cholinesterase (16), and then incubated at room temperature. During incubation there was a mean release of 2.4 pmol ACh/hr from the muscle into the medium (Table 1). This is similar to the value of 1.8 pmol/hr reported by Miledi et al., (16), who pointed out that only 2-3% of this amount can be accounted for by spontaneous release in the form of quanta. What is the source of the remaining 97-98% of the spontaneously released ACh? At any rate an important part of it cannot originate from the nerve terminals because after denervation there still was a release of about 1.5 pmol ACh/hr (Table 1). Because the release was decreased by about 1 pmol/hr by denervation, this amount presumably originates from the terminals; it is still some 20 times higher than can be accounted for by

TABLE 1: Spontaneous release of ACh ($pmol \cdot hr^{-1}$ per muscle).

	n	INNERVATED	DENERVATED
Frog sartorius (20°C)			
ChE inactivated	11	2.4 ± 0.31*	1.5 ± 0.31*
no ChE inhibition	4	0.8 ± 0.28	0.3 ± 0.32
Rat hemidiaphragm (37°C)			
ChE inhibited	6	85 ± 32	18 ± 7

Experiments were performed 16–20 days after denervation of frog
sartorius muscle and 8 days after denervation of left hemidiaphragm
of the rat. The weight of the sartorius muscles was 50–80 mg, that
of the hemidiaphragms 100–130 mg.
* Significantly different, $P_2 < 0.05$ (Student's t-test).

TABLE 2: Distribution of ACh and CAT in skeletal muscle.

	RAT DIAPHRAGM	FROG SARTORIUS
Total ACh (femtomol/endplate)	13	45
Total CAT (pmol/hr/endplate)	4	0.12
Nerve terminals: ACh	75%	75%
: CAT	70%	80–100%
Myelinated fibers: ACh	<10%	<10%
: CAT	30%	<20%
Muscle fibers: ACh	25%	25%
: CAT	∿ 3%	<20%

10^4 endplates per hemidiaphragm (about 10^5 per g); 10^3 endplates
per sartorius (about 1.6×10^4 per g).

TABLE 3: Origin of spontaneous ACh release from skeletal muscle.

	RAT DIAPHRAGM 37°C	FROG SARTORIUS 20°C
Total release	8.5	4.0
Molecular leakage from nerve terminals	6.7	1.5
Leakage from muscle fibers	1.8	2.5

ACh release in femtomol/hr/fiber. The diaphragm contains about 10^4
fibers and the sartorius about 600.

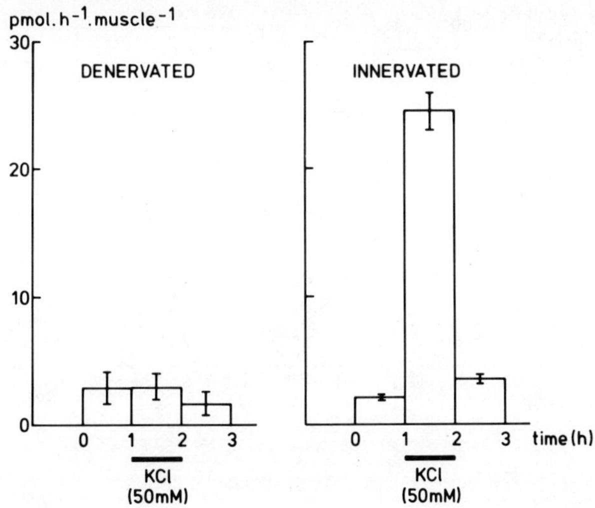

FIGURE 1: ACh release from frog sartorius at 20°C. The cholinesterase was inhibited by a pretreatment with 10 µM DEPP.
When 50 mM KCl was included in the Ringer, NaCl was reduced in order to maintain isotonicity. Means ± S.E.M.
of 4 observations.

spontaneous quantal release. Apparently there is a non-quantal leak
of ACh from the terminals of about 1 pmol/hr. This is in agreement
with the results by Katz and Miledi (12) who observed that curare,
applied ionophoretically to the junctions, caused a slight hyperpolarization of the muscle membrane, from which it was concluded
that there is a continuous leakage of ACh at the junctions, estimated to a few pmol/hr.

Table 1 shows further that some ACh (0.8 pmol/hr) was collected
in the medium when the cholinesterase was not inhibited. It appears
unlikely that the ACh which escaped hydrolysis was released at the
neuromuscular junction since cholinesterase is highly concentrated
at this site (4). Moreover, results obtained with La^{+++} ions, which
cause ACh release from motor nerve terminals, have shown that all
ACh released by La^{+++} is hydrolyzed unless cholinesterase is fully
inactivated (16,17). ACh release in the absence of cholinesterase
inhibition seemed to be decreased after denervation, but this decrease was not significant.

The ACh released from denervated muscle probably originates
from the muscle fibers. Although the myelinated motor axons may not
yet have degenerated at the time when the release was measured

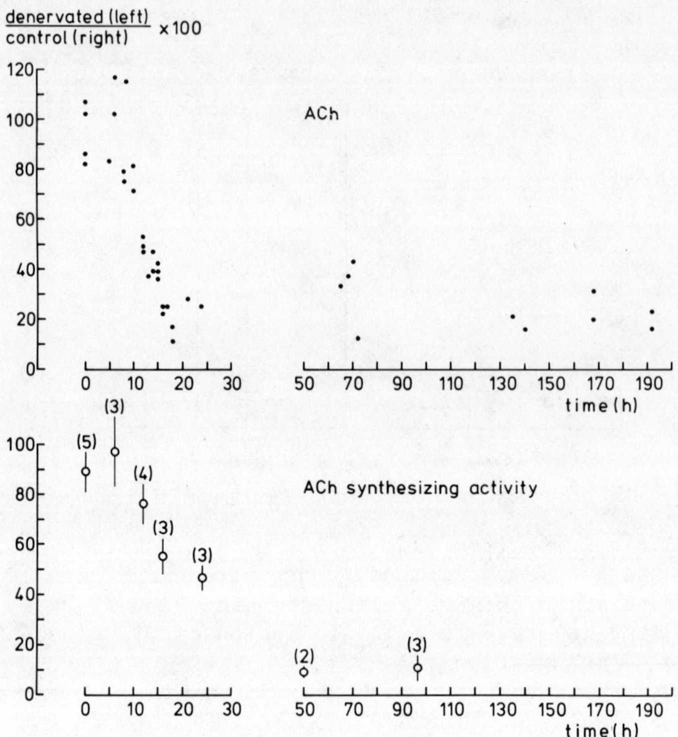

FIGURE 2: ACh and CAT in rat diaphragm after section of the phrenic
nerve. Single observations or means $\pm$ S.E.M. with number
of muscles in brackets.

(see 6,28) their ACh content is very small (15) and it is therefore
unlikely that they are the source of the observed release. Moreover, even from pieces of sciatic nerve, which contains many more
fibers and consequently much higher amounts of ACh (53 pmol/cm) than
the sartorius nerve(15), no significant ACh release was detected
during incubation (Molenaar, unpublished observations).

One might question the identity of the ACh collected in the
experiments of Table 1, especially since "ACh release" was observed
in the absence of a cholinesterase inhibitor. However, we found
that the electric eel acetylcholinesterase hydrolyzed practically
all ACh collected in the medium, confirming the ACh-like nature of
the substance collected.

Figure 1 shows that K^+ ions evoke a great release of ACh from the sartorius. On the other hand, no effect of K^+ was observed in denervated muscles. This excludes the possibility that some nervous structures, containing substantial amounts of ACh releasable by K^+, survive after denervation.

Acetylcholine in Rat Diaphragm

The ACh content in a hemidiaphragm was about 130 pmol, which is about 2 times less than previously found in muscles from older Wistar rats from a different source (21). About 90-95% of the ACh was found in the e.p. portion, indicating that in the rat diaphragm, as in the frog sartorius, the ACh is contained predominantly in the nerves.

Miledi and Slater (18) have reported that between 8 and 16 hr after section of the phrenic nerve, the terminals degenerate and neuromuscular transmission fails. The present results show that the ACh content decreased in this period by 80% (Fig. 2). This indicates that about 80% of the ACh is localized in the terminals. No substantial amount of ACh is contained in the intramuscular nerve branches, because they degenerate 24 to 48 hr after denervation (18), i.e. at a time when no further reduction of ACh took place. The remaining 20% of the ACh must therefore be located in the muscle fibers.

The decrease in ACh by denervation took place exclusively in the e.p. part of the muscle. The ACh concentration became equal to that of the non-e.p. part (about 200 pmol/g, similar to the value found in the non-e.p., in the frog sartorius [15]).

Acetylcholine Release from the Diaphragm

During incubation of the muscle at 37°C, there was a much greater spontaneous release of ACh than from the frog sartorius at 20°C. The release was reduced from 85 to 18 pmol/hr by denervation, that is, it was reduced to a larger extent than was the case in the sartorius. K^+ ions evoked a little release of ACh from denervated muscle, but this was only a few percent of the amount released by K^+ from innervated muscle. These results are in agreement with previous findings showing that ACh is released from the unstimulated denervated rat diaphragm (13,19,26). Just as in the frog, only a minute fraction of the ACh released from rat muscle at rest is quantal (7,9,19).

CAT in the Frog Sartorius

CAT activity in frog skeletal muscle is known to be very low.
Tucek et al. (28) found that the synthesis of ACh in a homogenate
of frog sartorius muscle was 30 nmol/g/hr and Braggaar-Schaap (3)
reported an activity as low as 3 nmol/g/hr. However, recent findings
(20) show that the actual CAT activity is even lower, because part
of the reported ACh synthesis was due to an enzyme different from
CAT.

It was already suspected by Tucek et al. (28) that part of the
synthesis of ACh in the frog sartorius was due to carnitine acetyltransferase (CarnAT) since CarnAT has been reported to be able to
acetylate Ch in mammalian muscle where it is present in large a-
mounts (25,30). Moreover, Tucek et al. (28) had noted that ACh
synthesis in homogenized sartorius was inhibited only incompletely
by 4-(1 -naphthylvinyl)-pyridine (NVP), a specific inhibitor of
mammalian CAT (29). On the other hand, these authors found that
NVP completely blocked CAT from frog sciatic nerve.

Using Ch-saturation kinetics we were able to distinguish CAT
from a different enzyme which might be identical to CarnAT, to
which we shall refer as psi-CAT. Whereas CAT in homogenized muscle
and sciatic nerve was saturated by exceptionally low Ch concentrations (Km 50 μM), psi-CAT showed no signs of saturation even with
Ch concentrations up to 10 mM.

The CAT activity in the sartorius was found to be present only
in e.p. segments, and in these it was superimposed upon the non-
saturable activity of psi-CAT, which was present in equal quantities
in non-e.p. and e.p. segments. With 0.15 mM Ch in the assay system,
a concentration by which CAT was nearly saturated while it was not
yet overshadowed by the activity of psi-CAT, we found that 17-22
days after denervation the ACh synthesizing activity in homogenized
e.p. parts had decreased from 3.18 ± 0.69 (5) to 1.62 ± 0.14 (5)
nmol/g/hr. In the non-e.p. parts the activity had not changed:
1.39 ± 0.20 (5) before and 1.35 ± 0.17 (5) nmol/g/hr after denervation was found. Moreover, no Ch saturable activity was observed
in denervated muscle (20).

These results suggest that in the frog sartorius practically
all CAT is located in the nerve terminals. However, due to the high
background of psi-CAT in muscle homogenates, we cannot exclude that
there is some CAT in the myelinated axons in the muscle or in the
Schwann cells of denervated muscle. It should be noted that the
minute amount of ACh in the Schwann cells (15) would require only
trace amounts of CAT for its synthesis.

CAT in the Rat Diaphragm

The CAT activity in the rat diaphragm is two orders of magnitude greater than in frog muscle. Hebb et al. (9) reported ACh synthesizing activities of about 600 nmol/g/hr, we found activities ranging from 200-500 nmol/g/hr with large variations between animals. The enzyme activity was 5-10 times higher in the e.p. part of the non-e.p. part of the diaphragm. When the ACh synthesizing activity was studied as a function of the Ch concentration again we could distinguish two enzymes: 1) a saturable enzyme (Km 500 μM), located mainly in the e.p. part of the diaphragm and 2) a psi-CAT not saturable by Ch, which was present in non-e.p. and e.p. parts.

The ACh synthesizing activity decreased by about 50% in the interval between 10 and 24 hr after denervation (Fig. 2). The disappearing enzyme activity is probably due to CAT localized in the motor nerve terminals, which degenerate during this period (18). In the interval between 24 and 48 hr after denervation, there was a further decrease of the enzyme activity, probably representing CAT localized in the myelinated fibers which degenerate during this period. The remaining activity (about 10% of the contralateral activity) is probably due to CarnAT (cf. 25,30), although a small amount of saturable activity (CAT) was found in homogenized non-e.p. segments of 24 and 96 hr denervated muscle. These findings agree well with those of Israel (10) who found that the ACh synthesizing enzyme activity in the e.p. region of the rat's sternomastoid muscle decreased in an interval between 20 and 50 hr after denervation, to the level of activity found in non-e.p. parts.

The results from rat and frog muscles are compared in Tables 2 and 3. It can be seen that both ACh and CAT have a very similar distribution in both muscles. However, the absolute amounts of ACh and CAT are very different; sartorius contains more ACh per endplate, and vastly less CAT, than the diaphragm. The resting ACh release from the muscle fibers is similar in both muscles, but the leakage from the terminals is about 4 times higher in the diaphragm.

ACKNOWLEDGEMENTS

We thank Mrs. P. Braggaar-Schaap, Miss J.W.M. Tas and Mr. A.L. v.d. Laaken for their excellent technical assistance. Financial support by FUNGO-ZWO (to P.C. Molenaar and R.L. Polak) and the MRC (to R. Miledi) is gratefully acknowledged.

REFERENCES

1. Bhatnagar, S.P. and MacIntosh, F.C. (1960): Proc. Canad. Fed. Biol. Soc. 3:12-13.

2. Birks, R., Katz, B. and Miledi, R. (1960): J. Physiol. Lond. 150:145-168.

3. Braggaar-Schaap, P. (1979): J. Neurochem. 33:389-392.

4. Couteaux, R. (1963): Proc. Roy. Soc. Lond. B. 158:457-480.

5. Dennis, M.J. and Miledi, R. (1974): J. Physiol. Lond. 237: 431-452.

6. Elul, R., Miledi, R. and Stefani, E. (1970): Acta Physiol. Latinoamer. 20:194-226.

7. Fletcher, P. and Forrester, T. (1975): J. Physiol. Lond. 251: 131-144.

8. Hebb, C.O. (1962): J. Physiol. Lond. 163:294-306.

9. Hebb, C.O., Krnjevic, K. and Silver, A. (1964): J. Physiol. Lond. 171:504-513.

10. Israel, M. (1970): Arch. d'Anat. Microsc. 59:68-98.

11. Katz, B. and Kuffler, S.W. (1941): J. Neurophysiol. 6:99-110.

12. Katz, B. and Miledi, R. (1977): Proc. Roy. Soc. Lond. B. 196:59-72.

13. Krnjevic, K. and Straughan, D.W. (1964): J. Physiol. Lond. 170:371-378.

14. Miledi, R. (1960): J. Physiol. Lond. 151:1-23.

15. Miledi, R., Molenaar, P.C. and Polak, R.L. (1977): Proc. Roy. Soc. Lond. B. 197:285-297.

16. Miledi, R., Molenaar, P.C. and Polak, R.L. (1977): In Cholinergic Mechanisms and Psychopharmacology, (ed) D.J. Jenden, Plenum Press, New York, pp. 377-386.

17. Miledi, R., Molenaar, P.C. and Pollak, R.L. (1980): J. Physiol. Lond. 309:199-214.

18. Miledi, R. and Slater, C.R. (1970): J. Physiol. Lond. 207: 507-528.

19. Mitchell, J.F. and Silver, A. (1963): J. Physiol. Lond. 165: 117-129.

20. Molenaar, P.C. and Polak, R.L. (1980): J. Neurochem. 35:1021-1025

21. Molenaar, P.C., Polak, R.L., Miledi, R., Alema, S., Vincent, A. and Newsom-Davis, J. (1979): Prog. Brain Res. 49:449-458.

22. Pezard, A. and May, R.M. (1937): C.R. Soc. Biol. Paris 124: 1081-1083.

23. Polak, R.L. and Molenaar, P.C. (1964): J. Neurochem. 23: 1295-1297.

24. Polak, R.L. and Molenaar, P.C. (1979): J. Neurochem. 32:407-412.

25. Roskoski, R., Mayer, H.E. and Schmidt, P.G. (1974): J. Neurochem. 23:1197-1200.

26. Straughan, D.W. (1960): J. Physiol. Lond. 170:371-378.

27. Titeca, J. (1935): Arch. Int. Physiol. 41:2-56.

28. Tucek, S., Zelena, J., Ge, I. and Vyskocil, F. (1978): Neuroscience 3:709-724.

29. White, H.L. and Cavallito, C.J. (1970): Biochim. Biophys. Acta 206:343-358.

30. White, H.L. and Wu, J.C. (1973): Biochemistry 12:841-846.

RELEASE OF ACETYLCHOLINE TRIGGERED BY THE VENOM

OF GLYCERA CONVOLUTA

R. Manaranche, M. Thieffry and M. Israel

Laboratoire de Neurobiologie Cellulaire
Centre National de la Recherche Scientifique
91190 Gif sur Yvette, France

INTRODUCTION

The polychaete annelid Glycera convoluta (3) has four small
glands, associated to its jaws, containing a toxic substance. When
injected into the coelomic cavity of crabs the venom rapidly killed
the animals. We therefore looked for a possible neurotoxicity of
this substance. We report here that the venom of Glycera (GV)
applied to frog neuromuscular preparations or Torpedo electric
organs, triggers an important increase in the frequency of miniature potentials, but it does not induce ultrastructural changes
in contrast to other presynaptic venoms (2).

RESULTS

Frog Neuromuscular Preparations

The effect of GV was studied on the cutaneous pectoris muscle
of the frog. The treatment with GV (0.05-0.2 gland/ml) led to an
important increase in the frequency of miniature endplate potentials
(mepp's). After a latency of 1-2 min after start of the venom
perfusion, the first effect appears as a short burst of mepp's
(Fig. 1B). Longer bursts occur in the following seconds until a
sustained activity of mepp's sets up (Fig. 1C). This activity (mean
mepp's frequency in the range 50-100 sec^{-1}) can be observed for
many hours. In an experiment in which the GV effects were continuously monitored in the same preparation for 19 hr, it was

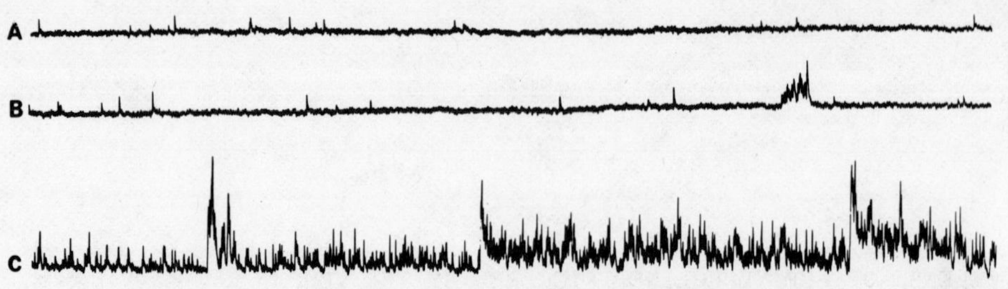

FIGURE 1: Effect of GV on spontaneous quantal release in a cutaneous
          pectoris fiber. A = control, b = isolated burst 85 sec
          after venom addition, C = sustained activity after 140 sec
          of GV action.

found in all fibers examined after 15 hr of venom action. After
19-20 hr, an abnormally high activity was still observed in some
fibers, other ones being almost silent. Bursts of high frequency
(up to 150 sec^{-1}) as well as periods of relatively low frequency
(20 sec^{-1}) occur in an unpredictable way. When the action of GV
was allowed to develop for several hours it was still possible to
obtain an almost complete reversal of the effect by washing with
venom free solution.

After 5 hr of GV action, leading to the release of about two
times the initial store of transmitter, no obvious change was
noticed in the 25 terminals which were examined (Fig. 2). The
number of synaptic vesicles was equivalent in both control and
treated preparations. In some cases their distribution seemed
slightly different from that of the control where vesicles appeared
more evenly distributed. However, in other terminals no change in
their distribution could be found. Neither mitochondria nor the
nerve terminal membrane nor the synaptic cleft were modified.

After 19 hr, most of the terminals are strongly modified. In
24 of the 26 terminals examined, the nerve terminal was invaded by
numerous membrane invaginations leading more or less to the frag-
mentation of the cytoplasm. Synaptic vesicles were still present,
but their size was much less homogenous. Many mitochondria were
swollen or even destroyed, and only identifiable by crest fragments.
However, in the other two terminals examined, the synaptic vesicle
population looked roughly normal, while some mitochondria were
extremely swollen. In all cases, no modification of the synaptic
cleft was observed.

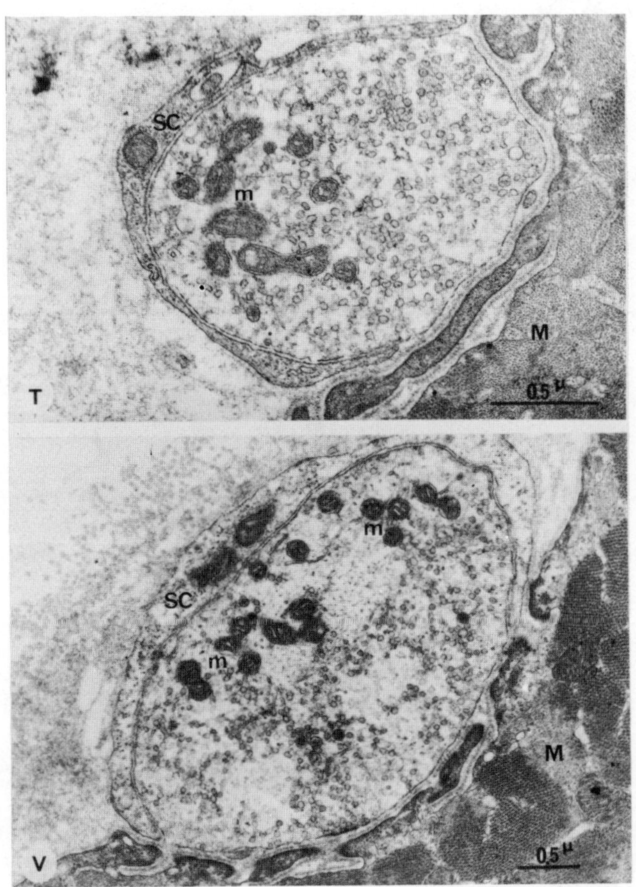

FIGURE 2: Cross sections of frog neuromuscular junctions. T = control preparation, V = preparation treated for 5 hr with GV (0.05 gland/ml), M = muscle, m = mitochondria, SC = Schwann cell.

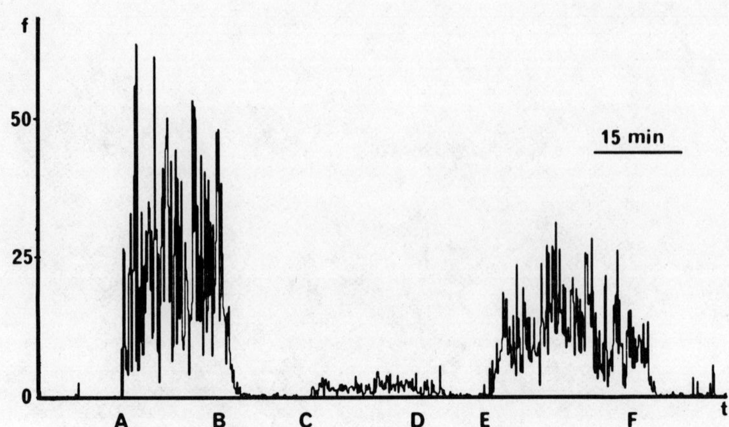

FIGURE 3: Blockade of GV action by D 600. Equal concentrations of
GV were applied between A and B, C and D, E and F. D 600
(10^{-4} M) was present in the solution between B and D.
(f = mean mepp's frequency during successive 10 sec
intervals.)

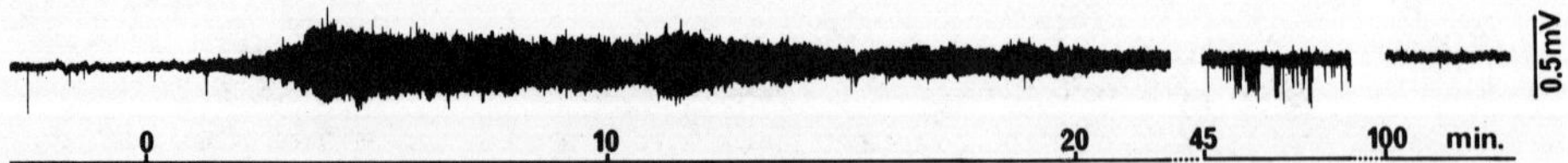

FIGURE 4: Effect of GV applied at t = 0 on the quantal release of
a nerve terminal of Torpedo electroplaque (AC record).

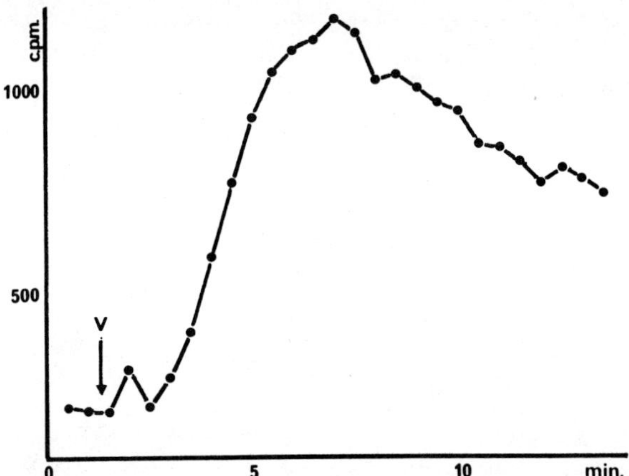

FIGURE 5: Release of ACh measured by the efflux of radioactive
          acetate from Torpedo electric organ prisms labelled with
          $(1-^{14}C)$ acetate. Venom applied at V in the superfusing
          solution.

The venom effect was not modified when GV was applied in the
presence of 12 µM tetrodotoxin following 20 min preincubation with
this agent. But when muscles were soaked in a solution where
calcium was substituted by 1 mM Mg Cl^2 addition of GV did not produce any detectable effect. Upon calcium addition, the venom effect
immediately developed; moreover, methoxy-verapamil (D 600) 10^{-4} M
blocked (reversibly) the venom action (Fig. 3).

Torpedo Nerve Electroplaques Preparations

After a latency of the order of 1 min, the perfusion of
electroplaques with a solution containing 0.1-0.2 gland/ml of GV
induces an important increase in the frequency of extracellularly
recorded miniature potentials. The activity reaches its maximum
after 4-5 min, then declines more slowly during the following 20-
40 min. After that time, some giant spontaneous potentials occur.
After about 100 min the effect is over (Fig. 4).

All the samples examined by electron microscopy were taken
from surface electroplaques. Preparations treated with venom were
kept for 1 to 3 hr in solution containing 0.15 gland/ml. The
number and distribution of synaptic vesicles were comparable in
control and treated preparations. No swelling of mitochondria and
no modifications of the presynaptic membrane and of the synaptic
cleft were observed.

With the <u>Torpedo</u> electroplaque preparation, it was possible
to verify that the release of transmitter estimated by the efflux
of labelled acetate paralleled the activity of the miniature potential. The time course of both rising and falling phases of the
radioactive efflux curve (Fig. 5) were comparable to those of the
electrically recorded phenomenon.

CONCLUSION

A common feature of many presynaptic toxins is to induce a
large increase in the frequency of miniature potentials while
important ultrastructural changes of nerve terminals were observed
(1). In view of the fact that the <u>Glycera</u> venom has a very low
phospholipase activity (C. Bon, personal communication) and does
not induce similar ultrastructural changes, it appears as an interesting potential tool for the study of neurotransmitter release.

REFERENCES

1.   Hurlbut, V.P. and Ceccarelli, B. (1979): In <u>Neurotexins:
     Tools in Neurobiology. Advances in Cytopharmacology</u>, Vol.3.
     (eds) B. Ceccarelli and F. Clementi, Raven Press, New York,
     pp. 87-127.
2.   Manaranche, R., Thieffry, M. and Israel, M. (1980): J. Cell
     Biol. <u>85</u>:446-448.
3.   Michel, C. (1966): Cah. Biol. Mar. <u>7</u>:367-373.

ELECTROPHYSIOLOGICAL AND MORPHOLOGICAL CORRELATES OF THE

RE-INNERVATION OF RAT NEUROMUSCULAR JUNCTION: IMPLICATIONS

ON THE ROLE OF MEMBRANE COMPONENTS SUCH AS GANGLIOSIDES IN

THE MOTOR NERVE SPROUTING

A. Gorio, G. Carmignoto, M. Finesso, A. Leon,
P. Marini, L. Tredese and R. Zanoni

Fidia Research Laboratories
Abano Terme, Italy

INTRODUCTION

The re-innervation process of a mammalian muscle can be divided
into several parts: a) axonal growth; b) recognition of the endplate with consequent reformation of the nerve-muscle contact;
c) maturation and stabilization of the new contact; and d) axonal
sprouting, which can be considered a consequence of both the regeneration process and the presence of an endogenous stimulus, perhaps coming from the muscle. Our goal is to study those critical
steps by means of histochemical and electrophysiological techniques.
Three points mentioned above (b,c,d) were studied extensively, using
frog neuromuscular preparations (6,7,18,24), and reviewed extensively by several authors (8,15). Although the studies performed on
mammalian muscle are numerous, only a few approached the problem
as a whole, trying to compare these different parameters (1) and
carefully control the re-establishment of the synapse and the
sprouting process itself.

Several years ago it was postulated that motoneuron sprouting
could be induced by a molecule called "neurocletin" (16,17). In
this paper we report that gangliosides may be such an inductive
agent. When applied exogenously, these membrane components stimulate cat nictitating membrane re-innervation (4) and neuromuscular
junction formation in vitro (22). Our results, suggesting a

sprouting inductive property of the gangliosides, may explain the above effects.

MATERIALS AND METHODS

Extensor Digitorum Longus (EDL) muscles were denervated by crushing the nerve with fine forceps just after the last gluteal nerve branches off.

Male, albino Sprague-Dawley rats (180-200 g) at the time of denervation were used. The animals were decapitated and the EDL muscle, along with a long piece of sciatic nerve, was quickly dissected out and mounted by usual methods. Electrophysiological experiments were performed at room temperature and begun 60 min after perfusion of the muscle with the bathing medium equilibrated with 95% O_2 and 5% CO_2. The composition of the standard Krebs solution was (in mM): NaCl, 113; $NaHCO_3$, 25; KCl, 4.7; KH_2PO_4, 1.2; glucose, 11; $MgSO_4$, 1.2 and $CaCl_2$, 2.5. The muscle fibers along the fine intramuscular branches of the nerve were impaled with KCl filled glass microelectrodes (resistance 10-20 M Ω). The electrical recordings were displayed on a dual beam oscilloscope. One beam was AC coupled at high gain to record miniature and endplate potentials (mepp's) and endplate potentials (epp's), and the other was DC coupled at low gain to record resting potentials. When changes in the ionic composition of the Krebs solution were studied, the chamber was perfused with the modified solutions. To study the effects of Black Widow Spider Venom (BWSV), perfusion was stopped and 50 µl of the venom was added. For more details on the preparation of BWSV see earlier publications (10,13). The scheme of the electrophysiological experiments comprised several parts and in each part at least 10 muscle fibers were impaled: a) at rest, to record resting potential and mepp frequency; b) 30 mM K^+ rich solution, to record mepp frequency; c) 10 mM Mg^{++} rich solution, to record type of innervation; and d) 50% hypertonic Krebs, to record mepp frequency.

Beef brain cortex gangliosides, purified by high pressure liquid chromatography, were injected daily (i.p.) at a dose of 50 mg/kg of body weight. The composition of the mixture was: GM_1 27%, GD_{1a} 39.7%, GD_{1b} 16%, GT 17%. The data in Tables 1 and 2 are shown as mean $\pm$ S.E.M.

For histochemical studies, EDL muscles were pinned on a wood board at resting length and frozen by immersion in liquid nitrogen. Longitudinal sections (30-50 µm thick) were cut with a Damon cryostat. The technique developed by Pestronk and Drackman (23) was

TABLE 1:   Time sequence for the recovery of K^+ stimulated synaptic activity in the denervated rat EDL muscle.

	Days from Denervation				
	14	15	18	25	35
Mepp's/sec (30 mM K^+)	1.5 ± 0.2	2.3 ± 0.1	18.0 ± 1.0	108.0 ± 6.3	200

TABLE 2:   Time sequence for the recovery of hyperosmolarity stimulated synaptic activity in the denervated rat EDL muscle.

	Days from Denervation			
	21	25	35	60
Mepp's/sec (50% hypertonic solution)	3.3 ± 0.2	8.5 ± 2.4	19.5 ± 3.0	49.0 ± 5.0

used to stain acetylcholinesterase (AChE) and nerves. The endplate region is demarcated by a blue precipitate and the nerves are black because of a gold silver precipitate.

RESULTS AND DISCUSSION

General Electrophysiological Features

Following nerve degeneration the neuromuscular transmission ceased and no nervous electrical activity could be recorded at suspected neuromuscular junctions. However, it has been reported that, although the endplate activity ceases in frog muscles after section and consequent degeneration of the nerve, small, spontaneous potentials are recorded and maintained throughout the period of denervation. These potentials are elicited by release of acetylcholine (ACh) packets from Schwann cells and therefore are referred

to as Schwann cell mepps. They are different from innervated endplate mepps, having a slower and more variable time course and a
very low frequency of 1 per min. In addition, their rate cannot
be increased by agents such as lanthanum, BWSV, ouabain, etc., which
normally have a marked effect (2).

In our experiments we were not able to record any spontaneous
potentials at the denervated EDL endplates, suggesting that Schwann
cells do not release packets of ACh in rat muscles. However, our
noise level was 80-100 µV and Schwann cell mepps might have been
lost in the noise level. Among the most dramatic changes induced
by the denervation were those of both the muscle membrane resting
potential and atrophy. The resting potential dropped from about
-72 mV to -55 mV and muscle weight fell to about 50% of the control
values.

Two weeks after crushing the nerve subthreshold epps could
be recorded at suspected endplate regions near the site of nerve
entrance, their amplitude being from 1 to 4 mV. Interestingly, the
epp grew rapidly and in less than 24 hr was over threshold and
capable of inducing muscle contraction. In contrast, mepps were
almost absent at first, 1% of the controlrate, but in about 10 days
the frequency became normal (3). The amplitude distribution was
skewed to the left showing a large prevalence of small mepps (for
details, see 20). It is worth noting that by the second day of
re-innervation the muscle had to be perfused with Krebs containing
10 mM Mg^{++} to avoid muscle contraction when the nerve was stimulated
and that the epps were 1 or 2 mV in amplitude. This finding suggests that the voltage dependent Ca^{++} channel was already well developed and able to evoke ACh release despite the high Mg^{++}
antagonism.

Accurate estimation of the resting potential was difficult because the fibers were atrophied and hard to impale. However,
measurements were fairly good and the variations small. As mentioned above, the muscle membrane resting potentials dropped following denervation and reached a lower limit of about -54.7 ± 1 mV.
The return toward normal resting potential was slow and lasted
about 10 days. (This is somewhat surprising because the muscle
activity was reinstated very quickly and the recovery of the membrane potential seemed to parallel the return toward a normal mepp
frequency, which would suggest a relationship between the two
phenomena.) Conversely, a few years ago, it was shown (19) that
denervated soleus muscles, when stimulated directly at a certain
rate, reached the normal resting potential in 10 days despite the
absence of the nerve.

Our impression is that during the re-innervation process
probably both mechanical activity and neural factors play an

effective role in recovery of the muscle membrane properties, and
that one does not exclude the other.

Miniature Endplate Potentials during Re-innervation

Extensive work on the electrophysiological characteristics of
muscle re-innervation has been done using frog sartorius and cutaneous pectoris muscles (6-8,21). It has been reported that the
first sign of re-established contact between nerve and muscle is
the recording of mepps due to ACh packets released spontaneously
from a regenerating nerve terminal and that stimulation of the
nerve, however, did not produce any epp. This was defined as the
"non-transmitting stage." The frequency was variable from animal
to animal and lower than normal, but interestingly, the response
to stimulation with K^+ rich solutions was normal. Only at a later
stage did the nerve stimulation evoke ACh release, at first originating a sub-threshold epp that subsequently grew to induce muscle
contraction. This picture is quite different in the rat EDL muscle.
In our experimental model we tried to follow very carefully the two
mechanisms of ACh release as they returned to the normal level.

Mepp frequency was very low at the early stages of re-innervation, even when preparations were stimulated with 30 mM K^+. This
concentration of K^+ is known to increase mepp rate about 100-fold
in normal muscles, from 2.5 to 250 mepp/sec. Accordingly, regenerating nerve terminals responded as well as normal ones to K^+
stimulation, although the frequency was much lower. Table 1 shows
the time sequence for the recovery of K^+ stimulated synaptic activity. The gradual return toward normal shows how this method of
stimulation can nicely define the stage of re-innervation of the
rat endplates.

In contrast, BWSV powerfully stimulated ACh release and raised
mepp frequency on any day of re-innervation, even on day 15. This
effect allowed us to determine the number of quanta contained in a
single nerve terminal. For instance, on day 16 the terminals contained about 16,000 quanta and the nerve evoked ACh release was
capable of inducing muscle contraction. The mepps, however, were
rare and the frequency in 30 mM K^+ rich solutions was 2.3/sec,
which is 1% of the rate in normally innervated muscles.

Another parameter which confirmed how gradual and slow the
recovery of the endplate function is during re-innervation was the
response to stimulation with 50% hypertonic solution. The perfusion of a neuromuscular preparation with the medium elicits a
marked increase of mepp frequency, which eventually leads to

depletion of synaptic vesicles (5). The mechanisms leading to
stimulation of transmitter release are quite intriguing because
they do not require Ca^{++} in the bathing solution. However, it was
found that surface receptor capping of lymphocytes or thymocytes
was induced by a hypertonic medium. This effect was strongly inhibited by concanavalin A, and the inhibition was reversed by colchicine (26). In addition, recent work on frog neuromuscular junction has led to the conclusion that this may be true for the frog
neuromuscular junction as well and that a similar phenomenon may
elicit a Ca^{++} independent transmitter release (11,12).

When the osmolarity of the medium was raised by 50%, the mepp
frequency of EDL muscles increased to 50/sec. During the re-innervation process, however, a detectable effect was observed
starting from day 21 (Table 2). It is obvious from these data that
the mechanisms underlying this type of release recover at a much
slower pace than the mechanisms stimulated by nerve action potential,
BWSV and K^+. This delay may reflect the time required in assembling
a functioning nerve terminal plasma membrane and related structures,
such as membrane components and sub-membrane cytoskeleton apparatus.
For instance, it is obvious that both the voltage dependent Ca^{++}
channel and the structures related to ACh release induced by nerve
activity mature rapidly. The K^+ stimulated release as well seems
to mature quite early, because it is capable of raising mepp frequency 100-fold as in normally innervated muscles. The effectiveness of this type of stimulation may reflect the quick recovery of
the voltage dependent Ca^{++} channel. However, the return toward
normal rates of release stimulated by K^+ is parallel to the recovery of spontaneous mepp frequency. This would suggest that spontaneous mepps and K^+ stimulated mepps are due to ACh packets released via the same pathway.

The slower time sequence of the recovery of ACh release in
hypertonic solutions again suggests a continuous rearrangement of
the nerve terminal membrane, which seems to be almost functionally
normal only 5 weeks after the first contact between nerve and muscle.
It is of interest to note that BWSV bypasses all these requirements
and stimulates release in the usual powerful way as soon as the
synapse is formed.

Sprouting and Ganglioside Treatment

Another parameter followed was the degree of nerve sprouting,
known to be a nerve response to damage. At the crush site, several
axons give rise to myelinated branches and the muscle will be invaded by a far larger number of axons. In addition, the motor
neurons near the target area or after innervation of an old endplate

may sprout again. The last case is called terminal "sprouting" and
the sprouts may be either myelinated or unmyelinated.

In normal muscles the endplates are innervated by axons which
leave intramuscular branches in the proximity of the endplate regions. There are as many axons leaving the nerve trunks as there
are endplates (Fig. 1). During the re-innervation process a few
very thin unmyelinated processes leave the almost empty nerve trunks
(Fig. 2) and travel parallel to the muscle fibers, sequentially
forming contact with several endplates. Often this result is
achieved by collateral sprouting (1,14,25). With time, an increasing number of motor neurons penetrate the muscle following this
characteristic type of growth. A consequence of the process is
the occurrence of polyneuronal innervation, precisely detected by
recording a multiple epp at the same muscle fiber (Fig. 3). The
measurement of the amount of polyneuronal innervation can be used
as an estimate of the sprouting during the innervation process.

The detection of polyinnervation was facilitated by the variable
and slow conduction velocity of the regenerating motoneurons. In
addition, they often had a different threshold for excitation. In
other words, it was possible to record the presence of several terminals at the same endplate even by stimulating the nerve at different intensities. At higher intensities more axons were recruited
and the epp recorded either peaked higher or was larger. The polyneuronal innervation became apparent on day 15 (about 20% of the
impaled fibers) and reached the maximum value of 66% between day 21
and 25. It then slowly declined, and 2-3 weeks later the superficial fibers were all monoinnervated (Fig. 4). The time course of
this phenomenon suggests that during the re-innervation process
there is a period of stimulated nerve growth induced by the inactive
muscle followed by a period characterized by the destabilization of
the redundant innervation; then a single nerve terminal will innervate the endplate.

When animals were treated with gangliosides there was a marked
stimulation of the sprouting. Precisely on the 14th day 40% of
the fibers were polyinnervated already, and a couple of days later
the maximum level was reached (9). This is a week earlier than the
untreated animals. Furthermore, the elimination of the excess innervation proceeded concomitantly for both treated and untreated
muscle. Therefore it seems that we are facing a dual phenomenon:
an endogenous stimulus, perhaps coming from the muscle, and a need
for membrane structures, such as gangliosides. The injected gangliosides, which are incorporated in the membrane of the regenerating nerves, may enforce the natural physiological phenomenon of
sprouting, which seems to be regulated by the denervated muscle.
The gangliosides alone cannot stimulate sprouting, this being suggested by the fact that the muscle contralateral to the denervated

LEGEND

FIGURE 1: Several axons leaving an intramuscular bundle to in-
          nervate endplates in normal EDL muscle.
          Scale bar = 100 µm.

FIGURE 2: Early stages of the re-innervation process. A very thin
          unmyelinated axon (2 small arrows) reaches the endplate
          (large arrow) with a faint AChE reaction.
          Scale bar = 10 µm.

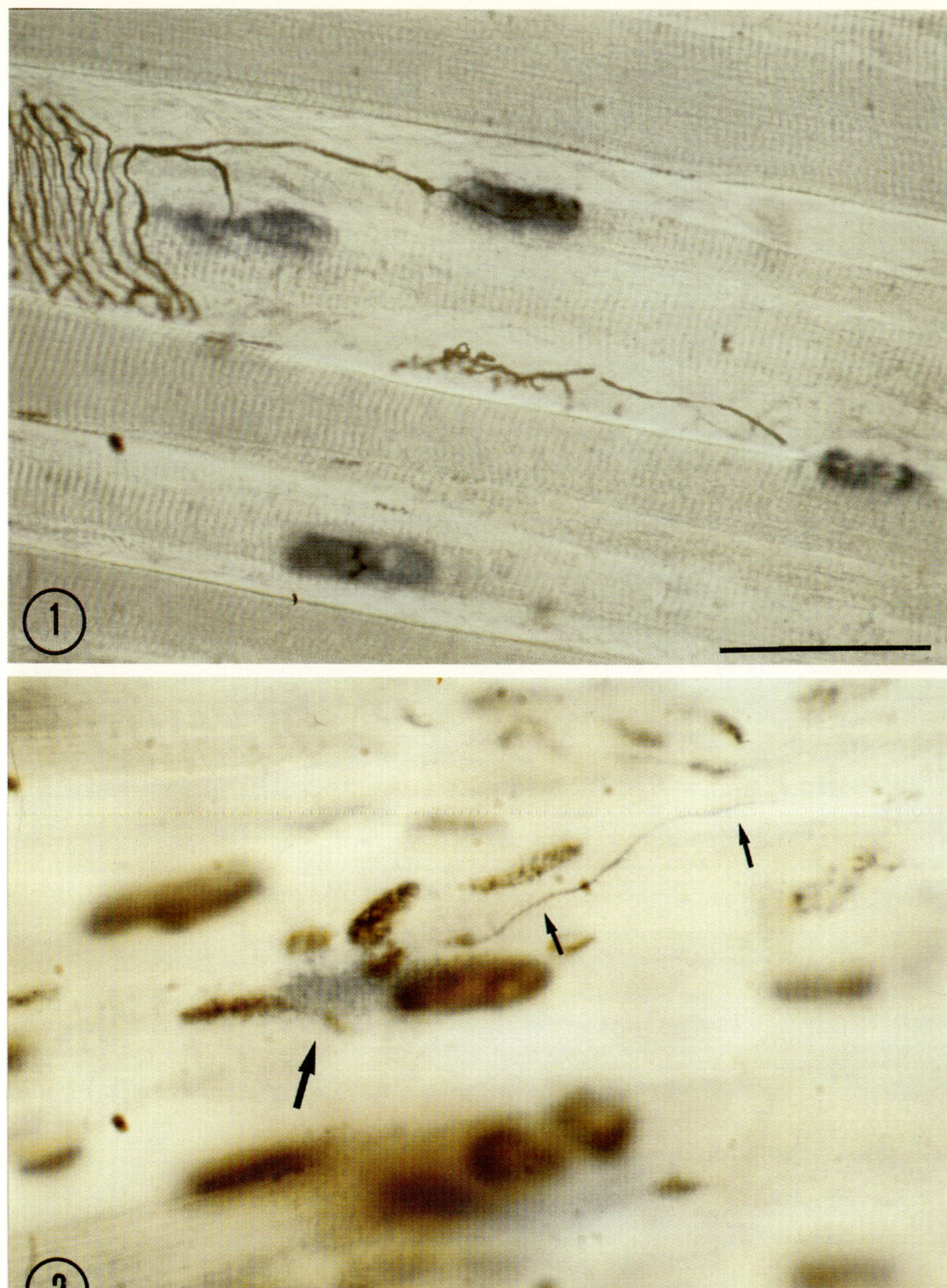

1
2

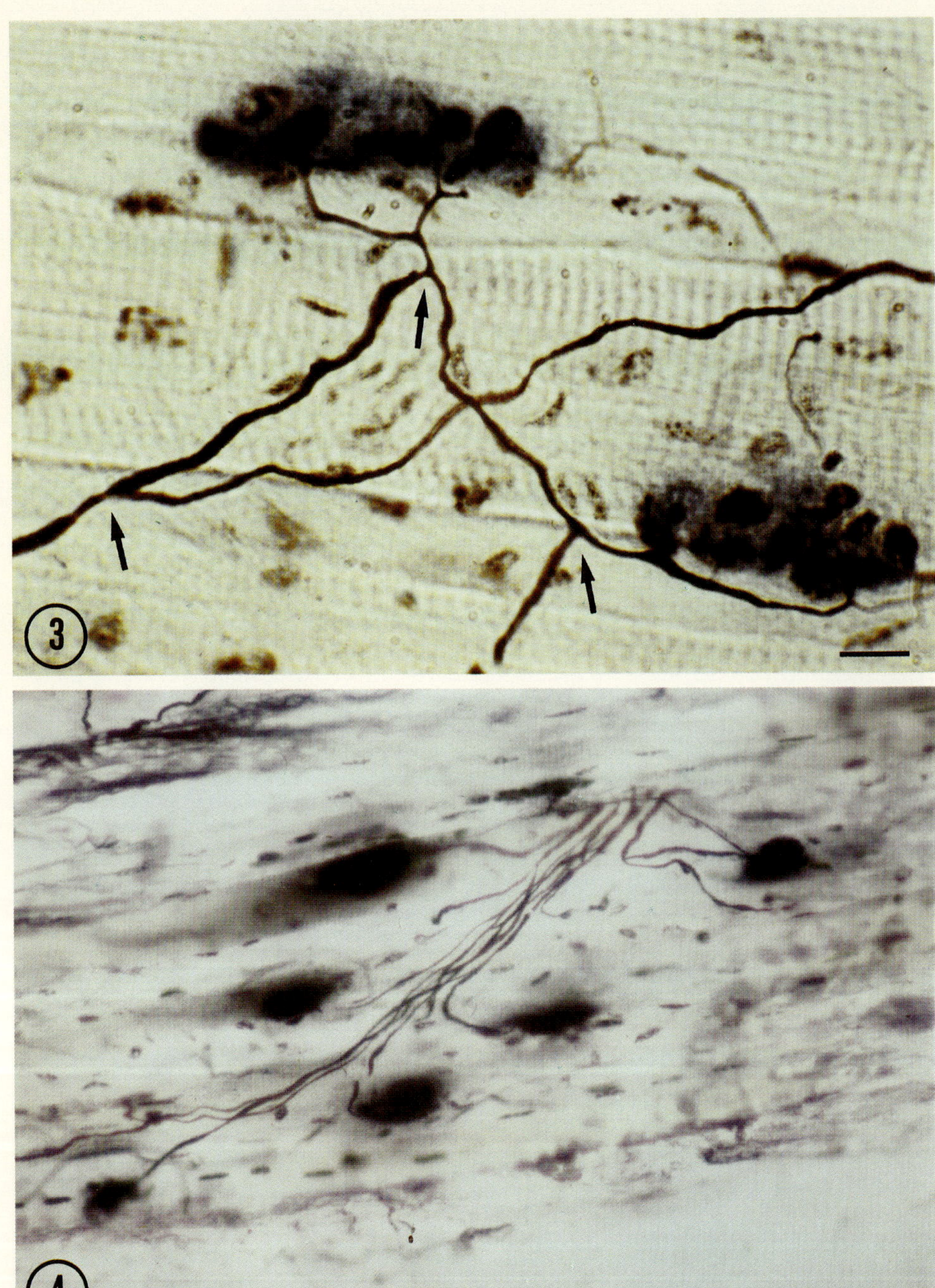

LEGEND

FIGURE 3: Single axon that innervates two endplates by collateral
         sprouting. Arrows indicate the points of sprouting.
         Scale bar = 10 µm.

FIGURE 4: Pattern of the muscle innervation several weeks after the
         first sign of endplate activity.
         Scale bar = 100 µm.

one does not show any sign of polyneuronal innervation even in
treated animals. Therefore the ganglioside treatment accelerates
a naturally occurring phenomenon which is present when there is a
request from the muscle, and when the requested ceases, the sprouting
process regresses, even if the animals are treated daily.

To conclude, the results of Ceccarelli (4) and Obata (22) can
be explained by the above results, suggesting that treatment increases the number of axons reaching the denervated area rather
than making them grow faster.

SUMMARY

In the process of muscle re-innervation several steps are involved. In our studies we have analyzed the recognition process,
the maturation of the synapse and the sprouting. Regarding the
first part, it can be concluded that the regenerated nerve muscle
contact occurs at the old endplate regions and in no instance did
we observe more than one endplate on the same muscle fiber.

The mechanisms of neurosecretion mature in two separate sets,
the first one dependent on the voltage dependent Ca^{++} channel being
effective at normal level from the early hours of contact. The
second one, regarding mepp frequency both at rest and stimulated
by hypertonic medium, is much slower and reaches the normal rate
several weeks later.

The degree of sprouting was assessed precisely by electrophysiological and histochemical means, and it has been shown that
membrane components such as gangliosides may be able to stimulate
and regulate motor neuron sprouting during the re-innervation
process.

REFERENCES

1.   Bennett, M.R., McLachlan, E.M. and Taylor, R.S. (1973): J.
        Physiol. 233:481-500.
2.   Bevan, S., Grampp, W. and Miledi, R. (1976):Proc. Roy. Soc.
        Lond. B 194:195-210.
3.   Carmignoto, G., Finesso, M., Tredese, L. and Gorio, A. (1981):
        IN Membranes, Molecules, Toxins and Cells. (eds) K. Block,
        L. Bolis and D.C. Tosteson, PSG Publications, Littleton
        (in press).

4. Ceccarelli, B., Aporti, F. and Finesso, M. (1976): Adv. Exp.
     Med. Biol. 71:275-293.
5. Clark, A.W., Mauro, A., Longenecker, H.E. Jr. and Hurlbut,
     W.P. (1970): Nature 225:703-705.
6. Dennis, M.J. and Miledi, R. (1974): J. Physiol. 239:553-570.
7. Dennis, M.J. and Miledi, R. (1974): J. Physiol. 239:571-594.
8. Gorio, A. (1980): IN A Multidisciplinary Approach to Brain
     Development (ed) C. Di Benedetta, Elsevier North-Holland,
     Amsterdam, pp. 439-452.
9. Gorio, A., Carmignoto, G., Facci, L. and Finesso, M. (1980):
     Brain Res. 197:236-241.
10. Gorio, A., Hurlbut, W.P. and Ceccarelli, B. (1978): J. Cell
     Biol. 78:716-733.
11. Gorio, A. and Mauro, A. (1979): Adv. Cytopharmacol. 3:129-140.
12. Gorio, A. and Mauro, A. (1979): J. Gen. Physiol. 73:245-263.
13. Gorio, A., Rubin, L.L. and Mauro, A. (1978): J. Neurocytol.
     7:193-205.
14. Gutmann, E. (1945): J. Anat. 79:1-8.
15. Harris, A.J. (1974): Ann. Rev. Physiol. 36:251-305.
16. Hoffman, H. (1950): Aust. J. Exp. Biol. Med. Sci. 28:383-397.
17. Hoffman, H. and Springell, P.H. (1951): Aust. J. Exp. Biol.
     Med. Sci. 29:417-424.
18. Letinsky, M.S., Fischbeck, K.H. and McMahan, U.J. (1976):
     J. Neurocytol. 5:691-718.
19. Lomo, T. and Westgaard, R.H. (1975): J. Physiol. 252:603-626.
20. McArdle, J.J. and Albuquerque, E.X. (1973): J. Gen. Physiol.
     61:1-23.
21. Miledi, R. (1960): J. Physiol. 154:190-205.
22. Obata, K., Oide, M. and Handa, S. (1977): Nature 266:369-371.
23. Pestronk, A. and Drachman, D.B. (1978): Muscle & Nerve 1:70-74.
24. Rotshenker, S. and McMahan, U.J. (1976): J. Neurocytol. 5:
     719-730.
25. Tello, F. (1970): Trab. Lab. Invest. Biol. Univ. Madrid 5:
     117-151.
26. Yahara, I. and Kakimoto-Sameshima, F. (1977): Proc. Nat. Acad.
     Sci. 74:4511-4515.

ISOLATION OF SYNAPTIC VESICLES FROM THE MYENTERIC PLEXUS OF

GUINEA PIG

G.H.C. Dowe, H. Kilbinger and V.P. Whittaker

Abteilung Neurochemie, Max-Planck-Institut fur
Biophysikalische Chemie, Gottingen, FR Germany

INTRODUCTION

The isolation of very pure cholinergic synaptic vesicles from
the electromotor terminals of Torpedo (2,5) has provided much information which would also be useful to have for cholinergic synaptic vesicles from the mammalian peripheral nervous system. The
only successful isolation of such vesicles is from ox superior
cervical ganglia (7). We have now obtained a highly purified fraction of vesicles rich in acetylcholine (ACh) from the myenteric
plexus-longitudinal muscle preparation (3) of guinea pig ileum.

METHODS

Longitudinal muscle with myenteric plexus attached was reduced
to small fragments (3) in a blender (Vertis 45) in 0.16 M NaCl containing 1 mM EGTA and 10 mM Tris HCl (pH 7.0). The supernatant and
washings from a once-washed pellet formed by centrifuging the 10%
dispersion at 1000 g for 20 min was combined (as fraction S_1) and
loaded onto a sigmoid sucrose-sodium chloride gradient formed in a
zonal rotor (T1-14) as described by Whittaker et al. (6). After
centrifuging at 48,000 rev/min for 3 hr, the gradient was unloaded
in the usual way and the fractions assayed for ACh. About 3-6 g
of tissue from 3-5 animals was used in one run.

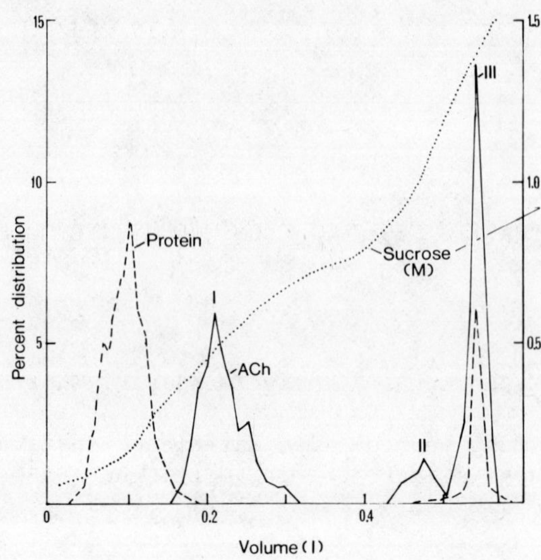

FIGURE 1: Separation of 3 fractions (I, II, III) of bound ACh from
soluble protein (SP) by the centrifugal density gradient
fractionation in a zonal rotor of a low speed supernatant
derived from a homogenate of guinea pig ileum myenteric
plexus-longitudinal muscle.

TABLE 1: Characteristics of ACh-rich peaks recovered after zonal
density gradient separation.

Parameter	Units	I	II	III
Density	M sucrose	0.48±0.02	0.96±0.02	1.47±0.02
ACh				
Peak Concentration	nmol/mg of protein	31 ±2	16 ±3	2 ±1
Distribution	% recovered	53 ±5	11 ±4	27 ±3
Recovery	% of S_1		94 ±7	
	% of homogenate		68 ±6	

Bound ACh in homogenate 0.5 ± 0.1(3) nmol/mg, 50 ± 5(8) % of total

Results are means of 5 experiments unless otherwise indicated in
parentheses.

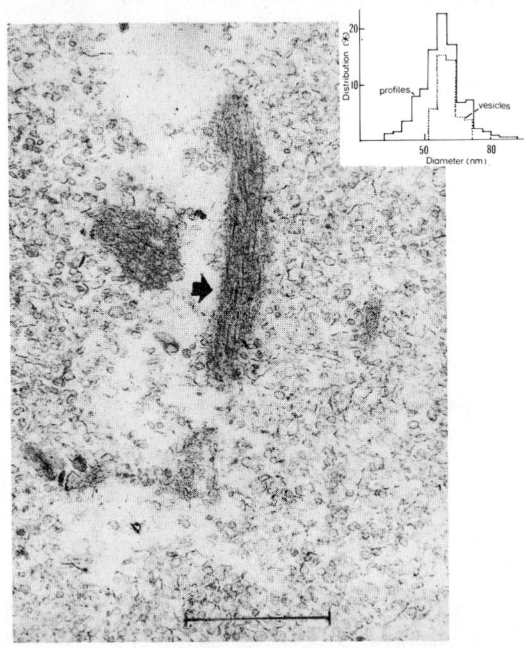

FIGURE 2: Electron micrograph of fraction I. Note presence of
synaptic vesicles of (insert) mean diameter 62 ± SD6
(108) nm as the sole lipoprotein membrane structures,
mingled with clumps of actomyosin fibrils (arrow). The
distribution of vesicle diameters shown in the insert
was obtained from the distribution of vesicle profile
diameters using a modified Wicksell procedure (2). The
bar is 1 μm.

TABLE 2: Acetylcholine content of mammalian particulate fractions.

Species	Tissue	Preparation	ACh (nmol/mg protein)	Reference
Ox	*	Synaptic vesicles	4	7
Guinea pig	**	Synaptic vesicles purified by glassbead column chromatography	14 ± 5(6)	1
Guinea pig	***	Synaptic vesicles	31 ± 2(5)	This report

* = superior cervical ganglion; ** = cortex, *** = myenteric plexus

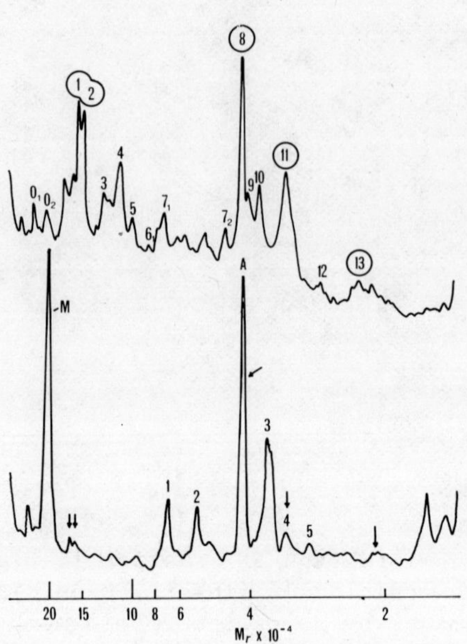

FIGURE 3: Densitometric scan of Coomassie blue stained SDS-PAGE of
(A: upper) Torpedo vesicle proteins, (B: lower) myenteric
plexus vesicle proteins. Components ringed in (A) are the
main specific vesicle proteins of Torpedo (4). These components all appear to be present in (B)(arrows). The components in (B) labelled 1-3,5 may be proteins of non-
cholinergic vesicles present in Peak I. Actomyosin (A,M)
is the main protein present, however, as expected from
Fig. 2.

RESULTS

After centrifuging, ACh was distributed in three peaks in the
gradient (Fig. 1). These peaks had the characteristics shown in
Table 1. The ACh was in the bound form since free ACh (A) was
rapidly destroyed by cholinesterases (B) would have been recovered,
if present, in the initial peak of soluble cytoplasmic protein (SP).

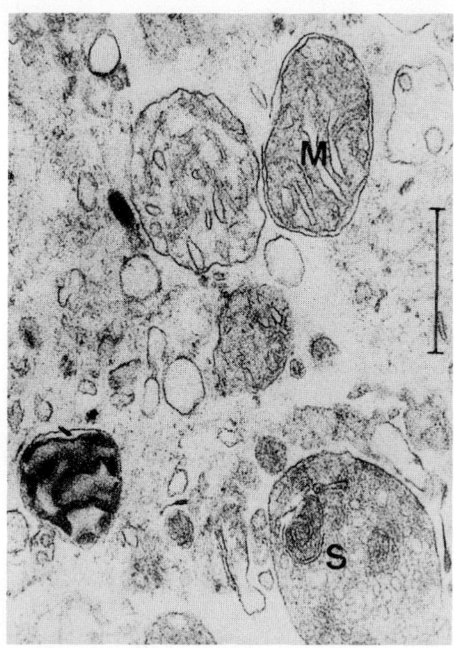

FIGURE 4: Electron micrograph of fraction III. Note presence of
         mitochondria (M) and a synaptosome (S). Bar is 0.5 μm.

Peak I had the highest ACh content (the highest so far reported for a mammalian particulate preparation; see Table 2). It
consisted (Fig. 2) of a homogeneous preparation of synaptic vesicles of diameter 62 $\pm$ 6 nm mingled with large clumps of actomyosin,
presumably derived from the longitudinal muscle of the ileum.
SDS-PAGE (Fig. 3) revealed two large peaks of actomyosin (A,M)
and several minor protein components, some of which (arrows) were
identical with the specific polypeptide subunits found in Torpedo
vesicles. The vesicles are estimated to be about 10% pure with
respect to protein on this basis, i.e. the limiting concentration
of ACh should be about 300–400 nmol/mg of protein. This would
correspond to a 60 nm vesicle filled with iso-osmotic (0.16 M) ACh.

Peak II provided insufficient material for examination but
may represent a dense vesicle peak like that found in Torpedo (VP_2).

Peak III was a mixed fraction (Fig. 4) containing membranes,
mitochondria (M) and a few synaptosomes (S).

REFERENCES

1. Nagy, A., Baker, R.R., Morris, S.J. and Whittaker, V.P. (1976):
    Brain Res. 109:285-309.
2. Ohsawa, K., Dowe, G.H.C., Morris, S.J. and Whittaker, V.P.
    (1978): Brain Res. 161:447-457.
3. Paton, W.D.M. and Zar, M.A. (1968): J. Physiol. 194:13-33.
4. Stadler, H. and Tashiro, T. (1979): Eur. J. Biochem. 101:
    171-178.
5. Tashiro, T. and Stadler, H. (1978): Eur. J. Biochem. 90:479-487.
6. Whittaker, V.P., Essman, W.B. and Dowe, G.H.C. (1972): Biochem.
    J. 128:833-846.
7. Wilson, W.S., Schultz, R.A. and Cooper, J.R. (1973): J. Neuro-
    chem. 20:659-667.

THE STRUCTURE OF CHOLINERGIC SYNAPTIC VESICLES

H. Stadler

Department of Neurochemistry
Max-Planck-Institute for Biophysical Chemistry
Gottingen, Free Republic of Germany

The secretory granules of cholinergic nerve terminals, the
synaptic vesicles, play a central role in synaptic transmission.
They store acetylcholine (ACh) and ATP and though it is widely
accepted now that vesicles "recycle" within the nerve terminal and
most probably release ACh directly into the synaptic cleft (15),
the sequence of events at the molecular level finally leading to
transmitter release is almost unknown.

Since it is now possible to isolate cholinergic synaptic vesicles in high purity (13) the aim of the investigations presented
here was to analyse their structure and chemical composition a
knowledge of which should lead to a better understanding of their
function and the molecular mechanisms involved in storage and
release of ACh at the synapse.

Vesicle Isolation and Chemical Composition

Cholinergic synaptic vesicles were isolated from the electric
organ of Torpedo marmorata by extracting the crushed and frozen
tissue with isoosmolar sodium chloride solutions in the presence
of EGTA and further purified by sucrose density gradient centrifugation first on a step gradient and second on shallow gradient in a
zonal rotor (13). The specific activity of ACh and ATP per mg
protein cannot be increased by further purification such as chromatography on glass beads (7) or flotation on sucrose gradients, indicating that a limiting composition had been reached. The ACh/mg
protein value is comparable with preparations isolated from Narcine
brasiliensis or Torpedo californica (3).

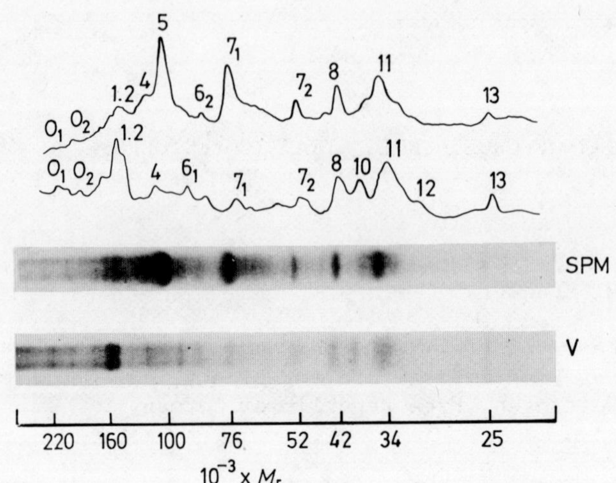

FIGURE 1: SDS gel electrophoresis of synaptic vesicles (V) and synaptosomal plasma membranes (SPM) from the electric organ of <u>Torpedo</u> <u>marmorata</u>. Electrophoresis was carried out in gradient gels (7.5–12.5%), staining with Coomassie Brilliant Blue. Bands 8 correspond to actin. The top shows densitometric tracings of the patterns.

TABLE 1: Chemical composition of cholinergic synaptic vesicles from <u>Torpedo</u> <u>marmorata</u>.

Constituent	Concentration (μmol/mg protein)	Percent Dry Weight
Protein	–	16
Acetylcholine	6.9 ± 0.63	16
ATP	1.0 ± 0.26	8
Phospholipid	3.82 ± 0.26	46
Cholesterol	1.60	10
Glycosaminoglycan	–	3

The final preparation is of high purity by morphological criteria, which is consistent with the very low activities found for
enzymes characteristic for plasma membranes such as AChE or Na^+-
K^+ ATPase; cytoplasmic markers are also present only in trace
amounts. The preparations are enriched in a Ca^{++}-Mg^{++}-activated
ATPase (1); these findings were recently confirmed by Rothlein and
Parsons (9).

The chemical composition of the fraction is shown in Table 1.
The vesicle membrane is relatively lipid-rich and protein poor. The
presence of an acidic glycosaminoglycan is noteworthy (11); it is
probably located inside the vesicles (unpublished observation).

If the number of vesicles in a sample is counted by quantitative electron microscopy and compared with the ACh content, the
number of ACh molecules per vesicle can be calculated. It was found
to be around 2 x 10^5 molecules per vesicle (8), which corresponds to
an accumulated ACh concentration of about 0.6-0.8 M. Unfortunately
absolute numbers of ACh molecules per vesicle have not yet been
compared to the size of the ACh quantum determined electrophysiologically but such studies are now being pursued in our laboratory.

Vesicle Proteins

The protein composition of the vesicles as analysed by one or
two dimensional gel electrophoresis is a sensitive test for vesicle
purity. As purification continues, some components are eliminated
and others are intensified. The final preparation has a simple
composition with only about five major bands (Fig. 1). The faint
bands in the pattern are, in contrast to the major bands, somewhat
unreproducible and are regarded as contaminations (10).

It was noted that band 8 (m.w. around 42,000) comigrates with
Torpedo muscle actin. Further characterization revealed that the
component is indeed an actin, but one slightly more basic than
muscle actin and the bulk actin of the electric organ (13,14). The
result indicates that vesicles contain a very special actin species
suggesting a specific association, and thereby excluding copurifi-
cation of contaminating amounts of filamenteous actin.

The protein composition does not change if vesicles are lysed
by hypoosmotic shock and the membranes recovered by centrifugation
and washed by various salt solutions of high or low ionic strength
(except that prolonged treatment with EGTA removes band 9, see
Fig. 1), a soluble protein present in the core can therefore be
excluded.

Recent progress in isolating synaptosomes from the electric organ (5) enabled us to isolate plasma membranes from them (10) and compare their protein pattern with that of vesicles (Fig. 1). The compositions are quite different. On two dimensional gel electrophoresis of a mixture of vesicle and plasma membrane protein only the vesicle actin co-migrates with a similar component of the synaptosomal plasma membrane. The prominent bands 5 and 7_1 (m.w. 100,000 and 76,000) in the plasma membrane are probably subunits of the Na^+-K^+ ATPase and the AChE; they are absent from vesicles or present only as trace contaminants. It seems, therefore, that the view that the vesicle membrane is a simplified version of the synaptic plasma membrane has to be revised (6). The difference in compositions requires a mechanism whereby, if vesicles fuse during transmitter release the vesicle membrane keeps its identity during exo- endocytotic cycles. It remains to be seen whether the characteristic vesicle proteins can be identified in synaptosomal plasma membranes isolated from electrically stimulated organs.

Proton - NMR - Analysis

The combined concentrations of ACh and ATP in vesicles are almost 1 M. This raises the question as to whether they are in a soluble form or restricted and osmotically inactive. An analysis of isolated vesicles by high resolution proton nuclear magnetic resonance was carried out by Stadler and Fuldner (12). The spectra are shown in Fig. 2. In Fig. 2a the spectrum shown is from vesicles treated with 50 µM paraoxon to inhibit AChE present in the fraction as a contamination. Figure 2b shows the spectrum after releasing the ACh from the vesicles by repeated freezing and thawing. In Fig. 2c the spectrum of a different preparation has been recorded. The esterase has not been blocked, the ACh was released from vesicles by freezing and thawing and has been hydrolysed to acetate and Ch (note that the methyl resonance of the acetyl group 3' is replaced by the resonance 2 corresponding to free acetate).

Essentially four observations must be noted: 1) The spectra are remarkably simple and indicate, apart from ACh the presence of only trace amounts of sucrose (used in preparing the vesicles) and trimethylamine oxide (a ubiquitous product in fish tissues). ATP does not give detectable resonances in this part of the spectrum. 2) The ACh resonances of intact vesicles are shifted compared to uncompartmented ACh (compare peaks 3 and 3'). This shift may be due to a different susceptibility in the vesicle core. 3) The integrals of the resonances before and after the lysis are identical, which means that all the vesicular ACh has been detected. 4) ACh

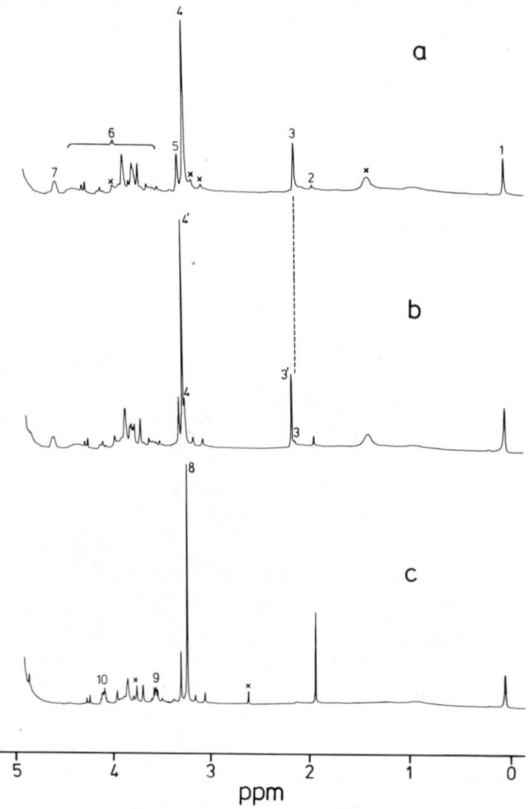

FIGURE 2: Proton NMR spectra of isolated cholinergic synaptic vesicles from _Torpedo_ _marmorata_. Vesicles were pelleted from
a sucrose density step gradient and dialysed against 0.4 m
NaCl, 20 mM phosphate pH 7.4 made in D_2O and the spectrum
recorded at 5°C. a) intact vesicles in the presence of
50 µM paraoxon (AChE blocker); b) the same preparation as
a), but after lysis of vesicles and release of ACh by repeated freezing and thawing. Dashed line indicates the
chemical shift between resonances 3 and 3'; c) another
vesicle preparation that has been lysed by freezing and
thawing in the absence of an AChE blocker. The ACh has
been hydrolysed by the esterase present in the sample.
Identified resonances: 1) 3-(trimethylsilyl)-propionic
acid (internal standard 1 mM); 2) free acetate; 3 and 3')
$-CH_3$ protons of the acetyl residue of intact ACh; 4 and
4') $-N-(CH_3)_3$ protons of ACh; 5) trimethylamine oxide;
6) resonances from residual sucrose and $-CH_2$-protons from
ACh; 7) $-CH_2$-protons of ACh adjacent to the acetyl group;
8) $-N-(CH_3)_3$ protons of Ch; 9 and 10) $-CH_2$-protons of Ch
superimposed on sugar resonances; x) unidentified
resonances.

present in vesicles shows line widths almost identical with those
of released ACh; therefore it may be concluded that ACh is essentially in free solution within the vesicles.

In contrast to other secretory granules where is seems that
low molecular weight secretion products are immobilized and probably osmotically inactive (4), ACh in synaptic vesicles is in
solution. The result is consistent with density measurements on
vesicles at different osmotic conditions suggesting that the
vesicular ACh is in an osmotically active state (2). It is clear
from Fig. 2c that the breakdown of ACh in any given sample (provided the concentration is high enough) can be observed easily and
quantitatively by proton NMR. In addition, the shift differences
mentioned, might enable one in future studies to follow the turnover of ACh pools and their breakdown in intact cholinergic tissue
or synaptosomes with this powerful non-perturbing method.

Altogether the results of the structural analysis presented
reveal a remarkable simplicity of the cholinergic synaptic vesicle
compared to other secretory granules. Few proteins, one of them
actin, are associated with a lipid rich membrane surrounding a
hydrated core filled with high concentrations of ACh and ATP, where
at least ACh is dissolved and in solution. The rest of the membrane proteins might represent carriers for ACh and ATP uptake or
the $Ca^{++}-Mg^{++}$ activated ATPase; their functional characterization
is in progress. The simplicity of the vesicle structure seems
promising in elucidating further studies, its precise function in
synaptic transmission. In addition, these vesicles might provide
a valuable model system to study phenomena of general interest such
as membrane fusion or the function of carriers.

REFERENCES

1.   Breer, H., Morris, S.J. and Whittaker, V.P. (1977): Eur. J.
        Biochem. 80:313-318.
2.   Breer, H., Morris, S.J. and Whittaker, V.P. (1978): Eur. J.
        Biochem. 81:453-458.
3.   Carlson, S.S., Wagner, J.A. and Kelly, R.B. (1978): Bio-
        chemistry 17:1188-1199.
4.   Giannattasio, G., Zanini, A. and Meldolesi, J. (1979): In
        Complex Carbohydrates of Nervous Tissue (eds) R.U. Margolis
        and R.K. Margolis, Plenum Press, New York, pp.327-345.
5.   Israel, M., Manaranche, R., Mastour-Frachon, P. and Morel, N.
        (1976): Biochem. J. 160:113-115.
6.   Mahler, H.R. (1977): Neurochem. Res. 2:119-247.

7.   Nagy, A., Baker, R.R., Morris, S.J. and Whittaker, V.P. (1976): Brain Res. 109:285-309.
8.   Ohsawa, K., Dowe, G.H.C., Morris, S.J. and Whittaker, V.P. (1980): Brain Res. 161:447-457.
9.   Rothlein, J.E. and Parsons, S.M. (1979): Biochem. Biophys. Res. Comm. 88:1069-1076.
10. Stadler, H. and Tashiro, T. (1979): Eur. J. Biochem. 101: 171-178.
11. Stadler, H. and Whittaker, V.P. (1978): Brain Res. 153:408-413.
12. Stadler, H. and Fuldner, H.H. (1980): Nature 286:293-294.
13. Tashiro, T. and Stadler, H. (1978): Eur. J. Biochem. 90: 479-487.
14. Zechel, K. and Stadler, H. (1979): H.-S., Z. Physiol. Chem. 360:409.
15. Zimmerman, H. (1979): Neuroscience 4:1773-1804.

INHIBITION OF MEMBRANE TRANSPORT SYSTEMS IN SYNAPTOSOMES

FROM <u>TORPEDO</u> ELECTRIC ORGAN BY SNAKE NEUROTOXINS

M.J. Dowdall and P. Fretten

Department of Biochemistry, University of Nottingham
Medical School, Queen's Medical Center
Nottingham NG7 2UH, United Kingdom

INTRODUCTION

Recently there has been considerable interest in the use of
naturally occurring neurotoxins as probes for elucidating the
molecular processes which underly chemical and electrical transmission in the nervous system. Of those neurotoxins which have
been both chemically and pharmacologically characterized a large
number occur in the venoms of poisonous snakes. According to Lee
(7),snake toxins can be broadly classified into five groups:
curaremimetic postsynaptic toxins, cardiotoxins, presynaptic neurotoxins, myonecrotic toxins and toxins affecting Na channels. As
tools, the curaremimetic neurotoxins (e.g. α-bungarotoxin) have
so far proved to be the most useful since they bind to nicotinic
cholinoceptors with high specificity in a quasi-irreversible
fashion. Pharmacological evidence suggests that toxins from the
other groups exhibit a similar degree of specificity at least at
the cellular level of organization. At the molecular level much
less is known about the nature of the target sites to which these
toxins are directed. The present chapter is concerned with studies
aimed at a better understanding of the target sites for the presynaptic neurotoxin group.

The chemistry and pharmacology of neurotoxins of this group
have recently been reviewed (7) and it suffices here to indicate
their general properties in summary form only. The group includes
the following toxins: notexin, <u>Notechis</u> II-5, β-bungarotoxin, taipoxin and crotoxin. It is now clear that all of these toxins have
intrinsic phospholipase A_2 activity and that this is critical for
neurotoxicity. In each case this catalytic activity resides in a
polypeptide chain of m.w. $\simeq$ 13,500 (the 13.5 K chain) which occurs

alone (e.g. notexin and <u>Notechis</u> II-5) or in combination with other
subunits which are either homologues of the 13.5 K **su**bunit (e.g.
taipoxin) or completely different (e.g. β-bungarotoxin and crotoxin).
All of these toxins produce neuromuscular blockade by acting primarily on motor nerve endings. Electrophysiological studies have
shown an initial stimulation of acetylcholine (ACh) release followed by slower but progressive inhibition of this process.

Using biochemical techniques there have been numerous studies
on the effect of β-bungarotoxin on mammalian brain synaptosomes. In
addition to reports which demonstrate effects on cholinergic synaptosomes (11) there are many others which show that synaptosomes containing other neurotransmitters are also affected by this toxin. It
has been suggested that a membrane depolarizing action could account
for these rather unselective actions (1,10,12). Subcellular particles, T-sacs, generated from the purely cholinergic nerve terminals of <u>Torpedo</u> electric organ have also been used as a test
preparation for studying the effects of a number of the presynaptic
neurotoxins including β-bungarotoxin (2). Potent inhibition of
high affinity choline (Ch) transport into T-sacs was reported (4)
and in a more detailed study with taipoxin a model for neurotoxicity
was proposed based on its specific recognition of the Ch transport
system (6). To test this hypothesis two other transport systems
(for acetate and adenosine) now known to be present in <u>Torpedo</u> nerve
terminals have been tested for their sensitivity to all the toxins
used previously. For measuring acetate and adenosine uptake synaptosomes (9) are more reliable than T-sacs and they have therefore been
used for the present study which compares the effects of all the
toxins on the uptake of Ch, acetate and adenosine.

MATERIALS AND METHODS

Toxins

Purified samples of notexin, <u>Notechis</u> II-5, <u>Enhydrina</u> myotoxin
and taipoxin were kindly provided by Dr. D. Eaker, Institute of
Biochemistry, Uppsala. A chemically modified taipoxin, PBP$_2$-tai-
poxin, was prepared by reaction with p-bromophenylacyl bromide as
described previously (6). β-Bungarotoxin was part of the first
batch made by Boehringer in 1976 (vials containing toxin from this
batch bear no batch number).

Synaptosome Preparation

Synaptosomes from the electric tissue of <u>Torpedo</u> <u>marmorata</u>

were prepared essentially by the procedure originally described by Morel et al. (9). The method involves forcing $\simeq$ 30 g of finely chopped electric tissue through stainless steel grids mounted on disposable 50 ml plastic syringes using a <u>Torpedo</u> physiological saline (osmotically fortified with 0.1 M sucrose and 0.3 M urea) as the dilution and washing medium. In the present experiments the physiological medium (for composition see 13) was saturated with 95% O_2, 5% CO_2 prior to use and grid sizes were (μm) 1140, 500, 308, 142, 62 and 39. The filtrate passing through the 39 μm mesh was diluted where necessary with physiological medium to a final volume of 4X the original weight of tissue (1 g $\equiv$ 1 ml) and then centrifuged at 1,000 g_{av} for 10 min. Synaptosomes in the supernatant fraction were collected as a pellet by centrifugation at 8,000 g_{av} for 30 min and after resuspension in physiological medium were further purified by density gradient centrifugation using a swing bucket rotor. The density gradient consisted of equal volumes of 0.3 M sucrose, 0.1 M urea and 0.6 M sucrose both in otherwise non-osmotically fortified physiological medium. After centrifugation at 69,000 g_{av} for 40 min synaptosomes were recovered at the 0.3 M - 0.6 M sucrose interface using a Pasteur pipette and retained at 0-4°C until use. Membrane transport studies were initiated as soon after preparation as possible with samples of synaptosomes which had been suitably diluted and equilibrated with ice-cold physiological medium.

Transport Assays

Synaptosome suspensions in physiological medium were allowed to equilibrate to room temperature (21-25°C) by a 30-45 min pre-incubation prior to mixing with ^{3}H-labelled substrates (acetate, adenosine or Ch) and neurotoxins. Incubation was initiated by adding 100 μl of the synaptosome suspension to 900 μl of physiological medium containing ^{3}H substrates and terminated 15 min later by addition of 3 ml ice-cold medium and rapid filtration through 25 mm diameter Whatman glass fiber filters (GF/F). After washing with 3 x 3 ml of ice-cold physiological medium filters were transferred to scintillation vials. Filter bound ^{3}H was released using a 10 min incubation with 1 ml H_2O (with or without further addition - see below) and determined by liquid scintillation using a Xylene: triton X-100 based scintillation fluid. For acetate uptake filters were presoaked in physiological medium containing 1 mM Na acetate. Synaptosomes derived from 0.8-1 g tissue were incubated with 2 μCi [3H]-acetate at a final concentration of 0.75 μM and incubations for zero time were used to correct for nonspecific adsorption of

[3H]-acetate to filters and synaptosomes. Elution of ^{3}H from filters
was with 1 mM Na acetate. Ch uptake was measured using synaptosomes from 0.4-0.5 g tissue and 1 μCi of N-[Me-^{3}H]-Ch at a final
concentration of 1 μM. Filters were presoaked in physiological
medium containing 1 mM Ch chloride. Nonspecific binding was determined using incubations in the presence of 0.5 mM hemicholinium-3
(HC-3) as previously described (3). Filter bound ^{3}H was eluted
with 1 mM Ch chloride. Adenosine uptake was measured again using
synaptosomes from 0.4-0.5 g tissue and 2.5 μCi [2-^{3}H]-adenosine at
a final concentration of 0.11 μM. Blank activities were determined
using 1 mM 5'AMP which blocks adenosine transport (8).

The effect of increasing concentrations of neurotoxins on each
transport system was tested using quadruplicate incubations for
each toxin concentration used. All transport studies were completed
within 6 hr of preparation and because of differential post-isola-
tion labilities of the three systems acetate uptake was always
tested first, Ch second and adenosine last.

RESULTS

In the present series of experiments mean values for the uptake of ^{3}H labelled acetate, Ch and adenosine were (pmol/15 min/mg
protein) 4.68, 84.1 and 18.0 respectively, at the external concentrations used. For Ch and adenosine these compare reasonably well
with uptake rates calculated from the kinetic parameters previously
reported (8,9,13). Thus for Ch the calculated uptake rate at 1 μM
using the results of Morel et al. (9) is 164 pmol/15 min/mg protein
and for adenosine at 0.11 μM calculated uptake rates are 11.3 and
21.6 pmol/15 min/mg protein using the results of Meunier and
Morel (8) and Zimmermann et al. (13), respectively. For acetate
uptake the present results are considerably lower than calculated
(35 pmol/15 min/mg protein) from those reported by Morel et al. (9).
The reason for the difference is not known.

Inhibitory Effect of Neurotoxins

Concentration dependent inhibition of all three transport
systems was seen with each of the neurotoxins studied. Inhibition
curves are shown in Figs. 1-3 where percent inhibition has been
plotted against the $\log_{10}$ of neurotoxin concentration. Concentrations of neurotoxin causing a 50% inhibition (IC$_{50}$s) were estimated
from these plots and are collected together in Table 1.

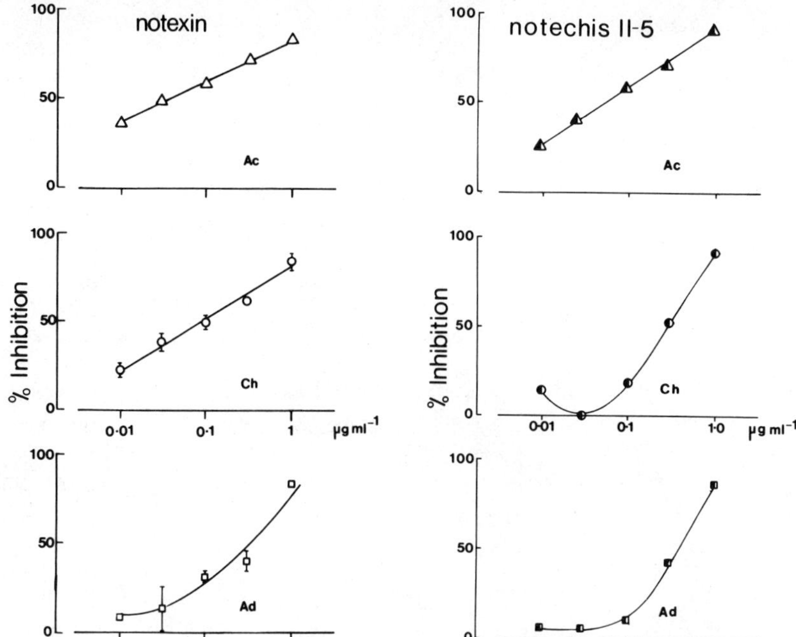

FIGURE 1: Dose-dependent inhibition of synaptosomal membrane transport systems by notexin and <u>Notechis</u> II-5. The uptake of
^{3}H labelled acetate (Ac), choline (Ch) and adenosine (Ad)
in the presence of varying concentrations of toxin was
measured as described in Materials and Methods. The
degree of inhibition is indicated as a function of the
$\log_{10}$ of toxin concentrations in µg/ml.

Inhibition curves for two of the single chain neurotoxins,
notexin and <u>Notechis</u> II-5, used in this study are shown in Fig. 1.
For both toxins significant inhibition of acetate uptake was observed at concentrations which had little effect on adenosine uptake. In addition to this difference the slopes of the inhibition
curves for both toxins were markedly steeper for the uptake of
adenosine than for acetate. Ch uptake inhibition followed the
"acetate-pattern" with notexin and the "adenosine pattern" with
<u>Notechis</u> II-5.

Figure 2 shows the inhibitory effect of another single chain
toxin, <u>Enhydrina</u> myotoxin, together with β-bungarotoxin, a two
chain toxin. As with the other single chain toxins, <u>Enhydrina</u>
myotoxin was more inhibitory towards acetate uptake than the other
two systems. The same was true for β-bungarotoxin which produced

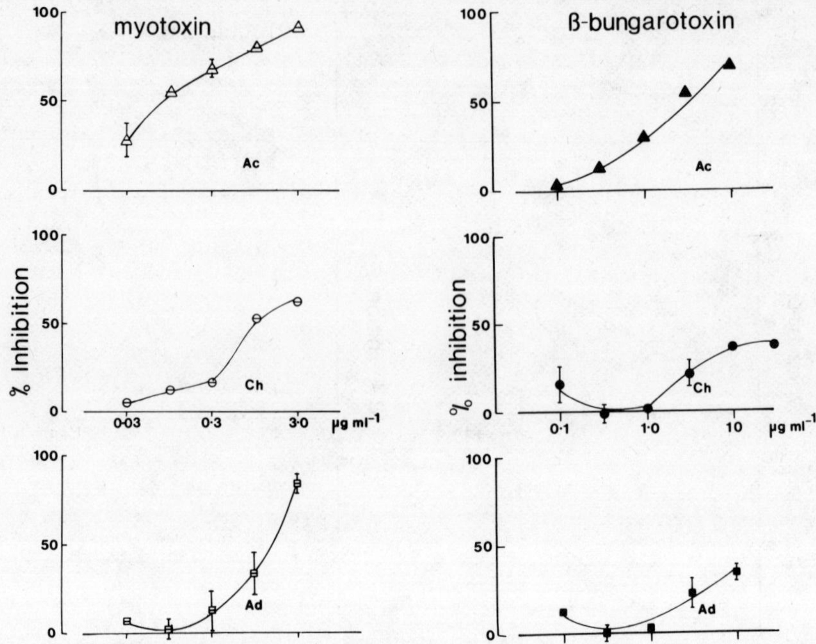

FIGURE 2: Dose-dependent inhibition of synaptosomal membrane trans-
port systems by _Enhydrina_ myotoxin and ß-bungarotoxin.
For details consult legend to Fig. 1 and Materials and
Methods.

only 30 and 40% inhibitions of adenosine and Ch respectively at
10 µg/ml but 70% inhibition of acetate uptake at the same concentration. For Ch uptake increasing the concentration of ß-bungarotoxin from 10 to 30 µg/ml did not increase inhibition and this suggests that a maximum possible blockade of Ch uptake by ß-bungarotoxin of 40%. This agrees with previous findings with _Torpedo_ T-
sacs (4). Similar considerations may also apply to adenosine although in this case concentrations greater than 10 µg/ml were not
tested. An adjusted IC_{50} figure for ß-bungarotoxin inhibition of
Ch uptake is 2.24 µg/ml which is very close to the IC_{50} figure of
2.5 µg/ml for inhibition of acetate uptake.

The effect of taipoxin and PBP_2-taipoxin on the three uptake
systems is shown in Fig. 3. By contrast to the preferential inhibition of acetate uptake seen with single chain toxins, taipoxin was
about equipotent towards all three uptake systems. Furthermore the
slopes of the inhibitory curves closely paralleled each other. The
chemically modified taipoxin derivative, PBP_2-taipoxin, which has

TABLE 1: Summary of the inhibitory effect of presynaptic snake neurotoxins on membrane transport systems of Torpedo electric organ synaptosomes.

Toxin	IC_{50} (µg/ml)		
	Acetate	Choline	Adenosine
Notexin	0.04	0.10	0.32
Notechis II-5	0.06	0.32	0.38
Enhydrina myotoxin	0.07	0.89	1.45
β-Bungarotoxin	2.50	2.24*	--
Taipoxin	0.66	0.69	1.32
PBP_2-Taipoxin	1.10	1.90	2.51

Concentrations of toxins causing 50% inhibition of each uptake system (IC_{50}) were determined graphically using the inhibition curves shown in Figs. 1-3.
* Concentrations causing 50% of maximum inhibition which was 40%.

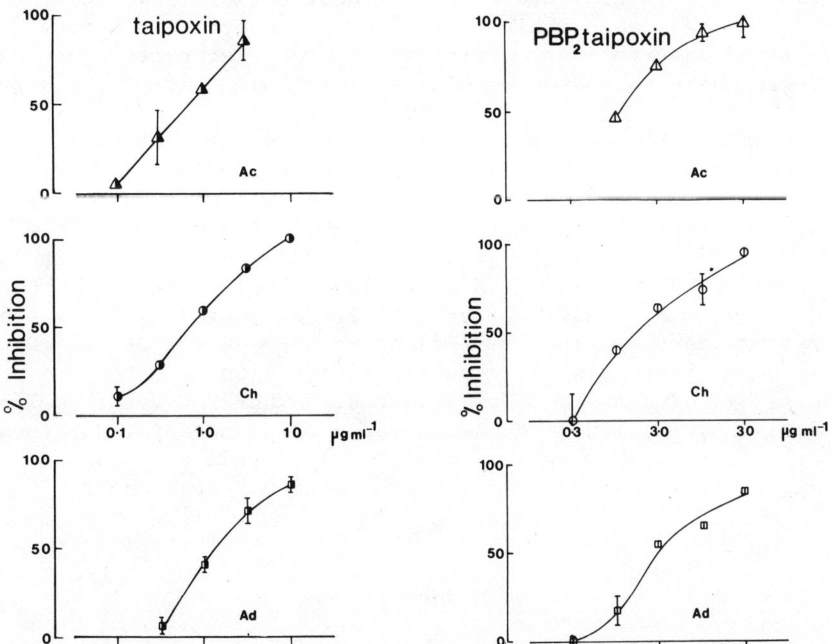

FIGURE 3: Dose-dependent inhibition of synaptosomal membrane transport systems by taipoxin and PBP_2-taipoxin. For details consult legend to Fig. 1 and Materials and Methods.

greatly reduced phospholipase A_2 activity and neurotoxicity (6) was
only slightly (2-3X) less inhibitory towards all three uptake systems. These observations agree with previous findings with PBP_2-
taipoxin and its effect on Ch uptake into Torpedo T-sacs (4).

The inhibitory potencies of all the toxins used in the present
study towards each of the uptake systems tested are given in Table 1.
Compared to earlier results with these toxins and their inhibitory
action on Ch uptake into T-sacs from Torpedo (4) some interesting
features emerge. The most striking difference is seen with the
single chain toxins - notexin, Notechis II-5 and Enhydrina myotoxin -
which are all significantly less effective on synaptosomes than on
T-sacs. These differences are 200X for notexin, 36X for Notechis
II-5 and 25X for Enhydrina myotoxin. By contrast taipoxin and
PBP_2-taipoxin were only 2X and β-bungarotoxin 4X less potent. The
reason for this marked shift in potency for the single chain toxins
is not clear. Possibly T-sacs are inherently more susceptible to
these toxins than synaptosomes. However, this would not explain why
the two different nerve terminal preparations behave with similar
sensitivities towards taipoxin. Another possibility is that the
inhibitory action on Ch uptake with the single chain toxins seen
previously was largely due to an indirect action. Lysolecithin and
fatty acids can cause inhibition of Ch transport into T-sacs (Dowdall,
unpublished observations) and these might be generated by the action
of single chain toxins on membranes other than that of the nerve
terminal in crude (P_2) T-sacs from electric tissue. A prime candidate would be the postsynaptic membrane (fragments of which contam-
inate the crude T-sac fraction) since both notexin and Enhydrina
myotoxin have postsynaptic myonecrotic activities. Further support
for an indirect action for the single chain toxins comes from observations on Ch uptake by gradient purified T-sacs. As with synaptosomes these purified T-sacs are less sensitive to single chain
toxins than crude T-sacs but essentially unchanged in their sensitivity towards taipoxin , PBP_2-taipoxin and β-bungarotoxin (Dowdall,
unpublished observations). Finally, inhibition mediated via phospholipid hydrolysis products would explain why PBP-Notechis II-5
and PBP-Enhydrina myotoxin have a markedly reduced potency towards
Ch uptake into T-sacs which parallels their reduced phospholipase
activity (Dowdall and Fohlman, unpublished observations).

DISCUSSION

Specificity of Action

The results of this study show that transport processes other
than that for Ch are inhibited by a number of different presynaptic
neurotoxins. Comparison of the concentration-dependency of these

inhibitory effects suggests that the neurotoxins can be divided into two groups; those which exert a preferential effect on acetate uptake (i.e. the single chain toxins) and those which show little discrimination between different uptake systems (i.e. taipoxin and PBP-taipoxin). The position of β-bungarotoxin in this classification is ambiguous because of a somewhat anomalous effect on Ch uptake (see Table 1). There are several possibilities which might explain this selective action of single chain toxins. The simplest of these is that this sub-group has a target site on the preterminal membrane which is different from that to which taipoxin and its derivative are directed. Alternatively the target site might be common to all these neurotoxins and the differential sensitivity of acetate might be due to subtle differences in toxin-target interactions or different degrees of phospholipid breakdown in the vicinity of the target by their intrinsic phospholipase A_2 activity. (The peculiar sensitivity of acetate uptake to these phospholipase A_2 neurotoxins might reflect the antagonist activity of released fatty acids towards an uptake system for their lowest homologue.)

Although both of these alternative possibilities provide theoretical grounds to explain the selective action of some neurotoxins towards acetate uptake neither can account for their "loss" of potency towards Ch uptake inhibition associated with purification of the nerve terminals. It is perhaps not just coincidence that those toxins which preferentially inhibit acetate uptake are the same as those which showed a marked "loss" of potency towards Ch uptake with purified synaptosomes. Conceivably even with purified synaptosomes trace amounts of contaminating membranes might provide phospholipid substrates for these toxins to generate fatty acids and lysophosphoglycerides in quantities sufficient to affect transport across the synaptosomal membrane. Any such effect might be superimposed on a more direct action at the presynaptic membrane. This could explain the peculiar sensitivity of acetate uptake seen with these toxins both in terms of inhibitory potency and the slope of the inhibition curves.

Inhibition of Choline Transport

It now seems unlikely that the toxin target sites constitute some part of the Ch transport system of the presynaptic membrane previously suggested (2,6). Recent observations of the relative insensitivity of Ch transport in synaptosomes from mammalian and cephalopod brain towards these toxins also support this conclusion (5). The results of the current study favor the view that there are major changes in membrane structure which result from toxin-target interactions and that these can be detected using any membrane transport process. Major changes in membrane fluidity have been shown to occur when taipoxin or PBP_2-taipoxin interact with T-sacs (5). Structural changes of this type might also provide some explanation for the effect of these toxins on the release of ACh from motor nerve terminals _in situ_. They might also be relevant to the idea that the toxins exert a depolarizing action on the pre-terminal membrane as recently proposed for the effect of β-bungaro-toxin on mammalian brain synaptosomes (1,10,12). We are currently testing this possibility by measuring membrane potentials in normal and toxin-treated _Torpedo_ synaptosomes.

ACKNOWLEDGEMENTS

Financial support by The Wellcome Trust and The Muscular Dystrophy Group of Great Britain is gratefully acknowledged. We also thank Professor C. Cazaux and his Staff at the Marine Station of Arcachon for their help in supplying _Torpedos_.

REFERENCES

1.   Dolly, J.O., Tse, C.K., Spokes, J.W. and Diniz, C.R. (1978): Biochem. Soc. Trans. 6:652-654.
2.   Dowdall, M.J. (1977): In _Cholinergic Mechanisms and Psycho-pharmacology_, (ed) D.J. Jenden, Plenum Press, New York, pp. 359-375.
3.   Dowdall, M.J., Barrantes, F.W., Stender, W. and Jovin, T.M. (1976): J. Neurochem. 26:1253-1255.
4.   Dowdall, M.J., Fohlman, J.P. and Eaker, D. (1977): Nature 269:700-702.
5.   Dowdall, M.J., Fohlman, J.P. and Watts, A. (1979): In _Advances in Cytopharmacology, Vol. 3: Neurotoxins: Tools in Neuro-biology_, (eds) B. Ceccarelli and F. Clementi, Raven Press, New York, pp. 63-76.
6.   Fohlman, J., Eaker, D., Dowdall, M.J., Lullmann-Rauch, R. Sjodin, T. and Leanders, S. (1979): Eur. J. Biochem. 94: 531-540.
7.   Lee, C.Y. (1979): In _Advances in Cytopharmacology, Vol. 3: Neurotoxins: Tools in Neurobiology_, (eds) B. Ceccarelli and F. Clementi, Raven Press, New York, pp. 1-16.
8.   Meunier, F.M. and Morel, N. (1978): J. Neurochem. 31:845-851.
9.   Morel, N., Israel, M., Manaranche, R. and Mastour-Frachon, F. (1977): J. Cell Biol. 75:43-55.
10. Ng, R.H. and Howard, B.D. (1978): Biochemistry 17:4978-4986.
11. Sen, I., Grantham, P.A. and Cooper, J.R. (1976): Proc. Nat. Acad. Sci. 73:2664-2668.

12. Smith, C.C.T., Bradford, H.F., Thompson, E.J. and MacDermot, J. (1980): J. Neurochem. 34:487-494.

13. Zimmermann, H., Dowdall, M.J. and Lane, D.A. (1979): Neuroscience 4:979-993.

BIOCHEMICAL AND BIOPHYSICAL ASPECTS OF STIMULATION INDUCED VESICLE

HETEROGENEITY IN CHOLINERGIC SYNAPTIC VESICLES

P.E. Giompres, H. Zimmermann and V.P. Whittaker

Abteilung Neurochemie, Max-Planck-Institut
fur Biophysikalische Chemie, Gottingen, FR Germany

In a series of previous experiments we provided evidence that
on stimulation-induced transmitter release synaptic vesicles in the
cholinergic nerve terminals of the <u>Torpedo</u> electric organ start to
undergo cycles of exo- and endocytosis. The recycled vesicles can
be shown to contain extracellular marker of high molecular weight
if it has been applied in high concentrations (1). They are smaller
and also denser and thus can be separated from vesicles which have
not yet released their stores of transmitter. If radiolabelled
precursors are applied the recycled vesicles contain both ACh (9)
and ATP (13) of high specific radioactivity. The evidence for this
is derived from density gradient centrifugation studies using sucrose gradients in a zonal rotor and from related electron micro-
scopical investigations.

Although it could be shown by biochemical methods that the ACh
in the denser vesicle fraction was contained in synaptic vesicles
only and not in particles containing entrapped cytoplasm, morphological analyses had revealed the presence of larger membrane particles in this fraction. Since a detailed analysis of the molecular
constituents of the types of vesicles generated on stimulation induced transmitter release is of great interest for the understanding
of the mechanism of vesicle recycling in cholinergic nerve terminals
we further purified the isolated vesicles by column chromatography
using porous glass beads. The reason for the changes in size and
density of recycled vesicles is not clear. We therefore measured
the changes in vesicle density, vesicle water space and its contents
in ACh and ATP during rest and during transmitter release. In order
to investigate to what extent the stimulation-induced alterations in
synaptic vesicles were reversible and whether the various parameters

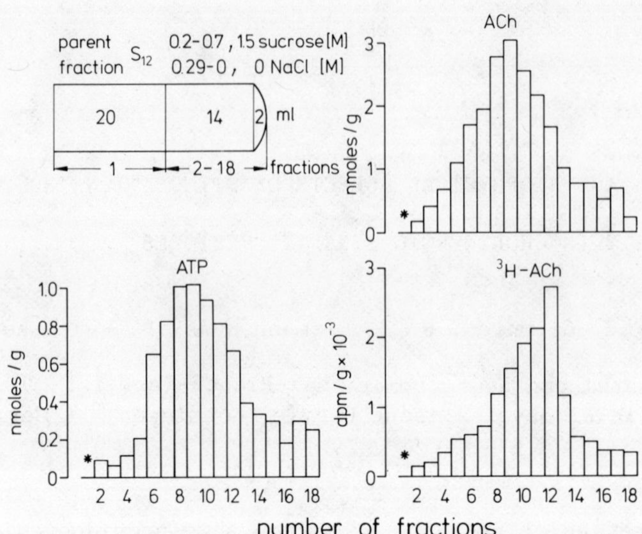

FIGURE 1: Density gradient separation of vesicles from stimulated tis-
         sue (0.1 Hz, 1800 pulses). Isolation of synaptic vesicles on
         an isoosmolar sucrose density gradient was as indicated. A
         block of electric tissue was perfused with [3H]-acetate and
         stimulated via the attached nerve. Tissue was homogeni-
         zed and the parent fraction prepared as described pre-
         viously (2). The amount of parent fraction layered onto
         the gradient corresponded to 20 g wet weight of tissue.
         * = not determined.

were reversed concomitantly we analyzed the same parameters during
various periods of rest following stimulation.

Isolation of Pure Small and Dense Vesicles

Vesicle heterogeneity revealed by sucrose density gradient
centrifugation in a swing-out rotor: In previous experiments
stimulation-induced separation of synaptic vesicles into two main
fractions was obtained by very shallow isoosmolar sucrose density
gradients in zonal rotors (6,12). A similar separation can be obtained if continuous sucrose gradients are used in a swing-out
rotor and the material is harvested from the gradient in small
fractions (Fig. 1). If blocks of electric tissue are perfused with
[3H]-acetate and stimulated via the attached nerve at a low

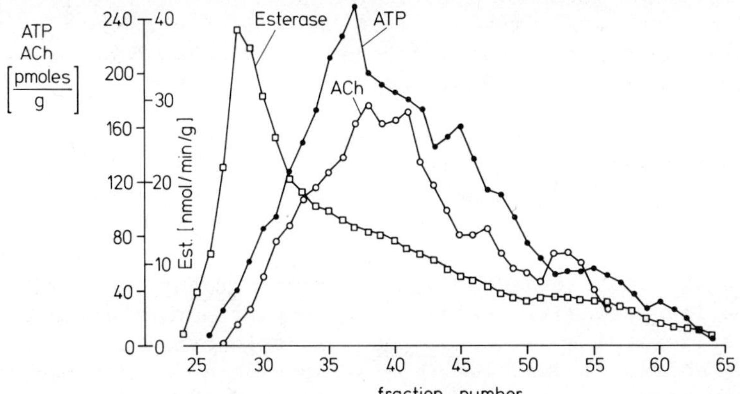

FIGURE 2: Purification of fractions containing synaptic vesicles ob-
          tained by density gradient centrifugation (fractions 7-12,
          Fig. 1) by column chromatography using porous glass beads
          (Electronucleonics, CPG-10, 300 nm pore size). The elu-
          tion pattern shows a clear separation of larger membrane
          particles in the void volume (peak of esterase activity)
          from the vesicle markers ACh and ATP.

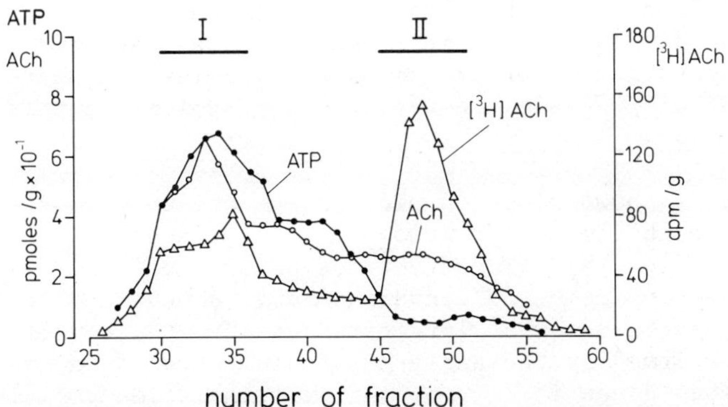

FIGURE 3: Stimulated, 0.1 Hz, 1800 pulses. Elution of prepurified
          synaptic vesicles from a glass bead column after previous
          perfusion of a block of electric tissue with [3H]-acetate
          and stimulation. The column separates between particles
          of lower (I) and higher (II) SRA of ACh.

frequency which does not reduce vesicle number (13) a peak of [3H]-
ACh can be separated from the main and coincident peaks of ACh and
ATP. This separation does not occur on fractionation of unstimulated control tissue. Analysis of enzyme markers (not shown)
further revealed that activity of lactate dehydrogenase as a marker
for cytoplasmic constituents and of cytochrome c oxidase as a marker
of mitochondrial membranes is absent from the peaks of ACh and [3H]-
ACh, whereas these fractions contain a mixture of membrane bound
and soluble cholinesterase.

Vesicle separation by column chromatography: Column chromatography using porous glass beads for further purification of synaptic
vesicles isolated from the Torpedo electric organ was first introduced by Morris (2) and led to a separation of larger membrane constituents from synaptic vesicles. Figure 2 shows the separation of
subcellular particles from a perfused and stimulated block of
electric tissue prepurified by density gradient centrifugation by
means of column chromatography. It can be seen that particles
associated with cholinesterase activity are eluted in the void
volume of the column, whereas the vesicle markers ACh and ATP are
delayed and appear as a coinciding peak separated from the esterase
containing particles. The trailing of esterase activity is likely
to be due to the presence of soluble esterases in the parent fraction. In a series of experiments nerve terminals were labelled with
[3H]-ACh and the distribution of particles containing the radiolabel was followed on column chromatography. As can be seen from
Fig. 3 no [3H]-ACh is eluted with the void volume of the column.
[3H]-ACh is, however, eluted together with the peaks of ACh and ATP
(I) as well as close to the salt volume of the column (II). This
latter pool of [3H]-ACh is also particle bound and represents a
transmitter pool of higher SRA than that contained in the main peak
of ACh and ATP. If, for comparison, vesicles isolated from unstimulated tissue are further fractionated by column chromatography a
second peak of [3H]-ACh close to the salt volume can also be
observed.

Figure 4 shows electron micrographs of subcellular particles
contained in the column fractions I and II (a and c) and the intermediate fraction (b). The fractions consist of pure synaptic vesicles without any membrane contamination; however, vesicles in a)
appear to have a somewhat larger radius than those in fraction c).
This is confirmed by the analysis of diameters of the vesicles of
the fractions given in Fig. 5. After stimulation vesicles isolated
from column fraction I (Fig. 5a) show a broader diameter distribution indicating the presence of both larger and smaller vesicles,
whereas those in column fraction II (Fig. 5c) are of smaller diameter only. If unstimulated tissue is used (Fig. 5, right column)
vesicles in fraction I comprise a homogeneously larger population,
whereas those in fraction II are again of smaller diameter. This

TABLE 1:   Specific radioactivity of ACh in the two peak fractions
           (I and II) of the column effluent.

Experimental Condition	SRA of ACh in Peak Fraction		Number of Experiments
	I	II	
Unstimulated control	269 ± 6	3013 ± 618	4
Stimulated (1800 pulses, 0.1 Hz)	876 ± 84	3690 ± 405	7

Values represent combined column fractions as indicated in Fig. 3
and are means ± S.E.M. of number of experiments as indicated above.

TABLE 2:   Changes in vesicle density and water space on low
           frequency stimulation.

Experimental Condition	Density (g/cm^3)	Water Space (percent of particle volume)	Number of experiments
Unstimulated control	1.046 ± 0.0003	65.2 ± 0.2	3
Stimulated (1800 pulses 0.1 Hz)	1.067 ± 0.0002	42.5 ± 4.0	2

Values represent the peak fraction of $[^3H]$-ACh on density gradients
and are means ± S.E.M. of three experiments or ± range of 2 experiments as indicated above.

is consistent with the previous fine structural observation (12)
that in unstimulated tissue only about 8% of the vesicles are of
the smaller diameter type, whereas on low frequency stimulation
(0.1 Hz, 1800 pulses) this percentage increases to 80% .

     Table 1 compares the SRA of ACh in the two main vesicle frac-
tions obtained by column chromatography from control and stimulated
tissue. SRA is always highest in column fraction II. The higher
SRA of ACh in fraction I after stimulation as compared to control
conditions is likely to be caused by the increase in the number of
small vesicles in fraction I after stimulation (compare Fig. 5).

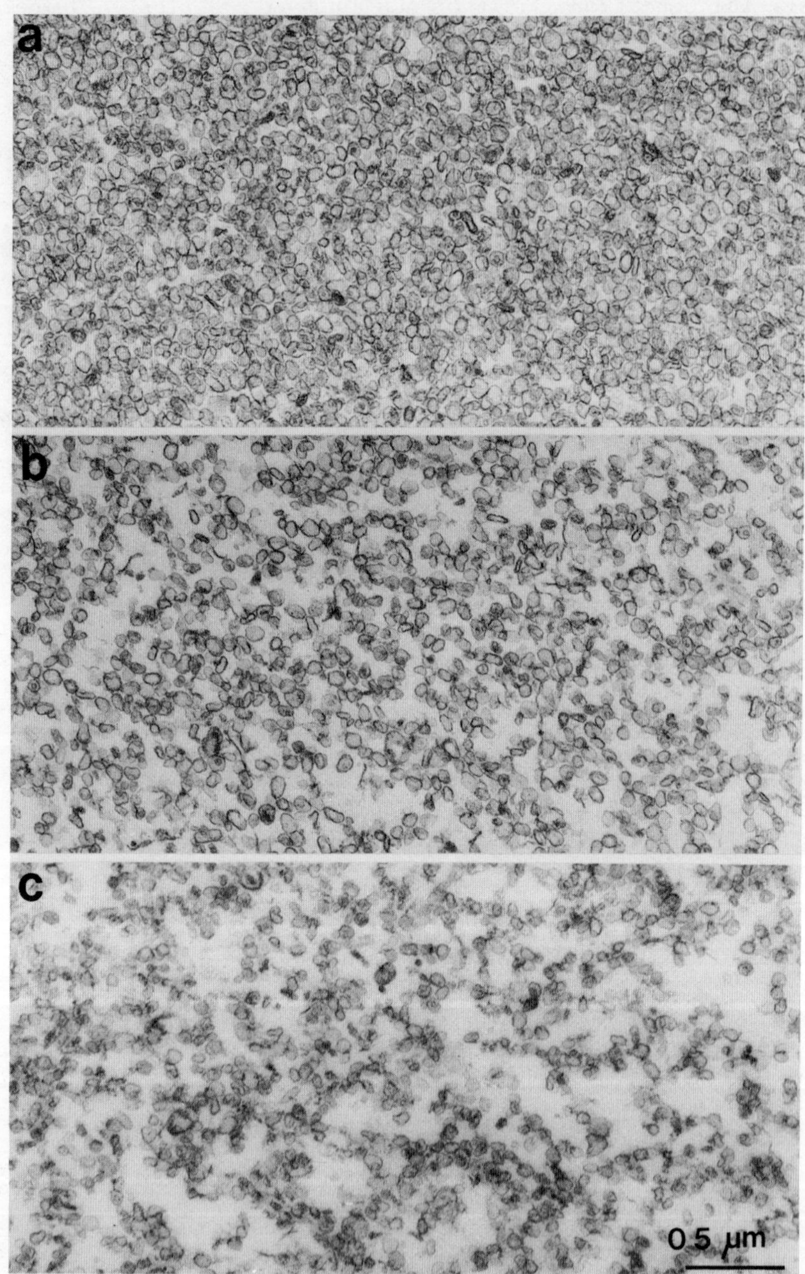
a
b
c
0.5 μm

LEGEND: FIGURE 4 – Electron micrographs of pooled fractions from
                  the column effluent after stimulation of tissue.
                  For comparison with Fig. 3, a) and c) represent
                  peaks I and II, respectively, whereas b) re-
                  presents the fractions between the two peaks.
                  All fractions contain highly purified vesicles
                  although these appear to differ in average
                  diameter.

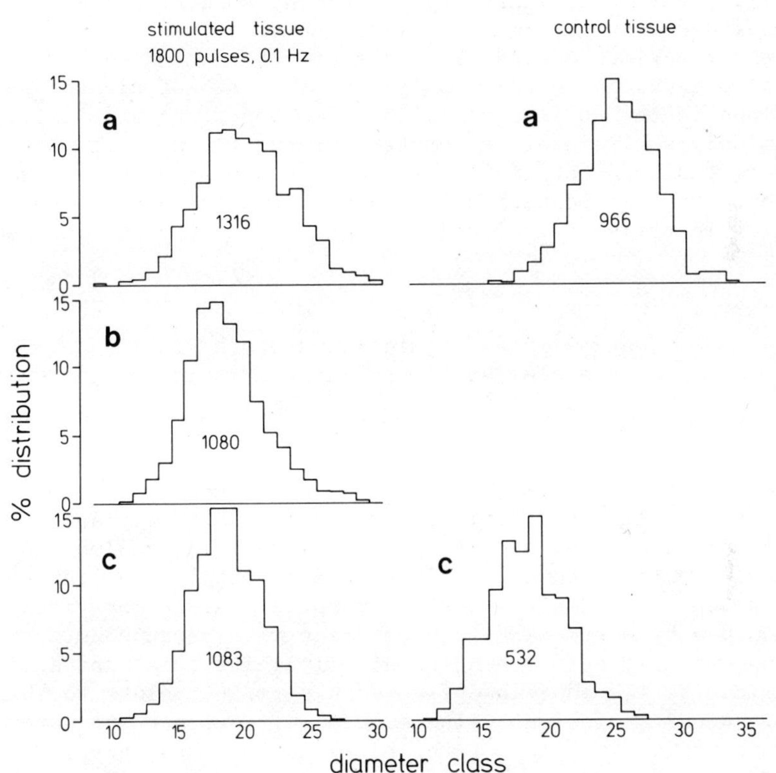

FIGURE 5: Normalized diameter distributions of synaptic vesicles
          recovered from the column effluent of the stimulation
          experiment shown in Fig. 4 and of a control experiment.
          a), b) and c) Represent fractions as in Fig. 4. The
          number of vesicles measured for each fraction is given in
          parentheses. Diameter classes are arbitrary units.

DISCUSSION

The results demonstrate that after application of radiolabelled precursor ([3H]-acetate) vesicles of different SRA of ACh can be separated by continuous sucrose density gradients using a swing out rotor similar to previous studies using zonal centrifugation (12). If these isolated vesicles are further chromatographed on porous glass beads radioactivity is recovered in two main fractions containing synaptic vesicles of larger and smaller diameter but not in the fraction of membranes containing high activity of esterases. Since the column elutes particles in the order of larger esterase containing particles, larger vesicles and smaller vesicles it may be concluded that a) radiolabelled ACh recovered in the density gradient fractions around the peaks of ACh and [3H]-ACh is contained in synaptic vesicles only; b) that ACh of highest SRA is contained in the smaller vesicles which appear on stimulation. A preliminary analysis of the protein of the two main fractions of synaptic vesicles obtained after column chromatography revealed that both of them possess all the main protein components detected normally in vesicles isolated from unstimulated electric tissue (7).

Changes in Biochemical and Biophysical
Vesicle Parameters on Stimulation
and During Subsequent Recovery

<u>Measurement of vesicle water space</u>: Since the volume of synaptic vesicles as determined from electron micrographs could have been affected by secondary changes due to fixation and embedding we determined as an additional physical parameter vesicular water space. If the cholinergic synaptic vesicle comprises a particle bounded by a semipermeable membrane with an aqueous core filled with osmotically active ACh and ATP (1,8) its water space may be determined by the addition of a dense permeant solute to the iso-osmotic density gradient. The solute will replace the water in the same volume fraction of the particle and thus increase its density. From the increase in density the water space of the particle may be calculated (3). Using glycerol (20%) as a dense membrane permeant solute the volume fraction of the vesicle in which water can be exchanged against glycerol was determined to be 65%. Table 2 shows the alterations of both vesicle density and water space induced by low frequency nerve stimulation. It can be seen that the increase in vesicle density (peak fraction of [3H]-ACh on a sucrose density gradient, compare Fig. 1) is accompanied by a decrease in the vesicular fraction which can be penetrated by glycerol by 35%. This would correspond to a decrease of about 20% which was what was observed in these experiments, compared to the 25% reported earlier (12).

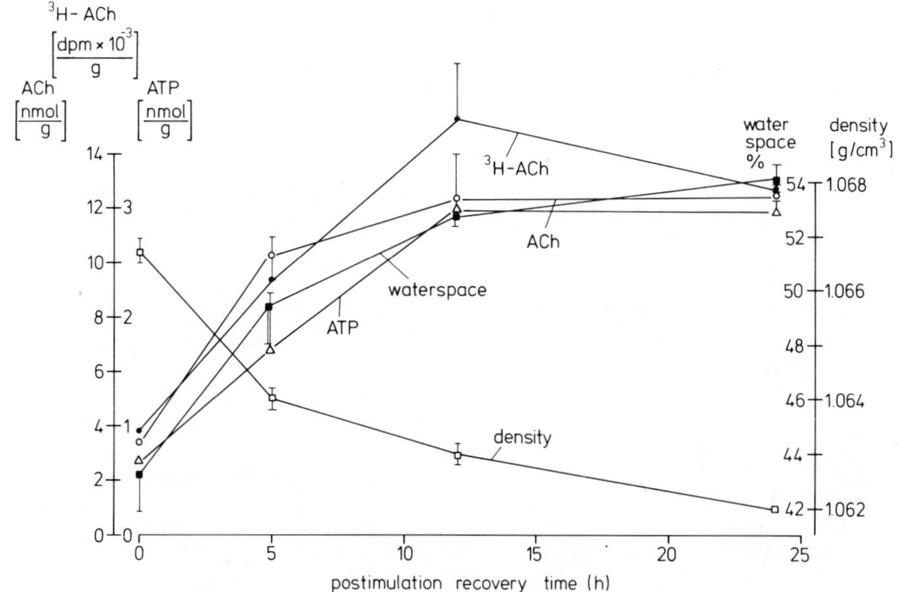

FIGURE 6: Poststimulation (1800 pulses, 0.1 Hz) recovery. Changes
in vesicle contents in ACh, ATP and [3H]-ACh as well as
vesicle density and water space during a poststimulation
recovery period. Perfusate containing [3H]-acetate was
recycled until the 12 hr time point and then exchanged
against label free medium. Only the [3H]-ACh peak of
density gradients is analyzed. Values are mean ± S.E.
range of two perfused blocks of electric tissue.

Recovery of chemical and physical parameters: Figure 6 shows
alterations in a number of vesicle parameters during a period of
rest following low frequency stimulation. All values refer to the
peak fraction of [3H]-ACh on sucrose density gradients. As can be
seen, all vesicle parameters become altered during recovery. The
recovery in vesicular ACh, ATP and [3H]-ACh is accompanied by an
increase in vesicular water space and a decrease in vesicle density.
The rate of change is highest at the beginning of the recovery
period and after 12 hr little further change is obtained. Within
24 hr of recovery vesicular density is reduced to 1.062 g/cm^3 corresponding to an increase in vesicular water space to 54%.

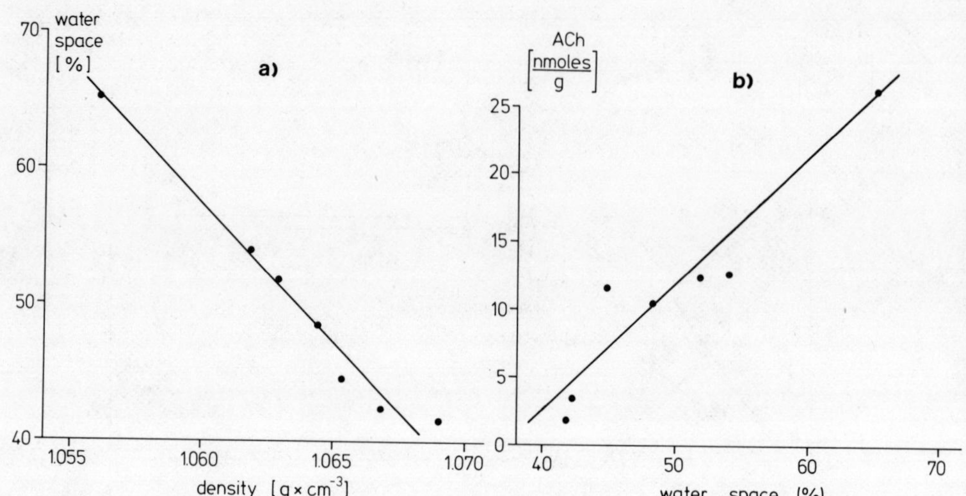

FIGURE 7: Correlation of vesicle density with water space (a) and
vesicular contents in ACh with vesicle water space (b).
Values represent the density gradient peak fraction of
[³H]-ACh which in control experiments coincides with the
peak fraction of ACh and ATP. The correlation coefficients were 0.986 (p < 0.001) for a) and 0.947
(0.01 < p < 0.001) for b).

Figure 7a shows that vesicle density is correlated with vesicular water space. This suggests that the water content of the
vesicle directly determines its density. Furthermore, the vesicular content in ACh is significantly correlated with vesicle water
space (Fig. 7b). This would suggest that the reloading with ACh
of the recycled vesicles is paralleled by an increase in vesicular
hydration volume which again decreases vesicle density. On the
other hand there was no significant correlation between vesicle
water space and vesicular ATP content. This became particularly
apparent in experiments in which the recycling of ACh was blocked
by HC-3 (which does not impair recycling of vesicular ATP [11]).
Thus ATP is likely to be stored in synaptic vesicles by a somewhat
different mechanism than ACh. No deterioration of the general fine
structural appearance could be observed on stimulation and during
24 hr of recovery.

Since the ACh content of the vesicles was correlated with
vesicle glycerol space and density it appears likely that a vesicle
newly formed after releasing its ACh by exocytosis has a core which
is hypoosmotic with respect to the surrounding cytoplasm. It is
then subjected to osmotic dehydration with an attendant reduction
in radius and in the glycerol space. If the vesicle membrane is
conserved, osmotic dehydration would also result in an increase in
density since the membrane has a density greater than that of water
and would now represent a larger proportion of the smaller vesicle.
Hypoosmotic conditions could be established in a newly formed vesicle if ACh were replaced by an electrochemical equivalent of another
ion (e.g. Ca^{++} [4]) which combined with a membrane or other vesicle
constituent to form an osmotically inactive complex. If such an ion
were during the recovery process to be exchanged for extra-vesicular
ACh, the original osmotic pressure of the vesicle core would become
reestablished and rehydration with the resumption of the original
vesicle volume and density would ensure.

It is noteworthy that the very slow process of vesicle reloading does not reflect the capacity of the ACh uptake mechanism
into synaptic vesicles but the generally low metabolic activity of
the stimulated tissue (low content of ATP, see Ref. 10). It has
been demonstrated (5) that the rate of vesicle refilling after
stimulation is limited by the very slow rate of overall resynthesis
of ACh in the nerve terminals and not by the rate of reformation of
the vesicles themselves. Thus the partial depletion of energy
resources and their slow restoration after simulation of this tissue
enables the experimenter to detect various discrete steps in the
process of vesicle recycling and reloading which might not be
observable in tissue with its full complement of energy resources
or in a tissue with generally higher metabolic rate.

In summary: 1) Synaptic vesicles were purified by column
chromatography using porous glass beads. Analysis of highly purified vesicles confirmed earlier observations from this laboratory
made by density gradient centrifugation, that the small and dense
vesicles formed on transmitter release contain ACh of high SRA if
a radiolabelled precursor is applied. 2) During a period of rest
following low frequency stimulation a concomitant reversal of the
stimulation induced depletion in vesicular contents in ACh and ATP,
the decrease in vesicular water space and radius and the increase
in vesicle density takes place. During this process significant
correlations were obtained between the changes in vesicular density
and water space as well as between vesicular contents in ACh and
vesicle water space. 3) Our results further support the mechanism
of transmitter release for cholinergic nerve terminals put forward
earlier (10). Vesicles release ACh by exocytosis and are immediately recycled. They are then refilled with ACh (and ATP). The
just recycled vesicles thus contain more newly synthesized ACh and

ATP than still resting vesicles (vesicle heterogeneity). The recycled vesicles have (due to their juxtaposition to the release
sites) a higher probability for renewed exocytosis than vesicles
which have not yet released their contents (preferential release
of newly synthesized transmitter).

REFERENCES

1.   Breer, H., Morris, S.J. and Whittaker, V.P. (1978): Eur. J.
        Biochem. 87:453-458.
2.   Morris, S.J. (1973): J. Neurochem. 21:713-715.
3.   Morris, S.J. and Schowanka, I. (1977): Biochim. Biophys. Acta
        464:53-64.
4.   Schmidt, R., Zimmermann, H. and Whittaker, V.P. (1980):
        Neuroscience 5: 625-638.
5.   Suszkiw, J.B. (1980): Neuroscience 5:1341-1349.
6.   Suszkiw, J.B., Zimmermann, H. and Whittaker, V.D. (1978): J.
        Neurochem. 30:1269-1280.
7.   Tashiro, T. and Stadler, H. (1978): Eur. J. Biochem. 90:479-487.
8.   Whittaker, V.P. and Stadler, H. (1980): IN Proteins of the
        Nervous System (eds) D.J. Schneider, R.H. Angeletti, R.A.
        Bradshaw, A. Grasso and B.W. Moore, Raven Press, New York.
9.   Zimmermann, H. (1978): Neuroscience 3:827-836.
10.   Zimmermann, H. (1979): Neuroscience 4:1773-1804.
11.   Zimmermann, H. and Bokor, J.T. (1979): Neurosci. Lett. 13:
        319-324.
12.   Zimmermann, H. and Denston, C.R. (1977): Neuroscience 2:695-714.
13.   Zimmermann, H. and Denston, C.R. (1977): Neuroscience 2:715-730.

ACETYLCHOLINE STORAGE AND CALCIUM CLEARANCE BY SYNAPTIC VESICLES

M. Israel

Department of Neurochemistry
Laboratory of Cellular Neurobiology; CNRS
91190, Gif-sur-Yvette, France

INTRODUCTION

The observation that recently synthesized transmitter is more
easily available for release than the rest of the store, when synaptic activity is induced by stimulation, is certainly a well accepted finding (11). It is also well established that up to 70% of
the transmitter is found in synaptic vesicles in a bound form, since
it is protected from the action of esterases. The isolation of
cholinergic synaptic vesicles extremely rich in acetylcholine (ACh)
from Torpedo electric organ has been achieved by several investigators (28-30,50).

With such an important store of ACh it seemed odd to have to
synthesize transmitter in order to feed the release mechanism. This
was particularly difficult to understand since vesicles   then
have to take up ACh rapidly from the cytoplasm where choline
acetylase had been localized (19,20,25,28,30), uptake and synthesis
being two additional steps before exocytosis. In Collier's experiments (11) the demonstration that recently synthesized ACh was readily available for release was based on a comparison of specific
radioactivities of the transmitter in the perfusate and in the tissue.
We decided to test this finding on Torpedo electric organ where ACh
compartments were directly measurable. Considering the localization of
choline acetylase one expected that the recently synthesized transmitter
would first appear on the cytosol, a compartment in which it should
be accessible to esterases after disrupting the nerve terminal membrane. When Torpedo electric organ prisms are incubated with a radioactive ACh precursor, the specific radioactivity of free ACh, the pool

hydrolyzed after homogenizing the tissue, is two to three times
higher than the specific radioactivity of bound ACh found in the
vesicles. After stimulation the specific radioactivity of free ACh
increases while no significant changes are found in the vesicular
bound ACh pool. Free ACh is not only utilized but renewed during
activity. This verified the view that recently synthesized ACh is
more available for release, but we did not expect to find that vesicles would neither lose ACh nor incorporate significant amounts of
radioactive ACh as a result of stimulation (15,16). These observations had a time resolution of about 150 impulses, they were then
scaled down to a few impulses (18,32), and finally to one impulse
after 4-aminopyridine (14; and This Volume). The ACh pools are here
determined immediately after brief stimulations, it is therefore
difficult to compare these results with data obtained after a long
fractionation (52,53).

In a recent work (12) it was shown that miniature endplate
potentials (mepps) generated by a false transmitter acetylmonoethylcholine have a more rapid decay time constant that mepps due
to ACh. This permitted us to test if evoked release would or would
not be a sum of independent packets of real and false transmitter
(40,41). The conclusion that the released transmitter came from a
pool where real and false transmitter were constantly mixed was drawn
from the observation that each packet was a mixture of ACh and
acetylmonoethylcholine, the proportion of false transmitter being
increasingly higher as a result of stimulation. This finding is in
full agreement with the biochemical data showing that the recently
synthesized transmitter appears in the cytoplasm where it is mixed
with the transmitter present. However, biochemical data give an
additional piece of information: in Torpedo electric organ it is
possible to exhaust the electrical discharge after a few minutes of
stimulation and still not find any significant change of the vesicular bound pool.

The conclusion that perhaps part of the vesicles were involved
while most of them remained unaffected was drawn (15) and it agreed
with a previous observation showing that synaptic vesicles were
heterogeneous, the heterogeneity being intrinsic to each vesicle,
or due to the existence of a subpopulation (42,43). We felt, however, that this view (active vesicles) was not the only possibility
and expressed the still less acceptable idea, that perhaps synaptic
vesicles were simply a store of ACh not directly involved in the
release process (15). After the impressive effects of the black
widow spider venom on the vesicular population, the massive disappearance of vesicles obtained by several investigators, even after
short stimulations of the electric organ (6,52,53), led us to question our results and try to find out the reason for such an important
discrepancy. We have found that the disappearance of vesicles and
the early fall of bound ACh is obtained when the electric organ is

not fully immersed in a physiological saline during electrical stimulation. In this condition the electric energy of the discharge cannot be dissipated in the saline. The high voltage of the discharge
then drives an abnormally high current through the tissue which is
altered. The then needs an extremely long time to recover. Such a
depletion of vesicles is not obtained in conditions of stimulation
in which the tissue is immersed in saline, even when stimulation is
prolonged until exhaustion of the electrical discharge (17,27). Recent data (see 17) confirm that the vesicles (fraction VP_1) do not
lose ACh if the tissue is stimulated immersed.

Returning to the point raised by Dunant et al. (15,16,18) and
Israel et al. (27) about the mechanism of the ACh release, we are
left with two possibilities. The first is, for example, the incorporation and release of ACh by "active vesicles," those of the active
zone; and the second being the non-involvement of vesicles. One
might be tempted to consider that the recent separation of a sub-
fraction of vesicles which have now been purified (Zimmerman, This
Volume) favors the first hypothesis, particularly if they have incorporated some of the very radioactive cytoplasmic ACh. We do know
the incorporation rate of radioactive ACh in the bound pool (31,35)
and during a long stimulation experiment, followed by a long fractionation, it can be found that some of the very radioactive free
ACh became incorporated in vesicles, particularly if these new vesicles formed by endocytosis were empty. In our experiments we are
measuring free and bound ACh immediately after stimulation or in the
course of stimulation, and no changes are detected in the bound pool.
Certainly in the long run newly formed vesicles will incorporate
cytoplasmic ACh; this has perhaps nothing to do with the mechanism
of ACh release. There is, however, a strong agreement favoring the
idea that "active" vesicles do the work.

The active vesicle hypothesis is certainly interesting, and
we consider with great interest the attempts to purify subpopulations
of vesicles. This should, however, not displace our attention from
the other possibility raised by Dunant et al. (15) which consider
that vesicles might simply not be involved in the release process.
I apologize for defending this possibility, which is provocative.
First I shall consider synaptic vesicles as a store and discuss the
conditions mobilizing this store, then briefly consider the release
mechanism.

The Vesicular ACh Store
Filled and Empty Vesicles

The concentration of vesicular ACh was, in previous reports,
estimated simply by dividing the ACh content by the vesicular volume
determined after counting them in a tissue section (26,27), or on

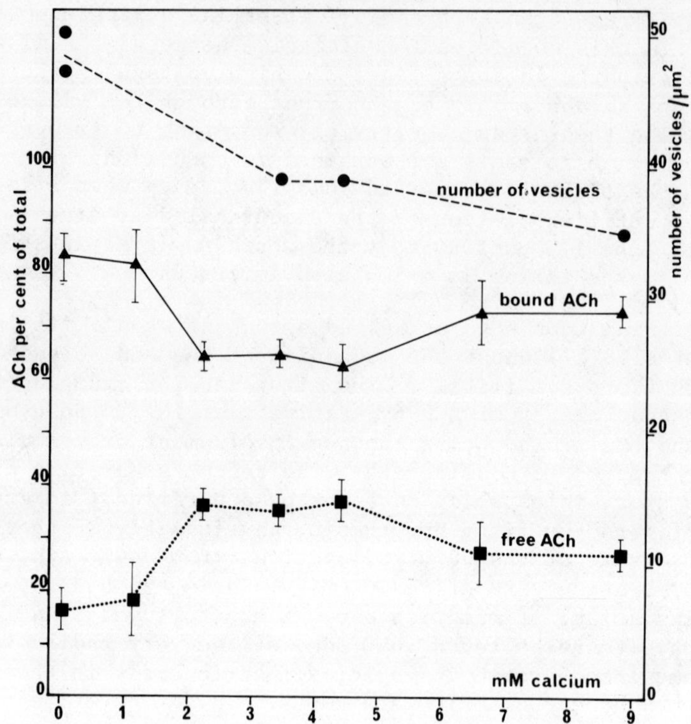

FIGURE 1: Effect of calcium on ACh compartmentation and on the number
of synaptic vesicles. Up to 85% of ACh is bound when the
tissue is incubated for about 90 min in a physiological
solution without calcium, while the number of synaptic vesicles is high (50 per μm^2). When calcium is increased the
bound pool falls to about 60% of the total but could not
be further decreased by raising the calcium concentration
to 9 mM. In contrast, the number of synaptic vesicles decreased to 35 per μm^2. Free ACh represents 15% of the
total when calcium is omitted and reaches a maximum of 40%
of the total between 2 and 4.5 mM calcium. Data from
P. Frachon-Mastour (diplome de l'Ecole Pratique des
Hautes Etudes, 1978).

freeze fractured synaptosomes. The concentration of 80 mM found
corresponds to about 10,000 molecules of ACh per vesicle. This value
is five to ten times lower than the concentration determined on synaptic vesicles isolated after fractionation (36,49). We therefore
have to consider the possibility that the fractionation procedure
purifies only synaptic vesicles full of ACh. This is not implausible

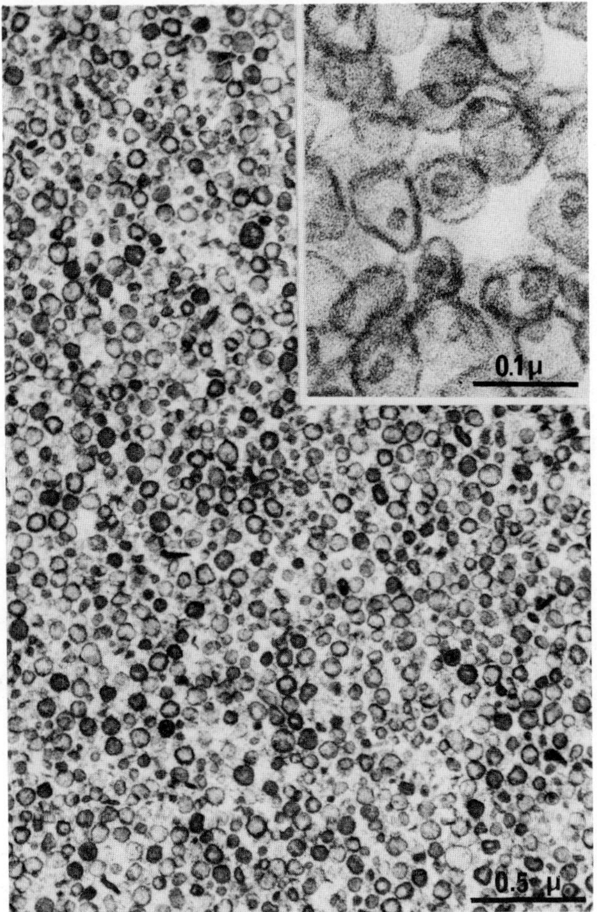

FIGURE 2: Isolated synaptic vesicles. Low magnification showing
the homogeneity of fraction 3. Insert: synaptic vesicles
contain a dense spot when fixed in presence of calcium.

since they are isolated according to their density, which might depend on their contents. Certainly the sources of errors in such
calculations are important. The possibility exists, however, that
only 20% of the vesicles counted _in situ_ contain ACh.

What are the factors controlling the size of the vesicular store?
As first shown by Birks (2), nerve terminals in the sympathetic ganglia fixed for electron microscopy in high Mg^{++} glutaraldehyde, are
considerably more rich in synaptic vesicles, than in usual fixations,
and it was found that an important number of synaptic vesicles

disappeared upon stimulation in conditions that do not change the ACh
content of the nerve terminals. If all the vesicles shown by this
technique had been incorporated in the presynaptic membrane as a
result of stimulation, the active zone would have been completely
disorganized and this was not apparent on the electron micrographs.
It is possible that many of the vesicles seen in high Mg^{++} glutaraldehyde fixation, in Birk's experiment, are empty ghosts which will
vanish upon stimulation (intracellular calcium being perhaps the trigger). The alternative interpretation would be that all the vesicles
were full of ACh while activity triggers the leakage of their contents in the cytoplasm leading to an increase of the cytoplasmic
ACh pool, and a reduction in the number of vesicles.

In Torpedo electric organ, nerve terminals contain more synaptic vesicles in tissues incubated in low calcium solutions before
fixation. In this condition (Fig. 1), the amount of bound ACh increases in parallel to the number of synaptic vesicles, while free
ACh fall to 12% of the total (Fig. 1). These observations are similar to those previously reported (1,32). This increase in bound ACh
is reversed when calcium is raised to its normal concentration (3 to
4 mM) in the physiological solution. For higher calcium concentration the effects were less evident, indicating that perhaps the
mobilization of bound ACh triggered by calcium occurs within a critical range of calcium concentrations. It was also shown (1) that
the late fall of bound ACh obtained on stimulated Torpedo electric
organ is accompanied by an increase in intracellular calcium. The
effects of calcium are probably not a direct action on the vesicle
membrane, since pure synaptic vesicles do not lose ACh when incubated
with calcium. Some investigators have advanced the idea that other
factors might be required, ATP, Mg, ascorbate (22,37,45). These results are interesting, but would need additional work to be understood and accepted. It is essential to avoid working on primary
fractions, which might contain many factors and enzymes.

Whatever the explanation for these results, it is probably that
divalent ions could influence the size of the bound ACh pool and
the number of synaptic vesicles. There are situations not yet very
clear where the number of vesicles follows very well variations of
bound ACh, while under other conditions the number of vesicles appears higher than expected. Perhaps the most convincing evidence
showing that the number of vesicles found in nerve terminal at a
given monment does not reflect the releasable ACh was obtained by
Ceccarelli and Hurlbut (8). The frog neuromuscular junction was stimulated until exhaustion in presence of hemicholinium to prevent ACh
synthesis, at this stage the terminal was full of vesicles but when
the black widow spider venom was applied the vesicles disappeared
while the usual avalanche of mepps was not obtained. This experiment
shows that if exocytosis of vesicular contents took place as a result
of venom action, that these vesicles did not contain ACh.

Uptake of Calcium by Synaptic Vesicles

Synaptic vesicles purified to morphological homogeneity (Fig. 2) often contain in their lumen a "dense spot" when fixed in glutaraldehyde containing calcium. This spot could be found even at low calcium concentration (100 μM) in the fixative (Fig. 2 insert). The dense spot was also observed _in situ_ by several investigators (5, 21,46). More recently this dense spot was obtained after a histochemical method for calcium (44). It was also found that nerve terminals contain a system other than ·mitochondria able to sequester calcium in the presence of ATP (3,4,47). It has been suggested that the smooth nerve terminal reticulum could be taking up that calcium, and this is certainly possible (3,4). But since synaptic vesicles might have common properties with the terminal cysternae of the reticulum (13), they might possess the property of concentrating calcium. In addition, synaptic vesicles are far more abundant in the terminal than the few cysternae. "Dense spot" found in vesicles can be removed by EGTA, and it is probable that we are dealing with a calcium precipitate. Thus led us to an attempt to find out if the purest synaptic vesicles we could get were able or not able to concentrate calcium in the presence of ATP and Mg.

In fractionation experiments the peak of ATP dependent calcium uptake coincides on the gradient with the peaks of ACh and ATP (Fig. 3a). Upon refractionation of pure vesicles on a continuous gradient by floatation, an extremely high specific activity for ACh, ATP and calcium uptake was found in fractions 11 and 12 (Fig. 3b). Whereas fractions 14 and 15 , which contained only small ACh and ATP shoulders, took up plenty of calcium, showing that vesicles emptied perhaps during the long fractionation procedure, were still able to concentrate calcium. Alternatively, we have seen that this situation might reflect a heterogeneity found _in vivo_. The basic finding is that synaptic vesicles identified by their ACh and ATP content are able to concentrate calcium. The vesicles keep this property after depletion. This was verified after a mild osmotic shock, which depleted most of their ACh content (Fig. 4). These findings are in accordance with the observations that synaptic vesicles contain a Ca-Mg dependent ATPase slightly stimulated by ACh (7).

The concentration of calcium within synaptic vesicles was estimated after measuring the calcium content of the fraction, and also from the uptake data. A value of the order of 15 mM could be determined. Since the cytoplasmic calcium concentrattion is of the order of 10^{-6} or 10^{-7} M, we can deduce that vesicles will concentrate calcium very efficiently. The K_m for calcium uptake is of the order of 5 μM, after correcting for ionized calcium, a value 1.5 times lower is found. Most of the kinetic parameters for this uptake have been described elsewhere (33). Certianly fractionation is not an absolute demonstration, but there is a whole set of observations (such as the dense spot which support the present finding.

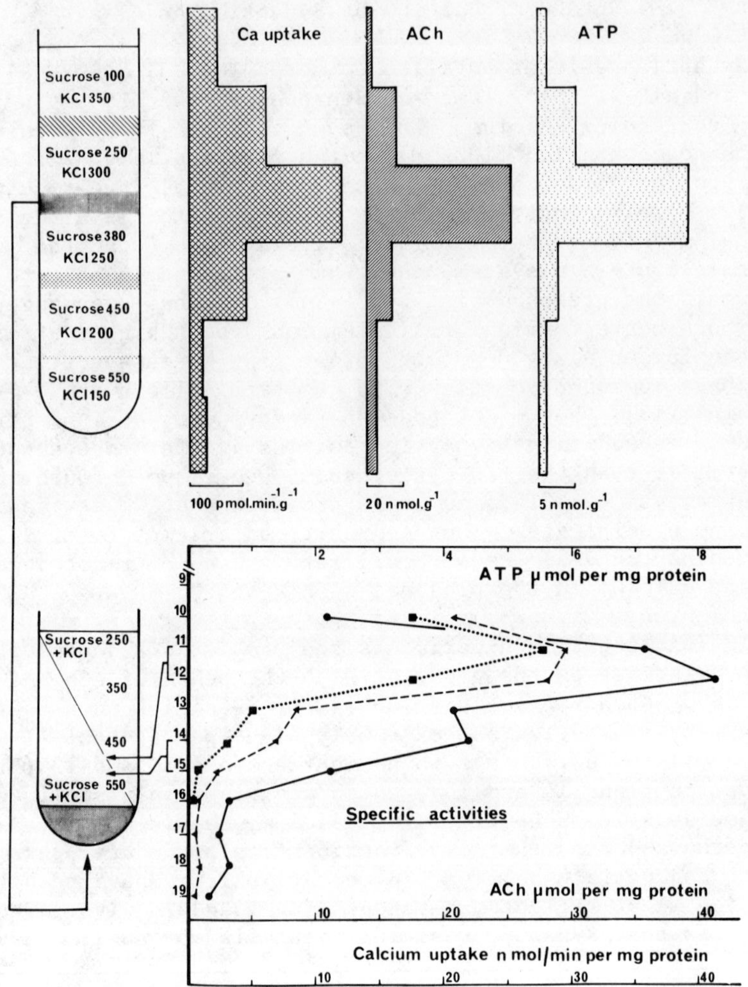

FIGURE 3: Refractionation of synaptic vesicles. The synaptic vesicles are isolated on a first discontinuous gradient (on top of 380 mM
sucrose-250 mM KCl, fraction 3). This fraction is the peak of ACh, ATP
and ATP-dependent calcium uptake (activities expressed per g original
tissue). The pure vesicles of fraction 3 (Fig. 2) are then refraction-
ed on a second continuous gradient by flotation. The specific activated peaks for ACh (■···■), ATP (▼---▼) and ATP-dependent calcium
uptake (●---●) were found in fractions 11 and 12, and could not be
separated. However, fractions 14 and 15 which take up calcium as well
have much less ACh and ATP. The sucrose molarity along the continuous
gradient is indicated, it expands from 550 mM to 250 mM sucrose, KCl
is added to keep the osmolarity at 850 mosm. This continuous gradient
was layered above fraction 3 to which sucrose was added to get 0.7 M
sucrose. All solutions were made in 10 mM Tris buffer (pH 7.2) and
gradients spun 150 min at 9500 g$_{av}$.

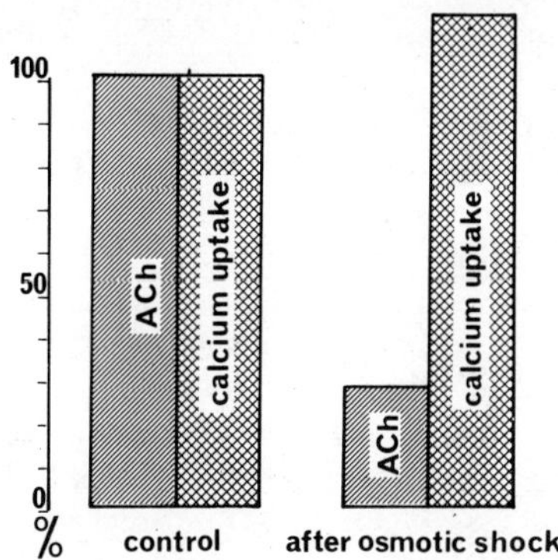

FIGURE 4: Uptake of calcium by depleted synaptic vesicles. After
          a mild osmotic shock, most of the vesicular ACh was re-
          leased, the uptake of calcium was unaffected, it was
          measured in an isoosmotic solution.

The idea that vesicles will be able to clear calcium from the
cytoplasm, may provide an interesting interpretation of the effects
of lanthanum on ACh release. If lanthanum is not taken up by vesicles as well as calcium by the ATP dependent process, then one might
expect that lanthanum will accelerate considerably the release of
ACh with little or no action on the vesicular population at least
in the first hours of lanthanum action.

The Endo-Exocytotic Cycle

As shown by many contributors, stimulation leads to the interior-
ization of extracellular markers in the vesicle lumen. This certainly demonstrates endocytosis (8-10, 23,51). It is probable that
at the frog neuromuscular junction, exocytosis of vesicular contents
takes place at specialized areas where one finds the double row of
vesicles facing the postsynaptic folds (23,24). The formation of
vesicles appears to be balanced by their utilization, explaining
that during stimulations, investigators have found either an increase or a decrease, or simply no change. The fact that it was
possible to trigger the vesicle cycle independently of quantal ACh
secretion is of great interest, as we have already pointed out
(see 8).

Stimulation in the presence of 4-aminopyridine triggers the
occurrence of many pits at the active zone (24), but these pits were
not found if stimulation took place in the absence of 4-aminopyridine.

Moreover, the pits appeared later than one would have expected from the latency and rising phase of the endplate potential. We should bear in mind that probably an important calcium entry took place in the presence of 4-aminopyridine. If vesicles close to the membrane have sequestered calcium, they might have to get rid of it perhaps by exocytosis. We certainly do not exclude the possibility that these vesicles also contain ACh, but from other experiments we have seen, that vesicles which presumably did not contain ACh could be utilized without any occurrence of mepps. This was indeed the case in the experiments using black widow spider venom after a stimulation in the presence of hemicholinium. The possibility that I would like to present is that the endoexocytotic cycle might reflect for its exocytotic part the removal of calcium sequestered in vesicles which might not contain ACh, these vesicles could either be new endocytotic vesicles or vesicles which have delivered their content to the cytoplasm. To switch off ACh release, cytoplasmic calcium has to be cleared, vesicles close to the presynaptic membrane might be particularly well situated to take up the entered calcium.

Evolution of Vesicles
in the Course of Synaptic Activity

Several sequences of events could be presented to envisage how the stores of vesicular ACh and ATP will contribute to the release process:

1.   Stimulation triggers the entry of calcium in the nerve terminal.

2.   Calcium activates the release mechanism plugged in the presynaptic membrane.

3.   Cytoplasmic transmitter is released.

4.   Ch and acetic acid also (in neuromuscular systems) are recycled.

5.   Synthesis of ACh occurs in the cytoplasmic pool at the expense of ATP.

6.   Calcium is also cleared at the expense of ATP in order to switch off the release process. The synaptic vesicles take up that calcium, being in a better   position to do so than the reticulum or mitochondria.

7.   For longer stimulations, calcium clearance and ACh synthesis which consume ATP, lead to the increase of calcium or of some other metabolite (H^+ ?) in the cytoplasm.

8.   Vesicular ACh and ATP now feed the cytoplasm. The release mechanism can go on, while ATP supports the synthesis of ACh but also the clearance of calcium.

9.   Vesicles have lost ACh and ATP but gained calcium, they will perhaps every now and then exocyte it, or it might diffuse out from the vesicle and be extruded by the membrane Na/Ca exchange

system.   The diffusion being slow, the cytoplasmic Ca concentration remains low.

10.   Metabolic recovery will fill the cytoplasmic ACh and ATP pools.

11.   New vesicles formed by endocytosis will concentration ACh and ATP by a process depending on the charge and composition of their matrix.

We speculated about an alternative physiological role for vesicles. Is it possible to envisage the mechanism of ACh release in non-vesicular terms? The present status of the vesicle hypothesis was discussed in a recent review (27).

The Mechanism of ACh Release

Should we envisage a single mechanism for quantal and non-quantal ACh release? Or is it more convenient to consider that ACh leakage (48) is independent from the secretion of packets? Certainly many discrepancies would be explained by the existence of two or more mechanisms for ACh release. One mechanism would let cytoplasmic ACh leak out, while the other would deliver packets of ACh from different populations of vesicles. Giant mepps, normal mepps or sub-mepps (39) resulted from different vesicular ACh contents.

If, on the other hand, we consider that it is premature to envisage two mechanisms of ACh release, then we are left with a complicated form of the vesicle hypothesis which is only able to explain transmitter leakage by the occurrence of permeability areas in the presynaptic membrane, due, perhaps, to the incorporation of the vesicle membrane when exocytosis takes place.

All this is certainly possible, but it is more simple to consider that since most of the vesicular pool remains stable while the free pool undergoes important changes during stimulation, that the mechanisms of ACh release is not directly dependent on vesicles. This stability of bound ACh is accepted, since under physiological conditions no fall of vesicular ACh can be elicited by stimulation.

"Carriers" releasing cytoplasmic ACh, either in molecular form or multimolecular packets, when polymerized would be an interesting possibility, but carriers are thought to be slow. Most probably protein molecules form under the effect of calcium transient channels through the membrane. These channels permit the release of ACh in molecular form from the cytoplasm, if oligomers of these proteins are formed, packets of transmitter will appear. A good example of such a mechanism is obtained when gramicidine molecules are included

in an atrificial membrane, and a quantal transfer of ions is elicited.
Certainly the polymorphic aspect of ACh packets would be explained
by the transient channel hypothesis. The objection about the non-
electrogenic nature of ACh release is not an important one since it
is possible to consider that a counter ion (why not ATP?) can be released with ACh. Perhaps the release of ACh is accompanied by the
influx of positive charges. The channel or gate hypothesis considered in parallel to the classical vesicular hypothesis was discarded
previously because of the high amounts of ACh associated with
synaptic vesicles, and perhaps we should carefully reconsider it.
The transient channel hypothesis would be most adequate if it could
be established beyond any doubt that free ACh is a real cytoplasmic
compartment.

A final comment should be made about the finding that upon
stimulation, ATP appears in the extracellular fluid. The release
of ATP is a consequence of pre- and postsynaptic ion fluxes or depolarization (see 34). Extracellular ATP probably reduces the entry
of calcium triggered by stimulation and contributes to regulation
of the release process.

From the very first contribution on release until the most
recent experiments, ATP is found in the shadow of ACh, and it is
probable that synaptic mechanisms would be greatly clarified if
such a relationship could be considered as an essential regulation.

REFERENCES

1.   Babel-Guerin, E. (1974): J. Neurochem. 23:525-532.
2.   Birks, R.I. (1974): J. Neurocytol. 3:133-160.
3.   Blaustein, M.P., Ratzlaff, R.W., Kendrick, N.C. and Schweitzer,
       E.S. (1978): J. Gen. Physiol. 72:15-41.
4.   Blaustein, M.P., Ratzlaff, R.W. and Schweitzer, E.S. (1978):
       J. Gen. Physiol. 72:43-66.
5.   Boyne, A.F., Bohan, T.P. and Williams, T.H. (1974): J. Cell
       Biol. 63:780-795.
6.   Boyne, A.F., Bohan, T.P. and Williams, T.H. (1975): J. Cell
       Biol. 67:814-825.
7.   Breer, H., Morris, J.J. and Whittaker, V.P. (1977): Eur. J.
       Biochem. 80:313-318.
8.   Ceccarelli, B. and Hurlbut, W.P. (1975): J. Physiol. (Lond)
       247:163-168.

9.   Ceccarelli, B., Hurlbut, W.P. and Mauro, A. (1972): J. Cell
     Biol. 54:30-38.
10. Ceccarelli, B., Hurlbut, W.P. and Mauro, A. (1973): J. Cell
     Biol. 57:499-524.
11. Collier, B. (1969): J. Physiol. (Lond) 205:341-352.
12. Colquhoun, D., Large, W.A. and Rang, H.P. (1977): J. Physiol.
     (Lond) 266:361-395.
13. Droz, B., Rambourg, A. and Koenig, H.L. (1975): Brain Res.
     93:1-13.
14. Dunant, Y., Eder, L. and Servetiadis-Hirt, L. (1980): J. Physiol.
     (Lond) 298:185-198.
15. Dunant, Y., Gautron, J., Israel, M., Lesbats, B. and Manaranche,
     R. (1972): J. Neurochem. 19:1987-2002.
16. Dunant, Y., Gautron, J., Israel, M., Lesbats, B. and Manaranche,
     R. (1974): J. Neurochem. 23:635-643.
17. Dunant, Y. and Israel, M. (1979): TINS:130-132.
18. Dunant, Y., Israel, M., Lesbats, B. and Manaranche, R. (1977):
     Brain Res. 125:123-140.
19. Fonnum, F. (1966): Biochem. Pharmacol. 15:1641-1643.
20. Fonnum, F. (1967): Biochem. J. 103:262-274.
21. Gautron, J. (1978): Biol. Cell. 31:31-44.
22. Hata, F., Kuo, C.H., Matsuda, T. and Yoshida, H. (1976): J.
     Neurochem. 27:139-152.
23. Heuser, J.E. and Reese, T.S. (1973): J. Cell Biol. 57:315-344.
24. Heuser, J.E., Reese, T.S., Dennis, M.J., Jan, L. and Evans, L.
     (1979): J. Cell Biol. 81:275-300.
25. Israël, M. (1970): Arch. Anat. Microsc. 59:67-98.
26. Israël, M. (1972): Actual. Pharmacol. 25:1-22.
27. Israël, M., Dunant, Y. and Manaranche, R. (1979): Prog. Neuro-
     biol. 13:237-247.
28. Israël, M. and Gautron, J. (1969): IN Symposium International
     Society for Cell Biology, Vol. 8, Academic Press, New York,
     pp. 137-152.
29. Israël, M., Gautron, J. and Lesbats, B. (1968): C.R.H. Acad.
     Sci, Paris 266:273-275.
30. Israël, M., Gautron, J. and Lesbats, B. (1970): J. Neurochem.
     17:1441-1450.
31. Israël, M., Hirt, L. and Mastour-Frachon, P. (1973): C.R.H.
     Acad. Sci. Paris 276:2725-2728.
32. Israël, M., Lesbats, B., Manaranche, R., Marsal, J., Mastour-
     Frachon, P. and Meunier, F.M. (1977): J. Neurochem. 28:1259-
     1267.
33. Israël, M., Manaranche, R., Marsal, J., Meunier, F.M., Morel,
     N., Lesbats, B. and Frachon, P. (1980): J. Memb. Biol. 54:
     115-120.
34. Israël, M. and Meunier, F.M. (1978): J. Physiol. (Paris) 74:
     485-490.
35. Israël, M. and Tucek, S. (1974): J. Neurochem. 22:487-493.
36. Kazuaki, O., Dowe, H.E., Morris, S.J. and Whittaker, V.P. (1979):
     Brain Res. 161:447-460.

37. Kuo, C.H., Hata, F., Yoshida, H., Yamatodami, A. and Hiroshi,
         W. (1979): Life Sci. $\underline{24}$:911-921.
38. Kriebel, M.E. (1978): Brain Res. $\underline{148}$:381-388.
39. Kriebel, M.E. and Gross, C.E. (1974): J. Gen. Physiol. $\underline{64}$:
         85-103.
40. Large, W.A. and Rang, H.P. (1978): J. Physiol. (Lond) $\underline{285}$:1-24.
41. Large, W.A. and Rang, H.P. (1978): J. Physiol. (Lond) $\underline{285}$:25-34.
42. Marchbanks, R.M. and Israel, M. (1971): J. Neurochem. $\underline{18}$:439-
         448.
43. Marchbanks, R.M. and Israel, M. (1972): Biochem. J. $\underline{129}$:1049-
         1061.
44. O'Hara, P.T., Wade, C.R. and Lieberman, A.R. (1979): J. Anat.
         $\underline{129}$:869-884.
45. Pinchasi, I. and Michaelson, D.M. (1979): FEBS Lett. $\underline{108}$:189-
         191.
46. Politoff, A.L., Rose, S. and Pappas, G.D. (1974): J. Cell Biol.
         $\underline{61}$:818-823.
47. Rahamimoff, H. and Abramovitz, E. (1978): FEBS Lett. $\underline{89}$:223-226.
48. Vizi, E.S. and Vyskocil, F. (1979): J. Physiol. (Lond) $\underline{286}$:1-14.
49. Wagner, J.A., Carlson, S.S. and Kelly, R.B. (1978): Biochem-
         istry $\underline{17}$:1199-1206.
50. Whittaker, V.P., Essman, W.B. and Dowe, G.H.C. (1972): Biochem.
         J. $\underline{128}$:833-846.
51. Zimmermann, H. and Denston, C.R. (1977): Neuroscience $\underline{2}$:695-714.
52. Zimmerman, H. and Whittaker, V.P. (1974): J. Neurochem. $\underline{22}$:
         435-450.
53. Zimmerman, H. and Whittaker, V.P. (1974): J. Neurochem. $\underline{22}$:
         1109-1114.

ACETYLCHOLINE CHANGES DURING TRANSMISSION

OF A SINGLE NERVE IMPULSE

Y. Dunant, J. Corthay, L. Eder and F. Loctin

Department of Pharmacology, C.M.U.
1211 Geneva 4, Switzerland

INTRODUCTION

Transmission of nerve impulses at cholinergic synapses takes place in a matter of milliseconds. This extreme rapidity has made the investigation of synaptic mechanisms by means of biochemical methods very difficult. The present study is an attempt to overcome this difficulty by looking for correlations between the electrophysiological and the biochemical analysis of transmission. All experiments were carried out on the electric organ of _Torpedo marmorata_.

MATERIALS AND METHODS

Acetylcholine Release

Slices of electric organ were excised from the anesthetized fish and small fragments containing one or two prisms (stacks of electroplaques) were carefully dissected. The acetylcholine (ACh) in the tissue was partially labelled with choline (Ch) and acetate at the same molar concentration but labelled with different isotopes. Under these conditions the two precursors were incorporated into ACh in a ratio 1 to 1. After a single electrical stimulus, or a brief burst of stimuli, the compound electroplaque potential (epp) was recorded and the radioactive Ch and/or acetate counted in the perfusion fluid, providing a sensitive assay for ACh release in the absence of anticholinesterase drugs (2).

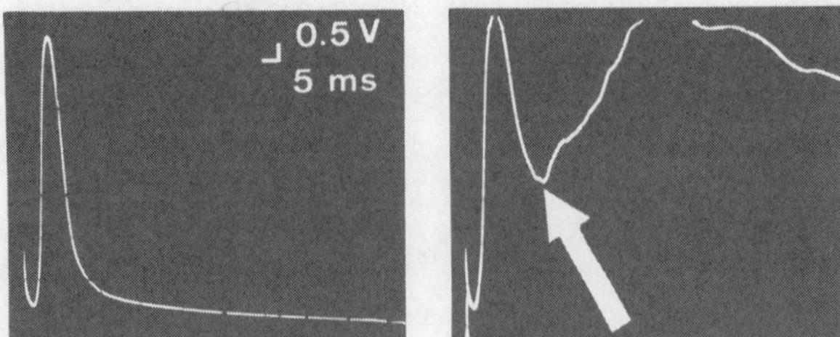

FIGURE 1: Electroplaque potential (epp) generated by an isolated prism in response to a single stimulus. The two records were taken before (left hand picture) and during (right hand picture) a freezing run. The arrow indicates the precise time of contact with the freezing fluid.

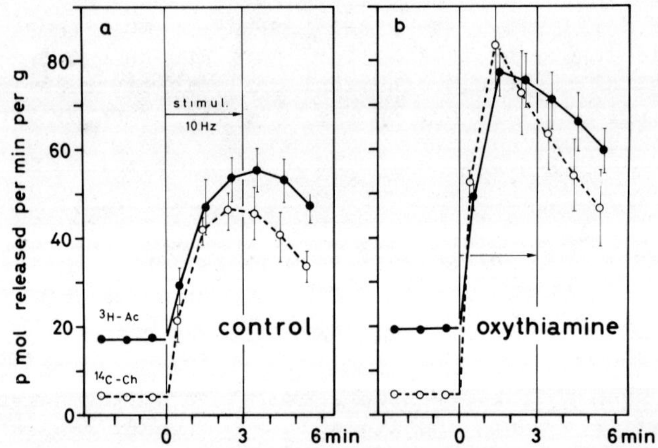

FIGURE 2: ACh release in response to a repetitive stimulation at 10 Hz for 3 min. The overflow of ^{3}H-acetate and ^{14}C-Ch was measured in the absence of anticholinesterase drug. In Fig. 2b, 10^{-4} M oxythiamine significantly potentiated the evoked ACh release

Measurements of Tissue Acetylcholine
at Short Time Intervals

We designed and used a machine which allows 24 tissue samples
to be stimulated with various patterns and quickly plunged into isopentane at −130°C. Up to 24 prisms excised from the electric organ
were placed on their respective electrodes on a tissue plunger.
After an initial test of the epp, the prisms were left for at least
1 hr at rest in the physiological medium to which drugs were added
when desired. The tissue plunger was then lifted upwards and
maintained by a magnetic lock. When the machine was started, stimulation was applied first to the sample number 1, and then succes-
sively to the others with a given time interval (10 msec to several
seconds). At the precise time desired, the magnetic lock liberated
the plunger which fell into isopentane. The stimulation pattern
continued during the descent. Usually the last samples were stimulated when already frozen and thus served as additional controls.
The epp of one of the 24 samples could be recorded during the run
and allowed for a precise measurement of the time to contact with
isopentane (see Fig. 1). The frozen samples were transferred into
liquid nitrogen and broken into two fragments. The first fragment
was finely pulverized in a ceramic motar with 5% (w/v) TCA for extraction of the total ACh. The second fragment served for the
determination of vesicle bound ACh. It was first pulverized in the
frozen state, and then homogenized at 5-10°C in the physiological
saline using a Potter device. Only at this stage TCA was added to
the isoosmotic homogenate. Synaptic vesicles were isolated by the
method of Israel at al. (8) with modification due to the fact that
we started with frozen tissue (see 20).

RESULTS AND DISCUSSION

Acetylcholine Release

Figure 2 gives an illustration of a typical experiment in which
the ACh in electric organ was labelled with ^{3}H-acetate and ^{14}C-Ch.
On stimulation, the two precursors were released in a roughly 1 to
1 ratio. In other experiments, the number of shocks was reduced in
the stimulation and it was possible to compare epp and ACh release
even in response to a single stimulus. A close correlation could
be established by this method between the biochemical and electrophysiological release in a number of conditions (calcium and temperature dependency, progressive depression of transmission in
series of successive stimuli, etc; see Ref. 2). In the present
study we want to emphasize the pharmacology of transmitter release
as studied in this purely cholinergic system.

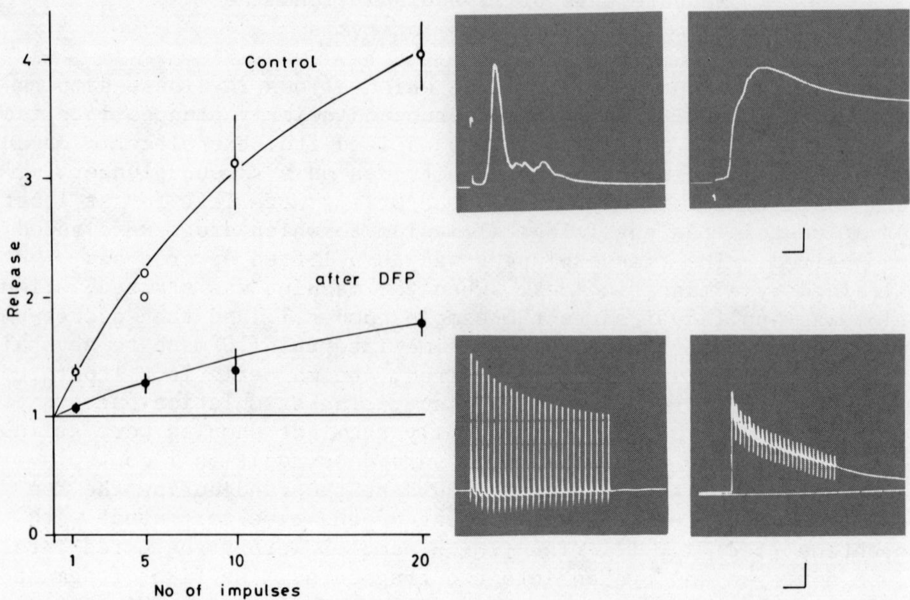

FIGURE 3: Effects of cholinesterase inhibition on transmitter release and epp. Pretreatment with DFP irreversibly abolished cholinesterase activity and resulted in a prolonga-
tion of the epp duration and a strong reduction of ACh
release, even in response to a single stimulus. The release function has been expressed as the ratio of the
peak of evoked release over the background radioactivity.
Calibration 0.5 V, and 5 ms (single impulse),200 ms
(20 impulses).

The most inexpected finding was that anticholinesterase drugs
greatly reduced evoked ACh release, chiefly in response to a single
stimulus (Fig. 3 and Table 1). Indeed, since the pioneer work of
Dale, Feldberg and coworkers, the release of ACh has been almost
always measured after complete inhibition of the tissue cholinesterases. ACh release was studied either in the presence of
eserine or after 1 hr of pretreatment with fluostigmine (DFP). We
determined that these drugs inhibited by more than 99%   the cholinesterase activity of the electrogenic tissue. As found by Feldberg and Fessard (6), anticholinesterases greatly prolonged the
duration of a single epp but reduced the ability of the organ to
sustain repetitive activity (Fig. 3, records). The release of
radiolabelled transmitter was strongly reduced by anticholinesterases.
The effect was prominent in the release due to a single stimulus.

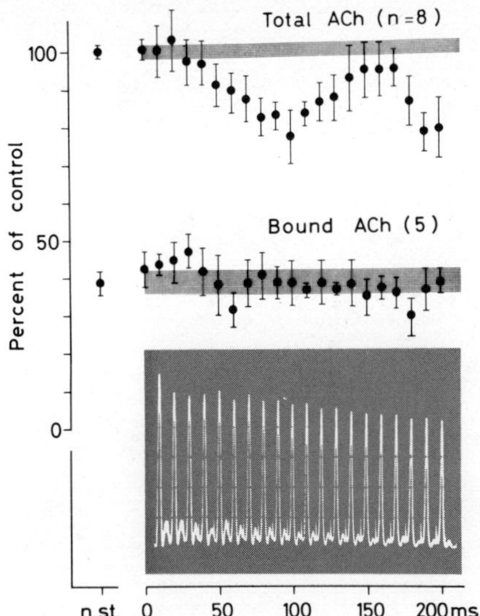

FIGURE 4: Changes in tissue ACh and giant epp observed in response
to a single nerve stimulus, given in the presence of
10^{-4} M 4-aminopyridine.

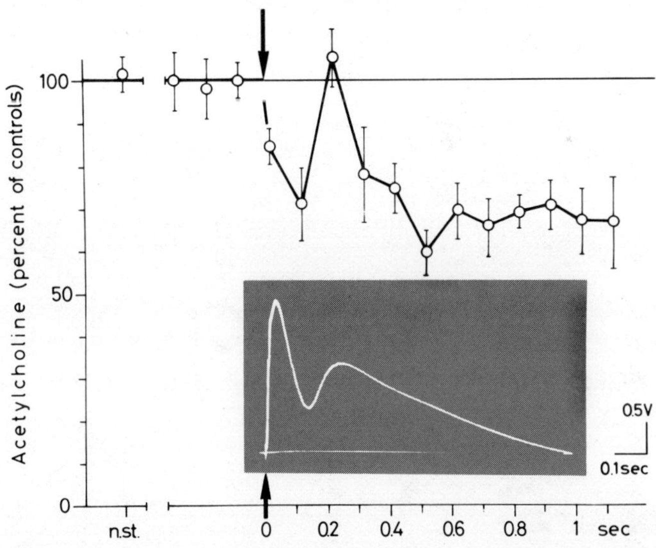

FIGURE 5: Changes in total ACh and vesicle-bound ACh during transmission of a brief burst of 20 impulses at 100 Hz. The
time intervals for ACh analysis was 10 ms, so that each
point in time corresponds to transmission of one more
impulse. No drug present.

TABLE 1: Pharmacology of ACh release in electric organ.

Substances	Precursors	Spontaneous Release	Number of Stimuli	Change in ACh Release (percent of control)	
Eserine (10^{-3})	^{3}H–Ac	+	20	7	(3)
(10^{-4})	^{3}H–Ac	0	20	25 ± 5	(9)
(10^{-4})	^{3}H–Ac	0	KCl	71	(3)
Eserine (10^{-4}) + α–bungarotoxin (7.10^{-6})	^{3}H–Ac + ^{14}C–Ch	0	20	23 ± 6	(6)
Eserine (10^{-4}) + tubocurarine* (10^{-4})	^{3}H–Ac	0	20	45	(3)
DFP (pretreatment 10^{-3})	^{3}H–Ac	0	1	17 ± 5	(6)
	^{3}H–Ac	0	5	25 ± 4	(6)
	^{3}H–Ac	0	10	18 ± 8	(6)
	^{3}H–Ac	0	20	25 ± 5	(10)
	^{14}C–Ac + ^{3}H–Ch	0	1800	52	(2)
	^{3}H–Ac	0	KCl	100	(1)
DFP (then tubocurarine* (10^{-4})	^{3}H–Ac	0	20	28 ± 5	(7)
Tubocurarine* (10^{-4})	^{3}H–Ac	0	20	82 ± 4	(3)
	^{14}C–Ch + ^{3}H–Ac	0	1800	89	(2)
Tubocurarine** (10^{-4})	^{3}H–Ac	0	20	31 ± 6	(8)
α–Bungarotoxin (7.10^{-6})	^{3}H–Ac	0	20	104	(3)
Carbachol (10^{-3})	^{3}H–Ac	++	20	9	(3)

Adenosine (10^{-3})	^{3}H–Ac	0	20	25 ± 4	(6)
	^{14}C–Ac + ^{3}H–Ch	0	1800	60 ± 7	(5)
	^{3}H–Ac	0	20	25 ± 4	(6)
Theophylline (10^{-3})	^{3}H–Ac	0	20	59	(3)
Tubercidine (10^{-4})	^{3}H–Ac	++	20	9	(2)
(10^{-3})	^{3}H–Ac	+++	20	25 ± 11	(5)
Adenosine (10^{-3}) + tubercidine (10^{-3})	^{3}H–AC	0	20	32	(3)
Adenosine (10^{-3}) + theophylline (10^{-3})	^{3}H–Ac	0	20	34	(3)
Thiamine (10^{-4})	^{3}H–Ch	0	1	130 ± 9	(4)
(10^{-3})	^{3}H–Ch	0	1	84 ± 8	(4)
(5.10^{-3})	^{3}H–Ch + ^{14}C–Ac	0	1800	65	(2)
Oxythiamine (10^{-4})	^{14}C–Ch	0	1	170 ± 10	(4)
(10^{-4})	^{3}H–Ch + ^{14}C–Ac	0	1800	164 ± 16	(5)
(10^{-3})	^{14}C–Ac	0	1800	300	(2)
Quinine (10^{-6})	^{3}H–Ac	0	20	103	(3)
(10^{-5})	^{3}H–Ac	0	20	67	(3)
(10^{-4})	^{3}H–Ac	+	20	4 ± 4	(7)
(10^{-4})	^{3}H–Ac	+	KCl	15	(3)
Quinidine (10^{-4})	^{3}H–Ac	+	20	0	(4)
Ouabaine (10^{-4})	^{3}H–Ac	++	20	0	(2)
4-Amino-pyridine					
(10^{-6})	^{3}H–Ac	0	1	197	(2)
(10^{-5})	^{3}H–Ac	0	1	1697 ± 303	(6)
(10^{-4})	^{3}H–Ac	0	1	3763 ± 853	(6)
(10^{-3})	^{3}H–Ac	0	1	6830	(3)

[FOR LEGEND SEE FOLLOWING PAGE]

LEGEND: TABLE 1
   The spontaneous releave of radioactivity is expressed: 0 =
   unchanged; + = slightly increased; +++ = greatly increased.
   Number of stimuli in 1 sec; 1800 = stimulation at 10 Hz for
   3 min; KCl = depolarization by 50 mM KCl.
   100% of control means no change in the stimulation evoked
   release. Means $\pm$ S.E.M. Number of samples in parentheses.
   Concentrations indicated are molarity.
   * = tubocurarine from Vifor; ** = tubocurarine from Sigma
   Ac = acetate; Ch = choline

TABLE 2:   Radioactivity of total ACh and vesicular ACh after 4-AP
           and rapid freezing.

	ACh (nmol/g)	Radioactive ACh (kBq/g)	Specific Radioactivity (Bq/nmol)
Total ACh	346 $\pm$ 13	11.8 $\pm$ 1.0	34
Bound ACh	264 $\pm$ 7	4.5 $\pm$ 0.3	18.4
Free (cytoplasmic)ACh	82	7.3	88.6
Synaptic vesicle ACh	9.15	0.17	18.6

When more impulses were delivered in the stimulation, anticholinesterases inhibited ACh release to a smaller extent. Only a
slight effect, if any, was observed when stimulation was made by
increasing the KCl concentration (Fig. 3 and Table 1).

The anticholinesterase effect was not due to some special
change in the radioactive pools of ACh, since neither the radioactive ACh nor the endogenous ACh were reduced in the tissue after
DFP treatment. A similar depression of ACh release by anticholinesterase drugs has been reported by Szerb and Somogyi (19),
working with mammalian brain slices. This effect of anticholinesterase drugs may perhaps be explained by the retro-inhibition of
transmission by ATP, recently proposed by Meunier et al. (14) and
Israel et al. (9,11). The ACh released at rest, being no longer
hydrolysed, would act on postsynaptic receptors, depolarizing the
electroplaques and causing the release of ATP. This postsynaptic

ATP, in turn, would act on the nerve endings by reducing further
release of ACh. In the work of Szerb and Somogyi (19), atropine
did restore the efficiency of the release process; this speaks in
favor of the above explanation, as did the recent experiments of
Miledi et al (16), who showed that the release of ACh in the DFP
treated rat disphragm was increased after treatment with α-bungarotoxin. We tried similar experiments with the electric organ of
Torpedo. Used alone, α-bungarotoxin (kindly supplied by Dr. Fulpius)
had no effect on ACh release at a concentration causing a complete
blockade of transmission. Apparently, this toxin had a purely
postsynaptic action. After treatment by α-bungarotoxin the depressive action of eserine was not significantly antagonized (Table
1). A similar conclusion was reached with curare. In fact, different effects were found when we used tubocurarine from different
sources. The drug provided by Vifor Company (Geneva) only caused
a 10-20% decrease of transmitter release, but tubocurarine of the
Sigma Company surprisingly caused a strong inhibition of release
(about 70% at the concentration which fully blocked transmission).
Thus this preparation of tubocurarine by itself had a marked presynaptic action and did not reverse depression due to anticholinesterases.

Adenosine and adenosine nucleotides also reduced ACh release
(see also 9,11). Theophylline had a depressive effect and did not
antagonize the adenosine action. Tubercidine by itself triggered
a spontaneous release of radiolabelled transmitter and this spontaneous release was antagonized by adenosine. Adenosine is transported into the nerve terminals of the Torpedo (15) and probably
acts on the "excitation-contraction" coupling of transmitter release. As for anticholinesterase, the adenosine effect was less
pronounced in prolonged stimulation and in KCl induced release.
These effects of adenosine and derivatives have also been studied
in great detail on other systems (see for example, Ref. 18 for
the brain).

Among the other drugs tested, some increased the spontaneous
ACh release but depressed the evoked ACh release (ouabain), others
mainly depressed the evoked release (quinine, quinidine). An
enormous potentiation was found in this tissue using 4-amino-
pyridine (4-AP). Also at the neuromuscular junction, the anti-
curare action of 4-AP was shown to be caused by a great potentiation
of the evoked "quantal" release (17). This drug is an interesting
tool which causes a quasi explosive liberation of transmitter (see
Ref. 7). The corresponding changes in tissue ACh will be described
in the next section.

Thiamine and its antivitamin oxythiamine also affected ACh
release. The action of oxythiamine was surprising since it caused
an approximately twofold increase of transmitter release (Fig. 2

and Table 1). This potentiation was seen in the epp which was not greatly increased in amplitude but presented a characteristic shoulder in its descending phase. Thiamine itself affected transmission in a more complicated manner, causing an increase in ACh release at 10^{-4} M and a decrease at higher concentrations (5). These findings may be of some functional interest since thiamine is very abundant in the electric organ, mainly in the form of phosphoric acid esters. The vitamin seems to be especially concentrated in the nerve terminals (3,4).

It may be concluded that this method enables one to closely compare electrophysiological and biochemical estimates of transmitter release, especially when the stimulation consists of a single nerve impulse.

Changes in Tissue Acetylcholine
Measured at Very Short Time Intervals

Our rapid freezing device was used to analyze changes in ACh and other components of the tissue in two conditions of stimulation which were 1) the giant potential elicited by a single stimulus in the presence of 4-AP and 2) a short burst of impulses at 100 Hz in the absence of any drug.

Figure 4 shows the giant epp due to 4-AP (10^{-4} M) and the corresponding changes of the total tissue ACh. The electrophysiological discharge was enormously enlarged. Soon after the unique stimulus (arrow), there was a first great peak that was followed by a late and prolonged rebound. The amount of total tissue ACh decreased by 25-30% during the first phase of the giant epp, that is, during the first 100 msec. Then, a surprising rapid and large ACh increase occurred transiently around 200 msec. This late ACh increase corresponded to and explained the rebound of the epp. The good agreement between the shape of the giant epp, and changes of tissue ACh was confirmed and analyzed in four additional experiments in which ACh was determined at shorter intervals (30 msec). The "rebound" of ACh was constantly found. In fact, it seemed to slightly precede in time the rebound of the epp.

Vesicular bound ACh was also measured in six experiments under the same conditions of stimulation (a single stimulus in the presence of 4-AP). The level of bound ACh did not exhibit any change parallel to those of total ACh but it remained stable during and after the giant epp, up to 1.5 sec after the stimulus. It was, therefore, the extravesicular pool of ACh (most probably cytoplasmic ACh) which was concerned in the early stage of activity.

Of course, the fact that the level of vesicular ACh remained constant might mean that cytoplasmic ACh very rapidly refills the vesicles. This was directly tested by labelling the ACh in the cytoplasm and the vesicles. As shown in Table 2, the specific activity of the vesicle-bound ACh was identical to that found in the vesicles isolated and purified by a density gradient centrifugation. In contrast, cytoplasmic ACh had a much higher radioactivity, so that any transfer from this pool to vesicular ACh could be easy to detect. Thus it was confirmed, using our conditions of labelling, 4-AP treatment and rapid freezing, that the bound ACh corresponds to the transmitter contained in synaptic vesicles (12). However, an unexplained finding was that the yield of total ACh and bound ACh, and the recovery of synaptic vesicles were lower after 4-AP treatment.

In four subsequent experiments, we measured radioactive and non-radioactive ACh at 0.1 sec intervals during the course of a giant 4-AP discharge. Changes in total ACh were similar to those described above, but vesicle bound ACh remained stable over 1.7 sec, with regard to both, the yield in nmol and the radioactivity. This result, although preliminary, clearly showed that no rapid transfer of cytoplasmic ACh occurred to vesicular ACh, nor to vesicular sub-fractions during the giant potential due to 4-AP.

In the second type of stimulation (Fig. 5), no drug was present and the tissue was given a short burst of impulses at 100 Hz, a pattern which resembles the normal discharge used by the fish in the sea for attack or defense. During the first 10 impulses of the burst, the level of total ACh decreased at a rate of about 2% of the inital store per impulse. Then a transient ACh resynthesis occurred with a surprising rapidity. These ACh changes, therefore, resembled those observed during the 4-AP giant potential. However, the difference was that no obvious correspondence was found in the 100 Hz burst between ACh changes and the size of the epp.

Vesicular bound ACh was also measured in the 100 Hz burst (Fig. 5). Once again, no significant decrease was observed from the control level over the 200 msec of the stimulation. Moreover, in three preliminary experiments, no sign of rapid transfer of cytoplasmic ACh to the vesicle population was found.

CONCLUSION

1.   It has been possible to detect significant changes in the
level of tissue ACh at very short time intervals during transmission
of nerve impulses. The limit of 10 msec, reached in the present
work for the time resolution of biochemical analysis, may theoretically still be improved, at the price, however, of greatly reducing
the size of tissue samples.

2.   The reason for observing such rapid changes of ACh stores
in this tissue is the extremely rapid action of acetylcholinesterase
(see 13). ACh is destroyed in a matter of milliseconds soon after
its release from nerve endings. The total amount of transmitter
released during a 4-AP giant potential can theoretically be hydrolysed in approximately 60 msec. Thus all the ACh found in the tissue when it is rapidly "quenched" for extraction can be regarded as
intracellular and most probably intraterminal.

3.   It was most surprising to find a net ACh resynthesis as
early as 200 msec after the onset of activity. This raised a difficult problem since the maximum velocity of choline acetyltransferase measured in electric organ homogenates is much too slow to
fit with this phenomenon. It may well be that the _in vitro_ kinetics
can hardlybe compared to the changes occurring _in situ_ because of
the spatial organization of enzymes in the synapses.

4.   Vesicular ACh remained stable in these very early stages
of synaptic activity. Moreover, there was no indication of any
rapid transfer to vesicular ACh. It can, therefore, be concluded
that it was the extravesicular pool of ACh, most probably cytoplasmic ACh, which was used and renewed during the course of the
giant epp and also during transmission of the short burst of impulses, a pattern very similar to the normal electrical discharge
of the fish. This conclusion confirmed and strengthened previous
studies in which the ACh pools were labelled by radioactive precursors and repetitively stimulated for several minutes (1).

5.   The mechanism of ACh release can, therefore, only be
explained by a membrane operator, able to bind the newly synthesized
axoplasmic ACh and to release it into the synaptic cleft. A functional definition of such a mechanism was recently proposed (10)
and discussed in the light of experimental data. The present results, obtained with a rapid freezing technique, are in good agreement with this proposal.

REFERENCES

1.   Dunant, Y., Gautron, J., Israel, M., Lesbats, B. and
     Manaranche, R. (1972): J. Neurochem. 19:1987-2002.
2.   Dunant, Y., Eder, L. and Servetiadis-Hirt, L. (1980): J. Phy-
     siol. Lond. 298:185-192.
3.   Eder, L., Dunant, Y. and Baumann, M. (1978): J. Neurocytol.
     7:637-645.
4.   Eder, L. and Dunant, Y. (1980): J. Neurochem. 35:1278-1286.
5.   Eder, L., Dunant, Y. and Loctin, F. (1980): J. Neurochem.
     35:1287-1296.
6.   Feldberg, W. and Fessard, A. (1942): J. Physiol. Lond. 101:
     200-216.
7.   Heuser, J.E., Reese, T.S., Dennis, M.J., Jan, Y., Jan, L. and
     Evans, L. (1979): J. Cell Biol. 81:275-300.
8.   Israel, M., Gautron, J. and Lesbats, B. (1968): C.R. Acad.
     Sci. Paris 266:273-275.
9.   Israel, M., Lesbats, B., Manaranche, R., Marsal, J., Mastour-
     Frachon, P. and Meunier, F.M. (1977): J. Neurochem. 28:
     1259-1267.
10. Israel, M., Dunant, Y. and Manaranche, R. (1979): Prog. Neuro-
     biol. 13:237-275.
11. Israel, M., Lesbats, B., Manaranche, R., Meunier, F.M. and
     Frachon, P. (1980): J. Neurochem. 34:923-932.
12. Marchbanks, R.M. and Israel, M. (1971): J. Neurochem. 18:
     439-448.
13. Marnay, A. (1937): C.R. Soc. Biol. Paris 126:573-574.
14. Meunier, F.M., Israel, M. and Lesbats, B. (1975): Nature 257:
     407-408.
15. Meunier, F.M. and Morel, N. (1978): J. Neurochem. 31:845-851.
16. Miledi, R., Molenaar, P.C. and Polak, R.I. (1978): Nature
     272:641-642.
17. Molgo, J., Lemeignan, M. and Lechat, P. (1975): C.R. Acad.
     Sci. Paris 281:1637-1638.
18. Phillis, J.W., Edstrom, J.P., Kostopoulos, G.K. and Kirkpatrick,
     J.R. (1979): Canad. J. Physiol. Pharmacol. 57:1289-1312.
19. Szerb, J.C. and Somogyi, G.T. (1973): Nature New Biol. 241:
     121-122.
20. Whittaker, V.P., Essman, W.B. and Dowe, G.H.C. (1972): Bio-
     chem. J. 128:833-846.

UPTAKE OF ACETYLCHOLINE INTO <u>TORPEDO</u> SYNAPTIC VESICLES <u>IN</u> <u>VITRO</u>

Y.A. Luqmani and P. Giompres

Abteilung Neurochemie, Max-Planck-Institut fur
Biophysikalische Chemie, Gottingen, FR Germany

INTRODUCTION

The discovery that brain synaptic vesicles contained acetylcholine (ACh)(6) seemed to confirm the earlier predictions of del
Castillo and Katz (3) concerning the quantized nature of ACh released at the neuromuscular junction. Subsequently, with improvements in subcellular fractionation techniques, estimates for intravesicular ACh concentration have been continually revised upwards,
and were recently reported to be as high as 0.8 M (8) in Torpedo
vesicles.

The absence of choline acetyltransferase (CAT) from synaptic
vesicle preparations suggests the cytoplasm as the source of vesicular ACh. But despite numerous demonstrations of <u>in situ</u> labelling of vesicular ACh (11), the general failure to observe a similar uptake by isolated vesicles, has given rise to much speculation
concerning the role of vesicles in transmitter release (7).

Recent evidence (2) and that presented in this paper, however,
contradicts previous studies, and shows that purified <u>Torpedo</u> vesicles can accumulate transmitter <u>in vitro</u>.

MATERIALS AND METHODS

Synaptic vesicles were isolated from <u>Torpedo marmorata</u> by a
modification of an earlier method (1) with purification steps involving a linear iso-osmotic glycine/dextran gradient and column

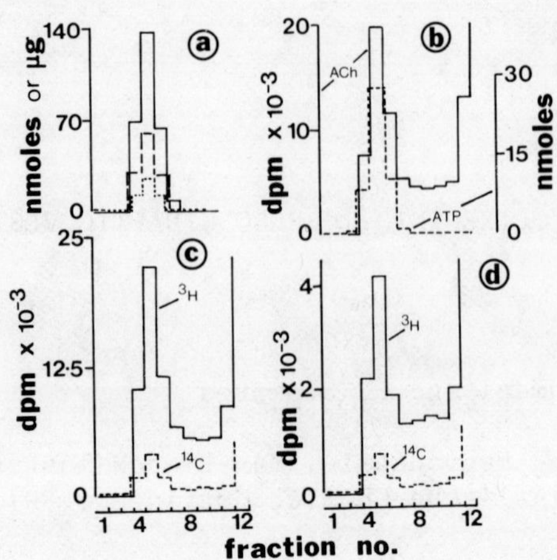

FIGURE 1: Separation of synaptic vesicles by gel filtration. Following incubation particulate material was separated on a 1.5 cm x 30 cm column of coarse Sephadex G-50, eluted with the same medium as used for incubation. (A) shows a typical elution profile with coincident peaks of protein (---), ATP (......) and bioassayable ACh (——) appearing in the void volume. Noe that the ratio ACh/ATP remains constant through fractions 4-7. (B) shows the elution profile of vesicles previously incubated with [^{3}H]-ACh. A peak of label, separated from the bulk of the radioactivity, appears synchronously with the endogenous ATP. Similarly, after a double labelling experiment, both [^{3}H]-ACh (——) and [^{14}C]-Ch (---) were eluted in the same fractions as containing vesicles (C), but the levels of both labels were reduced by about 80% when the column was eluted with water to cause hypo-osmotic lysis (D). The levels of endogenous ATP and bioassayable ACh were also considerably reduced, although protein content was unaltered (not shown, for clarity). In (A), ordinate units are nmol for ACh (——) and ATP (......) and µg (---) for protein.

---

chromatography using porous glass beads (CPG-10-3000, 120-300 mesh). Column eluted vesicles were concentrated by centrifugation on a 1 ml cushion of iso-osmotic sucrose, and used immediately for uptake studies. Incubation was normally for 1 hr at 26°C in 0.64 M sucrose, 5 mM Hepes (pH 7) in the presence of radiolabel, and was

terminated by passage through a column of Sephadex G-50. Fractions
(2 ml) were collected and vesicles located by their content of
adenosine 5'-triphosphate (ATP).

RESULTS

Vesicle Preparation: Purity and Stability

Electron microscopy showed a homogeneous vesicle population
with very little membrane contamination. ACh content was 2.35 µmol
mg^{-1} protein and the ACh/ATP ratio was 5.6 Incubation at 26°C for
1 hr did not result in significant loss of vesicular contents; ATP
was reduced by about 10% after 5 hr storage at 4°C.

Vesicular Uptake of Labelled ACh and Ch

Figure 1 shows the result of Sephadex gel filtration of vesicles under iso-osmotic conditions, eluting in the void volume, as
evinced by co-incident peaks of protein, ACh and ATP. Incubation
with [^{3}H]-ACh resulted in the appearance of label in the vesicle
peak, separated from the bulk of radioactivity. [^{14}C]-Ch was also
incorporated. Separation under hypo-osmotic conditions, showed
that the radioactivity was contained within an osmotically sensitive particle. The identity of vesicular ACh was verified by paper
chromatography, after ion exchange extraction.

Time Course of Uptake

The accumulation of both Ch and ACh was much slower than the
rapid uptake observed with catecholamines (9). Uptake of ACh was
still increasing after 2 hr incubation; for Ch, a more rapid initial
phase preceded a plateauing by 2 hr, and this was unaffected by
the presence of ACh.

Concentration Dependence of Uptake

Up to concentrations of 8 mM, uptake of ACh increased linearly
at 0°C. Only when this was subtracted from the uptake at 26°C was
a clear saturation observed, indicating that two processes were
involved at the higher temperature. Figure 2 shows the kinetics of

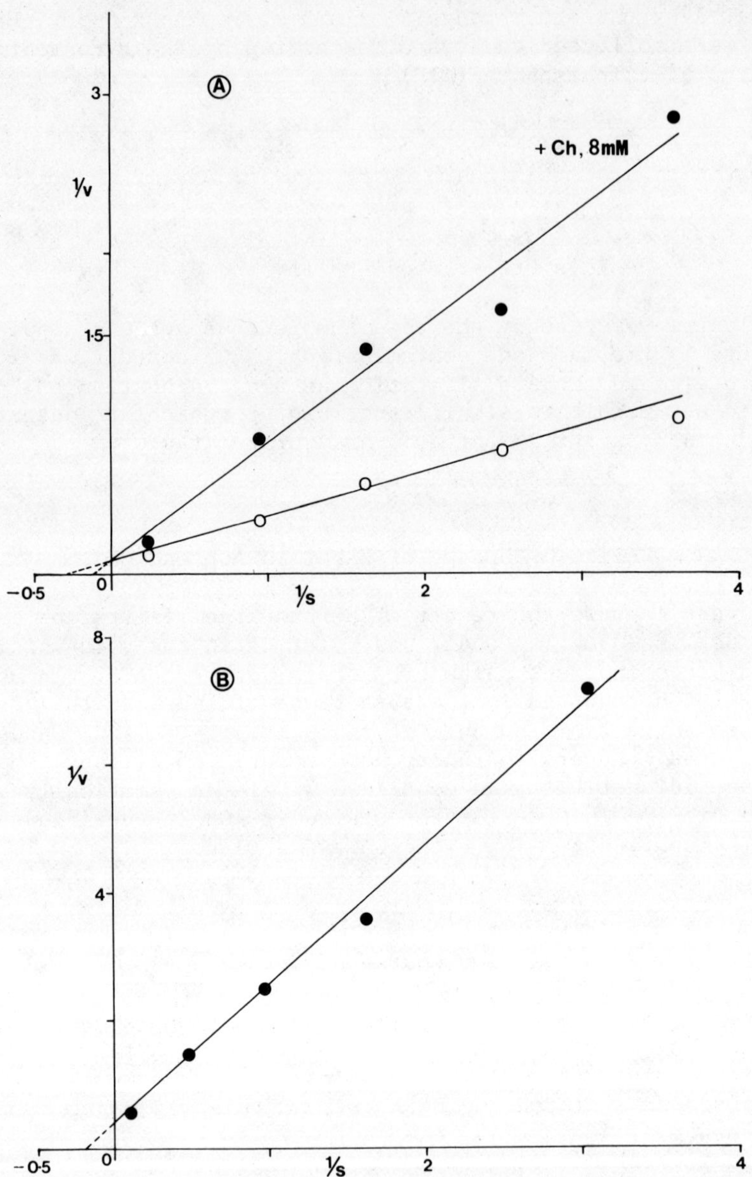

FIGURE 2: Effect of concentration on vesicular uptake of labelled
Ch and ACh. Vesicles (approximately 40-70 μg protein) were
incubated at 0°C or 26°C for 1 hr in the presence of a
fixed amount of [³H]-ACh or [³H]-Ch and the concentration
varied by addition of appropriate quantities of unlabelled
compound. Eserine (0.5 mM) was added to all samples containing ACh. Uptake at 0°C (similar to that at 26°C in
the presence of 0.5 mM HC-3) was subtracted from that

FIGURE 2 legend (continued)

at 26°C and the data transformed into a double reciprocal Lineweaver-Burke plot to determine the Michaelis-Menten kinetic constants. a) ACh was taken up with K_T 3.13 mM, V_m 30.86 nmol. mg protein^{-1}. hr^{-1} and competitively inhibited by Ch with K_{Tapp} 8.33 mM. b) Ch was also transported with K_T 9.99 mM, V_m 28.57 nmol. mg protein^{-1}. hr^{-1}. Abscissa are 1/M x 10^{-7} and coordinates 1/pmol. hr^{-1} x 10^{-7} (10^{-8} for a). Data are results of 4-5 experiments.

---

the saturable ACh uptake and its inhibition in a competitive manner by Ch. The latter was also transported with a similar V_{max} but a much higher K_T (Fig. 2). In comparison, uptake of glucose was 15- to 30-fold less, and remained linear with respect to concentration at both temperatures.

Hemicholinium as a Specific Inhibitor of Uptake

Hemicholinium-3 is a potent blocker of presynaptic Ch transport (4); it was also effective in blocking the vesicular process. Ch uptake was 45-55% less in the presence of the inhibitor (0.5 mM), whether it was added alone, with ACh, or with glucose. Similarly, uptake of ACh was also substantially inhibited (68%). The non-saturable uptake seen at 0°C, however, was not affected by hemicholinium-3. The accumulation of glucose was also not modified; this was true over the range of inhibitor concentrations used (up to 2 mM).

Do Vesicles Concentrate ACh?

The vesicular concentration of labelled substrates was calculated using theoretically derived values for the intra-vesicular volume (based on estimates of ATP concentration). From the proportion of radiolabelled substrate taken up, both Ch and ACh were accumulated to a level which was 3- to 6-fold above that in the medium. Similar calculations for intravesicular glucose and 2-deoxy-D-glucose showed that their accumulation did not exceed a quarter of their medium concentration over 1 hr.

CONCLUSIONS

1.   Isolated _Torpedo_ synaptic vesicles retain a saturable mechanism for the accumulation of ACh; Ch is a competitive substrate for this uptake process _in vitro_.

2.   The process has some similarities to the presynaptic Ch uptake system.

3.   Vesicle refilling could occur from a permanent or transient cytoplasmic pool of ACh, which may reach millimolar concentrations.

4.   The high K_T for ACh may explain why the recovery of vesicular ACh is extremely slow following stimulation induced tissue depletion in _Torpedo_ (10) despite much earlier and almost total recovery of vesicle numbers.

REFERENCES

1.   Breer, H., Morris, S.J. and Whittaker, V.P. (1977): Eur. J. Biochem. 80:313-318.

2.   Carpenter, R.S. and Parsons, S.M. (1978): J. Biol. Chem. 253: 326-329.

3.   Del Castillo and Katz, B. (1956): Prog. Biophys. Chem. 6: 121-170.

4.   Dowdall, M.J. (1977): IN The Biochemistry of Defined Neuronal Systems (ed) N.N. Osborne, Pergamon Press, Oxford.

5.   Fonnum, F. (1968): Biochem. J. 109:389-398.

6.   Hebb, C.O. and Whittaker, V.P. (1958): J. Physiol. 142:187-196.

7.   MacIntosh, F.C. and Collier, B. (1976): IN Handbook of Experimental Pharmacology (eds) E. Zaimis and J. MacLagen, Springer-Verlag, Berlin.

8.   Ohsawa, K., Dowe, G.H.C., Morris, S.J. and Whittaker, V.P. (1979): Brain Res. 161:447-457.

9.   Slotkin, T.A., Seidler, F.J., Whitmore, W.L., Lau, C., Salvaggio, M. and Kirksey, D. (1978): J. Neurochem. 31:961-968.

10.   Suszkiw, J.B. and Whittaker, V.P. (1979): Prog. Brain Res. 49: 153-162.

11.   Zimmermann,H. and Whittaker, V.P. (1977): Nature 267:633-635.

A VESICULAR SITE OF ORIGIN FOR THE RELEASE OF A CHOLINERGIC FALSE TRANSMITTER AT THE <u>TORPEDO</u> SYNAPSE

Y.A. Luqmani and V.P. Whittaker

Abteilung Neurochemie, Max-Planck-Institut fur
Biophysikalische Chemie, Gottingen, FR Germany

INTRODUCTION

The morphological studies of Ceccarelli, Heuser (2,6) and others
(for review see 14) have shown convincingly the increased occurrence of exocytotic profiles (of vesicles fusing with the presynaptic membrane) during induced activity at the neuromuscular
function. Vesicle recycling is apparent from the inclusion of extracellular markers, observed in a number of tissues (5,11,15), but
this does not necessarily prove that transmitter-containing vesicles
undergo a similar process. Similarly, biochemical studies involving
a comparison of the degree of radiolabelling of acetylcholine (ACh)
in subcellular pools with that of stimulus-released transmitter, has
also been subject to major difficulties of interpretation.

Some of these limitations can be overcome by a simultaneous
study of the distribution of ACh and of drugs which act as cholinergic false transmitters. An analysis of the ratio of true to false
transmitter in different subcellular fractions provides a sensitive
method of identifying the site of origin of released transmitter.
This also avoids the need to separate metabolically distinct vesicle
populations (12,15), the precise composition of which may be controversial. A number of recent studies have shown that at least 5 compounds act as precursors to cholinergic false transmitters (Table 1).

This report describes a study of the storage and release of
acetyltriethylcholine (acetyl TEC) as a false transmitter at the
<u>Torpedo</u> electromotor synapse.

TABLE 1:   Summary of work done on cholinergic false transmitters.

Precursor	Abbreviation	Preparations tested	References
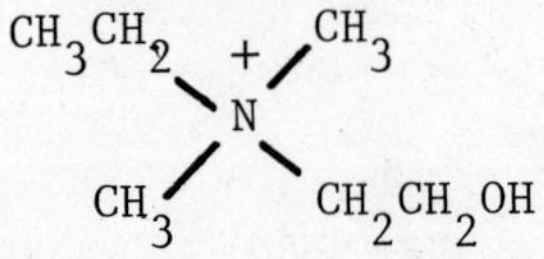 CH_3CH_2 CH_3 / N^+ / CH_3 CH_2CH_2OH	MEC	SCG EO D	3 Unpubl. Data 1,9
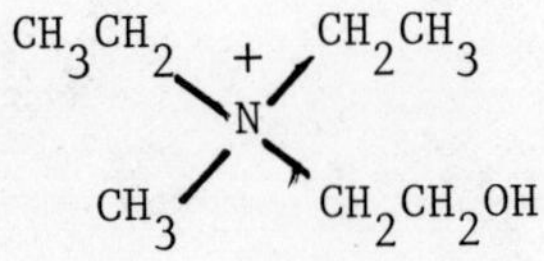 CH_3CH_2 CH_2CH_3 / N^+ / CH_3 CH_2CH_2OH	DEC	EO	Unpubl. Data
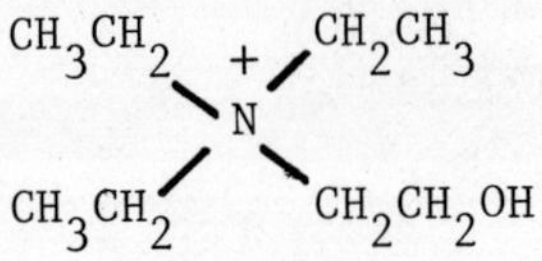 CH_3CH_2 CH_2CH_3 / N^+ / CH_3CH_2 CH_2CH_2OH	TEC	SCG EO	7 This paper
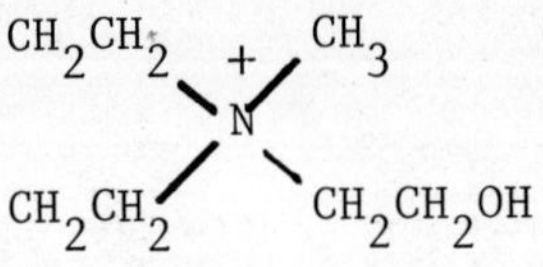 CH_2CH_2 CH_3 / N^+ / CH_2CH_2 CH_2CH_2OH	PC	SCG EO MP BC	3 16 8 12,13
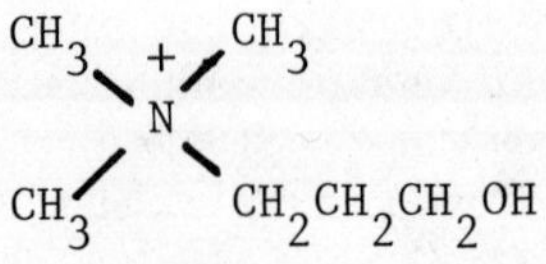 CH_3 CH_3 / N^+ / CH_3 $CH_2CH_2CH_2OH$	HC	SCG EO BC	6 10,13 12,13

Abbreviations of preparations tested: SCG = cat superior cervical
ganglion; BC = guinea pig brain cortex; EO = _Torpedo_ electric organ;
D = rat diaphragm; MP = myenteric plexus of small intestine of
guinea pig.

METHODS

All experiments were performed on isolated electric organs of
female _Torpedo marmorata_ supplied by the Institut de Biologie Marine,
Arcachon, France.   Innervated tissue blocks were perfused and stimulated essentially as described previously (10) and vesicles prepared
from them subsequently.   Experimental details are given in the
legend of Fig. 1.

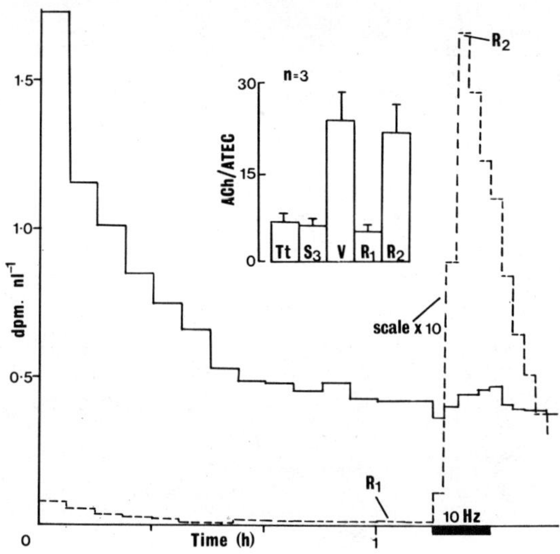

FIGURE 1: Release of acetyl TEC and ACh from an innervated, perfused block of electric tissue from <u>Torpedo</u> <u>marmorata</u>.
The tissue block corresponding to the territory of one
electromotor nerve and its accompanying blood vessels
was perfused with Torpedo Ringer solution containing TEC
(250 µM), Ch (25 µM) and [^{3}H]-acetate (16.7 µM, specific
radioactivity 300 Ci/mol) for 3 hr. Stimulation was then
applied at 0.1 Hz for 1 hr to promote incorporation of
newly synthesized transmitter into tissue stores. After
2 hr recovery in label and 70 min washout perfusion with
radioactivity-free Ringer solution containing paraoxon
(100 µM) to inhibit AChE the tissue was restimulated at
10 Hz for 10 min (release stimulus) while perfusion was
continued. The continuous line shows the total concentration of [^{3}H]-acetate (free and esterified) in the perfusate, the broken line, the concentration of [^{3}H]-ace-
tate (as esterified organic bases) passing into the organic phase when samples of perfusate were extracted
with allylcyanide containing sodium tetraphenylboride
(10 mg/ml). Note that the scale of extractable counts
is 10 times greater than that of total counts and that as
a result the concentration of acetylated bases in the
perfusate increased enormously from a very low baseline.
The insert diagram shows the ratio of ACh to acetyl TEC
(ATEC)in total tissue (Tt), a crude vesicle pellet (V),
in the perfusion luid before (R$_1$) and during (R$_2$) the
release stimulus and in the "cytoplasmic" fraction (S$_3$).
The extracted acetate esters were seperated by paper chromatography. Note the close agreement in the ratio of ACh
to ATEC in the perfusate after the stimulated release of
transmitter with that in vesicles, even though this ratio
is almost 5 times higher than found in whole tissue.

RESULTS

Triethylcholine: Precursor to a False Transmitter

Tissue blocks were perfused with [^{3}H]-choline (Ch) and [^{14}C]-TEC for 3 hr, and then given 360 pulses at 0.1 Hz, and allowed to recover for 2 hr, after which excess label was washed out in the presence of an anticholinesterase (paraoxon, 100 µM). Subsequent high frequency stimulation (10 Hz) resulted in rapid release of both labels; ^{3}H appeared predominantly as ACh and ^{14}C as acetyl TEC, showing that TEC has given rise to a probable false transmitter. A sharp increase in the ^{3}H to ^{14}C ratio indicated preferential release of the normal transmitter, probably due to its greater incorporation into the releasable transmitter pool, relative to acetyl TEC. The release of both compounds was shown to be Ca^{++}-dependent; inclusion of EGTA in the perfusion medium abolished response to a second stimulus.

Comparison of Tissue and Perfusate Ratos of True and False Transmitter

The tissue distribution of ACh and acetyl TEC was studied using [^{3}H]-acetate to label both transmitters; this method avoided complications arising from the additional consideration of labelled precursors. Experimental details are given in the legend to Fig. 1. Following washout with acetate free Ringer containing paraoxon, the organ block was restimulated at 10 Hz. The resulting increase in total radioactivity was due to the acetylated bases which were subsequently extracted from the perfusate using a liquid ion exchange system, and identified as a mixture of ACh and acetyl TEC. Thus, stimulation released relatively large amounts of transmitters previously present in the perfusion fluid in extremely low concentrations. Control blocks, not receiving the release stimulus, were used for subcellular fractionation. The inset in Fig. 1 shows that the molar ratio of ACh/acetyl TEC in the whole tissue (Tt) was very similar to that in the extra-vesicular "cytoplasm" fraction (S$_3$), and to that in the prestimulation perfusion fluid (R$_1$). The vesicle fraction (V), however, had a ratio several times greater, and was very close to that in the poststimulation perfusate (R$_2$).

Control Experiments

Blocks were perfused with intra- (2-deoxy-D-[1-^{3}H]-glucose) and extracellular ([^{14}C]-inulin, [^{14}C]-sorbitol) markers, and it was found that high frequency stimulation did not cause increased output of any of these substances, but did initiate the expected release of endogenous transmitter. In other experiments, utilizing 2-deoxy-D-[1-^{3}H]-glucose and [^{14}C]-Ch, vesicles isolated on a sucrose/ glycine linear gradient were found to contain the ^{14}C label but no ^{3}H.

CONCLUSIONS

1.   TEC is a precursor to a cholinergic false transmitter: acetyl TEC.
2.   It is likely that ACh and acetyl TEC are stored in the same vesicle population, but acetyl TEC is a much poorer substrate for vesicular uptake and also accumulates in the cytoplasm.
3.   Leakage of both transmitters from the cytoplasm probably accounts for their "resting release."
4.   The stimulus induced, Ca^{++}-dependent, release of both true and false transmitter occurs directly from a vesicular pool

It is concluded that vesicular storage is a prerequisite for the normal release of transmitter at the cholinergic terminals of Torpedo, and is likely to be true also for other synapses mediated by ACh.

REFERENCES

1.   Calquhoun, D., Large, W.A. and Rang, H.P. (1977): J. Physiol. 266:361-395.
2.   Ceccarelli, B., Hurlbut, W.P. and Mauro, A. (1972): J. Cell Biol. 57:499-524.
3.   Collier, B., Barker, L.A. and Mittag, T.W. (1976): Molec. Pharmacol. 12:340-344.
4.   Collier, B., Lovat, S., Ilson, D., Barker, L.A. and Mittag, T.W. (1977): J. Neurochem. 28:331-339.
5.   Fried, R.C. and Blaustein, M.P. (1976): Nature 261:255-256.
6.   Heuser, J. and Reese, T.S. (1973): J. Cell Biol. 57:315-344.
7.   Ilson, D. and Collier, B. (1976): Nature 254:618-620.
8.   Kilbinger, H. (1977): N.-S. Arch. Exp. Path. Pharmak. 296: 153-158.
9.   Large, W.A. and Rang, H.P. (1978): J. Physiol. 275:61-62P.
10. Luqmani, Y.A., Sudlow, G. and Whittaker, V.P. (1980): Neuro-science 5:153-160.
11. Model, P.G., Highstein, S.M. and Bennett, M.V.L. (1975): Brain Res. 98:209-228.
12. von Schwarzenfeld, I. (1979): Neuroscience 4:477-493.

13. von Schwarzenfeld, I., Sudlow, G. and Whittaker, V.P. (1979):
         Prog. Brain Res. 49:163-174.
14. Zimmermann, H. (1979): Neuroscience 4:1773-1804.
15. Zimmermann, H. and Denston, C.R. (1977): Neuroscience 2:715-730.
16. Zimmermann, H. and Dowdall, M.J. (1977): Neuroscience 2:731-739.

TRANSMEMBRANE POTENTIAL AND CATION DIFFUSION GRADIENTS IN ISOLATED
CHOLINERGIC SYNAPTIC VESICLES: POSSIBLE MODEL FOR ENERGIZATION OF
VESICULAR ACETYLCHOLINE UPTAKE

J.B. Suszkiw

Physiology Section, Biological Sciences Group
University of Connecticut, Storrs, Connecticut 06268 USA
and: Department of Neurochemistry, Max-Planck-Institut
fur Biophysikalische Chemie, Gottingen, FR Germany

INTRODUCTION

Recent experiments with Torpedo electroplaque synapses lend
support for the idea that in activated terminals synaptic vesicles
are mobilized into a pool of active vesicles which may undergo
several cycles of acetylcholine (ACh) release and reuptake (4,5,7).
The active but not the unmobilized depot vesicles become highly
charged with the newly synthesized ACh, suggesting that the exocytotic event somehow energizes these vesicles for subsequent uptake
of ACh from cytoplasm. A clue to the possible nature of this process was provided by the observation that gramicidin induces an
electrogenic influx of choline (Ch) or ACh into isolated vesicles
(1). This communication presents some experimental evidence consistent with the hypothesis that the exocytotic release of ACh and
the concomitant imposition of ΔNa^+ across the vesicle membrane due
to a rapid equilibration between the intravesicular and extracellular medium may be a precondition for energization of ACh uptake
into vesicles.

METHODS

Synaptic vesicles were isolated from Torpedo or Narcine electroplaques and were purified by two consecutive sucrose density step

TABLE 1: The relation between accumulation of $[^3H]$-Ch and $\Delta\Psi$ estimated from distribution of $[^3H]$-TPMP in isolated synaptic vesicles.

Vesicle Preparation	$\dfrac{[^3H]\text{-TPMP}_{in}}{[^3H]\text{-TPMP}_{out}}$	$\Delta\Psi = 58\ \log$	$\dfrac{[\text{TPMP}]_{in}}{[\text{TPMP}]_{out}}$	$\dfrac{[^3H]\text{-Ch}_{in}}{[^3H]\text{-Ch}_{out}}$
Fresh	7.1		49.4	8.3
Aged: 12 hr, 2°C or: fresh + 1 µM gramicidin	73.3		108.2	78

Synaptic vesicles (Narcine) were: a) extracted in 0.4 N NaCl-5 mM Hepes-1 mM EGTA, pH 7; b) purified in NaCl sucrose density gradients; c) concentrated by pelleting at 100,000 g; d) resuspended in 0.8 M Gly-5 mM hepes, pH 7.2. Steady state distribution of $[^3H]$-TPMP or $[^3H]$-Ch was determined as described in Methods. Vesicle protein: 15-30 µg/0.2 ml of incubation medium; $[^3H]$-TPMP$_{out}$: 7 µM; $[^3H]$-Ch$_{out}$: 0.12 µM. Fresh vesicles: uptake determined soon after resuspension of purified vesicles; aged vesicles: resuspended vesicles were left overnight at 2°C. Results are means of two or more determinations.

gradients, similar to the procedure used in (6). Three different types of extraction media were used: NaCl (Na vesicles), KCl (K vesicles) and Gly (Gly vesicles); all were 800 mOsM and were buffered with 5 mM Hepes, pH 7. The ATP content of purified vesicles ranged from 200 to 400 nmol/mg protein, and ACh ranged from 1000 to 2000 nmol/mg protein. Vesicles were concentrated by pelleting at 100,000 g and then were resuspended either in 0.8 M Gly or 0.4 M Lys HCl buffered with 5 mM Hepes-Tris.

Steady state distribution (1 hr incubation at 20°C) of $[^3H]$-triphenylmethylphosphonium·Br (TPMP), K$[^{14}C]$-SCN, $[^{14}C]$-methylamine·HCl, and $[^3H]$-Ch·Cl was determined by adaptation of the methodology described in (2), using ^3H_2O and $[^{14}C]$-dextran for determination of the intra- and extravesicular water space, respectively.

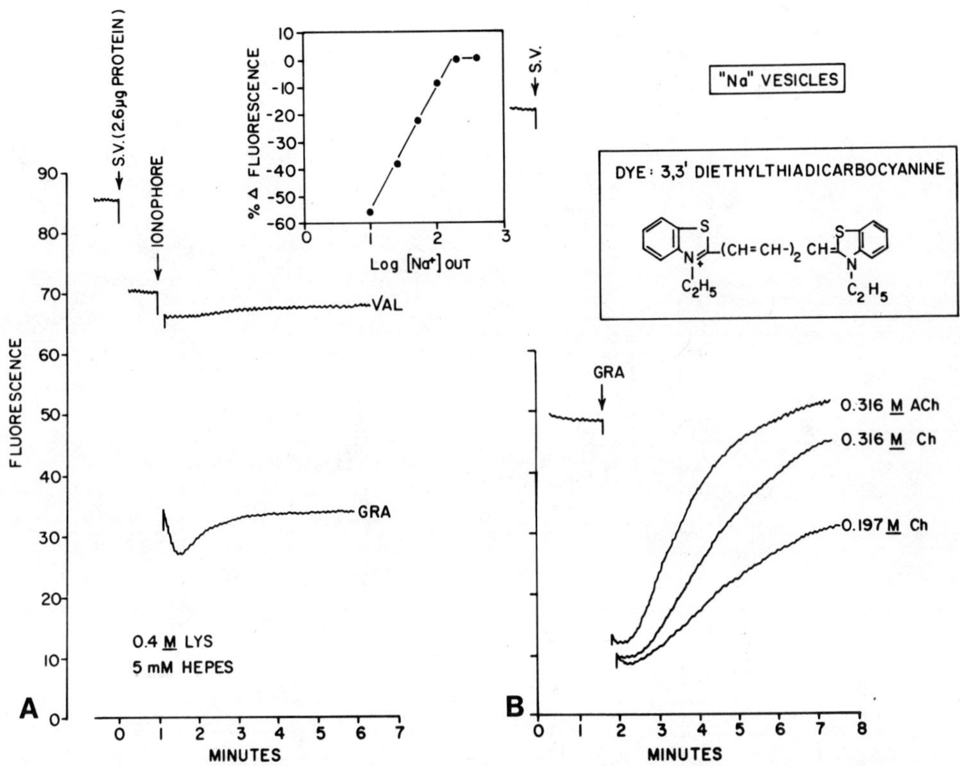

FIGURE 1: A: – Effect of gramicidin D and valinomycin on binding of
               diS-C$_2$[5] to synaptic vesicles (Torpedo) isolated in
               Na-rich media. Inset: gramicidin induced change in
               fluorescence as function of [Na$^+$]$_o$ (Results corrected
               for vesicle independent quenching of diS-C$_2$[5] fluo-
               resence by gramicidin).
           B: – Gramicidin induced electrogenic influx of Ch or ACh.
               Osmolarity and ionic strength in A and B were main-
               tained constant by replacing Lys-HCl with equivalent
               concentrations of NaCl, Ch·Cl, or ACh·Cl. diS-C$_2$[5],
               5 µM; gramicidin, 1.4 µg/ml; valinomycin, 0.6 µg/ml.

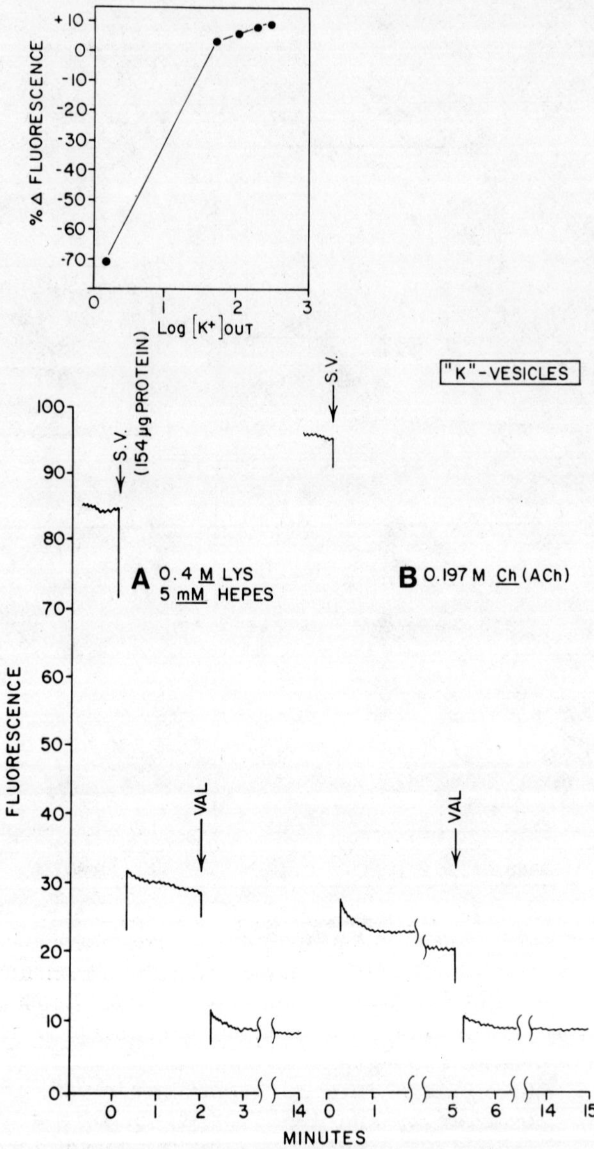

FIGURE 2: A: – Effect of valinomycin on binding of diS–C$_2$[5] to K-
vesicles (Torpedo). A small vesicle hyperpolariza-
tion is observed provided sufficiently concentrated
suspension of vesicles is used. Inset: valinomycin
induced change in fluorescence as function of [K$^+$]$_o$.
B: – Valinomycin induced vesicles hyperpolarization does
not appear to generate electrogenic influx of Ch
or ACh. diS–C$_2$[5], 5 μM; valinomycin, 0.6 μg/ml.

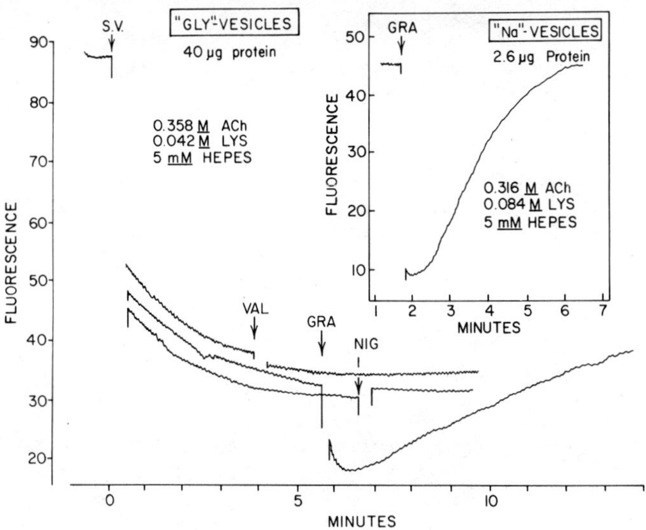

FIGURE 3: Interaction of diS-C$_2$[5] (5 µM) with Gly vesicles (Torpedo) in ACh, in presence of gramicidin (1.4 µg/ml), valinomycin (0.6 µg/ml) and nigericin (0.6 µg/ml).

$[\text{Tracer}]_{\text{in}}/[\text{Tracer}]_{\text{out}}$ was calculated from:

$$\frac{\dfrac{[\text{DPM Tracer}]_{\text{Pellet}}}{[\text{DPM Tracer}]_{\text{sup}}} - \dfrac{[\text{DPM Dext}]_{\text{pellet}}}{[\text{DPM Dext}]_{\text{sup}}}}{\dfrac{[\text{DPM }^3\text{H}_2\text{O}]_{\text{pellet}}}{[\text{DPM }^3\text{H}_2\text{O}]_{\text{sup}}} - \dfrac{[\text{DPM Dext}]_{\text{pellet}}}{[\text{DPM Dext}]_{\text{sup}}}}$$

Fluorescence measurements were done on Perkin-Elmer MPF-2 or MPF-3 spectrophotofluorometers. Pelleted vesicles were resuspended in 0.1 to 0.4 ml of 0.4 M lys HCl – 5 mM Hepes-Tris and were added to thermostated cuvettes (15°C) containing media of desired composition. The membrane potential was monitored with 3,3'diethyl-thiadicarbocyanine (diS-C$_2$[5]) and ΔpH was monitored with 9-amino-acridine.

RESULTS

Transmembrane potential estimated from the [^{3}H]-TPMP distribution (3) in freshly resuspended, isolated Na vesicles was approxi-

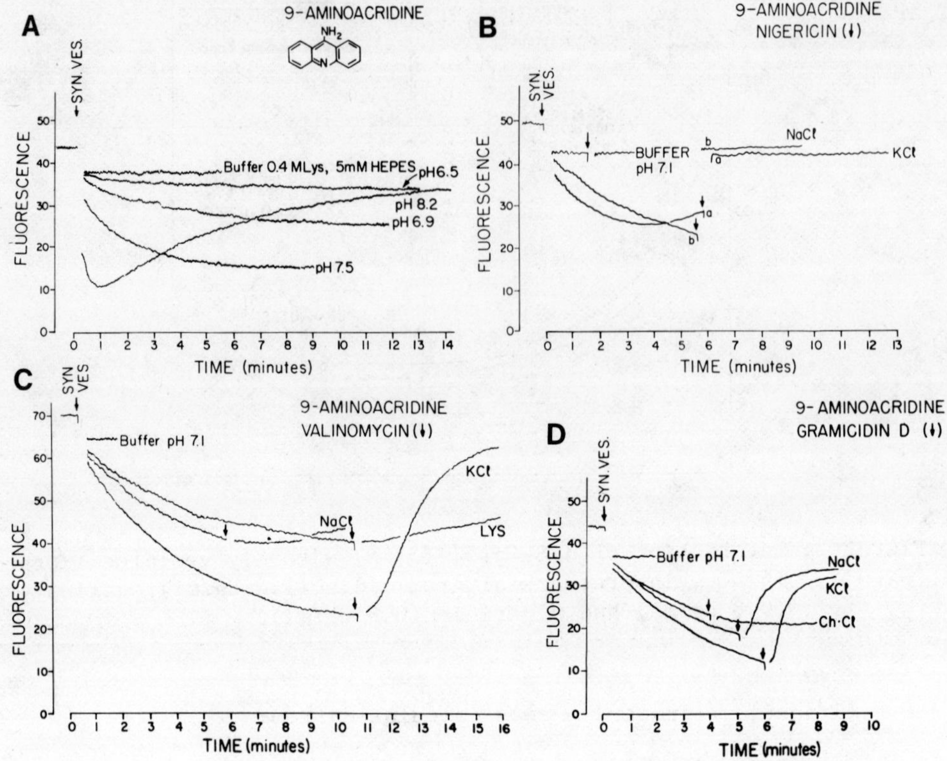

FIGURE 4: A: – pH sensitive binding of 9-aminoacridine (0.6 μM) to synaptic vesicles (3.8 μg protein/ml). Gly vesicles (Torpedo).
B: – Effect of nigericin (0.35 μg/ml) in NaCl or KCl medium.
C: – Effect of valinomycin (0.7 μg/ml) in NaCl or KCl medium.
D: – Effect of gramicidin (1.7 μg/ml) in NaCl, KCl or Ch·Cl.
All solutions 0.4 M 5 mM hepes.

---

mately 50 mV, negative inside, and increased to approximately 100 mV in vesicles left in suspension overnight at 2°C or when 1 μM gramicidin (but not valinomycin) was added to a suspension of fresh vesicles (Table 1), thus suggesting that loss of Na^+ was responsible for vesicle hyperpolarization. As $\Delta\psi$ increased, the accumulation of Ch also increased from approximately 8 x in fresh to approximately 80 x in aged vesicles. $[^{14}C]$-SCN^- was completely excluded from the vesicles.

In agreement with the steady state measurements, the addition of gramicidin to the Na vesicles caused their hyperpolarization, as

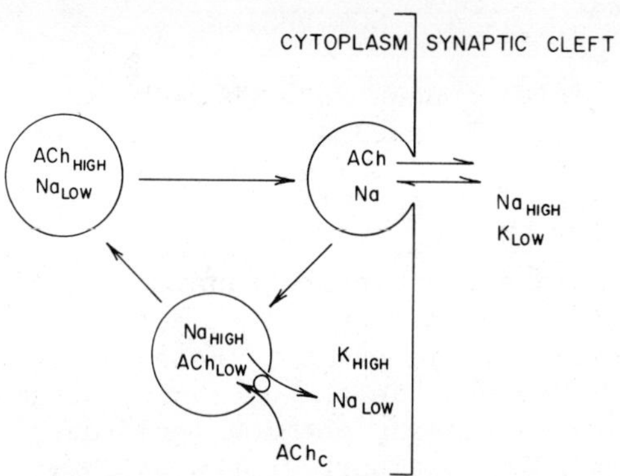

FIGURE 5: A model of _in situ_ activation of vesicular ACh uptake:
a) during exocytosis vesicles lose ACh and gain Na^+;
b) ΔNa^+ is established in the reformed, ACh depleted
vesicles; and c) dissipation of ΔNa^+ provides the driving
force for translocation of ACh against its chemical
gradient.

---

monitored by the instantaneous quenching of $diS-C_2[5]$ fluorescence.
This hyperpolarization was then followed by depolarization if Ch or
ACh were present in the medium (Fig. 1). Valinomycin had no effect
on the Na vesicles but caused hyperpolarization of the K vesicles
(Fig. 2). However, in this case the electrogenic   of Ch (ACh)
into vesicles was not detected, probably because the established K
diffusion potential was of insufficient magnitude to provide the
driving force for optically detectable translocation of Ch (ACh).
As expected, vesicles isolated in Gly (nominally Na^+-K^+-free medium)
did not respond to valinomycin, but gramicidin still induced an
influx of Ch (ACh) although the effect was greatly reduced relative
to Na vesicles (Fig. 3). Since the magnitude of vesicle hyperpolarization in this case was similar to that produced by valinomycin on K vesicles, the observed stimulation of Ch influx suggests a more specific role for Na^+.

The ΔpH across the vesicle membrane is small: 0.1 at pH_e 6.4,
0.19 at pH_e 6.7 and 0.4 at pH_e 7.7, as determined from $[^{14}C]$-methyl-
amine and nigericin sensitive 9-aminoacridine binding to vesicles
(Fig. 4). While ΔpH can be collapsed by the electroneutral exchange
of H^+ for Na^+ or K^+ in the presence of ionophore, it is unaffected
by gramicidin induced vesicle depolarization by Ch. The addition of
nigericin to the gramicidin-sensitive vesicles (Fig. 3) is likewise
without effect.

CONCLUSIONS

Synaptic vesicles concentrate Ch (ACh) as a function of vesicle membrane potential. Na^+ appears to be the principal mobile cation capable of generating the driving force for the concentrative translocation of Ch (ACh) into vesicles. The driving force can be conceptually separated into $\Delta\psi$ and ΔNa^+. It is proposed that _in situ_ ΔNa^+ is imposed across vesicle membrane as a consequence of transient contact formed between the intravesicular and extracellular space during the exocytotic event. The subsequent dissipation of the chemical ΔNa^+ may provide the necessary driving force for uptake of ACh against its increasing chemical gradient, possibly via the electroneutral Na^+/ACh^+ antiport system which is capable of discriminating between ACh and other cations in the cytoplasm (Fig. 5).

ACKNOWLEDGEMENTS

This research was supported by a USPHS grant (NS-15674) and by the Max-Planck Gesellschaft Summer Stipend.

REFERENCES

1. Carpenter, R.S. and Parsons, S.M. (1978): J. Biol. Chem. <u>253</u>: 326-329.
2. Johnson, R.G. and Scarpa, A. (1976): J. Gen. Physiol. <u>68</u>:601-631.
3. Schuldiner, S. and Kaback, H.R. (1975): Biochemistry <u>14</u>:5451-5461.
4. Suszkiw, J.B. and Whittaker, V.P. (1979): Prog. Brain Res. <u>49</u>:153-162.
5. Suszkiw, J.B., Zimmermann, H. and Whittaker, V.P. (1978): J. Neurochem. <u>30</u>:1269-1280.
6. Tashiro, T. and Stadler, H. (1978): Eur. J. Biochem. <u>90</u>:479-487.
7. Zimmerman, H. and Whittaker, V.P. (1977): Nature <u>267</u>:633-635.

INCORPORATION OF ACETATE INTO ACETYLCHOLINE AND OTHER COMPOUNDS

IN THE TORPEDO ELECTRIC ORGAN

J. Corthay, F. Loctin and Y. Dunant

Department of Pharmacology, C.M.U.
1211 Geneva 4, Switzerland

Acetylcholine (ACh) biosynthesis pathways are still unclear. In Torpedo electric organ the best precursor of acetyl moiety of ACh is acetate (2) as in the neuromuscular junction of the rat (1). Morot and Diebler (4) found an acetylCoA synthetase activity in the cytoplasm of the electric organ cells, but they could not separate any labelled acetylCoA from labelled acetate.

Slices of Torpedo electric organ are incubated with labelled acetate and the labelled compounds formed are separated by thin layer chromatography and high voltage electrophoresis. In addition to the labelled ACh, four other radioactive compounds are separated.

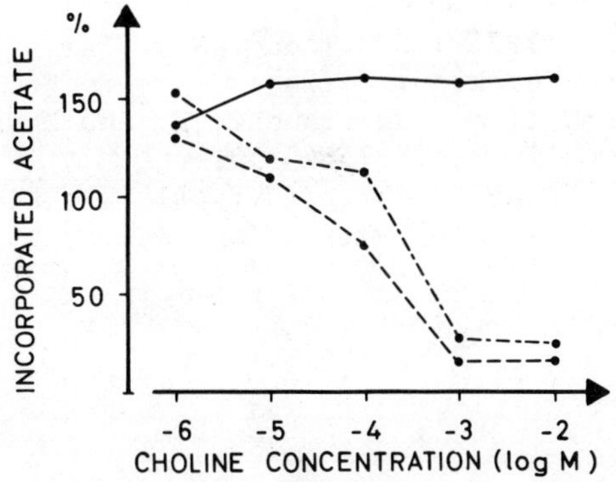

FIGURE 1: Effect of increasing Ch concentration on the incorporation of acetate in the metabolites. Values expressed as
a percentage of the results obtained without Ch in the
incubation medium.
Acetylcholine - ; Acetylcarnitine -- ; Aminoacids -.-

Their migration properties are compared to the reference compounds; three are aminoacids (glutamate, aspartate and glutamine) and the fourth one might be acetylcarnitine. A biochemical test was used to characterize presumed acetylcarnitine. Labelled AcCoA was formed with this substance which was used as a precursor in the presence of carnitine acetyltransferase.

The relative proportions of ACh, presumed acetylcarnitine and aminoacids (taken together) are, respectively, of 40%, 40% and 20% of the incorporated precursor with individual variations.

Figure 1 shows the effect of an increasing choline (Ch) concentration on the amount of labelled compounds. When Ch increases the incorporation of acetate into ACh, there is a concomitant decrease into the presumed acetylcarnitine and the aminoacids.

Isolated nerve endings from electric organ are prepared by the method of Morel et al. (3). These pure cholinergic synaptosomes are incubated with radioactive acetate; 90% of the incorporated radioactivity is found in ACh, the rest in aminoacids. Presumed acetylcarnitine is found in the pellet with the mitochondria and residual membranes. With other labelled precursors, glutamine or pyruvate, the incorporation is principally observed in the aminoacids and very weak amounts of labelled ACh or presumed acetylcarnitine are found. ACh is the major metabolite of acetate in nerve terminals of electric organ, but aminoacids and acetylcarnitine are other important ones.

REFERENCES

1.   Dreyfuss, P.M. (1975): C.R. Acad, Sci. Paris 280:1893-1898.
2.   Israel, M. and Tucek, S. (1974): J. Neurochem. 22:487-491.
3.   Morel, N., Israel, M., Manaranche, R. and Mastour-Franchon, P.
         (1977): J. Cell. Biol. 75:43-55.
4.   Morot-Gaudry, Y. and Diebler, M.F. (1979): Biochem. J. 180:
         297-301.

CHOLINERGIC RECEPTOR ISOLATION

W.H. Hopff, A.A. Hofmann, G. Riggio and P.G. Waser

Pharmacology Institute, University of Zurich
Gloriastrasse 32
CH 8006 Zurich, Switzerland

INTRODUCTION

The isolation of specific cholinergic bindin protein (acetyl-choline receptor = AChR) has been achieved successfully by numerous research groups (1-4,7-9,11,12). The ideal source for such preparations was the electric organ of several species of <u>Torpedo</u>. All successful preparations notwithstanding, there are still some serious problems waiting to be solved. The first problem is the stability of AChR. Compared to acetylcholinesterase (AChE) which can be kept at room temperature for several days without losing enzymatic activity, the AChR is less stable. Regarding stability as shown by immunological properties only, the AChR is remarkably stable. Even methods such as freeze drying of the electric organ or protein fractions during preparation, binding to animal toxins (mostly snake venoms) and application of drastic methods to release it from affinity columns do not interfere with the immunological properties. Regarding stability in terms of pharmacological properties, however, the AChR is very unstable and continuously loses its binding properties during isolation and over the course of time. It is still an open question whether a free isolated receptor in solution will ever behave like a receptor in its natural environment embedded in membranes. Some factors which might interact with this very sensitive protein are 1) ambient oxygen, 2) proteolytic enzymes (a number of proteolytic enzymes are liberated in the course of isolation), 3) microbes, 4) acetylcholine (ACh) (reasonable amounts of ACh are liberated during preparation), and 5) other agonist type molecules (small molecules with depolarizing ability have been used for affinity chromatography and/or elution from the affinity column.

EXPERIMENTAL

AChR was isolated employing the following preventive measures.

Ambient oxygen: Ambient oxygen was displaced by carbon dioxide. From the beginning to the end of isolation all solutions were saturated with carbon dioxide. When the electric organ of <u>Torpedo</u> is homogenized in the presence of a non-ionic detergent a dense foam is formed in the mixer and this foam ideally protects the solution from access by ambient oxygen.

Proteolytic enzymes: Phenylmethane sulfonyl fluoride (PMSF) has the advantage that most of the proteolytic enzymes in question are blocked without blocking AChE.

Microbes: All work was done under sterile conditions Sodium azide (0.02%) was added to all solutions.

Acetylcholine: As ACh is one of the best agonists of AChR it is obvious that after interaction with solubilized AChR the original conformation might be altered irreversibly. AChR can thus lose its pharmacological properties completely. For this reason ACh has to be removed immediately. It is an advantage that PMSF blocks only proteolytic enzymes and AChE is left active. Thus when the electric organ is homogenized and a great amount of ACh is liberated, a sufficient amount of AChE is simultaneously solubilized which immediately begins its catalytic action accelerating the hydrolysis of ACh. At the end of homogenization no ACh could be detected. The choline (Ch) formed during preparation was easily filtered off by ultrafiltration.

Other agonist type molecules: The same as for ACh, applies to small molecules of the agonist type. It has been long known that "leptocurares" react with the AChR in a depolarizing manner and "pachycurares" react in a non-depolarizing manner. In other words, small molecules or non-voluminous molecules with high affinity to the AChR also have high intrinsic activity and thus cause depolarization. Big or voluminous molecules called "pachycurares" with high affinity to the AChR do not have intrinsic activity. They block the AChR but leave the charge untouched, thus displaying their non-depolarizing ability. All ligands with intrinsic activity on the AChR were avoided. For this reason we abandoned affinity column preparation where the ligand was a depolarizing agent, as used in our eearlier work (10,13). In our opinion, the ligands from pachycurare type are best suited to affinity chromatography. These ligands possess high affinity and almost no intrinsic activity, and when they are present in solution the AChR is well protected against conformational changes. One of the substances best suited from our standpoint for the affinity chromatography of AChR is gallamine,

which was used earlier by Olsen et al. (9). These authors (9), however, used for linkage of this ligand to the solid support, one of
the sidechains of gallamine, thus sacrificing a considerable amount
of the non-depolarizing ability of the molecule. Furthermore, they
had to take into consideration a reasonable loss of the binding
property to the AChR. From this point of view we designed a
"gallamine" with three sidearms and a suitable coupling group in
the para-position to the center for optimum binding to the receptor.
The desired gallamine amide amine, however, cannot be easily synthesized. The synthesis of this compound was a time consuming
bottleneck in our isolation procedure. The isolation of AChR was
accomplished by the following procedure:

Protected homogenization and solubilization: All solutions
were CO_2 saturated to avoid access of ambient oxygen. To minimize
microbial contamination all work was done under sterile conditions
and sodium azide was added. For blockade of proteolytic enzymes,
PMSF, and for solubilization, cirrasol was added.

Centrifugation: For separation of non-solubilized material of
the electric organ, the mixture was centrifuged and the clear supernatant was used for further preparation.

Concentration: As the supernatant was rather voluminous we
concentrated the solution in a 15-plate fast ultrafiltration device
to a reasonable volume for gel filtration, followed by ultradialysis
to remove Ch and other small molecules from the solution.

Gel filtration: The crude protein mixture was prepurified by
conventional gel filtration using the "double reversed" technique.
After gel filtration, the solution again was concentrated by ultrafiltration.

Affinity chromatography: Gallamine amide amine was synthesized
and coupled to agarose separated by the well defined spacer we had
earlier used for affinity chromatography for AChE (5,6). With
receptor directed affinity chromatography we also solved the problem
of separation of AChR from AChE. This serious problem is not often
stressed in the literature. We have taken into consideration the
fact that a reasonable amount of AChE will be bound by the gallamine
amide amine inhibitor on the affinity column. To inactivate the
active site of AChE we blocked the enzyme with methylisopropyl-
fluorophosphate (MIFP) and thus could prevent binding of AChE to the
affinity column. When we applied the crude protein solution from
the preceding step containing a sufficient amount of MIFP to the
affinity column, AChE appeared with the bulk of non-specific proteins
at the front of the elution profile. The column can then be washed
for several days with buffer of considerable ionic strength, without
loss of detectable amounts of protein. For elution of AChR wc

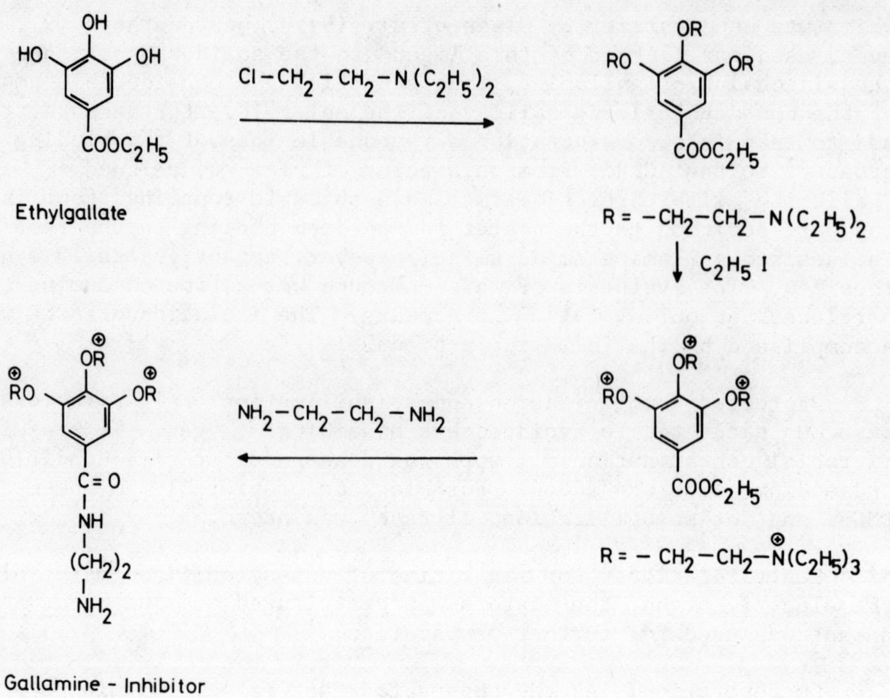

Ethylgallate

$R = -CH_2-CH_2-N(C_2H_5)_2$

$C_2H_5 I$

$R = -CH_2-CH_2-\overset{\oplus}{N}(C_2H_5)_3$

Gallamine - Inhibitor

FIGURE 1: Synthesis of gallamine amide amine inhibitor.

applied the same buffer solution with free gallamine added. After
a certain time the purified AChR detected by UV could be collected.
For storage in solution we kept the receptor in presence of gallamine. For the preparation of freeze-dried AChR free gallamine was
removed by ultrafiltration and by a final gel filtration. The clear
solution was dialyzed against distilled water and after freeze-
drying yielded a white powder.

MATERIALS AND METHODS

Our starting material for AChR isolation was the electric
organ of <u>Torpedo</u> <u>marmorata</u>; 2.5 kg of the electric organs was dissected and cut into small pieces. The organs were homogenized with
about one-third of their volume of buffer in a conventional kitchen
blender. This buffer contained 0.5 M NaCl, 0.02 M phosphate,
pH 7.4, 0.2% cirrasol ALN-WF, 0.01% phenylmethanesulfonyl fluoride
(PMSF), 0.02% sodium azide and was saturated with CO_2. After

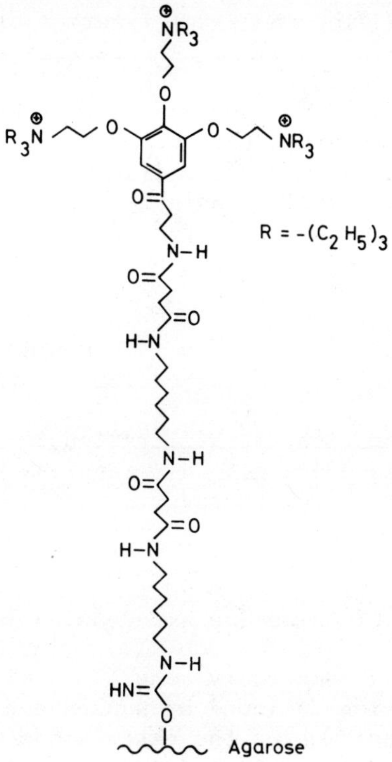

FIGURE 2: Gallamine amide amine inhibitor coupled to agarose.

homogenization the ACh content was estimated to be almost zero with the pH-stat using 0.01 M NaOH. The pH-stat was also applied for the estimation of AChE activity. AChE was inhibited by MIFP. The solution was made up with MIFP to 10^{-9} M. The pale reddish solution of about 3.5 liters was centrifuged at 7000 x g. The clear aqueous layer was concentrated by ultrafiltration (Sartorius filter SM 16520 with 15 SM 12134 filter plates). This ultrafiltration was carried out as fast as possible in order to filter off Ch and other organic or inorganic molecules. When the solution was concentrated to about 500 ml the crude protein solution was applied to a Sephadex G 100 K 100/100 column. Best results were obtained by our "double reversed" technique.

Double Reversed Technique

When protein solutions are applied at the top of a gel filtration column and run downwards, after a short time the preceding front of protein begins to broaden and irregular tailing impedes appropriate separation. This can be easily avoided by application of the protein to the bottom. The climbing front line stays sharp but in the course of time the final fraction of the run significantly broadens with irregular tailing. The results are somewhat better than after application to the top, but still are unsatisfactory. Therefore immediately after having applied the protein solution to the bottom we again reversed the column by inverting it. Now the pure buffer solution with less specific weight was layered over the protein solution. With this double reversed technique we could separate the faster running receptor-containing fraction in one step from the slower reddish fraction-containing proteins of lower molecular weight. For this technique Pharmacia columns are best suited. We used Pharmacia K 100/100 and K 215/100 columns because their frame is fixed at two points in the middle of the column and thus the column can easily be turned 180°. After gel filtration we again concentrated the protein fraction of higher molecular weights with the Sartorius ultrafiltration device. The protein was centrifuged at 30,000 x g. This material was the starting material for affinity chromatography.

Synthesis of Gallamine Amide Amine Inhibitor

Ethylgallate was completely etherified with an excess of di-ethylaminoethylchloride followed by subsequent quaternization with ethyliodide. Condensation of the ester group with ethylenediamine yielded the desired gallamine amide amine inhibitor, as shown in Fig. 1. The synthesis, with full details, will be published elsewhere.

Synthesis of the Affinity Gel

The affinity gel was synthesized according to our earlier work on AChE (5,6). After bromocyano activation agarose was treated with hexamethylenediamine and subsequently coupled with succinic anhydride. Then, with the aid of carbodiimide, hexamethylenedi-amine was coupled and again succinylated with succinic anhydride. At this point the gallamine amide amine inhibitor was coupled with the aid of carbodiimide. The structure of the spacer with the coupled gallamine amide amine inhibitor is shown in Fig. 2.

About 250 ml of the crude protein, collected with the double reversed technique was applied to the affinity gel in a K 50/100 Pharmacia column. We used standard buffer as previously described. Proteins in the effluent were detected using an LKB Uvicord II. From several preparations all eluates showed the same elution pro-file. Soon after application of the crude protein, the bulk of

TABLE 1: Aminoacid composition of AChR. Residues per 10^5 g protein.

ASP	74.8	ILE	45.9
THR	80.2	LEU	60.9
SER	77.6	TYR	21.0
GLU	115.0	PHE	33.8
PRO	50.5	LYS	53.2
GLY	49.2	HIS	22.9
ALA	54.6	ARG	30.2
VAL	49.2	TRP	8.6
MET	29.5		

TABLE 2: Carbonhydrate composition of AChR in percent weight.

Fucose	0.05
Mannose	0.39
Galactose	1.45
Glucose	0.21
NAc galactosamine	1.53
NAc glucosamine	0.85
NAc neuraminic acid	4.19
TOTAL PERCENT WEIGHT:	8.67

non-binding proteins together with AChE could be detected in one
large fraction. As expected, no AChE activity in this fraction
could be measured. Therefore in a separate run, we immediately
applied the crude protein after blockade with MIFP to the affinity
column and reactivated the eluate with obidoxim (Toxogonin: Merck)
We were then able to detect AChE activity, although for several
reasons concerning the problem of reactivability, we were unable
to quantify the exact amount of AChE.

Specific Elution From the Affinity Column

As stated in our earlier work (5,6) on AChE, our criteria for
affinity chromatography comprise not only specific adsorption on
the affinity column, but also a specific desorption, instead of a
more or less non-specific elution. This can be easily demonstrated.

When the first fraction of non-specific proteins has left the column
and no protein can be further detected from the effluent, at a certain time we add 0.02 M gallamine (Flaxedil; Specia, Paris) to the
elution buffer. After a definite time interval, following gallamine
application, the peak of AChR appears. In order to prove the accuracy of this elapsed time we washed the column for several days
with standard buffer in other runs. After addition of a gallamine
we always measured the same time interval from gallamine application
to the appearance of AChR.

The AChR fraction of about 100 ml was ultrafiltered in an
Amicon cell (model 402, filter PM 10) and finally applied to a
Pharmacia K 50/100 column of Sephadex G-100. One sample was dialyzed
against distilled water and then freeze dried to constant weight.
Based on this weight, the overall yield of AChR calculated on wet
weight of electric organ was 0.02%. Other samples were used for
conventional binding assays with α-bungarotoxin (11). For high affinity binding we found K_d 8 nM and for low affinity binding K_d
1700 nM. Other samples were analyzed for amino acid composition
and carbohydrate composition.

The amino acids were determined after hydrolysis with 6 N HCl
(24, 48 and 72 hr) and hydrolysis with p-toluenesulfonic acid for
the determination of tryptophan. Cysteine was not determined. The
carbohydrates were determined as earlier described by Werner et
al. (14).

DISCUSSION

Electric organs of various species of _Torpedo_ are the ideal
source for isolation of receptor protein (AChR). We used the electric organ of _Torpedo marmorata_. Contrary to the isolation of
AChE, the isolation of AChR demands much more care and effort. Our
main effort was to cut down the time of isolation to a minimum,
because of the instability of this protein with time. Further we
have taken into consideration some properties of AChR which might
interfere with its reactivity in the native conformation; oxidation
by ambient oxygen, proteolysis by concomitant enzymes, deleterious
action of contaminating microbes, stimulation with ACh and other
small molecules of agonist type during preparation. For affinity
chromatography we have therefore synthesized an inhibitor using a
ligand of the non-depolarizing type, a gallamine amide amine. Although affinity chromatography is a very specific method, prepuri-
fication under preventive measures cannot be omitted. When AChR in
its natural environment is stimulated, a considerable structural

alteration has to be taken into account. There one serious, but as
yet unsolved problem, is the detection of the activity of AChR in
solution. Approaches in this direction have been made by measuring
the different kinetics of low and high affinity binding. But binding
relates only to affinity. Intrinsic activity, however, can only be
measured in the animal or in whole organ preparations. Physicochemical methods have been applied to measure conformational changes
in membrane bound receptor preparations. Measuring intrinsic activity of AChR in its free state in solution with a simple and reliable
method is still the aim of numerous research groups. In spite of
much progress in receptor isolation and characterization, results
to date are incomplete and further work is required.

ACKNOWLEDGEMENTS

We gratefully acknowledge the financial help by Schweiz.
Nationalfonds grant Nr. 3,086,076; and especially for the Uvicord II
unit Stiftung fur wissenschaftliche Forschung der Universitat
Zurich.   We are also grateful to the following: the electric fish
supplied by the Institute Universitaire de Biologie Marine, Arcachon, France and to Dr. C. Cazaux and his staff for their generous
help; to Drs. M.C. Schaub and J. Watterson for the amino acid
analysis; to Drs. G. Werner and W. Wojnarowski for the carbohydrate
analysis; to Dr. G. Riggio for synthesizing MIPF in our laboratory;
to Dr. U. Quast for having introduced us to the elegant technique
of Schmidt and Raftery (11); to ICI (Zurich) for supplying cirrasol
ALN-WF. For his help and critical disucssion we would also like
to acknowledge Dr. J. Watterson; and for technical assistance to
Mr. G. Engler, Mrs. H. Meyer and Mrs. E. Schonenberger.

REFERENCES

1.   Biesecker, G. (1973): Biochemistry 12:4403-4409.
2.   Chang, H.W. and Bock, E. (1979): Biochemistry 18:172-179.
3.   Eldefrawi, M.E. and Eldefrawi, A.T. (1973): Arch. Biochem.
         Biophys. 159:362-373.
4.   Heilbronn, E. and Mattson, C. (1974): J. Neurochem. 22:315-317.
5.   Hopff. W.H. (1976): Nat. Forsch. Gesellsch. Zurich 121:223-260.
6.   Hopff, W.H., Riggio, G. and Waser, P.G. (1975): In Cholinergic
         Mechanisms (ed) P.G. Waser, Raven Press, New York.
7.   Karlsson, E., Heilbronn, E. and Widlund, L. (1972): FEBS Lett.
         28:107-111.
8.   Klett, R.P., Fulpius, B.W., Cooper, D., Smith, M., Reich, E.
         and Possani, L.D. (1973): J. Biol. Chem. 248:6841-6853.

9.   Olsen, R., Meunier, J.C., Weber, W. and Changeux, J.P. (1972):
     FEBS Lett. 28:96-100.
10. Riggio, G., Hopff, W.H. and Hofmann, A.A. (1979): Experientia
     35:587-588.
11. Schmidt, J. and Raftery, M.A. (1973): Biochemistry 12:852-856.
12. Schwyzer, R. and Frank, J. (1972): Helv. Chim. Acta 55:2678-2682.
13. Waser, P.G., Hopff, W. and Riggio, G. (1979): In Recent Advances
     in Receptor Chemistry,(eds) F. Gualtieri, M. Giannella, and
     C. Melchiorre, Elsevier, North Holland.
14. Werner, G., Wojnarowski, W., Hopff, W.H. and Waser, P.G. (1977):
     In Cholinergic Mechanisms and Psychopharmacology, (ed) D.J.
     Jenden, Plenum Press, New York.

NEW EVIDENCE FOR TRUE IMMUNOPHARMACOLOGIC BLOCKADE

IN MYASTHENIA GRAVIS

A.K. Lefvèrt*, S. Cuenoud and B.W. Fulpius

Department of Biochemistry, University of Geneva
Sciences II
CH-1211 Geneva 4, Switzerland

INTRODUCTION

Anti-acetylcholine receptor (anti-AChR) antibodies are generally considered to give rise to the neuromuscular deficiency observed
in myasthenia gravis (MG). Their precise mechanism of action is
still under discussion (for review, see 12).

One of the postulated mechanisms implies binding of these im-
munoglobins to the receptor located on the postsynaptic membrane
and complement mediated membrane destruction which leads to a reduction of the receptor density. On myasthenic motor endplates,
IgG and complement factors have been shown, bound to the postsynaptic region (6). Moreover, the morphological aspect of the endplate
is simplified and the postsynaptic area reduced. According to another proposed mechanism, reduction of AChR density could be caused
by antigenic modulation of AChR metabolism. This hypothesis rests
on observations made on human skeletal muscle grown in tissue culture in presence of myasthenic serum (4) but there has been no
direct demonstration of this pathogenic mechanism in myasthenic
muscles.

These two versions of MG pathogenesis are not fully satisfying.
Clinical observations suggest a third mechanism, an immunopharmaco-
logical blockade. As a matter of fact, the symptoms, especially
at early stages of the disease, resemble those of a transient

---

* Present address: Department of Clinical Chemistry, Huddinge
Hospital, Karolinska Institut, S-14186 Huddinge, Sweden.

curare-like neuromuscular block and usually they are promptly re-
verted by an adequate anticholinesterase medication. In addition,
at such early stages of the disease, the morphologic changes in the
endplate do not seem so prominent (17). Finally, one should remember the marked improvement which follows plasmapheresis (16)
and the even more pronounced effect of drainage of thoracic duct
lymph in this respect (3). Reciprocally, during retransfusion to
myasthenic patients of lymph or IgG fractions containing anti-AChR
antibodies, muscular weakness develops usually within 30 min and
disappears after a few hours, an effect which can be seen on elec-
tromyographic recordings using the decremental response to repetitive stimulation (10).

An immunopharmacologic blockade strongly suggests the existence
of anti-AChR antibodies with a direct and partly reversible effect
on acetylcholine (ACh) binding _in vivo_. These antibodies would
be expected to compete with cholinergic ligands and elapid neurotoxins for AChR binding. They would escape detection by the usual
assays which are based on binding of antibodies to a complex made
of AChR and radiolabelled neurotoxin.

Antibodies which interfere with toxin binding on AChR have
already been found by several authors, but the nature of the inhibition was not well characterized. They have been detected in different ways; Bender et al. (2) reported that 24 out of 32 myasthenic sera inhibited the binding of α-bungarotoxin (α-Btx) to
normal human motor endplates. Using detergent extracted denervated
rat muscle AChR, Almon and Appel (1) showed a non-competitive inhibition of toxin binding by myasthenic IgG. Mittag et al. (13)
used the same principle of assay and found an inhibition of 7% of
the myasthenic sera tested. Using a _Torpedo_ membrane fragments as
a source of AChR, Lefvert and Bergstrom (8) showed a partial inhibition of toxin binding by IgG fractions from 5 out of 6 patients
with severe MG. The inhibitory effect was found exclusively in
IgG subclass 3. A complete inhibition could not be demonstrated,
even with high IgG concentrations. Using detergent extracted
human skeletal muscle AcHR, Lindstrom (11) was not able to show
significant inhibition of toxin binding by myasthenic sera. With
a similar assay, Lefvert and Bergstrom (9) and Vincent and Newsom-
Davis (18) demonstrated that around 60% of myasthenic sera have
antibodies which interfere with toxin binding and 95% have antibodies directed against other sites than the ACh binding sites.
These authors did not find any concordance when they compared the
titers obtained with these two different methods.

The inhibition by myasthenic antibodies of toxin binding to
solubilized AChR does not necessarily prove the existence of anti-
AChR antibodies directed specifically against the ACh binding site.
Antibodies directed against sites other than the ACh binding site

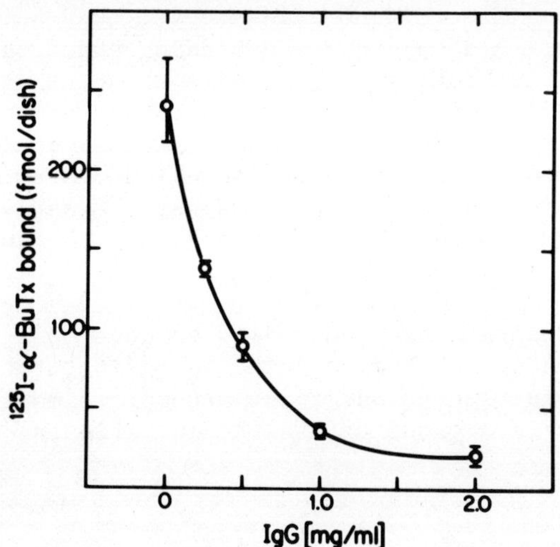

FIGURE 1: Labelling of chick muscle cell surface AChR by ^{125}I-α-Btx after preincubation with different concentrations of IgG prepared from the serum of a myasthenic patients. Cultures used were 8 days old.

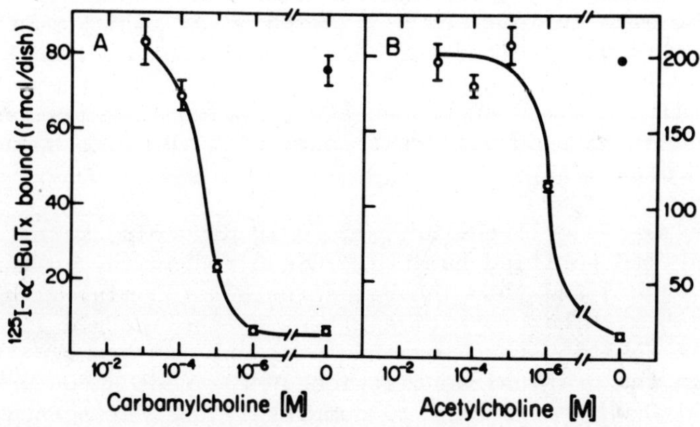

FIGURE 2: Effect of two cholinergic agonists; carbamylcholine (experiment A) and ACh (experiment B) on the inhibition of cell surface toxin labelling by IgG (2 mg/ml). Eight day (experiment A) and five day (experiment B) old cultures were preincubated for 120 min at 37°C. The controls (closed circles) were measured in the absence of human IgG. Experiment B was performed in the presence of physostigmine (10^{-4} M).

can also inhibit toxin fixation, either sterically (5) or allo-
sterically (19). There is no evidence yet that antibodies inhibiting toxin fixation do prevent ACh binding and lead to a curare-
like blockade. In order to test the inference that myasthenic IgG
block α-Btx binding by reacting with AChR determinants within the
ligand binding region implies the use of small cholinergic drugs
such as ACh, carbamylcholine and dimethyltubocurarine. These drugs
should protect in a dose-dependent manner the native AChR against
inactivation by myasthenic IgG to prove that the antibody molecules
are directed against the ACh binding site.

These considerations prompted us to develop a test in order
to detect anti-AChR antibodies directed against the ACh binding
site (7). We report here our first attempt to measure these specific antibodies in a group of myasthenic patients.

Method of Measuring Anti-AChR Antibodies
Directed Against the ACh Binding Site (7)

The test system used is based on the inhibition, by myasthenic
IgG, of ^{125}I-labelled α-Btx binding to AChR of chick embryo muscle
cells in culture. We have used AChR in its native environment,
since it is known that its extraction from the membrane exposes
hidden determinants and thus allows reaction with other antibodies
which might also inhibit toxin fixation (19). Chick muscle cells
are especially well suited for the detection of anti-AChR antibodies
directed against the ligand binding site since myasthenic immunoglobulins directed, on the human AChR, against other sites than the
ligand binding site do not cross-react with the corresponding
domain on chick AChR.

Chick muscle cell cultures, 5 to 10 days old, were incubated
for 45 min with purified human IgG dissolved in the culture medium.
They were washed and then incubated further (45 min) with ^{125}I-
labelled α-Btx. The protective effect of cholinergic ligands was
tested by incubating them together with the IgG fraction to be
tested. In the original description of the test a concentration of
1-2 mg/ml IgG was sufficient to completely prevent subsequent toxin
binding (Fig. 1). It should be noticed that, in the same experimental conditions, IgG did not displace α-Btx from pre-existing
toxin-receptor complexes. Low concentrations of the cholinergic
agonists ACh and carbamylcholine protected AChR in a dose-dependent
manner against the inactivation by IgG (Fig. 2). Cholinergic
antagonists such as d-tubocurarine, but not atropine, were also
effective in this respect (Fig. 3).

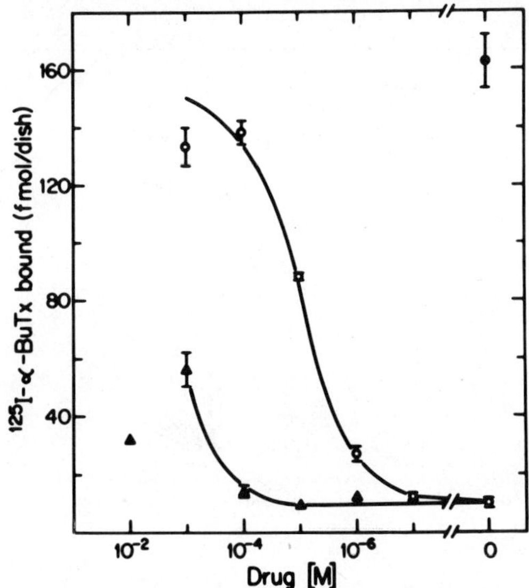

FIGURE 3: Effect of two cholinergic antagonists; dimethyltubocur-
          arine (open circles) and atropine (closed triangles) on
          the inhibition of cell surface toxin labelling by IgG
          (2 mg/ml). Six day old cultures were preincubated for
          120 min at 37°C. The control (closed circle) was measured
          in the absence of IgG.

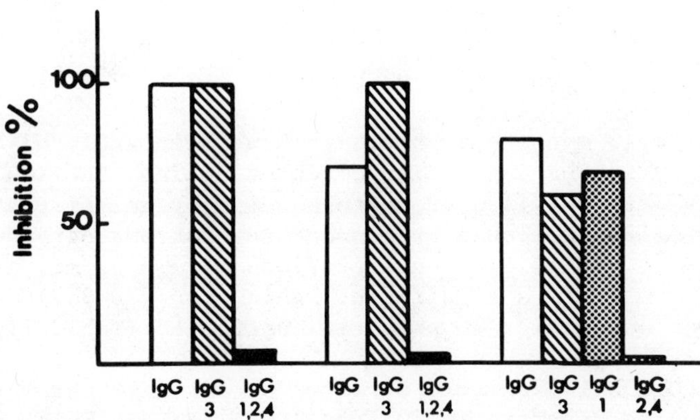

FIGURE 4: Inhibition of α-Btx binding to cell surface AChR after
          preincubation with the different IgG subclasses 1, 2, 3
          and 4. Each IgG concentration was calculated to corres-
          pond to 50% of the one existing in the serum, i.e. total
          IgG 5 mg/ml; IgG 3, 0.35 mg/ml, etc.

TABLE 1:   Anti-AChR antibodies in myasthenic patients using two
           assays.

Patient Number	Inhibition of Toxin binding to cell membrane AChR (percent)*	Binding to Toxin–AChR complex (pmol/ml)**
1	51	20.9
2	0	1.36
3	20	4.94
4	61	0.94
5	0	0
6	0	23.3
7	0	12.5
8	0	3.94
9	92	17.0
10	51	0.92
11	30	17.6
12	54	24.1
13	93	21.3

*   Inhibition of toxin binding to cell membrane AChR (chick
     embryo muscle cells).
**   Binding to detergent exctracted AChR-toxin complex (human
     skeletal muscle)(see, 14). IgG concentration in assay (1)
     was 3-5 mg/ml.

RESULTS

IgG from 13 myasthenic patients have been analyzed for their
content in anti-AChR antibodies directed against the ACh binding
site. They were purified from fresh serum by gel filtration
(Sephacryl G-300) followed by ion exchange chromatography (DEAE-
Sephadex A-50). With IgG concentrations of 3-5 mg/ml, it was possible to demonstrate a significant inhibition (20-98%) of toxin
binding to 8 out of 13 IgG preparations tested (Table 1).

The IgG subclasses were purified by affinity chromatography
on protein A to separate IgG 3 from IgG 1, 2, 4. This was followed by a restricted papain digestion, in the absence of reducing
agents, in order to digest IgG 1. IgG 2, 4 were then separated
from the papain digest by gel filtration (Sephacryl G-300). As
shown in Fig. 4, in 2 out of 3 patients tested, the antibody activity was found exclusively in the IgG subclass 3. In the third
patient, both subclasses 1 and 3 contained about equal amounts of

the antibody activity.   The IgG from one of the patients was
digested with papain and Fab and Fc fragments were isolated by
gel filtration (Sephacryl G-300) and ion exchange chromatography
(CM- and DEAE-Sephadex). Fc fragments were without any effect
on toxin binding. Fab fragments retained all the inhibitory properties of the native IgG (Fig. 5). This result confirms that the
interaction with AChR is characteristic of an antigen-antibody
reaction.

DISCUSSION

The method described above permits the detection of anti-AChR
antibodies interacting with the ACh binding site. It should be of
great use in the study of MG pathogenic mechanisms. First, of
course it remains to be shown that these antibodies have the same
effect on human skeletal muscle junctional AChR. However, this is
most likely, since the binding characteristics of other nicotinic
ligands are similar on AChR from different sources which suggests
that the protein structure and conformation of the ligand binding
site have been preserved during evolution.

Clinical evidence suggest that, at early stages of the disease, a direct inhibition of ACh binding by anti-AChR antibodies
might play a major role in the pathogenesis of myasthenic symptoms.

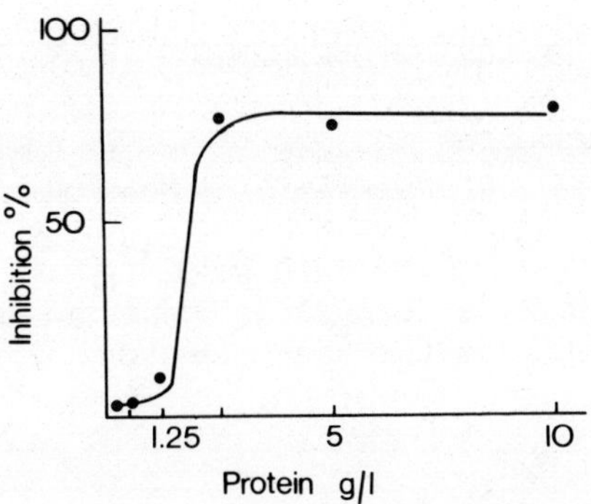

FIGURE 5: Inhibitory effect on toxin binding to cell surface AChR
           by preincubation with Fab fragments prepared from the
           IgG of a myasthenic patient.

In addition to its curare-like effect, the antibody will initiate, on binding to a receptor, activation of the complement reaction. This will eventually lead to structural damage of the postsynaptic membrane, liberation of membrane fragments containing AChR in the surrounding tissue and activation of the immune system with production of auto-antibodies to parts of AChR which are normally unexposed or inaccessible.

In early cases of MG it seems logical to assume that the concentration of antibodies competing for the ligand binding site might be more closely correlated to the severity of the clinical symptoms than that of the antibodies which are directed against other part of AChR than the ACh binding site. If this were true, this test should be of great value in diagnosing early cases of MG in which the detection of anti-AChR antibodies by the usual test is often negative.

The results reported here in analyzing IgG subclasses deserve further comment (for review, see 15). IgG subclasses differ in their capacity to form complexes,to aggregate and to bind complement. They differ also in their susceptibility to proteolysis. When compared to IgG 2 and 4, both IgG 1 and 3 bind much more strongly to complement to form complexes with complement factors. These properties are of importance when one considers the ability of an antibody to lead to tissue damage. In the context one should remember that, in another autoimmune disease, systemic lupus erythematodes, the occurrence of autoantibodies of subclasses 1 and 3 is well correlated to the development of severe clinical manifestations of the disease. It follows that an analysis of the subclasses of anti-AChR antibodies might have some prognostic value.

ACKNOWLEDGEMENTS

This work was supported in part by grant number 3,157,77 from the Swiss National Fund for Scientific Research, and by a grant from the Sandoz Foundation. The stay of A.K. Lefvert at the University of Geneva was made possible by a grant from the Swedish Medical Research Council, a fellowship from the European Science Foundation and a grant from the Roche Research Foundation.

REFERENCES

1.   Almon, R.R. and Appel, S.H. (1975): Biochim. Biophys. Acta
     393:66-77.
2.   Bender, A.N., Engel, W.K., Ringel, S.P., Daniels, M.P. and
     Vogel, Z. (1975): Lancet 1:607-609
3.   Bergstrom, K., Franksson, C., Matell, G., Nilsson, B.Y.,
     Persson, A., von Reis, G. and Stensman, R. (1975): Eur.
     Neurol. 13:19-30.
4.   Bevan, S., Kullberg, R.W. and Heinemann, S.F. (1977): Nature
     267:263-265.
5.   Dwyer, D.S., Bradley, R.J., Oh, S.J. and Kemp, G.E. (1979):
     Clin. Exp. Immunol. 37:448-451.
6.   Engel, A.G. (1979): Prog. Brain Res. 49:423-434.
7.   Fulpius, B.W., Miskin, R. and Reich, E. (1980): Proc. Nat.
     Acad. Sci.77:4326-4330.
8.   Lefvert, A.K. and Bergstrom, K. (1977): Eur. J. Clin. Invest.
     7:115-119.
9.   Lefvert, A.K. and Bergstrom, K. (1978): Scand. J. Immunol.
     8:525-533.
10. Lefvert, A.K. and Matell, G. (1979): In Plasmapheresis and the
     Immunobiology of Myasthenia Gravis (ed) P.C. Dau, Houghton
     Mifflin, Boston.
11. Lindstrom, J. (1977): Clin. Immunol. Immunopathol. 7:36-43.
12. Lindstrom, J. (1979): Adv. Immunol. 27:1-50.
13. Mittag, T., Kornfeld, P., Tormay, A. and Woo, C. (1976): New
     Eng. J. Med. 294:691-694.
14. Monnier, V.M. and Fulpius, B.W. (1977): Clin. Exp. Immunol.
     29:16-20.
15. Morell, A., Skvaril, F. and Barandun, S. (1975): IgG-Subklassen
     der menschlichen Immunoglobuline. Immunochemische, gene-
     tische, biologische und klinische Aspekte, S. Karger, Basel.
16. Pinching, A.J., Peters, D.K. and Newsom-Davis, J. (1976):
     Lancet 2:1373.
17. Tsujihata, M., Hazama, R., Ishii, N., Ide, Y., Mori, M. and
     Takamori, M. (1979): Neurology 29:654-661.
18. Vincent, A. and Newsom-Davis, J. (1979): In Advances in Cyto-
     pharmacology, Vol. 3, (eds) B. Ceccarelli and F. Clementi,
     Raven Press, New York.
19. Zurn, A. and Fulpius, B.W. (1979): Prog. Brain Res. 49:435-440.

IMMUNOHISTOCHEMICAL DEMONSTRATION OF IgG AND Fab FRAGMENTS

AT MOTOR ENDPLATES OF PASSIVELY TRANSFERRED MICE

B. Lowenadler, E. Heilbronn and K. Toyka*

Department of Neurochemistry and Neurotoxicology
University of Stockholm, c/o FOA Ursvik
S-104 50 Stockholm, Sweden, and
*Department of Neurology, University of Dusseldorf
Moorenstrasse 5
D 4000 Dusseldorf, West Germany

INTRODUCTION

Interaction between the circulating receptor antibodies and
the nicotinic acetylcholine receptor (nAChR) at the neuromuscular
junction is probably the main cause of the impaired neuromuscular
transmission in myasthenia gravis (MG) (for review see, e.g. 6).
It has been demonstrated that by passive transfer of human myasthenic immunoglobulins to recipient mice many of the features of
MG can be reproduced: reduced amplitudes of miniature endplate
potentials (MEPP), a decreased number of available nAChR:s (α-
bungarotoxin binding sites), decremental response to repetitive
nerve stimulation and occasionally clinical weakness. Increased
degradation of nAChR has also been observed in mice treated with
myasthenic immunoglobulins (10). Using muscle cell cultures,
evidence has been obtained that native receptor antibodies or bi-
valent $F(ab')_2$ fragments are needed to induce accelerated degradation of nAChR, since monovalent Fab fragments were inactive in
this respect (1).

In this study we have shown, using the mouse transfer model,
that both Fab fragments and IgG from myasthenic patients bind to
nAChR at the neuromuscular junction. Fab fragments alone do not
reproduce the myasthenic transmission defect but completely inhibit
the effect of IgG on MEPP amplitudes when transferred simultaneously.

MATERIALS AND METHODS

Mice of the C57 or BDF$_1$ strain were injected (i.p.) with the IgG and Fab fractions of myasthenic and control sera (see Table 1). The animals were treated for various times and then sacrificed and tested for neuromuscular transmission and presence of IgG and Fab at their motor endplates using conventional immunohistochemistry.

Miniature Endplate Potentials Measurements

MEPP measurements were carried out on the left hemidiaphragm maintained at 31°C in modified Liley's solution gassed with 95% O_2 and 5% CO_2. Intracellular recordings of MEPP were performed by standard microelectrode techniques as described earlier (14).

Light Microscopic Localization of IgG and Fab

Mice were sacrificed by decapitation. Tibialis anterior was excised and frozen by immersion in freon chilled in liquid nitrogen. Longitudinal cyrosections (10 μm thick) were cut on a microtome. The sections were fixed in acetone and rinsed in phosphate buffered saline pH 7.2 (PBS). For the localization of IgG, sections were incubated with protein A conjugated peroxidase (pA-HRPO) prepared according to Nakane and Kawoi (8) at a concentration of 25 μg/ml. After rinsing in PBS the sections were stained for peroxidase activity with H_2O_2 and diaminobenzidine (7). In the case of Fab localization, sections were incubated with anti-human Fab antibody prior to pA-HRPO and then processed as above.

Electron Microscopic Localization
of IgG at Motor Endplates

Thin strips of diaphragm muscle were processed for electron microscopic localization of IgG by the method of Engel et al. (3). After postfixation in OsO_4 and dehydration in alcohol, the muscle strips were embedded in Epon. Sections (1.5 μm semi-thin) were cut and studied under a light microscope for peroxidase stain. When endplates were found in a section, consecutive ultra-thin sections were cut for electron microscopy.

TABLE 1

Mouse Strain	Number of Animals	Treatment	Protein Received (mg/day)	MEPP (mV)
C 57		IgG - 3 days		
	2	myasthenic	11-12	not tested
	2	control	11-12	not tested
C 57		IgG - 30 days		
	2	myasthenic	11-12	not tested
	2	control	11-12	not tested
B6D2F$_1$/J		IgG - 10 days		
	2	myasthenic	12	0.37/0.32
	1	control	12	0.70
B6D2F$_1$/J	2	Fab - 9 days:myasthenic	8	0.80/0.95
	1	Fab - 7 days:control	8	0.69
	2	Fab + IgG - 9 days myasthenic	8+12/8+16	0.80/1.14

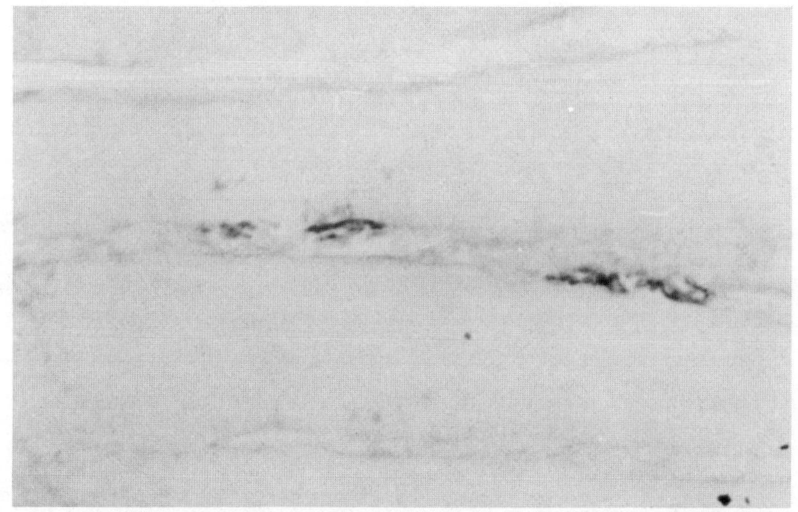

FIGURE 1: Cryosection of _tibialis anterior_ from mouse injected with myasthenic IgG for 10 days. Motor endplates stain for bound IgG (x 800).

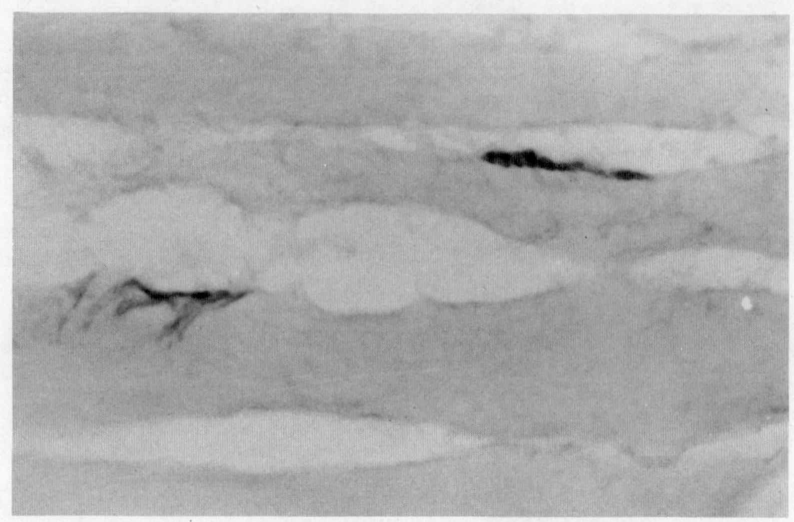

FIGURE 2: Cryosection of <u>tibialis</u> <u>anterior</u> from mouse injected with
          myasthenic Fab for 9 days. Motor endplates stain for
          bound Fab fragments (x 800).

FIGURE 3: Semithin Epon section of diaphragm muscle fiber from a
          mouse treated with myasthenic IgG for 3 days. Motor end-
          plate stain for bound IgG. A stained erythrocyte, due to
          endogenous peroxidase activity, can be seen at the upper
          left (x 1500).

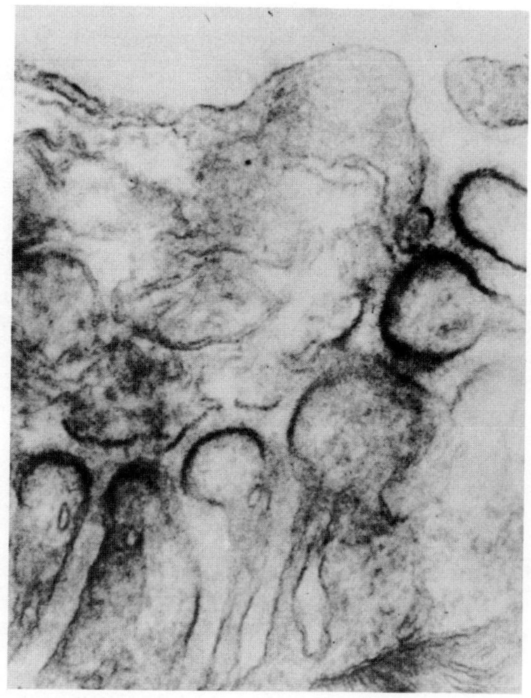

FIGURE 4: Electron micrograph of part of a motor endplate from
diaphragm muscle of a mouse receiving myasthenic IgG
for 30 days. Stain mainly at the tips of the postsynaptic folds which appear structurally normal (x 45,000).

RESULTS

Neuromuscular Transmission

Two of the mice injected with myasthenic IgG were tested with
MEPP measurements. Both these mice had significantly lowered MEPP
amplitudes, in agreement with earlier observations (14). The mean
MEPP amplitude in the diaphragm of the two mice injected with the
purified Fab fragments from two individual myasthenic patients were
within the normal range for control mice. Two mice injected with
both myasthenic IgG and Fab had normal MEPP amplitudes in clear
contrast to those injected with myasthenic IgG.

Immunohistochemistry

All mice receiving myasthenic IgG displayed peroxidase stain at their motor endplates, indicating the presence of IgG. In contrast no stained endplates were found in muscle from control animals. Motor endplate stain was also absent in specificity control sections (i.e. sections either preincubated with unlabelled protein A or with H_2O_2 omitted from the staining medium). Electron microscopic studies revealed the stain predominantly at the tips of the postsynaptic folds, in good agreement with the distribution of the AChR (5). Some stain was also seen at the presynaptic membrane.

Fab fragments were demonstrated at motor endplates in muscle cryosections of both mice treated with human myasthenic Fab by incubation with rabbit anti-Fab antibody followed by pA-HRPO. Sections from the same animals incubated with only pA-HRPO were negative. Fab could not be demonstrated at motor endplates of the one mouse receiving control Fab.

DISCUSSION

In this study we have demonstrated, by specific immunohistochemical staining, the presence of IgG and Fab fragments at the neuromuscular junction of mice injected with myasthenic immunoglobulin fractions but not in control mice.

No major ultrastructural changes were observed in animals treated with myasthenic IgG for 30 days. This is at variance with observations in human myasthenia gravis and its animal model, where widening of the synaptic cleft and a reduction of the postsynaptic surface area has been reported (2,11). A complement mediated lesion is likely to be the basic mechanism for these morphological changes since IgG and complement C3 and C9 were found in the synaptic cleft (4). Our neurophysiological findings suggest that monovalent Fab fragments obtained from myasthenic sera may be able to protect the neuromuscular junction from the effect of native myasthenic IgG, since in animals treated with myasthenic Fab + IgG or with myasthenic Fab alone no transmission defect could be observed. There are at least two possible reasons why Fab fragments do not impair neuromuscular transmission: the monovalent Fab is unable to activate complement (9) and also to cross-link receptors (1).

It is not known if Fab could have the same effect on motor endplates with already bound receptor antibodies. If so, treatment with homologous Fab may have some value as an immunotherapeutic agent.

REFERENCES

1. Drachman, D.B., Angus, C.W., Adams, R.N., Michelson, J.D. and Hoffman, G.J. (1978): New Eng. J. Med. 298:1116-1122.

2. Engel, A.G. and Santa, T. (1971): Ann. N.Y. Acad. Sci. 183: 46-63.

3. Engel, A.G., Lambert, E.H. and Howard, F.M. (1977): Mayo Clin. Proc. 52:267-280.

4. Engel, A.G., Sahashi, K., Lambert, E.H. and Howard, F.M. (1979): In Current Topics in Nerve and Muscle Research (eds) A.J. Aguayo and G. Karpati, Excerpta Medica, Amsterdam.

5. Fertuck, H.C. and Salpeter, M.M. (1974): Proc. Nat. Acad. Sci. 71:1376-1378.

6. Heilbronn, E. and Stalberg, E. (1978): J. Neurochem. 31:5-11.

7. Karnovsky, M.J. (1967): J. Cell Biol. 35:213-236.

8. Nakame, P.K. and Kawoi, A. (1974): J. Histochem. Cytochem. 22:1084-1091.

9. Ruddy, S., Gigli, I. and Austen, K.F. (1972): New Eng. J. Med. 287:489-495.

10. Stanley, E.F. and Drachman, D.B. (1978): Science 200:1285-1287.

11. Thornell, L.E., Sjostrom, M., Mattsson, C.H. and Heilbronn, E. (1976): J. Neurol. Sci. 29:389-410.

12. Toyka, K.V., Drachman, D.B., Griffin, D.E., Pestronk, A. and Kao, I. (1975): Science 190:397-399.

13. Toyka, K.V., Drachman, D.B., Griffin, D.E., Pestronk, A., Winkelstein, J.A., Fischbeck, K.H. and Kao, I. (1977): New Eng. J. Med. 296:125-131.

14. Toyka, K.V., Birnberger, K.L., Anzil, A.P., Schlegel, C., Besinger, U.A. and Struppler, A. (1978): J. Neurol. Neurosurg. Psychiat. 41:746-753.

ON THE EFFECTS OF ANTICHOLINERGIC AGENTS ON MUSCARINIC RECEPTORS

OF DIFFERENT LOCALIZATION

D.A. Kharkevich, A.P. Skoldinov, D.N. Samoilov and
V.A. Shorr

Department of Pharmacology, First Medical Institute and
Institute of Pharmacology, Academy of Medical Sciences
Moscow, USSR

INTRODUCTION

Dissimilarities in the sensitivity of muscarinic receptors of
different organs to agents of various pharmacological groups has
been noted (4-6,8-11,14,16-18). Thus, the search for compounds
which block muscarinic receptors of definite localization (heart,
exocrine glands, smooth muscles of visceral organs) is warranted
and will stimulate the synthesis of antimuscarinic agents of cardiotropic or broncholytic action and possibly of selective blockers
of glandular secretory activity, etc. In addition, particular
morphological and functional properties of muscarinic receptors of
various effector cells could be characterized with the use of such
selective drugs.

The present report gives an account of our results with agents
which prevalently affect muscarinic receptors of the heart.

RESULTS

Bis-quaternary ammonium derivatives
of diphenylcyclobutanedicarboxylic (truxillic) acids (I)

The compounds of this series belong to the group of active
antidepolarizing curare-like agents (6,7,11). These agents induce

tachycardia by blockade of muscarinic receptors of the heart (in
myoparalytic doses these compounds are devoid of ganglion-blocking
action). An important pecularity of their action lies in their
ability to block the bradycardia but not the hypotension induced
by acetylcholine (ACh). This datum testifies to the considerable
difference in the sensitivity of the muscarinic receptors of the
heart and blood vessels to the derivatives of truxillic acids.

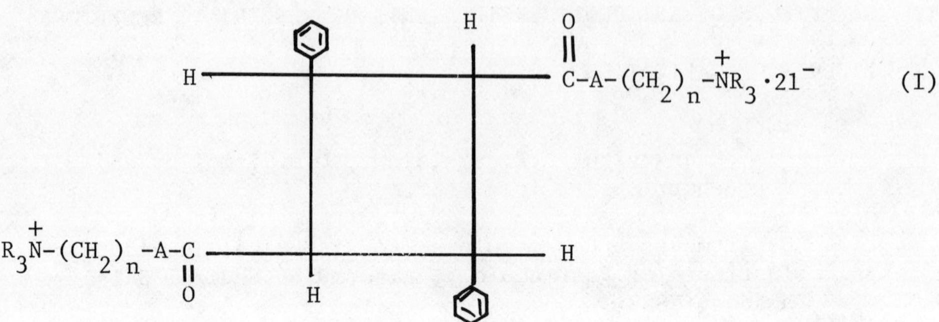

$$H-\underset{\underset{\underset{H}{|}}{|}}{\overset{\overset{\bigcirc}{|}}{C}}\qquad \overset{O}{\overset{\parallel}{C}}-A-(CH_2)_n-\overset{+}{N}R_3 \cdot 2I^- \qquad (I)$$

In accordance, the relationship between cardiotropic antimuscarinic activity and the structure of the truxillic acid derivatives was investigated. The following factors were considered:
a) the stereoisomerism of truxillic acids; b) the structure of
other components of the interonium part of a molecule; c) the
distance between cationic centers; d) the character of radicals
at quaternary nitrogen atoms.

To elucidate the role of steric configuration of truxillic
acids, the antimuscarinic activity of α, ε and γ truxillic acid
derivatives was compared (Table 1). It was found that the de-
rivatives of α and ε truxillic acid have higher antimuscarinic
activity than γ truxillic acid.

In the series of bis-esters of α truxillic acid (Table 2), the
highest antimuscarinic activity was observed at n = 4 and n = 3
(i.e. at 15 and 13 atoms between the quaternary nitrogen atoms).

The antimuscarinic activity of the compounds was considerably
affected by the radicals of the quaternary nitrogen atoms (Table 3).
Thus, at n = 3, the bis-trimethylammonium compounds showed lowest
activity. The successive replacement of the methyl groups by ethyl
groups increased cardiotropic antimuscarinic activity. The pronounced cardiotropic effect was a trait of the iodoethylates and
iodomethylates of bis-piperdine and bis-pyrrolidine derivatives
of α-truxillic acid.

It is worth noting that the replacement of ester groups by amide
groups evoked a sharp decrease in the cardiotropic antimuscarinic

TABLE 1: Significance of the spatial configuration of the central part of truxillic acid derivatives on anticholinergic activity.

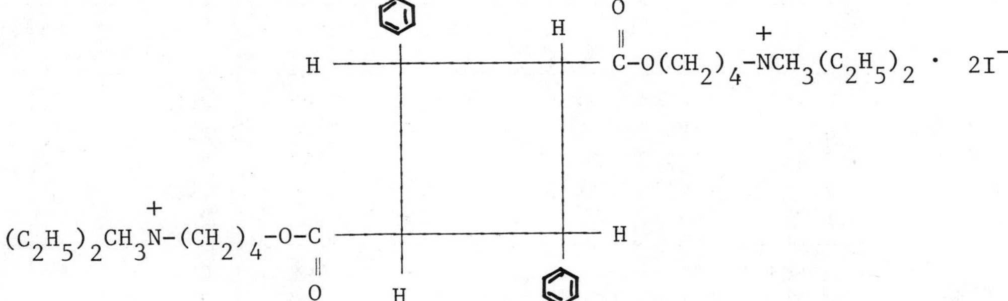

Compound Number	Spatial Isomerism of Truxillic Acids	Antimuscarinic Activity* (ED_{50})	Curare-like Activity	
			Head Drop in Rabbits (ED_{50})	Block of Neuromuscular Transmission in Cats (MED)
1	α	26 (19.1 – 32.9)	48 (41.3 – 55.7)	250–300
2	ε	28 (21.7 – 34.3)	130 (110.1 – 153.4)	400–500
3	γ	99 (53.2 – 144.8)	250 (236.9 – 263.7)	600–1000
Tubarine			137 (105 – 178)	180–230
Atropine Sulfate		1.3(0.5 – 2.1)		

ED_{50} = mean effective doses (μg/kg, i.v) and 95% confidence limits (n = 6–8).

MED = minimum effective dose (μg/kg, i.v) (n = 6–8).

* Inhibition of the negative chronotropic effect of ACh on cat heart.

TABLE 2: Significance of the distance between the quaternary nitrogen atoms of truxillic acid derivatives on anticholinergic activity.

$$R^1R^2R^3-\overset{+}{N}-(CH_2)_n-O-\overset{O}{\overset{\|}{C}} \qquad \overset{O}{\overset{\|}{C}}-O-(CH_2)_n-\overset{+}{N}R^1R^2R^3 \qquad 2I^-$$

Compound Number	n	$-\overset{+}{N}R^1R^2R^3$	Antimuscarinic Activity* (ED_{50})		Curare-like Activity ($\mu g/kg$)		Block of Neuromuscular transmission in cats (MED)
					Head-drop in Rabbits (ED_{50})		
4	2		184	(143.0 – 225.0)	170	(147.8 – 195.5)	600–800
5	3	C_2H_5	43	(26.5 – 59.5)	37.5	(32.4 – 43.3)	150–250
1	4	$-\overset{+}{N}-C_2H_5$	26	(19.1 – 32.9)	48	(41.3 – 55.7)	250–300
6	5		130	(105.0 – 155.0)	43.5	(37.2 – 49.6)	130–160
7	7	CH_3	145.5	(98.4 – 192.6)	98	(87.5 – 109.7)	500–600
8	2		50	(37.1 – 62.9)	73	(62.0 – 84.7)	600–700
9	3		30	(24.1 – 35.9)	22.5	(15.6 – 32.4)	110–150
10	4	$-\overset{+}{N}$	24	(20.5 – 27.5)	46	(43.3 – 48.7)	150–200
11	5		74	(53.7 – 94.3)	48	(42.8 – 53.7)	200–250
12	7	CH_3	352	(326.6 – 377.4)	140	(107.6 – 182.0)	1000–1200
Tubarine					137	(105 – 178)	180–230
Atropine sulfate			1.3	(0.5 – 2.2)			

FOR LEGENDS SEE TABLE 1.

TABLE 3: Significance of the radicals on the quaternary nitrogen atoms of truxillic acid derivatives on anticholinergic activity.

$$R^3R^2R^1N^+-(CH_2)_3-O-\overset{O}{\underset{\|}{C}}\cdots\overset{O}{\underset{\|}{C}}-O-(CH_2)_3-N^+R^1R^2R^3 \cdot 2I^-$$

Compound Number	$-N^+R^1R^2R^3$	Antimuscarinic Activity* (ED_{50})	Curare-like Activity	
			Head Drop in Rabbits (ED_{50})	Block of Neuromuscular Transmission in Cats (MED)
13	$-N^+(CH_3)_3$	155 (139.0 – 171.0)	255 (232.8 – 279.2)	600–800
14	$-N^+(CH_3)_2(C_2H_5)$	56 (26.1 – 85.9)	78 (55.7 – 109.2)	250–350
5	$-N^+(CH_3)(C_2H_5)_2$	43 (26.5 – 59.5)	37.5 (32.4 – 43.3)	150–250
15	$-N^+(C_2H_5)_3$	25 (25.4 – 33.6)	34 (31.6 – 36.5)	120–150
16	$-N^+$ (pyrrolidinium, CH_3)	44 (34.0 – 54.0)	31.5 (28.8 – 34.8)	130–170
9	$-N^+$ (piperidinium, CH_3)	30 (24.1 – 35.9)	22.5 (15.6 – 32.4)	110–150
17	$-N^+$ (piperidinium, C_2H_5)	19.8 (12.4 – 27.2)	21.7 (17.3 – 27.1)	130–160
18	$-N^+$ (morpholinium, CH_3)	74 (48.3 – 99.7)	81 (74.5 – 88.0)	500–600
Tubarine			137 (105.0 – 178.0)	180–230
Atropine sulfate		1.3 (0.5 – 2.1)		

TABLE 4: Muscarinic blocking and curare-like activity of drugs.

DRUG	Muscarinic blocking Activity* (ED_{50})	Curare-like Activity (MED)
Antratruxonium	19.8 (12.4 _ 27.2)	100–130
Cyclobutonium	43.0 (26.5 _ 59.5)	130–180
Decadonium	18.2 (16.5 _ 19.9)	250–300
Diadonium	26.1 (24.0 _ 28.2)	250–350
Tubocurarine	**	180–230
Decamethonium	***	30–40
Succinylcholine	****	60–80
Atropine	1.3 (0.5 _ 2.1)	

*	as in Table 1.
**	No effect at doses of 125 and 300 µg/kg.
***	No effect at doses of 30 and 120 µg/kg.
****	No effect at doses of 80 and 240 µg/kg.

activity. Thus, bis-esters of α-truxillic acid (see (I) where
A = -0-; n = 3; N^+R_3 = $N^+(C_2H_5)_2CH_3$) reduced ACh bradycardia in
anesthetized cats by 50% at a mean dose of 43 µg/kg i.v. For its
amide analog (see I) where A = -NH-; n = 3; N^+R_3 = $N^+(C_2H_5)_2CH_3$)
the ED_{50} was 444 µg/kg. The amide completely blocked transmission
from the sciatic nerve to the gastrocnemius muscle in anesthetized
cats in doses of 40-50 µg/kg and the ester in doses of 150-250 µg/kg.

A more detailed study was made on two of the α truxillic acid
derivatives (see (I) where anatruxonium n = 3; A = 0; $^+NR_3$ = $^+$N-ethyl-
piperidile) and cyclobutonium (where n = 3; A = 0; $^+NR_3$ = $N^+(C_2H_5)$
CH_3). Their anticholinergic effects were tested on the heart and
arterial pressure as well as on muscarinic receptors of the bronchi,
urinary bladder, ileum and salivary glands (Tables 4 and 5). Three or
four parameters were recorded simultaneously.

As mentioned above, both agents abolished the ACh induced
bradycardia but not hypotension. The initial antimuscarinic effect
appeared after the administration of 5-10 µg/kg of the compounds
and the complete elimination of bradycardia was obtained with 40-50
µg/kg of anatruxonium (Fig. 1) and 100-120 µg/kg of cyclobutonium.
Increasing the doses of these agents, even up to 500 and 1000 µg/kg,
did not reduce the hypotensive effect of ACh (Fig. 2). In most

---

NOTE: In the USSR anatruxonium and cyclobutonium are used in
anesthesiology as antidepolarizing curare-like agents.

TABLE 5: Comparative actions of curare-like drugs in myoparalytic doses on ACh evoked responses on anesthetized cats.

Response	DRUGS				
	Anatruxonium (150 µg/kg)	Cyclobutonium (200 µg/kg)	Decadonium (300 µg/kg)	Diadonium (350 µg/kg)	Atropine (50 µg/kg)
Bradycardia	CB	CB	CB	CB	CB
Bronchospasm	Decrease 49.5% (38.7–60.3)	Decrease 46.8% (36.0–57.6)	Increase 200% (50.4–349.6)	Decrease 25–50%(3) NS (4) Increase 25–100%(4)	CB
Hypotension	NS	NS	NS	NS	Decrease 57.9% (46.0–68.9)
Salivation	NS	NS	Increase 173.5% (102.1–244.9)	Increase 84.7% (50.5–118.9)	CB
Ileum contractions	NS	NS	Increase 53.7% (14.0–93.4)	NS (3) Increase 25–100% (3)	Decrease 91.6% (86.7–96.5)
Urinary bladder contractions	NS	NS	Increase 124.4% (48.4–200.4)	Increase 25–100% (5) NS (4)	Decrease 84.9% (78.0–91.8)

CB = complete block; NS = no significant change; mean with confidence limits (p = 0.05) estimated from six or more experiments; number of experiments in parentheses.

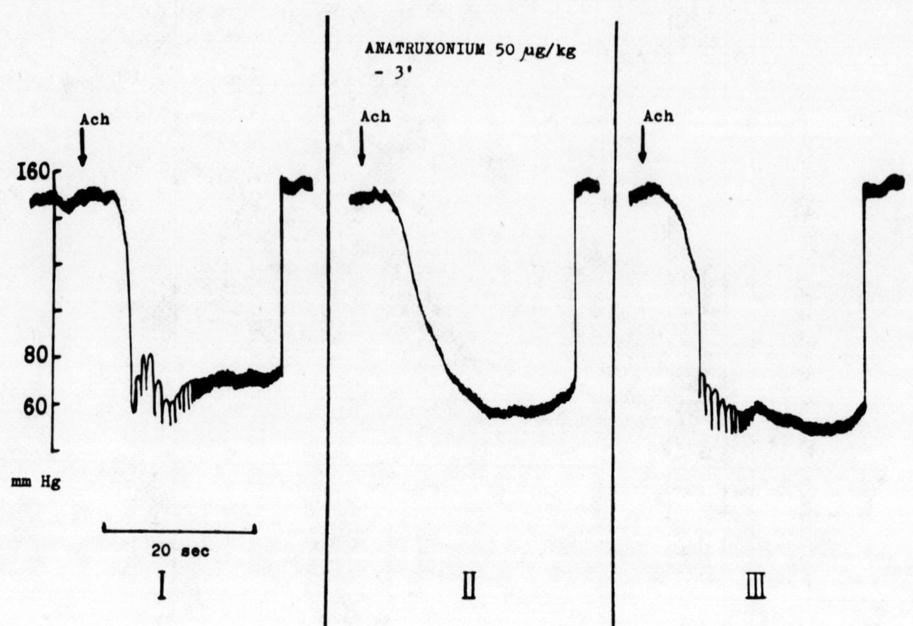

FIGURE 1: Effect of anatruxonium on ACh induced bradycardia and
           hypotension in an anesthetized cat.
           I - 12.15 - before the injection of anatruxonium.
           II - 12.30 - 3 min after the injection of anatruxonium
                        (50 µg/kg).
           III - 13.00 - 33 min after the injection of anatrixonium.
           ACh (20 µg/kg i.v.).

experiments, the drugs neither affected ACh induced salivation (Fig.
2) nor the contractions of the ileum and the urinary bladder. At
the same time, anatruxonium and cyclobutonium dose dependently
reduced, but did not abolish ACh action on the bronchi. Although
they considerably decreased bronchospasm, their blocking action
on the muscarinic receptors of the bronchi was less pronounced than
on the heart.

Anatruxonium induced a parallel shift to the right of the concentration response curves for carbachol on the isolated atrium and
ileum with no suppression of the maxium response. The pA_2 value
for atria was considerably greater than the pA_2 for the ileum
(Table 6).

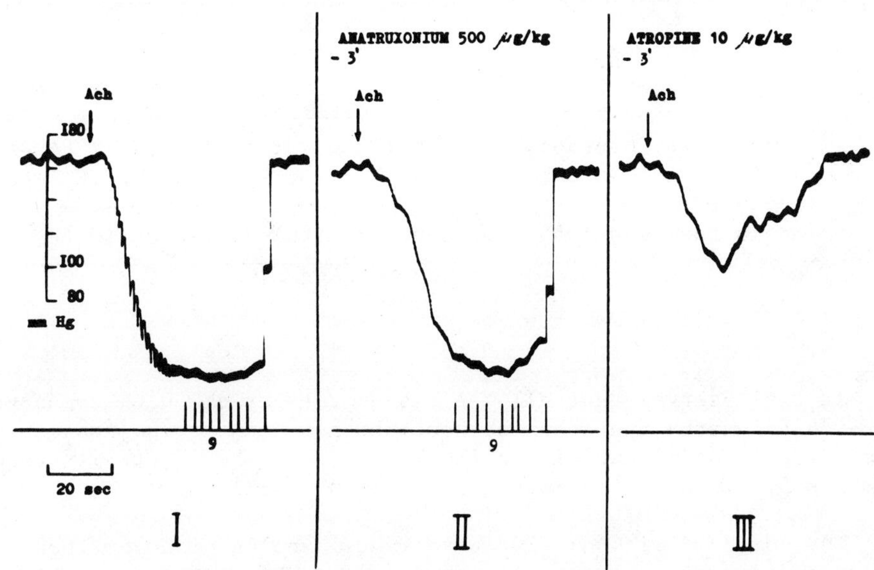

FIGURE 2: Effect of anatruxonium and atropine on ACh induced brady-
cardia, hypotension and salivation in an anesthetized
cat.
Downwards: arterial pressure, salivation from the
parotis in drops (number of drops indicated by vertical
lines and ciphers).
  I - 14.10 - before the injection of anatruxonium
 II - 14.25 - 3 min after the injection of anatruxonium
              (500 µg/kg)
III - 14.40 - 3 min after the injection of atropine
              (10 µg/kg)
ACh (20 µg/kg i.v.)

Bis-quaternary Ammonium Compounds
with Adamantyl Radicals

Earlier it was shown that the addition of a highly hydrophobic
adamantyl radical to different parts of the molecule of some nicotinic agonists transformed them into curare-like agents (12,13).
Some of the bis-quaternary ammonium compounds tested were shown to
possess antimuscarinic activity as well. It should be mentioned
that, like the truxillic acid derivatives, many compounds with
adamantyl radicals are characterized by their marked action on
cardiac muscarinic receptors.

TABLE 6:   Affinity of anatruxonium and atropine for muscarinic
           receptors of the isolated atrium and ileum of the rat.

Organ	Carbachol pD_2	Anatruxonium pA_2	Atropine pA_2
Atrium	6.34 (6.23–6.45)	7.50 (7.44–7.56)	8.75 (8.60–8.92)
Ileum	6.74 (6.64–6.84)	6.28 (6.17–6.39)	8.85 (8.74–8.96)

Each value is the mean (95% confidence limits) of 6–8 estimates.

It was observed that all the bis-quaternary ammonium compounds
with the polymethylene chain between the nitrogen atoms possess
cardiotropic antimuscarinic action (Table 7).   Their activity rose
from n = 5-6 to n = 7 and was maximum at n = 8-11.

The compound with n = 10 (N-adamantyl analog of decamethonium),
called decadonium, was examined in detail.   When administered in
small doses (50-60 µg/kg), decadonium abolished the negative chronotropic action of ACh without significantly changing its hypotensive
effect.   Higher doses (100-1000 µg/kg) did not alter, or increased
and prolonged the ACh hypotensive action.   The responses of the
bronchi, salivary glands, ileum and urinary bladder increased after
the administration of decadonium.   This may be accounted for by
the considerable anticholinesterase activity of decadonium.

Similar results were obtained for the N-adamantyl analog of
succinylcholine, diadonium (II)(9).

$$AD \ (CH_3)_2 N^+ - (CH_2)_2 - O - CO - (CH_2)_2 - CO - O - (CH_2)_2 N^+ (CH_3)_2 \ AD \quad 2I^- \qquad (II)$$

Like decadonium, this drug is characterized by its blocking action
on cardiac muscarinic receptors in doses of 80-90 µg/kg, while at
doses up to 2000 µg/kg it did not affect the hypotensive effect of
ACh.   Unlike decadonium, which increased the bronchospasm, diadonium
either reduced, increased or, in some cases, did not affect bronchospasm at all.   This variability of results can be explained as
being due to the dual counteractions of the antimuscarinic and
anticholinesterase properties of diadonium.   The anticholinesterase
activity of diadonium is lower than that of decadonium.   The K_i of
decadonium in human erythrocytes was $5.2 \cdot 10^{-7}$ M (non-competitive
antagonism).   Diadonium manifested a mixed type of action: for

TABLE 7:   Antimuscarinic activity of bis-quaternary ammonium compounds.

$$CH_3\overset{|}{\underset{R}{N}}-(CH_2)_n-\overset{|}{\underset{R}{N}}-CH_3 \cdot 2X$$

n	R	X	EC_{50}		
5	Ad	CH_3I	2.18 (1.59–2.78)	x 10^{-5}	M
6	Ad	CH_3I	2.13 (1.26–2.99)	x 10^{-5}	M*
7	Ad	CH_3I	8.36 (6.05–10.66)	x 10^{-6}	M
8	Ad	CH_3I	6.13 (3.60–8.66)	x 10^{-6}	M*
9	Ad	CH_3I	6.03 (4.86–7.20)	x 10^{-6}	M
10	Ad	CH_3I	4.75 (2.60–6.90)	x 10^{-6}	M*
10	(cyclohexyl)	CH_3I	4.55 (2.46–6.64)	x 10^{-5}	M
10	(phenyl)	CH_3I	7.67 (5.90–9.44)	x 10^{-6}	M
10	Ad	HCl	> 5	x 10^{-4}	M
11	Ad	CH_3I	3.65 (2.30–5.00)	x 10^{-6}	M

Concentrations of compounds inhibiting the negative iontropic effect of carbachol by 50% (means with 95% confidence limits). Experiment on isolated frog ventricle.
* In anesthetized cats, compounds (n = 6,8 and 10) completely block ACh induced bradycardia in doses of 1200–1500, 60–80 and 50–60 µg/kg, respectively, without affecting the hypotensive effect of ACh.

competitive inhibition, K_i = 3.9·10^{-6} M, for non-competitive inhibition, K_i = 1.7·10^{-5} M. Diadonium did not change, or increased, the stimulating effect of ACh on the muscarinic receptors of the ileum or urinary bladder, and in all experiments enhanced ACh induced salivation. Thus, decadonium and diadonium are characterized by their pronounced selectivity for the muscarinic receptors of the heart (Tables 4 and 5).

Compounds of Different Chemical Structure

A prevailing action on cardiac muscarinic receptors was found for some antihistamines and procaine. Mebhydroline hydrochloride abolished ACh induced bradycardia in cats in doses of 1-2 mg/kg without affecting its hypotensive action (Fig. 3). Even doses of 15-20 mg/kg failed to affect the latter.

On the spontaneously beating rat atrium, mebhydroline hydrochloride ($5 \cdot 10^{-7}$-10^{-6} M) antagonized the effect of carbachol. It evoked a parallel shift to the right of the concentration response curves without any change of the maximal response.

Diphenhyramine (5-10 mg/kg) and promethazine (0.5-3.5 mg/kg) showed effects similar to mebhydroline. Thus, all three antihistaminics, though quite different in their chemical structure, possess pronounced cardiotropic antimuscarinic activity with practically no effect on the muscarinic receptors of the blood vessels. Similar results were obtained for procaine (5-10 mg/kg) which eliminated bradycardia without changing the hypotensive effect of ACh and antagonized the action of carbachol on the rat atrium.

DISCUSSION

It is known that atropine is not uniformly potent in its effects on various organs. Similarly, the bis-quaternary ammonium salts of truxillic acids demonstrate a pronounced selectivity for cardiac muscarinic receptors. In cats, anatroxonium and cyclobutonium, the derivatives of α truxillic acid, effectively blocked the muscarinic receptors of the heart in small doses and failed to block ACh receptors of the ileum, urinary bladder, blood vessels and salivary glands. Their antimuscarinic action on the bronchi was weak. The selectivity of anatruxonium for the muscarinic receptors of the heart was confirmed in the experiments on isolated rat atrium and ileum. While the affinities of atropine for the muscarinic receptors of the atria and ileum were similar, those of anatruxonium differed considerably for these two tissues being markedly higher for the atrial muscarinic receptors.

A selective blocking action on cardiac muscarinic receptors was demonstrated by some bis-quaternary ammonium compounds with adamantyl radicals, including diadonium and decadonium. These compounds are particularly interesting as they contain hydrophobic adamantyl radicals which can conceivably form additional bonds with muscarinic receptors. In the derivatives of truxillic acids, the

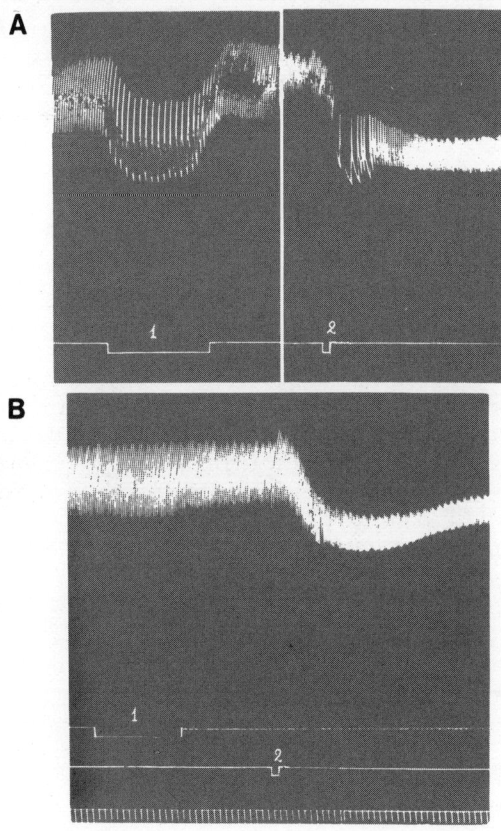

FIGURE 3: Effect of mebhydroline hydrochloride on the muscarinic
          receptors of the heart of a decerebrated cat.
          A – before the injection of mebhydroline hydrochloride
          B – 2 min after injection of mebhydroline hydrochloride
              (5 mg/kg i.v.)
          1 – stimulation of the peripheral end of the vagus
              nerve
          2 – an injection of ACh (5 μg/kg i.v.)
          Time scale: 1 division = 5 sec

central portion of the compounds (diphenylcyclobutanedicarboxylic
structure) is also highly hydrophobic. The hydrophobic interaction
(in the case of complementarity of hydrophobic structures to the
receptor) may be of considerable importance for the manifestation
of the antimuscarinic cardiotropic effect (10). Indeed, the analogs
without the hydrophobic structures (bis-quaternary ammonium salts of
aliphatic dicarboxylic acids and decamethonium) are devoid of these
properties.

The importance of hydrophobic interactions of muscarinic antagonists with the muscarinic receptors was also observed for 3-quinuclidyl benzilate (1). A significant decrease of the antimuscarinic cardiotropic activity of the α truxillic acid derivatives results from the replacemnt of carboxyl groups by amide groups. The same quality was shown by mono-quaternary ammonium derivatives of diphenylacetic and phenylacetic acids (2). The affinity of these compounds for the muscarinic receptors of the isolated guinea pig ileum decreased with the replacement of the ester groups by amide groups.   It is assumed that this peculiarity is due to the rigidity of amide bond which hinders the necessary contact of the agent with the receptor. On the other hand, in accordance with the existing concept, the less flexible compounds possess a higher selective activity. For bis-amides of α truxillic acid, this property is manifested in their high curare-like activity and feeble antimuscarinic activity when compared with the ester analogs. Thus, the selectivity of amides for the nicotinic receptors of the muscle endplate is much more pronounced than that of esters. One should keep in mind that the value for carbonyl oxygen in amides is smaller than in esters and the hydrophobibity of the former is somewhat lower. At the same time, the NH- group possess a mobile hydrogen atom which can participate in the intermolecular interactions and prevent the agent's binding with the muscarinic receptor.

In spite of its high anticholinesterase activity, decadonium, a bis-N-adamantyl compound, demonstrates a pronounced antimuscarinic action on the heart over a wide range of doses. Nevertheless, decadonium facilitates other effects of ACh, such as bronchospasm, contractions of the ileum and urinary bladder, and when used in larger doses, potentiates the hypotensive effect. These data demonstrate a high selective sensitivity of the muscarinic receptors of the heart for decadonium. Diadonium also possesses a pronounced selectivity for the muscarinic receptors of the heart.

In the bis-quaternary ammonium series of α-truxillic acid, changes in antimuscarinic activity paralleled the changes in the curare-like activity. Nevertheless, this does not mean that the structure of the muscarinic receptors of the heart is analogous to the ACh receptors of the endplate as was assumed by Saxena and Bonta (17). The present investigation, and some others, reveal that the pronounced affinity for the muscarinic receptors of the heart is peculiar to many non-curariform compounds, some antihistaminics and procaine (4,11,17). In addition, there are curare-like agents with no antimuscarinic activity in myoparalytic doses (decamethonium, succinylcholine, tubocurarine).

In conclusion, in accordance with these data, it would appear possible to design selectively acting antimuscarine cardiotropic

agents and broncholytic agents without blocking secretory activity
of exocrine glands.

REFERENCES

1. Aronstam, R.S., Abood, L.G. and Baumgold, J. (1977): Biochem.
   Pharmacol. 26:1689–1695.
2. Barlow, R.B., Bremner, J.B. and Soh, K.S. (1978): Brit. J.
   Pharmacol. 62:39–50.
3. Benfey, B.G., Yong, M.S., Belleau, B. and Melchiorre, C. (1979):
   Canad. J. Physiol. Pharmacol. 57:41–47.
4. Kharkevich, D.A. (1957): Farmakol. i Toksikol. 6:46–51.
5. Kharkevich, D.A. (1963):Abst. Sec. Int. Pharmacol. Meeting,
   Prague, 166, No.588.
6. Kharkevich, D.A. (1965): Farmakol. i Toksikol. 3:305–309.
   (Fed. Proc. 25,T521, 1966).
7. Kharkevich, D.A. (1966): Farmakol. i Toksikol. 1:47–53.
8. Kharkevich, D.A. (1970): Farmakol. i Toksikol. 4:395–399.
9. Kharkevich, D.A. (1970): Farmakol. i Toksikol. 5:531–536.
10. Kharkevich, D.A. (1974): Farmakol. i Toksikol. (Kiev, vyp)
    9:94–102.
11. Kharkevich, D.A. (1974): J. Pharm. Pharmacol. 26:153–165.
12. Kharkevich, D.A. (1975): Proc. Int. Congr. Pharmacol. 1:33–47.
13. Kharkevich, D.A. and Skoldinov, A.P. (1976): J. Union Mendel-
    eev's Chem. Soc. 21:124–129.
14. Richer, W.F. Jr. and Wescoe, W.C. (1951): Ann. NY Acad. Sci.
    54:373–394.
15. Rosen, J.M., Rauckman, E.J. and About-Donia,M.B. (1976): Bio-
    chem. Pharmacol. 25:2761–2764.
16. Samoilov, D.N. (1971): Farmakol. i Toksikol. 4:413–420.
17. Saxena, P.R. and Bonta, J.L. (1970): Eur. J. Pharmacol. 11:
    332–341.
18. Shorr, V.A. (1972): Farmakol. i Toksikol. 4:426–431.

LIGAND INTERACTIONS OF AXONAL MEMBRANES

H.C. Mautner, J.E. Jumblatt and J.K. Marquis

Department of Biochemistry and Pharmacology
Tufts University School of Medicine
Boston, Massachusetts   02111 USA

The postulate (18) that the acetylcholine (ACh) cycle plays
an essential part in controlling the cation permeability of electrically excitable postsynaptic membranes is accepted generally.

However, whether this cycle is involved in the permeability
changes of axonal membranes remains controversial. In general, cholinergic agonists and antagonists do not affect axonal function. However, in unmyelinated nerve fibers (1,21) and nodes of Ranvier (7),
where the structural barriers preventing access to trialkylammonium
compounds are minimal, physiological effects induced by cholinergic
ligands have been noted.

In studies of a series of local anesthetics in which one or
both of the oxygens of the ester group had been replaced with sulfur or with selenium, we noted that in intact squid axons the
blocking activity of these compounds was abolished if the dialkylamino group was quaternized. Ability to block axonal conduction
increased as the oxygens of the ester group were replaced by sulfur
or selenium (22) (Table 1). The relative abilities of analogous
esters, thiolesters, selenolesters, thionesters, thionothiolesters
and thionoselenolesters to block axonal conduction were paralleled
by their abilities to block the carbamylcholine induced depolarization of the electroplax preparation. However, in the electroplax
preparation, in which no structural barriers interfere with access
of onium compounds to their binding sites, the blocking actions
of tertiary amines and their trialkylammonium analogs were very
similar (Webb and Mautner, unpublished data) (Table 1).

367

TABLE 1:

$$\text{A} = \underset{\overset{\|}{\text{CBCH}_2\text{CH}_2}}{\overset{}{\bigcirc}}\overset{\overset{R}{\underset{|}{}}}{\underset{\underset{|}{\text{CH}_3}}{\text{N}}}{}^+\!\cdot\text{CH}_3$$

$$10^{-4}\ \underline{\text{M}}$$

A = O,S

B = O,S, Se

R = H (tertiary)

R = RH$_3$ (quaternary)

A	B	Electroplax Preparation: Percent Block of Carbamylcholine Response*		Squid Axon, Action Potential Percent Block of Electrical Activity**	
		Tertiary	Quaternary	Tertiary	Quaternary
O	O	10	15	0	0
O	S	40	58	25	0
O	Se	99	100	57	0
S	O	57	59	100	0
S	S	63	90	100	0
S	Se	97	91	100	0

*   Webb and Mautner (unpublished data)
** See ref. 22.

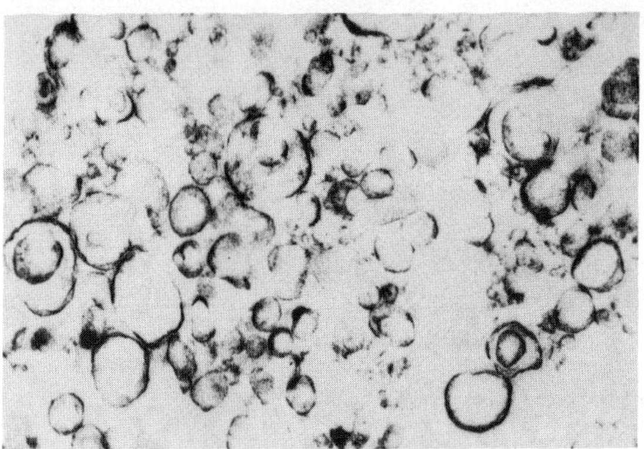

FIGURE 1: Lobster walking leg nerve membrane fraction after incubation in α-Btx-HRP conjugate and assay for peroxidase.
From Chester et al. (4).

The presence of "axonal cholinergic binding macromolecules" in
crustacean axonal membranes was reported several years ago (2,5,6,9).
In extending this work we compared the abilities of a series of
isologous local anesthetics carrying the ester, thiolester and
selenolester groupings to displace ^{3}H-nicotine from axonal plasma
membrane fragments derived from Homarus americanus (17). The relative abilities of oxo-, thio- and seleno- analogs to displace nicotine paralleled their relative blocking activities in axonal or
synaptic preparations. Quaternization of the dialkylamino groups
to form cationic trialkylammonium compounds, did not reduce the
abilities of the various analogs to displace nicotine, suggesting
that the barriers interferring with the binding of onium compounds
in intact axons were absent in the membrane fragments (17).

Electron microscopy showed that the membrane fragments formed
closed vesicles (4). Direct binding of ^{125}I-α-bungarotoxin (α-Btx)
to purified axon plasma membrane fragments and to solubilized membrane material could be demonstrated (14). Visualization of cholinergic binding sites on the electron microscopic level can be
achieved using a conjugate of α-Btx with horseradish peroxidase (12,
13). The binding of toxin to membrane vesicles derived from lobster walking leg nerves could be demonstrated using this technique
(4). The binding of the conjugate to vesicles could be abolished
by pretreatment with native toxin or with d-tubocurarine (Figs. 1,
2 and 3).

Electron microscopic localization of toxin binding sites could
also be accomplished using intact fibers from Homarus americanus or
the spider crab Libinia cmarginata (4). Where Schwann cell

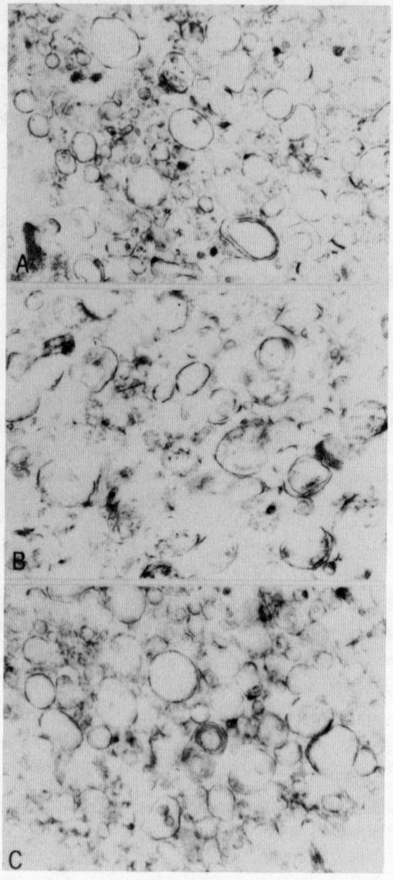

FIGURE 2: (A) Incubation in buffer alone. (B) Pretreatment with
native α-Btx and incubation in conjugate. (C) Pretreatment with d-tubocurarine and incubation in the conjugate. From Chester et al. (4).

---

membranes could be identified and distinguished from the axolemma,
these were generally unreactive. These findings are of interest
since previous studies of Rawlins and Villegas (20) showing binding
of ^{125}I-α-Btx at the axon-Schwann cell boundary could not distinguish on which surface binding actually occurred. Recently,
Freeman and Lentz (personal communication) have demonstrated the
binding of α-Btx horseradish peroxidase at the nodes of Ranvier of
rat sciatic nerve. Pretreatment of the nerves with native toxin
or with tubocurarine blocked the binding of the conjugate. Interestingly, if the myelin sheath was broken or separated by

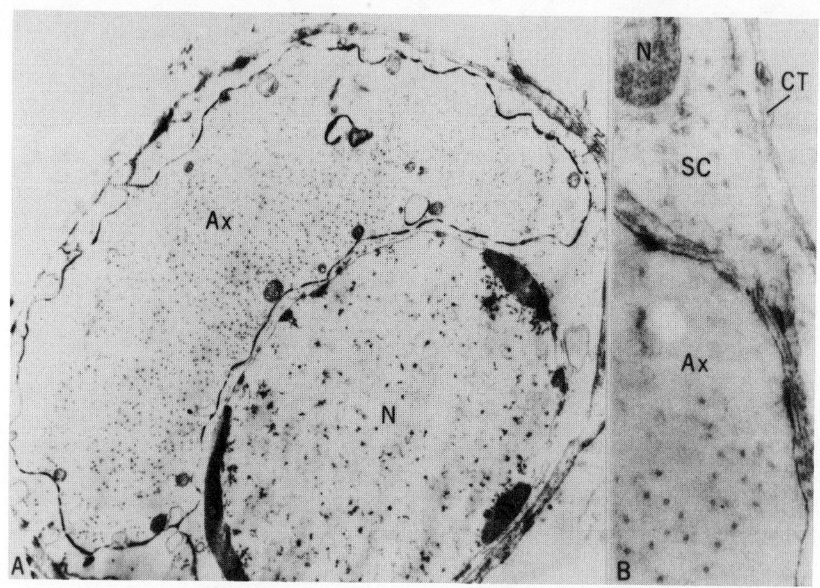

FIGURE 3: (A) Lobster axon (Ax) enveloped by Schwann cell (SC).
The nerve was incubated in conjugate and assayed for
peroxidase. The axon contains large numbers of micro-
tubules. N = Schwann cell nucleus. (B) Portion of spider crab axon (Ax) and adjacent Schwann cell (SC) identified by its nucleus (N) after incubation in conjugate
and assay for peroxidase. From Chester et al. (4).

enzymatic treatment, the axolemma in those regions was as reactive
as at the nodes of Ranvier.

It should be noted that while axonal membrane fragments and
solubilized material isolated from axonal membranes can bind ligands
believed to be specific labels of ACh receptor, and while the attachment of these ligands can be blocked by certain cholinergic
agonists and antagonists, major differences between the binding of
"cholinergic ligands" to synaptic and axonal membranes can be demonstrated. For instance, it was previously noted by Denburg et
al. (6) that procaine inhibits nicotine binding to axonal membranes
competitively, a finding we confirmed (4). On the other hand, synaptic membranes bind ACh much more tightly than axonal membranes
and competition between cholinergic agonists and local anesthetics
is more complex. In further contrast to synaptic membranes, the
axonal membranes bind α-Btx reversibly. While curare sensitive
toxin binding to axonal membranes is saturable, the binding has a
relatively low affinity (K_D 10^{-7} M), and is highly sensitive to ionic

TABLE 2:   Inhibition of binding of $[^{125}I]$-α-Btx (2.4×10^{-8} M) to
           axon plasma membrane fragment by various drugs.

DRUG (10^{-4} M)	PERCENT INHIBITION
d-Tubocurarine	69
Atropine	84
Nicotine	43
Carbamylcholine	15
Choline	17
Acetylcholine	10
($\pm$) Quinuclidinyl benzilate	0
Physostigmine	74
Neostigmine	16
Procaine	82
Tetracaine (3°C)	37
Tetracaine (4°C)	50
Lidocaine	29
GABA	0
Bicuculline	0
Octopamine	0
Tetrodotoxin	0
4-Aminopyridine	0
Veratridine	0

strength (Jumblatt and Mautner, unpublished data). Since the introduction of α-Btx as a "specific ligand" for labelling ACh receptor
(11), numerous studies of the affinities of this and related toxins
for membrane bound and for solubilized ACh receptor have been carried out using a variety of preparations. Comparisons of relative
affinities of toxin binding sites are complicated by the fact that
in many preparations the binding of α-Btx is very poorly reversible.
Furthermore, while the binding of cholinergic agonists and antagonists is blocked by α-Btx and related toxins, the opposite is not
necessarily the case (8). ACh receptor isolated from the electric
organ of electric fish has twice the number of α-Btx binding sites
as the number of binding sites for cholinergic agonists and antagonists (26). In the case of neurons, Patrick and Stallcup (19) and
Carbonetto et al.(3) have raised questions regarding reliability of
α-Btx as an indicator of the presence of nicotinic ACh receptor, although it is known that this toxin is bound to axons of cultured
chick sympathetic neurons.

    Table 2 summarizes the relative abilities of a series of ligands
to inhibit the binding of ^{125}I-α-Btx to lobster axon plasma membrane

fragments. Several of these ligands have been studied previously
for their abilities to displace ^{3}H-nicotine from axon plasma membrane fragments. Atropine proved to be more potent in displacing
α-Btx than d-tubocurarine, even though the former ligand is supposed to be a classical marker for muscarinic ACh receptor. Yet
quinuclidinyl benzilate, a molecule used widely as a label for muscarinic ACh receptor, proved unable to compete for the attachment
of α-Btx. Physostigmine, a tertiary amine was more potent than
neostigmine, a quaternary amine, in displacing α-Btx while procaine,
a local anesthetic of relatively low potency, was a better inhibitor
of toxin binding than the more potent derivative tetracaine or its
quaternary analog. Similarly, procaine is more effective in displacing labelled nicotine than is its physiologically more potent
analog tetracaine. Carbamylcholine and choline (Ch) show equal
ability to displace α-Btx, in contrast to the observation that the
former but not the latter compound was able to compete for the
attachment of ^{3}H-nicotine (17) in the same preparation. The high
ability of atropine to displace ^{125}I-α-Btx is paralleled by its
ability to displace ^{3}H-nicotine (6). These findings emphasize that
the axonal binding sites for either α-Btx or nicotine have little
resemblance to the binding sites of classical "nicotinic" or "muscarinic" ACh receptor. The findings also emphasize the lack of
importance of lipophilicity in determining access to the binding
sites on the axonal membrane vesicles.

Since labelling with α-Btx horseradish peroxidase conjugate
and the susceptibility of such labelling to pretreatment with d-
tubocurarine give parallel labelling of vesicles and of axolemma,
it can be assumed that the vesicles are "right side out." In support, solubilization of vesicles does not appear to increase the
levels of nicotine or α-Btx binding. The solubilization of "axonal
cholinergic binding material" using lysolecithin was reported by
Denburg (5) several years ago. Since then, several other solubil-
izing agents, including digitonin, butanol extraction and a series
of detergents (15), have been studied in our laboratory. Interestingly, 2 M NaCl can also be used for the partial solubilization
of this material.

Since it is known that nicotinic ACh receptor, as well as
AChE are glycoproteins (23,27) capable of being bound to concanavalin A-Sepharose columns, the con-A binding properties of solubilized axonal membrane fragments, as well as of AChE isolated from
lobster axon plasma membranes, were investigated. Elution of AChE
from con-A-sepharose with α-methyl mannoside yielded good purification of the enzyme (16), similar to the results achieved by Wiedmer
et al. (27) with electroplax enzyme. The columns retained only
about 30% of solubilized axonal membrane proteins. The fractions
retained, as well as the fractions not retained on the lectin
column were able to bind ^{125}I-α-Btx. However, only the con-A

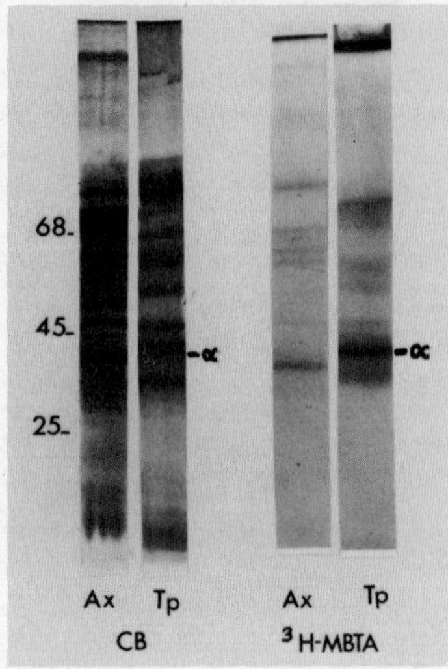

FIGURE 4: SDS slab gel electrophoresis of membrane fragments derived from lobster axons (Ax) or <u>Torpedo</u> <u>nobiliana</u>
electroplax (Tp). (CB) Polypeptides stained with Co-
massie Blue [³H]-MBTA labelled polypeptides visualized
by autofluorography.

binding fraction contained d-tubocurarine sensitive binding sites
(16). Thus, it is possible to separate curare sensitive and curare
insensitive α-Btx binding material. It should be remembered, however, that all the labelling of axolemma or of membrane vesicles
with α-Btx peroxidase is curare sensitive.

SDS gel electrophoresis was carried out on membrane material
derived from lobster walking leg nerve bundles or from the electric
organ of <u>Torpedo</u> <u>nobiliana.</u> ³H-4-(N-maleimido) benzyltrimethyl-
ammonium (MBTA), a ligand believed to be capable of selective
attachment to ACh receptor (10) and more specifically, to its α
subunit (26), was investigated for its ability to label synaptic
and axonal membrane material (Jumblatt and Mautner, unpublished
data). As expected, ³H-MBTA labelled a peptide band of 40,000
daltons, in <u>Torpedo</u> membranes corresponding to the α subunit, as
shown by SDS polyacrylamide gel electrophoresis. A peptide of
ca 37,000 daltons was labelled in a parallel experiment using
axonal membranes from <u>Homarus</u> <u>americanus.</u> Preliminary results
suggest that in contrast to synaptic material, MBTA labelling of
axonal membrane material is poorly inhibited by α-Btx or nicotine.

In conclusion, it appears that axonal membranes carry a constituent having some of the characteristics of ACh receptor. Fractionated axonal membranes form closed vesicles, the outer surface of which carries the material capable of binding cholinergic ligands. While the ligand specificity of these binding sites is quite different from that of either nicotinic or muscarinic ACh receptor, the presence of such sites along the axolemma and on vesicles derived from axons deserves further study, particularly since the axolemma is also the site of localization of AChE (24). The report that higher levels of ACh are found in surrounding Schwann cells than in the squid giant axon (25), raises further interest in the axolemmal cholinergic ligand binding material, and its possible interactions with the Schwann cell membrane facing it (25).

ACKNOWLEDGEMENT

We are grateful to the National Science Foundation for a grant (BNS-7722356) in support of this work.

REFERENCES

1. Armett, C.J. and Ritchie, J.M. (1961): J. Physiol. $\underline{155}$:372-384.
2. Balerna, M., Fosset, M., Chicheportiche, R., Romey, G. and Lazdunski, M. (1975): Biochemistry $\underline{14}$:5500-5511.
3. Carbonetto, S.T., Fambrough, D.M. and Muller, K.J. (1978): Proc. Nat. Acad. Sci. $\underline{75}$:1016-1020.
4. Chester, J., Lentz, T.L., Marquis, J.K. and Mautner, H.G. (1979): Proc. Nat. Acad. Sci. $\underline{76}$:3542-3546.
5. Denburg, J.L. (1973): Biochim. Biophys. Acta $\underline{298}$:967-972.
6. Denburg, J.L., Eldefrawi, M.E. and O'Brien, R.D. (1972): Proc. Nat. Acad. Sci. $\underline{69}$:177-181.
7. Dettbarn, W.D. (1960): Nature $\underline{186}$:891-892.
8. Heidmann, T. and Changeux, J.P. (1978): Ann. Rev. Biochem. $\underline{47}$:317-357.
9. Jones, S.W., Galasso, R.T. and O'Brien, R.D. (1977): J. Neurochem. $\underline{29}$:803-809.
10. Karlin, A. and Winnik, M. (1968): Proc. Nat. Acad. Sci. $\underline{60}$: 668-674.
11. Lee, C.Y. (1972): Ann. Rev. Pharmacol. $\underline{12}$:265-281.
12. Lentz, T.L. and Chester, J. (1977): J. Cell. Biol. $\underline{75}$:258-267.
13. Lentz, T.L., Mazurkiewicz, J.E. and Rosenthal, J. (1977): Brain Res. $\underline{132}$:423-442.

14. Marquis, J.K., Hilt, D.C. and Mautner, H.G. (1977): Biochem.
     Biophys. Res. Comm. 78:476-482.
15. Marquis, J.K., Hilt, D.C. and Mautner, H.G. (1979): Biochem.
     Pharmacol. 28:2094-2096.
16. Marquis, J.K., Hilt, D.C. and Mautner, H.G. (1980): J. Neuro-
     chem. 34:1071-1076.
17. Marquis, J.K., Hilt, D.C., Papadeas, V.A. and Mautner, H.G.
     (1977): Proc. Nat. Acad. Sci. 74:2278-2282.
18. Nachmansohn, D. (1955): Harvey Lectures 1953-1954:57-99.
19. Patrick, J. and Stallcup, B. (1977): J. Biol. Chem. 252:
     8629-8633.
20. Rawlins, F.A. and Villegas, J. (1978): J. Cell. Biol. 77:
     371-376.
21. Ritchie, J.M. (1967): Ann. NY Acad. Sci. 144:504-516.
22. Rosenberg, P. and Mautner, H.G. (1967): Science 155:1569-1571.
23. Salvaterra, P.M., Gurd, J.W. and Mahler, H.R. (1977): J. Neuro-
     chem. 29:345-348.
24. Villegas, G.M. and Villegas, J. (1974): J. Ultrastruc. Res.
     46:149-163.
25. Villegas, J. and Jenden, D.J.(1979): J.Neurochem. 32:761-766.
26. Weill, C.L., McNamee, M.G. and Karlin, A. (1974): Biochem.
     Biophys. Res. Comm. 61:997-1003.
27. Wiedmer, T., Gentinetta, R. and Brodbeck, U. (1974): FEBS
     Lett. 47:260-263.

AXONAL TRANSPORT OF CHOLINE PHOSPHOGLYCERIDES TO

CHOLINERGIC SYNAPSES

B. Droz, M. Brunetti, L. di Giamberardino, H.L. Koenig
and G. Porcellati

Departement de Biologie, Commissariat a l'Energie Atomique
C.E.N., Saclay, B.P.2, 91190 Gif-sur-Yvette, France,
Istituto di Chimica Biologica, Universite di Perugia, Italy
and Laboratoire de Cytologie, Universite Pierre et
Marie Curie, Paris, France

INTRODUCTION

Phospholipid molecules play an important role in synaptic function. They are, indeed, major components of synaptic vesicles and
presynaptic plasma membranes. In cholinergic synapses, choline (Ch)
phosphoglycerides could also act as possible donors of Ch since the
axons do not synthesize this nitrogenous base in notable amounts.

Ch phosphoglycerides of nerve endings may originate from a
triple source : a local synthesis in the axon terminal, an intercellular translocation from adjacent cells and an arrival of axonally transported phospholipids after their synthesis in the neuronal
cell body.

Local incorporation of labelled precursors of various classes
of phospholipid has been investigated in the giant cholinergic synapse of the chicken ciliary ganglion. When ciliary ganglia are
incubated in vitro for 15 min or 60 min in an oxygenated medium
containing $(2-^3H)$glycerol or (methyl-^3H)Ch, the amount of label
incorporated into phospholipids of the presynaptic axon terminal
and of the postsynaptic ganglion cell body can be determined by radioautography. While the ganglion cell perikarya actively incorporate
both precursors into phospholipids, the nerve endings do not contain
significant amounts of label when incubated with $(2-^3H)$glycerol. It

is, therefore, concluded that no net synthesis of phosphoglyceride occurs in these nerve endings (9). However, when ciliary ganglia were incubated with (methyl-^{3}H)Ch, a slight but definite incorporation of the radioactive tracer was observed in the cholinergic calyces. This low but clear incorporation of the tracer takes place in the absence of net biosynthesis of phosphoglycerides and probably results from an exchange reaction of the nitrogenous base with pre-existing phospholipids (18).

An intracellular translocation of phospholipids from the surrounding cells to the axon terminals was not seen in the chicken ciliary ganglion. It seems improbable that Ch phosphoglycerides made in satellite or ganglion cells are delivered in sufficient amounts to the nerve endings. Hence, the axonal transport of phospholipid would supply the nerve endings with Ch phosphoglycerides. This view is supported by biochemical investigations performed in the central and peripheral nervous system (1,5,14,16,20). The data have clearly shown that the movement of phospholipid all along the axons proceeds at the same rate as the fast axonal transport of protein and glycoprotein. Since our radioautographic studies have pointed to the axonal smooth endoplasmic reticulum as the organelle ensuring mainly the fast axonal transport of membrane proteins and glycoproteins to the cholinergic synapses of the chicken ciliary ganglion (8,10), we thought that it would be important to determine whether axonally transported phospholipids do, or do not, follow the same intra-axonal pathway.

MATERIAL AND METHODS

Two labelled precursors of phospholipid were chosen; (2-^{3}H)-glycerol, which is incorporated into newly formed phosphoglycerides, displays only a little recycling of the label (3,17); and Ch labelled in the N-methyl position with ^{3}H or ^{14}C is also incorporated into newly synthesized Ch lipids, but is largely recycled by exchange with another base or, to a greater extent, by incorporation into other phospholipid molecules (17).

In a series of radioautographic experiments, chickens received an injection of (2-^{3}H)glycerol or (methyl-^{3}H)Ch into the cerebral aqueduct. From the midbrain, newly formed radioactive phospholipids were transported by axonal flow along the preganglionic nerves to the axon terminals located in the ciliary ganglia. One ganglion was processed for high resolution radioautography of lipids and the other was homogenized to measure the radioactivity in the water soluble fraction and in the lipid extract (7).

In a series of biochemical determinations, chickens were simultaneously injected into the cerebral aqueduct with $(2-{}^3H)$glycerol and (methyl-^{14}C)Ch. Axonally transported labelled phospholipids were extracted from the ciliary ganglia. The main classes of labelled phospholipids were separated by thin layer chromatography and their specific radioactivity was estimated at time intervals ranging from 1 hr to 10 days. In these double labelling experiments the changes of isotopic ratio ^{14}C/^{3}H in Ch phosphoglycerides were followed to determine the extent of recycling of the nitrogenous base (14).

RESULTS AND DISCUSSION

Axonal Transport of Choline Phosphoglycerides

When (methyl-^{14}C)Ch and $(2-{}^3H)$glycerol were injected into the cerebral aqueduct, radioactive phospholipids started to appear in the ciliary ganglion after a 1 hr delay. They were found to consist mainly of Ch phosphoglycerides: 95% of (methyl-^{14}C)Ch and 65% of $(2-{}^3H)$glycerol labelled lipids recovered in the ciliary ganglia are represented by Ch phosphoglycerides at 20 hr after the injection. It is important to emphasize that from 6 hr to 10 days after the injection of $({}^{14}$C) and $({}^3$H) labelled precursors, the isotopic ratio ^{14}C/^{3}H in Ch phosphoglycerides increased from 0.44 to 3.62 in the ciliary ganglia. This progressive change of the isotopic ratio indicates that (methyl-^{14}C)Ch is extensively re-utilized for local synthesis of new Ch lipids in the ciliary ganglia (4).

With both (methyl-^{3}H)Ch or $(2-{}^3H)$glycerol, light microscope radioautographs showed that all the preganglionic axons were invaded by axonally transported labelled lipids which piled up in the cholinergic nerve endings of the chick ciliary ganglion. Examination of electron microscope radioautographs indicates that the axonal label was associated with profiles of the smooth endoplasmic reticulum (Fig. 1). Quantitative studies reveal that transported phospholipids were mainly concentrated in areas occupied by the profiles of the axonal smooth endoplasmic reticulum (Table 1).

The use of heavy metal impregnation, which permits the visualization of the axonal smooth endoplasmic reticulum in thick sections prepared for electron microscopy, reveals that this organelle extends as an elongated network of tubules channeling the axoplasm up to the presynaptic nerve endings (19). In the axon terminals thinner tubules display frequent relationships with vesicles resembling synaptic vesicles. It is highly probable that this network

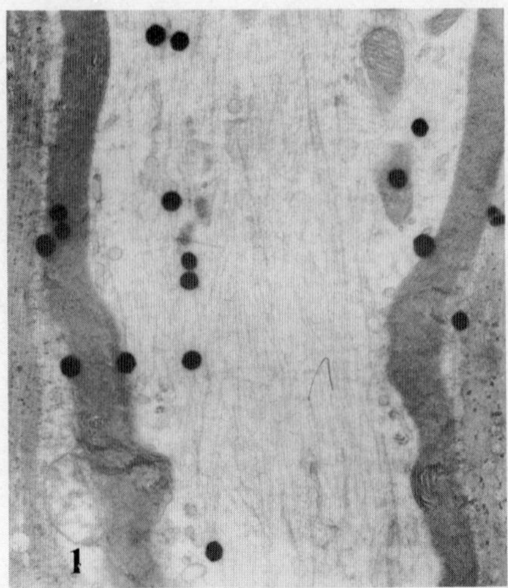

FIGURE 1: Electron microscope radioautographs of an axon after the
intracerebral injection of methyl-^{3}H-Ch. At 18 hr silver
grains are related to profiles of the axonal smooth endoplasmic reticulum and to the myelin sheath.

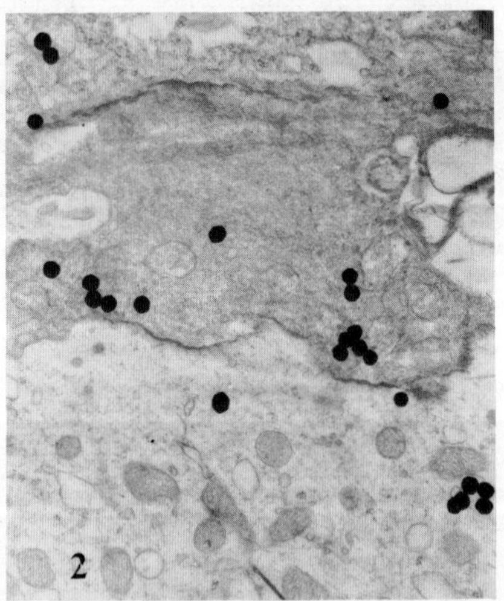

FIGURE 2: In a caliyciform nerve ending, the early labelling observed
at 6 hr after the intracerebral injection of methyl-^{3}H-Ch
points to the area occupied by the presynaptic plasma membrane and the closely apposited synaptic vesicles as the
most radioactive site of the cholinergic axon terminal.

is a dynamic structure continuously changing its configuration with
time, the membrane constituents of which flow continuously down to
the nerve endings. When axonal transport is interrupted, a simultaneous accumulation of labelled phospholipids and of profiles of
smooth endoplasmic reticulum occurs in the proximal segment (19).
It is, therefore, concluded that Ch phosphoglycerides as other membrane constituents of synapses are preferentially conveyed with the
smooth endoplasmic reticulum of the axon to the presynaptic nerve
endings in the cholinergic preganglionic nerves.

Distribution of Choline Phosphoglycerides
Transported to Presynaptic Axon Terminals

Electron microscope radioautographs performed at various time
intervals allowed us to study the distribution and kinetics of
phospholipids axonally transported to the calyciform nerve endings.
The smooth endoplasmic reticulum of the axon terminals shows the
highest concentration of label by 6 hr, the time of the arrival of
radioactive phospholipids (Fig. 3). This result suggests that Ch
phosphoglycerides are rapidly conveyed with the smooth endoplasmic
reticulum up to nerve endings (9,19). Later, the decrease in the
radioactivity in this organelle, while an increase is observed in
other structures, would be consistent with a transfer of labelled
phosphoglycerides from the smooth endoplasmic reticulum to other
membranous elements.

Presynaptic plasma membrane and synaptic vesicles closely apposited to this structure, exhibit clear differences in labelling,
depending on the injected precursor. Phospholipids labelled with
$(2\text{-}^3\text{H})$glycerol are rather scarce at 6 hr, whereas phospholipids
labelled with $(\text{methyl-}^3\text{H})$Ch are abundant at this early time interval (Fig. 2). This discrepancy, represented in Fig. 3, could be
due to the fact that, after the intracerebral injection of radioactive Ch, but not after glycerol, a definite peak of the water
soluble radioactivity is noted at 6-18 hr in the ciliary ganglion
(7). Thin layer chromatography revealed that this water soluble
radioactivity is mainly associated with compounds migrating with
Ch and acetylcholine (ACh)(4). To account for the early incorporation of Ch into the presynaptic plasma membrane and adjacent synaptic veiscles, a tentative theory could be advanced that radioactive Ch is exchanged <u>in situ</u> with unlabelled base of pre-existing
phospholipids by means of the rapid reaction of base-exchange (2).
If such a process could be confirmed as taking place in presynaptic
structures, a strategic localization could perhaps be ascribed to
the very restricted pool of phospholipids able to exchange base.

TABLE 1: Grain density counted over organelles of the preganglionic
          axons 6 hr after the injection of (2-^{3}H)glycerol or
          (methyl-^{3}H)Ch into the cerebral aqueduct.

	(2-^{3}H)glycerol	(methyl-^{3}H)Ch
Smooth endoplasmic reticulum	2.08 ± 1.73	5.03 ± 2.07
Axoplasm	0.03 ± 0.02	0.07 ± 0.02
Mitochondria	0.22 ± 0.19	0.27 ± 0.15
Axolemma	0.25 ± 0.17	0.31 ± 0.15

$$\text{grain density} = \frac{\text{percent grain distribution x grain concentration}}{\text{percent organelle volume}}$$

± S.D. (average of three chickens)

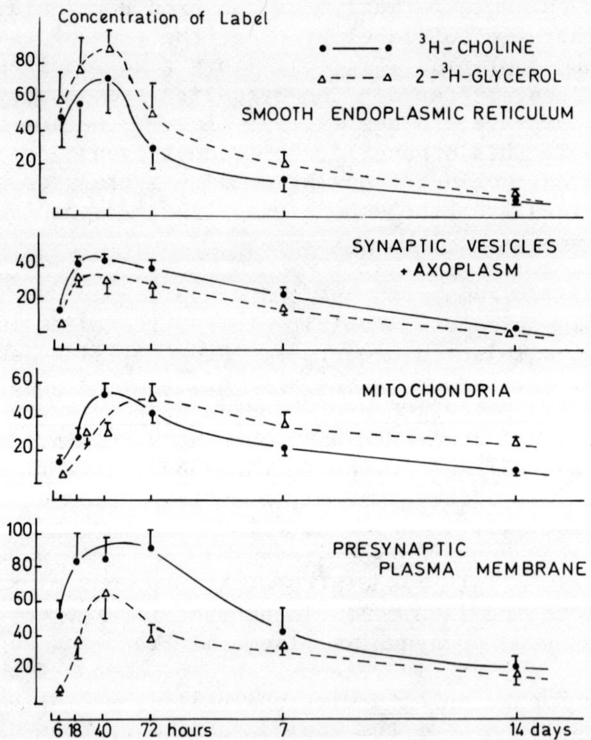

FIGURE 3: Time curves of label concentration in various components
           of the calyciform nerve endings in the chicken ciliary
           ganglion after injection of labelled precursors of phos-
           pholipids into the cerebral ventricle.

The great majority of the synaptic vesicle population follows
a similar pattern of labelling after the injection of radioactive
Ch or glycerol. It must be stressed that the radioautographic
technique used in this study preserves lipid but not water soluble
compounds such as free Ch or ACh, which are washed out. Thus the
rate of disappearance of the label reflects the turnover of the
labelled phospholipid constituent. By comparing the renewal of the
labelled phospholipid constituents with that of protein and glycoprotein in the synaptic vesicles of the calyciform nerve endings,
the kinetics indicate that synaptic vesicles are not renewed as a
whole entity but their constituents are probably recycled locally
(9).

Presynaptic mitochondria contain radioactive phospholipids
(Fig. 3). The presence of Ch phosphoglycerides in these organelles
would result from two mechanisms: first, mitochondria receive fast
transported Ch phosphoglycerides from smooth endoplasmic reticulum;
and second, mitochondria having incorporated radioactive phospholipids synthesized in the nerve cell body are slowly conveyed as
organelles along the preganglionic axons to nerve endings (Figs.3-5).
If we exclude a possible exchange of Ch at the level of the presynaptic membrane region (Figs. 2 and 3), Ch phosphoglycerides integrated within membranes of these cholinergic nerve endings are
provided with fast axonal transport which thus ensures a supply of
most synaptic Ch phosphoglycerides.

Release and Recycling of Axonal Choline

During axonal transport of phospholipids labelled with either
(2-^{3}H)glycerol or (methyl-^{3}H)Ch, the appearance of label in the
myelin sheath raises the question of whether there is an intercellular passage of labelled phospholipids from axon to myelinating
Schwann cell (6,7). Recently Brunetti (unpublished observation)
has definitely proved that a small amount of axonally transported
Ch phosphoglycerides is recovered in myelin fractions. Thus both
radioautography and cell fractionation point to a transneuronal
passage of Ch phosphoglycerides from the axon to its encompassing
myelin sheath (Fig. 1). In addition to this process, another mechanism is revealed when axonally transported phospholipids are
labelled with (methyl-^{3}H)Ch. Radioactive Ch of axonal origin,
probably freed from labelled Ch phosphoglycerides under the action
of a phospholipase activity, is indeed taken up by adjacent Schwann
cells and re-utilized for biosynthesis of new myelin Ch lipids (Fig.1).
This interpretation is supported by both radioautographic and biochemical data (4,6,7). Therefore axonal transport would participate
in the glial metabolism by providing myelin with base and Ch phosphoglyceride in the peripheral as well as in the central nervous
system (11,13).

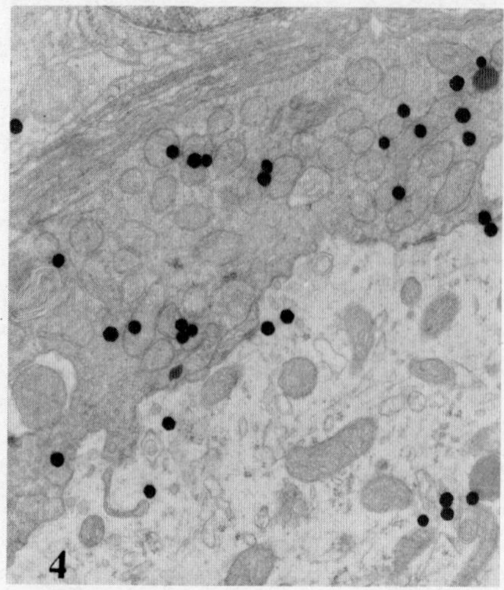

FIGURE 4: Electron microscope radioautograph of nerve endings in the ciliary ganglion 3 days after injection of (methyl-^{3}H)-Ch into the cerebral aqueduct of chickens. Note the labelling in the postsynaptic perikaryon.

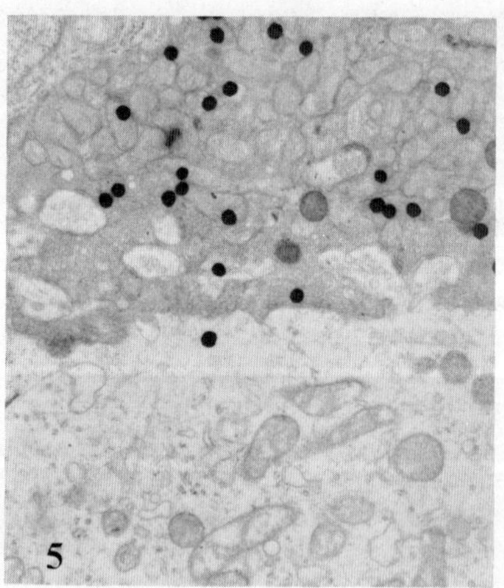

FIGURE 5: Electron microscope radioautograph of nerve endings in the ciliary ganglion 3 days after injection of (2-^{3}H)-glycerol into the cerebral aqueduct of chickens. Note the lack of labelling on the postsynaptic perikaryon.

In the axon terminals of the ciliary ganglion, the Ch phosphoglycerides labelled with (2-^{3}H)glycerol do not display any significant transfer of entire molecules of phospholipids from pre- to postsynaptic elements. In Fig. 5 only stray silver grains are found over the postsynaptic ganglion cell body. In contrast, Figs.2 and 4, obtained after injection of (methyl-^{3}H)Ch, show numerous silver grains located over the postsynaptic perikaryon and the satellite cells. Hence, although no transfer of Ch phosphoglyceride occurs, the presence of incorporated radioactive Ch can only result from recycling of Ch released from the axon terminal either as such, or as ACh. The reincorporation of labelled Ch in the ciliary ganglion is also supported by the results of double labelling experiments, which show a gradual increase of the isotopic ratio ^{14}C/^{3}H in Ch phosphoglycerides. In the same ganglion, a re-utilization of labelled amino acids released from broken down radioactive proteins has been observed (8). Thus the local catabolism of radioactive Ch phosphoglycerides in the cholinergic calyces would produce a release of free labelled Ch or a derivative in the narrow cleft situated between the nerve ending and the adjacent cells. The reuptake of the base by the postsynaptic ganglion cell for incorporation into newly synthesized phospholipids as well as by the cholinergic axon terminal itself for synthesis of its neurotransmitter reflects the importance of Ch recycling in the neuronal economy.

REFERENCES

1.   Abe, T., Haga, T. and Kurokawa, M. (1973): Biochem. J. 136: 731-740.
2.   Arienti, G., Brunetti, M., Gaiti, A., Orlando, P. and Porcellati, G. (1976): Adv. Exp. Med. Biol. 72:62-78.
3.   Benjamins, J.A. and McKhann, G.M. (1973): J. Neurochem. 20: 1111-1120.
4.   Brunetti, M., Droz, B., Di Giamberardino, L. and Porcellatti, G. (1978): Proc. Eur. Soc. Neurochem. 1:484.
5.   Currie, J.R., Grafstein, B., Whitnall, M.H. and Alpert, R. (1978): Neurochem. Res. 3:479-492.
6.   Droz, B., Brunetti, M., Di Giamberardino, L., Koenig, H.L. and Porcellati, G. (1979): Soc. Neurosci. Symp. 4:344-360.
7.   Droz, B., Di Giamberardino, L., Koenig, H.L., Boyenval, J. and Hassig, R. (1978): Brain Res. 155:347-353.
8.   Droz, B., Koenig, H.L. and Di Giamberardino, L. (1973): Brain Res. 60:93-127.
9.   Droz, B., Koenig, H.L., Di Giamberardino, L., Couraud, J.Y., Chretien, M. and Souyri, F. (1979): Prog. Brain Res. 49: 23-44.

10. Droz, B., Rambourg, A. and Koenig, H.L. (1975): Brain Res. 93:1-13.
11. Gould, R.M., Sinatra, R.S., Spivack, W., Berti, M., Wisniewski, H.M., Lindquist, T. and Ingoglia, N. (1979): Soc. Neurosci. Abst. 5:60.
12. Grafstein, B., Miller, J.A., Ledeen, R.W., Haley, J. and Specht, S.C. (1975): Exp. Neurol. 46:261-281.
13. Haley, J.E. and Ledeen, R.W. (1979): J. Neurochem. 32:735-742.
14. Haley, J.E., Tirri, L.J. and Ledeen, R.W. (1979): J. Neurochem. 32:727-734.
15. Markov, D., Rambourg, A. and Droz, B. (1976): J. Microsc. Biol. Cell. 25:57-60.
16. Miani, N. (1963): J. Neurochem. 10:859-874.
17. Miller, S.L., Benjamins, J.A. and Morell, P. (1977): J. Biol. Chem. 252:4025-4037.
18. Porcellati, G., Arienti, G., Pirotta, M. and Giorgini, D. (1971): J. Neurochem. 18:1395-1417.
19. Rambourg, A. and Droz, B. (1980): J. Neurochem. 35:16-25.
20. Toews, A.D., Goodrum, J.F. and Morell, P. (1979): J. Neurochem. 32:1165-1173.

ACETYLCHOLINESTERASE AND BUTYRYLCHOLINESTERASE: SIMILARITIES

IN NORMAL AND DENERVATED MUSCLES, DIFFERENCES IN AXONAL TRANSPORT

L. di Giamberardino and J.Y. Couraud

Department of Biology
C.E.A. de Saclay
BPN #2
91190 Gif-Sur-Yvette, France

INTRODUCTION

Acetylcholinesterase (AChE: EC 3.1.1.7) and butyrylcholinesterase (BuChE: EC 3.1.1.8) are two enzymes distinguishable by
substrate specificity (12,14), inhibitors sensitivity (14) and,
more recently, by immunological properties (16) and thermal stability (8,16). In spite of many distinct properties, the two enzymes are often found together (3) and show analogous structure
(11,16) and similar behavior in some instances (13).

In this study we investigated the molecular forms of both enzymes in normal and denervated skeletal muscles and in sciatic
nerve of the chicken. We report here that the two enzymes exhibit
parallel variations in developing and denervated muscles, but, contrary to AChE (1,4), BuChE is not conveyed by rapid axonal transport in the sciatic nerve.

MATERIALS AND METHODS

Muscles and nerves were taken from Leghorn chickens raised in
our laboratory. Denervation of anterior latissimus dorsi (ALD) and
posterior latissimus dorsi (PLD) was done by cutting the brachial
nerve next to the brachial plexus 7 days before analysis. Homogenization, extraction and enzyme assay were as in (13). Rapid

axonal transport was monitored by cutting the sciatic nerve and
measuring the amount of enzyme accumulated in the 2 mm tip of the
proximal stump 4 hr after nerve cut (1,4). Sucrose gradient sedimentation analysis was as in (4). AChE activity is expressed in
International Units (IU), 1 IU corresponding to the hydrolysis of
1 micromole of substrate per minute. Sedimentation coefficents
are in Svedberg units ($S_{20,w}$).

RESULTS AND DISCUSSION

The physiological function of BuChE and its relations with
AChE are still open to debate (8,13,14,16). In this regard it
seemed relevant to compare the variations of the two enzymes, and
their molecular forms, in developing and denervated muscles, and to
ascertain whether BuChE, like AChE (1,4), is conveyed along the
axon by rapid axonal transport.

As reported elsewhere, a set of 4 major molecular forms of
AChE are found in chicken muscles (5,10,13,15) and nerves (1,4,5,
15). The forms are labelled L_1, L_2, M and H, respectively (for
light, medium and heavy), according to their position along the
sucrose gradient (4). In this study we found that BuChE displays
an analogous set of molecular forms, with only slightly lower sedimentation coefficient values with respect to AChE (Table 1)(see
also, 11, 16).

The H form has often been referred to as H_2 (10,11,13) to
distinguish it from a 14.7 S form, labelled H_1 (11), which is not
always resolved in the sedimentation profile presented here (Figs.
1 and 2). As regards variations of level and forms of cholinesterases (ChE) in developing and denervated muscles, both AChE and
BuChE concentrations in the PLD and ALD decrease between the 9th
and 28th day of hatching (Table 2)(see also 10,11). During the
same time lapse the muscle weight increases (Table 2), so that the
total enzyme level per muscle remains approximately constant. The
molecular form pattern of AChE of either muscle does not change
appreciably in 9 or 28 day old chickens. The form variations reported in ALD by Lyles and Barnard (10) are probably due to differences among chicken varieties (Leghorn vs Davis-412 strain).

In denervated muscles both enzyme concentrations rise dramatically, particularly in 28 day old chickens (Table 2), while the
molecular form pattern shifts towards the light forms with almost
complete disappearance of the H form (Fig. 1).

Given the preferential cytoplasmic localization of the L forms
(2), their drastic increase could indicate that denervation enhances

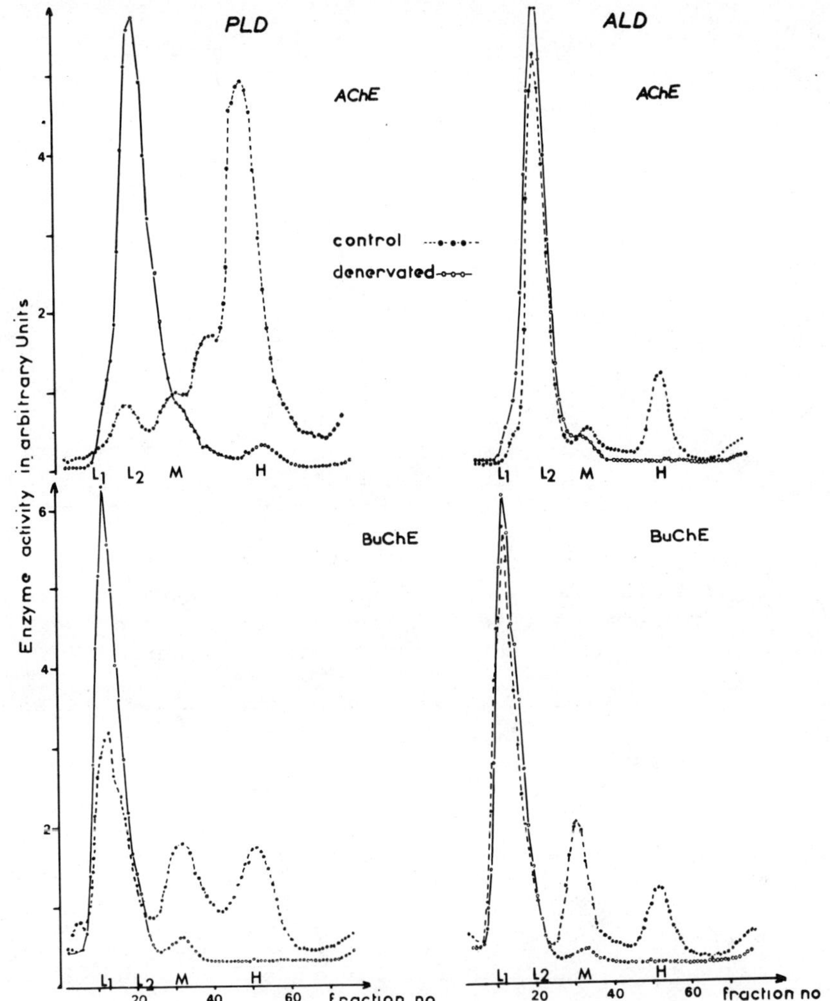

FIGURE 1: Sedimentation patterns of AChE and BuChE, after centrifugation in a 5-20% sucrose gradient. PLD and ALD were
taken from 28 day old chicken, denervated at 21 days.
Note the presence of 14.7 form (H_1 in ref. 11) between
M and H in the AChE pattern of control PLD.

TABLE 1:   Sedimentation coefficients of the AChE and BuChe
molecular forms.

	L_1	L_2	M	H
AChE	4.8	7.2	11.6	20.0
BuChE	4.4	6.0	11.2	19.1

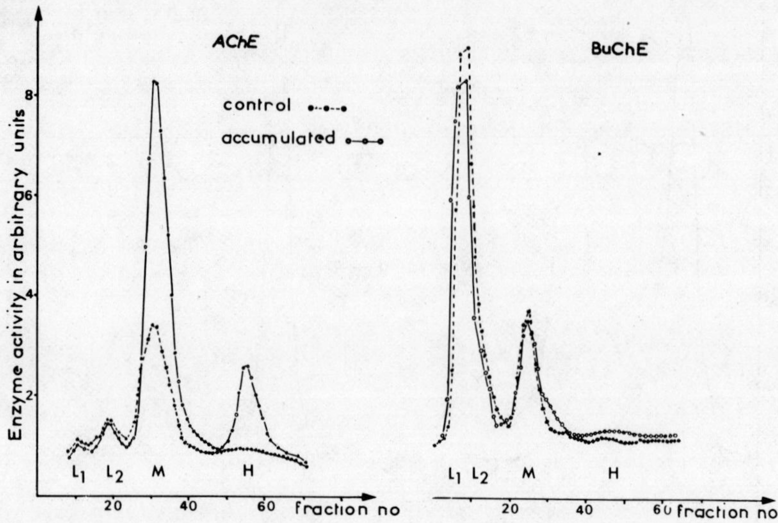

FIGURE 2: Sedimentation patterns of AChE and BuChE after centrifugation in a 5–20% sucrose gradient of extracts from 2 mm
nerve segments taken from control sciatic nerve and from
the tip of the central stump 4 hr after nerve cut
(accumulated).

TABLE 2: AChE and BuChE concentration (IU g^{-1}) and muscle weight
of control and denervated PLD and ALD of 9 and 28 day
old chickens.

	AChE		BuChE		Muscle Weight (mg)	
	C	D	C	D	C	D
PLD*	3.6	23.0	0.35	2.3	27.0	26.0
	±0.8	±4.0	±0.05	±0.4	±5	±5
PLD**	1.0	20.0	0.12	3.1	141	197
	±0.3	±5.0	±0.06	±0.9	±17	±13
ALD*	11.0	14.8	0.43	0.56	14.6	26.4
	±3.0	±0.4	±0.1	±0.18	±0.6	±2.4
ALD**	4.7	8.4	0.19	0.51	107	126
	±1.4	±0.5	±0.01	±0.10	±9	±12

* = 9 days; ** = 28 days; C = control; D = denervated
Mean ± S.D. of at least 3 determinations.

protein synthesis in these muscles. The loss of AChE H form after
denervation has been taken as an indication of its synaptic localization (2,7,15,17); the question can be raised as to whether the
parallel loss of the H form of BuChE implies a synaptic role for
this enzyme. The structural resemblance and the analogous variations of levels and forms of AChE and BuChE during development (10)
and upon muscle denervation (11) certainly do speak in favor of
some functional similarities in these two enzymes, as proposed by
Vigny et al. (16).

Axonal flow of AChE and BuChE was measured by the accumulation
of their activities at a severed sciatic nerve. In the 2 mm segment at the end of the central stump, AChE activity rises from
8.1 to 31.3 mIU 4 hr after nerve cut. The activity increase is
entirely due to the accumulation of the M and H forms (Fig. 2)(1,4).
On the contrary, the activity of BuChE (2 mIU) and the pattern of
its molecular forms remains unchanged (Fig. 2), indicating that
none of the BuChE forms are conveyed by rapid axonal transport in
sciatic nerve.

As we have shown elsewhere (1), only a fraction (about 23%)
of nerve AChE is conveyed by rapid axonal transport, the rest of
it being either stationary or slowly moving. The absence of rapidly moving BuChE under our conditions means that this enzyme is
either extra-axonal in the sciatic nerve or, if located inside the
axon, is stationary or slowly moving.

The physiological meaning of the rapid axonal transport of
the AChE M and H forms is not quite understood. Its bearing on
endplate AChE, purported by some (6) and excluded by others (9)
can be considered a conceivable working hypothesis (1). On the
contrary, in view of our results, a possible contribution of the
nerve to neuromuscular junction BuChE, if any, can be ruled out.

In conclusion, we confirm here the structural homologies of
AChE and BuChE molecular forms (11,16) and their parallel variations in chicken skeletal muscles. These similarities might be
the expression of homologous function in muscles. However, the
homology is not maintained in sciatic nerve since, contrary to
AChE, none of the BuChE form is conveyed by rapid axonal transport.

ACKNOWLEDGMENTS

This study was supported in part by INSERM grant (C.R.L No.
79-4-316-6). DiG. is affiliated with INSERM U.153, directed by
Dr. M. Fardeau. We thank Mrs. R. Hassig for her greatly appreciated technical assistance.

REFERENCES

1.   Couraud, J.Y. and Di Giamberardino, L. (1980): J. Neurochem. 35:1063-1066.
2.   Couraud, J.Y., Koenig, M. and Di. Giamberardino, L. (1980): J. Neurochem. 34:1209-1218.
3.   Davis, R. and Koelle, G. (1978): J. Cell Biol. 78:785-809.
4.   Di Giamberardino, L. and Couraud, J.Y. (1978): Nature 271: 170-172.
5.   Di Giamberardino, L., Couraud, J.Y. and Barnard, E.A. (1979): Brain Res. 160:196-202.
6.   Fernandez, H.L., Duell, M.J. and Festoff, B.W. (1980): J. Neurochem. 11:31-39.
7.   Hall, Z. (1973): J. Neurobiol. 4:346-361.
8.   Koelle, G.B., Kitto Rickard, K. and Ruch, G.A. (1979): Proc. Nat. Acad. Sci. 76:6012-6016.
9.   Lomo, T. (1980): Trends Neurosci. 23:126-129.
10. Lyles, J.M. and Barnard, E.A. (1980): FEBS Lett. 109:9-12.
11. Lyles, J.M., Silman, I. and Barnard, E.A. (1979): J. Neurochem. 33:727-738.
12. Meyers, D.K. (1953): Biochem. J. 55:67-79.
13. Silman, I., Di Giamberardino, L., Lyles, J., Couraud, J.Y. and Barnard, E.A. (1979): Nature 280:160-162.
14. Silver, A. (1974): The Biology of Cholinesterases, North-Holland, Amsterdam.
15. Vigny, M., Di Giamberardino, L., Couraud, J.Y., Rieger, F. and Koenig, J. (1976): FEBS Lett. 69:277-280.
16. Vigny, M., Gisiger, V. and Massoulie, J. (1978): Proc. Nat. Acad. Sci. 75:2588-2592.
17. Vigny, M., Koenig, J. and Rieger, F. (1976): J. Neurochem. 27:1347-1353.

THE ACTIVITY OF GLYCEROPHOSPHOCHOLINE PHOSPHODIESTERASE

IN BRAIN TISSUE

G.B. Ansell and S. Spanner

Department of Pharmacology
University of Birmingham Medical School
Birmingham, B15 2TJ, England

INTRODUCTION

Considerable interest has been generated in recent years about
the supply of choline (Ch) for acetylcholine (ACh) synthesis in the
brain (for short review see reference 4). This interest commenced
with observations made over a decade ago that the capacity of brain
tissue to synthesize Ch was non-existent, though this conclusion
may have to be modified in view of more recent findings (23). Thus
the brain has to assimilate Ch from the blood. There are some
doubts whether the free Ch in the blood is the most immediate source
of Ch for ACh synthesis and interest has therefore focussed on two
problems. Is the Ch brought to the brain in a lipid-bound form and
releases some Ch or is the lipid bound form assimilated into the
pool of brain lipid-bound Ch from which Ch is released? The fact
that there is in brain tissue a Ch-generating system (11,12,18),
suggests that there is a mechanism for releasing Ch from Ch lipids
in brain tissue which is enzymic in nature. The possible mechanisms
have been discussed recently by Ansell and Spanner (4). Illingworth
and Portman (14) measured the activity of a number of enzymes catal-
yzing the hydrolysis of phosphatidylcholine and sphingomyelin (the
major Ch lipids in brain) in homogenates of monkey brain, and concluded that Ch could derive either from sphingomyelin by reactions
(I) and (II):

    (I)   Sphingomyelin → N-acylsphingosine + phosphocholine
    (II) Phosphocholine → Ch + organophosphate

TABLE 1: Reported values for the activity of GPC-diesterase in nervous tissue.

Tissue	pH	Added Cation	Buffer	μmol/g tissue^{-1},h^{-1}	Reference Number
Rat forebrain	8.9	None	25 mM bicarbonate	36	21
Rat spinal cord	8.9	None	25 mM bicarbonate	58	21
Rat cerebellum	8.9	None	25 mM bicarbonate	25	21
Hen forebrain	8.9	None	25 mM bicarbonate	13	21
Hen spinal cord	8.9	None	25 mM bicarbonate	10	21
Hen cerebellum	8.9	None	25 mM bicarbonate	3	21
Human spinal cord	8.9	None	25 mM bicarbonate	9	21
Human cerebellum	8.9	None	25 mM bicarbonate	3	21
Monkey cerebral cortex	8.6	5 mM Ca^{++}	0.4 mM Tris	0.7*	14
Monkey brain stem	8.6	5 mM Ca^{++}	0.4 mM Tris	3.3*	14
Rat cerebrum	8.0	mM Mg^{++}	30 mM aminoethylpropandiol	30	17
Rat spinal cord	8.0	mM Mg^{++}	30 mM aminoethylpropandiol	24	17
Rat cerebellum	8.0	mM Mg^{++}	30 mM aminoethylpropandiol	29	17
Rat brain	8.5	10 mM Ca^{++}	100 mM glycine	43	10

* Calculated from the authors' results on the assumption that the protein content is 10% of the wet weight of the tissue.

or from phosphatidylcholine (sn-1,2-diacylglycero-3-phosphocholine)
by reactions (III), (IV), (V):

(III)   Phosphatidylcholine → fatty acid + lysophosphatidyl-
         choline
(IV)   Lysophosphatidylcholine → glycerophosphocholine +
         fatty acid
(V)   Glycerophosphocholine → glycerophosphate + choline

Reaction (II), catalyzed by alkaline phosphatase (EC 3.1.3.1),
is feeble at physiological pH in vitro (19), but Illingworth and
Portman (14) found significant activity at pH 7.4 in their preparations. Phosphocholine is readily hydrolyzed by brain tissue in
vivo (2). Clearly the rates of reactions measured in vitro may not
be very relevant to rates in vivo. For example, assays of phospholipase activities are usually done in the presence of detergents,
and assays of alkaline phosphatase which hydrolyzes phosphocholine
are usually done in the presence of concentrations of phosphocholine
which do not exist in vivo. If phosphatidylcholine is a major source
of Ch, then the terminal reaction, i.e. reaction (V) catalyzed by the
enzyme glycerophosphocholine phosphodiesterase (GPC-diesterase,
EC 3.1.4.2) is very important. In fact, the recent experiments of
Jope and Jenden (15) strongly suggest that glycerophosphocholine is
a precursor of free Ch in the brain in vivo.

This enzyme has not been studied in great detail and there are
relatively few reports of its activity in brain and only one study
of its subcellular distribution. Most of the available information
for whole brain tissue is summarized in Table 1. There are some
variations in activity between brain areas and considerable species
variation. According to Webster et al. (21) and Illingworth and
Portman (14) there is more activity in areas rich in white matter
(spinal cord and brain stem, though not cortical white), but this
is not apparent from the study by Mann (17) on spinal cord. The
optimum activity is in the region of pH 9, though there is significant activity at physiological pH (21) and the liver enzyme is
optimally active at pH 7.5 according to Dawson (7); though it is
9.3 according to Baldwin and Cornatzer (5). In the important respect
of cation requirement, the concensus of opinion is that neither
homogenates nor subcellular fractions require the addition of metallic divalent cations for optimal activity though all activity is
abolished by mM ethylene diamine tetra-acetic acid (EDTA). Most
authors have used additional magnesium as the cation in the assay
and one may assume that a metallic cation is required, though presumably in low concentration.

METHODS AND RESULTS

Tissues

Rat forebrain and striatum were used for most of these studies.
Subcellular fractions were prepared by the method of Gray and
Whittaker (13) as modified by Spanner and Ansell (19). For some
experiments the caudate nucleus of the ox was subfractionated by
the method of Gregg, Spanner and Ansell (unpublished observations).

Assay of GPC-Diesterase

Glycerophosphocholine as its cadmium chloride complex, obtained
from Sigma, was dissolved in water at a concentration of approximately 15 mg per ml and passed through a column of mixed bed resin consisting of 2 ml of Amberlite IRC 50 (H+) + 4 ml of Amberlite IR 45
(OH-)(20). In this system the phosphate ester passes through the
column but cadmium and chloride ions are retained. After washing
with 2 x 5 ml of water, the effluent was neutralized with a little
sodium bicarbonate, taken to dryness, and diluted to give a concentration of 40 μmol/ml after the addition of glycerophospho[1,2 -^{14}C]-
choline, obtained from ICN Pharmaceuticals (Irvine, California) to
give a specific radioactivity of approximately 1400 dpm/μmol. The
incubation system finally adopted consisted of 0.5 ml of 50 mM
glycylglycine buffer (pH 8.6) containing 2 mM $MgCl_2$ and 0.25 ml of
a microsomal suspension equivalent to 50 mg of original tissue weight
to which was added at zero time 0.25 ml of labelled glycerophosphocholine to give a final substrate concentration in the system of
10 mM. Incubation was carried out at 37°C for 30 min, and the reaction terminated by heating the tubes in a boiling water bath for
10 min. After the suspension had been centrifuged and the pellet
washed with 1 ml water the pooled cloudy supernatants were passed
through a column (0.9 x 3.5 cm) of Zeolite 225 (Na$^+$)(52-100 mesh),
and the column washed through with 20 ml of water which eluted
glycerophosphocholine and phosphocholine. Ch was eluted with 15 ml
of 0.5 M KCl, the eluate counted and the amount of hydrolysis calculated for the radioactivity. This separation of Ch from glycerophosphocholine was devised by Mann (17). Zeolite 225 is a cation
exchange resin comparable to Dowex 50 and obtained from the Permutit
Company Ltd. (England). No evidence was obtained in this study for
the release of phosphocholine from glycerophosphocholine, though
Abra and Quinn (1) reported the presence of this reaction in brain
tissue.

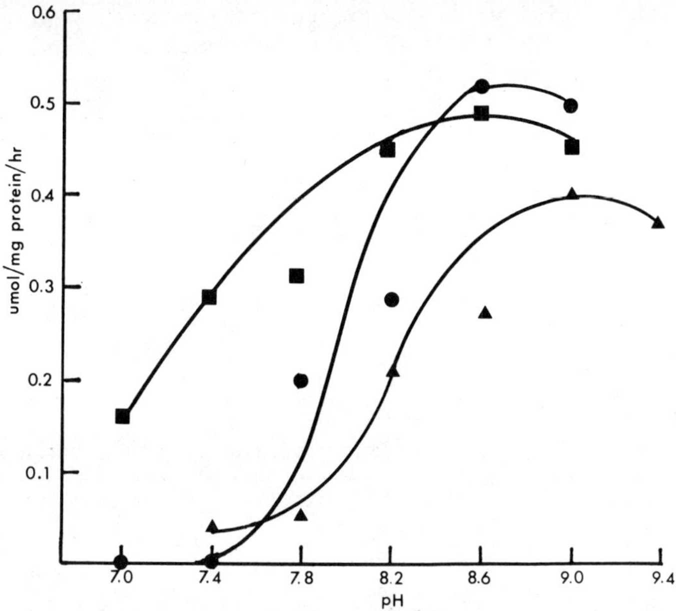

FIGURE 1: The activity of brain GPC-diesterase in the microsomal
fraction from rat forebrain as a function of pH.
▲ = 50 mM Tris buffer; ■ = 25 mM glyclglycine buffer;
● = 20 mM veronal buffer.

Effect of pH

In Fig. 1 is shown the activity of the enzyme in the microsomal
fraction of rat forebrain as a function of pH with three different
buffers: glycylglycine, Tris and veronal. The optimum was found to
be in the region of 8.6 with all the buffers though in 25 mM glycylglycine buffer there was also a 50% activity at pH 7.4 (cf. reference
21). With the other two buffers the activity was almost non-existent
at physiological pH and with Tris buffer the optimum activity was
lower than with glycylglycine. For all subsequent experiments a
final concentration of 25 mM glycylglycine buffer (pH 8.6) was used.
At this pH the activity in the presence of 10 mM glycerophosphocholine was linear with increasing amounts of tissue protein up to
2 mg.

TABLE 2: The activity of GPC-diesterase in subcellular fractions*
          of rat striatum and ox caudate nucleus.

	nmol, mg protein^{-1},h^{-1}	
	Rat Striatum	Ox Caudate
Homogenate	150	130
P_2	N.D.	N.D. (2)
P_2A	N.D.	–
P_2B	N.D.	N.D. (4)
P_2C	N.D.	–
P_3	280	290 $\pm$ 150(5)**
S_3	N.D.	N.D.

N.D. = not detected.
* = Nomenclature of Gray and Whittaker (13).
** = Represents 62% of the activity of the original homogenate.

Subcellular Distribution

The object of examining striatum and caudate was to see if brain
areas rich in ACh and the enzymes that metabolize it, contain more
of an enzyme capable of releasing Ch. The results obtained with
rat striatum and ox caudate are shown in Table 2. Although significant amounts of GPC-diesterase were found in the P_2 (synaptosomal)
fraction of rat forebrain, when rat striatum and ox caudate were
examined all the measurable activity was found in the P_3 (microsomal)
fraction. About 62% of the activity in the homogenate of caudate
was recovered in the microsomal fraction (Table 2). However, the
activity of rat striatum was low compared with whole forebrain which
was 246 $\pm$ 55(8) nmol, mg protein^{-1}, h^{-1}; ox caudate also contained a
low level of activity. Webster <u>et al</u> (21) found low activity in
human caudate.

Effect of Cations

In common with other reports on this enzyme it was found that
neither the tissue homogenates nor the microsomal fraction required
the addition of divalent metal cations though mM Mg^{++} was routinely
added (the activity was rather lower in the presence of mM Ca^{++}).
Concentrations in excess of 10 mM Mg^{++} at pH 8.6 caused inhibition.
That divalent metallic cations are necessary was confirmed by carrying out incubations in the presence of 2 mM EDTA or 2 mM ethylene
glycol-bis (β-aminoethyl ether) N,N-tetraacetic acid (EGTA) when

TABLE 3: The activity of GPC-diesterase (nmol, mg protein^{-1}, h^{-1}) in the microsomal fraction of rat forebrain in the presence of Mg^{++}, Ca^{++} and chelating agents.

Chelating Agent	mM Mg^{++}	mM Ca^{++}	No added cation*	No added cation* (preincubated**)
2 mM EDTA	36	0	0	0
0.2 mM EDTA	225	406	16	110
2 mM EGTA	10	33	20	0
0.2 mM EGTA	122	314	52	122
---	454 $\pm$ 46(9)	341	484(3)	–

Values are the mean of two determinations, except where indicated.

* The concentration of Ca^{++} measured in the absence of chelating agent was approximately 14 µM (calcium electrode).

** The microsomal preparation, buffer and chelating agent were incubated for 8 min before the substrate was added.

TABLE 4:   The activity of GPC-diesterase in the microsomal fraction
           of rat forebrain prepared in mM EDTA and washed twice in
           0.25 M sucrose.

Concentration of added cation (μM)	Activity as percentage of control*	
	Mg^{++}	Ca^{++}
10	131	114
1	129	114
0.5	106	132

The microsomal fraction was incubated with substrate and buffer as
described in the text.
* 465 nmol, mg protein^{-1}, h^{-1}.

activity was almost abolished (Table 3). It was also noted that the
activity of the enzyme was partially inhibited when 0.2 mM EDTA or
0.2 mM EGTA was used, when mM Mg^{++} was added (Table 3) or in the
absence of added cations when the tissue had been preincubated with
the chelating agent. When the microsomal fraction from rat forebrain was prepared in 0.25 mM sucrose containing 2 mM EDTA, the
activity of this fraction could be restored simply by washing it
twice in 0.25 mM sucrose; no added Ca^{++} or Mg^{++} was necessary, even
at concentrations as low as 0.5 μM (Table 4).

It might be implied that the enzyme does require a divalent
cation, and that these experiments are imperfect in that concentrations of cation of 10^{-5} M or lower are difficult to produce without
the use of metal ion buffers (22) since the contribution of cations
at this level from reagents was not controlled. The Ca^{++} ion concentration in the reagents was about 14 μM.

Effect of Inhibitors

Since GPC-diesterase is a phosphodiesterase, inhibitors of the
phosphodiesterase which hydrolyse cyclic-AMP were tested. Theophylline inhibited the enzyme when activity was measured in whole
tissues to the extent of 46% at 3 mM concentration, and had negligible effect at 300 μM. 3-Isobutyl-1-methylxanthine inhibited the
enzyme to the extent of 50% at 3 mM and 28% at 300 μM and may therefore be an inhibitor of some use. Hemicholinium-3 (the bishemi-
acetal form of α'-dimethylethanolamino-4,4'-biacetophenone), which

inhibits Ch kinase (3) also inhibited GPC-diesterase to the extent
of 54% at 40 μM and is therefore relatively powerful as in inhibitor.
In this experiment the microsomal fraction from forebrain was used.

DISCUSSION

The preliminary findings reported here confirm that, in rat
brain tissue the enzyme GPC-diesterase is quite active and that most
of its activity is found in the microsomal fraction. In the only
other study of the subcellular distribution of this enzyme in brain
tissue (17) a lot of activity was found in the synaptosomal fraction.
Although, in the present study some activity was found in the synaptosomal fraction of forebrain, the microsomal fraction had the highest
relative specific activity (3.1 compared with total particulate fraction). Furthermore, there was no detectable activity in the synaptosomal fraction, or indeed the P_2 fraction as a whole, from rat
striatum and ox caudate (Table 2). The absence of GPC-diesterase
from the synaptosomal fraction of caudate of striatum strongly suggests that in this area at least, the enzyme is not associated with
neurones. Since no activity was found in the P_2A fraction, which
contains plasma membranes and myelin, it might be concluded that the
enzyme is absent from plasma membranes in this brain area, though it
was in the P_2A fraction of forebrain. In this respect, therefore,
the subcellular distribution in striatum differs from that of rat
liver where, in a detailed study by Lloyd-Davies et al. (16), the
enzyme was found to be distributed between the plasma membranes and
the cytosol. In a study on rat kidney, Baldwin and Cornatzer (5)
found the enzyme to be largely microsomal. It is possible to suggest
therefore, that in brain tissue the enzyme may be located in glial
endoplasmic reticulum.

In the presence of 2 mM EDTA or 2 mM EGTA the activity of GPC-
diesterase in the microsomal fraction of brain was abolished (Table 3)
as observed by other authors (7,16,21). However, no added cations
are necessary for maximal activity as was originally noted by Webster
et al. (21). No explanation can at present be offered for two further
observations noted in Table 3. First, preincubation of microsomal
fraction with 0.2 mM EDTA or 0.2 mM EGTA caused a slight stimulation
of activity. Second, the same concentration of EDTA or EGTA, though
having little effect on the activity in the presence of mM Ca^{++} greatly affected this activity in the presence of mM Mg^{++}.

When the microsomal fraction of rat forebrain was prepared in
0.25 mM sucrose containing mM EDTA and then washed twice with 0.25 M
sucrose alone, the activity was restored to normal and even very low
concentrations of Mg^{++} or Ca^{++} did not greatly increase activity

(Table 4). These observations suggest that the cation requirement can be restored or replaced by the low concentration of Ca^{++} (ca. 14 μM) usually present in our reagents. It might be pointed out that for the enzyme from kidney, Zn^{++} has been proposed as an essential part of the enzyme (6); though the studies of Dawson (7) on liver and Webster et al. (21) on brain, would not suggest a role for Zn^{++} in these tissues. Further investigation on the role of metallic cations on the activity of GPC-diesterase in brain are very necessary.

Hemicholinium-3 would appear to be a good inhibitor of GPC-diesterase in brain (I_{50} ca. 40 μM) and may be a useful tool in further studies on the role of the enzyme in the release of Ch. It is interesting that hemicholinium-3 inhibits Ch kinase (3) and acetyl-cholinesterase (9), but not Ch acetyltransferase (8). Potent organo-phosphorous inhibitors of acetylcholinesterase such as di-isopropyl-fluorophosphonate have no effect on GPC-diesterase according to Webster et al. (21), which is perhaps not surprising since glycero-phosphocholine is unlikely to bind to the esteratic site of acetyl-cholinesterase.

Although GPC-diesterase was not found in high concentration in an area of the brain where release of Ch for ACh synthesis would be expected to be important, further investigations are called for. Some very direct evidence has been produced in this paper for its localization in a glial microsomal fraction which could explain why destruction of cholinergic neurones _in vivo_ failed to prevent the postmortem enzymic release of Ch (12).

ACKNOWLEDGEMENT

The interest of Professor P.B. Bradley is appreciated.

REFERENCES

1.   Abra, R.M. and Quinn, P.J. (1975): Biochim. Biophys, Acta 380:436-441.
2.   Ansell, G.B. and Chojnacki, T. (1963): IN Biochemical Problems of Lipids (ed) A.C. Frazer, Elsevier, Amsterdam, pp. 425-432.
3.   Ansell, G.B. and Spanner, S. (1974): J. Neurochem. 22:1153-1155.
4.   Ansell, G.B. and Spanner, S. (1979): IN Nutrition and the Brain: Choline and Lecithin in Brain Disorders, Vol. 5 (eds) A. Barbeau, J.H. Growdon and R.J. Wurtman, Raven Press, New York, pp. 35-46.

5. Baldwin, J.J. and Cornatzer, W.E. (1968): Biochim. Biophys. Acta 164:195-204.

6. Baldwin, J.J., Lanes, P. and Cornatzer, W.E. (1969): Arch. Biochem. Biophys. 133:224-232.

7. Dawson, R.M.C. (1956): Biochem. J. 62:689-693.

8. Diamond, I. and Milfay, D. (1972): J. Neurochem. 19:1899-1909.

9. Domino, E.F., Schellenberger, M.K. and Frappier, J. (1968): Arch. Int. Pharmacodyn. Ther. 176:42-49.

10. Dross, K. (1975): J. Neurochem. 24:701-706.

11. Dross, K. and Kewitz, H. (1972): N.-S. Arch. Pharmacol. 274: 91-106.

12. Freeman, J.J. and Jenden, D.J. (1976): Life Sci. 19:949-962.

13. Gray, E.G. and Whittaker, V.P. (1962): J. Anat. (Lond) 96:79-88.

14. Illingworth, D.R. and Portman, O.W. (1973): Physiol. Chem. Phys. 5:365-373.

15. Jope, R.S. and Jenden, D.J. (1979): J. Neurosci. Res. 4:69-82.

16. Lloyd-Davies, K.A., Michell, R.H. and Coleman, R. (1972): Biochem. J. 127:357-368.

17. Mann, S.P. (1975): Experientia 31:1256-1257.

18. Schuberth, J., Sparf, B. and Sundwall, A. (1970): J. Neurochem. 17:461-468.

19. Spanner, S. and Ansell, G.B. (1979): Biochem. J. 178:753-760.

20. Tattrie, N.H. and McArthur, C.S. (1958): Biochem. Preps. 6:16-19.

21. Webster, G.R., Marples, E.A. and Thompson, R.H.S. (1957): Biochem. J. 65:374-377.

22. Wolf, H.U. (1973): Experientia 29:241-249.

23. Zeisel, S.H., Blusztajn, J.K. and Wurtman, R.J. (1979): IN Nutrition and the Brain: Choline and Lecithin in Brain Disorders, Vol. 5 (eds) A. Barbeau, J.H. Growdon and R.J. Wurtman, Raven Press, New York, pp. 47-55.

SYNTHESIS OF CHOLINE IN THE BRAIN

H. Kewitz and O. Pleul

Department of Clinical Pharmacology
Free University of Berlin, Hindenburgdamm 30
D-1000 Berlin 45, West Germany

Specific radioactivities of choline (Ch), phosphorylcholine
(PCh), lecithin, lysolecithin and glycerylphosphorylcholine (GPCH)
have been measured in blood, liver, muscle, and including acetylcholine (ACh) in brain after injection (i.v) of three labelled
precursors, Ch, the methyl group of methionine and ethanolamine.
In relation to the specific activity of Ch in blood there was significantly more radioactivity in the Ch of brain after labelled
methyl groups and labelled ethanolamine than after labelled Ch.

Since the Ch moiety of lipids, which returns to the Ch pool
contained less radioactivity after labelled methyl groups or labelled ethanolamine than after labelled Ch, it is a most likely interpretation of the data that Ch in brain can be formed by the
methylation of free ethanolamine. Data from liver confirm the
formation of lecithin in this tissue by the methylation of phosphatidylethanolamine. No indication was found for the synthesis of
Ch in muscle.

Kinetics of Ch in brain have been calculated (nmol x g^{-1} x
min^{-1}): turnover rate of Ch, 36; rate of synthesis of Ch by
methylation and net loss of Ch into the blood stream, 6; inflow
from the blood, 6; outflow into the blood, 12; transfer into lipids
and vice versa, 20; transfer to ACh and vice versa, 4.

INTRODUCTION

Leakage of almost 6 nmol of Ch per min from the brain into the
venous blood flow has been concluded from data on the arteriovenous

difference by Dross and Kewitz (4) confirmed by Freeman et al. (5), Aquilonius et al. (1) and Spanner et al. (11). This means that the brain needs per minute supplementation of 25% of its free Ch or 17% of the turnover rate. On principle two pathways may be considered for supplementation, either de novo synthesis of Ch within the brain or the influx of Ch containing intermediates from the arterial blood supply. De novo synthesis of Ch in the brain has been thought un-likely as yet.

With regard to supplementation by Ch-containing intermediates, Illingworth and Portman (6) provided some data supporting the as-sumption that lysophosphatidylcholine would be the vehicle to trans-port Ch into the brain. The test performed to differentiate between these alternatives is based on the observation that after i.v. in-jection of labelled Ch, the specific radioactivity of free Ch in the brain is much lower than in the blood, as shown by Dross and Kewitz (3) which is due to the dilution by unlabelled or less label-led free Ch formed within the brain, most of it by hydrolysis of GPCh.

The dilution of labelled Ch and, therefore, also the difference between specific radioactivities (SRAs) of Ch in the blood and in the brain should be less if the Ch synthetized or liberated within the brain were to carry radioactivity. This can be achieved by the application of labelled methyl groups or labelled ethanolamine and their expected incorporation into Ch and Ch-containing intermediates at any site in the body. Following the SRAs of the various inter-mediates in blood, brain, liver and skeletal muscle, it should be possible to recognize the entrance compartment of the radioactive label in each organ and during the rising phase of SRA, precursor-successor relations as well.

METHODS

Three precursors have been used [methyl-^{14}C]-Ch, [methyl-^{14}C]-methionine and [1,2-^{14}C]-ethanolamine. Animals were sacrificed 3 and 6 hr after injection by i.v. After extraction, various inter-mediates were carefully separated by repeated paper chromatography, hydrolysis if suitable, and ion pair extraction. Radioactivity was estimated in a scintillation counter (Packard) and Ch by the di-picrylamine photometric method. Details of the procedure have been published by Kewitz and Pleul (7), where the standard deviations of the figures can also be found.

TABLE 1: Specific radioactivity (dpm x nmol^{-1}) of Ch and intermediates after administration (i.v.) of [methyl-^{14}C]-Ch, [methyl-^{14}C]-methionine and [1,2-^{14}C]-ethanolamine.

Tissue	After Injection (hr)	Ch	ACh	PCh	Lecithin	Lysolecithin	GPCh	Sphingomyelin
[methyl-^{14}C]-Ch								
Blood	3	223	–	165	26	62	24	8
	6	137	–	81	41	53	15	13
Brain	3	33	31	40	3.2	4.2	4.5	0.09
	6	31	19	25	3.5	5.1	6.6	0.21
Liver	3	117	–	202	121	49	51	12
	6	74	–	79	69	55	66	30
Skeletal	3	146	–	–	66	88	51	17
muscle	6	95	–	–	62	89	112	16
[methyl-^{14}C]-methionine								
Blood	3	12	–	2.2	28	86	5.4	13
	6	22	–	8.0	30	64	13	14
Brain	3	16	12	8.3	0.35	2.1	0.7	–
	6	24	13	11	0.73	0.95	1.4	–
Liver	3	41	–	32	129	70	50	23
	6	43	–	33	108	63	64	41
Skeletal	3	14	–	17	2.2	–	4.7	–
muscle	6	9	–	13	1.4	–	9.5	–
[1,2-^{14}C]-ethanolamine								
Blood	3	1.9	–	–	10	18	0.71	1.7
	6	7.5	–	5.1	24	56	4.6	8.3
Brain	3	2.6	1.2	1.7	0.068	–	0.12	–
	6	9.8	8.9	6.3	0.28	–	0.38	–
Liver	3	20	–	6.3	45	29	17	6.1
	6	33	–	21	86	58	40	37
Skeletal	3	0.98	–	–	0.39	–	–	–
muscle	6	4.8	–	4.2	0.36	–	–	–

TABLE 2: Specific radioactivity of Ch and glycerylphosphoryl-
          choline (GPCh) in rat brain 3 and 6 hr after injection
          by i.v. of labelled Ch, methionine or ethanolamine.

	Ch		GPCh		Ch/GPCh	
Precursor	3 hr	6 hr	3 hr	6 hr	3 hr	6 hr
[methyl-^{14}C]- Ch	33	31	4.5	6.6	7.3	4.7
(methyl-^{14}C]- methionine	16	24	0.7	1.4	22.9	17.1
(1,2-^{14}C]- ethanolamine	2.6	9.8	0.38	0.38	21.6	25.8

TABLE 3: Specific radioactivity (dpm x nmol^{-1}) of Ch in blood and
          various organs, 3 and 6 hr after injection by i.v. of
          labelled precursors.

	[methyl-^{14}C] Ch		[methyl-^{14}C] Methionine		[1,2-^{14}C] Ethanolamine	
Organ	3 hr	6 hr	3 hr	6 hr	3 hr	6 hr
Blood	233	137	12	22	1.9	7.5
Brain	33	31	16	24	2.6	9.8
Liver	117	74	41	43	20	33
Muscle	146	95	14	9.5	0.98	4.7

RESULTS

After injection of labelled Ch, the highest specific activity
was found in the free Ch of the blood that was 7 and 4 times higher
than in the brain (Table 1). Whereas the labelling of ACh and PCh
in the brain was almost in the same range as Ch, the specific activity of phosphatidylcholine, due to the tremendous pool size appeared much lower. Lysophosphatidylcholine and GPCh had a somewhat
higher and almost equal SRA, respectively, relative to their precursor phosphatidylcholine. This may indicate that parts of the
phosphatidylcholine pool have higher turnover rates than the rest.
We interpret the lowering of the SRA of Ch arriving to the brain
by the dilution with Ch derived from the hydrolysis of the low
labelled GPCh. According to the SRA of GPCh, the dilution of the
radioactivity was less in the liver and absent in skeletal muscle.

TABLE 4:   Ratios of specific radioactivity between Ch in blood and
           organs after injection by i.v. of labelled precursors.

Precursor	hr	Blood/Brain	Blood/Liver	Blood/Muscle
[methyl-^{14}C]-	3	6.8	2.0	1.6
Ch	6	4.4	1.6	1.2
[methyl-^{14}C]-	3	0.67	0.32	1.0
methionine	6	0.98	0.64	2.9
[1,2-^{14}C]-	3	0.79	0.08	1.4
ethanolamine	6	0.76	0.26	1.9

TABLE 5:   Ratios of specific radioactivity between Ch and PCh or
           lecithin (lec) in rats after injection (i.v.) of labelled
           precursors.

		Brain		Liver		Muscle	
Precursor	hr	Ch/ PCh	Ch/ lec	Ch/ PCh	Ch/ lec	Ch/ PCh	Ch/ lec
[methyl-^{14}]-	3	0.8	10	0.6	1.0	–	2.3
Ch	6	1.3	9	0.9	1.0	–	1.5
[methyl-^{14}]-	3	2.0	52	1.6	0.4	0.9	7.5
methionine	6	2.2	40	1.4	0.4	0.8	6.7
[1,2-^{14}C]-	3	1.6	57	3.1	0.4	–	3.4
ethanolamine	6	1.6	36	1.7	0.4	1.4	14.0

After the application of [methyl-^{14}C]-methionine (Table 1),
the most striking difference was the higher SRA of free Ch in the
brain than in the blood, although the labelling of GPCh in the
brain was relatively lower than after injection of labelled Ch.
Thus the dilution of the SRA of free Ch by the less labelled Ch
derived from the hydrolysis of GPCh was ineffective when methionine,
labelled in the methyl group, was given (Table 2).   Results with
[1,2-^{14}C]-ethanolamine (Table 1) were quite similar but on a lower
scale of SRAs.

With both precursors, methionine labelled in the methyl group
and chain labelled ethanolamine, the highest SRA was found in the
lecithin fraction of liver.   Skeletal muscle deposits almost 50%

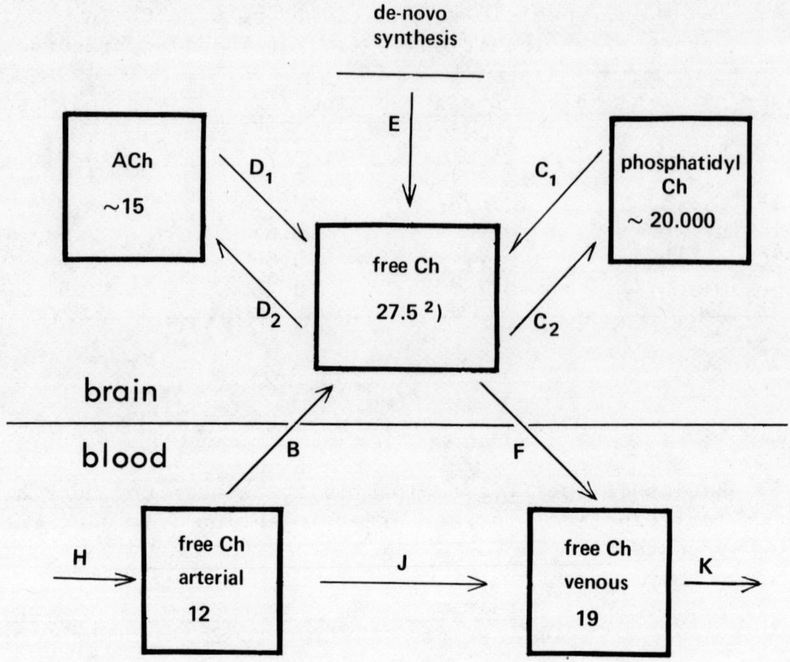

Numbers inside of the squares = pool size (n moles x g⁻¹ or ml⁻¹)

FIGURE 1

of the body's free Ch. In this large compartment, the highest
SRA appeared as the free Ch and, on a similar level, as PCh.

DISCUSSION

Based on the hypothesis, which should be tested, it has been
predicted that the difference between SRAs of Ch in blood and brain
would be reduced if Ch could be labelled within the brain. Results
show that we succeeded in labelling free Ch in the brain by the
application of labelled methionine or ethanolamine. The predicted
reduction of the difference between specific activities of Ch in
the blood and brain can be seen in Tables 3 and 4.

The surplus of labelled Ch could hardly be generated from
labelled PCh or one of the two successors, because they had

relatively low SRAs. One serious argument against this interpretation mentioned by Tucek (12) is based on the assumption that the
phosphatidylcholine pool might not be homogeneous and only a small
proportion might turnover at a rapid rate. Although this argument
is of some fascination it would require a different interpretation
of results with the application of labelled Ch. If it is true that
the low labelled Ch from the hydrolyzed GPCh dilutes the highly
labelled free Ch after the application of labelled Ch, this should
also happen after the application of the two other precursors.
Therefore we don not have any indication to support this argument.

On the basis of our data we are not able, however, to rule
out for certain the possibility that a very small fraction of the
phosphatidylcholine may be synthesized in the brain by methylation
of unlabelled phosphatidylethanolamine with labelled methyl groups
from methionine or by the methylation of labelled phsophatidyl-
ethanolamine. This conceivable small fraction with a high SRA
should be metabolized by phospholipase D preferentially rather than
by phospholipase A. Although not very likely, theoretically this
pathway may exist. This assumption gets support since phospholipase D has been found in the brain by Saito and Kanfer (10).
Furthermore, Blusztajn et al. (2) and, independantly, Mozzi and
Porcellati (8) have also recently described in vitro synthesis of
lecithin by methylation of phosphatidylethanolamine in a synaptosomal prepartion of bovine brain and in rat brain cortex,
respectively.

The second serious argument against our interpretation stems
from experiments of Illingworth and Portman (6), who studied the
uptake of lysophosphatidylcholine from plasma into the brain in
squirrel monkeys. They found overall ratios in the brain of tritium
labelled Ch to C-14 labelled fatty acids between 2.75 and 4.3. This
they considered close enough to 3.75 (that of the injected lysophosphatidylcholine) to assume the uptake of the intact molecule into the
brain. With regard to the enormous variation of the ratio it might
be reasonable, however, to consider also secondary distribution of
the labelled components, and the more so since only 1% of the radioactivity appeared in the brain.

The ratios of tritium to C-14 in lysophosphatidylcholine or
phosphatidylcholine has not been specified in the paper of Illingworth and Portman (6), although 70% of the label at the fatty acid
was found in PCh and more than 20% in triglycerides and free fatty
acids. The comparison of these data with our own results is also
difficult because of the species difference and of the high un-
physiological dose of lysophosphatidylcholine which Illingworth
and Portman have used. In our experiments we did not find any elevation of the SRA of lysophosphatidylcholine in the brain which would
designate this compartment as the entrance of labelled Ch.

Nevertheless, Nordberg (9) has made a very reasonable point, namely, that in our experiment the lysophosphatidylcholine in the blood carried a high SRA and may have delivered the labelled Ch to the brain. This interpretation would require the assumption that only a very small fraction of the lysophosphatidylcholine, namely 1/50 of the pool size, or 2 nmol/g, would have been labelled and that this small compartment would have to have a very short turnover time of almost 20 sec. Such a compartment has not yet been found, but would be of great interest if it can be shown.

According to Nordberg's argument (9), one would also expect that the lysophosphatidylcholine would be formed in the liver before transport to the brain. Thus the level of SRA of lysophosphatidylcholine in the brain should be related to that in the liver. This relation cannot be found in our data. SRA of lysophosphatidylcholine in the brain was higher after the application of labelled Ch than after labelled methionine, although we found the reverse relation in the liver.

Furthermore, if Nordberg (9) is correct, the difference between the specific activity of Ch and lysophosphatidylcholine should be smaller in the experiments with labelled methionine or ethanolamine than in that with labelled Ch. According to Illingworth and Portman's (6) finding, the difference should be reduced, expecially between Ch and PCh, because they found 70% of the label in that fraction. Both these differences were not smaller, and actually were larger after labelled methionine or ethanolamine than after labelled Ch, as shown with lecithin in Table 5. This is also a strong argument against the speculation that a small fraction of lecithin with high metabolic activity might be the entrance for the labelled Ch into the brain.

In conclusion, the arguments given in the literature do not disprove the interpretation of the data that free Ch may be synthesized in the brain of rats by the methylation of ethanolamine. Unfortunately, no confirmation of our results by _in vivo_ experiments are available. Experiments in our laboratory have not been conclusive so far. There is a short note in the recent paper by Blusztajn et al. (2) that considerable amounts of free Ch were formed by methylation on a synaptosomal preparation from rat brain. We look forward to learning the details of this work.

In our view, Ch keeps a key position by interconnecting the metabolism of phospholipids and transmitter in the brain (Fig. 1). We calculated a turnover rate of almost 36 nmol x min^{-1} x g^{-1}, which means a turnover time of 0.75 min. The pool may get its supply from four different sources. Twenty nmol may be contributed from lipid metabolism, 4 nmol from ACh hydrolysis, 6 nmol should arrive from the bloodstream, and we consider 6 nmol to be synthesized _de novo_ within the brain.

Whether the supply from the different sources may compensate for each other is not known. A diminution of the Ch pool, however, is supposed to be of crucial significance for brain function. The Ch pool could be a vulnerable component in the metabolism of brain which needs careful regulation. It will be the work of the future to elucidate this regulatory mechanisms.

REFERENCES

1.   Aquilonius, S.-M., Coder, G., Lying-Tunell, U., Malmlund, H.D. and Schubert, J.(1975): Brain Res. 99:430-433.

2.   Blusztajn, J.K., Zeisel, S.H. and Wurtman, R.J. (1979): Brain Res. 179:319-327.

3.   Dross, K. and Kewitz, H. (1966): N.-S. Arch. Pharmakol. 255: 10.

4.   Dross, K. and Kewitz, H. (1972): N.-S. Arch. Pharmakol. 274: 91-106.

5.   Freeman, J.J., Choi, R.L. and Jenden, D.J. (1975): J. Neurochem. 24:729-734.

6.   Illingworth, D.R. and Portman, O.W. (1972): Biochem. J. 557: 557-567.

7.   Kewitz, H. and Pleul, O. (1977): Proc. Nat. Acad. Sci. 73: 2181-2185.

8.   Mozzi, R. and Porcellati, G. (1979): FEBS Lett. 100:363-366.

9.   Nordberg, A. (1977): Acta Physiol. Scand. Suppl. 445.

10. Saito, M. and Kanfer, J. (1973): Biochem. Biophys. Res. Comm. 53:391-398.

11. Spanner, S., Hall, R.C. and Ansell, G.B. (1976): Biochem. J. 154:133-140.

12. Tucek, S. (1978): Acetylcholine Synthesis in Neurons, Chapman and Hall, London.

ORIGIN OF ACETYL GROUPS OF ACETYLCHOLINE IN THE BRAIN AND

THE ROLE OF ACETYLCOENZYME A IN THE CONTROL OF ITS SYNTHESIS

S. Tucek, V. Dolezal and J. Ricny
Institute of Physiology, Czechoslovak Academy of Science
14220 Prague, Czechoslovakia

INTRODUCTION

Problems concerning the origin of acetyl groups of acetylcholine (ACh) in the brain have been reviewed recently (19,27). It
is the prevailing view that the acetylcoenzyme A (acetylCoA) used
for the synthesis of ACh in the brain is derived from pyruvate (12,
16,28,29) and is formed by pyruvate dehydrogenase, which is localized in the mitochondrial matrix. The suggestion that there is
extramitochondrial pyruvate dehydrogenase present in the brain (11)
is not supported by observations made on subcellular fractions of
brain homogenates (25).

A major unresolved problem is the way in which the acetyl
groups from the intramitochondrial acetylCoA are supplied to the
extramitochondrial space, since the inner mitochondrial membrane
is believed to be impermeable to acetylCoA and molecules of a
similar size and charge (5; however, see Refs. 17 and 25 for data
suggesting that this need not be an absolute rule). It is assumed
that some compounds capable of crossing the inner mitochondrial
membrane act as carriers of acetyl groups to the extramitochondrial
space (5,27). Several pieces of indirect evidence (7,23,24) support the view that citrate acts as the carrier of acetyl groups
for the synthesis of ACh. At the same time, however, very small
incorporation of radioactive label from citrate into ACh has been
found in experiments in which citrate was administered into the
cerebrospinal fluid (28,29) or included into the incubation medium
in experiments with brain slices (1,16) and synaptosomes (13).

Acetate, glutamate and N-acetylated amino acids have also been
considered as candidates for the role of carriers of acetyl groups

in cholinergic neurons, but their participation in the synthesis of
ACh appears unlikely (for review, see Refs. 19 and 27). The possibility that acetylcarnitine is involved in the transport of acetyl
groups for the synthesis of ACh has been investigated only indirectly. Lefresne et al. (13) incubated brain synaptosomes in the
presence of labelled pyruvate and measured the incorporation of
label into ACh. When they added unlabelled acetylcarnitine to the
incubation medium, no dilution of radioactivity incorporated into
ACh occurred. From this result it appeared unlikely that acetylcarnitine is utilized as a source of acetyl groups for the synthesis
of extramitochondrial acetylCoA and ACh.

Comparison of the Utilization of Different Substrates
for the Synthesis of Acetylcholine
in Slices of Rat Caudate Nuclei

In our experiments the utilization of several labelled substrates for the synthesis of ACh was compared with the utilization
of unlabelled glucose. Slices of rat caudate nuclei were incubated
in the presence of 5 mM glucose and of 0.1 mM $[1\text{-}^{14}C]$-acetate,
0.1 mM $[2\text{-}^{14}C]$-pyruvate, 0.1 mM$[1,5\text{-}^{14}C]$-citrate, or 1 mM $[1\text{-}^{14}C\text{-Ac}]$-
acetylcarnitine. In some experiments 5 mM $[U\text{-}^{14}C]$-glucose was the
only substrate added to the incubation medium. After 60 min incubation, the total content of ACh in the tissue and medium was measured
by bioassay on the guinea pig ileum. To measure the synthesis of
radiolabelled ACh, samples treated and non-treated with acetylcholinesterase were extracted with tetraphenylboron (4) and the
amount of label present in the acetyl groups of ACh was calculated
from the difference in the extracted radioactivity.

Under the conditions used, ACh was simultaneously synthesized
from the unlabelled glucose and from the labelled substrate. In
order to compare the utilization of both, we introduced the term
"relative specific preference" (RP) of the tissue for individual
substrates. RP was calculated from the formula:

$$RP = \frac{\dfrac{\text{synthesis of } [^{14}C]\text{-ACh}}{\text{total synthesis of ACh}}}{\dfrac{\text{concentration of } [^{14}C]\text{-substrate}}{\text{total concentration of substrates}}}$$

The concentration of glucose was multiplied by 2 for the calculations since 1 mole of glucose yields 2 moles of acetylCoA. The
assumption was that, if the tissue were using a substrate for the
synthesis of ACh with the same avidity as glucose, the proportion

TABLE 1:   Values of "relative preference" for different labelled
           substrates, calculated from measurements of ACh synthesis
           in slices of caudate nuclei incubated in the presence of
           glucose and labelled precursors.

SUBSTRATE	RELATIVE PREFERENCE ($\pm$ S.E.M.)
$[2-^{14}C]$-pyruvate	1.46 $\pm$ 0.08
$[U-^{14}C]$-glucose	1.17 $\pm$ 0.05
$[1-^{14}C]$-acetylcarnitine	0.63 $\pm$ 0.05
$[1,5-^{14}C]$-citrate	0.27 $\pm$ 0.03
$[1-^{14}C]$-acetate	0.18 $\pm$ 0.04

For definition of relative preference see text.

TABLE 2:   Effect of 2.5 mM (-)-hydroxycitrate on the synthesis of
           $[^{14}C]$-ACh from $[^{14}C]$-labelled substrates in slices of
           rat caudate nuclei.

SUBSTRATE	Synthesis in the presence of (-)-hydroxycitrate (percent of synthesis in its absence)
$[U-^{14}C]$-glucose	67.4
$[2-^{14}C]$-pyruvate	75.6
$[1-^{14}C]$-acetylcartine	91.6
$[1,5-^{14}C]$-citrate	13.9

of labelled ACh in the total synthesis of ACh would correspond to
the proportion which the concentration of the labelled substrate
represented in the total concentration of substrates. The value
of RP would then equal unity.

     Values of RP actually calculated from the results of our exper-
iments are summarized in Table 1. Apparently the substrates were
utilized for the synthesis of ACh in the following order of preference: pyruvate > glucose > acetylcarnitine > citrate > acetate.
The higher RP found for pyruvate in comparison with glucose may
depend on the fact that the metabolic pathway from pyruvate to

acetylCoA is shorter than that from glucose. The value of RP for
[U-^{14}C]-glucose was expected to equal unity; the reason why it was
higher has not been clarified. The value of RP for [1-^{14}C]-acetyl-
carnitine was much higher than expected in view of the earlier negative findings by Lefresne et al. (13), referred to in the Introduction. It is apparent from our direct measurements of the incorporation of label that acetylcarnitine may be efficiently utilized for the synthesis of ACh <u>in vitro</u> and that its possible role
in the supply of acetyl groups for the synthesis of ACh <u>in vivo</u>
will have to be given serious consideration. The values of RP for
[1,5-^{14}C]-citrate and [1-^{14}C]-acetate were low. It is worth noting
however, that the utilization of citrate was higher in our experiments than in comparable experiments described by several other
authors (1,13,16). Calculations regarding citrate have been made
on the assumption that the [^{14}C]-atom from position 5 only is incorporated into the acetylCoA and ACh arising from citrate.

Effect of (−)-hydroxycitrate
on the Synthesis of Acetylcholine
in Slices of Rat Caudate Nuclei

(−)-Hydroxycitrate is a specific inhibitor of ATP-citrate
lyase, the enzyme responsible for the extramitochondrial (26) formation of acetylCoA from citrate. In view of data supporting the
idea that citrate is the carrier of acetyl groups for the synthesis
of ACh (7,23,24), it appeared of interest to investigate the effect
of the inhibition of ATP-citrate lyase upon the incorporation of
labelled atoms from various precursors into ACh. The experiments
were performed using a technique similar to that detailed in the
previous section. However, 1 mM Na-EGTA was substituted for CaCl$_2$
in the incubation medium. Unlabelled 5 mM glucose, together with
0.1 mM [1,5-^{14}C]-citrate, 0.1 mM [2-^{14}C]-pyruvate, or 1 mM [1-^{14}C]-
acetylcarnitine, or 5 mM [U-^{14}C]-glucose alone served as substrates.

In experiments with [1,5-^{14}C]-citrate, the synthesis of labelled ACh had been diminished by 64%, 82% and 87% in the presence of
0.5 mM, 2.5 mM and 10 mM (−)-hydroxycitrate, respectively. A
2.5 mM concentration of the inhibitor was then used in experiments
with other labelled precursors of ACh (Table 2). It is apparent
from Table 2 that the utilization of pyruvate and glucose for the
synthesis of ACh was diminished by only one-quarter to one-third
under conditions when the utilization of citrate was depressed by
more than 80%. The formation of CO$_2$ from the substrates tested was
not diminished by 2.5 mM (−)-hydroxycitrate.

The inhibition of ACh formation from glucose observed in our
experiments is similar to that described by Sterling and O'Neill

(24). It is apparent from the present observations, however, that
the effect of (-)-hydroxycitrate could not have been due to the
chelation of Ca^{++} ions in the incubation medium, and that the concentration of the inhibitor used was sufficient to suppress the
enzymatic cleavage of citrate. The most likely interpretation of
the results with (-)-hydroxycitrate seems to be that ATP-citrate
lyase is responsible for the supply of only one-quarter to one-
third of the acetylCoA which is used for the synthesis of ACh from
glucose or pyruvate. Alternative interpretations are possible,
however. Need for caution in drawing definitive conclusions is
particularly apparent from the observation that citrate itself can
inhibit the synthesis of ACh (see next section).

The Inhibitory Effect of Citrate
on the Synthesis of Acetylcholine

When 2.5 mM sodium citrate was added to the incubation medium
containing 5 mM glucose, the synthesis of ACh (measured by bioassay)
in slices of rat caudate nuclei was inhibited by 55%, compared to
that observed during incubations with glucose alone. An incubation
medium without Ca^{++} ions and with 1 mM EGTA was used in these experiments. The formation of CO_2 from glucose was not changed by the
addition of citrate. The mechanism of the inhibitory effect of
citrate on the synthesis of ACh (noted earlier, see Ref. 15) is
difficult to explain. An inhibition of phosphofructokinase or of
pyruvate dehydrogenase by citrate does not appear to be responsible,
since citrate was found to inhibit the synthesis of ACh also in
experiments with 10 mM pyruvate as the sole substrate, without affecting the rate of formation of $^{14}CO_2$ from [1-^{14}C]-pyruvate. It
has been shown earlier (17,25) that an increase in the concentration
of Ca^{++} in the medium in which brain mitochondria are suspended
brings about an increase in the output (leakage?) of acetylCoA from
the mitochondria. The question arises whether the decrease in the
synthesis of ACh caused by citrate might depend on an opposite
phenomenon, i.e. on a decrease in the output of acetylCoA from the
mitochondria when the low intracellular concentration of Ca^{++} is
further diminished by the citrate entering the cells from the
incubation medium.

Relation Between the Availability of Acetylcoenzyme A
and the Content and Synthesis of Acetylcholine
in Slices of Caudate Nuclei

The mechanism responsible for the maintenance of a stable concentration of ACh in cholinergic neurons has not been clarified with
certainty; relevant experimental findings and their interpretations

have been discussed recently (27). It appears likely that the concentration of ACh in the compartment of its synthesis approximates the concentration that is expected when the reaction catalysed by choline acetyltransferase is at equilibrium (18,27). If this is so, the concentration of ACh should reflect changes in the availability of Ch and of acetylCoA. Kuntscherova (10) discovered in 1972 that the administration of a massive dose of Ch brings about an increase in the concentration of ACh in the brain, heart atria and gut. Her observations have been confirmed in other laboratories (2,6; for a critical review see also Ref. 14). Conditions interfering with the transport of Ch across nerve cell membranes have been found to decrease the level of ACh in the brain (3,8). Only few and contradictory data are available, however, with regard to the relation between the levels of acetylCoA and ACh in the brain (9,20).

In the present experiments, changes in the concentration of acetylCoA were elicited in slices of rat caudate nuclei by changing the concentration of glucose in the incubation medium or by adding metabolic inhibitors into the medium. Data on the content of acetyl-CoA, ACh and Ch in the tissue and of ACh in the incubation medium (21,22) obtained after 60 min of incubation in the presence of a comparatively high concentration of Ch (0.2 mM), paraoxon and of three different concentrations of glucose are summarized in Table 3. The concentration of acetylCoA in the tissue was directly related to the concentration of glucose in the medium. The content of ACh in the tissue and, in experiments with a depolarizing (30 mM) concentration of K^+ ions, the amount of ACh released into the medium was directly related to the content of acetylCoA in the slices. Under the conditions of experiments in Table 3, the relation between the concentration of acetylCoA and of ACh in the tissue was linear and could be expressed, within the range of concentrations examined, by the equations:

$$(a, \text{ at } 5 \text{ mM } K^+) \qquad [ACh] = 47.3 \, [acetylCoA] + 56.2$$
$$(b, \text{ at } 30 \text{ mM } K^+) \qquad [ACh] = 58.5 \, [acetylCoA] + 31.6.$$

In experiments with metabolic inhibitors, the concentration of glucose in the medium was 10 mM. Bromopyruvate (0.25–1 mM), an inhibitor of pyruvate dehydrogenase, and 2,4-dinitrophenol (0.05–0.2 mM), an uncoupler of oxidative phosphorylation, produced a dose-dependent decrease in the concentration of both acetylCoA and ACh in the slices. Sodium cyanide caused an increase in tissue acetyl-CoA and ACh content when used at a 50 μM concentration. Higher concentrations of sodium cyanide brought about a decrease of acetylCoA and ACh levels.

TABLE 3: Concentrations of acetylCoA, ACh and Ch in the tissue and the content of ACh in the medium after incubations (60 min) of slices of rat caudate nuclei in the presence of different concentrations of glucose.

	CONCENTRATION OF GLUCOSE		
	0.5 mM	2.0 mM	10.0 mM
A: Medium with 5 mM K$^+$			
AcetylCoA in the tissue	1.30 ± 0.17	1.85 ± 0.15	4.10 ± 0.21
ACh in the tissue	83.25 ± 8.81	154.33 ± 7.11	267.16 ± 18.28
ACh in the medium	84.78 ± 5.48	190.07 ± 17.33	141.62 ± 14.51
Ch in the tissue	860.50 ± 36.24	833.50 ± 48.92	742.50 ± 79.93
B: Medium with 30 mM K$^+$			
AcetylCoA in the tissue	0.56 ± 0.09	1.65 ± 0.18	3.18 ± 0.20
ACh in the tissue	63.00 ± 6.02	122.53 ± 5.27	224.32 ± 10.14
ACh in the medium	126.53 ± 7.85	406.70 ± 39.11	863.18 ± 86.17
Ch in the tissue	814.83 ± 34.04	724.33 ± 38.91	563.17 ± 39.51

All results are expressed as nmol/g fresh weight (mean ± S.E.M.)

Simultaneous measurements of the concentrations of acetylCoA
and ACh in the tissue performed under several different conditions
thus reveal a close correlation between the two values. It is worth
noting that, in experiments with different concentrations of glucose
(Table 3), changes in the concentration of ACh were not accompanied
by corresponding changes in the concentration of Ch in the tissue.
The observed correlation between acetylCoA and ACh appears to be
best explained as a direct consequence of the law of mass action,
i.e. on the assumption that the reaction of ACh synthesis is close
to equilibrium and that the level of ACh in the compartment of synthesis depends on the levels of both substrates (Ch and acetylCoA)
and of the second product (CoA). Presumably all mechanisms responsible for the homeostasis of intracellular acetylCoA, Ch and CoA will
have to be taken into account in attempts to explain the control of
ACh concentration and of ACh synthesis in cholinergic neurons.

SUMMARY

The way in which acetyl groups from the intramitochondrial
acetylCoA are supplied for the extramitochondrial synthesis of ACh
still remains unclear. A comparatively efficient utilization of
$[1-^{14}C]$-acetylcarnitine for the synthesis of $[^{14}C]$-ACh revealed in
experiments with slices of caudate nuclei points to acetylcarnitine
as a possible carrier of acetyl groups for the synthesis of ACh.
Experiments with (-)-hydroxycitrate, an inhibitor of ACh-citrate
lyase, did not provide support for the view that this enzyme is
responsible for the supply of most of acetylCoA for the synthesis
of ACh. Direct correlation between tissue levels of acetylCoA and
of ACh, and in experiments with 30 mM K^+, also between the level
of acetylCoA in the tissue and the amount of ACh released into the
medium, was discovered in experiments in which slices of caudate
nuclei were incubated in the presence of varying concentrations of
glucose. AcetylCoA and ACh levels were also directly related in
the slices that had been incubated in the presence of metabolic
inhibitors (bromopyruvate, 2,4-dinitrophenol, NaCN). These observations accord with the view that the reaction of ACh synthesis is
close to equilibrium in cholinergic neurons and that the level of
ACh in the compartment of its synthesis depends on the supply of
both substrates and removal of both products of the reaction
catalyzed by choline acetyltransferase.

ACKNOWLEDGEMENTS

The authors thank Drs. A. Silver, A. Sullivan, B. Collier, and
the Wellcome Trust for assistance in providing biochemical reagents.

REFERENCES

1. Cheng, S.-C. and Brunner, E.A. (1978): J. Neurochem. 30:
    1421-1430.
2. Cohen, E.L. and Wurtman, R.J. (1975): Life Sci. 16:1095-1102.
3. Collier, B., Poon, P. and Salehmoghaddam, S. (1972): J.
    Neurochem. 19:51-60.
4. Fonnum, F. (1969): Biochem. J. 113:291-298.
5. Greville, G.D. (1969): In Citric Acid Cycle (ed) J.M. Lowen-
    stein, M. Dekker, New York, pp. 1-136.
6. Haubrich, D.R., Wedeking, P.W. and Wang, P.F.L. (1974): Life
    Sci. 14:921-927.
7. Hayashi, H. and Kato, T. (1978): J. Neurochem. 31:861-869.
8. Hebb, C.O., Ling, G.M., McGeer, E.G., McGeer, P.L. and
    Perkins, D. (1964): Nature (Lond) 204:1309-1311.
9. Heinrich, C.P., Stadler, H. and Weiser, H. (1973): J. Neuro-
    chem. 21:1273-1281.
10. Kuntscherova, J. (1972): Physiol. Bohemoslov. 21:655-660.
11. Lefresne, P., Beaujouan, J.C. and Glowinski, J. (1978):
    Nature (Lond) 274:497-500.
12. Lefresne, P., Guyenet, P. and Glowinski, J. (1973): J. Neuro-
    chem. 20:1083-1097.
13. Lefresne, P., Hamon, M. Beaujouan, J.C. and Glowinski, J.
    (1977): Biochimie 59:197-215.
14. MacIntosh, F.C. (1979): In Nutrition and the Brain, Vol. 5
    (eds) A. Barbeau, J.H. Growden and R.J. Wurtman, Raven Press,
    New York, pp. 201-217.
15. McLennan, H. and Elliott, K.A.C. (1950): Amer. J. Physiol.
    163:605-613.
16. Nakamura, R., Cheng, S.-C. and Naruse, H. (1970): Biochem. J.
    118:443-450.
17. Polak, R.L., Molenaar, P.C. and Braggaar-Schaap, P. (1977):
    In Cholinergic Mechanisms and Psychopharmacology (ed) D.J.
    Jenden, Plenum Press, New York, pp.511-524.
18. Potter, L.T., Glover, V.A.S. and Saelens, J.K. (1968): J. Biol.
    Chem. 243:3864-3870.
19. Quastel, J.H. (1977): In Cholinergic Mechanisms and Psycho-
    pharmacology (ed) D.J. Jenden, Plenum Press, New York,
    pp.411-430.
20. Reynolds, S.F. and Blass, J.P. (1975): J. Neurochem. 24:185-
    186.
21. Ricny, J. and Tucek, S. (1980): Anal. Biochem. (in press).
22. Ricny, J. and Tucek, S. (1980): Biochem. J. (in press).
23. Sollenberg, J. and Sorbo, B. (1970): J. Neurochem. 17:201-207.
24. Sterling, G.H. and O'Neill, J.J. (1978): J. Neurochem. 11:
    525-530.
25. Tucek, S. (1967): Biochem. J. 104:749-756.

26. Tucek, S. (1967): J. Neurochem. 14:531-545.
27. Tucek, S. (1978): Acetylcholine Synthesis in Neurons.
       Chapman and Hall, London.
28. Tucek, S. and Cheng, S.-C. (1970): Biochim. Biophys. Acta
       208:538-540.
29. Tucek, S. and Cheng, S.-C. (1974): J. Neurochem. 22:893-914.

CHOLINE AVAILABILITY AND THE SYNTHESIS OF ACETYLCHOLINE

D.R. Haubrich, N.H. Gerber, A.B. Pflueger and W.J. Pouch

Merck Institute for Therapeutic Research
Department of Pharmacology
Neuropsychopharmacology Section
West Point, Pennsylvania 19486 USA

Most of the work on transmitter replacement therapy by means
of precursor loading has been directed toward manipulation of neurotransmitters synthesized from aromatic amino acids (47). However,
beginning in the mid 1970's, clinical investigators became interested in the use of choline (Ch) to treat disorders in which too
little acetylcholine (ACh) was formed in the brain, or in which the
formation of an excess of the neurotransmitter was believed to be
desirable. This interest in Ch as a therapeutic agent was triggered by basic research showing that the administration of Ch to
laboratory animals could cause an increase in the level of ACh in
the brain, a phenomenon first described in 1972 by Kuntscherova
(32). Although the beneficial effects of Ch in some disorders such
as tardive dyskinesia (see 2) may reside in other properties of the
molecule (e.g. a direct postsynaptic muscarinic action of Ch has
been demonstrated (9,20,31)), there is evidence that part of the
effect of Ch is elicited through stimulation of ACh synthesis and
release. This evidence is reviewed in the first part of this
report.

In the second part of this report, results of some of our current studies on the metabolic pathways for Ch in the periphery are
presented and the possible role of these pathways in regulation of
the amount of Ch available within cholinergic neurons for synthesis
of ACh is discussed. These metabolic pathways are particularly
important in Ch therapy because one of the major problems associated
with use of Ch as a thereapeutic agent is that extremely high doses
must be administered in order to attain a sustained increase in

blood levels, presumably because the compound is rapidly oxidized
and phosphorylated in tissues. Furthermore, when a high dose of Ch
is administered orally, most is converted in the gut to trimethylamine with a resultant "fishy" odor as the major side effect.

Evidence that Choline is a Rate Limiting Substrate
in the Synthesis of Acetylcholine

With the introduction of methods for measurement of the rate
of ACh synthesis _in vivo_, it became evident that the values observed
were considerably lower than those obtained _in vitro_. For example, a
rate of ACh synthesis greater than 8000 nmol/g/hr can be achieved
when brain homogenates are prepared such that all the choline
acetyltransferase (CAT) is solubilized and assays are performed in
a medium of physiologic ionic strength containing an optimum concentration of Ch (27). In contrast to this high rate of _in vitro_
synthesis of ACh, relatively low rates of less than 1200 nmol/g/hr
occur _in vivo_ in brains of rats (23). This discrepancy between the
maximal rate of synthesis that can be achieved _in vitro_ (i.e. the
V_{max} rate) and the _in vivo_ rate of synthesis would occur if the
enzyme were not saturated _in vivo_ with its substrate (i.e. the substrate concentration _in vivo_ is below its K_m).

The K_m of CAT for Ch is greater than 400 μM (45). Although
accurate estimates of the level of Ch within cholinergic nerve
terminals have not been possible, in regions of rat brain the concentration of Ch is 35 μM or less (23). Lesioning of cholinergic
neuronal tracts causes a decrease in the level of ACh (14) with
little or no change in levels of Ch (14,39), which indicates that
Ch is not concentrated within the cells. Therefore, its concentration in cholinergic neurons is probably no greater than in the surrounding tissue. The lowest level of free Ch in any tissue studied
occurs in cerebrospinal fluid where it is less than 4 μM (5). Thus,
it seems likely that the level of Ch within cholinergic neurons is
below the K_m of CAT for this substrate. This hypothesis is further
supported by the fact that Ch for the synthesis of ACh must be
transported from outside cholinergic neurons, as indicated by the
fact that hemicholinium-3 (HC-3), which inhibits the transport of
Ch across nerve membranes, prevents the synthesis and release of
ACh, and this inhibition is overcome by Ch (4,6).

Based upon the above theoretical considerations, we decided to
test the hypothesis that Ch is a rate limiting substrate in synthesis of ACh _in vivo_. The precursor was administered intravenously to guinea pigs in the hope that an elevation of tissue concentration of Ch would stimulate the rate of synthesis of ACh by a

mass action effect and cause an increase in the concentration of
the neurotransmitter. In this study, Ch administration induced a
marked increase in levels of ACh in several peripheral tissues of
guinea pigs including the adrenal glands and heart (24,26). As
shown in Fig. 1, the increase in the concentration of both Ch and
ACh in the adrenal gland was proportional to the dose of Ch. In
this tissue, the increase in ACh concentration was greater than
fivefold when the highest dose of Ch was given. In contrast to the
adrenal gland, the greatest increase in concentration of ACh which
occurred in the heart was only 75%, and this occurred with all doses
of Ch, even though the concentration of Ch in the heart continued
to increase with dose. This difference between the two tissues in
their response to Ch may be related to differences in the kinetics
of ACh synthesis, the ability of neurons to store newly synthesized
transmitters, or the rate of release of ACh, although these possibilities remain to be investigated.

In our initial studies of the effect of Ch on tissue ACh levels,
we were unable to demonstrate any change in brain levels of ACh
measured 2 min after intravenous administration of Ch (26). In
subsequent studies the precursor was given by different routes in an
effort to raise brain ACh levels and was found to induce such an effect when given intraventricularly or intra-arterially (22,24). In
the meantime, studies by Racagni et al. (37) had shown that Ch administered intravenously did cause an increase in brain ACh levels
but that the change could be observed only when the rats were sacrificed 0.5 min after the injection. As shown in Fig. 2, intraperitoneal administration of Ch induced a modest increase in the concentration of Ch and ACh in the corpus striatum of rat brain. Similar results have been reported by Cohen and Wurtman (8). The Ch-
induced increase in brain ACh is blocked by HC-3,which inhibits
neuronal transport of Ch (17). Furthermore, Ch administration does
not increase adrenal ACh when the splanchnic nerve is sectioned
(17). In addition, Ch administration has been shown to augment the
increase in ACh concentration which occurs in the corpus striatum
of rats treated with paraoxon to inhibit cholinesterase (42). These
latter findings suggest that the Ch induced increase in concentration of ACh is caused by stimulation of ACh synthesis within cholinergic neurons. Including our own work, an increase in brain ACh
induced by administration of either Ch or lecithin has been demonstrated in at least eight different laboratories (see Table 1).
However, the increase in brain Ch or ACh is never large when Ch is
given parenterally (approximately 20% for ACh; Fig. 2 and Table 1),
especially when compared to the effect on adrenal glands (see Fig.1),
and requires the administration of a large dose of Ch.

Investigators in several laboratories have been unable to confirm the finding that Ch administration raises brain ACh levels in
normal animals (13,33,36,44), whereas some have shown an effect only

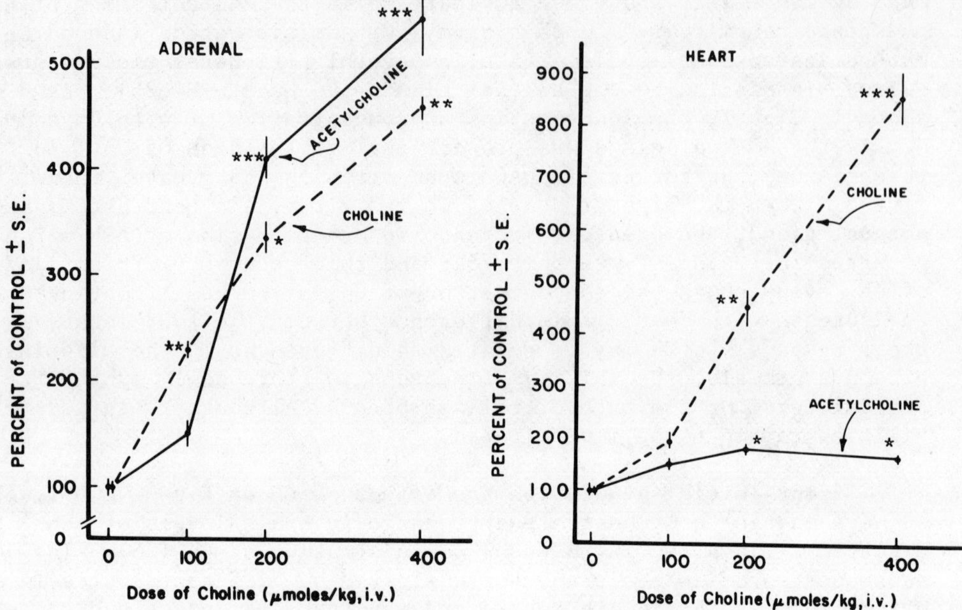

FIGURE 1: Effect of Ch administration on Ch and ACh concentration in guinea pig heart and adrenals. Ch chloride was administered intravenously to guinea pigs anesthetized with pentobarbital sodium (35 mg/kg i.p.) and the animals were sacrificed 2 min after injection of Ch. Values are the averages of 4-6 determinations. Asterisk indicates that values are significantly different from control (P < 0.05; Student's t-test).

when the steady-state level of ACh was reduced by dietary deficiency of Ch (44) or by lesioning striatal neurons with kainic acid (33). The discrepancy does not seem to be related to the method of assay as suggested by Pedata et al. (36). The Ch induced increase in brain ACh has been detected using bioassay (33), gas chromatographic (GC) methods (37,42,44) and enzymatic/radioisotopic assays (8,12, 22). Furthermore, the change is not simply an artifact caused by interference of Ch in the assay for ACh, as indicated by the fact that there is no correlation between the increase in Ch and that of ACh in the heart after Ch is administered (Fig. 1). The possibility exists that, in studies in which no change in ACh was observed after Ch administration, the rate of release of ACh may also have been stimulated to the same degree as its rate of synthesis such that no net change in steady state levels of ACh had occurred.

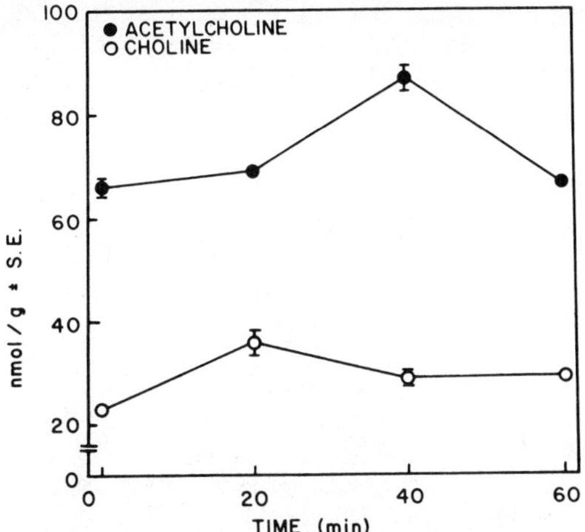

FIGURE 2: Effect of Ch administration on Ch and ACh in corpus striatum
of rats. Rats were killed by microwave irradiation of the
head after intraperitoneal administration of Ch (0.7 mmol/
kg) (data from Ref. 22).

To evaluate the possibility that Ch elicits a central effect
by stimulation of the rate of synthesis and release of ACh, we have
utilized an indirect measure of central cholinergic function. Cholinomimetic agents when administered to laboratory animals stimulate
the rate of turnover of dopamine in the corpus striatum (18), presumably because they activate cholinoceptive neurons which indirectly impinge upon dopaminergic fibers.

The effect of Ch on the rate of synthesis of dopamine was assessed _in vivo_ after inhibition of dopa decarboxylase with NSD-1015,
as described previously (7). As shown in Table 2, oral administration of Ch augmented by 46% the increase in concentration of dopa
induced by treatment with NSD-1015. This indicates that Ch administration significantly enhanced the rate of synthesis of dopamine.

In order to further evaluate Ch induced changes in dopamine
metabolism, we measured changes in levels of homovanillic acid (HVA),
a metabolite of dopamine (19). As shown in Table 3, oral administration of Ch caused a significant increase in the level of HVA in
the corpus striatum. Repeated administration of low doses of Ch to
fasted rats was found to be more effective in stimulation of dopamine metabolism than treatment with a single large dose. The data
in Table 3 also show that an increase in the level of plasma Ch of
less than twofold was sufficient to stimulate functional activity

TABLE 1:   Effects of Ch administration on tissue Ch and ACh.

Dose µmol/kg	Route	Species	Time min	Tissue	Assay[*]	Percentage Increase[**]		
						Choline	ACh	Ref.
1400	s.c.	rat	80	Brain	B	n.m.	20	33
			60	Atria	B	n.m.	34	33
			60	Ileum	B	n.m.	33	33
			20	Plasma	C	1100	n.m.	33
200	i.v.	guinea pig	2	Adrenal	E	235	311	16
			2	Heart	E	352	80	16
			2	Brain	E	n.c.	n.c.	16
$10 \cdot min^{-1}$	arterial perfusion	guinea pig	15	Brain	E	49	30	17
429,858	i.p.	rat	20	Brain	E	123	n.c.	8
			40	Brain	E	40	22	8
200	i.v. (pulse)	rat	0.5	Corpus Striatum	GC/MS	146	25	38
			2	Corpus Striatum	GC/MS	n.c.	n.c.	38
5	i.v.	rat	6	Plasma	GC/MS	76	n.m.	38
			6	Occipital Cortex	GC/MS	46	17	38
25	i.v.	rat	6	Corpus Striatum	GC/MS	74	20	38
0.5 (total)	i.vent.	rat	15	Corpus Striatum	E	n.m.	60	20
2.5 (total)	i.vent.	rat	15	Hippocampus	E	n.m.	30	20,23
50 (total)	i.vent.	rat	15	Cerebral Cortex	E	n.m.	51	20
700	i.p.	rat	20	Corpus Striatum	E	50	n.c.	20
			40	Corpus Striatum	E	n.c.	20	20
			20	Plasma	E	500	n.m.	20
714	i.p.	rat	40	Kainate-lesioned Striatum	E	224	95	34
429	i.p.	rat	20	Hippocampus	E	59	92	28
$15 \cdot min^{-1}$	i.v. infusion	rat	15	Cortex	E	156	n.c.	11
			15	Striatum	E	61	11	11
429	i.p.	rat	60	Paraoxon-Treated Striatum	GC	26	28	44
500	i.p.	rat	40	Choline-Deficient Striatum	GC	n.m.	33	45
433	i.p.	rat	20,40	Corpus Striatum	E	20	n.c.	9
1299	i.p.	rat	20,40	Corpus Striatum	E	144,73	20-25	9[+]

* B = bioassay; C = colorimetric; E = enzymatic; GC = gas chromatographic; MS = mass speectrometric;   ** n.m. = not measured; n.c. = no change. Where a number is shown, the change was statistically significant.
+ Also increased with i.v. infusion.

of central cholinergic neurons. Fasting of the rats prior to treatment with Ch was found to be necessary to elicit the response, possibly because of variable rates of absorption of Ch in fed animals. The finding of a Ch induced increase in dopamine metabolism agrees

TABLE 2:   Effect of Ch on the rate of synthesis of dopamine.

Treatment	Concentration of Dopa (nmol/g $\pm$ S.D.)
NSD 1015	3.9 ± 0.6
NSD 1015 + Ch	5.7 ± 1.0*

The rate of synthesis was assessed by measuring the accumulation of Dopa induced by administration of the Dopa decarboxylase inhibitor NSD-1015 (7). Three doses of Ch chloride (1 mmol/kg each) were given orally at 1 hr intervals. NSD-1015 (100 mg/kg) was administered i.p. with the last dose of Ch, and the rats were decapitated 1 hr later for fluorometric measurement of Dopa in the corpus striatum. * Different from NSD-1015 alone ($P < 0.01$; Student's t-test).

TABLE 3:   Effect of Ch on rat striatal homovanillic acid and serum Ch.

Dose (mmol/kg)	Number of Doses	Homovanillic Acid (percent control $\pm$ S.D.)	Serum Ch (percent control $\pm$ S.D.)
10	1	129 ± 27	–
10	2	191 ± 20*	2154 ± 645*
5	2	166 ± 35*	627 ± 345*
1	2	139 ± 40*	209 ± 54*
1	3	186 ± 68*	327 ± 118*
0.5	3	153 ± 15*	190 ± 64*
0.25	3	102 ± 22	–

Rats fasted for previous 20 hr were treated by oral intubation with Ch chloride. Doses of Ch were administered 60 min apart and rats were decapitated 30 min after last dose. Homovanillic acid data are from 3 separate experiments. Statistical analysis (Student's t-test) were performed using vehicle treated control group for each experiment. Values calculated from averages of 6 determinations, each of which consisted of pooled corpora striata from 2 rats. Ch was measured in a separate experiment using serum taken from the orbital sinus of the anesthetized (chloral hydrate, 350 mg/kg i.p.) rats. Average concentration of homovanillic acid, measured fluorometrically, was 1.6 ± 0.3 nmol/g and that of Ch, measured enzymatically, was 11.6 ± 6 nmol/ml. * Different from control ($P < 0.05$) (From Ref. 19).

with the results of Ulus and Wurtman (41) showing that treatment of
rats with Ch stimulated the activity of tyrosine hydroxylase measured in vitro.

To assess the role of cholinergic receptors in the increase in
dopamine metabolism elecited by treatment with Ch, rats were given
atropine to block the response to cholinergic stimulation. As shown
in Table 4, treatment with atropine blocked by about 50% the Ch
induced increase in the level of the metabolite. This finding indicates that the Ch induced increase in the concentration of HVA is
mediated by activation of cholinergic receptors. This conclusion
is consistent with results of another study which showed that administration of Ch prevented the atropine induced decrease in brain
ACh concentration (43).

To determine if Ch augments the metabolism of dopamine by increasing the rate of synthesis and release of ACh, rats were treated
with 4-(1-naphthylvinyl)pyridine (NVP) prior to administration of
either Ch or physostigmine. NVP is a potent in vitro inhibitor of
CAT (40). The in vivo rate of synthesis of ACh is also inhibited
when NVP is administered to animals (21,30). The results in Fig. 3
show that oral administration of either Ch or physostigmine caused
a significant increase in the concentration of HVA in brain and that
pretreatment with NVP completely prevented the increase induced by
administration of Ch. In contrast to its effect on the Ch induced
increase in HVA, pretreatment of rats with NVP did not prevent the
increase induced by administration of physostigmine (Fig. 3), which
indicates that NVP does not block the response elicited by stimulation of cholinergic receptors. As previously shown (18), NVP caused
a decrease in the concentration of HVA possibly because it inhibits
cholinergic function by inhibition of ACh synthesis. Because the
increase in dopamine metabolism produced by Ch is blocked by an
inhibitor of ACh synthesis, whereas the effect of the acetylcholinesterase (AChE) inhibitor is not, we conclude that Ch and physostig-
mind produce their central cholinergic effects by different mechanisms and that Ch does so by the stimulation of ACh synthesis and
release from its presynaptic terminals.

Regulation of Choline
Available for Synthesis of Acetylcholine

The results presented in the previous section support the
hypothesis that Ch is a rate limiting substrate for the synthesis
of ACh. They indicate that raising the extracellular concentration
of Ch will lead to an increase in the concentration of Ch within
cholinergic neurons and this will result in an enhancement of the
rate of synthesis and release of ACh. Because the Ch that is converted to ACh in brain is derived ultimately from free Ch in the

TABLE 4:   Effect of atropine on the Ch induced increase in brain
           homovanillic acid.

Treatment	Homovanillic Acid (percent control $\pm$ S.D.)
Vehicle	100 ± 16
Atropine	96 ± 19
Ch	146 ± 14*
Atropine and Ch	122 ± 16**

Atropine sulfate (2 µmol/kg) was administered intraperitoneally to
rats which had been fasted for the previous 20 hr; 10 min later
Ch chloride (10 mmol/kg) was administered orally, followed 1 hr
later by a second dose of Ch. Rats were decapitated 20 min after
the second dose of Ch for fluorometric measurement of homovanillic
acid in the corpus striatum. Values calculated from average of 6
determinations, each consisting of pooled corpora striata from two
rats. Data are from Ref. 19. * = Different from vehicle treated,
P < 0.01 (Student's t-test); ** = Different from Ch alone, P < 0.05
(Student's t-test).

periphery (25), some of the factors that might regulate the steady
state concentration of peripheral Ch are now being investigated in
our laboratory.

The major pathways for metabolism of Ch are well known and include phosphorylation and oxidation in mammalian tissue. Ch kinase
(EC 2.7.1.32) catalyzes the phosphorylation of Ch in a reaction in
which ATP serves as phosphate donor. This enzyme was purified from
yeast more than 25 years ago (46) and has since been shown to occur
in a soluble form in a variety of tissues.

The oxidation of Ch, first demonstrated during the 1930's (3,
34) is known to proceed by two separate reactions. The first is
catalyzed by Ch dehydrogenase (EC 1.1.99.1), a mitochondrial enzyme
which converts Ch to betaine aldehyde. The second enzyme, betaine
aldehyde dehydrogenase (EC 1.2.1.8), is a soluble enzyme which converts the aldehyde to betaine. The sequence of reactions is
illustrated below:

$$(CH_3)_3N^+CH_2CH_2OH \longrightarrow (CH_3)_3N^+CH_2CHO \longrightarrow (CH_3)_3N^+CH_2COOH$$

Choline   Betaine aldehyde   Betaine

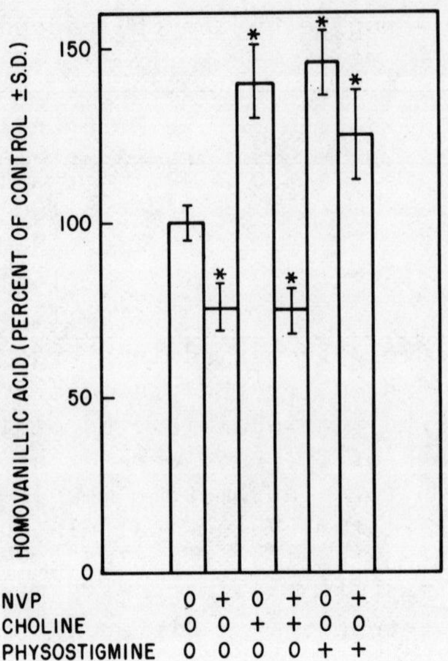

FIGURE 3: Effect of inhibition of ACh synthesis on the increase in
homovanillic acid (HVA) induced by Ch or physostigmine.
An inhibitor of ACh synthesis, 4-(1-naphthylvinyl)pyridine
(NVP; 0.37 mmol/kg i.p.) was administered to rats which
had been fasted for the previous 20 hr; 10 min later
either Ch chloride (10 mmol/kg) or physostigmine sulfate
(2.5 μmol/kg) was administered orally, followed 1 hr later
by a second oral dose of each compound. Rats were decapitated for measurement of striatal HVA 20 min after the
second treatment with physostigmine or Ch. Values are
average of 6 determinations, each consisting of pooled
corpora striata from two rats. * Significantly different
(P < 0.01; Student's t-test) from that of control.

The tissue and species distribution of the enzymes for Ch
metabolism are illustrated in Table 5.   Ch dehydrogenase is present predominantly in liver and kidney. Ch kinase has a more widespread occurrence with high activity in liver and kidney as well as
in lung, spleen, brain and duodenum.

In order to assess the possible importance of these two enzymes
in regulation of the metabolism of Ch in vivo, we have utilized two
analogs of Ch: dimethylaminoethanol (deanol) and 2-amino-2-methyl-
propanol (2-AMP). As shown in Table 6, deanol and 2-AMP are competitive inhibitors of Ch dehydrogenase in vitro, with K_i values in

TABLE 5:   Enzymes of Ch metabolism.

| Species | Tissue | Enzyme Activity (nmol/min·mg $\pm$ S.D.) | |
		Ch Dehydrogenase	Ch Kinase
Rat	Liver	22 $\pm$ 0.3	1.6 $\pm$ 0.2
	Kidney	17 $\pm$ 1.8	0.8 $\pm$ 0.1
	Lung	0.2 $\pm$ 0.06	0.7 $\pm$ 0.3
	Brain	0.06 $\pm$ 0.01	1.1 $\pm$ 0.1
	Blood	N.D.	N.D.
	Spleen	N.D.	1.5 $\pm$ 0.3
	Heart	N.D.	0.5 $\pm$ 0.7
	Muscle	N.D.	0.03 $\pm$ 0.03
	Fat	N.D.	0.03 $\pm$ 0.003
Guinea pig	Liver	0.15 $\pm$ 0.1	0.2 $\pm$ 0.03
	Kidney	1.8 $\pm$ 0.3	1.2 $\pm$ 0.07
	Lung	N.D.	1.1 $\pm$ 0.07
	Brain	N.D.	0.6 $\pm$ 0.05
	Blood	0.06 $\pm$ 0.05	N.D.
	Spleen	0.02 $\pm$ 0.02	1.03 $\pm$ 0.1
	Heart	0.02 $\pm$ 0.02	0.1 $\pm$ 0.02
	Muscle	N.D.	0.02 $\pm$ 0.02
	Fat	N.D.	0.78 $\pm$ 0.9
	Duodenum	0.38 $\pm$ 0.01	1.1 $\pm$ 0.05

Enzyme activities were measured radioisotopically. Values averages
from 3 animals.   N.D. = not detectable.

TABLE 6:   Inhibition _in vitro_ of Ch dehydrogenase.

Inhibitor	Inhibitory Constant (K_i)
Dimethylaminoethanol	1.8
2-Amino-2-methylpropanol	1.4

Inhibitory constants were determined using a Dixon plot.   Ch dehydrogenase was measured in liver mitochondria using ^{3}H-Ch as substrate (Haubrich and Gerber, Biochem. Pharmacol., in press).

the lower millimolar range. Furthermore, deanol is a substrate for
Ch kinase _in vitro_ (16) and therefore would be expected to compete
with Ch for phosphorylation. To determine if these agents inhibit
Ch metabolism _in vivo_, mice were treated with high doses of each
compound when Ch-methyl-^{3}H was administered intravenously. The
kidneys were removed 3 min later and analyzed for endogenous free
Ch and for radiolabelled Ch, betaine aldehyde, betaine and phosphorylcholine.

As shown in Table 7, radiolabelled Ch was converted rapidly
in the kidney to both phosphorylcholine and to the oxidation products. Administration of either deanol or 2-AMP induced a significant increase in the concentration of free Ch in the tissue. In
addition, the treatment also led to an increase in the concentration
of radiolabelled Ch in kidney, suggesting a decrease in its rate of
turnover, and caused a decrease in the amount of radiolabelled
metabolites. When the specific activity of Ch is taken into account and the rate of metabolism calculated, the results indicate
that deanol or 2-AMP almost completely inhibit both the oxidation
and phosphorylation of Ch _in vivo_ (Table 7). These results show
that Ch kinase and Ch dehydrogenase are important enzymes in regulation of the concentration of free Ch in the periphery, and suggest
that activity of these enzymes might regulate cholinergic function
indirectly through control of Ch availability. Deanol administration is known to induce an increase in levels of ACh in brain (22)
and to elicit central cholinergic effects in humans (see 38). The
present finding suggests that part of the mechanism of action of
deanol may involve elevation of peripheral levels of free Ch through
inhibition of Ch metabolism, which then leads to mass action induced stimulation of the rate of synthesis and release of ACh in
cholinergic neurons of the brain.

Another metabolic pathway that may be important in control of
free Ch in cholinergic neurons is the deamination which is catalyzed
by an enzyme present in intestinal bacteria (35). When large doses
of Ch are given orally, more than 60% of the administered dose is
converted in the gut to trimethylamine, which causes an undesirable
"fishy" odor associated with therapeutic administration of Ch (11).
In order to assess the possible role of intestinal metabolism of
Ch in regulation of the steady state level of Ch in the body, we
investigated the effect of treatment with antibiotics on the levels
of Ch in blood. Rats were given a combination of antibiotics as
indicated in Table 8 which, together with fasting, reduced to nearly undetectable levels the titre of both aerobic and anaerobic
microorganisms in the intestine. As shown in Table 8, sterilization
of the gut significantly increased the concentration of blood Ch.
In addition, oral administration of Ch caused a significant increase in blood levels of Ch in sterilized rats but not in the controls measured 15 min after treatment (Table 8). At intervals of

TABLE 7: Metabolism of [^{3}H]-Ch in mouse kidney and the effect of Ch analogs.

| Endogenous Ch (nmol/g ± S.D.) | Radioactivity (dpm/mg ± S.D.) | | | Rate of | |
	Choline	Phosphorylcholine	Betaine + Bet. aldehyde	Phosphorylation	Oxidation
Treatment: None					
118 ± 28	225 ± 98	678 ± 119	1178 ± 469	102 ± 22	181 ± 81
Treatment: Dimethylaminoethanol					
411 ± 96*	1919 ± 1249*	13 ± 10*	217 ± 204*	1.3 ± 0.3*	10 ± 7*
Treatment: 2-Amino-2-methylpropanol					
465 ± 103*	3361 ± 1911*	50 ± 18*	658 ± 514*	2 ± 0.3*	27 ± 15*

Female mice were treated intraperitoneally with 5 mmol/kg of each compound; 2 hr later, ^{3}H-Ch-methyl was administered intravenously and mice were decapitated 3 min later. The kidneys were immediately frozen and Ch and metabolites were isolated by electrophoresis and chromatography and measured as described previously (25). Rates of phosphorylation and oxidation were calculated from the equation:

$$nmol/g/min = \left[\frac{^3H\text{-product}}{^3H\text{-choline}} \right] \times endogenous\ choline \div 3\ min$$

* Different from control (P < 0.05; Student's t-test).

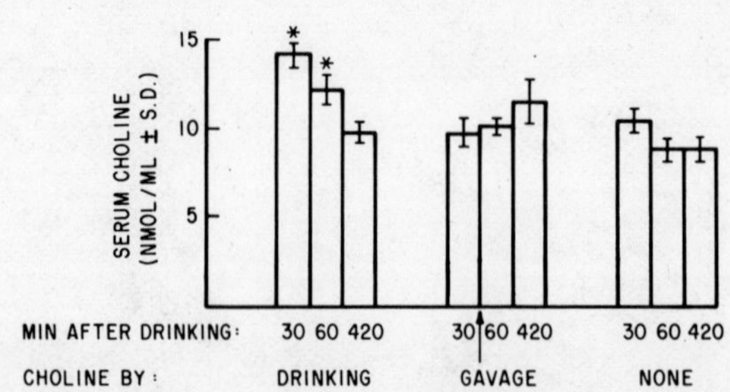

FIGURE 4: Comparison of serum Ch in rats given Ch by gavage or in
           the diet. Rats deprived of drinking water overnight were
           allowed free access to water with Ch for 2 hr. During
           this period rats consumed 1.12 mmol/kg of Ch. This dose
           of Ch was then given by gavage to rats that had drunk
           water without Ch. Control rats received both vehicles.
           * Values different from control animals that did not
           receive Ch (n = 6 rats per group).

TABLE 8:   Effect of gut sterization on Ch absorption.

Min after Ch	Serum Ch (nmol/ml ± S.D.)		
(0.5 mmol/kg p.o.)	Control	Sterilized	Statistical Significance vs Control (P)
0	13.7 ± 3.9	18.5 ± 5.7	< 0.05
15	18.5 ± 7.7	27.0 ± 9.7*	< 0.05
30	24.4 ± 12.4*	28.3 ± 8.2*	> 0.1
45	23.9 ± 10.8*	23.7 ± 5.6*	> 0.1
60	24.6 ± 9.1*	24.7 ± 7.0*	> 0.1
120	16.6 ± 6.0	21.6 ± 5.1	< 0.05
240	14.1 ± 4.5	16.7 ± 2.3	< 0.1

Ch was administered to rats which had been fasted for 2 days and
treated for 3 days with neomycin sulfate (100 mg b.i.d.), clinda-
mycin palmitate HCl (50 mg b.i.d.) and amoxicillin (50 mg/kg s.i.d.).
* Different from zero time (P < 0.05; Student's t-test). (n = 6-
12 rats/group).

30, 45 and 60 min after administration of Ch, its concentration was
increased significantly, and to the same extent, in blood of both
control and sterilized rats. Two hours after treatment Ch levels
had returned essentially to normal in the blood of both groups.
These results suggest that absorption of Ch from the gut may be
limited to some extent by bacterial metabolism.

In another study we compared the effect on blood Ch levels of
a single dose of Ch given by gavage with that of an equal amount
of Ch ingested in drinking water over a 2 hr period. These results,
shown in Fig. 4, indicate that Ch given over the longer period was
more effective in raising the concentration of blood Ch than a
single large dose. The reason for the difference in activity of
Ch given by gavage as compared to ingestion is not clear from this
study but may be related to differences in metabolism of Ch within
the gut. Nevertheless, this finding does have potential clinical
applications. In most clinical studies of Ch large oral doses are
given at relatively infrequent intervals (10,15,29). The present
results suggest that consumption of smaller amounts of Ch at more
frequent intervals might be a more effective means of raising plasma levels of Ch. Such treatment might also diminish the amount
of trimethylamine formed since this seems to be associated primarily
with administration of high doses of Ch.

SUMMARY

The following lines of evidence suggest that Ch is a rate
limiting substrate in the synthesis of ACh: 1) the concentration
of Ch within cholinergic neurons is below the K_m of CAT for the
substrate, 2) raising the extracellular concentration of Ch causes
an increase in the concentration of ACh in brain and peripheral
tissues, 3) the Ch-induced increase in concentration of ACh is
blocked by HC-3, 4) administration of Ch elicits a central cholinergic effect (i.e. stimulation of dopamine metabolism) which
depends upon stimulation of the rate of synthesis of ACh.

The possibility that alternate metabolic pathways for Ch in
the periphery can regulate the amount of Ch available for acetylation within cholinergic neurons is now being investigated. Our
studies and those of others indicate that the major metabolic pathways for Ch in mammals are phosphorylation and oxidation. Compounds that inhibit these pathways were found to cause accumulation
of Ch in peripheral tissues. In addition to oxidation and phosphorylation, Ch is also known to be converted to trimethylamine by
bacteria in the gut. Sterilization of the intestines was found to
increase the rate of absorption of Ch administered orally. These

findings suggest that it may be possible to influence central cholinergic function by modification of Ch metabolism in the periphery.

REFERENCES

1. Aquilonius, S.M. and Eckernas, S.A. (1975): New Eng. J. Med. 293:1105-1106.
2. Barbeau, A., Growdon, J.H. and Wurtman, R.J. (eds) (1979): Nutrition and the Brain, Vol. 5, Raven Press, New York.
3. Bernheim, F. and Bernheim, M.C.C. (1938): Amer. J. Physiol. 121:55-60.
4. Birks, R. and MacIntosh, F.C. (1961): Canad. J. Biochem. Physiol. 39:787-827.
5. Bowers, M.B. Jr. (1967): Life Sci. 6:1927-1933.
6. Brown, L. and Feldberg, W. (1936): J. Physiol. 88:265-283.
7. Carlsson, A. and Lindqvist, M. (1973): J. Neural. Trans. 34: 79-91.
8. Cohen, E.L. and Wurtman, R.J. (1975): Life Sci. 16:1095-1102.
9. Consolo, S., Ladinsky, H. and Gomeni, R. (1979): Pharmacol. Res. Comm. 11:903-918.
10. Davis, K.L., Hollister, L.E., Barchas, J.D. and Berger, P.A. (1976): Life Sci. 19:1507-1516.
11. DeLaHuerga, J. and Popper, H. (1951): J. Clin. Invest. 30: 463-470.
12. Eckernas, S.A. (1977): Acta Physiol. Scand. 449 (Suppl.).
13. Flentge, F. and Vanderberg, C.J. (1979): J. Neurochem. 32: 1331-1333.
14. Friesen, A.J.D., Ling, G.M. and Nagai, M. (1967): Nature 214: 722-724.
15. Growdon, J.H., Hirsch, M.J., Wurtman, R.J. and Weiner, W. (1977): New Eng. J. Med. 297:524-527.
16. Haubrich, D.R. (1973): J. Neurochem. 21:315-328.
17. Haubrich, D.R. and Chippendale, T.J. (1977): Life Sci. 20: 1465-1478.
18. Haubrich, D.R. and Goldberg, M.E. (1975): Neuropharmacology 14:211-214.
19. Haubrich, D.R. and Pflueger, A.B. (1979): Life Sci. 24:1083-1090.
20. Haubrich, D.R., Risley, E.A. and Williams, M. (1979): Biochem. Pharmacol. 28:3673-3674.
21. Haubrich, D.R. and Wang, P.F.L. (1976): Biochem. Pharmacol. 25:669-672.
22. Haubrich, D.R., Wang, P.F.L., Clody, D.E. and Wedeking, P.W. (1975): 17:975-980.
23. Haubrich, D.R., Wang, P.F.L., Herman, R.L. and Clody, D.E. (1975): Life Sci. 17:739-748.

24. Haubrich, D.R., Wang, P.F.L. and Wedeking, P. (1974): Fed. Proc. 33:477.

25. Haubrich, D.R., Wang, P.F.W. and Wedeking, P.W. (1975): J. Pharmacol. Exp. Ther. 193:246-255.

26. Haubrich, D.R., Wedeking, P.W. and Wang, P.F.L. (1974): Life Sci. 14:921-927.

27. Hebb, C. (1972): Physiol. Rev. 52:918-957.

28. Hirsch, M.J., Growdon, J.H. and Wurtman, R.J. (1977): Brain Res. 125:383-385.

29. Hollister, L.E., Jenden, D.J., Amaral, J.R.D., Barchas, J.D., Davis, K.L. and Berger, P.A. (1978): Life Sci. 23:17-22.

30. Krell, R.D. and Goldberg, A.M. (1975): Biochem. Pharmacol. 24:391-396.

31. Krnjevic, K. and Reinhardt, W. (1979): Science 206:1321-1323.

32. Kuntscherova, J. (1972): Physiol. Bohem. 21:655-660.

33. London, E.D. and Coyle, J.T. (1978): Biochem. Pharmacol. 27: 2962-2965.

34. Mann, P.J.G. and Quastel, J.H. (1937): Biochem. J. 31:869-878.

35. Neill, A.R., Grime, D.W. and Dawson, R.M.C. (1978): 170:529-535.

36. Pedata, F., Wieraszko, A. and Pepeu, G. (1977): Pharmacol. Res. Comm. 9:755-761.

37. Racagni, G., Trabucchi, M. and Cheney, D.L. (1975): N.-S. Arch. Pharmacol. 290:99-105.

38. Re., O. (1974): Curr. Ther. Res. 16:1238-1242.

39. Sethy, V.H., Kuhar, M.J., Roth, R.H., Van Woert, M.H. and Aghajanian, G.K. (1973): 55:481-484.

40. Smith, J.C., Cavallito, C.J. and Foldes, F.F. (1967): Biochem. Pharmacol. 16:2438-2441.

41. Ulus, I.H. and Wurtman, R.J. (1976): Science 194:1060-1061.

42. Wecker, L. and Dettbarn, W.D. (1979): J. Neurochem. 32:961-967.

43. Wecker, L., Dettbarn, W.D. and Schmidt, D. (1978): Science 199:86-87.

44. Wecker, L. and Schmidt, D.E. (1978): Soc. Neurosci. Abst. 4:436.

45. White, H.L. and Wu, J.C. (1973): J. Neurochem. 20:297-307.

46. Wittenberg, J. and Kornberg, A. (1953): J. Biol. Chem. 202: 431-444.

47. Wurtman, R.J. (1979): IN Nutrition and the Brain, Vol. 5 (eds) A. Barbeau, J.H. Growdon and R.J. Wurtman, Raven Press, New York, pp. 1-12.

ACETYLCHOLINE SYNTHESIS AND GLUCOSE OXIDATION WITH VARIOUS OXYGEN

LEVELS IN VIVO AND IN VITRO

G.E. Gibson, H.J. Ksiezak and T.E. Duffy

Department of Neurology, Cornell Medical College, and
The Burke Rehabilitation Center
White Plains, New York 10605   USA

Normal brain function is dependent upon a continual supply of
oxygen. For example, memory and the ability to learn a complex task
are impaired when the arterial oxygen tension (PaO$_2$) is reduced from
a normal of approximately 90 mm Hg to about 60 mm Hg (13).   The biochemical basis of the brain's sensitivity to a decrease in oxygen
(hypoxic hypoxia) is unknown, but it is apparently not due to an inability to maintain the level of energy metabolites (5,14). Consequently, the focus in the search for the molecular mechanism in
hypoxia has shifted to those neurotransmitters whose synthesis depends upon oxygen. The rate-limiting step in the formation of dopamine, norepinephrine and serotonin requires molecular oxygen;
furthermore, their turnover is decreased by hypoxia. However, the
importance of decreased catecholamine and serotonin synthesis in
the development of symptoms of hypoxia remains questionable since
the turnover of these neurotransmitters is not reduced when hypoxia
is accompanied by stress (2).

Impaired acetylcholine (ACh) metabolism is intriguing as a
mechanism which causes the behavioral symptoms of hypoxia. Cholinergic transmission in the superior cervical ganglion is blocked at
levels of oxygen which do not alter axonal conduction, postsynaptic
sensitivity to ACh (3) or energy levels in the synaptic region (12).
The anticholinergic drug, scopolamine, produces symptoms similar to
those observed during hypoxia (e.g. loss of recent memory [4]). Impaired oxidation of glucose, pyruvate or ketone bodies leads to a
decrease in ACh synthesis (9), even though less than 1% of the

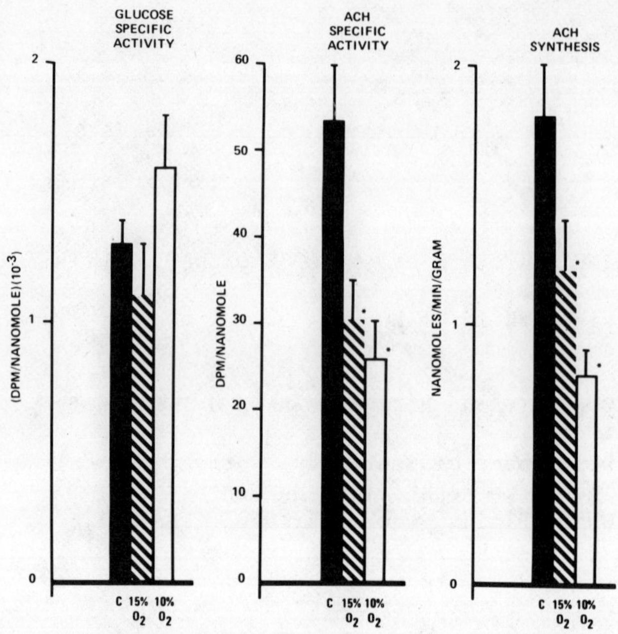

FIGURE 1: ACh synthesis measured from [U-^{14}C]-glucose with decreasing oxygen tensions in vivo. One minute before
sacrifice by "brain blowing," animals were injected with
[U-^{14}C]-glucose (1 mCi/kg). The incorporation into ACh
and the rate of synthesis were determined as previously
described (11). * = $p < 0.05$.

oxidized substrate is normally converted to ACh. In vivo, even
mild anemic hypoxia (NaNO$_2$) or histotoxic hypoxia (KCN) impairs
the synthesis of ACh (7). However, studies of chemically induced
hypoxia may not be directly relevant to hypoxic hypoxia, since KCN
or NaNO$_2$ may alter ACh turnover by some mechanism other than reduced oxygen availability. Furthermore, most studies which relate
metabolism to brain function in man and animals have used arterial
oxygen tension as the measure of hypoxia. It seemed logical to
determine whether the degrees of systemic hypoxic hypoxia that are
known to cause neurological impairment in man also lead to impaired
ACh synthesis.

The effect of mild hypoxic hypoxia was examined in vivo with
rats by decreasing the percent oxygen in the inspired air from a
control of 30% to 15% or 10% oxygen for 15 min. Consequently the
PaO$_2$ in the blood decreased from a control of 120 $\pm$ 2 mm Hg to 57 $\pm$
1   or 42 $\pm$ 1 mm Hg, respectively. ACh synthesis was studied by
labelling both the acetyl and choline (Ch) moieties by simultaneous
injection of [U-^{14}C]-glucose and [^{2}H$_4$]-Ch. The incorporation of

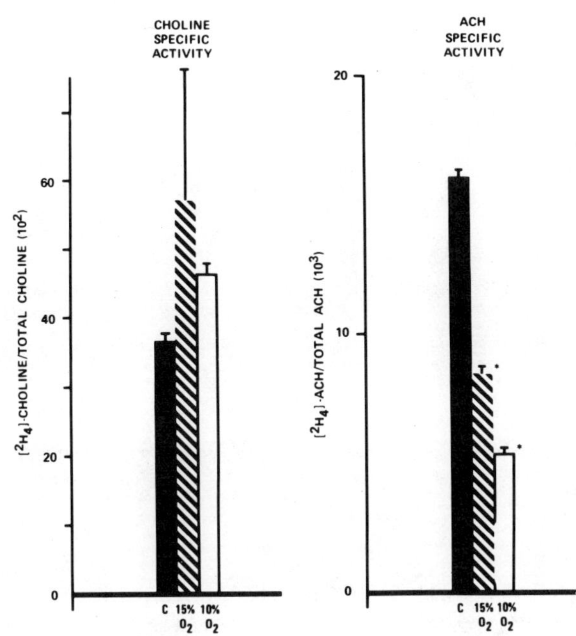

FIGURE 2: Specific activity of $[^2H_4]$-Ch and $[^2H_4]$-ACh with decreasing oxygen tensions <u>in vivo</u>. Animals were sacrificed
by "brain blowing" 1 min after injection of $[^2H_4]$-Ch
(20 μmol/kg). The specific activities of $[^2H_4]$-Ch and
$[^2H_4]$-ACh were determined by gas chromatography mass
spectrometry (6).

[U-^{14}C]-glucose into ACh was reduced by 43% and 52% of control with
15% and 10% oxygen, respectively (Fig. 1). Similarly, the incorporation of $[^2H_4]$-Ch into ACh was reduced by 48% and 55% of control
by 15% and 10% oxygen, respectively (Fig. 2). The specific activity
of the precursor pools (glucose and Ch) did not decrease (Figs. 1
and 2). The reduced incorporation of either precursor into ACh
was not due to a dilution of the labels if one assumes uniform
mixing of the pools. Therefore levels of hypoxic hypoxia which
impair mental function in man also decrease the synthesis of ACh
in rats.

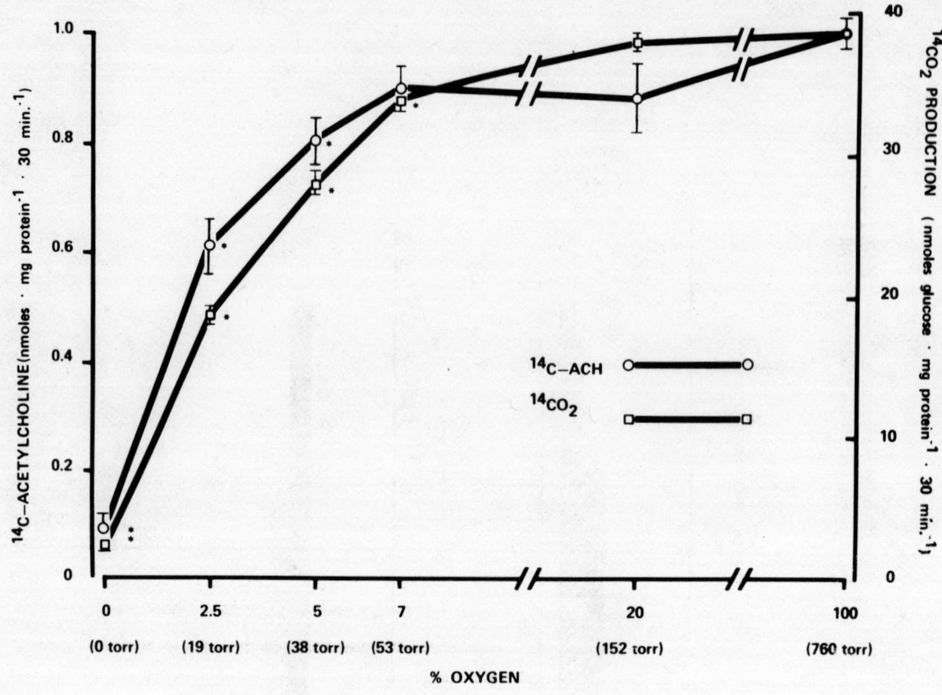

FIGURE 3:   ACh synthesis and CO_2 production in rat brain slices with
varying oxygen tensions. Slices were prepared and incubated with 1.5 mM [U-^{14}C]-glucose (5 μCi/ml)(8). Each
flask was flushed for 10 min with a gas having the percent O_2 shown on the horizontal axis. The oxygen content in the incubation medium was monitored with a microelectrode and found to be within theoretical limits for
the duration of the incubation. Incorporation of [U-^{14}C]-
glucose into ACh and the release of $^{14}CO_2$ was determined
as previously described (10).

The reduction of ACh synthesis by hypoxic hypoxia _in vivo_
could have been the result of a generalized decrease in neuronal
firing rather than a primary effect on ACh metabolism. Therefore,
we studied ACh metabolism _in vitro_ where the secondary effect of
these pathways is minimized. In rat brain slices, a reduction in
the synthesis of ACh occurred at oxygen levels similar to those
which inhibited ACh turnover _in vivo_ (Fig. 3). When the oxygen

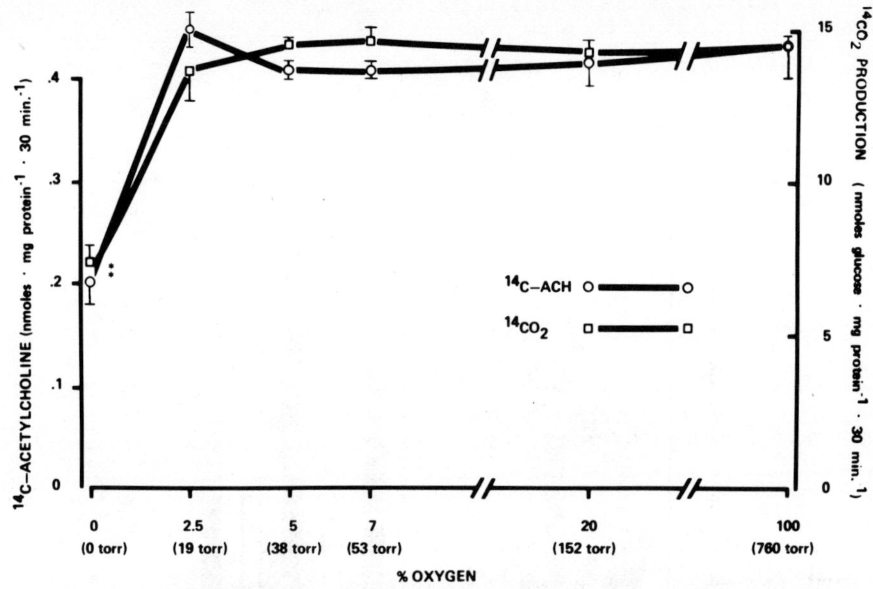

FIGURE 4:   ACh synthesis and CO_2 production by synaptosomes
             incubated with various oxygen tensions. Synaptosomes were
             prepared by the method of Booth and Clark (1). Incubation
             procedures were exactly as described in Fig. 3 for slices.

tension was decreased below 53 mm Hg, a sharp decline in ACh production from glucose was observed (Fig. 3).   The reduction paralleled: 1) the decrease in overall glucose metabolism as measured
with [U-^{14}C]-glucose, 2) the decrease in pyruvate decarboxylation
as measured with [3,4-^{14}C]-glucose, and 3) the activity of the citric acid cycle as measured with [2-^{14}C]-glucose. ACh synthesis and
metabolic flux through all the pathways of glucose oxidation that
we examined were reduced by more than 90% with incubations under
N_2 (pO_2 < 1 mm Hg) or with antimycin A, an inhibitor of oxidative
phosphorylation (Fig. 5).

     In synaptosomes, the relation between glucose oxidation and
ACh synthesis was different than that observed with slices or in
vivo. ACh metabolism in synaptosomes was not as sensitive to decreases in oxygen. The production of ACh was maintained at the
control level with a pO_2 even as low as 19 mm Hg (Fig. 4). When
the incubations were conducted under anaerobic conditions, ACh

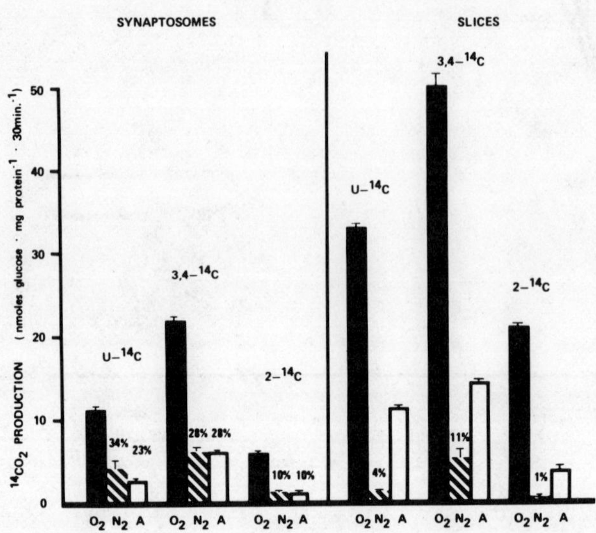

FIGURE 5: $^{14}CO_2$ production from either [U-^{14}C]-, [3,4-^{14}C]- or
[2-^{14}C]-glucose in the presence of 100% O_2, 100% N_2 or
2 µM antimycin A. Incubations were as described in Fig.3.
Increasing the concentration of antimycin to 20 µM had no
additional effect on $^{14}CO_2$ production by synaptosomes but
inhibited $^{14}CO_2$ production from [2-^{14}C]-glucose in slices
to 4% of control.

synthesis was at least 30% of the control rate. As with slices,
a parallel inhibition of ACh synthesis with overall glucose oxidation ([U-^{14}C]-glucose decarboxylation) and pyruvate decarboxylation
([3,4-^{14}C]-glucose decarboxylation) was apparent (Fig. 5).

However, ACh synthesis and citric acid cycle oxidation ([2-
^{14}C]-glucose decarboxylation) were not inhibited to the same extent.
During incubation with N_2,citricacid cycle activity was inhibited by
greater than 90% while ACh synthesis declined less than 70%.The
results also suggest that under N_2 the coupling between pyruvate
dehydrogenase and the citric acid cycle is not as "tight" in synaptosomes as in slices.

In conclusion, ACh synthesis *in vitro* seems to parallel pyruvate decarboxylation ($[3,4-^{14}C]$-glucose oxidation) more closely than citric acid cycle activity ($[2-^{14}C]$-glucose oxidation). This link may partially explain the relation between decreased oxygen levels and reduced ACh synthesis *in vivo*.

ACKNOWLEDGEMENTS

The work reported here was supported by USPHS NIH grants NS-03346 and NS-16997, the Winifred Masterson Burke Relief Foundation and the Will Rogers Institute

REFERENCES

1. Booth, R.F.G. and Clark, J.B. (1978): Biochem. J. 176:365-370.
2. Davis, J.N., Giron, L.T., Stanton, E. and Maury, W. (1979): Adv. Neurol. 26:219-223.
3. Dolivo, M. (1974): Fed. Proc. Amer. Soc. Exp. Biol. 33:1043-1048.
4. Drachman, D.A. (1978): IN Psychopharmacology: A Generation of Progress (eds) M.A. Lipton, A. DiMascio and K.F. Killam, Raven Press, New York, pp. 651-662.
5. Duffy, T.E., Nelson, S.R. and Lowry, O.H. (1972): J. Neurochem. 19:959-977.
6. Freeman, J.J., Choi, R. and Jenden, D.J. (1975): J. Neurochem. 24:729-734.
7. Gibson, G.E. and Blass, J.P. (1976): J. Neurochem. 27:37-42.
8. Gibson, G.E. and Blass, J.P. (1976): J. Neurochem. 26:1073-1078.
9. Gibson, G.E. and Blass, J.P. (1979): Biochem. Pharmacol. 28:133-140.
10. Gibson, G.E. and Shimada, M. (1980): Biochem. Pharmacol. 29:167-174.
11. Gibson, G.E., Shimada, M. and Blass, J.P. (1978): J. Neurochem. 31:757-760.
12. Harkonen, M.H.A., Passonneau, J.V. and Lowry, O.H. (1969): J. Neurochem. 16:1439-1450.
13. Luft, U.C. (1975): IN Handbook of Physiology, Sect. 3: Respiration (eds) W.O. Fenn and H. Rahn, American Physiological Society, Washington, pp. 1099-1145.
14. Siesjo, B.K. and Nilsson, L. (1971): Scand. J. Clin. Lab. Invest. 27:83-96.

REGULATION OF ACETYLCHOLINE RELEASE DURING INCREASED

NEURONAL ACTIVITY

L. Wecker and A.M. Goldberg*

Department of Pharmacology, Louisiana State University
Medical Center, New Orleans, Louisiana 70112 and
*Department of Environmental Health Sciences
Division of Toxicology, The Johns Hopkins University
School of Hygiene and Public Health
Baltimore, Maryland   21205

INTRODUCTION

The availability of choline (Ch) in the microenvironment of
the nerve terminal and its transport across the neuronal membrane
may play a significant role in the dynamic regulation of acetylcholine (ACh) metabolism. However, the precise nature of, or
factors involved in, the regulatory processes have not been un-
equivocally determined. In particular, relatively little attention
has been focussed on the possibility that the mechanisms regulating
the metabolism of ACh can be modified or directly influenced by the
state of neuronal activity. Plasticity, the ability to alter or
adapt with the immediate environment, is characteristic of numerous
biological systems. Hence, it is possible that the regulatory
mechanisms controlling the metabolism of ACh may vary according to
the demands imposed on the neuron. In order to characterize the
nature of the regulatory process(es) and elucidate the determinant
effects of the availability and transport of Ch on the metabolism
of ACh, we have taken into account the physiological status of the
neuron. Consideration of both the supply to and the demand of the
system for Ch will undoubtedly clarify the relationship between
neuronal activity and possible regulatory mechanisms.

Precursor Availability

Although free Ch can be generated from Ch containing compounds
in the brain (6,9,13,16,27), central cholinergic neurons lack the
ability to synthesize Ch de novo (1). Hence, brain Ch available
for the synthesis of ACh appears to be derived from sources outside
the cholinergic neuron and include: 1) Ch derived from phospholipids,
2) free Ch in plasma, and 3) Ch generated from the hydrolysis of ACh.
Under "normal" physiological and biochemical conditions, the steady-
state concentration of ACh in brain is maintained within relatively
narrow limits and there is a net efflux of Ch from brain into the
venous return (9,13,16). Therefore, evidence suggests that economic
stability exists in the "normal (quiescent)" situation and that the
supply of Ch for the synthesis of ACh is adequate to meet the demand.

Radiolabelled Ch, administered as an intravenous pulse, through
injection, or by dietary supplementation, is rapidly incorporated
into brain ACh (9,19,39,40). Hence, exogenous sources of Ch may be
utilized for the synthesis of ACh. If the supply of Ch were regulatory for the metabolism of ACh, then one may expect enhanced synthesis of ACh to result from an increased availability of Ch. When
rats were injected with Ch iodide (60-120 mg/kg, free base, i.p.)
and sacrificed 15-90 min following injection, the concentration of
free Ch in the striatum and hippocampus increased significantly
15 min after injection to 132% and 151% of controls, respectively
(48). Ch levels returned to control by 30 min and remained constant
thereafter. Despite the increased Ch supply, no change in the
steady-state concentration of ACh was detected in any brain region
at any time following the acute administration of Ch (48). Although
discrepant results have been reported concerning the effects of administration of Ch on the synthesis and steady-state concentration
of ACh in brain (10,15,21,22,35,38,49), date from this laboratory
indicate that under "normal" demands, exogenous Ch does not modify
levels of ACh.

In order to determine whether the exogenous availability of
Ch plays a significant role in the synthesis of ACh when economic
stability is altered, either the endogenous supply of Ch can be
limited or the demand for Ch can be increased. Since the dietary
intake of Ch accounts for a significant fraction of plasma Ch levels
(19,23), endogenous Ch availability was restricted by maintaining
rats on a Ch deficient diet for 2 weeks (47). This dietary deficiency of Ch significantly decreased the concentration of ACh in the
striatum from control levels of 91.3 to 78.1 nmol ACh/g tissue and
this 14% decrease was reversed by the acute administration of Ch (47).
Hence, brain levels of ACh were increased by Ch in compromised animals, but not in normal animals, indicating that the exogenous
availability of Ch is important when the endogenous supply of Ch
is limited.

TABLE 1:   Effects of choline administration on paraoxon induced
           alterations in acetylcholine levels in rat striatum.

Treatment	Acetylcholine Concentration	
	nmol/g Tissue	Percent Control
Saline	73.7 ± 2.64(8)	---
Ch chloride	77.6 ± 4.82(6)	105
(60 mg/kg, free base)		
Paraoxon	109.7 ± 2.06(6)*	149
(0.23 mg/kg)		
Paraoxon	130.1 ± 7.42(5)*,**	177
after Ch pretreatment		

Rats were injected (i.p.) with Ch 60 min prior to the administration of paraoxon (sc) and sacrificed 15 min following the second
injection by head-focussed microwave irradiation. ACh levels were
quantitated by pyrolysis gas chromatography. Each value is the
mean ± S.E.M. The number of animals is in parentheses.
* Significantly different from saline or Ch injected animals
   P < 0.05.
** Significantly different from paraoxon alone, P < 0.05.
Adapted from Wecker and Dettbarn (46).

     Since a significant portion of the Ch used for the synthesis
of ACh is generated from the hydrolysis of ACh (11,20,36,37), the
endogenous availability of Ch can also be limited by acetylcholinesterase inhibition. When rats were injected with the cholinesterase inhibitor paraoxon, at a dose that inhibited the brain
enzyme greater than 90%, the levels of ACh in the striatum increased
from 73.7 to 109.7 nmol ACh/g tissue (Table 1). Pretreatment with Ch,
60 min prior to the administration of paraoxon, significantly potentiated the effect of paraoxon (46) and increased the levels of ACh to
130.1 nmol ACh/g tissue (Table 1). Hence, the administration of Ch,
which has no significant effect on the levels of ACh under eco-
nomically stable conditions, is utilized for the synthesis of ACh
in the striatum when the endogenous supply of Ch is limited.

     In addition to decreasing supply, the economic stability of
the system may be altered by increasing the neuronal demand for Ch.
The antimuscarinic agent atropine, increases neuronal activity in
cholinergic neurons (14,31,44), thereby increasing the demand for
Ch necessary to maintain the synthesis of ACh. Apparently the
endogenous availability of Ch is insufficient to support synthesis
during atropine induced demand and a reduced level of ACh results
(28,48). When rats were injected with 5 mg/kg atropine sulfate,

TABLE 2: Utilization of exogenous choline for acetylcholine synthesis in rat striatum during pharmacologically induced increased neuronal demand.

| DRUG TREATMENT | ACETYLCHOLINE CONCENTRATION | | | |
| | Drug Treatment Alone | | After Choline Pretreatment | |
	nmol/g Tissue	Percent Control	nmol/g Tissue	Percent Control
Saline	77.8 ± 4.65(6)	--	77.6 ± 4.82(6)	100
Atropine	56.5 ± 7.34(6)*	73	73.4 ± 3.67(6)	94
(5 mg/kg)				
Saline	77.8 ± 3.32(8)	--	79.7 ± 3.10(8)	102
Fluphenazine	67.8 ± 2.63(8)*	87	73.5 ± 5.51(8)	94
(0.5 mg/kg)				

Rats were injected (i.p.) with saline or Ch iodide 60 min prior to the administration of atropine sulfate or 45 min prior to the administration of fluphenazine and sacrificed by head focussed microwave irradiation 30 min or 45 min after the second injection, respectively. ACh levels were quantitated by pyrolysis gas chromatography. Each value is the mean $\pm$ S.E.M. The number of animals is indicated in parentheses.
* Significantly different from corresponding saline injected animals, $P < 0.05$.

TABLE 3: Effects of supplemental choline on acetylcholine release in the perfused phrenic nerve-hemidiaphragm preparation.

| | Acetylcholine Release | | |
| | HEPES | HEPES buffer with 30 µM Ch | |
Conditions	pmol ACh/min	pmol ACh/min	Percent Control
Unstimulated (spontaneous)	0.84 ± 0.07(6)	0.95 ± 0.08(14)	113
Stimulated (7 Hz, 0.2 msec, 0.5-0.8 V)	5.13 ± 0.62(4)	6.75 ± 0.07(5)*	132

ACh release was measured after 60-70 min in aliquots of the perfusate by radioisotopic techniques. Adapted from Bierkamper and Goldberg (4).
* Significantly different from stimulated preparations in HEPES buffer, P < 0.01.

the levels of ACh in striatum decreased to 73% of control values (Table 2). Pretreatment with Ch, 60 min prior to atropine, significantly prevented the atropine induced depletion of ACh and the levels of ACh following the combined administration of Ch and atropine were 94% of control values. Similarly, when cholinergic neuronal firing in the striatum was enhanced indirectly by the dopamine receptor blocking agent, fluphenazine (0.5 mg/kg), the levels of ACh decreased to 87% of control (Table 2) and this effect was prevented by the prior administration of Ch (45). Therefore, exogenous Ch is utilized for the synthesis of ACh in the striatum under conditions of pharmacologically induced increased demand.

In addition to the utilization of exogenous Ch by the brain, the effects of Ch availability on the release of ACh in the isolated perfused phrenic nerve-diaphragm preparation were investigated (4). When Ch (30 µM) was added to the perfusate, the spontaneous release of ACh was not significantly different from control values of 0.84 pmol ACh/min/ hemidiaphragm (Table 3), similar to results obtained in the superior cervical ganglion preparation (30). However, when the phrenic nerve was stimulated at 7 Hz (0.2 msec, 0.5-0.8 V) for 60-70 min, Ch in the perfusate significantly increased the evoked release of ACh to 132% of control values, from 5.13 to 6.75 pmol ACh/min/ hemidiaphragm. Therefore, the results indicate that under conditions of either increased neuronal demand or decreased neuronal supply, the exogenous availability of Ch is a significant factor involved in regulating the synthesis of ACh.

Transport Mechanisms

The pool of Ch available for the synthesis of ACh in the brain is extracellular to the site of synthesis (5,7,17) and Ch must be transported across the neuronal membrane for utilization (2). Two transport systems for Ch have been characterized in synaptosomes and brain minces, one with a high affinity (HA, K_m < 10 µM) and one with a low affinity (LA, K_m >10 µM) for Ch (8,12,18,51). Evidence indicates that the rate of Ch transport by the HA carrier in synaptosomes directly reflects changes in neuronal activity (34,42) and is inversely related to the endogenous concentration of ACh (25,50). While much evidence supports the premise that the sodium dependent HA system is intimately involved with supplying Ch for utilization in the synthesis of ACh (3,18,51), studies have shown that ACh synthesis can be inhibited without altering the HA system (26) and pharmacologically induced depletion of ACh is not always accompanied by a change in HA uptake (41). Furthermore, Ch transported by the LA system can also be acetylated (8,43) and, in brain cell cultures, most of the Ch utilized for the synthesis of ACh is transported by the LA uptake system (52). Therefore, Ch transported by either system can be utilized for the synthesis of ACh.

In order to determine the relative contributions of the HA and LA uptake systems in supplying Ch for the synthesis of ACh during various states of neuronal demand, Ch accumulation was measured in cortical minces in the presence of either low (5 mM) or high (35 mM) potassium (8). In 5 mM potassium, both HA and LA transport were observed, but in 35 mM potassium, only one component was present with a K_m of 68.3 µM and a maximal velocity of 21.4 nmol/g/2 min, similar to the parameters characteristic of the LA uptake system. Hence, the velocity of transport by the HA system in cortical minces may be insufficient to support the demands imposed by potassium induced depolarization. Similar results have been obtained in the superior cervical ganglion preparation of the cat (11). In superfused retinal preparations incubated with labelled Ch, physiological stimulation (light) did not alter the HA uptake system, but did increase both the rate of Ch incorporation into ACh, as well as the synthesis and release of ACh (32). Hence, it has been suggested that during high demand situations, extracellular Ch may be transported by a LA carrier into the nerve terminal where it is acetylated and enters a releasable pool of ACh (8).

Recent studies (28) with brain synaptosomal preparations incubated in either 5 mM or 35 mM potassium indicate the presence of both transport systems (Table 4). However, when chloride was omitted from the incubation medium and replaced by propionate, only the LA component of transport was observed, supporting data that chloride is essential for the HA uptake system (29). Furthermore, at 35 mM potassium (chloride-free), the velocity of transport of the LA system

TABLE 4: Kinetic parameters of choline transport systems in synaptosomes in chloride free medium.

Medium K^+ Concentration (mM)	Anion	High Affinity		Low Affinity	
		K_m (μM)	V_{max} (pmol/mg protein/4 min)	K_m (μM)	V_{max} (pmol/mg protein/4 min)
5	Cl$^-$	3.6 ± 0.4	138 ± 5	18.2 ± 2.8	249 ± 17
35	Cl$^-$	8.4 ± 3.2	87 ± 28	59.0 ± 3.7*	220 ± 16
5	Pr$^-$	Undetectable		31.5 ± 1.9**	206 ± 29
35	Pr$^-$	Undetectable		47.9 ± 9.8	112 ± 18***

Each value is the mean $\pm$ S.E.M. of determinations from 4 experiments. Adapted from Ksiezak and Goldberg (28).

* Significantly different from corresponding 5 mM K$^+$, P < 0.001.
** Significantly different from Cl$^-$ present 5 mM K$^+$, P < 0.01.
*** Significantly different from corresponding 5 mM K$^+$, P < 0.05,

was significantly reduced. When the synthesis and release of ACh
were measured in synaptosomes incubated in chloride-containing
buffer with 5 mM potassium, the concentration of ACh in the pellet
increased during a 30 min incubation period from 211 to 357 pmol
ACh/mg protein. In 35 mM potassium, ACh was released into the
incubation medium and the concentration of ACh in the pellet decreased significantly to 247 pmol ACh/mg protein (28). Therefore,
even though both components of transport were present, depolarized
tissue apparently could not synthesize enough ACh to support the
demands imposed by release. When chloride was replaced by propionate in the incubation medium, with only the LA transport component present, the synthesis and release of ACh were maintained
in preparations incubated with 5 mM potassium. In 35 mM potassium,
where there was a significant decrease in the velocity of transport
of the LA system (Table 4), the concentration of ACh in the pellet
was 110 pmol ACh/mg protein with a significant decrease in the
release of ACh, indicative of the absence of ACh synthesis. Therefore, under these conditions, the LA transport process could support
the synthesis of ACh in normal potassium containing buffer, but with
the decrease in the velocity of transport, it could not supply Ch
for the replenishment of tissue stores.

CONCLUSION

Results from the present studies indicate that the mechanisms
involved in the regulation of the metabolism of ACh are directly
influenced by neuronal activity. During non-depolarized physiological states (5 mM potassium), there is apparently a sufficient
supply of Ch to meet the demands imposed on the cholinergic neuron.
However, when the economic stability of the system is altered, by
either increasing the demand for, or decreasing the supply of Ch,
then the regulatory mechanisms may adapt to the needs of the immediate environment. Although there are no known states of dietary
Ch deficiency in humans, it is possible that endogenous Ch availability may be limited as a result of pharmacological alterations
or pathological conditions. Similarly, increases in the demand
for Ch may occur consequent to physiological or pharmacological
impositions.

Recent data have shown that morphine administration increases
the efflux of dopamine from striatal neurons with a low firing rate,
but not from neurons with a normal or increased firing rate (33).
Hence, the concept that neuronal activity can modify the "normal"
intrinsic mechanisms regulating the dynamics of neurotransmitter
metabolism is probably not unique for the cholinergic neuron and
may apply to all neurotransmitter systems.

In conclusion, evidence supports the premise that neurotransmitter regulatory mechanisms are highly dependent on the physiological status of the neuron and that neuronal activity or pathological conditions must be considered for valid evaluation of the primary determinants regulating the dynamics of neurotransmitter metabolism.

ACKNOWLEDGEMENTS

These studies were supported in part by a research grant from the National Institute of Mental Health (MS-33443) and from the National Institute of Environmental Health Sciences (EHS-00034 and EHS-00454). The authors would like to thank D. Stanga, E. Blanc, D. Doerfler and M. Blake for the typing of the manuscript.

REFERENCES

1. Ansell, G.B. and Spanner, S. (1975): In Cholinergic Mechanisms, (ed) P.G. Waser, Raven Press, New York, pp. 117-129.

2. Barker, L.A. (1976): In Biology of Cholinergic Function, (eds) A.M. Goldberg and I. Hanin, Raven Press, New York, pp. 203-238.

3. Barker, L.A. and Mittag, T. (1975): J. Pharmacol. Exp. Ther. 192:86-94.

4. Bierkamper, G.G. and Goldberg, A.M. (1979): In Nutrition and the Brain, Vol. 5 (eds) A. Barbeau, J.H. Growdon and R.J. Wurtman, Raven Press, New York, pp. 243-251.

5. Birks, R. and MacIntosh, F.C. (1961): Canad. J. Biochem. Physiol. 39:787-827.

6. Browning, E.T. (1976): In Biology of Cholinergic Function, (eds) A.M. Goldberg and I. Hanin, Raven Press, New York, pp. 187-201.

7. Browning, E.T. and Schulman, M.P. (1968): J. Neurochem. 15: 1391-1405.

8. Carroll, P.T. and Goldberg, A.M. (1975): J. Neurochem. 25: 523-527.

9. Choi, R.L., Freeman, J.J. and Jenden, D.J. (1975): J. Neurochem. 24:735-741.

10. Cohen, E. and Wurtman, R.J. (1975): Life Sci. 16:1095-1102.

11. Collier, B. and MacIntosh, F.C. (1969): Canad. J. Physiol. Pharmacol. 47:127-135.

12. Dowdall, M.J. and Simon, E.J. (1973): J. Neurochem. 21:969-982.

13. Dross, K. and Kewitz, H. (1972): Naun.-Schmid. Arch. Pharmacol. Exp. Pathol. 274:91-106.

14. Dudar, J.D. and Szerb, J.C. (1969): J. Physiol. 203:741-762.

15. Eckernas, S.A., Sahlstrom, L. and Aquilonius, S.M. (1977):
     Acta Physiol. Scand. 101:404-410.
16. Freeman, J.J., Choi, R.L. and Jenden, D.J. (1975): J. Neuro-
     chem. 24:729-734.
17. Guyenet, P., Lefresne, P., Rossier, J., Beaujouan, J.C. and
     Glowinski, J. (1973): Molec. Pharmacol. 9:630-639.
18. Haga, T. and Noda, H. (1973): Biochem. Biophys, Acta 291:
     564-575.
19. Hanin, I. and Schuberth, J. (1974): J. Neurochem. 23:819-824.
20. Hanin, I., Massarelli, R. and Costa, E. (1972): Adv. Biochem.
     Pharmacol. 6:181-203.
21. Haubrich, D.R., Wedeking, P.W. and Wang, P.F.L. (1974): Life
     Sci. 14:921-927.
22. Haubrich, D.R., Wang, P.F.L., Clody, D.E. and Wedeking, P.W.
     (1975): Life Sci. 17:975-980.
23. Haubrich, D.R., Wang, P.F.L., Chippendale, T. and Procter, E.
     (1976): J. Neurochem. 27:1305-1313.
24. Holmstedt, B. (1967): Ann. N.Y. Acad. Sci. 144:433-458.
25. Jenden, D.J., Jope, R.S. and Weiler, M.H. (1976): Science
     194:635-637.
26. Jope, R.S. and Jenden, D.J. (1977): Life Sci. 20:1389-1393.
27. Kewitz, H., Dross, K. and Pleul, O. (1973): In Central Nervous
     System: Studies on Metabolic Regulation and Function, (eds)
     E. Genazzani and H. Herken, Springer-Verlag, New York, pp.21-32.
28. Ksiezak, H. and Goldberg, A.M. (1979): Neurosci. Abst. 5:591.
29. Kuhar, M.J. and Zarbin, M.A. (1978): J. Neurochem. 31:251-256.
30. Lovat, S., Millington, W. and Collier, B. (1978): Proc. Canad.
     Fed. Biol. Sci. 21:2.
31. Lundholm, B. and Sparf, B. (1975): Eur. J. Pharmacol. 32:
     287-292.
32. Masland, R.H. and Livingstone, C.J. (1976): J. Neurophysiol.
     39:1210-1219.
33. Moleman, P. and Bruinvels, J. (1979); Nature 281:686-687.
34. Murrin, L.C., DeHaven, R.N. and Kuhar, M.J. (1977): J. Neuro-
     chem. 29:681-687.
35. Pedata, F., Wieraszko, A. and Pepeu, G. (1977): Pharmacol.
     Res. Commun. 9:755-761.
36. Perry, W.L.M. (1953): J. Physiol. 119:439-454.
37. Potter, L.T. (1970): J. Physiol. 206:145-166.
38. Racagni, G., Trabucchi, M. and Cheney, D.L. (1975): Naun.-
     Schmid. Arch. Pharmacol. 290:99-105.
39. Schuberth, J., Sparf, B. and Sundwall, A. (1969): J. Neurochem.
     16:695-700.
40. Schuberth, J., Sparf, B. and Sundwall, A. (1970): J. Neurochem.
     17:461-468.
41. Sherman, K.A., Hanin, I. and Zigmond, M.J. (1978): J. Pharmacol.
     Exp. Ther. 206:677-686.
42. Simon, J.R., Atweh, S. and Kuhar, M.J. (1976); J. Neurochem.
     26:909-922.

43. Suszkiw, J.B., Beach, R.L. and Pilar, G.R. (1976): J. Neuro-
       chem. 26:1123-1131.
44. Szerb, J.C., Malik, H. and Hunter, E.G. (1970): Canad. J.
       Physiol. Pharmacol. 48:780-790.
45. Trommer, B.A., Schmidt, D.E. and Wecker, L. (1980): Fed. Proc.
       39:412.
46. Wecker, L. and Dettbarn, W -D. (1979): J. Neurochem. 32:
       961-967.
47. Wecker, L. and Schmidt, D.E. (1979): Life Sci. 25:375-384.
48. Wecker, L. and Schmidt, D.E. (1980): Brain Res. 184:234-238.
49. Wecker, L., Dettbarn, W -D. and Schmidt, D.E. (1978): Science
       199:86-87.
50. Weiler, M.H., Jope, R.S. and Jenden, D.J. (1978): J. Neuro-
       chem. 31:789-796.
51. Yamamura, H.I. and Snyder, S.H. (1973): J. Neurochem. 21:
       1355-1374.
52. Yavin, E. (1976): J. Biol. Chem. 251:1392-1397.

THE UPTAKE AND ACETYLATION OF ^{3}H-CHOLINE IN BRAIN OF INBRED

STRAINS OF MICE

R. Reynolds and A.K. Prince

University of London, King's College
The Strand, London WC2R 2LS England

INTRODUCTION

In a study of inbred strains of mice it was reported that whole
brain ChAT activities varied in the following order: Balb/c > DBA/2 >
C57Bl/6 (6). ChAT activities in both the temporal and frontal cortex of the DBA/2 strain of mice were reported to be greater than in
the C57Bl/6 strain (5). It was suggested that these differences
may be correlated with differences in learning ability. The in
vivo turnover rates of acetylcholine (ACh) in frontal-parietal cortex of these strains were in agreement with ChAT activities in this
order (2). Thus greater ChAT activity may be necessary where turnover rate is greatest.

The availability of choline (Ch), however, may be a major factor
in regulation of the synthesis of ACh and the possibility that this
may also vary between strains has not been investigated. We, therefore, investigated the uptake of ^{3}H-Ch and its acetylation in small
slices of mouse frontal cortex from four inbred strains of mice.

METHODS

^{3}H-Ch Uptake: Adult male mice (8-10 weeks, Balb/c, DBA/2,
C57Bl/10 and C57Bl/6 strains) were decapitated and frontal cortex
immediately dissected and sliced (0°C, 0.05 x 0.05 x approximately
1 mm). The slices were suspended in sodium free Tris-Krebs solution
(made iso-osmolar with sucrose, 236 mM; 0°C; pH 7.4). Tissue (20 mg)
was immediately assayed (37°C, 10 min) for Ch uptake in sodium free

medium and in Tris-Krebs solution containing NaCl (118 mM), using
0.1 µM ^{3}H-Ch (1 µCi). Tissue slices were recovered by rapid centrifugation, washed twice, solubilized with Soluene-350 (Packard) and
counted for ^{3}H. Alternatively, 20 mg aliquots were incubated with
0.6 µM ^{3}H-Ch (12.5 µCi). The tissue was recovered by centrifugation,
extracted in formic acid-acetone and the extract analyzed for ^{3}H-Ch
and derivatives by high voltage electrophoresis (7).

ChAT Assay: Tissues were homogenized in 2 mM sodium phosphate
(pH 7.0), containing 1 mM EDTA. Homogenates were incubated with
Triton X-100 (0.05%; 2 hr; 0°C), diluted fivefold, and assayed
essentially as described previously (3; 2 mM Ch; 100 µM 1-^{14}C-
acetylCoA; 150 mM NaCl; pH 7.0; 37°C).

RESULTS

ChAT activities in frontal cortex of Balb/c, DBA/2 and C57Bl/10
strains of mice were not significantly different. However, activity
in the C57Bl/6 strain was 15% smaller than in the other strains
(P < 0.02; Table 1).

^{3}H-Ch uptake was greatest in C57Bl/6 strain mice and decreased
in the following order: C57Bl/6 > C57Bl/10 > DBA/2 > Balb/c (Table 1).
These differences in ^{3}H-Ch uptake were apparent when values were
compared either as uptake in medium containing sodium, or as sodium
dependent uptake, calculated by subtracting values for uptake in the
absence of sodium from those in the presence of sodium.

The proportion of ^{3}H-Ch that was acetylated following uptake in
medium containing sodium was 69-72% in all except in the C57Bl/6
strain which was significantly different at 81% (Table 2). If correction is made for uptake and acetylation in sodium free medium
these conversions become 90-93% with no significant differences
between strains.

DISCUSSION

The ChAT activities were in broad agreement with those previously reported (5), suggesting the largest activities are required
where turnover rates are greatest (2). However, the present results
also show the ratio of ChAT activity to Ch uptake is in the order
of 1000/1 in all strains. Thus the potential for kinetic coupling
of uptake and acetylation is apparent (1). Kinetic coupling requires
that the rates of synthesis of ACh from extracellular Ch are determined by the rates of uptake and are independent of ChAT activity.

TABLE 1: ^{3}H-Choline uptake and ChAT activity in slices of mouse frontal cortex.

Strain	ChAT Activity (μmol/hr/g tissue)	^{3}H-choline Uptake (nmol/hr/g tissue)		
		In medium Containing 118 mM sodium	In sodium-free medium	Calculated sodium-dependent uptake
Balb/c	7.359 ± 0.270(9)	3.660 ± 0.325(10)****	1.159 ± 0.131***	2.501 ± 0.276****a
DBA/2	7.265 ± 0.276(9)	3.985 ± 0.175(11)****	1.061 ± 0.043****	2.924 ± 0.190****
C57BL/10	7.408 ± 0.191(9)	5.227 ± 0.314(9)***	1.185 ± 0.100****	4.042 ± 0.296**
C57BL/6	6.274 ± 0.247(6)*	6.651 ± 0.250(10)	1.674 ± 0.071	4.977 ± 0.262

Values are means ± S.E.M. n = numbers in parentheses. Statistical evaluation was by Student's t-test.
a = The S.E.M. was calculated using the equation: var (x−y) = var(x) + (y) −2. covar (x,y).
ChAT activity: * P < 0.02.
^{3}H-Ch uptake, different from C57BL/6: ** P < 0.05; *** P < 0.01; **** P < 0.001.

TABLE 2: Metabolism of ^{3}H-choline in slices of mouse frontal cortex.

^{3}H-label recovered as:

	In medium containing sodium (118 mM)			In sodium free medium			Calculated sodium dependent uptake		
	^{3}H-PCh	^{3}H-ACh	^{3}H-Ch	^{3}H-PCh	^{3}H-ACh	^{3}H-Ch	^{3}H-PCh	^{3}H-ACh	^{3}H-Ch
Balb/c (13)	4.8±0.8	69.3±2.0	26.0±1.7	14.7±2.4	9.3±1.2	76.0±2.5	0.8±0.3	92.0±1.5	7.2±1.6
DBA/2 (12)	5.7±1.1	71.9±1.0	22.7±0.9	21.5±3.9	10.3±0.8	68.2±3.7	0.6±0.3	93.4±1.0	6.0±1.0
C57BL/10 (13)	5.2±0.8	71.0±1.0	23.8±0.8	18.7±2.4	9.0±0.8	70.8±2.3	0.7±0.3	90.6±1.2	8.4±1.0
C57BL/6 (5)	2.6±0.2	81.1±1.8*	16.1±1.5	10.3±0.8	7.3±1.3	82.4±2.0	1.4±0.1	93.2±1.0	5.4±0.9

^{3}H-Metabolite expressed as the mean ± S.E.M. of the percentages it represented of the total recovered ^{3}H-label.

n = numbers in parentheses. Statistical evaluation by Student's t-test. *   P < 0.001.

However, the present results show Ch uptake to be smallest in strains
of mice where ACh turnover rates, and ChAT activities, are largest.
Thus a simple model involving kinetic coupling may not represent
the true role of the uptake of exogenous Ch in the synthesis and
turnover of ACh. Intracellular sources of Ch may play a vital role
(4).

The present results (Table 1) and those of Durkin et al. (2)
indicate that the largest steady state concentrations of Ch in vivo
are found in cortex of strains of mice showing the smallest ChAT
activities and the largest rates of uptake of ^{3}H–Ch. This observation is consistent with the kinetic coupling of Ch uptake and
acetylation, which predicts steady state concentrations of Ch proportional to the ratio:rate uptake/V_{max} ChAT.

^{3}H–ACh synthesis by kinetically coupled uptake and acetylation
requires that the fraction of ^{3}H–Ch acetylated, following uptake,
is determined solely by ChAT activity and is independent of Ch uptake. The present results show that on balance acetylations remain
unchanged between strains of mice, the only discrepancy being C57B1/
6 strain which shows a greater fraction of ^{3}H–Ch acetylated in
medium containing sodium, even though it possesses significantly
less ChAT than the other strains. However, this difference is not
apparent after correction for ^{3}H–Ch acetylated in the absence of
sodium.

Thus, although kinetic coupling explains some aspects of the
regulation of synthesis of ACh in mouse frontal cortex, the relationship between Ch uptake and the turnover rate of ACh, is not adequately explained by this model.

ACKNOWLEDGEMENTS

We thank the Science Research Council and Sterling Winthrop
Group Ltd (Europe) for support.

REFERENCES

1.   Burgess, E.J., Atterwill, C.K. and Prince, A.K. (1978): J.
        Neurochem. 31:1027–1033.
2.   Durkin, T., Ayad, G., Ebel, A. and Mandel, P. (1977): Brain
        Res. 136:475–486.
3.   Fonnum, F. (1975): J. Neurochem. 24:407–409.
4.   Kessler, P.D. and Marchbanks, R.M. (1979): Nature 279:542–544.
5.   Mandel, P., Ayad, G., Hermetet, J.C. and Ebel, A. (1974): Brain
        Res. 72:65–70.
6.   Tunnicliff, G., Wimer, C.C. and Wimer, R.E. (1973): Brain
        Res. 61:428–434.
7.   Wong, P.T-H. and Prince, A.K. (1979): Neuropharmacology 18:511–513.

LEAST SQUARES ANALYSES OF THE KINETICS OF CHOLINE UPTAKE

IN RAT BRAIN SYNAPTOSOMES

A.M. Kotas and A.K. Prince

University of London, King's College
The Strand, London WC2R 2LS England

INTRODUCTION

One assumption implicit to least squares analyses is that any
variation in the error of the dependent variable is known, so that
the correct weighting can be applied (3). However, there has been
no discussion of the errors associated with measurements of choline
(Ch) uptake. Cleland (2) defined weights for two different distributions, in which errors were either constant (homoscedastic) or
proportional to velocity (heteroscedastic).

We investigated which of these distributions was most representative of our measurements of Ch uptake. Least squares analyses
were then used to assess the significance of the deviations from
the Michaelis-Menten function, and to investigate which of several
classical models of Ch uptake, involving discrete mechanisms, best
fit the data. Both weighting schemes were used, and the results of
the analyses compared.

EXPERIMENTAL AND STATISTICAL

Synaptosomes (P_2 fraction) rat cerebral cortex were
prepared and used as described previously (1). Assays of uptake
were in quadruplicate at 10 to 12 Ch concentrations. Ten independent experiments were made, each using freshly prepared synaptosomes.

The significance of deviations from the Michaelis-Menten equation was tested by the ratio, F, of the deviation to the residual

mean squares, applied to a weighted double-reciprocal transformation. For i groups of observations (i different substrate concentrations), ni values of v in group i, and total number of velocities n, the deviation and residual mean squares (MS) were calculated from the sums of squares (SS) defined by:

$$\text{Deviation MS} = (\text{Groups SS} - \text{Regression SS})/(i-2)$$
$$\text{Residual MS} = (\text{Total SS} - \text{Groups SS})/(n-i)$$

Where $\overset{i}{\Sigma}$, $\overset{ni}{\Sigma}$, and $\overset{n}{\Sigma}$, represent summation over i groups, ni replicates within the i'th group, and n total observations:

$$\text{Groups SS} = \left[\overset{i}{\Sigma}\ \frac{(\overset{ni}{\Sigma}\ wy)^2}{\overset{ni}{\Sigma}\ w}\right] - \frac{(\overset{n}{\Sigma}wy)^2}{\overset{n}{\Sigma}w}$$

$$\text{Regression SS} = \frac{(\overset{n}{\Sigma}w.\overset{n}{\Sigma}wxy - \overset{n}{\Sigma}wx.\overset{n}{\Sigma}wy)^2}{\overset{n}{\Sigma}w\left[\overset{n}{\Sigma}wx^2.\overset{n}{\Sigma}w - (\overset{n}{\Sigma}wx)^2\right]}$$

$$\text{Total SS} = \overset{n}{\Sigma}wy^2 - \frac{(\overset{n}{\Sigma}wy)^2}{\overset{n}{\Sigma}w} \qquad \text{(see Sokal and Rohlf, 1969)}$$

$x = 1/s$, $y = 1/v$, w = weight (v^4 for homoscedastic errors, v^2 for heteroscedastic errors (2)). This procedure was used in a one-tailed test of significance with i-2 and n-i degrees of freedom. The kinetic parameters were calculated from the formulae given by Wilkinson (6), but modified here to incorporate either weighting scheme.

Model fitting used a weighted, iterative, non-linear least squares routine in Fortran IV (BMDX85) minimizing the weighted error mean square: $\Sigma w(v-v')^2 \cdot 1/(n-p)$: n, total number of velocity measurements (458); p, number of kinetic parameters (1 to 5); w, weight 1 and $1/v^2$, homoscedastic and heteroscedastic errors respectively (2); v, observed velocity; v', predicted velocity.

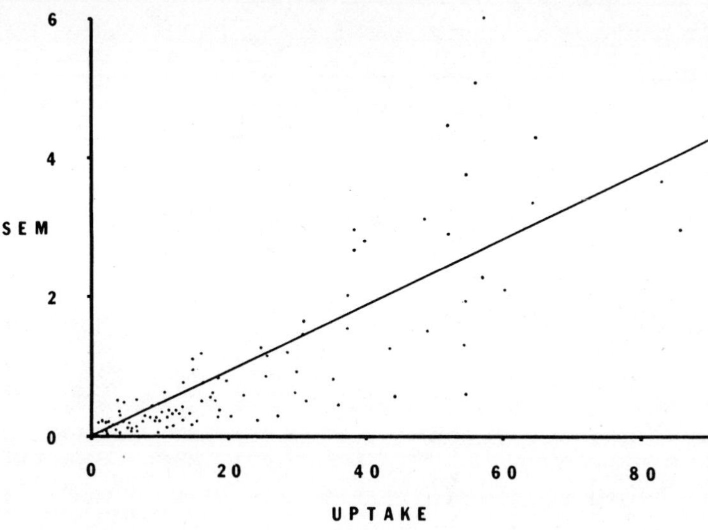

FIGURE 1: Distribution of errors in ^{3}H–Ch uptake. The relationship
is shown between the standard error (in each of 116 quadruplicate assays) and mean assay value. The solid line
represents the theoretical relationship: S.E.M. = 0.04724·v

RESULTS

Standard error (S.E.M.) in velocity increased with velocity
(Fig. 1), and was better represented by y = mv + c (c = -1.57 x
10^{-2}; EMS = 0.418), than by y = mv (EMS = 0.426), or y = c (EMS =
1.295); y = S.E.M., m = gradient; c = intercept (constant error).

There was no evidence of deviations from Michaelis–Menten kinetics at Ch concentrations of 0.1 to 2.75 μM, using either weighting
scheme (Table 1). In the range approximately 0.1 to 50 μM Ch, however, deviations were apparent with both weighting schemes, unless
groups of velocities at concentrations between approximately 5 and
50 μM were omitted from the analyses.

Using either weighting scheme, uptake was best described by
models incorporating one hyperbolic and one first order term, two
hyperbolic, or two hyperbolic and one first order term (models 3,
4, 5; Table 2). Models involving one hyperbolic or one first order
term showed progressively poorer fits (models 2, 1).

TABLE 1: Analysis of deviations from Michaelis–Menten kinetics.

Choline concn. range (no. of concns.)	Homoscedastic weighting $w = v^4$				Heteroscedastic weighting $w = v^2$			
	Largest concn. used in analysis (μM)	F	Vu	Ku (μM)	Largest concn. used in analysis (μM)	F	Vu	Ku (μM)
1) 0.125–50 (12)	50	7.09[a]	72.05	6.36	50	17.3[a]	60.85	3.51
	5	1.98[d]	37.60	1.95	5	1.49[d]	37.22	1.92
2) 0.33–40 (11)	40	23.5[a]	92.71	9.09	40	15.8[a]	71.04	4.74
	20	0.67[g]	50.67	3.27	20	0.80[f]	48.37	2.93
3) 0.33–40 (10)	40	59.6[a]	58.55	6.48	40	8.89[a]	47.83	3.83
	5	1.46[d]	32.26	2.34	20	1.71[d]	37.65	2.83
4) 0.33–40 (11)	40	10.6[a]	63.95	8.53	40	14.4[a]	46.84	4.18
	20	2.18[c]	35.08	3.15	10	1.67[d]	29.88	2.36
5) 0.10–40 (12)	40	4.61[a]	59.31	5.18	40	5.17[a]	49.85	3.04
	10	0.66[f]	36.18	2.08	20	2.04[c]	39.91	2.34
6) 0.125–50 (12)	50	6.86[a]	60.20	6.30	50	5.18[a]	52.27	3.78
	2	1.69[d]	27.15	1.55	20	1.86[d]	39.84	2.72
7) 0.125–50 (12)	50	3.81[b]	63.99	5.99	50	5.14[a]	54.59	3.70
	20	1.57[d]	44.16	3.04	10	4.10[h]	38.65	2.42
8) 0.1–1.2 (12)	1.2	0.22[g]	26.44	1.56	1.2	0.19[g]	28.35	1.76
9) 0.25–2.75 (12)	2.75	1.08[e]	31.20	1.68	2.75	1.51[d]	32.12	1.82
10) 0.25–2.75 (12)	2.75	0.19[g]	30.38	1.50	2.75	0.16[g]	31.38	1.66

a: $P < 0.001$; b: $P < 0.005$; c: $0.1 > P > 0.05$; d: $0.25 > P > 0.1$; e: $0.5 > P > 0.25$; f: $0.75 > P > 0.5$; g: $P > 0.75$; h: $0.01 > P > 0.005$. Further deviations were to smaller velocities. Minimum acceptable level of significance: $P < 0.05$. V_{max} units are pmol/hr/mg tissue.

DISCUSSION

The errors in Ch uptake are best described as heteroscedastic (Fig. 1). A mixed error distribution gives a small and negative constant error (c), which is difficult to rationalize (5).

Uptake at 0.1 to 2.75 μM Ch was consistent with a high affinity mechanism obeying Michaelis–Menten kinetics. Deviations at larger concentrations of Ch were greater than predicted to uptake by this relationship (Table 1).

TABLE 2: Fit of uptake data to different kinetic models.

Model		Best estimates of parameters, resultant error mean square*	
		$w = 1$ [†]	$w = 1/v^2$ [‡]
1) $v = K.S$	K;	1.58	2.16
	EMS;	116.54	0.51914
2) $v = Vu.S/(S+Ku)$	Vu;	61.89	46.29
	Ku;	6.63	3.43
	EMS;	31.294	0.04017
3) $v = Vh.S/(S+Kh) + K.S$	Vh;	28.41	30.86
	Kh;	1.80	2.19
	K;	0.740	0.583
	EMS;	20.963	0.03209
4) $v = Vh.S/(S+Kh) + Vl.S/(S+Kl)$	Vh;	24.14	30.85
	Kh;	1.44	2.19
	Vl;	145.4	2.97×10^4
	Kl;	133.2	5.09×10^3
	EMS;	20.763	0.03216
5) $v = Vh.S/(S+Kh) + Vl.S/(S+Kl) + K.S$	Vh;	24.14	30.86
	Kh;	1.44	2.19
	Vl;	145.3	2.60×10^7
	Kl;	133.2	4.02×10^{16}
	K;	0.0	0.583
	EMS;	20.808	0.03223

* Error mean square; EMS; appropriately weighted value. Units are
Vu, Vh, Vl: pmol/hr/mg tissue; Ku, Kh, Kl, S: µM; K: 10^{-6}/hr/mg
tissue. [†] homoscedastic error, [‡] heteroscedastic error.

Assuming the deviations arise from mechanisms unrelated to
high affinity uptake, they may be accounted for by a first order
process, possibly diffusion, or low affinity uptake (models 3, 4).
Contributions from both additional mechanisms (model 5) seem unlikely; this model reduces to model 3 or 4, depending on the
weighting scheme used. The error mean squares for models 3, 4 and
5 are too similar for any further conclusions to be drawn.

That these conclusions also apply where weighting appropriate
to homoscedastic errors was used should be interpreted with caution;
the observed sensitivity of the overall kinetic parameters to the
different weighting schemes (Table 1 and 2) suggests that conclusions drawn from these parameters should rely on appropriately
weighted analyses.

REFERENCES

1.   Cleaver, G., Kotas, A.M., Prince, A.K., Reynolds, R. and Wong,
        P.T-H. (1980): Brit. J. Pharmacol. 69:337P-339P.
2.   Cleland, W.W. (1967): Adv. Enzymol. 29:1-32.
3.   Henderson, P.J.F. (1978): IN Techniques in Protein and Enzyme
        Biochemistry B113 (ed) H.L. Kornberg, Elsevier, Amsterdam,
        pp. 1-43.
4.   Sokal, R.R. and Rohlf, F.J. (1969): Biometry, W.H. Freeman &
        Co., San Francisco, pp.428-436.
5.   Storer, A.C., Darlison, M.G. and Cornish-Bowden, A. (1975):
        Biochem. J. 151:361-367.
6.   Wilkinson, D.D. (1961): Biochem. J. 80:324-332.

INFLUX AND EFFLUX OF N,N,N-TRIMETHYL-N-PROP-2-YNYLAMMONIUM

BY A RAT BRAIN SYNAPTOSOME PREPARATION

L.A. Barker

Department of Pharmacology, Mt. Sinai School of Medicine
City University of New York
New York, New York   10029 USA

INTRODUCTION

Several analogs of choline (Ch) have been studied as substrates
for the Ch uptake systems present in synaptosomes or brain slices
(4-7,9,12,13,15). These analogs are monoethylcholine, diethyl-
choline, triethylcholine, pyrrolidinecholine, homocholine, sulfo-
choline, 1,1-dimethyl-3-hydroxypiperidine, N-(4-hydroxylbutyl)-N,N,
N-trimethylammonium (TMHBA) and N,N-dimethylaminoethanol (DMAE), and
all have a free hydroxyl group. With the exceptions of TMHBA (16)
and DMAE (15) they are alternative substrates for the high affinity
Ch uptake system as well as precursors to false cholinergic transmitters. These findings and the results of studies on Ch analogs
as inhibitors of the high affinity uptake of Ch (22) define some
of the structural requirements for alternative substrates. To be
recognized as a substrate, it appears that the Ch analog must be a
simple alkyl ammonium or sulfonium that has no more than three
methylene groups between the quaternary head and terminal hydroxyl
group. The observation that acetylcholine (ACh) is not transported
by the high affinity Ch uptake system suggests that a free hydroxyl
group also is required for a Ch analog to be a substrate for this
transport (17,18). To test this notion, a Ch analog devoid of a
hydroxyl group, N,N,N-trimethyl-N-prop-2-ynylammonium (TMPYA), was
evaluated as a substrate for the Ch high affinity uptake system.
TMPYA was chosen on the basis that the electrostatic potential it
generates is quite similar to that of Ch (25). On that basis, TMPYA
should be recognized by and interact with binding sites for Ch (see
24 for review). Further, the high affinity transport of Ch, monoethylcholine, pyrrolidinecholine, and homocholine by rat brain
synaptosomes is so closely linked to acetylation (3,7,28), it is

difficult to study this uptake system separately from the associated
metabolism of the substrates. Thus it is desireable to have a non-
metabolizable analog of the natural substrate for transport studies.
The preliminary results of studies on one such analog, TMPYA, are
reported here.

Metabolism Experiments

Experiments were carried out to determine if TMPYA is a substrate for choline acetyltransferase (CAT) and if it is metabolically altered upon incubation with a rat brain synaptosomal preparation,
P_2 fraction. TMPYA up to 50 mM is neither a substrate nor an inhibitor of rat brain CAT. Synaptosomes were incubated with [N-methyl-
^{3}H]-TMPYA and the quaternary ammonium compounds subsequently extracted and separated by thin layer chromatography (2). The chromatographic behavior of [^{3}H]-containing material extracted from
synaptosomes incubated at 0°C or at 37°C for 10 min in the presence
or absence of 10 μM hemicholinium-3 (HC-3) was indistinguishable
from that of authentic [^{3}H]-TMPYA. Thus, TMPYA does not appear to
be metabolized under these conditions.

Influx Studies

As expected, TMPYA was found to be an inhibitor of the synaptosomal uptake of Ch. At concentrations less than 9 μM, TMPYA inhibits the high affinity uptake of Ch (S = 0.5 to 5 μM) in an apparently competitive fashion with a K_i of 3.9 $\pm$ 0.7 μM (n = 7). TMPYA at
18 μM and higher concentrations produces a non-surmountable inhibition of Ch uptake. Based on these observations, the highest concentrations of TMPYA used in studies on the kinetics of uptake
was 8 μM.

The uptake of [^{3}H]-TMPYA by whole rat brain synaptosomal preparations was studied over the range of 0.5 to 8 μM. The apparent
K_T and V_{max} values for the uptake of TMPYA are 4.6 $\pm$ 0.8 μM (n = 6)
and 14.3 $\pm$ 4.0 pmol/mg P_2 protein/min (n = 6), respectively. The
apparent K_T value for the uptake of TMPYA is not significantly different (p > 0.1) from its K_i value for the inhibition of Ch uptake.
Similarly, Ch, which has a K_T of 2.5 μM (4), inhibits the uptake of
TMPYA; the calculated K_i (average of two experiments), is 3.4 μM.
In addition, as is seen with Ch (3,4,11) the uptake of TMPYA is
inhibited biphasically by HC-3 (Fig. 1). This observation suggests
the uptake of TMPYA is by both the high and low affinity Ch carriers.
The apparent K_i values for the inhibition of the high affinity up-
take of TMPYA and Ch by HC-3 are 0.05 μM and 0.06 μM, respectively.
The results of the studies on TMPYA inhibition of Ch uptake and

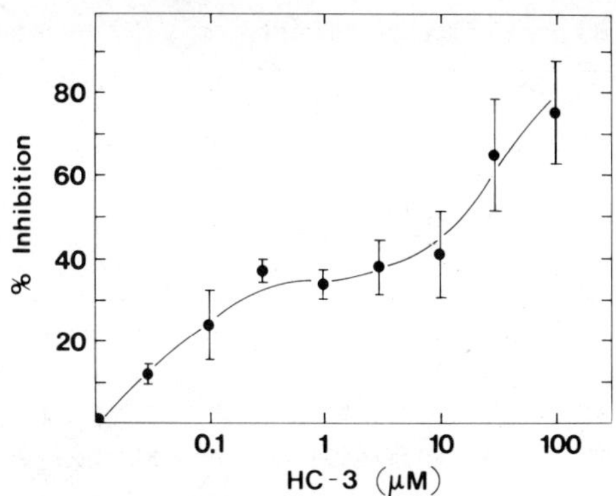

FIGURE 1: Log dose response curve for HC-3 inhibition of TMPYA
         (1 μM) uptake. Each point and bar represents the mean
         and S.E.M. for 3-4 experiments.

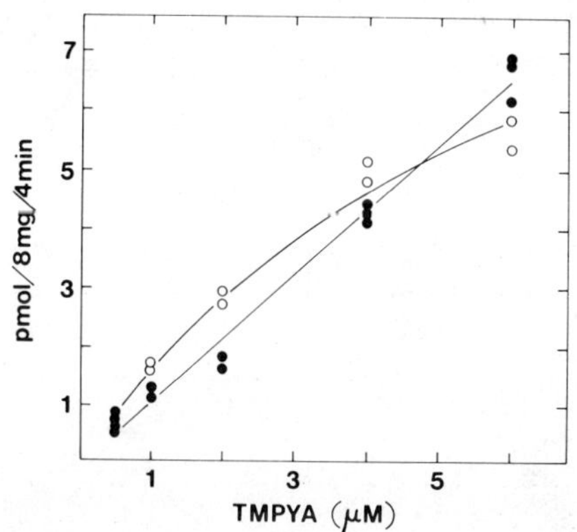

FIGURE 2: Uptake of TMPYA by a synaptosomal preparation incubated
         in regular KRP (O) or 5 mM Na$^+$ KR buffers (O). The
         uptake in KRP is best described by Michaelis-Menton kine-
         tics where

$$V = \frac{10.3 \text{ pmol/8 mg/min x S}}{4.5 \text{ μM + S}}$$

         (r = 0.9523, p < 0.003). The uptake in low Na$^+$ is best
         described by apparent first order kinetics where V =
         $(V_m/K_T) \cdot S$, the equation for the best fit is 1.08 (r =
         0.9893, p < 0.001). The results of one experiment are
         shown.

TABLE 1:   High affinity uptake of TMPYA and choline.

COMPOUND	K_T (μM)	K_i (μM)	HC-3 K_i (μM)
TMPYA	$4.6 \pm 0.8(6)$	$3.9 \pm 0.7(7)$	$0.05 \pm 0.02(3)$
Choline	$2.5 \pm 0.2(10)$	$2.7, 4.1$	$0.06 \pm 0.02(3)$

Values are mean $\pm$ S.E.M. (n). The K_i for TMPYA is with respect to choline and the K_i for choline is with respect to TMPYA and was calculated from the expression $K_i = IC_{50}/[1 + (S)/K_T]$. The K_i values for HC-3 are for TMPYA or choline as substrates and calculated from IC_{50} values.

TABLE 2:   Effect of HC-3 and low [Na^+] on uptake of choline and TMPYA.

SUBSTRATE	[μM]	TREATMENT (percent of control uptake)	
		HC-3 ($0.1\ \mu M$)	5 mM Na^+
Choline	0.1	$22 \pm 3(6)$	$26 \pm 4(7)$
TMPYA	0.1	$38 \pm 6(4)$	$129 \pm 18(4)$

Values are mean $\pm$ S.E.M. (n).

vice versa and the studies on the inhibition of Ch and TMPYA uptake by HC-3 (Table 1) show that TMPYA is an alternative substrate for the synaptosomal high affinity Ch uptake system.

Along with a marked sensitivity to inhibition by sub-micromolar concentration HC-3, another important characteristic of the high affinity Ch uptake system, in contrast to the low affinity system, is its dependency upon the concentration of Na^+-ions (17). In Table 2 are the results of experiments comparing the uptake of $0.1\ \mu M$ Ch or TMPYA in 5 mM Na^+ Krebs Ringer phosphate (KRP) buffer or in regular KRP in the presence of $0.1\ \mu M$ HC-3. As expected, both treatments reduced Ch uptake. Unexpectedly, the uptake of TMPYA was not decreased, but slightly increased, by lowering the Na^+ ion concentration of the buffer. The results of further experimentation showed that the kinetics of TMPYA uptake (S = 0.5 to 8.0 μM) are

TABLE 3:   Effect of high K^+ or 3-bromopyruvate pretreatment on the uptake of choline and TMPYA.

PRETREATMENT	UPTAKE (Percent of Control)	
	Choline	TMPYA
35 mM K^+ KRP	144 ± 7.7(6)	143, 200
500 µM 3-Bromopyruvate	37.9 ± 6.8(6)	24.8 ± 6.7(5)
500 µM 3-Bromopyruvate (0°C)	105.0 ± 10.0(6)	

Pretreatments were for 10 min at 37°C, following which the P_2 suspension was diluted 5x with KRP buffer and re-isolated by centrifugation. After resuspension in KRP, the uptake of ^{3}H-choline (0.5 µM) or ^{3}H-TMPYA (0.5 µM) was determined. Values are mean ± S.E.M. (n).

TABLE 4:   Effect of Na^+ ion removal on choline stimulated efflux of ^{3}H-TMPYA.

EFFLUX BUFFER	Efflux**	Choline stimulated Efflux**
KRP	14762 ± 81	
KRP + 25 µM choline	15450 ± 205*	688 ± 123
Na^+-free KR	14544 ± 407	
Na^+-free KR + 25 µM choline	17348 ± 823*	2803 ± 641*

Synaptosomes were loaded with ^{3}H-TMPYA (4.2 µM) by incubation at 37°C for 5 min. The loaded synaptosomes were re-isolated by centrifugation and following a surface wash, resuspended in the desired buffer. Samples were transferred to tubes and then re-incubated for 5 min at 37°C. Efflux was terminated by transfer to an ice bath and ultrafiltration. Values are mean ± S.E.M. (n = 3).
* Values are significantly different from appropriate controls (p < 0.05, paired t-test). ** cpm/5 min.

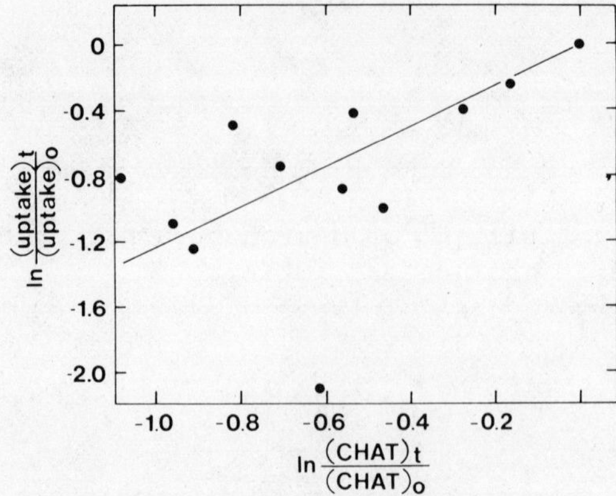

FIGURE 3: Correlation between time dependent inhibitions of synaptosomal ChAT and synaptosomal choline uptake by 3-bromo-
pyruvate. The equation for the best line forced through
the origin is ln (uptake)$_t$/(uptake)$_o$ = 1.238· ln (ChAT)$_t$/
(ChAT)$_o$ (r = 0.8681, p = 0001).

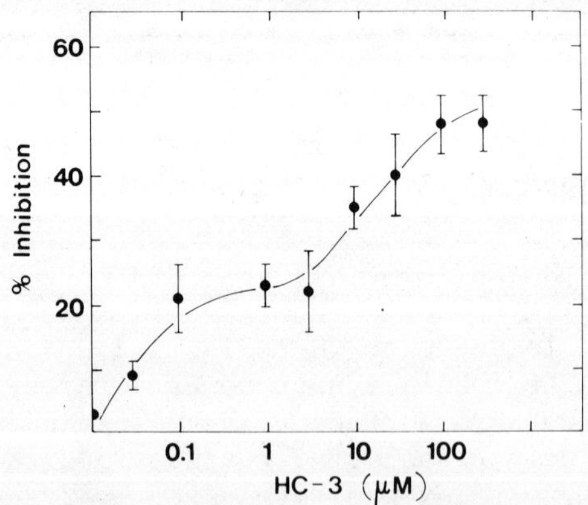

FIGURE 4: Trans-inhibition of TMPYA efflux by HC-3. Results are
mean and S.E.M. for 3-4 experiments.

dependent upon the concentration of Na^+ ions. Namely, the uptake
velocity of TMPYA in low Na^+ KRP is linear with respect to the concentration of TMPYA. On the other hand, in regular KRP the rate of
uptake of TMPYA is best described by Michaelis-Menton kinetics
(Fig. 2). By analogy with Ch, it is assumed that at a low Na^+ ion
concentration the uptake of TMPYA is mediated by a low affinity uptake process. Thus, the apparent linear dependence of the uptake
volocity on the concentration of TMPYA simply means that the concentrations of substrate used here are less than $0.1 \cdot K_T$.

In other experiments TMPYA and Ch uptake were determined following pretreatment with 35 mM K^+-KRP, a procedure known to stimulate Ch uptake (1,20,21,23); or 0.5 mM 3-bromopyruvate (3-BP), a
procedure known to impair ACh synthesis by brain slices (10) and
synaptosomes (14,16) as a result of its inhibiting acetyl-CoA formation from pyruvate. The results (Table 3) show that the uptake of
a low (0.5 μM) concentration of TMPYA like that of Ch is stimulated
by prior depolarization with 35 mM K^+-KRP. This observation provides
further support to the contention that TMPYA is an alternative substrate for the Ch high affinity uptake system.

The results of the studies on 3-BP by Jope and coworkers (16)
gave results that suggested 3-BP had no effect on Ch uptake _per se_,
but reduced tracer accumulation secondarily to the inhibition of
ACh synthesis. In this study (16) the accumulation of tracer $[^2H_4]$-
Ch, as measured by the synaptosomal content of $[^2H_4]$-Ch and $[^2H_4]$-
ACh following exposure to 0.5 mM 3-BP was reduced about 50%. The
decrease was due entirely to the near total inhibtion of $[^2H_4]$-ACh
synthesis; no change in $[^2H_4]$-Ch content was seen. Since the effects of 3-BP appeared to be related to Ch metabolism as opposed
to Ch uptake, it was of interest to determine the effect of 3-BP on
the uptake of TMPYA. As noted in Table 3, the uptake of both TMPYA
and Ch is inhibited by pretreatment with 3-BP. Further, it should
be noted that the inhibition is seen following the removal of 3-BP
and re-incubation of the synaptosomes with labelled substrate.
Since TMPYA does not appear to undergo metabolic conversion under
the conditions used here, it can be concluded that in addition to
its well demonstrated effects on ACh synthesis, 3-BP directly inhibits Ch transport. That the inhibition is observed after the
removal of 3-BP suggests that it is due to the alkylation of a
nucleophic group present at the Ch recognition site.

Further studies with 3-BP at 0.5 mM showed that the inhibition
of Ch uptake is temperature (Table 3) as well as time dependent.
The time dependent inhibition of uptake appears to follow first
order kinetics; the pseudo first order rate constant is $0.110 \pm$
0.013 min^{-1} (r = 0.9000, p = 0.001). CAT was measured in samples
of the same synaptosomal preparations and also was found to be inhibited in a time and temperature dependent manner by 3-BP.

Inhibition at 0°C after 20 min incubation was nil, 1 $\pm$ 6%. The
inhibition of synaptomal CAT also appeared to follow first order
kinetics. The pseudo first order rate constant is 0.084 $\pm$ 0.005 min^{-1}
(r = 0.9742, p = 0.0001). The inhibitions of Ch uptake and synaptosomal CAT by 3-BP are highly correlated (Fig. 3) (r = 0.8681, p =
0.0001; for the best line forced through the origin: y = 1.238·X).
It is very tempting to conclude that this observation demonstrates
a causal relationship between the inhibition of CAT and the inhibition of uptake as would be expected were high affinity Ch uptake
and acetylation coupled (4). Alternatively, this correlation could
obtain if nucleophilic groups of comparable reactivity are present
at the Ch binding sites on CAT and the transport system. It should
be noted that the apparent first order rate constant for 0.5 mM
3-BP inhibition of purified pyruvate dehydrogenase complex is
0.077 min^{-1} (19), a value not too dissimilar from those observed
here for the inhibition of Ch uptake and synaptosomal CAT.

Efflux Studies

The efflux of TMPYA, like the influx, does not occur at 0°C.
At 37°C the efflux is reasonably linear through 5 min, at which
time 34.8 $\pm$ 2.3% (n = 7) of the total accumulated [^{3}H]-TMPYA is
released to the medium. The efflux of TMPYA, like the influx, is
inhibited biphasically by HC-3 (Fig. 4) with IC$_{50}$ values of 0.03 μM
and 20 μM. The maximal inhibition of efflux by HC-3 varied between
40 and 60%. Thus, the efflux of TMPYA appears to be mediated by
three pathways. One route is insensitive to inhibition by HC-3 and
hence not part of the Ch uptake systems. The remaining two efflux
pathways are sensitive to inhibition by HC-3 in a fashion analogous
to the inhibition of the high and low affinity Ch uptake system.
Depending upon the experiment, 20 to 40% of the HC-3 sensitive efflux could be antagonized selectively by low concentrations of HC-3.
This is evidenced by the plateau between 0.1 and 3 μM in the inhibition log-dose response curve (Fig. 4). Thus the efflux of TMPYA via
the high affinity route is between 1 and 3% of the total accumulated
TMPYA/min.

Similar experiments were carried out with synaptosomes preloaded by incubation with 0.5 μM [^{3}H]-Ch. The release of [^{3}H] during
the second incubation was 17 $\pm$ 8% (n = 4) of the total accumulated.
Less than 10% of this release was inhibited by 10 μM HC-3, a concentration of inhibitor which produces 99% occupancy of the high
affinity carrier. The small magnitude of inhibition by 10 μM HC-3
suggests that [^{3}H]-Ch taken up during the loading incubation is not
available for efflux by the high affinity route. This is in agreement with the observations of Yamamura and Snyder (28) that virtually
all of the Ch taken up by the high affinity carrier is converted to
ACh which is an uptake inhibitor. Thus, it would seem that studies

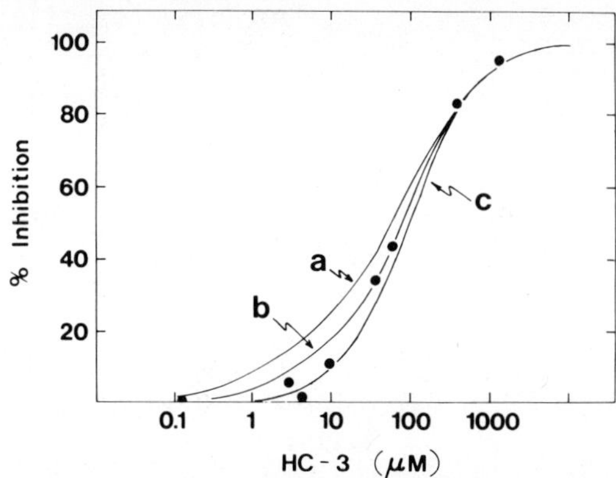

FIGURE 5: Log–dose   response curve for trans-inhibition by HC-3 of
TMPYA efflux stimulated by 100 μM choline added to trans
compartment.   Data points are means of triplicate determinations.   The solid lines are the results from computer
simulations of HC-3 inhibition of choline stimulated TMPYA
efflux.   The efflux is assumed to be mediated by both the
high and the low affinity choline carriers. Simulations
are shown for the cases where the high affinity route
mediates 20% (a), 10% (b), and 0% (c) of the total HC-3
sensitive efflux.   The inhibition of efflux produced by
HC-3 is assumed to be a function if its occupancy of the
high and low affinity choline recognition sites where
choline is viewed as a competitive ligand, and is described by the expression: $r = (Rm_1 \times S)/(S + K_1(1 + I/K_{i1}))$
$+ (Rm_2 \times S)/(S + K_2(1 + I/K_{i2})$ where r is the response and
Rm_1 and Rm_2 the maximal contributions due to the high and
low affinity sites, respectively.   S and I are the concentrations of HC-3 and choline, respectively. K_1 and K_2 are
the apparent K_i values for HC-3 inhibition of high and low
affinity choline uptake, 0.05 μM and 50 μM, respectively.
K_{i1} and K_{i2} are the apparent Michaelis constants for high
and low affinity choline uptake, 2.5 μM and 100 μm,
respectively.

---

on the kinetics of the high affinity uptake of [3H]-Ch or [3H]-Ch
analogs that are acetylated will be subject to minimal artefact due
to exchange reactions when low concentrations of substrate and short
incubation periods are used.   Because free Ch is formed from the
breakdown of Ch-containing compounds when synaptosomes are incubated
at 37°C (16), it is also advisable to avoid preincubation of synaptosomes prior to the addition of substrate (see also 26).

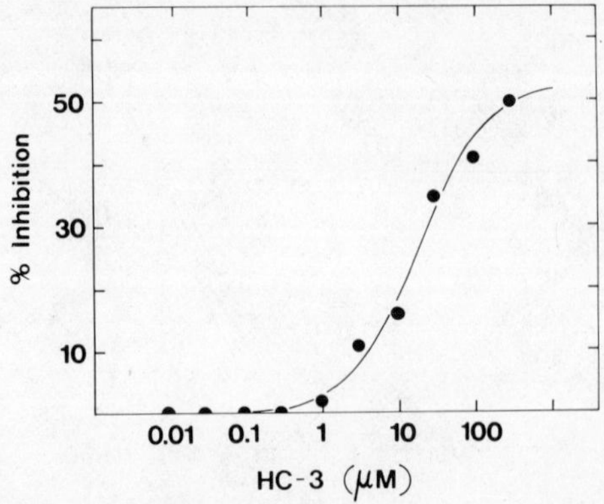

FIGURE 6: Log-dose response curve for HC-3 inhibition of basal
TMPYA efflux in a Na^+ free buffer. Values are mean of
triplicate determinations. The solid line is a computer
fit for the equation $r = R_{max}/1 + (K/A)^N$ where r is the
response (inhibition of efflux); R_{max} the maximal inhibition of total efflux by HC-3; K, the concentration of
HC-3 producing half maximal inhibition; A the concentration of HC-3 and N, the Hill coefficient. The parameters
of best fit are $R_{max} = 68$, $K = 65$ μM and $N = 1.04$.

The efflux of TMPYA is stimulated by added Ch; 10 μM produces
a 10% and 100 μM Ch produces a 37% increase in basal efflux. However, HC-3 inhibition of Ch (100 μM) stimulated efflux is not biphasic, but monotonic with an IC_{50} 65 μM (Fig. 5). This suggests
that the trans-stimulation of TMPYA efflux by Ch is mainly by the
low affinity Ch uptake route, i.e. that pathway for the lease sensitivity to HC-3. The results of computer simulation studies (Fig.
5) support this notion. Were Ch stimulated efflux only by the low
affinity route, reduction of external Na^+-ion concentration should
have little or no effect on the ability of Ch to produce trans-
stimulation of TMPYA efflux. The results of such experiments are
shown in Table 4. Here, it is seen that the basal efflux rates of
TMPYA in regular KRP and in a Na^+-free buffer are equal. Further,
the removal of Na^+ ions does not reduce the trans-stimulated efflux
produced by Ch, as would be expected if the high affinity transport
of Ch were necessary for the observed stimulation. In, fact, the
opposite is seen, Na^+ ion removal enhances Ch stimulated TMPYA efflux. This observation is in agreement with that of Diamond and
Kennedy (8) who showed that Na^+ ions inhibit the synaptosomal low
affinity uptake of Ch. It is concluded that the trans-stimulation

of TMPYA efflux by Ch occurs mainly if not exclusively by the low affinity uptake system.

In other experiments, the effects of Na^+ ion removal of the ability of HC-3 to inhibit efflux of TMPYA has been studied. The results of these studies, shown in Fig. 6, clearly indicate that Na^+ ion removal reduces the affinity of HC-3 for the high affinity Ch recognition site. Namely, no trans-inhibition of efflux is seen at a concentration of HC-3 (1 µM) which under normal conditions gives greater than 90% occupancy of the Ch high affinity site. Based on this observation, it can be inferred that the role of Na^+ ions in the high affinity uptake of Ch is one of modulating substrate affinity. The results of the experiments on Na^+ ion removal on the trans effects of Ch on TMPYA efflux are in agreement with this conclusion. Were Na^+ ions required only for substrate translocation but not for substrate binding as has been proposed (27), then removal of Na^+ ions should result in a trans-inhibition of TMPYA efflux by Ch. Such is not the case (Table 4). While these results do provide experimental support for the role of Na^+ ion regulation of Ch binding to the high affinity site, they do not preclude a role of Na^+ in translocation. Since efflux of TMPYA by the high affinity system originates from a compartment low in Na^+, substrate translocation by the high affinity route would seem to be not absolutely dependent upon Na^+ ions, as has been suggested (27).

SUMMARY

The results of these studies show that an analog of Ch need not have a free hydroxyl group to be an alternative substrate for the synaptosomal high affinity Ch uptake system. Instead, they suggest that for an analog of Ch to be an alternative substrate it must generate an electrostatic field potential similar to that generated by Ch in order to be recognized by the high affinity site. The importance of the electrostatic field potential formed by small molecules in drug-receptor interactions has been well documented (24); here the ability of TMPYA to mimic Ch provides yet another example. TMPYA should prove to be a very useful tool for elucidating the mechanism of high affinity Ch transport and the relationship between transport and ACh release.

ACKNOWLEDGEMENTS

L.A. Barker is the recipient of an NINCDS Career Development Award (1K04-NS00274). The research carried out in the author's laboratory was supported in part by a grant from NIMH (MH-24949). Curve fitting and computer simulation studies were carried out by programs available on the PROPHET system computer, a national resource supported by the Chemical-Biological Information Handling Program, Division of Research Resources (NIH).

REFERENCES

1.   Barker, L.A. (1976): Life Sci. 18:725-732.
2.   Barker, L.A., Dowdall, M.J. and Whittaker, V.P. (1972): Biochem. J. 130:1063-1080.
3.   Barker, L.A. and Mittag, T.W. (1973): FEBS Lett. 35:141-144.
4.   Barker, L.A. and Mittag, T.W. (1975): J. Pharmacol. Exp. Ther. 192:86-94.
5.   Chiou, C.Y. (1974): 14:1721-1733.
6.   Collier, B., Boksa, P. and Lovat, S. (1979): Prog. Brain Res. 49:107-121.
7.   Collier, B., Lovat, S., Ilson, D., Barker, L.A. and Mittag, T. W. (1977): J. Neurochem. 28:331-340.
8.   Diamond, I. and Kennedy, E.T. (1969): J. Biol. Chem. 244:3258-3263.
9.   Frankenberg, L., Heimburger, G., Nilson, G. and Sorbo, B. (1973): Eur. J. Pharmacol. 23:37.
10. Gibson, G.E., Jope, R. and Blass, J.P. (1975): Biochem. J. 148:17-23.
11. Haga, T. and Noda, H. (1973): Biochem. Biophys. Acta 291:564-575.
12. Hemsworth, B.A., Shreeve, S.M. and Veitch, G.B.A. (1979): Brit. J. Pharmacol. 66:465P.
13. Howard-Butcher, S., Cho, A.K. and Schaeffer, J.C. (1974): Fed. Proc. 33:1660.
14. Jope, R.S. and Jenden, D.J. (1977): Life Sci. 20:1389-1393.
15. Jope, R.S. and Jenden, D.J. (1979): J. Pharmacol. Exp. Ther. 211:472-479.
16. Jope, R.S., Weiler, M.H. and Jenden, D.J. (1978): Neurochemistry 30:949-954.
17. Kuhar, M.J. (1978): IN Cholinergic Mechanisms and Psychopharmacology (ed) D.J. Jenden, Plenum Press, New York, pp. 447-456.
18. Kuhar, M.J. and Simon, J.R. (1974): J. Neurochem. 22:1135-1137.
19. Maldonado, M.E., Oh, K.-J. and Frey, P.A. (1972): J. Biol. Chem. 247:2711-2716.
20. Murrin, L.C. and Kuhar, M.J. (1976): Molec. Pharmacol. 12:1082-1090.
21. Roskoski, R. (1978): J. Neurochem. 30:1357-1361.

22.   Simon, J.R., Mittag, T.W. and Kuhar, M.J. (1975): Biochem.
        Pharmacol. 24:1139-1142.
23.   Weiler, M., Jope, R.S. and Jenden, D.J. (1978): J. Neurochem.
        31:789-796.
24.   Weinstein, H. (1975): Int. J. Quant. Chem. QBS2:59-69.
25.   Weinstein, H., Maayani, S., Srebrenik, S., Cohen, S. and
        Sokolovsky, M. (1973): Molec. Pharmacol. 9:820-834.
26.   Wheeler, D.D. (1978): J. Neurochem. 31:109-120.
27.   Wheeler, D.D. (1979): J. Neurochem. 32:1197-1213.
28.   Yamamura, H.I. and Snyder, S.H. (1973): J. Neurochem. 21:
        1255-1374.

INTERACTION OF CHOLINE TRANSPORT WITH ACETYLCHOLINE RELEASE AND
SYNTHESIS

R.M. Marchbanks

Department of Biochemistry, Institute of Psychiatry
De Crespigny Park, London SE5 8AF United Kingdom

It has been reported from several laboratories (2,9,10) that
choline (Ch) uptake in the synaptic region is stimulated by acetylcholine (ACh) release. This would seem to be a possible control
mechanism for ACh synthesis because Ch uptake and extra mitochondrial AcCoA concentrations are known to be rate limiting. The
mechanism of this control has been supposed (10) to lie in a close
association of the Ch transport process with ACh synthesis. This
association had been previously invoked and defined by Barker and
Mittag (1) to explain their results on the transport and acetylation of various analogues of Ch. A simple explanation of the
equivalence of half saturating concentration for transport and
acetylation that they observed would be that the transport process
is rate controlling but they proposed that "transport and acetylation are coupled," i.e. Ch transported into the synaptosome has
no "free" existence and is immediately acetylated by choline acetyltransferase (ChAT). The coupling hypothesis of Barker and Mittag (1)
might seem a useful basis for a control mechanism but it flies in
the face of evidence (3,7) that at the high ionic strengths prevalent inside the cell ChAT is a soluble enzyme not associated with
any particulate or membranous material.

In this report I would like to outline the evidence that Ch
transport and ACh synthesis are not coupled, at least in the sense
proposed by Barker and Mittag (1). This caveat is necessary because the term "kinetic coupling" has been introduced by Suszkiw
and Pilar (11) and Jope _et al_. (4) to describe the fact that Ch
transport is one of the rate controlling steps in ACh synthesis.
Such usage assists neither clarity nor logic but is not an issue
insofar as it expresses the relative speeds of transport and acetylation. If the coupling hypothesis is discarded it is necessary to

seek another explanation for the well known activation of Ch transport by the release of ACh. Evidence that ACh acts as a transinhibitor of the Ch carrier is presented and it is proposed that the stimulation of Ch influx consequent upon depolarization is caused by the relief of ACh-induced trans-inhibition.

An unequivocal test of the coupling hypothesis cannot be made by measuring the conversion of radiolabelled Ch to ACh under various conditions (as for example in reference 4). This is because there is no way of distinguishing the proportion of Ch that has gone into structures that do not contain ChAT. Virtually all the preparations in use contain an unknown degree of contamination with structures that do not synthesize ACh so the overall percentage conversion will not be interpretable in terms of coupling. An alternative strategy is to apply pulses of Ch of varying specific radioactivity outside the membrane; if only the Ch that is transported is acetylated (as the coupling hypothesis supposes) there should be a one to one correspondence between the specific radioactivity of Ch outside and that of the recently synthesized ACh inside the terminal (coupling ratio = 1). If, however, there is mixing with intraterminal pools of Ch the ratio of specific activities $S_{ACh}/S_{(Ch\ outside)}$ will be less than 1; the greater the mixing, the smaller the ratio. This experiment gets over the problem that not all synaptosomes are cholinergic; since conversion is not measured Ch that has entered non-cholinergic synaptosomes will not affect the results.

ACh is known to exist in two compartments inside the synaptic terminal; the cytoplasmic compartment that contains most of the recently synthesized ACh, and the vesicular compartment that contains preformed ACh. The two only exchange relatively slowly with each other. In order that the ratio of specific activities accurately reflects the degree of mixing it is important that only recently synthesized ACh is considered. To do this Kessler and I (6) decided to label the acetyl moiety by preincubation with [14C]glucose. The amount of recently synthesized ACh could then be measured as the increment of [14C]ACh formed during the [3H]Ch pulse divided by the specific radioactivity during the pulse of the [14C]acetyl CoA formed from glucose.

The results of these experiments (6) are summarized in the top part of Table 1. Somewhatsurprisingly the coupling ratio was much less than 1 and also increased as the Ch concentration was increased above the so called high affinity region. Because of the assumptions about ACh exchangeability one might not expect the actual value of the coupling ratio to be very accurate, but it is quite inconsistent with the coupling hypothesis that it should rise as the Ch concentration is raised out of the high affinity range.

TABLE 1:   Coupling of choline transport to acetylcholine synthesis
           by different methods and under various conditions.

Method	Conditions	Ch Concentration (μM)	Coupling Ratio
Prelabelling with [^{14}C]-glucose**	glucose (8 μM)	0.2	0.013
		0.8	0.029
		10.0	0.360
	glucose (0.5 mM)	0.2	0.031 ± 0.06
		0.5	0.039 ± 0.001
		10.0	0.485 ± 0.029
Prelabelling with [^{14}C]-Ch	glucose (10 mM)	0.6*	0.17 ± 0.03
		1.2*	0.25 ± 0.04
		13 *	0.3
		51 *	0.43 ± 0.18
	low tissue concentration 0.3 mg protein/ml	1.9*(1.0)	0.07 ± 0.004
		51 *(50)	0.191
	no glucose	1.6*	0.183 ± 0.025
		51 *	0.651 ± 0.33

The coupling of Ch transport to ACh synthesis was determined under
various conditions as described in the text.   Except where indicated, the tissue concentration was 1-2 mg synaptosomal protein
per ml incubation medium.   Values of the Ch concentration marked
with one asterisk have been corrected for the maximal isotopic
dilution caused by the total efflux of Ch during the pulse.   This
figure is not very accurate in the case of experiments with low
tissue concentrations and exogenous Ch concentration is given in
parentheses.   The variability shown is the S.E.M., n = 4.
**   (see reference 6)

        A problem with these experiments concerns the glucose concen-
tration.   In order to avoid isotopic dilution it was found necessary to use very low glucose concentrations and although raising
the glucose concentration from 8 to 500 μM had little effect it
was obviously desirable to carry out the experiments in the physiological range of glucose concentration.   Another problems concerns
isotopic dilution of Ch extrasynaptosomally.   Our studies of Ch
efflux (see below) had suggested that over the course of the [^{3}H]-
Ch pulse its specific activity had approximately halved at added
Ch concentrations less than 1 μM but not much changed above this.
Because dilution occurs over the time course of the pulse it is not
clear how the gradual decrease of Ch specific activity should be
corrected for and it would not in any case affect the coupling

ratio determined at higher Ch concentrations or the general pattern
of the results.

While we were studying the kinetics of Ch acetylation a way
out of these difficulties suggested itself. We found that the
degree of acetylation of [^{14}C]Ch taken up into synaptosomes was
independent of time, i.e. the percentage conversion to [^{14}C]ACh
was the same at the earliest time intervals (20 sec) as the longest
(30 min) even though the amount of intrasynaptosomal [^{14}C]Ch increases some ten- to twentyfold during this period. This is a
reflection of the fact that ChAT is in such excess (12) that the
acetylation reaction is at thermodynamic equilibrium. It follows
then that the specific radioactivities of Ch and recently synthesized ACh in cholinergic terminals will be equal. These considerations allow the determination of the coupling ratio in a
completely different way. The synaptosomes were preloaded for
30 min with [^{14}C]Ch (2.0 μM), this was then removed by washing and
the [^{3}H]Ch pulse added. At the end of the 10 min pulse the synaptosomes were suparated and the ratio:

$$\frac{[^{3}\mathrm{H}]\mathrm{ACh}}{[^{14}\mathrm{C}]\mathrm{ACh}} \cdot \frac{S_{(\text{synaptosomal } [^{14}\mathrm{C}]\mathrm{Ch})}}{S_{(\text{Ch outside})}}$$

was determined.

This ratio equals the coupling ratio and the results are
shown in the lower part of Table 1. It can be seen that the values
are much the same as when [^{14}C]glucose was used as a precursor.
The [^{14}C]Ch prelabelling allows a test of the effect of glucose
on the coupling ratio. The presence or absence of glucose had
little effect on the coupling ratio. An addional advantage of the
[^{14}C]Ch prelabelling method is that efflux of [^{14}C]Ch into the
medium can be measured simultaneously with the pulse of [^{3}H]Ch. It
is not clear exactly how correction for isotopic dilution should
be applied since the full extent of dilution only occurs at the
end of the pulse. To check this the coupling ratio was determined
with a tissue:medium ratio reduced by dilution 4 times. The
coupling in the diluted samples lies in between the coupling
determined with minimum and maximum corrections for isotopic
dilution. The estimations of the coupling ratio by two different
methods produced values of about 0.1 at Ch concentrations of 1.0 μM
rising to 0.2–0.6 at higher Ch concentrations 10–50 μM. This means
that recently transported Ch has no privileged access to ChAT and
appears to mix with the endogenous Ch pool. Indeed, the extent of
coupling appears to be roughly equal to the contribution of transported Ch to total endogenous Ch as one would expect if transport
and acetylation are independent.

TABLE 2: Effect of different depolarizing stimuli and divalent
          cations on choline uptake.

| | Depolarizing Conditions: Percent Stimulation of Choline Uptake | |
	25 mM K^+	70 µM Veratrine
Control	25.9	22.5
		16.5
No $CaCl_2$	0	12.4
	0.6	0
$MgCl_2$ (18 mM)	10.2	0
	0	0

Synaptosomes (P_2B) were preincubated in normal, high K^+ or vera-
trine containing medium for 30 min at 37°C. The controls contained
2 mM $CaCl_2$ and 2 mM $MgCl_2$. These cations were varied in the test
samples as indicated. They were then washed twice and incubated
with [^{3}H]Ch (2 µM, 0.5 µCi/ml) for 4 min.

TABLE 3: Effect of acetylcholine and hemicholinium-3 on choline
          efflux.

External Concentration (mM) of Ch bases and HC-3			[^{3}H]Ch efflux (pmol min^{-1}.mg^{-1})
Ch	ACh	HC-3	
0			13.09 ± 1.06
0.02			21.44 ± 1.7
0.02	20		14.47 ± 1.04
0.02		0.5	13.12 ± 1.1
0.02	20	0.5	12.80 ± 1.25

Synaptosomes (P_2B) were preincubated with [^{3}H]Ch (2 µM, 2 µCi/ml)
for 30 min at 37°C. They were then washed 3 times and resuspended
in normal medium containing the additions as shown and efflux of
[^{3}H]Ch measured over a 4 min period. The figures are the mean ±
S.E.M. of 4 determinations.

     If the coupling hypothesis is discared we need to find an
explanation of the effect that ACh release has on the activation
of Ch transport. In our hands this activation is dependent on the
prior depolarization in the presence of Ca^{++}, while high concentrations of Mg^{++} prevent the activation (Table 2). The necessity

for Ca^{++} and the inhibitory effect of magnesium suggest that the
release of ACh (rather than depolarization _per se_) is the causative
step.

Our explanation of the activation of Ch transport follows from
the properties of the Ch carrier. Influx of $[^3H]$Ch into synaptosomes was studied simultaneously with the efflux of $[^{14}C]$Ch, mine
and Wonnacott's results (8) may be summarized:

i.   Influx of Ch is saturable and stimulated by high
      internal (trans) concentrations of Ch.

ii.   Efflux of Ch is saturable and also stimulated by
      trans (outside) concentrations of Ch.

iii. Influx and efflux rates are not the same but at
      saturating trans concentrations approach a
      common maximal velocity.

The most likely explanation for these results is that the Ch is
transported inwards by a reciprocating carrier which must be returned to the outside (or its outside facing conformation) if influx is to proceed, and _vice versa_ for efflux. Transport in the
opposite direction is facilitated by combination with Ch but since
influx is not in general equal to efflux, it is clear that the
carrier can traverse the membrane either alone or in combination
with some other substance, as yet unknown. It may also be that
sodium co-transport is important in this respect.

The kinetics of such a system will be rather complicated, it
is clear that the simple application of the Michaelis-Menten treatment is quite incorrect. As an example, consider the possible effects of compounds that compete for binding to the carrier site.
If they are transported they will act as transactivators as well
as cis-inhibitors, but if they are not carried across the membrane
they will inhibit from both the cis and trans directions.

ACh is a weak competitor for the binding site but is itself
hardly transported. In this respect it behaves similarly to the
drug hemicholinium-3 (HC-3). It would be predicted that both
compounds produce trans-inhibition of Ch transport. This is demonstrated in Table 3, which shows the effect of outside concentrations
on ACh and HC-3 on Ch efflux.

Ch shows trans-activation of influx, HC-3 shows trans-inhibition, and a similar effect, but at much higher concentrations, is
shown by ACh. There is a technical reason why this effect of ACh
is difficult to demonstrate experimentally. ACh hydrolyses spontaneously so that there will always be some 1-5% Ch present in
solutions of the ester no matter how effectively the esterase is

blocked. Ch produces trans-activation at much lower concentrations
than ACh causes inhibition. Consequently only a small hydrolysis
of ACh will be required to produce sufficient Ch to mask the inhibitory effect of the ester.

In fact the cytoplasmic concentration of ACh at the neuromuscular junction is thought to be in the mM range (5) so that the
concentrations found to cause trans-inhibition are physiologically
relevant. Trans-inhibition of Ch influx by ACh is more difficult
to demonstrate because preloading with ACh results in high internal
concentrations of Ch due to reversal of the reaction catalyzed by
ChAT. This swamps the effect of ACh, nevertheless, a statistical
analysis by Marchbanks et al. (13) of the effects of increased
internal ACh concentration show that it does inhibit Ch influx.

In summary, we propose that depolarization induced activation
of Ch uptake is mediated by ACh release from the synaptosomes. The
binding of ACh without concomitant transport has the effect of im-
mobilizing the carrier on the side of the high ACh concentration.
It will not, therefore, be available for Ch transport from the
other side of the membrane. If the ACh concentration is reduced
by the release process the immobilization of the carrier will be
relieved, thus enabling it to return to the outside of the membrane and mediate Ch influx. That it is cytoplasmic ACh that interacts with the carrier need not disconcert those who wish to believe
in release by ACh by exocytosis. The mechanism would work as well
whatever was the immediate source of released ACh provided the
cytoplasmic concentration was eventually lowered. Synaptosomes
are artificial particles and it is pertinent to ask how relevant
the phenomena is to the physiological control of ACh levels _in
vivo_. One of the most noticeable artificial features of synaptosomes is that a rapid breakdown of phospholipids leads to intrasynaptosomal concentrations of Ch considerably higher than _in vivo_.
A high intraterminal Ch concentration will tend to nullify (by
competition) the effect of ACh in inhibiting influx. The lower
tissue Ch concentrations found _in vivo_ will therefore allow a much
more dramatic activation of transport as a result of ACh release
than we have been able to measure.

ACKNOWLEDGEMENT

I am grateful to the MRC for support of these studies with
grants G974/907/N, G976/298/N and G979/507/N.

REFERENCES

1. Barker, L.A. and Mittag, T.W. (1975): J. Pharmacol. Exp. Ther. 192:86-94.
2. Collier, B. and Ilson, D. (1977): J. Physiol. 264:489-509.
3. Fonnum, F. (1967): Biochem. J. 103:262-270.
4. Jope, R.S., Weiler, M.H. and Jenden, D.J. (1978): J. Neurochem. 30:949-954.
5. Katz, B. and Miledi, R. (1977): Proc. Roy. Soc. Lond. B, 196:59-72.
6. Kessler, P.D. and Marchbanks, R.M. (1979): Nature (Lond) 279:542-544.
7. Marchbanks, R.M. and Israel, M. (1972): Biochem. J. 129:1049-1061.
8. Marchbanks, R.M. and Wonnacott, S. (1979): Prog. Brain Res. 49:77-87.
9. Polak, R.L., Molenaar, P.C. and van Gelder, M. (1977): J. Neurochem. 29:477-485.
10. Simon, J.R., Atweh, S. and Kuhar, M.J. (1976): J. Neurochem. 26:909-922.
11. Suszkiw, J.B. and Pilar, G. (1976): J. Neurochem. 26:1133-1138.
12. Tucek, S. (1978): In Acetylcholine Synthesis in Neurons, Chapman and Hall, London.

ADDED IN PROOF:

13. Marchbanks, R.M., Wonnacott, S. and Rubio, M.A. (1981): J. Neurochem. 36:379-393.

CHOLINE TRANSPORT AND THE REGULATION OF ACETYLCHOLINE SYNTHESIS

IN SYNAPTOSOMES

R.S. Jope and D.J. Jenden

Department of Pharmacology, School of Medicine, and
The Brain Research Institute, University of California
Los Angeles, California 90024 USA

In recent years evidence has accumulated supporting the proposal
that the transport of choline (Ch) into nerve endings by the high
affinity transport system may be the rate limiting step in acetylcholine (ACh) synthesis (for reviews see 6,9). We have investigated
the interconnections between Ch transport and ACh synthesis with
primary interest in: 1) the nature of the coupling between the high
affinity transport of Ch and its acetylation catalyzed by choline
acetyltransferase (ChAT), 2) in testing the possible existence of an
intracellular pool of free Ch available for acetylation in nerve
endings, and 3) in investigating the effects of varying extracellular
concentrations of Ch on the synthesis and release of ACh.

The development of a sensitive method for measuring stable isotopic derivatives of Ch and ACh, namely, combined gas chromatography
mass spectrometry (GCMS), has provided the means to measure endogenous
and tracer concentrations of Ch and ACh simultaneously in single,
small samples (5). We have utilized this system to measure the high
affinity transport and acetylation of deuterium labelled Ch in synaptosomes derived from rat brains (8). Synaptosomes (P_2 fraction) were
incubated at 37°C and 0°C for 4 min with $[^2H_4]$-Ch (2 μM) and the tissue and medium concentrations of ACh and Ch were measured by GCMS.
Average values from eight experiments of this type are described in
Fig. 1. After subtraction of the $[^2H_4]$-Ch accumulation obtained at
0°C, approximately 60% of the transported $[^2H_4]$-Ch was acetylated.
This method also allows for the simultaneous measurement of endogenous and released $[^2H_0]$-ACh and $[^2H_0]$-Ch. Although the $[^2H_0]$-ACh
values specifically reflect cholinergic synaptosomes the $[^2H_0]$-Ch
values do not, since $[^2H_0]$-Ch is contained in other particles in
this fraction and is released by phospholipid catabolism in all
membranes.

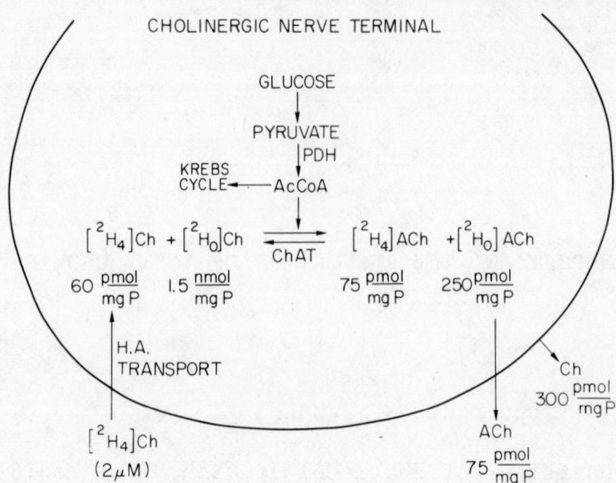

FIGURE 1: Representative GCMS measurements of Ch utilization. Synaptosomes (P_2 fraction) from rat brain were incubated in
Krebs-Ringer phosphate buffer containing $[^2H_4]$-Ch (2 µM),
glucose (10 mM) and paraoxon (40 µM) for 4 min at 37°C and
0°C. The samples were centrifuged at 10,000 x g at 2°C
for 15 min. The tissue and supernatant concentrations of
endogenous and labelled Ch and ACh were determined by GCMS
using $[^2H_9]$-ACh and $[^2H_9]$-Ch as internal standards (5).
Uptake of $[^2H_4]$-Ch at 0°C was subtracted from that obtained at 37°C; no $[^2H_4]$-ACh was synthesized at 0°C. Values
are averages of 8 experiments, described in detail in
Ref. 8.

Kinetic analysis of high affinity Ch transport using GCMS provides results similar to those reported by other investigators using
radiolabelled tracers, as well as providing data on the endogenous
levels of ACh and Ch. For example, Fig. 2 shows a Lineweaver-Burke
plot of high affinity Ch transport in striatal synaptosomes measured
in Krebs-Ringer phosphate buffer containing 5 or 35 mM KCl (in all
studies reported in this paper, kinetic parameters were estimated
by unweighted, nonlinear regression using an iterative Gauss-Newton
procedure). This result is of particular interest because there is
currently disagreement in the literature as to the effect of high
K^+ concentration on Ch transport. Carroll and Goldberg (2) reported
that the apparent K_T for Ch was increased tenfold (from 6 to 68 µM)
in the presence of 35 mM K^+, indicating a disappearance of the high
affinity component of Ch transport. On the other hand, Simon and
Kuhar (12) reported that the presence of 35 mM K^+ increased the apparent

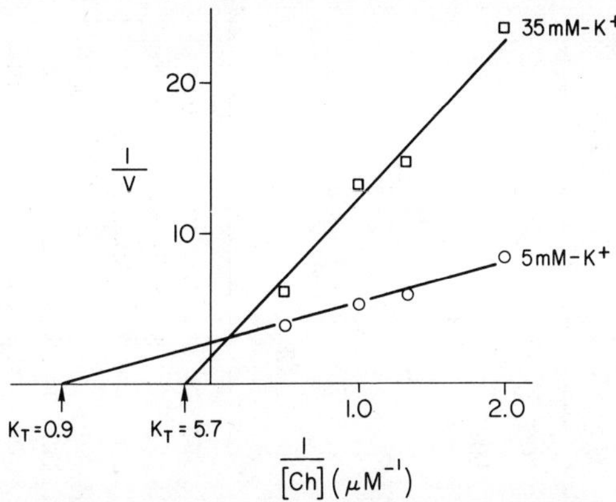

FIGURE 2: Lineweaver–Burke plot of high affinity Ch transport
measured in striatal synaptosomes in the presence of 5
or 35 mM K^+. High affinity Ch transport was calculated
as the tissue concentration of $[^2H_4]$-ACh plus $[^2H_4]$-Ch
(less that obtained at 0°C) plus $[^2H_4]$-ACh released during
a 4 min incubation at 37°C in Krebs–Ringer phosphate buffer containing $[^2H_4]$-Ch, glucose (10 mM), and paraoxon
(40 μM) (n = 4). V = nmol/mg protein/4 min.

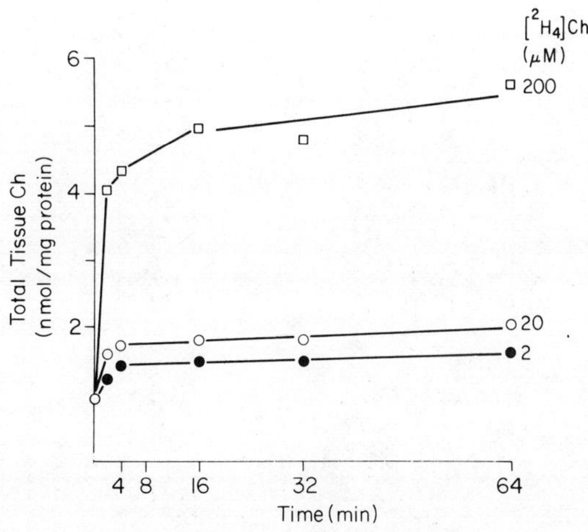

FIGURE 3: Rat brain synaptosomes were incubated at 37°C in Krebs–
Ringer phosphate buffer containing paraoxon (40 μM) and
2, 20 or 200 μM $[^2H_4]$-Ch. Synaptosomal concentrations of
Ch were determined by GCMS (n = 4).

TABLE 1: Summary of different methods of calculating the kinetic parameters of high affinity Ch transport.

METHOD	5 mM K^+		35 mM K^+		$\dfrac{K_T \text{(high } K^+)}{K_T \text{(low } K^+)}$	$\dfrac{V_{max} \text{(high } K^+)}{V_{max} \text{(low } K^+)}$
	V_{max}	K_T	V_{max}	K_T		
1. Tissue $[^2H_4]$-ACh + $[^2H_4]$-Ch	262	0.78	549	8.35	10.7	2.1
2. Method 1/Ch S.A.	241	0.32	292	2.97	9.3	1.2
3. Tissue $[^2H_4]$-Ch + tissue & medium $[^2H_4]$-ACh	368	0.92	543	5.71	6.2	1.5
4. Method 3/Ch S.A.	340	0.44	329	2.32	5.1	1.0

Rat brain synaptosomes (P_2 fraction) were incubated in Krebs–Ringer phosphate buffer containing 5 or 35 mM K^+, paraoxon (40 µM) and $[^2H_4]$-Ch. In all methods the uptake of $[^2H_4]$-Ch at 0°C was subtracted from the values obtained after a 4 min incubation at 37°C. The mean specific activity (S.A.) of Ch in the medium was obtained by averaging the values obtained at 0°C and 37°C for each concentration of $[^2H_4]$-Ch. Division by the specific activity of Ch in the medium corrects for varying degrees of dilution of the added labelled Ch (dependent on the concentration added) and the results indicate the total (labelled and unlabelled) Ch transported.

K_T = µM; V_{max} = pmol/mg protein/4 min (n = 4).

TABLE 2:   Acetylation of Ch after high affinity Ch transport.

Preincubation	CONCENTRATION (pmol/mg protein)	
	After Preincubation with $[^2H_4]$-Ch	After Incubation with HC-3 and Pyruvate
Pyruvate		
$[^2H_0]$-ACh	75.6 ± 3.9	67.9 ± 18.1
$[^2H_4]$-ACh	20.9 ± 2.2	27.6 ± 5.7
No Pyruvate		
$[^2H_0]$-ACh	45.0 ± 6.4	67.7 ± 6.7*
$[^2H_4]$-ACh	6.7 ± 2.0	21.3 ± 2.0*

Hippocampal synaptosomes were preincubated at 37°C with $[^2H_4]$-Ch
(2 µM) with or without pyruvate (5 mM). After 10 min, HC-3 (10 µM)
and pyruvate (5 mM) were added and the incubation continued for 5
min. The samples were poured into ice-cold centrifuge tubes and
centrifuged at 10,000 at 4°C for 10 min. ACh was determined by
GCMS and values are means ± S.E. of three incubations for each
treatment.   * P < 0.05 vs after preincubation.

K_T for Ch much less (from 0.5 to 0.8 µM). The results in Fig. 2 are
intermediate between these values as there was a sixfold increase in
the apparent K_T for Ch (from 0.9 to 5.7 µM), similar to the magnitude
of the change reported by Carroll and Goldberg (2), while the apparent K_T was still in the high affinity range in agreement with
Simon and Kuhar (12). The released $[^2H_4]$-ACh, which amounted to 17
to 31% and 28 to 35% of the total $[^2H_4]$-ACh synthesized in 5 and
35 mM K^+, respectively, was included in the calculations of high affinity $[^2H_4]$-Ch transport reported in Fig. 2. A variety of methods
can be used to calculate high affinity Ch transport, as shown in
Table 1. Usually calculations using Method 1 have been reported.
However, this method does not take into account the released labelled
ACh, which is derived from Ch transported during the incubation. This
is a substantial proportion of the newly synthesized ACh and may be
altered by treatments under study. Method 1 also does not take into
account the specific activity of Ch in the medium, which is obviously
increasingly affected as the concentration of labelled Ch is reduced.
In the case of high K^+, although the kinetic parameters vary depending on the calculation method used, its effect on the apparent
K_T relative to that of low K^+ is similar in all cases (5 to tenfold

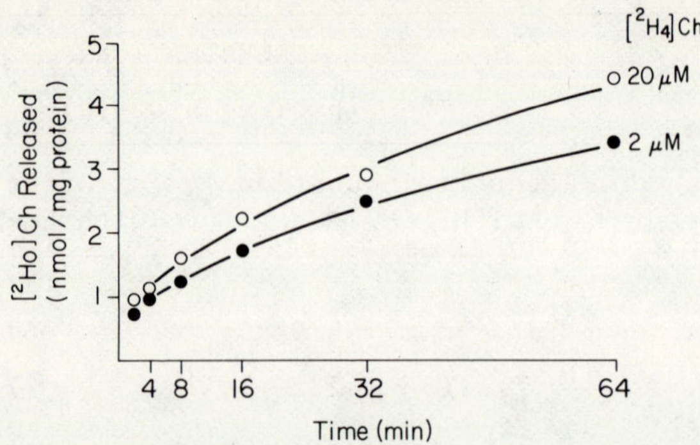

FIGURE 4: Rat brain synaptosomes were incubated at 37°C in Krebs-Ringer phosphate buffer containing paraoxon (40 μM) and 2 or 20 μM [²H₄]-Ch. Medium concentrations of Ch were determined by GCMS (n = 4).

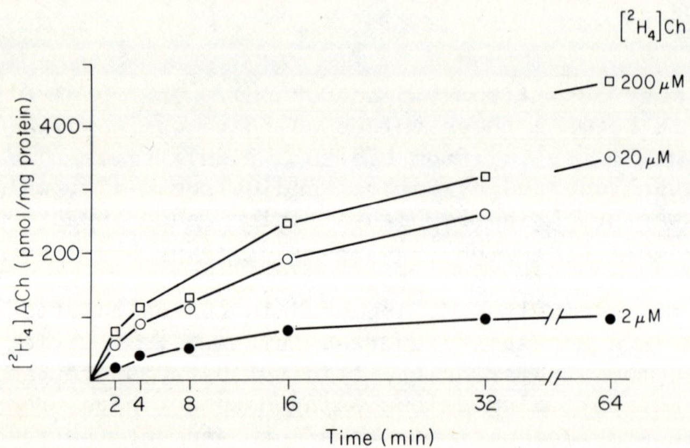

FIGURE 5: Rat brain synaptosomes were incubated at 37°C in Krebs-Ringer phosphate buffer containing paraoxon (40 μM) and 2, 20 or 200 μM [²H₄]-Ch. Following incubation the samples were centrifuged for 12 min at 10,000 g at 2°C and concentrations of ACh and Ch were determined by GCMS. Standard errors were smaller than the symbols (n = 2 at 8 and 64 min; n = 4 at other time points).

increase) while its effect on V_{max} depends on the method used. This change could be due to a decrease in membrane potential, which is likely to reduce the apparent affinity of an external cation for its carrier. The effects of these corrections on kinetic parameter estimates with other treatments should also be investigated, especially when the release of ACh or Ch is altered.

The existence of a free pool of Ch in cholinergic nerve endings was deduced from the results of two types of experiments. We have previously reported (7,8) that when AcCoA production is impaired there is reduced synthesis of ACh, thus lowering the rate of high affinity Ch transport into the nerve ending. Treatment with bromopyruvate, which inhibits pyruvate dehydrogenase and thus reduces the availability of AcCoA resulted in an increased synaptosomal concentration of $[^2H_4]$-Ch, while that of $[^2H_4]$-ACh was greatly decreased with increasing concentrations of bromopyruvate. We concluded from these and similar experiments that the high affinity transport and acetylation of Ch are distinct steps coupled kinetically, rather than directly as proposed by Barker and Mittag (1). Inhibition of Ch transport by bromopyruvate was competitive, as the apparent K_T for Ch was increased (1.57 vs 0.73 μM) while the V_{max} was unaltered. Although the mechanism of this inhibition is not known, one possible explanation for this effect is that the intracellular concentration of free Ch increased because acetylation was blocked, and the smaller electrochemical gradient of Ch reduced the rate of its transport into synaptosomes. Thus perhaps the intracellular concentrations of both Ch and ACh determine the rate of high affinity Ch transport, as we have previously proposed (4,6).

The implication that there may be a pool of free Ch in cholinergic nerve endings led us to attempt to demonstrate acetylation of Ch in this compartment (Table 2). To allow high affinity Ch transport to proceed with minimal acetylation, synaptosomes were first incubated with $[^2H_4]$-Ch (2 μM) for 10 min in the absence of glucose (7,8) since endogenous sources of AcCoA are minimal (10). Hemicholinium-3 (HC-3; 10 μM) was then added to inhibit further Ch transport and pyruvate (5 mM) was added to provide a source of AcCoA. It was observed that $[^2H_4]$-Ch that had been transported during the preincubation was acetylated during the second incubation. Complete inhibition of high affinity Ch transport during the second incubation was confirmed by addition of $[^2H_{13}]$-Ch, which was not transported or acetylated. These results indicate that a pool of free Ch can exist in cholinergic nerve endings and is available for acetylation. They also indicate that coupling of transport with acetylation of Ch is not obligatory (1,8).

We were interested in further characterization of this pool of Ch, particularly the possibility of altering its size by experimental manipulations, including Ch loading, and the effects of these changes

on the synthesis and release of ACh. Initially, the time course of
the utilization of $[^2H_4]$-Ch over a 100-fold concentration range (2
to 200 µM) was measured. Figure 3 shows that the tissue Ch quickly equilibrated with the medium Ch concentration. However, the concentration of $[^2H_0]$-Ch in the tissue remained constant throughout
the time course of the experiment and was independent of the concentration of $[^2H_4]$-Ch n the medium. Much of the accumulated $[^2H_4]$-Ch
was also bound at 0°C, and it cannot therefore be ascertained from
the data how much of this $[^2H_4]$-Ch was actually transported into
synaptosomes. For example, at 4 min, the accumulation at 0°C was
40, 62 and 84% of that at 37°C with 2, 20 and 200 µM $[^2H_4]$-Ch,
respectively. Figure 4 shows that the rate of release of $[^2H_0]$-Ch
into the media was approximately the same in the presence of 2 and
20 µM $[^2H_4]$-Ch (these measurements were not made with 200 µM $[^2H_4]$-
Ch). Much of the released Ch is probably due to the breakdown of
Ch containing lipids, though the contribution of the efflux of Ch
cannot be determined. However, it is clear that if the efflux of
Ch makes up a significant proportion of the total Ch released, it
is independent of the concentration of Ch in the medium.   This is
in direct contrast to the suggestion of Marchbanks and Wonnacut (11)
that Ch efflux is determined by the concentration of Ch in the medium.

    The incorporation of $[^2H_4]$-Ch into tissue ACh is shown in Fig.5.
Incubation with $[^2H_4]$-Ch led to increasing specific activity of
$[^2H_4]$-ACh in the tissue. However, the total concentration of ACh
($[^2H_0]$ plus $[^2H_4]$) in the tissue was independent of the concentration of $[^2H_4]$-Ch in the medium (Fig. 6). Thus the apparently different rates of synthesis of synaptosomal $[^2H_4]$-ACh indicated in
Fig. 5 are only due to the varying precursor specific activity and
not the concentration of Ch in the medium.   If ChAT maintains an
equilibrium between Ch and ACh by the law of mass action this result
indicates that the intracellular concentration of Ch is maintained
constant under these conditions.   The rate of release of ACh from
synaptosomes in the presence of 2, 20 and 200 µM $[^2H_4]$-Ch in the
medium is shown in Fig. 7.   Although the initial amount of ACh
in the medium varied with Ch concentrations, the rate of release
(i.e. the slope of the lines) was constant through the time course
and was independent of the concentration of $[^2H_4]$-Ch in the medium.
These results (Figs. 5-7) indicate that the rate of synthesis and
release of ACh in synaptosomes is not altered by increasing the
concentration of Ch in the medium from 2 to 20 or 200 µM. These
results support the proposal that the high affinity Ch transport
system, but not the low affinity system, is entirely responsible
for supporting ACh synthesis in synaptosomes. The results also
indicate that increased Ch concentration does not alter the turnover
of ACh in synaptosomes, in agreement with the _in vivo_ findings of
Eckernas _et al_. (3).

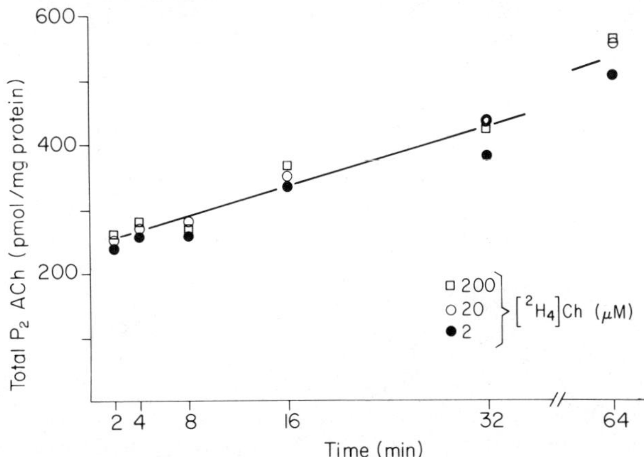

FIGURE 6: Rat brain synaptosomes were incubated at 37°C in Krebs-
Ringer phosphate buffer containing paraoxon (40 μM) and 2,
20 or 200 μM [^{2}H$_4$]-Ch. Concentrations of ACh and Ch were
determined by GCMS. Standard errors were smaller than the
symbols (n = 2 at 8 min; n = 4 at other time points). Total
ACh = [^{2}H$_0$]-ACh and [^{2}H$_4$]-ACh in final pellet.

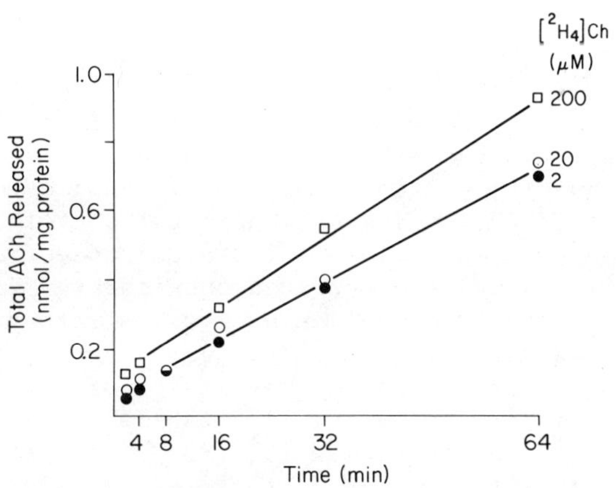

FIGURE 7: Rat brain synaptosomes were incubated at 37°C in Krebs-
Ringer phosphate buffer containing paraoxon (40 μM) and 2,
20 or 200 μM [^{2}H$_4$]-Ch. Concentrations of ACh and Ch were
determined by GCMS. Standard errors were smaller than the
symbols (n = 2 at 8 min; n = 4 at other time points). Total
ACh = [^{2}H$_0$]-ACh and [^{2}H$_4$]-ACh in the final supernatant.

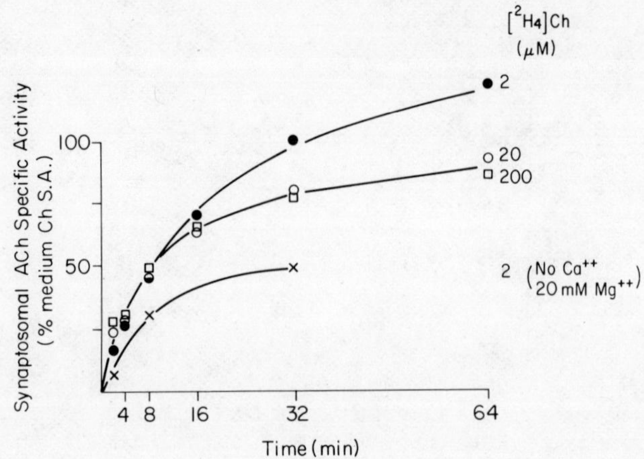

FIGURE 8: Equilibration of synaptosomal ACh with Ch in the medium.
Synaptosomes were treated as described in the Fig. 7 legend.
The specific activity of synaptosomal ACh is given as a
percentage of the specific activity of Ch in the medium to
indicate the rate at which the synaptosomal ACh equilibrates with Ch in the medium. In some samples (x), Ca^{++} was
omitted from the medium and $MgCl_2$ (20 mM) was added to
reduce ACh release.

The relationship between the specific activity of synaptosomal
ACh to that of Ch in the medium is presented in Fig. 8. The rate at
which the ACh specific activity approached that of Ch in the medium
was independent of the medium Ch concentration. Therefore the turnover rate of ACh was independent of the Ch concentration present.
The entire synaptosomal ACh pool had equilibrated with Ch in the
medium after 32 min of incubation. This was rather surprising considering the slow rate of turnover reported for vesicular ACh (13).
This finding was further examined by measuring the concentrations
and specific activities of ACh in the labile bound and stable bound
fractions of ACh in synaptosomes (Fig. 9). In this experiment synaptosomes were depleted of 40% of their ACh by a preincubation for
10 min at 37°C in a Na^+-free phosphate buffer containing 35 mM KCl.
Following centrifugation at 10,000 g for 12 min the synaptosomes
were incubated for 15 min at 37°C in Krebs-Ringer phosphate buffer
containing 2, 10, 50 or 400 μM $[^2H_4]$-Ch. The synaptosomes were
centrifuged and the pellet resuspended in 2 ml H_2O containing paraoxon (40 μM) and EGTA (0.2 mM). The labile bound and stable bound

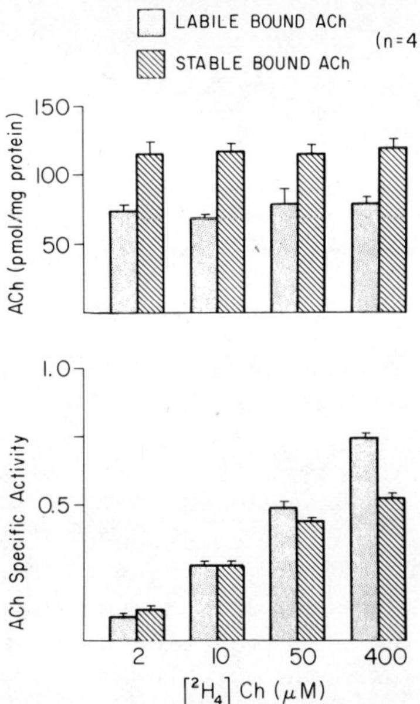

FIGURE 9: Rat brain synaptosomes were preincubated for 10 min at
37°C in a Na$^+$-free phosphate buffer containing 35 mM KCl.
This treatment depleted both labile bound and stable bound
ACh by approximately 40%. Following centrifugation at
10,000 g at 2°C for 12 min the synaptosomes were incubated
for 15 min at 37°C in Krebs-Ringer phosphate buffer containing 2, 10, 50 or 400 μM [²H₄]-Ch (no anticholinesterase
was added). The synaptosomes were centrifuged and the
pellet was resuspended in 2 ml H$_2$O containing paraoxon
(40 μM) and EGTA (0.2 mM). The labile bound and stable
bound fractions of ACh were separated by centrifugation
at 100,000 g for 60 min. No labile bound ACh was detected
when paraoxon was omitted from the final medium. ACh concentrations were determined by GCMS (n = 4).

fractions of ACh were separated by centrifuging at 100,000 g for
60 min. As in whole synaptosomes, the concentration of ACh in both
subcellular compartments was independent of the concentration of
[²H₄]-Ch in the medium. In addition, the specific activities of
labile bound and stable bound ACh were the same in all except the
highest concentration of [²H₄]-Ch. This indicates that the ACh in
these two compartments equilibrated during the 15 min incubation
period.

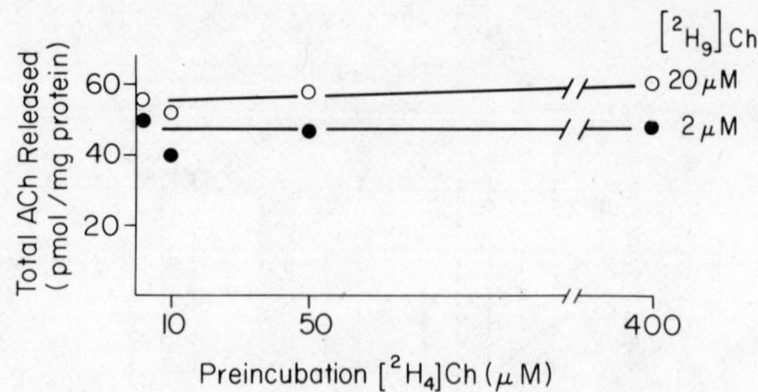

FIGURE 10: Rat brain synaptosomes were preincubated for 10 min at 37°C in Krebs-Ringer phosphate buffer containing 2, 10, 50 or 400 μM $[^2H_4]$-Ch (no anticholinesterase was added). The synaptosomes were then centrifuged at 10,000 g for 12 min at 2°C, resuspended in 10 ml buffer and recentrifuged. The pellet was resuspended and incubated for 4 min at 37°C or 0°C in Krebs-Ringer phosphate buffer containing paraoxon (40 μM) and 2 or 20 μM $[^2H_9]$-Ch. The incubation was stopped by centrifugation and the tissue and supernatant concentrations of ACh and Ch were determined by GCMS using $[^2H_{13}]$-ACh and $[^2H_{13}]$-Ch as internal standards. Standard errors were smaller than the symbols (n = 6-8). Total ACh = $[^2H_0]$-ACh + $[^2H_4]$-ACh + $[^2H_9]$-ACh.

Since Ch is presumably accumulated by noncholinergic synaptosomes and glial elements present in the preparations (presumably for lipid synthesis), it is not immediately obvious whether or not the intracellular concentration of Ch in cholinergic nerve terminals is altered by the concentration of Ch in the medium. To examine this problem we preincubated synaptosomes for 10 min at 37°C in 2, 10, 50 and 400 μM $[^2H_4]$-Ch. The synaptosomes were then centrifuged at 10,000 g at 2°C for 12 min, resuspended in buffer, recentrifuged and resuspended. The synaptosomes were then incubated for 4 min at 37°C in buffer containing paraoxon (40 μM) and 2 or 20 μM $[^2H_9]$-Ch on the premise that altered $[^2H_4]$-Ch concentration in cholinergic synaptosomes might alter the subsequent transport or acetylation of $[^2H_9]$-Ch. As in the results shown in Fig. 6, the ACh content of synaptosomes was unaffected by the concentration of Ch in the medium. There was also no effect of the preincubation concentration of Ch in the medium on the release of ACh during the second incubation (Fig. 10). The transport and acetylation of 2 μM $[^2H_9]$-Ch following preincubation with various concentrations of $[^2H_4]$-Ch are shown in Fig. 11. Upon first glance at the data it may appear that the transport of $[^2H_9]$-Ch and synthesis of $[^2H_9]$-ACh were reduced at increasing

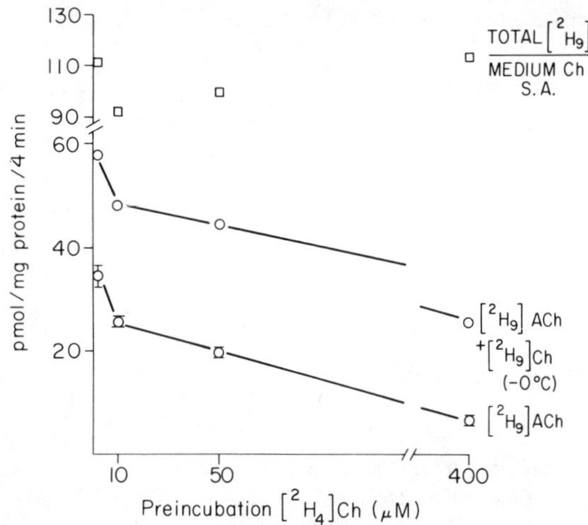

FIGURE 11: The effect of varying $[^2H_4]$-Ch concentrations in a preincubation on $[^2H_9]$-ACh synthesis and high affinity Ch
transport using 2 μM $[^2H_9]$-Ch. Synaptosomes were treated
as described in the Fig. 12 legend. High affinity Ch transport is the sum of the tissue concentrations of $[^2H_9]$-ACh
and $[^2H_9]$-Ch, less the $[^2H_9]$-Ch accumulated at 0°C (no
$[^2H_9]$-ACh was synthesized at 0°C). For the total transport calculation, the tissue values were divided by the
mean of the specific activity of Ch in the medium after
4 min at 37°C and 0°C.

concentrations of $[^2H_4]$-Ch in the preincubation. However, this
apparent reduction was caused by dilution of the $[^2H_9]$-Ch by $[^2H_4]$-
Ch that was present in the medium of the resuspended synaptosomes.
There was no apparent effect of the concentration of $[^2H_4]$-Ch in the
preincubation medium on the total high affinity Ch transport when
dilution of added $[^2H_9]$-Ch was corrected for by dividing by the
measured specific activity of $[^2H_9]$-Ch in the medium. Therefore
the intracellular pool of Ch in cholinergic synaptosomes is limited
in size and cannot be detectably altered by changes in the extracellular concentration of Ch under these conditions.

As a general working model of cholinergic metabolism we propose
the following model to describe the metabolism of Ch and ACh in a
cholinergic nerve ending. The intracellular Ch equilibrates with
the specific activity of Ch in the medium. Intracellular Ch and
ACh are equilibrated by ChAT by mass action and labile bound and

stable bound ACh in turn equilibrate. When ACh is released the cytoplasmic concentration of ACh is reduced. This drives ACh synthesis
by mass action which utilizes intracellular Ch. Utilization of intracellular Ch reduces its concentration and results in increased net Ch
transport and acetylation in the nerve ending. It is implicit in
this model that AcCoA becomes available at a sufficient rate for the
ACh concentration to be maintained. The production of AcCoA in the
cytoplasm could determine the kinetics of the entire system and itself be subject to regulatory control (6).

ACKNOWLEDGEMENTS

The authors thank Esther Shaw for technical assistance and Flo
Comes for editorial assistance. This work was supported by MH-17691.

REFERENCES

1.   Barker, L.A. and Mittag, T.W. (1975): J. Pharmacol. Exp. Ther.
       192:86-99.
2.   Carroll, P.T. and Goldberg, A.M. (1975): J. Neurochem. 27:93-99.
3.   Eckernas, S.A., Sahlstrom, L. and Aquilonius, S.M. (1977): J.
       Neurochem. 20:1083-1097.
4.   Jenden, D.J., Jope, R.S. and Weiler, M.H. (1976): Science 194:
       645-637.
5.   Jenden, D.J., Roch, M. and Booth, R.A. (1973): Anal. Biochem.
       55:438-448.
6.   Jope, R.S. (1979): Brain Res. Rev. 1:313-344.
7.   Jope, R.S. and Jenden, D.J. (1977): Life Sci. 20:1389-1392.
8.   Jope, R.S., Weiler, M.H. and Jenden, D.J. (1978): J. Neurochem.
       30:949-954:
9.   Kuhar, M.J. and Murrin, L.C. (1978): J. Neurochem. 30:15-21.
10.   Lefresne, P., Guyenet, P. and Glowinski, J. (1973): J. Neuro-
       chem. 20:1083-1097.
11.   Marchbanks, R.M. and Wonnacott, S. (1979): IN The Cholinergic
       Synapse (ed) S. Tucek, Elsevier Sci. Publ. Co., New York,
       pp. 77-88.
12.   Simon, J.R. and Kuhar, M.J. (1976): J. Neurochem. 27:93-99.
13.   Suszkiw, J.B., Zimmerman, H. and Whittaker, V.P. (1978): J.
       Neurochem. 30:1269-1280.

CHOLINE UPTAKE IN NERVE CULTURES AND IN SYNAPTOSOMAL PREPARATION

IS REGULATED BY THE ENDOGENOUS POOL OF CHOLINE

R. Massarelli and T.Y. Wong

Centre de Neurochimie du CNRS
11 rue Humann
67085 Strasbourg, Cedex, France

INTRODUCTION

Several experimental reports have, in the past, indicated that choline (Ch) is not synthesized in nerve cells through the mechanisms which are present in the liver, i.e. methylation of free ethanolamine or of ethanolamine containing phospholipids (for a review see 3). However, in 1972 Dross and Kewitz (2) showed that Ch may be produced in the brain.

At the moment it is not clear where this Ch comes from, whether it represents real endocellular synthesis or hydrolysis from lecithin; nor is it known where this phenomenon takes place, in neurons, in glial cells, or in both. Ch transport thus represents the only mechanism which, at the moment, is relatively well known and may act as a source for the cellular needs of Ch.

In 1973 Yamamura and Snyder (12), and later on several other authors (see 4), pointed out the possibility that the transport of Ch might be mediated by two or three mechanisms (including passive diffusion) with different affinities for the substrate. The essential features of the transport of Ch obtained with crude synaptosomal preparations may be summarized as follows: 1) the high affinity mechanism ($K_m = 10^{-6}$ M, HAM) seems to be specific for cholinergic neurons; 2) the HAM seems to be directed towards the synthesis of acetylcholine (ACh) through a mechanism which may require obligatory or kinetic coupling of the carrier with Choline acetyltransferase (CAT: EC 2.3.1.6); 3) the HAM is dependent upon the concentration of external Na^+ ions and Cl^- ions; 4) the HAM is inhibited by ouabain and other metabolic inhibitors and its activation energy is rather high (24 Kcal/mol) (for a review see 4).

In the past few years studies on Ch uptake performed in our own and several other laboratories have shown that a certain number of discrepancies might lead to different conclusions. Notably, it has been observed that: 1) the HAM is not specific for cholinergic neurons but may be observed in all neurons, in glia, in fibroblasts, in hepatocytes, in erythrocytes and in bacteria (11), irrespective of the presence or absence of an ACh compartment (see 5 for review); 2) the HAM of Ch may change with respect to affinity, depending on the concentration ranges of substrate which are employed. This has also been recently observed in synaptosomes where some authors (9) have found a K_m of = 4 x 10^{-7} M (the usual K_m was 2-6 x 10^{-6} M) with a concentration of Ch lower than usual (0.24-2.5 µM). It is not clear whether these authors consider the 10^{-7} M value as a sort of "super" HAM of Ch uptake; 3) the HAM of Ch uptake in nerve cells is not directed towards the synthesis of ACh and actually the essential metabolite of Ch in nerve cell bodies is phosphorylcholine (PCh) and this is so regardless of the external concentration of Ch (6); 4) the dependence upon external Na^+ ions concentrations, the effect of metabolic inhibitors and the study of the energy requirements of Ch uptake suggest that the transport of Ch (at all external concentrations) is of the facilitated diffusion type (5); 5) the inhibition of the HAM by hemicholinium-3 (HC-3) is usually reported as a typical characteristic of this mechanism. However, the low affinity mechanism (LAM) of Ch uptake is also inhibited by HC-3 (1,8); 6) it was shown that incubation of nerve cell cultures with HAM concentrations of Ch less than 1 µM dramatically lowers the endogenous compartment of free Ch and it was observed that only when the Ch pool is in non-physiological conditions a HAM of Ch transport may be observed (5).

On the basis of these findings, the hypothesis was advanced that Ch transport might be regulated by its endocellular metabolism and that only one carrier might be involved which would change of conformation with changes in the metabolism of the Ch pool (6). The present experiments were performed in order to prove or disprove this hypothesis.

MATERIALS AND METHODS

Neuronal cell cultures were prepared according to previously published procedures and the uptake and metabolism of Ch in nerve cell cultures were performed as previously reported (7,12).

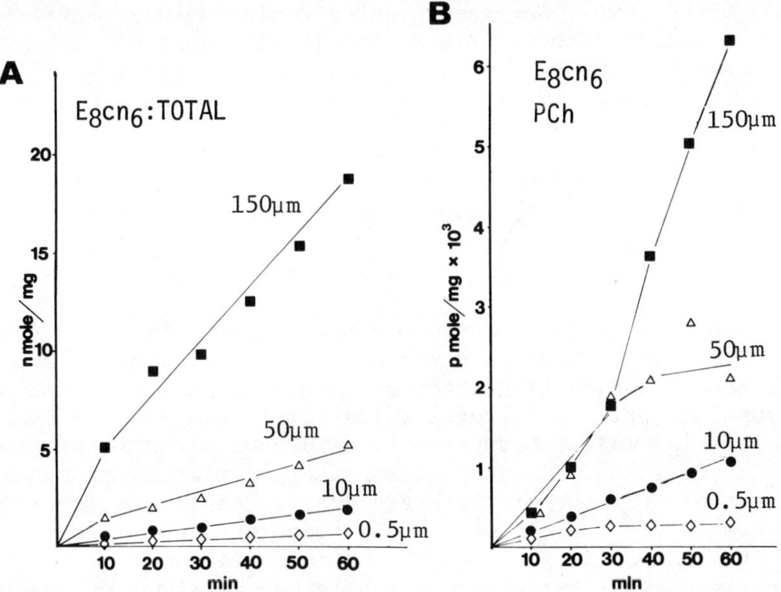

FIGURE 1: Incorporation and distribution of radioactive Ch (0.5, 10, 50 and 150 µM) in the homogenate (total, A) and in phosphorylcholine (PCh, B). In neuronal cultures from chick embryo brain (E$_8$cn$_6$).

Crude synaptosomal (crude mitochondrial, CM) fractions were obtained with standard procedures. The uptake and metabolism of Ch in crude synaptosomal preparations as function of time were studied by preincubating the organelles with or without 10 µM cold Ch and incubating with 1 or 100 µM [^{3}H] Me Ch (specific activity 8 Ci/mmol: Amersham). At the end of the incubation aliquots (0.5 mg protein) were carefully layered on 5 ml of 0.32 M sucrose and centrifuged. The pellet was digested with 500 µl 0.5 N HCl and lyophilyzed. The dry material was dissolved in 40 µl of 50% ethanol containing 10 mM of cold carriers and the Ch containing compounds were separated by thin layer chromatography as already described. The kinetics of Ch uptake in CM fractions was measured as follows (all procedures performed at 37°C). The organelles were preincubated with or without cold Ch on Sartorius filters (SM 11106 pore size 0.45 µ) in the pit of a Millipore filtration apparatus (model 1225) for 10 min, vacuum was then applied and the tissue washed with prewarmed (37°C) Krebs Ringer solution. Incubation with various concentrations of [^{14}C] Me Ch (1-50 µM, specific

activity 59 mCi/mmol, Amersham) lasted 2 min and was followed by
repetitive washed with 0.147 M NaCl. In all cases of preincubation
in the presence of cold Ch (in cells or in synaptosomes) the final
values were corrected by the remaining cold Ch (which in any case
did not account for more than 0.01 µM).

RESULTS AND DISCUSSION

When primary pure neurons were preincubated in the absence of
Ch (20 min) and incubated for various times with 0.5 µM [^{14}C]-Ch,
the pattern of radioactive incorporation resembled the one already
found in mixed primary cultures (5). Total incorporation was linear
with time (Fig. 1A), PCh apparently saturated after 20-30 min
(Fig. 1B), while Ch, ACh and betaine (Bet) compartments showed bell
shaped curves of incorporation (Fig. 2A). Considering that the
cells were in constant infusion conditions and that even after
60 min the radioactive content in the incubation medium did not
practically change, these results could be explained in terms of
non-steady state conditions of the Ch pool (similar curves were
indeed proved to be due to a non-steady state of Ch in primary mixed
cultures and in neuroblastoma cells)(5).

It should be noted that similar to what was already found in
mixed cultures of rat brain (13) a low concentration of Ch(0.5 µM) did
not label the ACh compartment as in synaptosomes, but seemed rather
to preferentially label the PCh compartment. When Ch concentrations
were increased to 10 µM, the shape of the incorporation curve
changed. Total uptake was still linear (Fig. 1A) and reflected the
linear incorporation into PCh while Ch compartment was saturated
at 10 min (Fig. 1B) and the incorporation into ACh and Bet was
linear with time (Fig. 2B).

In the next experiment the neurons were incubated with 50 µM
Ch. This concentration represents, roughly, the content of Ch in
the growth medium (Eagle's modified Dulbecco containing 20% fetal
calf serum) 1 day after the change of the medium. We may then consider 50 µM Ch as the physiological concentration for the cells
since we used them 24 hr after the change of the growth medium.
As shown in Fig. 1A, total uptake was linear while PCh incorporation
approached saturation (Fig. 1B), free Ch compartment was saturated
earlier than 10 min, while ACh and Bet compartments were approaching
saturation around 20-30 min (Fig. 2C).

What would have happened, especially in the synthesis of ACh,
if the external concentration of Ch was brought above the physiological level? When neurons were incubated with 150 µM Ch total

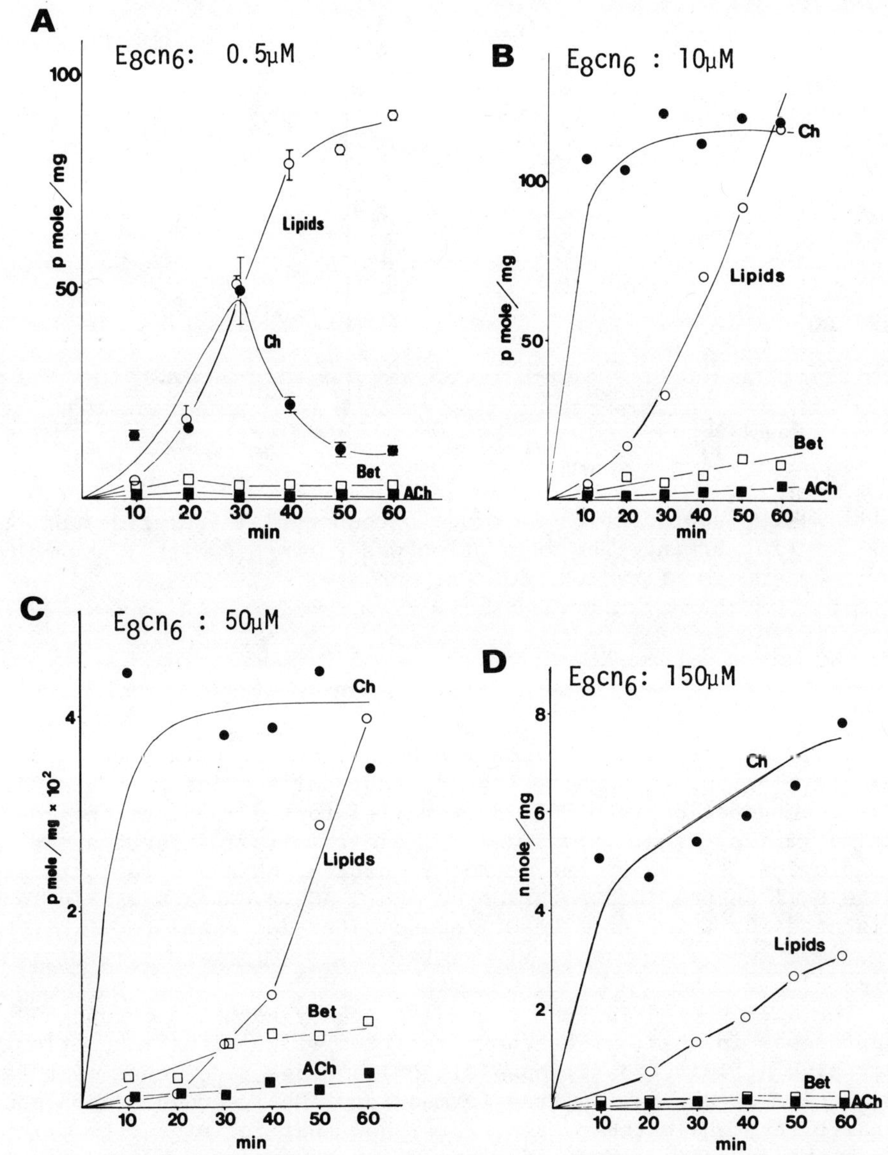

FIGURE 2: Ch radioactive distribution in exclusively neuronal cultures from the brain of chick embroy (E_8cn_6). Incubation with 0.5 μM (A), 10 μM (B), 50 μM (C), and 150 μM (D) radioactive Ch.

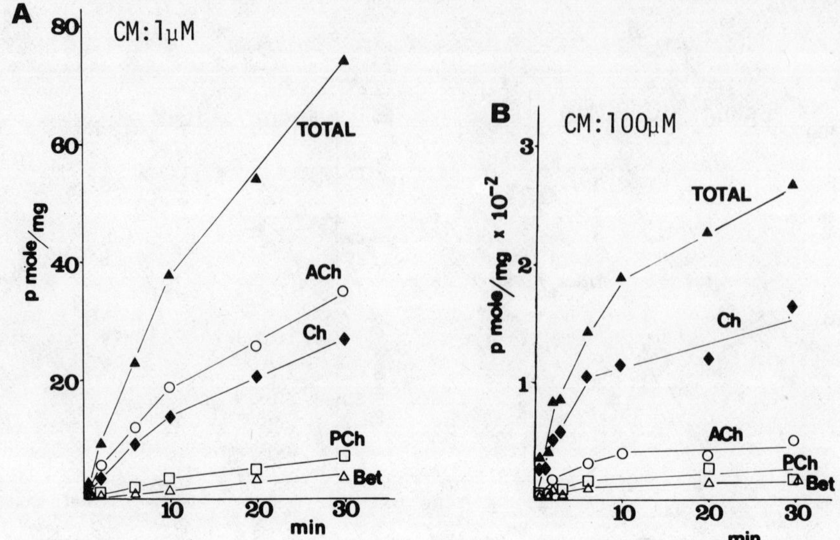

FIGURE 3: Ch incorporation in crude mitochondrial (CM) fractions of
         rat brain. The incubation was performed with 1 µM and
         100 µM of radioactive Ch.

uptake was linear (Fig. 2A); Ch was highly labelled (higher than
PCh)(Fig. 1B) but through a clear diffusional component (Fig. 2D),
while incorporation into PCh was parallel to the one into Ch containing lipids (Figs. 1B and 2D). These results indicated a non-
physiological status of the metabolic pool of Ch and were confirmed
by the bell shaped incorporation of ACh. ACh synthesis was in fact
stimulated for the first 30-40 min and later decreased dramatically
(Fig. 2D).

The data showed an important discrepancy between neurons and
synaptosomes in that, as has been reported (13), after incubation
under high affinity conditions (0.5 µM) a large percentage of total
radioactive uptake may be transformed into ACh in crude synaptosomal
preparations, while the present data show that cells need to maintain a physiological level of Ch in order to properly synthesize ACh.
In order to understand the reasons for this discrepancy crude
synaptosomal fractions were incubated with 1 µM and 100 µM [^{3}H]-Ch
as a function of time and the radioactivity was measured in each
[^{3}H]-Ch containing compartment. Preincubation of these fractions
without Ch and incubation with 1 µM [^{3}H]-Ch showed a linear incorporation in total radioactivity with an apparent breaking point at
10 min (Fig. 3A), the ACh and Ch compartments showed a similar

TABLE 1:   Kinetic parameters of Ch uptake measured after preincubation with or without cold Ch.

Tissue	Ch Preincubation (μM)	K_m (μM)	V_{max} (pmol/mg/2min)
E_8cn_6	0	H 4.5	113.6
		L 33.3	212.8
	50	14.3	322.6
E_8c_{10}	0	H 0.8	44.8
		L 14.3	333.3
	50	14.3	545.5
$E_{10}cg_{16}$	0	H 0.3	12.2
		L 20.0	222.2
	50	25.0	454.5
CM	0	H 6.4	19.0
		L 35.7	66.7
	100	26.3	66.7

H = high affinities; L = low affinities.

Similar data were observed in cells and synaptosomes.

incorporation.   The apparent bending of the line at 10 min may reflect dilution of the medium radioactivity by cold Ch exiting from the organelles; otherwise, it may simply be due to the beginning of saturation.   The results were in accordance with previously published data since, almost at each time point 45% of the total radioactivity was incorporated into ACh, 40% into Ch, 3% into PCh and 2% into Bet.

When the fractions were preincubated with 100 μM cold Ch and incubated with 100 μM [^{3}H]-Ch, all compartments were saturated (with the exception of total and free Ch compartments where a diffusional component accounted for the last points of the curve)(Fig. 3B). ACh compartment reached a plateau of about 40 pmol/mg at 10 min and at 8 min more than 80% of this compartment was already saturated.   It should be noted that with 1 μM Ch the ACh compartment reached a value of 40 pmol/mg only after 30 min of incubation.

On the basis of this data we suggest that: 1) LAM concentrations of Ch saturate faster ACh compartment compared to HAM concentrations; 2) in nerve cells the first synthesized metabolic

TABLE 2:   Effect of neuraminidase (0.5 U/ml) on Ch transport and
           metabolism in CM fractions.

	Control	NEUase
Kinetic Parameters		
K_m (μM)	H 5.1 L 35.7	20.2
V_{max} (pmol/mg/min)	H 44.0 L 286.0	119.0
Percent Total Uptake		
Ch	34.0 ± 0.3	36.0 ± 2.0
ACh	59.5 ± 2.3	60.0 ± 3.3
Bet	2.8 ± 0.3	2.6 ± 0.1
PCh	3.5 ± 0.4	3.3 ± 0.7

Total [^{14}C]-Ch uptake in NEUase (A. ureafaciens, Nakarai) was reduced by 45% after 1 min incubations. H = high affinities; L =
low affinities.

compartment is PCh, while in synaptosomes it is ACh. This may be
explained by the different concentrations of Ch kinase (ChK)(higher
in cell bodies) and CAT (higher in nerve terminals). Tentatively
we might conclude that the HAM of the transport of Ch is coupled
somehow to these two enzymes in cell bodies and in synaptosomes.

Our hypothesis, expressed as a correlation between transport
and metabolism of Ch (5), may indicate a coupling phenomenon between
Ch transport and CAT and ChK activity; however, our data never suggested the presence of two separate mechanisms of transport.

In order to show whether the kinetic parameters of Ch transport might be influenced by the metabolic status of the Ch pool,
neuronal cultures (E_8cn_6), mixed cultures (E_8c_{10}), glia cultures
($E_{10}cg_{16}$) and CM fractions were preincubated in the absence and in
the presence of 50 μM or 100 μM Ch (in CM) prior to kinetic studies.
After thorough washing of cultures and synaptosomes, and incubations
with various concentrations of radioactive Ch (usually 0.25–100 μM),
the data showed (Table 1) that after preincubation with Ch the HAM
disappeared.

This result was obtained when the cells were in physiological (50 μM) conditions would confirm the correlation between Ch transport and Ch metabolism and would indicate that only one mechanism is involved in the transport.

The uptake of Ch may thus be regulated by the endogenous metabolism of the Ch. pool. Moreover, the kinetic experiments showed that: 1) this regulation (or coupling) takes place inside the cell membrane (the only differences between pretreated and control cultures was in the levels of intracellular Ch since the amount of exogenous radioactive Ch was, after correction, the same) and 2) we had further confirmation that non-nervous cells (such as $E_{10}C_{816}$) do show a HAM (actually with a higher K_m value than neurons).

Our next step was to find a system by which we might influence the transport of Ch by acting at the outside face of the cell membrane without influencing Ch metabolism. We had previously shown that incubation of nerve cell cultures with neuraminidase decreased the total uptake of 1 μM Ch (10). CM fractions were then preincubated in the presence of 0.5 U neuraminidase per ml of preincubating medium for 15 min, and incubated afterwards for 1 min with various concentrations of Ch in order to measure the kinetic constants. Under these conditions we could observe that the HAM was totally abolished (Table 2).

If the HAM system was then coupled to the synthesis of ACh, then, in the presence of sialidase, we should have observed a reduction of ACh synthesis. This, however, was not the case, as is shown in Table 2. Transport of Ch was reduced, but in percentage there was no difference in radioactive distribution of Ch between treated and control fractions and, more specifically, there was no reduction in ACh synthesis after the disappearance of the HAM.

CONCLUSIONS

From the present data, it would appear probable that the transport of Ch is mediated, in cells as well as in crude synaptosomal fractions, by a single transport mechanism which is correlated (coupled) with, and kinetically regulated by, the endogenous Ch pool metabolism. However, no specific correlation can be observed between the transport of low concentrations of Ch and the synthesis of ACh.

ACKNOWLEDGEMENTS

The experiments with sialidase have been performed in collaboration with Drs. H. Dreyfus and S. Harth. We would like to thank Ms. C. Orphanides for her excellent secretarial help.

REFERENCES

1.   Diamond, I. and Milfay, D. (1972): J. Neurochem. 19:1899-1909.
2.   Dross, K. and Kewitz, H. (1972): N.S. Arch. Pharmacol. 274: 91-106.
3.   Freeman, J.J. and Jenden, D.J. (1976): Life Sci. 19:949-962.
4.   Kuhar, M.J. and Murrin, L.C. (1978): J. Neurochem. 30:15-17.
5.   Massarelli, R. (1977): IN Cholinergic Mechanisms and Psychopharmacology. (ed) D.J. Jenden, Plenum Press, New York, pp. 539-550.
6.   Massarelli, R., Wong, T.Y., Froissart, C. and Robert, J. (1979): IN Progress in Brain Research. (ed) S. Tucek, Elsevier, Amsterdam, pp. 89-96.
7.   Pettmann, B., Louis, J.C. and Sensenbrenner, M. (1979): Nature 281:378-380.
8.   Richtie, A.K. and Goldberg, A.M. (1970): Science 169:489-490.
9.   Simon, J.R. and Kuhar, M.J. (1976): J. Neurochem. 7:93-99.
10. Stefanovic, V., Massarelli, R., Mandel, P. and Rosenberg, A. (1975): Biochem. Pharmacol. 24:1923-1928.
11. Thomas, A.M., Lambert, P.A. and Poxton, I.R. (1978): J. Gen. Microbiol. 109:313-319.
12. Yamamura, H.I. and Snyder, S.H. (1973): J. Neurochem. 21: 1355-1374.
13. Yavin, E. (1976): J. Biol. Chem. 251:1392-1397.

INTERRELATIONSHIP BETWEEN ACETYLCHOLINE RELEASE FROM SYNAPTOSOMES

AND Na-K ATPase ACTIVITY

J.R. Cooper and E.M. Meyer

Department of Pharmacology, School of Medicine
Yale University
New Haven, Connecticut 06510 USA

INTRODUCTION

Since the mechanism of neurotransmitter release remains unclear,
we have recently considered the possibility that the terminal membrane bound Na-K ATPase is involved with this process (17). Specific inhibitors of this enzyme, such as the cardiac glycoside ouabain,
release neuronal acetylcholine (ACh)(19,27) and catecholamines (12,
22), even in the absence of extracellular calcium. These observations suggest that Na-K ATPase inhibition might bypass a calcium
dependent step in neurotransmitter release. Since calcium ions also
inhibit this enzyme (21), we have proposed that treatments which
promote calcium influx indirectly inhibit Na-K ATPase activity and
increase neurotransmitter efflux in a manner similar to that seen
with ouabain in calcium-free media.

In this paper we show that several direct inhibitors of Na-K
ATPase activity (as monitored in intact synaptosomes with ouabain-
sensitive rubidium uptake)(30) concomitantly release ACh via a
calcium independent process. Veratridine, electrical field stimulation, and A23187 also inhibit this enzyme mediated ion pump activity
and simultaneously release ACh but these three treatments act via
calcium dependent mechanisms.

MATERIALS AND METHODS

Male adult Sprague-Dawley albino rats (Charles River Laboratories: Wilmington, MA) were used in all of the studies. Synaptosomes were prepared from cerebral cortices as described previously (21) and suspended in ice-cold Krebs Ringer (KR) buffer.

Simultaneous Estimation of [³H]ACh Release and Ouabain Sensitive ^{86}Rb Uptake

Synaptosomes were loaded with 10 μM [³H]choline (Ch) (0.2 Ci/mmol) to label ACh as described previously (23). After several washes, the synaptosomes were suspended in KR or calcium-free KR (plus 1 mM EGTA) at 37°C, and [³H]ACh efflux and ouabain sensitive ^{86}Rb uptake were determined concomitantly by dividing each sample into 2 aliquots, plus or minus a 2 min pretreatment with 500 μM ouabain (17). Drugs or electrical field stimulation were applied to the synaptosomes immediately after adding a tracer amount of ^{86}Rb (1 μCi/g). Field stimulations were applied as described previously (17) by a Grass stimulator. Use of a Millipore vacuum filtration system permitted ^{86}Rb uptake measurements within 30 sec of treatment. The [³H]ACh released into the medium by synaptosomes was separated from labelled Ch, according to Nemeth and Cooper (18).

Synaptosomal membrane potential: Changes in the membrane potentials of synaptosomal preparations were monitored with the dye, 3,3'-dipentyl 2,2'-oxacarbocyanine (6,24). Synaptosome suspensions were illuminated at 475 nm, and fluorescence emission was measured at 500 nm using a Farrand spectrofluorometer.

Synaptosomal ^{45}Ca-efflux: Synaptosomes were prelabelled with a tracer quantity of ^{45}Ca in KR at 26°C or 37°C for 30 min. Samples were then diluted 20/1 with calcium-free KR containing ether saline diluent, veratridine or ouabain. Immediately before and at various times after dilution, synaptosomal ^{45}Ca levels were measured via vacuum filtration.

Synaptosomal protein determinations: Protein was estimated by the method of Lowry et al. (16).

Drugs: Veratridine, A23187, ouabain, strophanthidin and p-chloromercuribenzene sulfonate were purchased from Aldrich Chemical Company (Metuchen, NJ). Ch kinase was purchased from Sigma (St. Louis, MO). Dr. L. Cohen generously provided us with the 3,3'-dipentyl 2,2'-oxacarbocyanine dye. [³H]Ch and ^{86}Rb were purchased from New England Nuclear (Boston, MA).

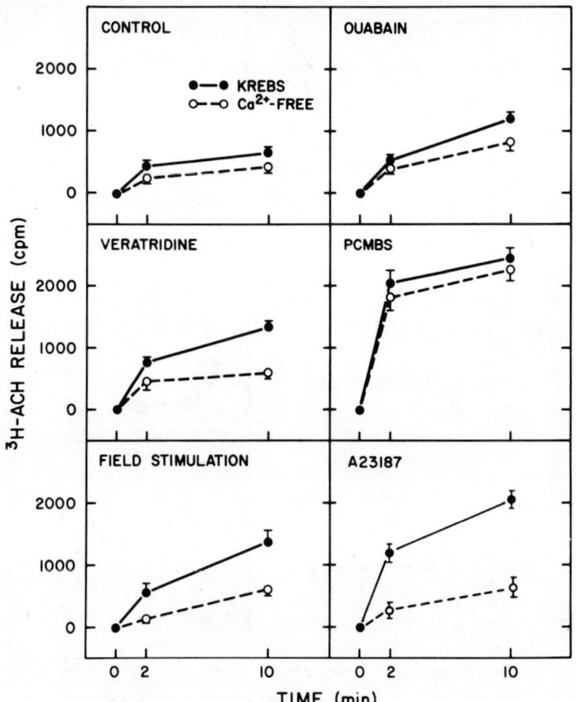

FIGURE 1: Synaptosomal [³H]ACh release in the presence or absence
of extracellular calcium. Following a 2 min preincubation
at 37°C, synaptosomes were treated with veratridine (10 μM),
A23187 (10 μg/ml), field stimulation (60 V, 60 Hz, 5 msec
pulse duration), ouabain (500 μM) or PCMBS (500 μM). At
several times thereafter, [³H]ACh released into KR (●──●)
or calcium free KR plus 1 mM EGTA. (O--O) was determined
and expressed as mean ± S.E.M. (n = 5-6/group).

RESULTS

Synaptosomal [³H]ACh release and ouabain-sensitive ⁸⁶Rb uptake
are shown in Figs. 1 and 2, respectively. Control ACh efflux was
slightly higher in KR than in calcium free KR but this calcium dependent release was greatly enhanced by veratridine, A23187, or
electrical field stimulation within 2 min. Ouabain and p-chloromercuribenzene sulfonate (PCMBS) also released [³H]ACh significantly
more than did KR or calcium-free KR controls.

Of the total ⁸⁶Rb uptake, about 50% was ouabain-sensitive. None
of the agents tested affected the ouabain-insensitive uptake but all

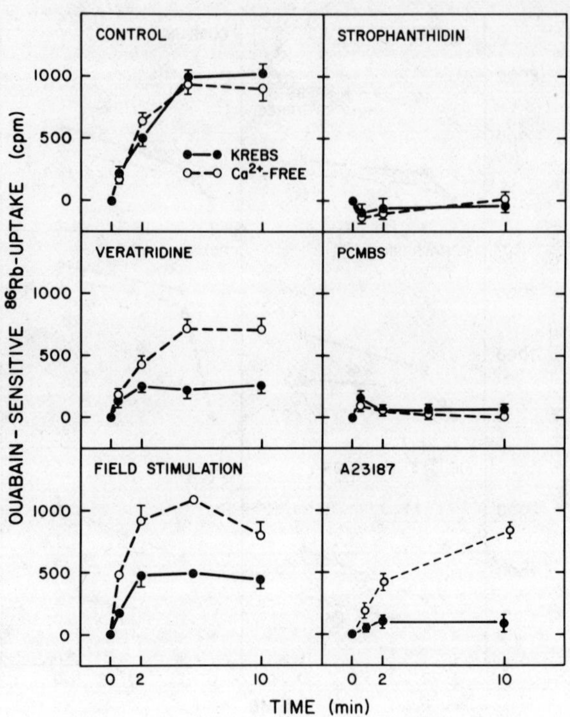

FIGURE 2: Synaptosomal ouabain sensitive [^{86}Rb]-uptake. Immediately
after the synaptosomes described in Fig. 1 were treated,
^{86}Rb was also added and its uptake was determined at
several times afterwards. The difference between [^{86}Rb]-
uptake plus or minus a 2 min ouabain (500 µM) pretreatment
was taken as the ouabain sensitive fraction in KR (   )
or calcium free KR (O--O) Mean $\pm$ S.E.M. (n = 5-6/group).

of them markedly inhibited the ouabain-sensitive ^{86}Rb uptake. PCMBS
and strophanthidin attenuated ouabain-sensitive ^{86}Rb uptake within
15-30 sec in the absence or presence of extracellular calcium.
Veratridine, electrical field stimulation, and A23187 significantly
decreased ouabain-sensitive ^{86}Rb uptake as well, but via a calcium
dependent process.   Further, in dose-response experiments, both with
veratridine and field stimulation, a complete parallelism was observed between ACh release and inhibition of rubidium uptake (data
not shown).

The membrane potentials of the purified synaptosomal preparations were surprisingly only moderately decreased by veratridine,

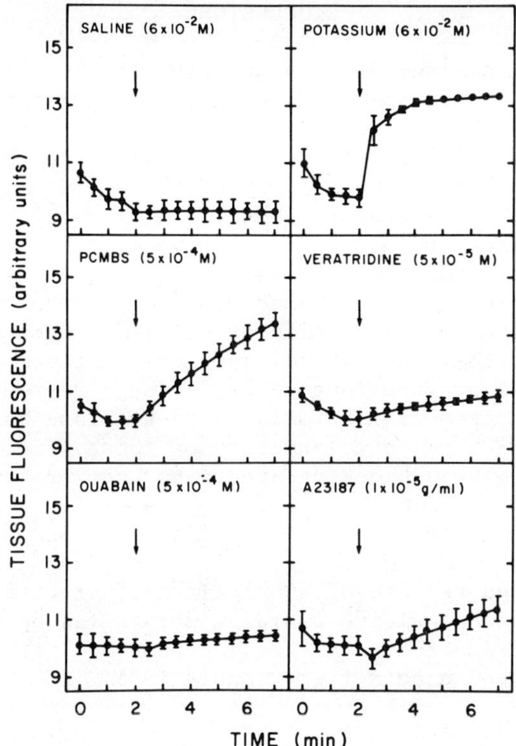

FIGURE 3: Synaptosomal fluorescence following treatments with Na-K
ATPase inhibitors. After a 10 min preincubation (time = 0)
in KR at 37°C, synaptosomes (about 1 mg protein) were
added to a cuvette containing 5 µg of 3,3'-dipentyl-2,2'-
oxacarbocyanine dye in a total volume of 3 ml KR. Two min
later (marked with an arrow), agents were added and the
resultant changes in fluorescence recorded (all values
are mean ± S.E.M. of 4 experiments).

ouabain, or A23187 as measured by changes in voltage-sensitive cy-
anine dye fluorescence (Fig. 3), suggesting only a slight depolarization by these agents. PCMBS elicited a pronounced depolarization,
although this change took several minutes, as contrasted with an almost immediate depolarization with 60 mM K⁺.

Treatment with either 10 µM veratridine or 500 µM ouabain had
no significant effect on ^{45}Ca efflux from preloaded synaptosomes at
either 26°C or 37°C compared to contols for at least 5 min after
dilution with calcium-free KR (data not shown).

DISCUSSION

It is becoming increasingly apparent from the literature that the process of neurotransmitter release may be a non-exocytotic one, especially in the brain where transmitters are released within 300 μsec of depolarization-induced calcium influx (for reviews see 10 and 15). Although the mechanism by which calcium influx triggers transmitter release is obscure, results from our laboratory strongly support a modulatory role for Na-K ATPase in the process. This possibility was first suggested by Paton et al. (19), who showed that inhibitors of Na-K ATPase released ACh from the guinea pig ileum, even in the absence of extracellular calcium. Numerous studies subsequently showed that ouabain releases transmitters from a variety of other neuronal preparations as well (for a review, see 27 or 28). Conversely, norepinephrine stimulates Na-K ATPase activity and inhibits neuronal ACh release, and both of these effects are blocked by pretreatment with α-blocking agents but not propanolol (4,13).

Using a direct measure of Na-K ATPase ion pump activity, we have been able to show for the first time a concomitant inhibition of this enzyme and release of ACh in intact nerve terminals (17). The use of ouabain-sensitive rubidium uptake (as opposed to potassium uptake) to approximate Na-K ATPase ion pump activity in intact cells has been described in detail previously (30).

A more fundamental question than whether direct, pharmacological inhibitors of Na-K ATPase can release transmitters remains. Does increased calcium influx promote neurotransmitter release by inhibiting this enzyme? Calcium is certainly a potent inhibitor of Na-K ATPase when exposed to intracellular sites (2,17,21) and our results indicate that veratridine, field stimulation and A23187 rapidly inhibit this enzyme via a calcium dependent process as they release ACh. Since none of these calcium dependent treatments decreased Na-K ATPase activity in osmotically lysed synaptosomes (17), it seems likely that they exerted their action on this enzyme in intact nerve terminals by increasing calcium levels intracellularly.

Veratridine's principal site of action is to prevent voltage sensitive sodium channels from closing (1,8) and treatment with this agent can increase synaptosomal calcium levels in at least three ways: 1) by depolarizing the nerve terminal and opening voltage-sensitive calcium channels (5), 2) by promoting calcium influx through sodium channels (14), and 3) by inhibiting a transmembrane Na-Ca exchange pump (3). Although we find that veratridine increases synaptosomal sodium levels (not shown), it does not appreciably depolarize this preparation. This finding probably reflects the relatively depolarized state of synaptosomes (with respect to sodium ions) suspended in KR buffer. Sodium concentrations can reach over 100 mM under these conditions (7). It therefore appears that veratridine's calcium dependent action on neurotransmitter release is not mediated entirely by voltage-sensitive calcium channels. Interestingly, we also find that tetrodotoxin, which blocks sodium channels, attenuates veratridine's actions on both ACh release and ouabain-sensitive Rb uptake (17). Along similar lines, electrical field stimulation increases calcium influx by opening voltage-sensitive calcium channels (30), while A23187 acts as an ionophore for this divalent cation (20).

How might changes in Na-K ATPase activity modulate ACh release from nerve terminals? One possibility is that inhibition of this enzyme indirectly releases ACh by increasing intracellular sodium levels. Elevated sodium concentrations have been shown to increase calcium efflux from synaptosomal mitochondria (25), and this may effect neurotransmitter release. We observed that ouabain, PCMBS, and calcium-dependent inhibitors of ouabain-sensitive rubidium uptake increase synaptosomal sodium levels. However, veratridine and A23187 increased synaptosomal sodium concentrations in a calcium-free medium to a greater extent than did ouabain, but did not release ACh or inhibit Na-K ATPase activity under these conditions (Meyer and Cooper, in preparation). Further, one would expect that treatments which promoted mitochondrial calcium secretion and elevated intrasynaptosomal levels of free calcium, should also increase the rate of calcium secretion, but we detected no significant effect of veratridine or ouabain on the release of ^{45}Ca from synaptosomes preloaded with this ion. Finally, it is interesting to note that the maximum stimulation of calcium secretion from synaptosomal mitochondria is observed with less than 100 mM sodium (25), which, as noted above, is about the concentration of this ion observed in tissue maintained in KR buffer (7). Therefore, it would appear that increased sodium levels may be insufficient to account for neurotransmitter release in our preparations.

Since Na-K ATPase may be an electrogenic pump in some tissues (21), we also examined whether its inhibition might depolarize synaptosomes, and thereby promote ACh release. However, we found little effect of ouabain, A23187 or veratridine on dye monitored membrane potential. These synaptosomes were apparently intact and able to maintain potassium ion gradients, since 60 mM K$^+$ increased tissue fluorescence, indicating a depolarization.

Thus far we have been unable to account for Na-K ATPase modulation of ACh release on the basis of increased sodium accumulation, calcium efflux from intracellular binding sites, or membrane depolarization, so we have hypothesized that this enzyme's rapid production of protons (upon ATP hydrolysis) may be involved. The influence of internal pH on neurotransmitter release has been overlooked in the literature, presumably because it is generally assumed

that the intracellular buffering capacity for this ion obviates even
short term variations in the pH of the nerve terminal. However, the
proximity of Na-K ATPase produced protons to the intracellular membrane may allow a local direct action of protons on an ACh gating
mechanism. According to this hypothesis, normal production of
protons would keep the ACh gate closed. Inhibition of Na-K ATPase
activity by calcium entry would decrease proton production, transiently increasing the pH at localized sites in the internal side of the
terminal membrane. This decrease in local hydrogen ion concentration
would then open a gate for the neurotransmitter. After calcium is
removed by sequestration or a transmembrane pump, normal Na-K ATPase
activity would decrease the pH and close the ACh gate. Although
this hypothesis could easily be accommodated into the non-quantal
release of transmitter, it could also account for voltage-dependent
quantal release if calcium influx proportionately inhibited the
enzyme and if ACh were assumed to be localized at the terminal membrane where it was released as a bolus when the gate opened. In
support of this proton-sensitive gate hypothesis, elevating the
extracellular hydrogen ion concentration has been shown to reduce
the ACh released per volley in mammalian neuromuscular preparations
(9,11). Further, hydrogen ions block other ionic channels such as
the voltage-dependent sodium channel (29). The possibility that
changes in intracellular proton production might block ACh channels
by a similar mechanism remains an intriguing possibility.

Finally, we must ask whether it is plausible that one enzyme
could regulate several ion gradients in addition to the rapidly
ongoing neurotransmitter release process. Along these lines, it
has recently been shown that the brain contains two distinct Na-K
ATPases in each of several mammalian species studied to date (26).
One of these Na-K ATPase is found associated only with neuronal
plasma membranes (designated the alpha (+) form). Perhaps this
neuronal Na-K ATPase has a specific role in the release of brain
neurotransmitters, while the other Na-K ATPase (the alpha form),
which is distributed throughout the body, is more directly responsible for the regulation of sodium-potassium exchange.

REFERENCES

1.   Abita, J.P., Chicheportiche, R., Schweitz, H. and Lazdunski, M.
        (1977): Biochemistry 16:1838-1844.
2.   Akera, T. and Brody, T.M. (1970): J. Pharmacol. Exp. Ther.
        176:545-557.
3.   Baker, P.F., Blaustein, M.P., Hodgkin, A.K. and Stienhard, R.A.
        (1969): J. Physiol. (Lond) 200:431-459.
4.   Beani, L., Bianchi, C., Giacomelli, A. and Tamberi, F. (1978):
        Eur. J. Pharmacol. 48:179-193.

5.   Blaustein, M.P. (1975): J. Physiol. (Lond) 247:617-655.
6.   Blaustein, M.P. and Goldring, J.N. (1975): J. Physiol. (Lond) 247:589-615.
7.   Campbell, C.W.B. (1976): Brain Res. 101:594-599.
8.   Catterall, W.A. (1975): J. Biol. Chem. 250:4053-4059.
9.   Cohen, I. and Van der Kloot, W. (1976): J. Physiol. (Lond) 262:401-414.
10. Cooper, J.R. (1977): IN Neurotransmitter Function: Basic and Clinical Aspects (ed) W.S. Field, Symposia Specialists, New York, pp. 135-141.
11. Del Castillo, J., Nelson, T.E. and Sanchez, V. (1962): J. Cell Comp. Physiol. 59:35-44.
12. Garcia, A.G. and Kirpekar, S.M. (1973): Brit. J. Pharmacol. 47:729-741.
13. Gilbert, J.C., Wylie, M.G. and Davison, D.V. (1975): Nature 255:237-238.
14. Hille, B. (1972): J. Gen. Physiol. 59:637-653.
15. Israel, M. and Dunant, Y. (1979): Prog. Brain Res. 49:125-139.
16. Lowry, O.H., Rosebrough, N.J., Farr, A.L. and Randall, R.J. (1951): J. Biol. Chem. 193:265-275.
17. Meyer, E.M. and Cooper, J.R. (1980): J. Neurochem. (in press).
18. Nemeth, E.F. and Cooper, J.R. (1978): Brain Res. 165:166-170.
19. Paton, W.D.M., Vizi, E.S. and Aboo Zar, M. (1971): J. Physiol. 215:819-848.
20. Pfeiffer, D.R., Taylor, R.W. and Lardy, H.A. (1978): Ann. New York Acad. Sci. 307:402-421.
21. Robinson, J.D. and Flashner, M.S. (1979): Biochim. Biophys. Acta 549:145-176.
22. Rutledge, C.O. (1978): Biochem. Pharmacol. 47:729-741.
23. Sen, I., Grantham, P.A. and Cooper, J.R. (1976): Proc. Nat. Acad. Sci. USA 73:2664-2668.
24. Sgaragli, G.P., Sen, I., Baba, A., Schulz, R.A. and Cooper, J.R. (1977): Brain Res. 134:113-123.
25. Silbergeld, E.K. (1977): Biochem. Biophys. Res. Commun. 77:464-499.
26. Sweadner, K.J. (1979): J. Biol. Chem. 254:6060-6067.
27. Vizi, E.S. (1972): J. Physiol. (Lond) 226:95-117.
28. Vizi, E.S. (1978): Neuroscience 3:367-384.
29. Woodhull, A.M. (1973): J. Gen. Physiol. 61:687-708.
30. Yamamoto, S., Akera, T. and Brody, T.M. (1979): Biochem. Biophys. Acta 555:270-284.

ACETYLCHOLINE METABOLISM IN RAT NEOSTRIATAL SLICES

M.H. Weiler, D.J. Jenden, I.J. Bak* and U. Misgeld**

*Departments of Pharmacology and Neurology
University of California, Los Angeles, California 90024
**Max-Planck-Institut fur Hirnforschung, Department of
Neurobiology, Frankfurt-Main, Federal Republic of Germany

INTRODUCTION

Brain slice preparations have been used for the in vitro study
of acetylcholine (ACh) metabolism since the 1930's (15). More recently, the brain slice preparation as described by McIlwain and
Rodnight (11) has been recognized as a useful tool for studying
electrophysiological characteristics of brain tissue (21). An intrinsic cholinergic network in rat neostriatal slices has been described both electrophysiologically (13) and pharmacologically (12)
and morphological studies showed a good correlation between cell
preservation and the recording of electrical activity (1).

In such slices the electrophysiological integrity of the cholinergic synapse is maintained for more than 10 hr after slice preparation. We report here biochemical studies of ACh metabolism in
slices prepared in a manner similar to that for electrophysiological
studies. The purpose of these studies was 1) to examine ACh
metabolism, including release, in the neostriatal slices under normal and potassium depolarized conditions and 2) to observe effects
of choline (Ch) on ACh metabolism in both normal and potassium depolarized slices.

METHODS

Slices (200-400 μm) were hand sliced from trimmed neostriatal
tissue blocks at the level of the anterior commissure using a glass

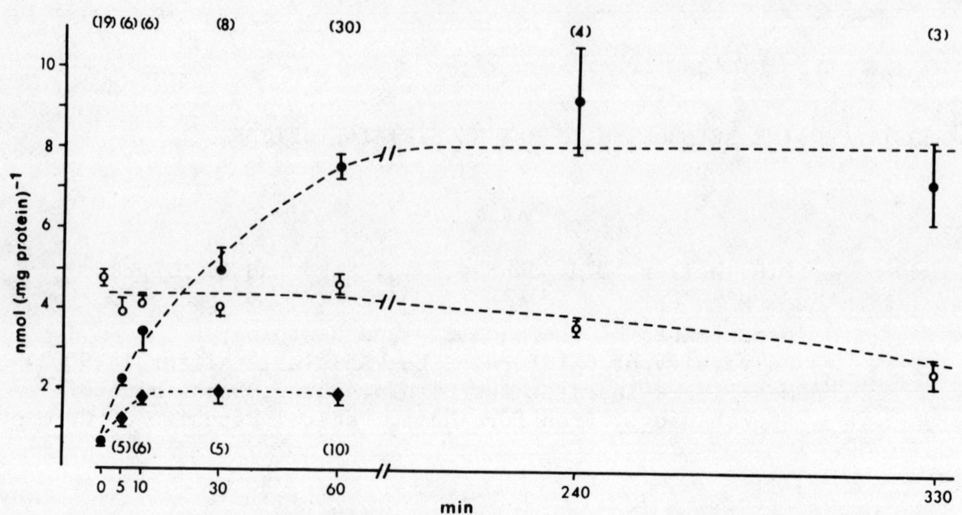

FIGURE 1:   Time course of ACh accumulation and Ch levels in rat
            neostriatal slices. The slices were incubated at 36°C
            in oxygenated Krebs-Ringer bicarbonate solution (mM con-
            centrations: NaCl 124; KCl 5.1; $MgSO_4$ 1.3; KH_2PO_4 1.22;
            $NaHCO_3$ 25.5; $CaCl_2$ 2.3, and glucose 10.2) containing
            $[^2H_4]$Ch (10 μm). The number of determinations for ACh
            and Ch is indicated in parentheses above. The number of
            determinations for ACh accumulation in the presence of
            HC-3 (5 μM) is indicated in parentheses below. Each value
            in nmol/mg protein is the mean $\pm$ S.E. of total ACh
            (●——●), total Ch (○——○), and total ACh in the
            presence of HC-3 (◆——◆)

guide (11). While slicing, the brain was continually moistened with
oxygenated Krebs-Ringer bicarbonate buffer (see Legend, Fig. 1) at
36°C. Immediately after slicing, the tissue was either transferred
to a low K^+ (5.1 mM) or high K^+ (25 mM) buffer containing $[^2H]$Ch or
equilibrated for 60 min in normal (low K^+) buffer before exposure
to potassium or $[^2H]$Ch. Tetram (10 μM) was used to allow determination of ACh release. At designated time points slices were fixed
in 15% aqueous formic acid, 85% acetone and prepared for analysis
by combined gas chromatography mass spectrometry (GCMS) (7,9).

TABLE 1:   ACh synthesis during recovery period (10 μM [^{2}H$_4$]Ch).

INCUBATION TIME	[KCl]		A	B	C	SYNTHESIS (A+B-C)
10 min	5.1 mM	Total ACh	0.25 ± 0.05(4)	4.91 ± 0.63(8)	1.07 ± 0.07(10)	4.09 ± 0.75
		[^{2}H$_4$]ACh	0.06 ± 0.02(4)	1.08 ± 0.27(8)	0.07 ± 0.03(10)	1.07 ± 0.32
	25 mM	Total ACh	0.61 ± 0.14(4)*	2.96 ± 0.20(8)**	1.07 ± 0.07(10)	2.50 ± 0.27
		[^{2}H$_4$]ACh	0.10 ± 0.03(4)	0.48 ± 0.06(8)*	0.07 ± 0.03(10)	0.51 ± 0.09
60 min	5.1 mM	Total ACh	1.84 ± 0.32(4)	9.01 ± 1.01(8)	1.07 ± 0.07(10)	9.78 ± 1.21
		[^{2}H$_4$]ACh	0.80 ± 0.17(4)	3.00 ± 0.49(8)	0.07 ± 0.03(10)	3.73 ± 0.59
	25 mM	Total ACh	5.18 ± 0.73(4)***	5.42 ± 0.44(8)***	1.07 ± 0.07(10)	9.53 ± 0.66
		[^{2}H$_4$]ACh	1.68 ± 0.19(4)**	1.97 ± 0.15(8)	0.07 ± 0.03(10)	3.58 ± 0.21

ACh synthesis during recovery period (10 and 60 min) in 5.1 mM and 25 mM KCl Krebs-Ringer bicarbonate buffer at 36°C. Media contained tetram (10 μM) and [^{2}H$_4$]Ch (10 μM). All values are expressed as the mean ± S.E. nmol (mg protein)$^{-1}$. The number of determinations are in parentheses. The values in the column labelled synthesis represent the net synthesis of ACh, which was calculated by adding the amount released (column A) to the change in slice content from the beginning (column C) to the end (column B) of the designated incubation time. Asterisks denote significant difference from 5.1 mM KCl: * $p < 0.05$; ** $p < 0.02$; *** $p < 0.01$. In this and subsequent tables the standard error for the synthesis estimate was calculated by pooling the data variance estimate(s^2) from the corresponding entries in columns A, B and C. The variance of (A+B-C) was assumed equal to: $s^2 \left(\dfrac{1}{n_A} + \dfrac{1}{n_B} + \dfrac{1}{n_C} \right)$.

RESULTS

We have previously shown (20; and Fig. 1) that 1) immediately after preparation the slices accumulate ACh at an initial rate which is 1.3 to tenfold higher than most in vivo estimates of ACh turnover; and 2) after a recovery period (1 hr after preparation) neostriatal slice ACh and Ch levels are about tenfold higher than those observed when brain tissue is optimally fixed in vivo by fast microwave fixation (5,14,18); and 3) when the medium is supplemented with $[^2H]$Ch (10 μM) the mole ratio of $[^2H]$ACh indicates that the Ch supplied externally does not mix readily with the endogenous Ch. In the present experiments we have examined the synthetic capacity of the slices under depolarized conditions (25 mM KCl) and in the presence of two different amounts of Ch.

High K^+ During Recovery Period

In 60 min the net synthesis of total and $[^2H]$ACh during the recovery period was not affected by high K^+ (Table 1). In the presence of high K^+ the neostriatal slices accumulated 40% less ACh, and ACh release was stimulated 2.8-fold. However, when the net ACh synthesis is calculated by adding the amount released to the change in slice content during 60 min, the rate of synthesis was no different from that seen in low K^+.

In contrast, the initial rate of synthesis of ACh was markedly affected in the presence of high K^+ (Table 1). In the first 10 min the net synthesis of both $[^2H]$ACh and total ACh was less than that of control. Initial ACh release was stimulated 2.4-fold, but the tissue accumulation rate was reduced by 49%.

There was a substantial efflux of $[^2H_0]$Ch from the neostriatal slices at a rate of 130 pmol $mg^{-1}min^{-1}$ between 10 and 60 min (data not shown), in agreement with other studies of brain both in vivo (6) and in vitro (2,3). However, the Ch content changed less than 1 nmol mg^{-1} during the 60 min incubation, confirming that the efflux is generated chemically rather than by translocation of preformed Ch.

High K^+ Following Recovery Period

If the neostriatal slices were allowed to recover for 60 min the net synthesis during the second 60 min was substantially less than that synthesized during the recovery period. This would be expected if steady state conditions were approached by the end of the first hour. Exposure to high K^+ during the second 60 min period resulted in a twofold stimulation of synthesis of both total and $[^2H]$ACh (Table 2). Tissue ACh levels did not change significantly from the beginning of incubation, and the increase in synthesis was, therefore, accounted for by the sevenfold increase in ACh release in high K^+ medium.

As before, initial rates of net ACh synthesis were not increased in slices exposed to high K^+, and synthesis of $[^2H]ACh$ was significantly reduced. A significant increase in ACh release was more than offset by a decrease in tissue levels.

$[^2H_0]Ch$ efflux between 10 and 60 min was 66 pmol $mg^{-1}min^{-1}$ and 40 pmol $mg^{-1}min^{-1}$ under normal and potassium depolarized conditions, respectively. The tissue Ch content changed on the average less than 1 nmol mg^{-1} under both conditions.

Effect of Ch Under Normal and Depolarized Conditions

Because previous studies (4,20) indicated that the supply of Ch for ACh synthesis in brain slices is from a site different from that where ACh is synthesized, we examined the effects of a fivefold greater concentration of Ch (50 μM vs 10 μM) on ACh metabolism in the neostriatal slices.

Under normal and depolarized conditions there was a twofold increase in net ACh synthesis during the initial 10 min when the medium was supplemented with more $[^2H_4]Ch$ (compare: Tables 2 and 3), although the differences were significant only for the $[^2H]ACh$ (p < 0.01 and p < 0.001 for 5.1 and 25 mM K^+, respectively). However, the initial rate of ACh synthesis in high K^+ was still less than the rate in the low K^+ buffer when both contained $[^2H]Ch$, although the differences are not significant.

After 60 min incubation in 50 μM Ch, the synthesis of ACh was significantly greater under depolarizing conditions than in 5.1 mM K^+ (Table 3), and the net synthesis could in both conditions be quantitatively accounted for by ACh formed from $[^2H]Ch$ supplied externally. The final steady state level of total ACh was not different from that seen following incubation in 10 μM Ch (Tables 2 and 3), column B), but the amount of ACh synthesized from $[^2H]Ch$ was significantly (p < 0.001) increased by Ch loading only when the slice was depolarized. This increase was reflected both in the released ACh (p < 0.01) and in that retained in the slice (p < 0.01).

DISCUSSION

The data presented demonstrate that the effect of high K^+ on ACh synthesis in rat neostriatal slices depends upon the length of time the slices are exposed to high K^+. Initially, net ACh synthesis

TABLE 2:   ACh synthesis following 60 min recovery period (10 μM [^{2}H$_4$]Ch).

INCUBATION TIME	[KCl]		A	B	C	SYNTHESIS (A+B−C)
10 min	5.1 mM	Total ACh	0.17 ± 0.04(5)	8.38 ± 0.89(10)	7.22 ± 0.35(18)	1.33 ± 1.14
		[^{2}H$_4$]ACh	0.02 ± 0.01(5)	0.41 ± 0.05(10)	0	0.43 ± 0.07
	25 mM	Total ACh	1.10 ± 0.08(6)***	6.62 ± 0.41(8)	7.22 ± 0.35(18)	0.50 ± 0.79
		[^{2}H$_4$]ACh	0.00 ± 0.01(6)	0.19 ± 0.01(8)**	0	0.19 ± 0.01**
60 min	5.1 mM	Total ACh	0.92 ± 0.19(5)	8.71 ± 1.02(7)	7.22 ± 0.35(18)	2.41 ± 1.25
		[^{2}H$_4$]ACh	0.18 ± 0.04(5)	1.41 ± 0.18(7)	0	1.59 ± 0.22
	25 mM	Total ACh	6.67 ± 0.50(6)***	7.02 ± 0.51(8)	7.22 ± 0.35(18)	6.47 ± 0.88*
		[^{2}H$_4$]ACh	1.08 ± 0.14(6)***	1.83 ± 0.20(8)	0	2.91 ± 0.26***

ACh synthesis (10 and 60 min) in 5.1 and 25 mM KCl buffer following a 60 min recovery period. Media contained tetram (10 μM) and [^{2}H$_4$]Ch (10 μM). Values expressed similarly to those in Table 1. Asterisks denote significant difference from 5.1 mM KCl: * p < 0.02; ** p < 0.005; *** p < 0.001.

TABLE 3: ACh synthesis following 60 min recovery period (50 µM $[^2H_4]$Ch).

INCUBATION TIME	[KCl]		A	B	C	SYNTHESIS (A+B-C)
10 min	5.1 mM	Total ACh	$0.16 \pm 0.04(7)$	$9.65 \pm 1.49(\ 9)$	$7.22 \pm 0.35(18)$	2.59 ± 1.40
		$[^2H_4]$ACh	$0.01 \pm 0.03(7)$	$0.84 \pm 0.15(\ 9)$	0	0.85 ± 0.17
	25 mM	Total ACh	$1.50 \pm 0.25(8)**$	$6.61 \pm 0.71(10)$	$7.22 \pm 0.35(18)$	0.89 ± 0.86
		$[^2H_4]$ACh	$0.02 \pm 0.02(8)$	$0.63 \pm 0.08(10)$	0	0.65 ± 0.09
60 min	5.1 mM	Total ACh	$1.11 \pm 0.26(6)$	$8.39 \pm 1.24(\ 7)$	$7.22 \pm 0.35(18)$	2.28 ± 1.16
		$[^2H_4]$ACh	$0.28 \pm 0.10(6)$	$2.09 \pm 0.46(\ 7)$	0	2.37 ± 0.51
	25 mM	Total ACh	$6.29 \pm 0.58(8)**$	$6.48 \pm 0.56(10)$	$7.22 \pm 0.35(18)$	$5.55 \pm 0.85*$
		$[^2H_4]$ACh	$2.00 \pm 0.24(8)**$	$3.33 \pm 0.34(10)*$	0	$5.33 \pm 0.44**$

Same as Table 2 except the $[^2H_4]$Ch concentration was 50 µM. Asterisks denote significant difference from 5.1 mM KCl: * $p < 0.05$; ** $p < 0.001$.

is depressed, but prolonged exposure to high K^+ may stimulate net
synthesis. These results are consistent with the observations of
Simon and Kuhar (17) in which high concentrations of K^+ decreased
Ch uptake into synaptosomes and with observations of Vaca and Pilar
(19) in which K^+ depolarization depressed initial rates of ACh synthesis in chick ciliary ganglia. The effects of K^+ depolarization
on ACh synthesis over prolonged periods are also consistent with
the suggestion of Vaca and Pilar (19) that ACh synthesis is stimulated over prolonged periods of depolarization because a new steady
state is achieved between ACh synthesis and release. As a result of
a rapidly adjusting equilibrium between the four reactants of Ch
acetylase a change in flux across the neuronal membrane would tend
to compensate for ACh depletion.

We also conclude that under certain circumstances, and particularly in the depolarized state, additional Ch may play a permis-
sive role in increasing ACh metabolism in rat neostriatal slices.
Initial net synthesis rates were increased by Ch, while net synthesis during a 60 min exposure was not significantly affected.
These results are compatible with a quasi-equilibrium model of ACh
synthesis regulation which predicts that extracellular Ch and intracellular ACh levels are directly related to each other if the acetyl-
coenzyme A supply remains unchanged (cf ref. 8).

ACKNOWLEDGEMENTS

This work was supported by USPHS grant MH 17691 and the Scott
Fund. The authors are grateful for the editorial assistance of Flo
Comes and the exceptional technical assistance of Esther Shaw.

REFERENCES

1.   Bak, I.J., Misgeld, U., Weiler, M. and Morgan E. (1980): Brain
        Res. 197:341-353.
2.   Bhatnagar, S.P. and MacIntosh, F.C. (1967): Canad. J. Physiol.
        Pharmacol. 45:249-266.
3.   Browing, E.T. (1971): Biochem. Biophys. Res. Comm. 45:1586-1590.
4.   Browning, E.T. and Schulman, M.P. (1968): J. Neurochem. 13:
        515-524.
5.   Butcher, S.H., Butcher, L.L., Harms, M.S. and Jenden, D.J.
        (1976): J. Microwave Power 11:61-65.
6.   Dross, K. and Kewitz, H. (1972): N.-S. Arch. Pharmacol. 274:
        91-106.
7.   Freeman, J.J., Choi, R.L. and Jenden, D.J. (1975): J. Neuro-
        chem. 24:729-734.

8.   Jenden, D.J. (1979): IN Brain Acetylcholine and Neuropsychia-
     tric Disease (eds) K.C. Davis and P.A. Berger, Plenum Press,
     New York, pp. 483-513.

9.   Jenden, D.J., Roch, M. and Booth, R. (1973): Anal. Biochem.
     55:438-448.

10. Lefresne, P., Guyenet, P. and Glowinski, J. (1973): J. Neuro-
     chem. 20:1083-1097.

11. McIlwain, H. and Rodnight, R. (eds)(1962): IN Practical Neuro-
     chemistry, Churchill-Livingstone, London, pp. 109-133.

12. Misgeld, U. and Bak, I.J. (1979): Neurosci. Lett. 12:277-282.

13. Misgeld, U., Okada, Y. and Hassler, R. (1979): Exp. Brain Res.
     34:575-590.

14. Nordberg, A. and Sundwall, A. (1976): Acta Physiol. Scand. 93:
     307-317.

15. Quastel, J.H., Tennenbaum, M. and Wheatley, A.H.M. (1936):
     Biochem. J. 30:1668-1681.

16. Sattin, A. (1966): J. Neurochem. 13:515-524.

17. Simon, J.R. and Kuhar, M.J. (1976): J. Neurochem. 27:93-99.

18. Stavinoha, W.B., Frazer, J. and Modak, A.T. (1977): IN Chol-
     inergic Mechanisms and Psychopharmacology (ed) D.J. Jenden,
     Plenum Press, New York, pp. 169-180.

19. Vaca, K. and Pilar, G. (1979): J. Gen. Physiol. 73:605-628.

20. Weiler, M.H., Misgeld, U., Bak, I.J. and Jenden, D.J. (1979):
     Brain Res. 176:401-406.

21. Yamamoto, C. and McIlwain, H. (1966): J. Neurochem. 13:1333-
     1343.

THE CHOLINERGIC SYSTEMS OF THE EYE

T.W. Mittag

Department of Pharmacology
Mount Sinai School of Medicine
CUNY
New York, New York 10029 USA

INTRODUCTION

Tissues of the eye possess several interesting features which
make them particularly useful for certain types of experimental
studies on transmitter systems. For example, ocular structures
can be readily denervated or given chronic drug treatments with
little trauma or systemic drug effect to the experimental animal.
Some of these advantages are illustrated by studies discussed here
on acetylcholine (ACh) and the cholinergic system present in cornea,
and studies on the dynamics of ACh receptors in the intraocular
smooth muscle of iris and ciliary body.

METHODS

Eyes were obtained from adult female Dutch Belt (pigmented)
or New Zealand (nonpigmented) rabbits and female mongrel cats,
sacrificed by parenteral pentobarbital.

Choline acetyltransferase (CAT) was determined on tissue
homogenates by the radiometric assay previously described (14).
Acetylcholinesterase (AChE) and butyrylcholinesterase (BuChE) were
determined by radiometric assay of the "functional" surface enzyme
and total homogenate activity as previously described (18,19). The
drug sensitivity of cat irides was determined from complete cumula-
tive dose-response curves to carbaymlcholine and to acetyl-β-methylcholine, with the tissues set up as isolated organ bath preparations

(9). Cat iris was "denervated" by unilateral retrobulbar injection
of 0.2 ml absolute ethanol (12) and then both irides from the same
animal were compared with respect to drug induced contractile responses. Responses were quantitated as the pD_O value = neg log of
the intercept on the dose axis of the extrapolated linear part of
the dose-response curve. ACh receptors (AChR) were determined by
binding of I^{125}-α-bungarotoxin (α-Btx) to 1% Triton X-100 extracts
of tissues for nicotinic AChR (20), and by the binding of H^3 $\pm$
quinuclidinyl benzilate (QNB) to tissue homogenates (22). Uptake
of H^3-choline (Ch) and the separation of labelled ACh by thin layer
chromatography were done as previously described (1,2).

ACh in the Cornea

The high ACh content of the cornea has been known for over
30 years (7). It is found only in the epithelium, which is an
avascular layer 5-8 cells thick. The outer cell layers are constantly shed and replaced by cell division of the basal layer. The
epithelium has no known efferent effector nervous system, the fine
network of nerves which is present appears to be exclusively sensory
in function (23) but may also be involved in growth and maturation
of the epithelial cells (6,7). Over the past few years we have attempted to define the characteristics and function(s) of the cholinergic system in this tissue but no clear understanding of its
role(s) has yet emerged.

Table 1 summarizes the CAT activity of the corneal epithelium
from several species in comparison with other ocular structures (14).
The CAT specific activity in cornea varies widely among species.
It is very low in feline cornes, but in the rabbit is usually higher
than other ocular structures such as the iris + ciliary body or
the retina, both of which contain cholinergic neuroeffector systems.
Most of our studies have been done on rabbit corneas which show a
wide range of activities from 2-40 nmol ACh/hr/mg protein. However,
the two corneas of an individual animal have closely similar enzyme
levels suggesting that CAT level in cornea might be genetically
determined. A heritable component in corneal CAT is indicated by
the finding that animals from an inbred rabbit colony and litter-
mates showed a much closer range of activities than randomly bred
animals from the same supplier (15).

The source of Ch for synthesis of ACh by epithelial cells could
be either from the aqueous humor or from the tear film. The latter
seems more likely since the permeability of the corneal endothelium
and stroma to cations would be rather low unless a specific transport system was involved. Table 2 shows that the cornea is able
to synthesize ACh from exogenous radioactive Ch. Whole corneas and,

TABLE 1: Mean CAT activity of ocular structures of various species.

Tissue	Feline	Rabbit	Bovine	Human
		μmol ACh/hr/g protein		
Cornea	0.2	22.5	16.0	11.0
Ciliary body	4.0	5.8	2.8	20.3
Iris	3.1		4.3	1.8
Retina	0.9	2.1	11.2	11.0

TABLE 2: Uptake of Ch into whole rabbit cornea and whole retina
incubated in Krebs-Ringer for 30 min. The percentage of
the total radioactive Ch content termed transported Ch
was determined by correcting for the Ch imbibed by the
tissues in 30 min at 4°C. The acetylated fraction was
determined by separation by thin layer chromatography of
the tissue extracts (2).

Incubation Medium (μM Ch)	Cornea (percent)		Retina (percent)	
	Transported	Acetylated	Transported	Acetylated
0.5	39	14.0	75	(< 3)*
1.0	41	8.5	75	(< 3)
10.0	53	(3.0)*	84	(< 3)

* The limit of chromatographic resolution for ACh from Ch under
these conditions was determined as 3%).

for comparison, whole retinae, were incubated in labelled Ch, briefly rinsed, and analyzed for ACh and Ch content (2). When whole
tissues were incubated in this way, large amounts of Ch were imbibed nonspecifically by the tissue so that the proportion of
labelled ACh is relatively small (Table 2). This nonspecific Ch
component can be estimated from the amount of Ch imbibed by incubating the tissue at 4°C, and represents 45-60% of the Ch content
of tissues incubated at 30°C. The additional uptake of label is
"transported" Ch and ACh. It is clear that under these conditions
cornea utilizes exogenous Ch for ACh synthesis. However, no detectable ACh is formed by retina under these conditions although
there is a large Ch uptake. ACh synthesis by retina requires

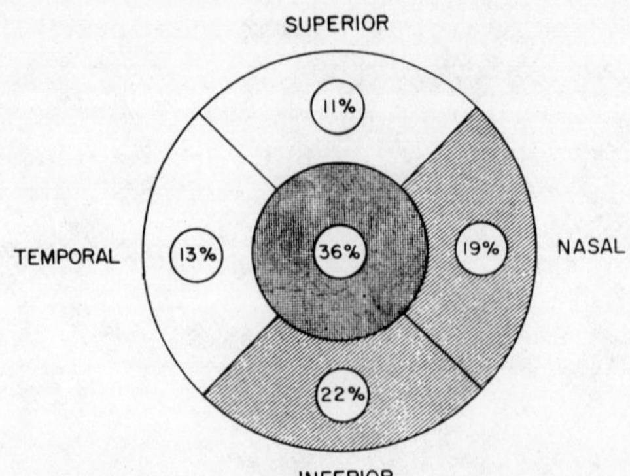

FIGURE 1: Regional distribution of CAT activity in rabbit corneas.
          (n = 11). Central cornea had significantly higher
          (p < 0.05) enzyme activity than nasal or inferior cornea
          which, in turn, had higher enzyme activity (p < 0.05)
          than superior or temporal cornea. Percentages designate
          each segment's proportion of total corneal enzyme activity.

functional activity (e.g. on/off light flashes) which was not done
for these experiments (13).

The high CAT enzyme levels and high ACh content of cornea suggest that the fine nerve fiber network cannot account for these
specific activities. The non-neuronal epithelial cell locus of
CAT is confirmed by the presence of the enzyme in epithelial cells
grown in tissue culture from corneal explants (8). However, cultured epithelial cells cannot be used to study the ACh system because the level of CAT declines with the age of the cultures
(Mittag, unpublished observation). This suggested that the level
of CAT in these cells may be affected by environmental and trophic
factors. Absence of a presumed "trophic" influence from the sensory
nerves could account for the rapid loss of enzyme activity in cultured cells. We have obtained some experimental evidence suggesting
an influence of innervation and environmental factors on CAT in the
cornea in situ. Figure 1 shows the distribution of CAT specific
activity in the quadrants and central area of rabbit corneas. The
five segments are of equal area and the percentages designate the
proportion of the total enzyme activity. The central cornea consistently has a much higher CAT specific activity, about threefold
that of the adjacent superior quadrant. This may be correlated
with the density of sensory nerves, since sensory activity may follow a roughly similar distribution (4).

Environmental factors also affect corneal CAT. Closing of the eyelids causes a rapid loss of CAT with a maximal decline of about 80% by seven days (16), without any obvious microscopic changes in morphology or in other enzymes of the epithelial cells. New epithelial cells which grow out from the corneal margins in 3-4 days to cover denuded areas of stroma have very low CAT activity. However, after four weeks of regrowth the mature epithelial cells have CAT activities comparable to the control eye. These findings correlate with results obtained with corneal explants in tissue culture cited above, and can be explained by a "trophic" influence of sensory innervation on CAT, which becomes manifest at later stages of epithelial regrowth. The available evidence shows that the expression of ACh synthesis in corneal epithelial cells is highly variable. It appears to be correlated with cell age and may be a feature of the senescent population of epithelial cells which do not proliferate. Thus one of the possible functions of corneal ACh, in addition to a possible role in water or ion transport mechanisms   . (21), might involve growth and maturation of cells.

The mechanism of action of corneal ACh is still completely unknown. Attempts to demonstrate classical types of ACh receptors in epithelial tissue have not been successful (17). No net binding of QNB indicative of muscarinic AChR or of α-Btx indicative of nicotinic AChR can be demonstrated at appropriate concentrations of these specific receptor ligands (up to 10^{-7} M). Thus it must be concluded that if corneal ACh effects are receptor mediated, then these receptors do not bind the characteristic ligands.

Although a great deal of new information has been obtained about the corneal cholinergic system, its role(s) in this tissue is almost as uncertain today as 30 years ago when Hellauer suggested a sensory function for the ACh in the cornea.

Receptor Regulation in Iris and Ciliary Muscle

The iris and ciliary body represent smooth muscle structures which are readily accessible for experimental studies. These structures can be decentralized by appropriate sympathetic or parasympathetic ganglionectomy or even by retrobulbar ethanol injection (ciliary denervation). They can also be chronically drug treated by the topical route or by intravitreal or anterior chamber injection. Finally, they provide smooth muscle tissues which contain nicotinic AChR in amphibia, reptiles and birds, whereas in mammals these structures contain only muscarinic AChR.

Binding studies with specific receptor ligands have provided tools to correlate tissue response with quantitative measurements

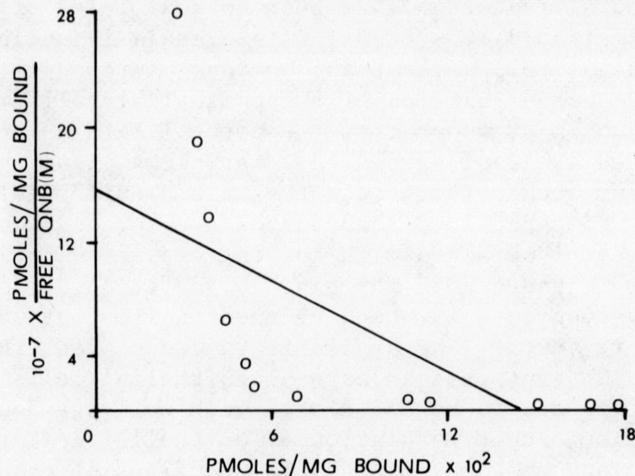

FIGURE 2: Scatchard plot of QNB binding to normal rabbit iris + ciliary body homogenates. Line drawn is best fit to the data points obtained by PROPHET ligand binding analysis when programmed for ONE set of binding sites.

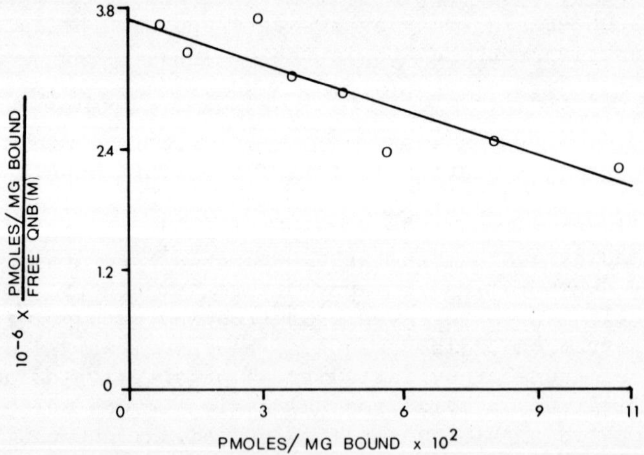

FIGURE 3: Scatchard plot of QNB binding to homogenates of sub-sensitive iris (phospholine[R] treated) fitted to a single set of binding sites.

of receptor sites. These techniques have been applied to the iris
of rabbits and cats to study adaptive changes in drug response at
the receptor level.

Topical treatment of the rabbit eye with echothiophate (Phospholine[R]) causes rapid miosis. However, the miotic response declines on repeated treatment, so that by the third day the treated
eye becomes unresponsive and is in a subsensitive state (3). The
change in receptor parameters associated with the subsensitive state
can be studied with the QNB binding. The binding isotherm for QNB
in control eyes clearly has two components as shown in Fig. 2,
indicating two sets of binding sites. By comparison, the subsensitive iris + ciliary body shows only a single set of binding sites
in the Scatchard plot (Fig. 3), corresponding to the low affinity,
high capacity set of QNB binding sites also present in normal tissues. Thus the cholinergic subsensitivity is associated with drop-
out of the high affinity sites. These data are summarized in
Table 3 showing binding data analyzed by use of PROPHET computer
programs for analysis of ligand binding (5).

The occurrence of two sets of QNB binding sites is in itself
an interesting finding. The sites were further investigated in
normal iris + ciliary body tissues by competition binding studies
of atropine and oxotremorine against QNB binding (Figs. 4 and 5).
High affinity sites, presumably those which mediate ACh responses
in normal iris + ciliary body show a Hill coefficient close to 1
with antagonists, e.g. atropine, but significantly less than 1 with
agonists, e.g. oxotremorine (Table 4). The low affinity set of QNB
binding sites gave much smaller Hill coefficients (between 0.3 and
0.4) for both agonists and antagonists. These data durther distinguish the two sets of QNB binding sites and suggest that low
affinity sites interact with negative co-operativity. The significance of the low affinity set of QNB binding sites remains to be
established.

Cholinergic supersensitivity, the functional opposite of the
subsensitive state, can also be studied in the iris. The cat iris
can be functionally denervated by retrobulbar injection of ethanol
(12). This treatment damages the ciliary nerve at the point of
entry into the globe and the consequent degeneration of cholinergic
nerve terminals in the iris can be confirmed by the loss of the CAT
enzyme (Table 5). However, the sensitivity to cholinergic agonists
of the in vitro tissue bath preparation of such irides showed no
significant difference from the control irides (Table 6). Thus,
denervation did not produce any evident changes at the receptor
level, based on drug response. Direct binding studies have confirmed this, because ciliary ganglionectomy did not cause any significant change in the number and affinity of the high affinity set
of QNB binding sites (11).

TABLE 3: QNB binding to iris + ciliary body homogenates.

Data Plot	Control Eyes		Untreated Eyes		Treated Eyes	
	Kd (nM)	Bmax (pmol/mg)	Kd (nM)	Bmax (pmol/mg)	Kd (nM)	Bmax (pmol/mg)
A: High affinity binding range (1–10 nM QNB)						
Hyperbolic	1.01	2.00	1.12	1.14	–	–
Reciprocal	0.98(R.99)*	2.12	1.03(R.99)	1.17	–	–
Scatchard	0.74(R.91)	1.95	1.07(R.96)	1.22	–	–
B. Low affinity binding range (10–50 nM QNB)						
Hyperbolic	17.1	11.1	54	5.85	82	5.66
Reciprocal	12.2(R.99)	9.3	63(R.99)	6.64	62(R.99)	4.58
Scatchard	15.4(R.97)	10.0	52(R.95)	5.76	65(R.89)	4.84

* R = regression coefficient

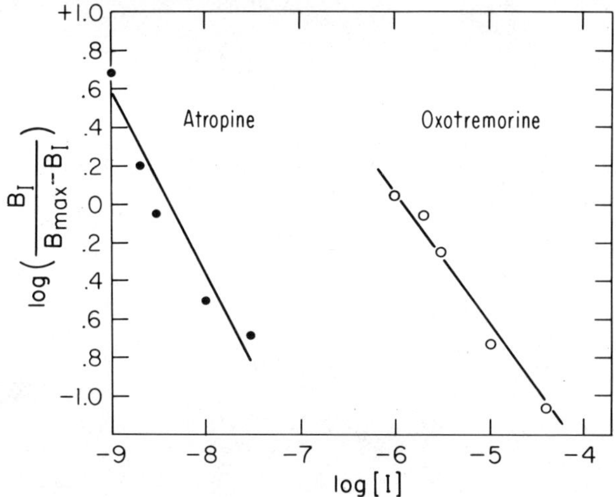

FIGURE 4: Hill plots of competition binding data for atropine and oxotremorine against the high affinity set of QNB binding sites. QNB was used at a fixed concentration of 0.5 nM.

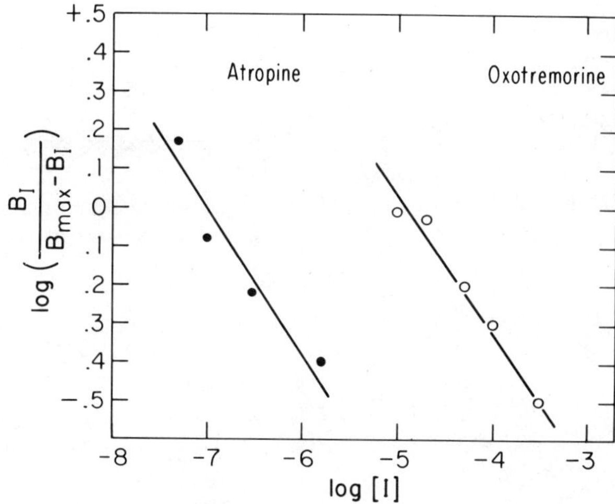

FIGURE 5: Hill plots of the competition binding data for atropine and oxotremorine against the low affinity set of QNB binding sites. QNB was used at a fixed concentration of 20 nM.

TABLE 4: Ligand competition for QNB binding sites in iris +
          ciliary body homogenates.

Ligand	K_I	Hill Coefficient
A. Low Affinity QNB binding sites		
Atropine	35 nM	0.38
Oxotremorine	4.2 µM	0.32
B. High affinity QNB binding sites		
Atropine	0.86 nM	0.96
Oxotremorine	0.24 µM	0.68

TABLE 5: Effect of retrobublar ethanol on ACh synthesis activity
          of the cat iris.

| Time | pmol ACh/hr/mg Tissue | |
(post-treatment)	Control Eye	Treated Eye
1 day	468	266
	310	181
	531	498
	342	363
7-14 days	610	2
	147	6
	354	8

TABLE 6:   The effect of retrobulbar ethanol on the contractile
           response of the cat iris to cholinergic agonists _in vitro_.

Time	Carbamylcholine pD_O		Acetyl-β-methylcholine pD_O	
(post-treatment)	Control	Treated	Control	Treated
1 day*	7.78	8.13	6.14	6.17
1 week	7.67	7.91	6.31	7.26
2 weeks	7.83	8.27	6.14	6.64
Untreated eyes from normal cats	8.02 ± 0.2 (n = 15)		6.31 ± 0.52 (n = 18)	

* 3 animals in each group.

TABLE 7:   Effect of retrobulbar ethanol denervation on AChE and
           BuChE activity* of cat iris, 7-14 days post-treatment.
           Total enzyme activities determined on homogenates and
           surface enzyme activity on whole tissues.

Enzyme Activity	Iris			Percent Change	SD
	Normal	Control (n = 12)	Treated (n = 12)		
Total AChE	–	6.0 ± 3.9	2.87 ± 2.8	NS	p< 3
Total BuChE	–	56.1 ± 10.7	29.3 ± 8.2	NS	p< 2
Surface AChE	2.1 ± 0.1	1.54 ± 1.4	1.00 ± 0.4	-52%	p< 0.05
Surface BuChE	8.1 ± 0.2	6.6 ± 0.2	9.6 ± 1.0	-48%	p< 0.05

*   Enzyme activities are expressed as the pseudo first order rate
    constant k (1, min^{-1}, g $tissue^{-1}$) for hydrolysis of BuChE and
    acetyl-β-methylcholine (AChE) and are not absolute enzyme
    activities (see, ref. 19). All iris figures are k ± S.D.
NS = not significant; SD = statistical difference

The ethanol induced denervation did cause a 50% loss of both
specific AChE and nonspecific BuChE in iris homogenates (Table 7).
If the AChE change involved a loss of the functional fraction of
the enzyme, then the response pattern of the denervated iris should
show apparent increased sensitivity to ACh. The response pattern
expected would be similar to that found for anticholinesterase
treated iris preparations, that is, a shift to the left of the dose
response curve to agonists that are AChE substrates (acetyl-β-methyl
choline) but little change in response to nonsubstrates such as
carbamyl Ch (9,10). Since the denervation did not produce any significant change in response to either agonist, the "functional"
fraction of tissue AChE appears to be unchanged by denervation in
spite of an overall loss of 50% of the enzyme. The functional fraction of the enzyme can be estimated in thin tissues by assay of the
"surface" AChE, defined as that fraction of the enzyme which interacts with the agonist/substrate as it diffuses from the bulk phase
to the receptor sites mediating the response of the in vitro bath
tissue preparation. Assays of the surface cholinesterases support
the results inferred from the dose reaponse data that denervation
causes little apparent change in the functional pools of enzyme
(Table 7).

The phenomenon of supersensitivity to topical cholinergic
agonists, which is shown by the denervated iris sphincter in vivo
cannot be satisfactorily accounted for by changes in muscarinic
receptor sensitivity or in cholinesterase enzymes. Other possible
mechanisms must therefore be considered. Supersensitivity could
arise from adaptive changes beyond the receptor level such as in
receptor mediated ion channels; from increased penetration of cornea
and iris tissue by cationic agents such as topically applied cholinergic agonists; or from mechanical changes in the normal antagonistic balance between iris sphincter and dilator muscle fibers.
The latter possibility deserves some serious consideration, because a relatively small degree of adaptive relaxation in the tone
of the sympathetically innervated dilator muscle could produce an
apparent supersensitivity in the parasympathetic sphincter muscle.
This raises the general possibility that some "supersensitivities"
may be indirect and actually arise from adaptive changes in inhibitory or antagonistic mechanisms which operate on the cell or
tissue in question.

ACKNOWLEDGEMENTS

This research was supported by USPHS grants EY-01243 and
EY-02619 from the National Eye Institute.

REFERENCES

1. Barker, L.A. and Mittag, T.W. (1973): FEBS Lett. 35:141-144.
2. Barker, L.A. and Mittag, T.W. (1975): J. Pharmacol. Exp. Ther. 192:86-94.
3. Bito, L.Z., Hyslop, K. and Hyndman, J. (1967): J. Pharmacol. Exp. Ther. 157:159-169.
4. Boberg-Ans, J. (1955): Brit. J. Opthal. 39:705-726.
5. Bolt, Beranek & Newman, Inc. (1979): Public Procedures Notebook: A Program Exchange for PROPHET Users. The Chem. Biol. Info. Handling Prog. NIH.
6. Brucke, H. von (1938): Klin. Wochenschrift. 17:420-422.
7. Brucke, H. von, Hellauer, H.F. and Umrath, K. (1949): Ophthalmologia 117:19-35.
8. Gnadinger, M.C., Heiman, R. and Markstein, R. (1973): Exp. Eye Res. 15:395-399.
9. Harris, L.S., Mittag, T.W., Denmark, L.W., Cohn, K. and Galin, M.A. (1972): Exp. Eye Res. 13:1-8.
10. Harris, L.S., Shimmyo, M. and Mittag, T.W. (1973): Arch. Ophthalmol. 89:49-51.
11. Korczyn, A.D., Kloog, Y., Heron, D.S., Sachs, D.I. and Sokolovsky, M. (1979): Invest. Ophthal. Vis. Sci. 18: Suppl. 189.
12. Kornblueth, W. (1949): Amer. J. Opthalmol. 32:781-790.
13. Masland, R.H. and Livingstone, C.J. (1976): J. Neurophysiol. 39:1210-1219.
14. Mindel, J.S. and Mittag, T.W. (1976): Invest. Ophthalmol. 15:808-814.
15. Mindel, J.S. and Mittag, T.W. (1977): Exp. Eye Res. 24:25-33.
16. Mindel, J.S. and Mittag, T.W. (1978): Exp. Eye Res. 27:359-364.
17. Mittag, T.W. (1979): Invest. Ophthal. Vis. Sci. 18:Suppl. 189.
18. Mittag, T.W., Ehrenpreis, S. and Hehir, R.M. (1971): Biochem. Pharmacol. 20:2263-2273.
19. Mittag, T.W., Ehrenpreis, S. and Patrick, P. (1971): Arch. Int. Pharmacodyn. Ther. 191:270-278.
20. Mittag, T.W., Tormay, A. and Massa, T. (1978): Mol. Pharmacol. 14:60-68.
21. Williams, J.D. and Cooper, J.R. (1965): Biochem. Pharmacol. 14:1286-1289.
22. Yamamura, H.I. and Snyder, S.H. (1974): Mol. Pharmacol. 10: 861-868.
23. Zander, E. and Weddell, G. (1951): J. Anat. 85:68-99.

LIGHT EVOKED RELEASE OF ACETYLCHOLINE FROM THE RABBIT RETINA IN VIVO

M.J. Neal and S.C. Massey

Department of Pharmacology, The School of Pharmacy
University of London
Brunswick Square, London WC1N 1AX, England
*Present Address: Department of Neurobiology and Anatomy
The Medical School, University of Texas Health Sciences
Houston, Texas 77030 USA

INTRODUCTION

Acetylcholine (ACh) is probably a synaptic transmitter substance in the retina of most, if not all, vertebrates (17,35). However, in spite of a wealth of biochemical (3,22,37,44,45), pharmacological (16,25,38,43,49,50) and histochemical (4,27,39,46) evidence which supports a transmitter role for ACh in the retina, there has only been one demonstration of ACh release in response to flashes of light (26).

The release of a substance in response to appropriate physiological stimulation is an important criterion in establishing a neurotransmitter, and with this in mind, we have studied the effect of light on the release of ACh from the retina of the anesthetized rabbit. In addition, since GABA is also thought to be a retinal transmitter (36), we have attempted to study the interaction of the presumed GABAergic and cholinergic systems by examining the effect of GABA and GABA antagonists on the light evoked release of ACh from the retina. Preliminary reports of this work have been published (28,29).

METHODS

Eye-Cup Preparation

Rabbits were anesthetized with urethane (1.5 g/kg i.p.) and the
head was clamped so that one eye faced upwards. This eye was sutured
to a ring for support and then the cornea, iris and lens were removed. The vitreous was removed by gentle traction with tissue
"wicks" using a dissection microscope. Any over-vigorous traction
caused immediate retinal detachment and preparations in which detachment or hemorrhage occurred were rejected. The eye-cup was
filled with Krebs bicarbonate Ringer (500 µl) containing [^{3}H]-choline
(Ch) (10 µM) and this solution was left in contact with the retina
for 30 min. The medium containing [^{3}H]-Ch was removed, and the
retina was washed for 60 min by continuous irrigation with fresh
medium containing eserine sulphate (30 µM). Eserinized medium
(500 µl) was then placed in the eye-cup using a syringe mounted on
a micro-manipulator and the medium was replaced at 10 min intervals.
The [^{3}H]-ACh and [^{3}H]-Ch in the resulting samples were separated by
high voltage electrophoresis and measured by liquid scintillation
counting (42).

Electroretinography

The ERG was recorded in all experiments and preparations in
which a satisfactory ERG could not be obtained were abandoned. Low
resistance electrodes (balsa wicks soaked in saline/agar) were connected via silver/silver chloride electrodes to an amplifier and
the ERG could be monitored on an oscilloscope or averaged and drawn
out by a potentiometric pen recorder.

Light Stimulation

Most experiments were performed in dim red light and the retina
was dark adapted for 60 min before samples were collected from the
eye-cup. The retina was stimulated for 10 min periods by flashes
of light from a xenon lamp (Devices photic stimulator, 3182). The
retinal illuminance was increased twice during the period of stimulation in an attempt to obtain the maximum firing rate of ganglionic
cells (25). The characteristics of the Devices photic stimulator
result in a frequency dependent illuminance. At 3 Hz, which was the
frequency used most often, the retinal illuminance was started at
1.6 lux, this was increased to 5.2 lux at 3 min and to 16 lux at
6 min. In more recent experiments, a quartz halogen lamp was used

and flashes (25% duty cycle) were produced by an electronically
controlled shutter.

RESULTS

Effect of Light Stimulation
on the Release of [^{3}H]-Ch and [^{3}H]-ACh

In the dark adapted retina there was a steady spontaneous
resting release of radioactivity. Electrophoretic analysis of the
samples revealed that about 80% of the spontaneous release was due
to [^{3}H]-Ch and only 20% was [^{3}H]-ACh (Fig. 1); no other radioactive
metabolites, such as betaine or phosphorylcholine were detected.

Light stimulation of the retina at a frequency of 3 Hz caused
a prompt increase in the release of radioactivity (approximately
2 times the resting release) which was almost entirely accounted
for by an increase in the efflux of [^{3}H]-ACh (Fig. 1). Thus, light
stimulation increased the release of [^{3}H]-ACh by 4.10 $\pm$ 0.29 (mean
$\pm$ S.E.M.; n = 20) times the spontaneous resting release while the
efflux of [^{3}H]-Ch increased by only 1.02 $\pm$ 0.02 (mean $\pm$ S.E.M.;
n = 20) times the resting release.

This light evoked release was not apparent if an ERG could not
be recorded, but in a functional eye-cup preparation it could be
repeated as many as 8 times after a 30 min recovery period. In the
subsequent experiments only the results with [^{3}H]-ACh are described
since no significant changes in the efflux of [^{3}H]-Ch occurred.

Effect of Calcium Free Medium
on the Light Evoked Release of [^{3}H]-ACh

The light evoked release of [^{3}H]-ACh was calcium dependent.
Thus, when the medium in the eye-cup was replaced with calcium free
medium containing 20 mM magnesium chloride, the evoked release of
[^{3}H]-ACh was abolished being only 0.78 $\pm$ 0.05 (mean $\pm$ S.E.M.; n =
4) times the resting release (Fig. 1). On return to normal medium,
the light evoked response recovered.

Exposure of the retina to calcium free, high magnesium medium
produced striking changes in the ERG. At the lowest light intensity

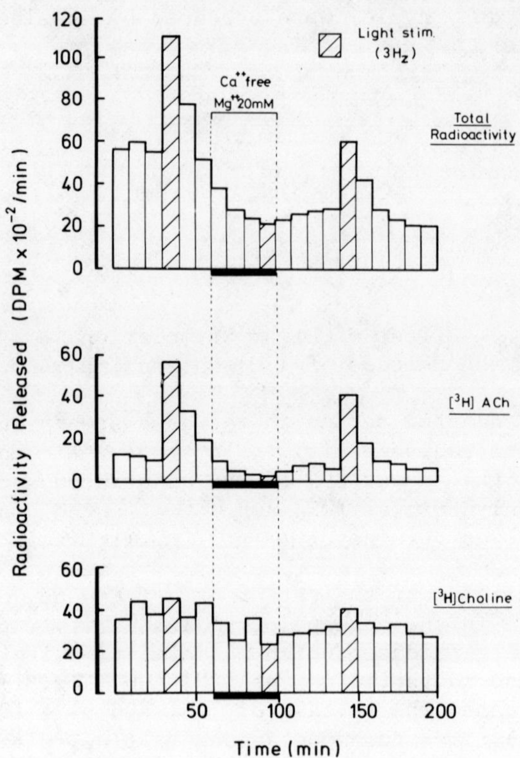

FIGURE 1: Effect of light flashes (3 Hz, hatched columns) on the
release of radioactivity, [³H]-ACh and [³H]-Ch from the
rabbit retina in vivo. In Ca-free medium the light
evoked release of ACh was abolished.

the ERG was completely abolished while at the two higher intensities
the isolated a-wave clearly remained. On returning to ordinary
Krebs medium, a normal series of ERGs could again be recorded.

Effect of Stimulus Frequency
on Light Evoked [³H]-ACh Release

The effect of different stimulation rates on the release of
[³H]-ACh is illustrated in Fig. 2 and summarized in Fig. 3. The
light evoked release of [³H]-ACh increased with frequency from 0.5
to 3 Hz but no further increase was observed at 10 Hz, while at
30 Hz the evoked release was not significantly different from the

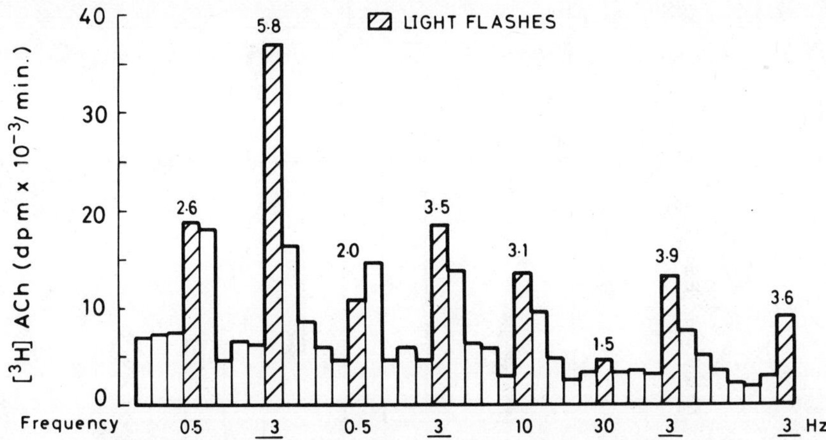

FIGURE 2: Effect of frequency on light evoked release of [3H]-ACh.
The numbers under the hatched columns represent the light
stimulation frequency. When the same frequency was repeated in an experiment the mean was calculated and used
as one result in Fig. 3. The figures above the columns
show the peak release/mean resting release of the two
preceding samples.

spontaneous resting release. ERG recordings revealed that up to
2 Hz the shape of b-wave was unchanged but, except for the first
response, a reduction in amplitude was evident at 4 Hz. As the
stimulus frequency was increased, the amplitude of the b-wave was
further reduced until it was barely visible at 16 Hz and absent at
30 Hz. These two higher frequencies still produced a normal first
response as well as a small negative off response.

Effect of Light Adaptation
on Evoked [3H]-ACh Release

In four separate experiments, the spontaneous resting release
of [3H]-ACh was apparently not affected by continuous background
illumination of 1 to 5 log units greater than the flashing light
stimulus. However, the light evoked release of [3H]-ACh produced
by flashes (3 Hz) was reduced by 34% (mean of 2 experiments), when
the background illumination was approximately 1 log unit greater
than the stimulus, and was abolished when the background illumination

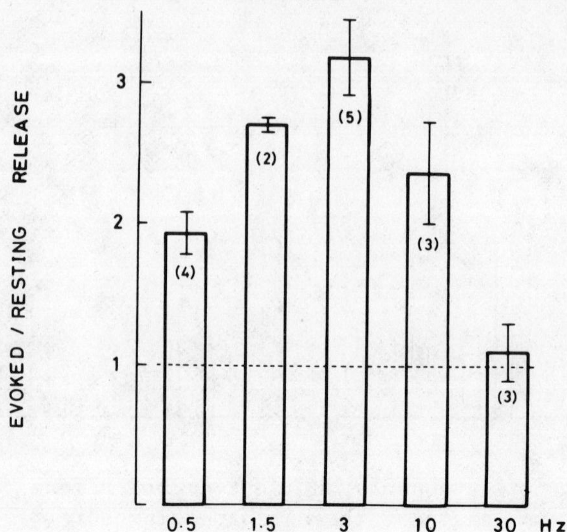

FIGURE 3: Effect of frequency on light evoked release of $[^2H]$-ACh.
The results are expressed as the relative efflux coefficient (peak release/mean release of preceding two
samples). Each result is the mean ± S.E. mean of 2 to
4 experiments and the result for each experiment was itself the mean of 1 to 4 determinations.

was raised a further 4 log units. In the latter case, 3 Hz light
evoked release returned after 30 min dark adaptation.

Effect of GABA on Evoked $[^3H]$-ACh Release

There is much inferential evidence that GABA may be a transmitter in the retina (36) and as GABA may be involved in presynaptic inhibition (10) for which there is strong anatomical evidence
in the retina (11,47), we have examined the effects of GABA and
GABA antagonists on the release of $[^3H]$-ACh.

Exposure of the dark adapted retina to GABA (1 mM) had little
effect on the spontaneous resting release of ACh, but the light
evoked release (3 Hz) was reduced by more than 50% (Fig. 4). Thus,
the evoked release in controls = 4.62 ± 0.569, while in the presence

of GABA the evoked release = 2.18 $\pm$ 0.274 times the resting release, mean $\pm$ S.E.M. of 4 separate experiments, p < 0.05. The normal response returned when the medium containing GABA was replaced with normal medium (Fig. 4).

In contrast to its inhibitory effect on ACh release, GABA produced a striking increase in the amplitude of the ERG. This was particularly noticeable at lower light intensities when the amplitude of the b-wave was sometimes double that seen in the absence of GABA.

Effect of GABA Antagonists
on Evoked [³H]-ACh Release

Exposure of the dark adapted retina to picrotoxin (20 μM) (Fig. 5) or bicuculline (5 μM)(not illustrated) caused striking increases in the spontaneous resting release of [³H]-ACh to 4.1 times and 5.0 times the previous resting releases, respectively (each result the average of 3 experiments). The drugs did not increase the release of [³H-]-Ch from the retina.

The light evoked release (3 Hz) of [³H]-ACh in the presence of either bicuculline or picrotoxin was approximately the same as that seen in the absence of the drugs, i.e. about 3 times greater than that seen in the two collection periods immediately before stimulation. However, compared with the spontaneous release in normal medium containing no GABA antagonist, the light evoked release of [³H]-ACh was up to 12 times the resting release, the highest seen in this series of experiments.

Picrotoxin decreased the amplitude of the ERG. Initially, this trace appeared qualitatively similar to the ERG recorded in calcium free medium as the a-wave was relatively unaffected. However, the c-wave persisted and at high intensities a small b-wave could be seen. Paradoxically, bicuculline did not alter the ERG.

In three experiments, 1 mM GABA was applied to the retina in the presence of 20 μM picrotoxin. Even at this low concentration, picrotoxin abolished the inhibitory effect of GABA on the light evoked release of [³H]-ACh (Fig. 5). The effect of GABA on the ERG was also abolished.

In contrast to the GABA antagonists, the glycine antagonist, strychnine (1 μM) had no effect on either the spontaneous or light evoked release of [³H]-ACh.

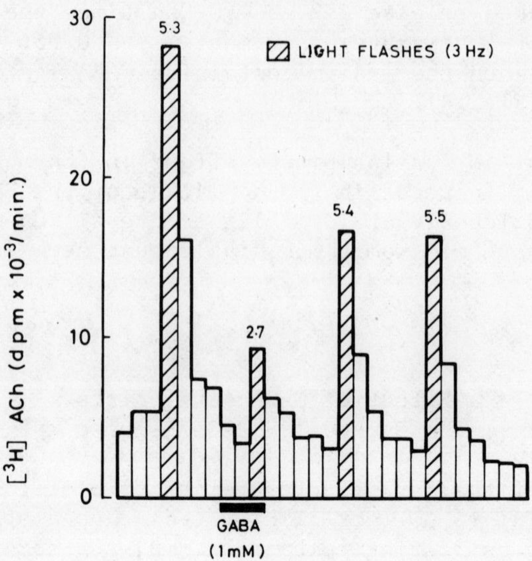

FIGURE 4: Experiment showing the effect on [^{3}H]–ACh release of exposing the retina to medium containing GABA (1 mM).

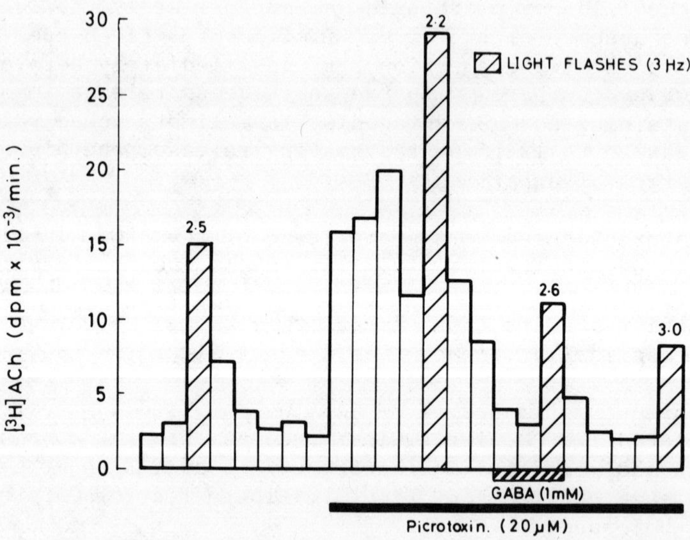

FIGURE 5: Experiment showing effect of picrotoxin on [^{3}H]–GABA release. The GABA antagonist greatly increased the spontaneous resting release of [^{3}H]–ACh and abolished the inhibitory effect of GABA on the light evoked release.

DISCUSSION

The present experiments, which confirm and extend the work of
Masland and Livingstone (26) show that stimulation of the rabbit
retina by flashes of light causes a calcium dependent increase in
ACh release. Exposure of the retina to calcium free medium also
abolished the ERG, except the a-wave which is thought to be a photoreceptor response (5,12,41), indicating that inhibition of synaptic
transmission across the retina (8,13) is associated with the aboli-
tion of the light evoked release of ACh.

The light evoked release of ACh increased with frequency in
the range 0.5 to 3 Hz but above 10 Hz, the evoked release diminished
in parallel with the amplitude of the b-wave, and at 30 Hz no evoked
release of ACh was observed. Although this reduction in evoked
release at higher frequencies may, at least partly, involve changes
in adaptation, more recent experiments using a photic stimulator
with a 25% duty cycle revealed similar changes in the evoked release of ACh with frequency. The failure of high frequencies to
elicit ACh release is more likely to be due to effects on transmission at neurones more distal to those which are believed to be
cholinergic in the rabbit retina.

The present study does allow certain tentative conclusions in
regard to the identity of the cholinergic neurones. The fact that
the efflux of ACh from the retina is increased at all implies that
the cholinergic neurones are depolarized by light. As the distal
neurones of the retina, i.e., the photoreceptors, horizontal and
some bipolar cells, respond to light stimulation with hyperpolari-
zations (19,51) and ganglion cells are always postsynaptic (14),
this strongly suggests that the cholinergic neurones are amacrine
cells or depolarizing bipolars. Furthermore, it appears that a
flashing stimulus is necessary for the light evoked release of ACh
from the retina (as for dopamine release [20]), suggesting that the
cholinergic neurones release ACh in response to the ON and/or OFF
portion of the stimulus.

Cells giving a sustained (depolarizing) response to light should
exhibit a prolonged release of neurotransmitter during continuous
light stimulation, whereas transient cells, which give a brief
response and then return to near their resting potential after the
onset of continuous light, should release little neurotransmitter.
However, a flashing stimulus, causing repeated depolarizations
should evoke a maximum release from transient cells, as observed in
these experiments, indicating that the cholinergic neurones may be
transiently responding cells.

Photoreceptor, horizontal and both classes of bipolar cells
give sustained responses to light but some amacrine and ganglion

cells respond with transient depolarizations (19,51). Although the
sustained release of [3H]-ACh from a small population of depolarizing bipolar cells may have gone unnoticed against the background
release during continuous light stimulation, it seems probable that
the majority of cholinergic neurones may be transiently responding
amacrine cells, which respond best to the onset and termination of
a stimulus (18).

The conclusion that certain amacrine cells in the rabbit retina
are cholinergic is consistent with many of the studies referenced
in the Introduction. In particular, acetylcholinesterase has been
localized in amacrine cells (39) and [3H]-Ch has been shown by autoradiography to be accumulated by certain amacrine cells of rabbit
retinae incubated under ACh conserving conditions but not by retinae
exposed to a flashing light (27).

In certain non-mammalian species such as the turtle, mudpuppy
and goldfish, there is evidence that ACh may be a transmitter in
some photoreceptors (22,45,46). However, in the carp retina, ACh
and eserine are without effect on horizontal cells (33).

The finding that the light evoked release of ACh was significantly reduced by GABA, suggests the possibility of inhibitory
GABAergic neurones presynaptic to the cholinergic neurones in the
retina. The possibility that GABA merely blocked transmission in
the outer plexiform layer is unlikely since the amplitude of the
ERG was enhanced by GABA. The suggestion that GABA interacts with
cholinergic neurones in the inner plexiform layer is consistent
with much evidence suggesting that GABA may be an amacrine cell
transmitter substance (36). However, it should be noted that species
variation may be important, because although in the rabbit retina
the GABAergic neurones appear to be mostly restricted to the proximal half of the retina, in other non-mammalian retinae, GABA may
also be a transmiter at horizontal cell synapses (23,24). In the
cat [3H]-GABA is accumulated by interplexiform cells as well as
amacrine cells (34).

GABA is thought to be a major inhibitory transmitter in the
central nervous system (21) and may have a presynaptic as well as
postsynaptic effect (10). Although there is strong anatomical
evidence for presynaptic inhibition in the inner plexiform layer
of the retina (11,47) our present experiments cannot differentiate
between these two possibilities.

The GABA antagonists, picrotoxin and bicuculline, both abolished
the inhibitory effect of GABA on the light evoked release of [3H]-
ACh, but only picrotoxin antagonized the enhancing effect of GABA
on the ERG, possibly reflecting some difference in the mechanism of
the two antagonists. More importantly, both antagonists caused a

large increase in the spontaneous resting efflux of $[^3H]$-ACh in the absence of light stimulation implying that there is a constant dark release of GABA which inhibits cholinergic neurones in the inner plexiform layer.

This suggestion is supported by intracellular recording from amacrine cells in the mudpuppy whose transient responses to light were enhanced by bicuculline and picrotoxin (31). Some amacrine cells receive input from both types of bipolar cells (30) and hyperpolarizing bipolar cells are thought to maximally release an excitatory transmitter in the dark (9,15). Thus a possible mechanism for the dark release of GABA could be the dark depolarization of GABAergic amacrine cells receiving input from hyperpolarizing bipolar cells (31).

Further evidence for amacrine cell modulation by GABA comes from studies of the proximal negative response (PNR), thought to reflect the activity of transient amacrine cells (6). In frog retinae, this response was depressed by GABA, but enhanced by picrotoxin and bicuculline, which also removed the suppressive effect of a larger flashed spot, suggesting that the surround inhibition may be GABA mediated (7,32). From the effects of bicuculline on the intensity/response curve of the PNR, Mooney (32) has suggested that GABA may be involved in the transformation from sustained to transient signals which occurs in the inner retina.

GABA is inhibitory to ganglion cells (1,40,48), while picrotoxin and bicuculline enhanced their response (31) but abolished their directional sensitivity, thought to result from asymmetrical lateral inhibition by amacrine cells (12), suggesting a role for GABA in this mechanism (52). Physostigmine also excited, but abolished the directional sensitivity of rabbit ganglion cells, which then responded to stimulation in the null direction, in a manner indistinguishable from the effect of picrotoxin (2). The suggestion that retinal cholinergic neurones are tonically inhibited by GABA and thus excited by picrotoxin provides a simple explanation for this last observation, but the same effect could arise postsynaptically at the ganglion cells where ACh and GABA are excitatory and inhibitory respectively (25,31). These experiments cannot differentiate between the two mechanisms although it is possible that they may be complimentary. Glycine may also be an amacrine cell transmitter (36), but since strychnine did not affect the release of ACh, it seems unlikely that glycinergic neurones have an inhibitory input to the cholinergic neurones.

ACKNOWLEDGEMENTS

A part of this work was supported by the SRC, to whom we are grateful. S.C. Massey was a SRC Scholar.

REFERENCES

1.   Ames, A. and Pollen, D.A. (1969): J. Neurophysiol. 32:424-444.
2.   Ariel, M. and Daw, N.W. (1978): Neurosciences 4:619.
3.   Atterwill, C. and Neal, M.J. (1978): Exp. Eye Res. 27:659-672.
4.   Baughman, R.W. and Bader, C. (1977): Brain Res. 138:469-485.
5.   Brown, K.T. (1968): Vision Res. 8:633-677.
6.   Burkhardt, D.A. (1970): J. Neurophysiol. 33:405-420.
7.   Birkhardt, D.A. (1972): Brain Res. 43:246-249.
8.   Cervetto, L. and Piccolino, M. (1974): Science 183:417-418.
9.   Dacheux, R.F., Frumkes, T.E. and Miller, R.F. (1979): Brain Res. 161:1-12.
10. Davidson, N. and Southwick, C.A.P. (1971): J. Physiol. 219:689-708.
11. Dowling, J.E. (1968): Proc. Roy. Soc. Lond. B 170:205-228.
12. Dowling, J.E. (1970): Invest. Ophthalmol. 9:655-680.
13. Dowling, J.E. and Ripps, H. (1973): Nature 242:101-103.
14. Dubin, M.W. (1970): J. Comp. Neurol. 140:479-506.
15. Frumkes, T.E. and Miller, R.F. (1979): Brain Res. 161:13-24.
16. Glickman, R.D. and Adolph, A.R. (1979): Invest. Ophthalmol. (Suppl)18:p.33.
17. Graham, L.T. (1974): IN The Eye Vol. 6 (eds) H. Davson and L.T. Graham, Academic Press, New York,, pp. 283-342.
18. Hedden, W.L. and Dowling, J.E. (1978): Proc. Roy. Soc. Lond. B 201:27-55.
19. Kaneko, A. (1970): J. Physiol. 207:623-633.
20. Kramer, S.G. (1971): Invest. Ophthalmol. 10:438-452.
21. Krnjevic, K. (1974): Physiol. Rev. 54:418-540.
22. Lam, D.M.K. (1972): Proc. Nat. Acad. Sci. 69:1987-1991.
23. Lam, D.M.K. and Steinman, L. (1971): Proc. Nat. Acad. Sci. 68:2777-2781.
24. Marc, R.E., Stell, W.K., Bok, D. and Lam, D.M.K. (1978): J. Comp. Neurol. 182:221-246.
25. Masland, R.H. and Ames, A. (1976): J. Neurophysiol. 39:1220-1235.
26. Masland, R.H. and Livingstone, C.J. (1976): J. Neurophysiol. 39:1210-1219.
27. Masland, R.H. and Mills, J.W. (1979): J. Cell Biol. 83:159-178.
28. Massey, S.C. and Neal, M.J. (1978): J. Physiol. (Lond) 280:51-52p.
29. Massey, S.C. and Neal, M.J. (1979): J. Neurochem. 32:1327-1329.
30. Miller, R.F. and Dacheux, R.F. (1976): J. Gen. Physiol. 67:639-659.
31. Miller, R.F., Dacheux, R.F. and Frumkes, T.E. (1977): Science 198:748-750.
32. Mooney, R.D. (1978): Brain Res. 145:97-115.
33. Murakami, M., Ohtsu, K. and Ohtsuka, T. (1972): J. Physiol. 227:899-913.
34. Nakamura, Y., McGuire, B.A. and Sterling, P. (1980): Proc. Nat. Acad. Sci. 77:658-661.

35. Neal, M.J. (1976): IN Transmitters in the Visual Process.
       (ed) S.L. Bonting, Pergamon Press, Oxford, pp. 127-143.
36. Neal, M.J. (1976): Gen. Pharmacol. 7:321-332.
37. Neal, M.J. and Gilroy, J. (1975): Brain Res. 93:548-551.
38. Negishi, K., Kato, S., Teranishi, T. and Laufer, M. (1978):
       Brain Res. 148:85-93.
39. Nichols, C.W. and Koelle, G.B. (1968): J. Comp. Neurol. 133:
       1-15.
40. Noell, W.K. (1959): Amer. J. Ophthalmol. 48:347-370.
41. Penn, R.D. and Hagins, W.A. (1969): Nature 223:201-205.
42. Potter, L.T. and Murphy, W. (1967): Biochem. Pharmacol. 16:
       1386-1388.
43. Pourcho, R.G. (1979): Vis. Res. 19:287-292.
44. Ross, C.D. and McDougal, D.B. (1976): J. Neurochem. 26:521-526.
45. Sarthy, P.V. and Lam, D.M.K. (1979): J. Neurochem. 32:455-461.
46. Schwartz, I.R. and Bok, D. (1979): J. Neurocyt. 8:53-66.
47. Sosula, L. and Glow, P.H. (1970): J. Comp. Neurol. 140:
       439-478.
48. Straschill, M. and Perwein, J. (1969): Pflug. Arch. 312:45-54.
49. Straschill, M. and Perwein, J. (1969): Pflug. Arch. 339:
       289-298.
50. Vogel, Z., Maloney, G.J., Long, A. and Daniels, M.P. (1977):
       Proc. Nat. Acad. Sci. 74:3268-3272.
51. Werblin, F.S. and Dowling, J.E. (1969): J. Neurophysiol. 32:
       339-355.
52. Wyatt, H.J. and Daw, N.W. (1975): Science 191:204-205.

BIOCHEMICAL AND PUPILLOGRAPHIC STUDIES OF GUINEA PIG IRIS

DURING ACETYLCHOLINESTERASE INHIBITION

F. Fonnum, N.E. Søli, P.K. Opstad, M. Opsahl and
R. Lund-Karlsen

Norwegian Defence Research Establishment
Division for Toxicology, P.O. Box 25
N-2007 Kjeller, Norway

INTRODUCTION

The effect of cholinergic drugs on the iris function is well
known. Acetylcholinesterase (AChE) inhibitors such as soman leads
to miosis, and atropine-like compounds give myadriasis. In the
present communication we describe the miotic effect of single and
multiple doses of soman to the eye. It will be show that transmitters other than acetylcholine (ACh) must be involved in the
fine regulation of the pupillary function.

METHODS AND MATERIALS

These experiments were carried out on albino guinea pig,
weighing 300-400 g.   Soman was dissolved in 0.9% NaCl and instil-
led into the conjunctival sac with a microsyringe (Hamilton) in a
volume from 1 to 5 µl. Systemic intoxication of animals was obtained by subcutaneous injection of soman (1-5 µg/100 g body weight).
External and total AChE was assayed as described by Lund-Karlsen
and Fonnum (8).

Muscarinic receptor binding was carried out with quinuclio-
dinyl/benzylate (QNB), at a final concentration of 5 nM, as described by Lund-Karlsen (7). The catecholamines were determined
by a radiochemical method (3).

The pupillography was carried out in darkness, and after 30 min dark adaptation, the eye was illuminated with infra-red light. The pupillary diameter was studied with a Grundig TV camera (model FA-70) connected to a Sony video tape recorder (Model AV-3670 CE) and an ordinary Sony TV monitor. The pupillary diameter is given in mm measured on the TV screen and is not the true diameter size. The method is described in detail by Lund-Karlsen and Søli (9). If not otherwise specified, the pupillary diameter was measured 1 hr after application of soman.

RESULTS

Neurotransmitter Interaction in Iris

The iris muscle is composed of two parts, the sphincter and the dilator muscle (Fig. 1). The sphincter muscle is responsible for the contraction of the iris when the eye is exposed to light, and is mainly innervated by cholinergic fibers from the ciliary ganglion (4). In agreement with this, the highest levels of choline acetyltransferase (CAT) and of the muscarinic receptor are found in this part of the iris (Table 1). In addition, both histo-fluorescent (11) and biochemical (Table 1) studies have shown the presence of noradrenergic fibers from superior cervical ganglion to the sphincter muscle.

The main proportion of the noradrenergic fibers ends in the dilator part of the muscle (11)(Table 1). In some cases it was found that the same axon projected both to the dilator and the sphincter (10). The dilator muscle also contains some CAT, but it is not known to what extent this activity is due to cholinergic axons traversing the tissue on its way to the dilator (Table 1). In discussing the effect of AChE inhibitors on the function of the iris, it is important to remember that both the cholinergic input from the ciliary ganglion and the noradrenergic input from the superior cervical ganglion are under cholinergic control (Fig. 1).

Total and External AChE Activities

During this study we observed the different behavior of the total AChE, and the so-called external AChE (8). The contribution of buturylcholinesterase (BuChE) was excluded through the use of the specific inhibitor iproniazid. Total AChE activity is defined as the rate of hydrolysis of 1.3 mM ACh by an iris homogenate. The

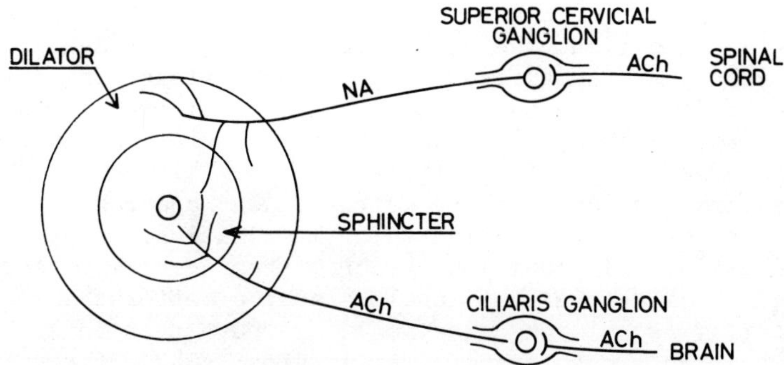

FIGURE 1: The localization of fibers modulating the iris function.
NA = noradrenaline; ACh = acetylcholine.

TABLE 1:   Distribution of transmitter parameters in iris.

	Sphincter	Dilatator
CAT	1.6	0.46
AChE	74	56
Muscarinic Receptor	+++	0
DA	0.7	0.7
NA	35	50
A	0.1	0.1

Activities per mg net wt.

external ACh activity is the rate of hydrolysis of 1.3 mM ACh by
the intact tissue. In a normal iris the external AChE activity
accounts for about 15% of the total AChE activity. The external
AChE activity in smooth muscle is linear both with regard to time
and tissue weight. The external AChE is present both in the sphincter and dilator part of the muscle and is not affected by removal
of the ciliary ganglion, whereas the total AChE is reduced by 40%.
This indicates that the external AChE is more closely associated
with the activity of the muscle membrane than with the presynaptic
structures; topically applied soman inhibits total and external
AChE to the same extent (15). The spontaneous recovery of external
AChE is faster than the recovery of total AChE and resembles more
closely the recovery of function (8). In addition, after PAM-in-
duced recovery of iris inhibited by DFP, the recovery of external
AChE followed the recovery of muscle function better than total
AChE (Mittag, personal communication). It has, therefore, been
suggested that in smooth muscle, the external AChE reflects in
some way the functional part of AChE (5,8,12).

Correlation between Pupillary Function and AChE Inhibition

    Drug effect on pupillary function can be studied by measuring
the pupillary diameter. In these studies the parameter was measured by means of infra-red television recording. Topical application of soman, an irreversible AChE inhibitor, to the eye, leads
to a striking correlation between the reduction in pupillary diameter and AChE inhibition (Fig. 2); both external and total AChE
were inhibited to the same extent (15). It was also surprising to
detect a significant reduction of pupillary size as soon as 20%
inhibition of AChE; this contrasts with the situation in striated
muscle, where contractions are not affected before 80% of the
activity is inhibited (1).

However, we found several situations showing that the pupillary function is not only controlled by the cholinergic fibers to
the iris during AChE inhibition but: 1) although AChE was completely, or almost completely inhibited, light flashes produced a
further decrease in iris contraction (2.4 mm pupil compared to
1.2 mm); 2) daily application of soman, which was always accompanied by complete AChE inhibition, did lead to less pupillary
contraction (2.4 mm pupillary diameter at day 1, 3.6 mm at day 6);
3) subcutaneous injection of lethal doses of soman did not result
in pupillary contraction, although both the total and external
AChE were inhibited by more than 90%. This phenomenon has also
been observed for humans (14); and 4) the contribution of the noradrenergic fibers to the control of pupillary diameter was studied

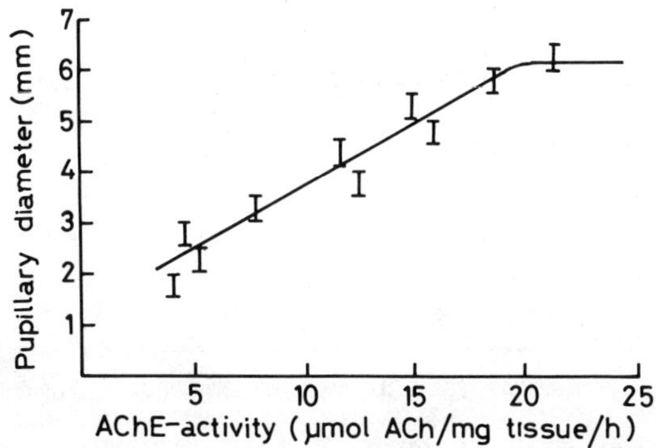

FIGURE 2: The correlation between pupillary diameter and total AChE activity.

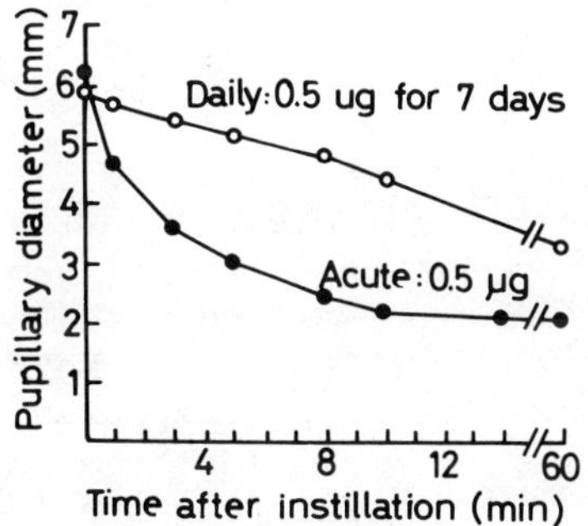

FIGURE 3: The dynamic response to iris after topical application of soman to the eye. Single instillation (●); daily administration for 1 week (O)

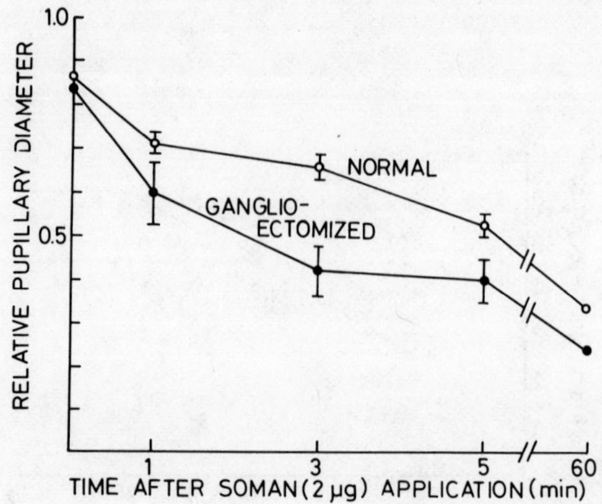

FIGURE 4:   The dynamic response to iris after topical application
of soman to the eye. Normal iris (O); iris ganglio-
ectomized 1 week before (●).

by superior cervical ganglionectomy. After degeneration of the
noradrenergic input the pupillary diameter was 8.5 ± 0.2 mm compared to 9.0 ± 0.1 mm in normal iris. After complete inhibition
of AChE the diameter of this pupil was reduced to 2.5 ± 10.2 mm
instead of 3.8 ± 0.3 mm in unoperated animals.

Dynamic Responses of Pupillary Function

The different miotic responses to similar degrees of AChE
inhibition achieved through different experimental models (single
or repeated topical instillation of soman, subcutaneous injection
of soman, the response of a normal and noradrenaline-deprived iris
to soman intoxication) was reflected even more markedly in the time
course of pupil constriction. Thus, repeated daily instillation of
a small dose of soman in the eye leads to a much slower response
than a single instillation of the same dose (0.5 μg) (Fig. 3).
Single application (1 μg) of soman in the gangliectomized iris
leads to a faster contraction than in unoperated iris (Fig. 4).

The return of the pupillary diameter to normal size does not
correspond well with the return of total AChE activity of this
iris. After a single dose of soman the animals recovered to normal
pupillary size within 24 hr; the total AChE recovered only from
about 5 to 20% in 24 hr. The external AChE corresponded somewhat

better since it recovered from less than 5% to about 40% in 24 hr
(8). Although repeated instillation with soman did not change the
rate of recovery of either the external or total AChE, the pupillary
diameter recovered more rapidly than after repeated instillations
than after a single instillation (6 hr as compared to 24 hr).

Receptor Binding

The QNB binding in iris was studied in order to see whether
changes in the muscarinic receptor could be responsible for some
of the changes in pupillary function. The number of QNB binding
sites in iris were not changed after denervation of the cholinergic
input (7). An increase in the number of muscarinic receptors can,
therefore, not be the explanation for denervation supersensitivity
in this preparation (6,13). The number of QNB binding (5 nM) sites
did not change as a result of single and multiple instillations
of soman in the eye.

DISCUSSION

The present paper shows that there exists a strict correlation
between AChE activity of the iris and the pupillary diameter in the
dark shortly after topical application of an anticholinesterase
such as soman. Even at almost total AChE inhibition, however, a
light flash will promote a further contraction of the pupil. It is
hard to believe that this light reflex may be caused by a further
ACh release during inhibition of iridial AChE. A better explanation may be an active relaxation of the dilator muscle driven by
noradrenaline release. Support for this hypothesis was gained by
the finding of a smaller pupillary diameter during AChE inhibition
in an iris deprived of its noradrenergic input.

When soman was administered systematically, iridial AChE was
inhibited without any effect on the pupillar size. Similarly, when
large topical doses of soman, accompanied by systemic effects, were
administered, then the pupillary diameter was reduced less than
expected. Systemic inhibition of AChE affects the pupillary function in several ways: 1) inhibition of iris AChE; 2) inhibition
of AChE in ciliary ganglion, and 3) inhibition of AChE in superior
cervical ganglion. Thus, both the cholinergic and noradrenergic
cells projecting to the iris are affected. At present we cannot
differentiate between the contributions from these alternatives.

After inhibition of iridial AChE some compensatory mechanisms
affected the pupillary response. The pupillary diameter returned
more rapidly to normal than both the total and external AChE. The
relaxation of the pupil is too rapid to be accounted for by the
recovery of AChE alone, but could be due to changes in ACh turnover
of the iris synapse, changes in the muscarinic receptor, increases
in activity of noradrenaline synapse, or changes in the co-existing
peptide transmitters.

With repeated topical instillations of soman to the eye, the
miotic response of the compound (both final pupillary diameter and
rate of contraction of pupil) were considerably reduced. Pupillary
size, however, recovered more rapidly. This phenomenon was first
detected after three days of treatment and was constant for many
days; similar findings have been observed on other species such as
the dog and the rabbit (2). In the latter cases the possibility
of systemic intexication was not excluded. Separate experiments
determined that these results were not caused by a reduced AChE
inhibition in the iris by soman or by an increased de novo synthesis
of the external or total AChE, or in a change in the number in
muscarinic receptors. However, it has recently been demonstrated
(Mittage, personal communication) that the so-called high affinity
QNB binding measured at 0.1 nM shows a decline; whereas QNB binding
at 5 nM is unaffected.

It is, in addition, possible that peptidergic fibers or co-
existence of peptides in the cholinergic or noradrenergic fibers
to the iris may affect its function. However, this possibility
has not yet been examined.

REFERENCES

1.   Barstad, J.A.B. (1960): Arch. Int. Pharmacodyn. 128:143-168.
2.   Bito, L.Z. and Banks, N. (1969): Arch. Ophthal. 82:681-686.
3.   De Prada, M. and Zurcher, G. (1976): Life Sci. 19:1161-1174.
4.   Ehinger, B. and Falck, B. (1966): Acta Physiol. Scand. 67:
         201-209.
5.   Harris, L.S., Shimmoy, M. and Mittag, T.W. (1973): Arch.
         Ophthal. 89:49-51.
6.   Kloog, Y., Sachs, D.I., Korczyn, A.D. Heron, D.S. and
         Sokolovsky, M. (1979): Biochem. Pharmacol. 28:1505-1512.
7.   Lund-Karlsen, R. (1978): Exp. Eye Res. 27:577-581.
8.   Lund-Karlsen, R. and Fonnum, F. (1977): J. Neurochem. 29:
         577-581.
9.   Lund-Karlsen, R. and Søli, N. (1979): Acta Opthal. 57:41-47.

10.   Malmfors, T. (1965): Acta Physiol. Scand. 64:377-382.
11.   Malmfors, T. (1967): Circ. Res. 20 & 21(Suppl.3):25-42.
12.   Mittag, T.W., Ehrenprise, S. and Hehir, R.M. (1971): Biochem.
          Pharmacol. 20:2263-2273.
13.   Sachs, D.I., Kloog, Y., Korczyn, A.D., Heron, D.S. and
          Sokolovsky, M. (1979): Biochem. Pharmacol. 28:1513-1518.
14.   Sim, V.M. (1975): IN Cholinergic Mechanisms (ed) P.G. Waser,
          Raven Press, New York, pp. 395-398.
15.   Søli, N.E., Lund-Karlsen, R., Opsahl, M. and Fonnum, F.
          (1980): J. Neurochem. 35:723-728.

ON THE IONIC MECHANISM OF PRESYNAPTIC MUSCARINIC RECEPTOR
ACTION IN RAT HIPPOCAMPUS

O. Nordstrom, A. Westlind, B. Hedlund, A. Unden
and T. Bartfai

Department of Biochemistry, Arrhenius Laboratory
Stockholm, Sweden

INTRODUCTION

Occupancy of the muscarinic acetylcholine receptor (mAChR)
with an agonist may trigger a number of different responses dependent on the localization of the receptor. The mAChR mediates depolarizing actions in Renshaw cells (10-12), in sympathetic ganglion (13) and hyperpolarizing actions in the heart (21). It activates synthesis of cGMP in a number of tissues (1) and inhibits
the β-receptor activated adenylate cyclase in heart (25). Some
of these actions take place on the presynaptic side and some on
the postsynaptic side of the synapse.

Despite recent advancement in our knowledge of the ligand
binding properties and distribution of the mAChR (2,7) we still
lack information on the ionic basis of its action. The aim of
this study is to approach this question by examining the muscarinic
inhibition of ^{3}H-acetylcholine (ACh) release from hippocampal
synaptosomes and slices (17,19,20).

Presynaptic Muscarinic Receptors

Release of ^{3}H-ACh from synaptosomes prepared from rat hippocampi is inhibited by muscarinic agonists that act on autoreceptors
localized on the same nerve ending that releases ^{3}H-ACh (17). This
conclusion was reached by studies on the effects of exogenous muscarinic ligands on release of ^{3}H-ACh from a dilute suspension of

nerve endings where the concentration of the endogenously released neurotransmitter was kept lower than 10^{-8} M. In such a preparation it is assumed that the studied pharmacological action originates from the exogenous ligands. Under these conditions carbachol and oxotremorine depress, and atropine enhances release of ^{3}H–ACh evoked by depolarization with K^+ (25–35 mM). This release is tetrodotoxin (TTX, 10^{-4} M) insensitive, indicating that impulse propagation is not involved in the presynaptic feedback regulation by muscarinic receptors that were consequently termed autoreceptors.

In order to obtain information on the ionic mechanism whereby the muscarinic autoreceptor influences release we studied release of ^{3}H–ACh as evoked by K^+, veratridine, ouabain, and the Ca^{++} iono-phore A23187. These agents induce release from synaptosomes or slices of the hippocampus by different mechanisms as indicated in Fig. 1.

It is known that K^+, veratridine or electrical nerve stimula-tion evoked release of ^{3}H–ACh require Ca^{++} influx, wherease the ouabain evoked release may be Ca^{++} independent. We re-examined the effect of these agents on release from hippocampal slices. Rat hippocampal slices, 0.4 mm thick, were loaded with 8 µM choline (Ch) with ^{3}H–Ch as a tracer at 37°C for 40 min in Krebs-Ringer buffer in the presence of 10 µM eserine to inhibit acetylcholinesterase (AChE) The slices were then placed in fresh medium and spontaneous efflux of ^{3}H–ACh and ^{3}H–Ch was studied for 15 min at 37°C. The drugs were then introduced and release examined by sampling medium at 5 min intervals for a period of 15 min. In some experiments ^{3}H–Ch and ^{3}H–ACh were separated by high voltage paper electrophoresis and in such cases ^{3}H–Ch accounts for about 20% of the evoked release. Data in Table 2 are expressed as the sum of $[^3H]$–ACh and $[^3H]$–Ch.

Table 1 shows the dependency of the release evoked by K^+, oua-bain and veratridine on the extracellular Ca^{++} concentration. As expected, K^+, veratridine and A23187 (data not shown) evoked re-lease was dependent on extracellular Ca^{++} while ouabain induced release was not fully abolished in the absence of Ca^{++} in the medium. In other systems, such as mouse diaphragm, Vizi (24) found that ouabain induced release was independent of extracellular Ca^{++} and he used this finding as an argument for a direct involvement of Na^+K^+-ATPase in the release (route [8] in Fig. 1) (23). After a 60 min incubation period with Ca^{++} free medium (+ EGTA, 1 mM) we could not fully verify this assumption and indeed in cortical slices Vizi (22) found some 20–30% of the release of ACh evoked by ouabain is Ca^{++} dependent.

Figure 2 summarizes our data on the activity of the presynaptic receptor under different release conditions. Atropine enhanced $[^3H]$–ACh + $[^3H]$–Ch efflux from hippocampal slices when release was

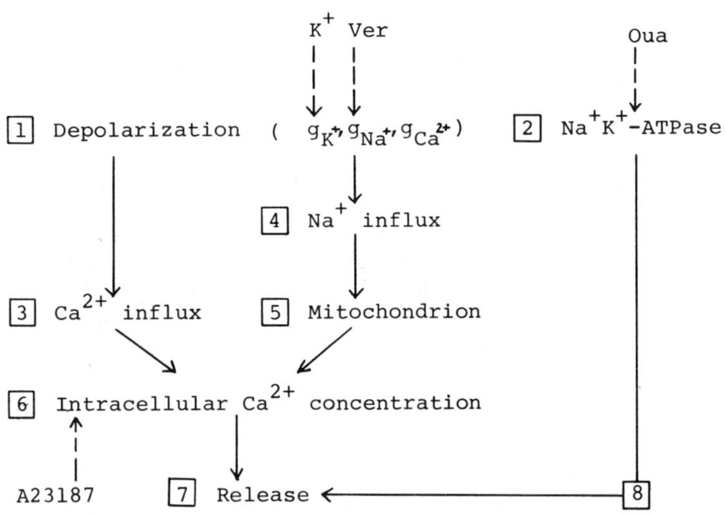

FIGURE 1: Schematic diagram of the ionic mechanisms of release. Release can be evoked by depolarization [1] at the arrival of an action potential. This depolarization (by K^+ or veratridine [Ver]) is known to involve changes in g_{K^+}, g_{Na^+} and subsequently leads to opening of the "late voltage dependent Ca^{++} channel," [3]. Influx of Ca^{++} as well as influx of Na^+ [4], that is known to trigger the liberation of Ca^{++} from the mitochondria [5], and leads to increased intraterminal Ca^{++} concentration [6], which triggers release of neurotransmitter [7]. Inhibition of Na^+-K^+-ATPase (by ouabain [Oua]) [2] also leads to release of neurotransmitters. This action is likely to be due to depolarization of the terminal based on passive Na^+ leakage when the pumping of Na^+ by the ATPase is blocked. Depolarization could then trigger events leading to release. However, numerous investigators believe that the activity of the enzyme may directly (vis. bypassing the events in the above scheme) influence neurotransmitter release (cf. for review see ref. 23). Arrows with broken lines indicate presumed site of action of drugs used to evoke release while studying presynaptic receptor mechanisms.

TABLE 1: Ca^{++} dependency of the $[^3H\text{-}ACh + {}^3H\text{-}Ch]$ release from rat hippocampal slices.

| | Percent of Basal Release | |
Agent	2.0 mM Ca^{++}	No Ca^{++} 1 mM EGTA
K^+ (25 mM)	270 ± 25(8)*	160 ± 30(8)
Veratridine (10^{-4} M)	304 ± 28(4)*	92 ± 24(4)
Ouabain (10^{-4} M)	342 ± 47(4)*	181 ± 24(4)*

* = $p < 0.01$ different from control (basal release: 2500 cpm/mg protein/10 min). Number in parentheses indicates number of experimental points.

TABLE 2: Interference of release evoking agents with binding of the muscarinic antagonists (QNB) to the receptor in membranes from rat forebrain.

Drug Added	Percent Specific Binding of ^3H-QNB (5 nM)
None	100
Veratridine (5 x 10^{-6} M)	34 ± 6*
Veratridine (5 x 10^{-6} M) + TTX (10^{-7} M)	51 ± 7*
Ouabain (10^{-4} M)	42 ± 7*
A23187 (30 µg/ml)	97 ±10
K^+ (25 mM)	105 ± 8

Binding was measured in a Krebs-Ringer buffer at 25°C for 60 min as described in Ref. 6. 100% Binding correspondes to 1.2 pmol ^3H-QNB/mg protein.   * = $p < 0.01$.

evoked by K^+ (25 or 35 mM). Under these circumstances carbachol (10^{-3} M) inhibited the release. Thus the presynaptic receptor was fully active. Its site of action may be on the regulation of Ca^{++} influx [3], of Na^+ influx [4], or anywhere later in the depolarization-release coupling sequence.

It is known that muscarinic agonists enhance K^+ efflux from smooth muscle (4). This would, in the case of the nerve terminal, lead to hyperpolarization and thereby reduce Ca^{++} influx. However, when depolarization of the slices or synaptosomes is induced by raising K^+ concentration, we effectively inhibit K^+ influx which makes involvement of g_K less likely.

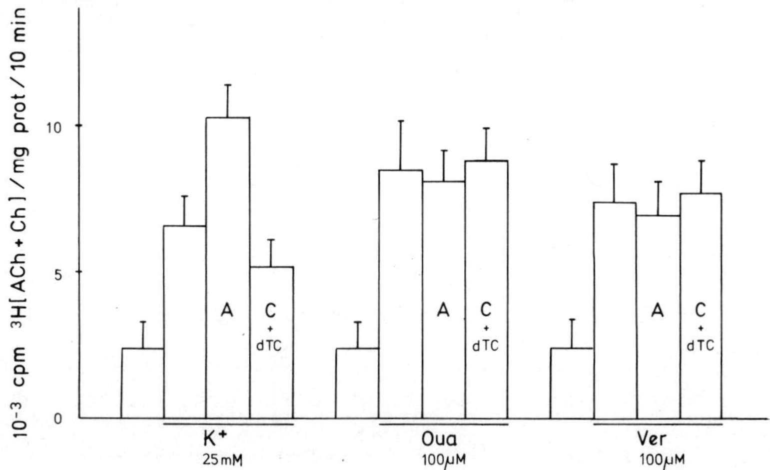

FIGURE 2: The effect of muscarinic agents, atropine (A, 10^{-6} M) and
carbachol (C, 10^{-3} M) on [^{3}H-ACh + ^{3}H-Ch] release evoked
by different agents. d-Tubocurarine (dTC, 10^{-5} M) was
added together with carbachol to block its nicotinic effects. The data are presented as mean $\pm$ S.D. for at
least 4 determinations.

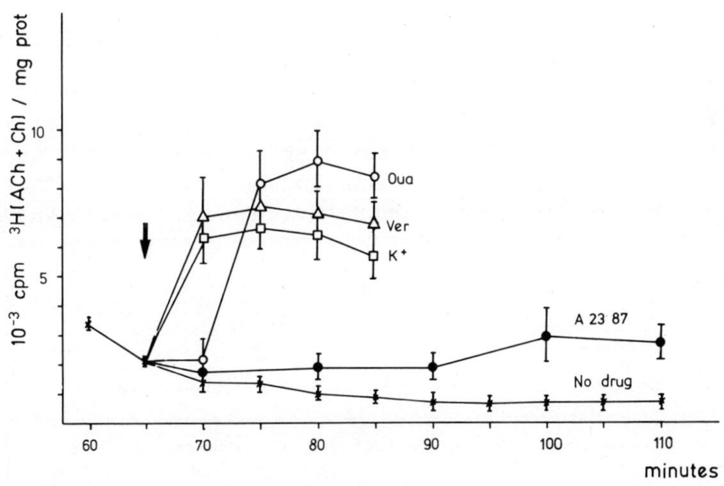

FIGURE 3: The time course of the [^{3}H-ACh + ^{3}H-Ch] release evoked by
different agents. Conditions used for loading and re-
lease are the same as those described in Table 1.

Changes of Na^+ conductance ($g_{Na}+$) as the ionic basis of muscarinic inhibition of release are possible. However, Na^+ ions do not seem to play any role in the muscarinic actions such as muscarinic receptor mediated increase of cGMP levels in NG10815 cells (Bartfai, unpublished results). Regulation of Ca^{++} channel or pump by the muscarinic receptor has already been suggested and may be operative in this context (18).

To examine involvement of Na^+ fluxes in the action of the presynaptic muscarinic receptor we studied veratridine (10^{-4} M) evoked release. This release is Ca^{++} dependent and rapid, similar to K^+-evoked release (cf. Fig. 3). Atropine or carbachol were ineffective in modulating the veratridine evoked release (Fig. 2). Before using this finding as an argument on the mechanism of release we examined our findings to see whether veratridine directly interacts with the muscarinic receptors. It was found that veratridine (5 μM) produces an 80% inhibition of the antagonist binding studied by ^{3}H-QNB (Table 2). This effect is only partly reversed by TTX (10^{-7} M) and probably due to irreversible damage of the membrane during the 60 min incubation with the alkaloid. However, for this long incubation we chose a twentyfold lower concentration (5 μM) than that used for the shorter (15 min) release experiments where 10^{-4} M veratridine was used. Inhibition of antagonist binding by veratridine was also observed by Luqumani et al.(14).

While direct interaction of the muscarinic receptor with veratridine prevents us from conclusively excluding the involvement of Na^+ channels in presynaptic muscarinic receptor action, the fact that both atropine and carbachol modulated K^+ evoked release in the presence of TTX (10^{-4} M) that would fully block Na^+ channels, argues against involvement of Na^+ channels.

Thus we turned to ouabain, a cardiac glycoside that inhibits Na^+K^+-ATPase, to study the release. Study of the dose response relationship indicated that ouabain induces maximal release of ^{3}H-ACh at 10^{-4} M concentration. Ouabain (10^{-4} M) induced release was much slower than that evoked by K^+ or veratridine (Fig. 3). This release was not modulated by atropine up to 10^{-5} M concentration. Carbachol was also ineffective in suppressing the release of ^{3}H-ACh + ^{3}H-Ch from hippocampal slices (Fig. 2).

We re-examined our findings to see whether direct interactions between the receptor and ouabain took place, thereby preventing detection of any presynaptic modulation. Table 2 shows that binding of ^{3}H-quinuclidinyl benzylate is inhibited by ouabain (10^{-4} M) to 60% since it binds to the receptor and may prevent agonist binding. Thus the ouabain evoked release would already be upward modulated. Alternatively, the release cannot be modulated by the muscarinic receptor since this is rendered inaccessible for both antagonists and agonists. The finding (Fig. 2) that carbachol (10^{-3} M) had no effect on ouabain evoked release argues for a blockade of the receptors toward the agonist site also.

Our results suggest that absence of muscarinic modulation of ouabain evoked release can be explained by the trivial finding that ouabain blocks this receptor. Thus these results do not yield evidence for ACh activated Na^+K^+-ATPase (9) in the regulation of release.

The use of Ca^{++} ionophore A23187 presents the possibility of evoking release (5,8,16) by bypassing changes in g_{Na^+}, g_{K^+}, $g_{Ca^{++}}$ and Na^+K^+-ATPase activity. This release, in our hands, was slow (cf. Fig. 3) and rather small (120–150%) even at high ionophore concentration (30 µg/ml). The ionophore evoked release was not enhanced by atropine (10^{-6} M). This finding is in agreement with data on ionophore evoked release of ACh in cortical slices (5). The data in Table 2 indicates that the ionophore , unlike ouabain and veratridine, does not inhibit ligand binding to the receptor. Therefore we interpret the lack of effect of muscarinic ligand on release evoked by A23187 as an indication that the regulation of intracellular Ca^{++} concentration is a step beyond muscarinic receptor control. The muscarinic receptor on the cholinergic nerve endings may regulate an event that leads to changes in intracellular Ca^{++} concentrations, but once A23187 is used to change this parameter, the effects of the receptor are bypassed. In a non-vertebrate preparation of synaptosomes from <u>Torpedo ocellata</u>, Michaelson et al. (15) indicated that oxotremorine inhibited the K^+, as well as the A23187 evoked release of ACh, and influenced a Ca^{++} dependent phosphorylation that is believed to be involved in the release. Thus, in that preparation, Ca^{++} influx may precede the action of the muscarinic autoreceptor (15).

In studies on ion movements in smooth muscle K^+ and Ca^{++} were suggested for muscarinic receptor action (3,4). Studies on muscarinic receptor mediated increases in cGMP levels have also implicated regulation of Ca^{++} movements (18). The present study provides no conclusive clue as to the identity of the ion movements involved in presynaptic receptor action, but allows for the possibility that the principal ion involved is Ca^{++}.

ACKNOWLEDGEMENTS

This work was supported by USPHS grants from NIMH and by the Swedish Medical Research Council.

REFERENCES

1. Bartfai, T., Study, R.E. and Greengard, P. (1977): IN Chol-
    inergic Mechanisms and Psychopharmacology (ed) D.J. Jenden,
    Plenum Press, New York, pp. 285-295.
2. Birdsall, N.J.M. and Hulme, E.C. (1976): J. Neurochem. 27:7-16.
3. Burgen, A.S.V. and Spero, L. (1968): Brit. J. Pharmacol. 34:
    99-115.
4. Burgen, A.S.V. and Spero, L. (1970): Brit. J. Pharmacol. 40:
    492-500.
5. Casamenti, F., Mantovani, P. and Pepeu, G. (1978): Brit. J.
    Pharmacol. 63:259-265.
6. Hedlund, B. and Bartfai, T. (1979): Mol. Pharmacol. 15:531-544.
7. Heilbronn, E. and Bartfai, T. (1978): Prog. Neurobiol. 11:
    171-188.
8. Holz, R.W. (1975): Biochim. Biophys. Acta 375:138-152.
9. Kometiani, Z.P. and Jariashvill, T.J. (1975): Proc. Acad. Sci.
    Georg. SSR, Biol. Ser. 1:190-198.
10. Krnjevic, K. and Phillis, J.W. (1963): J. Physiol. 165:274-304.
11. Krnjevic, K. and Phillis, J.W. (1963): J. Physiol. 166:296-327.
12. Krnjevic, K. and Phillis, J.W. (1963): J. Physiol. 166:328-350.
13. Kuba, K. and Koketsu, K. (1978): Prog. Neurobiol. 11:77-169.
14. Luqumani,Y.A., Bradford, H.F., Birdsall, N.J.M. and Hulme,
    E.C. (1979): Nature 277:481-482.
15. Michaelson, D.M., Avissar, S., Kloog, Y. and Sokolovsky, M.
    (1979): Proc. Nat. Acad. Sci. 76:6336-6358.
16. Michaelson, D.M. and Sokolovsky, M. (1978): J. Neurochem. 30:
    217-230.
17. Nordstrom, O. and Bartfai, T. (1980): Acta Physiol. Scand.
    108:347-383.
18. Study, R.E., Breakefield, X.O., Bartfai, T. and Greengard, P.
    (1978): Proc. Nat. Acad. Sci. 75:6295-6329.
19. Szerb, J.C. (1977): IN Cholinergic Mechanisms and Psychophar-
    macology (ed) D.J. Jenden, Plenum Press, New York, pp.49-60.
20. Szerb, J.C., Hadhazy, P. and Dudar, J.D. (1977): Brain Res.
    128:285-291.
21. Trautwein, W., Kuffler, S.W. and Edwards, C. (1956): J. Gen.
    Physiol. 40:135-145.
22. Vizi, E.S. (1972): J. Physiol. 226:95-117.
23. Vizi, E.S. (1975): IN Cholinergic Mechanisms (ed) P.G. Waser,
    Raven Press, New York, pp. 199-211.
24. Vizi, E.S. and Vyskocil, F. (1979): J. Physiol. 286:1-14.
25. Watanabe, A.M., McConnaughey, M.M., Strawbridge, R.A., Fleming,
    J.W., Jones, L.R. and Besch, H.R. (1978): J. Biol. Chem.
    253:4833-4839.

MUSCARINIC CHOLINERGIC RECEPTOR LOCALIZATION

BY RADIOHISTOCHEMISTRY

J.K. Wamsley, M.A. Zarbin, N.J.M. Birdsall* and M.J. Kuhar

Departments of Pharmacology and Experimental Therapeutics
and Psychiatry and the Behavioral Sciences
The Johns Hopkins University School of Medicine
Baltimore, Maryland 21205, and
*National Institute for Medical Research
The Ridgeway, Mill Hill, London NW7 1AA, England

INTRODUCTION

Following the development of binding methods for studying
receptors in vitro (18), it was possible to develop radiohistochemical methods in parallel for the light microscopic localization of receptors (9,19). While one can examine the anatomical
distribution of receptors by simple dissection and in vitro biochemical studies, the autoradiographic microscopic analyses provide several advantages over this latter approach. The two main
advantages are that the histochemical procedures provide a great
sensitivity in the measurement of receptors and an excellent level
of anatomical resolution in their localization. Thus, these studies provide an accurate, quantitative, high resolution assessment
of the distribution of receptors throughout the brain or other
tissues. These investigations are a basis for understanding the
widespread drug action in brain and elsewhere and are valuable
complements to other histochemical methods for mapping functional
neurotransmitter pathways.

Our laboratory first began its studies on the autoradiographic, light microscopic distribution of muscarinic receptors by utilizing in vivo injections of radiolabelled quinuclidinyl benzilate (QNB). Before beginning microscopic analysis we showed, in some detail, that intravenous administration of small doses of ^{3}H-QNB into rats resulted in a fairly selective localization of ^{3}H-QNB to muscarinic receptors in the brain in vivo (19). After completing these bio-chemical studies, it was possible to localize the binding sites with light microscopic autoradiography (10-12), utilizing techniques that prevent or minimize diffusion of drug from binding site (16). Fol-lowing these initial investigations on localizing the muscarinic receptor, we were able to extend our methodology to the localization of other receptors including opiate and dopamine receptors (5,7,8, 13).

While these studies, which relied on in vivo administration of provided valuable information and demonstrated the usefulness of this approach, there were several methodological limitations. First of all, the procedure involved administering radio label-led drugs systemically. Following the delivery of the drug to the brain by the circulation, there was a relatively good spe-cific binding of ligand to receptor. The problem is that this only worked for drugs that had a high affinity for the receptor. Thus, we could not examine the distribution of a wide variety of recep-tors because suitable ligands were not available. Because of this limitation, we have been spending a significant amount of time developing other methods that would permit greater flexibility. Recently we have developed a method which involves the binding of ligands to receptors in slide mounted tissue sections and the gen-eration of autoradiograms by the apposition of emulsion coated coverslips (20). By using this method we found a distribution of opiate receptors that was the same as that observed with the older method involving in vivo labelling of receptors (20). This in vitro labelling method has several important advantages. It is possible by washing to reduce nonspecific binding to low levels (by the in vivo procedure it was not possible to wash). One can use ligands that are not entirely suitable for in vivo labelling (for example, one can use tritiated peptides which do not usually cross the blood-brain barrier). It is more economical because one does not have to load an entire animal with radioactive ligand. One can use postmortem tissues including that from humans. One can perform a variety of autoradiographic studies under a wide variety of ligand binding conditions, with small pieces of tissue, to ex-amine variations of receptor densities with a high degree of ana-tomical resolution. This report describes our results on the localization of receptors in the brain at the light microscopic level.

Light Microscopic Studies

Receptor Distribution in Rat Brain Utilizing ^{3}H-QNB: As mentioned above, we had previously examined the distribution of muscarinic receptors in the rat forebrain following the systemic administration of ^{3}H-QNB (19). More recently, we have utilized the
new in vitro labelling procedure (20) to examine the distribution
of ^{3}H-QNB binding in the rat hindbrain (Wamsley, in preparation).
Before beginning autoradiography, we performed detailed biochemical
studies showing that the characteristics of ^{3}H-QNB binding in our
slide mounted tissue sections was the same as that found for QNB
binding to muscarinic receptors in tissue homogenates. The binding
was of a high affinity, had the same regional distribution and was
displaced only by cholinergic muscarinic agents and not by unrelated drugs. Because of the high affinity of QNB for the receptor,
it was possible to perform extensive washing and to have sections
with very high specific to nonspecific ratios of about 70 to 1.
Thus, the autoradiographs provided a relatively clear and detailed
picture of the distribution of muscarinic receptors in the rat
brain (Fig. 1).

While it is not possible to present all of the details of the
anatomical studies (Wamsley: in preparation), some of the results
can be summarized. We observed high densities of receptors in the
pontine nuclei, the superior colliculus, the floor of the fourth
ventricle (including the locus ceruleus), central gray, parabrachial
nuclei and the nucleus of nerve V. Other areas showing high densities in more caudal regions included the nucleus of nerve VII,
nucleus ambiguus, and the nucleus of nerve XII. We also observed
high densities in parts of the spinal cord including the substantia
gelatinosa and the ventral horn. Areas showing elevated but lower
levels of autoradiographic grains included the dorsal medial spinal
trigeminal nucleus, the cochlear nuclei, the vestibular nuclei,
nucleus tractus solitarii and nucleus gracilis and cuneatus. In
addition, some reticular nuclei in the medulla showed elevated
levels of autoradiographic grains. Our results with QNB are in
good agreement with those of Rotter and coworkers (14,15), who
utilized an irreversible ligand for mapping muscarinic receptors.

Light Microscopic Localization of Multiple Agonist Binding
Sites: Binding studies involving the use of agonists have revealed
that there are probably three populations of muscarinic binding
sites which recognize agonists with different affinities. In the
cerebral cortex there appears to be a minor population of binding
sites with a very high affinity for agonists, another larger population with a high affinity for agonists and a third population
with a relatively low affinity. Interestingly, radiolabelled
antagonists recognize all of these sites with about the same high
affinity (1-4). Biochemical studies on the distributions of these

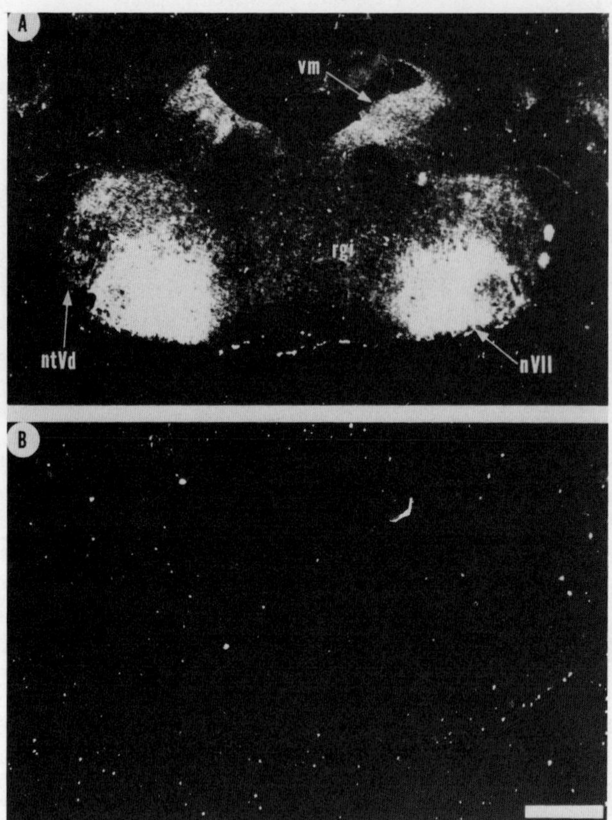

FIGURE 1: A – Darkfield photomicrograph demonstrating autoradiographic grain densities (on an emulsion-coated cover-
slip) over a section through the caudal pons. The tissue
section was immediately below the plane of focus (although it cannot be seen in this darkfield photomicrograph) and was in register with the identified nuclei.
The tissue's muscarinic receptors had previously been
labelled with ^{3}H-QNB and thus the autoradiographic grains
(appearing white with darkfield illumination) demonstrate
areas which contain muscarinic cholinergic receptors.
B – Photomicrograph of an adjacent section to that shown
in A, which had been treated with atropine (1 µM) along
with the ^{3}H-QNB. Atropine blocks the binding of the ^{3}H-
QNB to specific muscarinic cholinergic receptors. Thus any
autoradiographic grains seen in this photomicrograph represent nonspecific (background) binding. Bar = 1 mm.
ntVd = Pars dorsalmedialis of the spinal trigeminal nucleus;
nVII = nucleus of nerve VII (facial); rgi = reticular
nucleus; vm = medial vestibular nucleus.

different sites have revealed that some areas of brain contain a
relatively large concentration of the high affinity agonist binding
site, whereas other areas do not. This differential distribution
is a strong indication that the various populations of agonist
binding sites represent distinct entities with different distributions in the brain (3). We have recently completed preliminary
studies indicating that it is possible to localize these different
agonist binding sites by light microscopic autoradiography (17).

   The procedures are stated briefly as follows. Slide mounted
tissue sections are incubated with ^{3}H-N-methylscopolamine (^{3}H-NMS).
This drug labels all muscarinic cholinergic receptors with the same
high affinity (1,4). Some sections are incubated in the presence
of unlabelled carbachol as well as ^{3}H-NMS. The concentration of
carbachol is selected so that it occupies most of the high affinity
agonist sites while leaving most of the low affinity agonist sites
unoccupied. Therefore, under these precise conditions, ^{3}H-NMS is
preferentially labelling, with a high degree of selectivity, the
low affinity agonist sites. Without carbachol, NMS has the same
relative occupancy of all the different classes of muscarinic cholinergic agonist sites. However, in the presence of carbachol, NMS
has an approximately 45-fold greater occupancy of the low affinity
sites compared to the high affinity agonist sites. Hence, by comparing slides incubated with ^{3}H-NMS alone and slides incubated with
carbachol and ^{3}H-NMS, one can observe the distribution of the different classes of muscarinic receptors (17). Because only one concentration of carbachol was used, we include both of the high affinity binding sites as one class.

   Autoradiographs reveal that certain brain regions had high
concentrations of high affinity muscarinic cholinergic agonist
binding sites. These areas include lamina IV of the cerebral cortex (Fig. 2), nucleus tractus diagonalis, some thalamic nuclei,
the zona incerta and the dorsal lateral geniculate body (17). A
more detailed examination of this distribution is in progress.

Significance of the Light Microscopic Studies

   As mentioned above, the main advantages of these studies is
that one can obtain a very sensitive measurement of receptor in
tissue with a high degree of anatomical discrimination. These results can be used in many ways. They can be used by other investigators in other disciplines as maps pointing out sites where high
densities of receptors occur and that would be useful for a further
exploration of the function of muscarinic cholinergic receptors _in
situ_. For example, electrophysiological studies could be directed

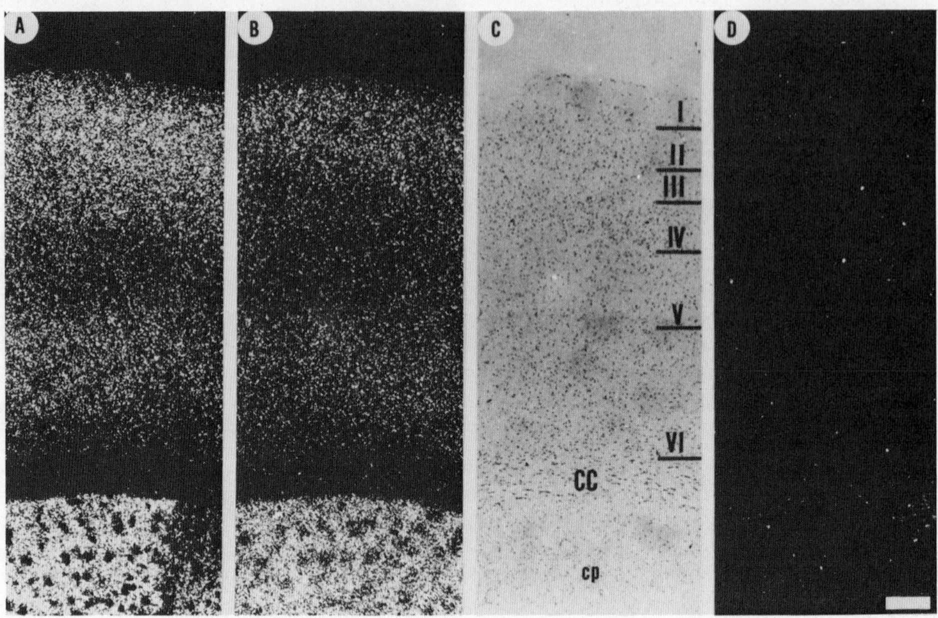

FIGURE 2: <u>A</u> - Darkfield photomicrograph of autoradiographic grain
densities over rat parietal cortical tissue labelled with
^{3}H-NMS. This muscarinic antagonist labels all muscarinic
sites and reflects the localization of muscarinic cholinergic receptors in the cerebral cortex.
<u>B</u> - Photomicrograph of the grain distribution over an ad-
jacent section which was treated with 10^{-4} M carbachol
along with the ^{3}H-NMS. The autoradiographic grains in
this photomicrograph indicate the low affinity muscarinic
agonist cholinergic binding sites. Note the relatively
large displacement of binding from laminae IV indicating
a relatively high concentration of high affinity binding
sites in this area. See text for further details.
<u>C</u> - Brightfield photomicrograph showing cellular distri-
bution in the tissue as it appears beneath the autoradiographic grains. The Roman numerals indicate approximate
deliminations of the six layers of cortex as seen at this
level (area #2 of Krieg). CC = corpus callosum; cp =
caudate-putamen.
<u>D</u> - Darkfield photomicrograph of an adjacent tissue sec-
tion which was treated with atropine to displace specific
muscarinic cholinergic binding. Bar = 200 μ.

to the various brain areas showing high densities of different classes of agonist receptors in an attempt to discern the significance of both high and low affinity agonist sites. At this point, there is no clear indication why separate high and low affinity agonist sites exist and what their relative function or importance might be.

At present, histochemical methods for localizing cholinergic neurons are in a state of development (6). These mainly involve the immunohistochemical localization of the enzyme choline acetyltransferase. Our radiohistochemical investigations of receptor localization would be important complements to the immunohistochemical studies by furthering the description of functional cholinergic pathways in various brain regions.

Another important use of these studies is that they provide "road maps" of cholinergic drug action in brain. It is known that administration of anticholinergic drugs results in a variety of behavioral changes in animals. These receptor localizations can be used to help understand some of these behavioral changes. For example, anticholinergic drugs are useful in the treatment of motion sickness. Motion sickness appears to be due to an over-stimulation of the vestibular labyrinthine system. Related to this, we found high densities of muscarinic receptors in the vestibular nuclei which receive neuronal fibers from the sensory specializations in the membraneous labyrinth of the inner ear. Hence, it is likely or at least possible, that drug action in the vestibular nuclei accounts for the antimotion sickness properties of these drugs. It is also well known that anticholinergic drugs provide a mild vagal excitation. Our results provide insights into the mechanism of this excitation as we observed high densities of receptors in the various nuclei associated with the vagus nerve. These include the dorsal motor nucleus of the vagus, nucleus ambiguus, nucleus tractus solitarii and nucleus commissuralis. In summary, the radiohistochemical localization of muscarinic receptors at the light microscopic level is feasible and provides additional insight into cholinergic mechanisms.

ACKNOWLEDGEMENTS

The authors acknowledge the technical assistance of N. Taylor, the clerical assistance of D. Weimer and M. Flutka, grant support from the McKnight Foundation and USPHS grant MH-25951. MJK is the recipient of a RCDA Type II award (MH-00053) and JKW of postdoctoral fellowship (HD-05739).

REFERENCES

1. Birdsall, N.J.M., Burgen, A.S.V. and Hulme, E.C. (1978): Molec. Pharmacol. 14:723-736.
2. Birdsall, N.J.M. and Hulme, E.C. (1976): J. Neurochem. 24:7-16.
3. Birdsall, N.J.M., Hulme, E.C. and Burgen, A.S.V. (1980): Proc. Rev. Soc. Lond. B 207:1-12.
4. Hulme, E.C., Birdsall, N.J.M., Burgen, A.S.V. and Mehta, P. (1978): Molec. Pharmacol. 14:737-750.
5. Klemm, N., Murrin, L.C. and Kuhar, M.J. (1979): Brain Res. 169:1-9.
6. Kuhar, M.J. (1976): IN Biology of Cholinergic Function (eds) A.M. Goldberg and I. Hanin, Raven Press, New York, pp. 3-28.
7. Kuhar, M.J. (1978): Ann. NY Acad. Sci. 311:35-48.
8. Kuhar, M.J. (1978): Fed. Proc. 37:153-157.
9. Kuhar, M.J. (1978): IN Neurotransmitter Receptor Binding (eds) H.I. Yamamura, S.J. Enna and M.J. Kuhar, Raven Press, New York, pp. 113-126.
10. Kuhar, M.J. and Yamamura, H.I. (1974): Proc. Soc. Neurosci. 4:294.
11. Kuhar, M.J. and Yamamura, H.I. (1975): Nature 253:560-561.
12. Kuhar, M.J. and Yamamura, H.I. (1976): Brain Res. 110:229-243.
13. Murrin, L.C. and Kuhar, M.J. (1979): Brain Res. 177:279-285.
14. Rotter, A., Birdsall, N.J.M., Burgen, A.S.V., Field, P.M., Hulme, E.C. and Raisman, G. (1979): Brain Res. Rev. 1:141-166.
15. Rotter, A., Birdsall, N.J.M., Field, P.M. and Raisman, G. (1979): Brain Res. Rev. 1:167-184.
16. Stumpf, W.E. and Roth, L.G. (1966): J. Histochem. Cytochem. 14:274-287.
17. Wamsley, J.K., Zarbin, M., Birdsall, N. and Kuhar, M.J. (1980): Brain Res. (in press).
18. Yamamura, H.I., Enna, S.J. and Kuhar, M.J. (eds)(1978): Neurotransmitter Receptor Binding, Raven Press, New York.
19. Yamamura, H.I., Kuhar, M.J. and Snyder, S.H. (1974): Brain Res. 80:170-176.
20. Young, W.S. III and Kuhar, M.J. (1979): Brain Res. 179:255-270.

"BINDING STATES" OF MUSCARINIC RECEPTORS

N.J.M. Birdsall, C.P. Berrie, A.S.V. Burgen, R. Hammer,
E.C. Hulme and J. Stockton

Division of Molecular Pharmacology
National Institute for Medical Research
Mill Hill, London NW7 1AA, United Kingdom

An examination of the binding of antagonists to the receptors
in rat cerebral cortex has shown that this occurs to an apparently
uniform set of sites (8). Furthermore, the binding constants so
obtained are in close agreement with those estimated by antagonism
of the contraction of the longitudinal smooth muscle of guinea pig
ileum. However, it has been reported by Barlow et al. (1) that the
activity of some antagonists on guinea pig heart was not identical
to that on the ileum.

This appears to be the case for binding to rat heart homogenates. N-Methylscopolamine (NMS), propylbenzilycholine (PrBCh)
and atropine all bind more weakly in the heart (0.16–0.8 times)
and this could be confirmed for the antagonist for which Barlow et
al. (1) found the greatest discrepancy, 4-diphenylacetyl-N-methyl-
piperidine methiodide (DI-4)(Fig. 1), where it can be seen that
Di-4 has an eightfold lower affinity in the heart. By contrast,
only small differences in affinity constant have been found between
different areas of brain.

We have recently reported (6) that a new antagonist, pirenzepine, does not accord with the general pattern for antagonists and
indeed not only shows a range of apparent binding affinities in
different tissues, but in several tissues the binding curve obtained by competition with tritiated antagonists such as [^{3}H]-NMS
or [^{3}H]-PrBCh is not mass action. For instance, in the cortex the
Hill coefficient, nH, is 0.73 and in the sublingual gland, nH = 0.72.

Using the curve analysis procedures described previously (3),
it is possible to resolve these flat curves into three populations

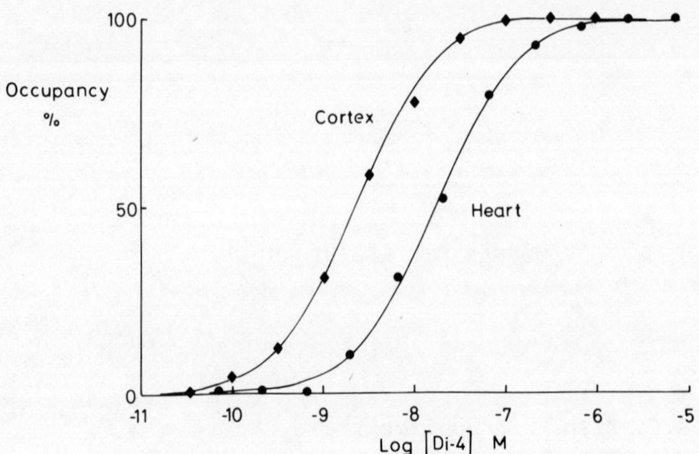

FIGURE 1: Occupancy concentration curves for the binding of Di-4 to muscarinic receptors in homogenates of a) rat heart (●) and b) rat cerebral cortex (◆) by competition with a tritiated antagonist and using methods described previously (2,8). The curves are best fit simple mass action binding curves.

TABLE 1: Comparison of fractional populations of pirenzepine and agonist sites in different tissues of the rat.

TISSUE	AGONIST SITES			PIRENZEPINE SITES		
	SH	H	L	A	B	C
Lacrimal gland	0		100	0	100	0
Atria	73	19	8	0	0	100
Hippocampus	4	17	79	79	21	0
Medulla-pons	30	42	28	0	100	0

SH, H and L are the superhigh, high and low affinity agonist sites. A, B and C are the three putative pirenzepine sites as defined in the text.

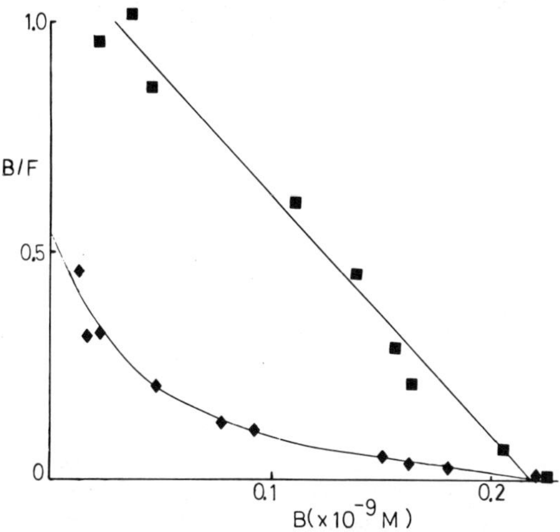

FIGURE 2: Scatchard plot of the binding of [^{3}H]-N-methylscopolamine
to muscarinic receptors on heart membranes in a 20 mM
Tris-Hepes buffer (pH 7.5) at 30°C in the absence (◆)
and presence (■) of 10^{-4} M 5'-guanylylimidodiphosphate.

of sites (A, B, C) with affinities 4 x 10^7 M^{-1}, 4 x 10^6 M^{-1} and
1 x 10^6 M^{-1}, respectively. As with the analysis of agonist binding
(3,5), it appears that differing proportions of these sites account
for the dispersed I_{50} values for pirenzepine in different tissues.

It is possible that this variation in affinity accounts for the
pharmacological selectivity of pirenzepine which apparently antago-
nises gastric secretion at lower concentrations than are required
to block smooth muscle contraction or the vagal slowing of the
heart (9,10,12). However, the I_{50} of pirenzepine for the receptors
in guinea pig gastric mucosa (2 x 10^{-7} M) is only a factor of four
lower than that found in the heart (8 x 10^{-7} M). As the nH is less
than one in the fundus mucosa, perhaps it is the fraction of high
affinity receptors present in this tissue and some of the other
exocrine glands that is somehow correlated with sensitivity to the
drug.

Pirenzepine shows no agonist properties, and the binding
heterogeneity is not affected by guanine nucleotides. The question
therefore arises of whether the pirenzepine heterogeneity is just
the agonist heterogeneity detected by an unusual antagonist. The
study of a number of tissues shows that there is no correlation

between the heterogeneity tested in these two ways in either central or peripheral tissues (Table 1). Pirenzepine is thus uncovering a type of heterogeneity distinct from that detected by agonists.

Guanine nucleotides inhibit agonist binding to muscarinic receptors in the rat heart (2,7,11,13). Thus GTP (10^{-3} M) increases the I_{50} for carbachol from 1.1 x 10^{-6} M to 2.3 x 10^{-5} M (2). Analysis of the data suggests that this is attributable to a conversion of a major fraction of the SH and H sites into L or L-like sites (7). In brain a similar but much smaller effect is seen which is most pronounced in the medulla-pons (7). Guanine nucleotides also produce modest changes in the binding of antagonists to muscarinic receptors in the heart (but not the brain). For instance, the affinity for NMS is increased 1.8-fold to give a value which is closer to the affinity in the medulla. It seems as if the "coupling" of GTP binding sites to receptors is very nearly complete in the heart, but is only to a subpopulation of receptors in the brain.

When antagonist binding to the heart is examined under conditions of low ionic strength, the binding curve deviates markedly from a simple mass action curve. This is illustrated in Fig. 2 by a Scatchard plot of [^{3}H]-NMS binding. Analysis of the binding curve by a two-site model reveals that NMS is binding to a major fraction (0.8) of sites, affinity constant 3 x 10^8 M, and a minor fraction (0.2), affinity constant 7 x 10^9 M. In the presence of 5-guanylylimidodiphosphate (10^{-4} M) homogeneity is restored and the affinity becomes 5 x 10^9 M.

DISCUSSION

The results described here reveal an increasing complexity in the "binding states" that the muscarinic receptor can assume. We have previously postulated that the receptors can exist in four states, a ground state to which antagonists bind preferentially and three agonist binding states (SH, H, L) (3,5). The results with GTP indicate that, at least for the SH and H forms of the receptor, there can exist GTP sensitive (G_+) and GTP insensitive (G_-) forms. It is probably adequate to assume that the G_+ forms are also responsible for the small but reproducible increase in affinity of antagonists in the presence of guanine nucleotides in the heart. The fact that in this tissue the proportion of receptors switched into a low affinity state for antagonists by low ionic strength and restored to high affinity by GTP corresponds with the sum of SH and H sites is probably coincidental (7). It could be that the low ionic strength changes the association of the receptor with another protein which could be the G protein. It is also interesting that in tissues in which the GTP effect is small the effect of low ionic strength was not detected (4). In fact, in the brain low ionic strength leads to an <u>increase</u> in affinity of muscarinic antagonists without the induction of receptor heterogeneity (4).

The differences in antagonist binding to cardiac muscarinic receptors must be due to a difference in the ground state of the cardiac receptor compared with that from other tissues. We have no evidence that enables us to distinguish between a difference in primary structure or in conformation – both are compatible with the changes in structure-activity for antagonists found. The subtypes detected by pirenzepine could also fit either possibility, but the pirenzepine structure must sense a different set of structural differences from those detected by agonists.

REFERENCES

1.   Barlow, R.B., Berry, K.J., Glenton, P.A.M., Nicolaou, N.M. and Soh, K.S. (1976): Brit. J. Pharmacol. <u>56</u>:613-620.
2.   Berrie, C.P., Birdsall, N.J.M., Burgen, A.S.V. and Hulme, E.C. (1979): Biochem. Biophys. Res. Commun. <u>87</u>:1000-1005.
3.   Birdsall, N.J.M., Burgen, A.S.V. and Hulme, E.C. (1978): Molec. Pharmacol. <u>14</u>:723-736.
4.   Birdsall, N.J.M., Burgen, A.S.V., Hulme, E.C. and Wells, J.W. (1979): Brit. J. Pharmacol. <u>67</u>:371-377.
5.   Birdsall, N.J.M., Hulme, E.C. and Burgen, A.S.V. (1980): Proc. Roy Soc. B. <u>207</u>:1-13.
6.   Hammer, R., Berrie, C.P., Birdsall, N.J.M., Burgen, A.S.V. and Hulme, E.C. (1980): Nature (Lond) <u>283</u>:90-92.
7.   Hulme, E.C., Birdsall, N.J.M., Berrie, C.P. and Burgen, A.S.V. (1980): Proc. EMBO Workshop: Drug receptors in central nervous system (in press).
8.   Hulme, E.C., Birdsall, N.J.M., Burgen, A.S.V. and Mehta, P. (1978): Molec. Pharmacol. <u>14</u>:737-750.
9.   Jennewein, H.M. (1979): In <u>Behandlung des Ulcus Pepticum mit Pirenzepin</u> (eds) A.L. Blum and R. Hammer, Karl Demeter, Munich, pp. 41-48.
10. Matsuo, Y. and Seki, A. (1979): Arznemittelforschung <u>29</u>: 1028-1035.
11. Rosenberger, L.B., Roeske, W.R. and Yamamura, H.I. (1979): Eur. J. Pharmacol. <u>56</u>:179-180.
12. Sachs, G., Kasbeker, D.K. and Berglindh, T. (1979): In <u>Die Behandlung des Ulcus Pepticum mit Pirenzepin</u> (eds) A.L. Blum and R. Hammer, Karl Demeter, Munich, pp. 18-25.
13. Wei, J.-W. and Sulakhe, P.V. (1979): Eur. J. Pharmacol. <u>58</u>: 91-92.

INVESTIGATION ON THE EFFECT OF CHOLINERGIC AGONISTS AND

ANTAGONISTS ON CYCLIC NUCLEOTIDE LEVELS IN VIVO:

EFFECT OF LYSO-PHOSPHATIDYLSERINE

S. Teolato, G. Calderini, A.C. Bonetti, A. Battistella
and G. Toffano

Fidia Research Laboratories
Abano Terme, Italy

INTRODUCTION

Muscarinic receptor stimulation has been shown to increase cGMP levels in nervous tissue, both in vivo (25) and in vitro (1, 11,18), suggesting that such a nucleotide, by analogy with the role postulated for cAMP, mediates the muscarinic actions of acetylcholine (ACh)(5,21). Moreover, it has been proposed (14,15) that cGMP and cAMP, by acting in opposite direction, provide the mechanism for bi-directional control of intracellular processes.

However, attempts to show coupling between the muscarinic receptor and guanylate cyclase have been unsuccessful (15) and only recently (18) it has been suggested that the occupancy of muscarinic receptors may activate a soluble intracellular guanylate cyclase indirectly, with calcium acting as the intracellular mediator. A good correlation between the regional distribution of guanylate cyclase and muscarinic receptor (30) in the brain is, however, not yet apparent.

This work examines some of the evidence for or against the hypothesis that cGMP is the intracellular mediator for the actions of ACh in vivo. We have investigated the effect of cholinergic (physostigmine, choline [Ch]) and anticholinergic (scopolamine) drugs on both cAMP and cGMP levels in different brain areas and compared the results with those obtained by administering lysophosphatidylserine (lyso-PS), a pharmacological active phospholipid (4)

TABLE 1: Effect of scopolamine, physostigmine and Ch on cGMP and cAMP in different brain areas of rat.

Drug (i.p.)	Dose (mg/kg i.p.)	Hippocampus*		Cerebral Cortex		Cerebellum	
		cGMP	cAMP	cGMP	cAMP	cGMP	cAMP
None		0.28±0.010	3.81±0.17	0.71±0.025	8.60±0.35	7.33±0.32	7.00±0.50
Scopolamine (30 min)	1	0.30±0.018	4.20±0.20	0.75±0.05	8.21±0.40	7.30±0.52	7.25±0.35
Physostigmine (30 min)	1	0.27±0.015	3.91±0.17	0.60±0.04	8.91±0.22	6.80±0.53	6.40±0.40
Choline (60 min)	300	0.26±0.02	4.15±0.21	0.67±0.03	8.33±0.30	8.20±0.62	7.42±0.36

* Values are expressed as pmol/mg protein ± S.E.M.

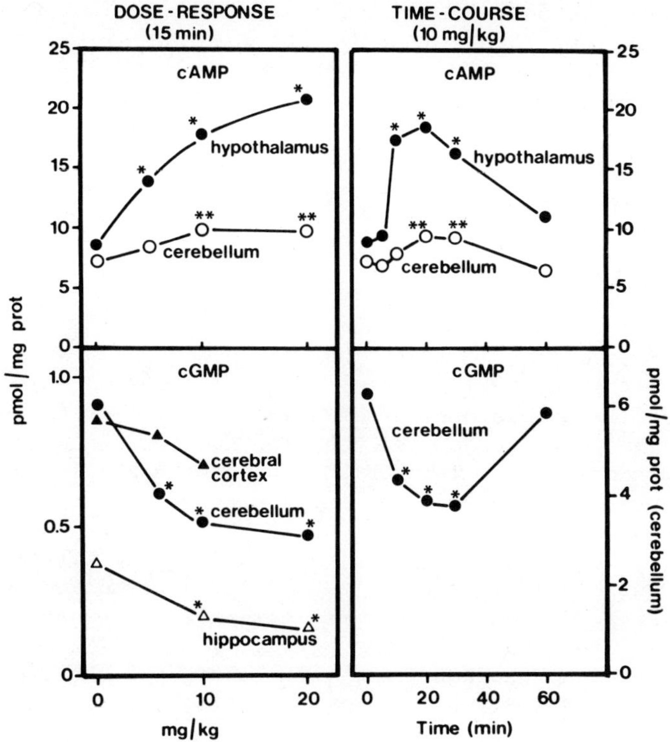

FIGURE 1: Effect of LYSO-PS on cerebral cyclic nucleotides. Values are the mean of at least 8-10 determinations, the S.E.M. is less than 10%.

MATERIALS AND METHODS

Scopolamine, physostigmine and Ch chloride were purchased from commercial sources. Lyso-PS, prepared and purified from bovine brain PS, as previously reported (4), was dissolved in 0.05 M Tris·HCl buffer (pH 7.8) and injected intravenously to 110-120 g Sprague-Dawley rats.

Animals were sacrificed by exposing their heads to focussed microwave irradiation (2.6 Kw x 5 sec) at various times after injection (6). After dissection of the brain regions, cGMP was determined by a radioimmunoassay (29) and cAMP by the protein binding procedure (13). Differences in the amount of cyclic nucleotides

whether determined on the total supernatant or as isolated compounds after column chromatography (24) were negligible.

Protein was determined according to the method of Lowry et al.
(22). Statistical analysis was performed by a two-way factorial
experiment and the Duncan new multiple range test.

RESULTS

The effect of scopolamine, physostigmine and Ch at doses capable
of eliciting pharmacological effect (30) on cAMP and cGMP levels in
different brain regions are reported in Table 1. At those doses
scopolamine induces stereotyped behavior and depresses the spon-
teneous alternation in a T-maze (20), physostigmine increases ACh
content in the brain (9,19), and antagonizes the effect of scopolamine on spontaneous alternation, while Ch increases the ACh precursor level in the brain (8,12), antagonizes the effect of scopolamine on spontaneous alternation and possibly stimulates muscarinic
receptors (28). In the hippocampus, cerebral cortex and cerebellum
all drugs fail to change the basal level of both cAMP and cGMP. The
lack of effect was also confirmed when we studied the time course
or when, in the case of Ch, we injected rats daily for a week.

In contrast, lyso-PS is capable of modifying cAMP and cGMP
levels in specific brain regions (Fig. 1). Lyso-PS induces a dose
dependent increase of hypothalamic cAMP, while its effect on cerebellar cAMP is less marked. No effect in the striatum, hippocampus
and cerebral cortex is apparent. In the same animals, these concomitantly occurs a significant decrease of cGMP in all brain areas
investigated, with the hypothalamus being the only exception. The
maximum effect is achieved with doses ranging from 10 to 20 mg/kg
within 15-30 min of the injection.   Since several neurotransmitters
are associated with the cAMP and cGMP systems (16), we have investigated what neurotransmitter could be involved in the lyso-PS effect.

The increase of hypothalamic cAMP induced by lyso-PS is completely prevented by reserpine treatment (10 mg/kg, given 3 hr
before lyso-PS), and partially antagonized by haloperidol, a DA-
receptor blocking agent, given at 1 mg/kg 30 min before lyso-PS
(Table 2). Of the two anticholinergic drugs tested, atropine (20 mg/
kg i.p., 30 min before lyso-PS) is completely ineffective, while
scopolamine (1 mg/kg i.p., 30 min before lyso-PS) has a minimal
effect.

Anticholinergic drugs have a similar effect on cerebellar
cGMP level. However, haloperiodol _per se_ decreases cGMP in the

TABLE 2: Action of antiadrenergic and anticholinergic drugs on the effect of lysophosphatidylserine (Lyso-PS) on hypothalamic cAMP and cerebellar cGMP.

Treatment	Hypothalamic cAMP	Cerebellar cGMP
None	8.54 ± 0.40	8.30 ± 1.08
Lyso-PS (10 mg/kg)	$24.66 \pm 1.85*$	$4.08 \pm 0.46*$
Reserpine (10 mg/kg)	11.06 ± 0.55	6.87 ± 1.31
Reserpine + Lyso-PS	$11.13 \pm 1.60**$	$3.54 \pm 0.45*$
Haloperidol (1 mg/kg)	10.28 ± 0.27	$4.38 \pm 0.34*$
Haloperidol + Lyso-PS	$16.69 \pm 1.02**$	$3.24 \pm 0.33*$
Scopolamine (1 mg/kg)	10.02 ± 0.78	11.34 ± 1.06
Scopolamine + Lyso-PS	$18.50 \pm 1.98**\#$	$7.77 \pm 1.29***$
Atropine (20 mg/kg)	9.36 ± 0.31	7.06 ± 1.07
Atropine + Lyso-PS	$23.10 \pm 2.31*$	$4.53 \pm 0.53*$

Values are expressed as pmol/mg protein $\pm$ S.E.M.   Animals were sacrificed 15 min after lyso-PS. Reserpine was injected 3 hr and the other drugs 30 min before lyso-PS.
*   Statistically significant difference with respect to control P < 0.01.
**   Statistically significant difference with respect to lyso-PS P < 0.01; **# P < 0.05.
***   Statistically significant difference with respect to scopolamine, P < 0.01.

TABLE 3: Action of GABA agonist and antagonist on the effect of lyso-PS con cGMP.

Treatment	Brain Region		
	Cerebellum	Hippocampus	Cerebral Cortex
None	8.02 ± 0.56	0.34 ± 0.04	0.87 ± 0.02
Lyso-PS (10 mg/kg)	$4.10 \pm 0.25*$	$0.19 \pm 0.01*$	0.75 ± 0.03
Isoniazid (450 mg/kg)	$17.05 \pm 2.27*$	0.46 ± 0.04	0.85 ± 0.02
Isoniazid + lyso-PS	9.5 ± 1.00	0.36 ± 0.02	0.88 ± 0.05
AOAA (50 mg/kg)	$3.02 \pm 0.47*$	$0.19 \pm 0.02*$	$0.53 \pm 0.05*$
AOAA + lyso-PS	$3.28 \pm 0.31*$	$0.21 \pm 0.01*$	$0.51 \pm 0.03*$

Values are expressed as pmol/mg protein $\pm$ S.E.M. Isoniazid (s.c.) and AOAA (i.p.) were injected 15 and 250 min before lyso-PS. Animals were sacrificed 15 min after lyso-PS. * = Statistically significant difference with respect to control, P < 0.01.

cerebellum (2,3) and its effect is potentiated by lyso-PS. These
data indicate that the mechanism of action of lyso-PS differs from
that of haloperidol and probably does not involve cholinergic
mechanisms.

Table 3 shows the effect of a GABA agonist and antagonist on
the effect of lyso-PS on cGMP in different brain regions. The GABA
agonist AOAA, decreases per se cGMP level in all brain areas investigated, confirming the general effect of drugs which increase
GABAergic transmission (17,26). The effect of AOAA on cGMP parallels but is not additive to that reported for lyso-PS. In contrast
isoniazid, a GABA antagonist, increases cGMP levels and restores
the basal nucleotide value in association with lyso-PS.

DISCUSSION AND CONCLUSION

Scopolamine, physostigmine and Ch at doses capable of eliciting
pharmacological effect (8,9,12,19,20,28) do not induce any significant changes in cAMP and cGMP levels in the hippocampus, cerebellum
and cerebral cortex. This suggests that an apparent cGMP modification in the above areas is not involved in at least some of the
pharmacological actions of these drugs. In contrast, lyso-PS modifies the levels of both cyclic nucleotides in specific brain areas.
Lyso-PS is the active form of PS (4). It has been shown that soni-
cated dispersion of PS, given intravenously, increases ACh release
from the cerebral cortex of urethane anesthetized rats (7,23). Since
this effect is prevented by pimozide treatment or by a septal lesion,
it was inferred that PS increases cortical ACh output by acting on
a cholinergic pathway originating from or passing through the septum,
with an indirect mechanism possibly involving a monoaminergic link
(23).   On the other hand, the cholinergic property of this drug is
confirmed by its capability of antagonizing the effect of scopolamine on the spontaneous alternation in a T-maze (27). This instinc-
tive exploratory behavior is disrupted by anticholinergic drugs,
probably by acting on the hippocampus (10).

In studying the molecular basis supporting the central effect
of PS, we recently provided evidence that PS can be activated in
vivo by a phospholipase A_2 mediated metabolic conversion. When
the resulting purified derivative, lyso-PS was used, high biological
activity was detected at remarkably low doses (0.5-1.0 mg/kg), giving
conclusive support for its involvement in the pharmacological effect
of PS (4). Similar to its precursor, lyso-PS antagonizes the effect
of scopolamine on the spontaneous alternation. However, this supposed cholinomimetic effect is not related to cyclic nucleotide
changes. In fact, induced lyso-PS increase of cAMP is counteracted

by antiadrenergic agents, whereas its effect on cGMP seems to involve the GABAergic system.

In general it has been reported that excitatory drugs increase
cGMP in the cerebellum, while depressive drugs have an opposite
effect (2,6,17,26). These observations lead to the suggestion that
cGMP may be involved in a more general sense in the fast adaptive
processes of the central nervous system independent of the function
of a specific neurotransmitter (17). Lyso-PS depresses spontaneous
motor activity and we can include it in the family of depressive
drugs which decreases cGMP, possibly through a GABAergic mechanism.

REFERENCES

1.   Bartfai, T., Study, R.E. and Greengard, P. (1977): IN Chol-
      inergic Mechanisms and Psychopharmacology (ed) D.J. Jenden,
      Plenum Press, New York, pp. 285-295.
2.   Biggio, G., Costa, E. and Guidotti, A. (1977): J. Pharmacol.
      Exp. Ther. 200:207-215.
3.   Biggio, G. and Guidotti, A. (1977): Nature 265:240-242.
4.   Bigon, E., Boarato, E., Bruni, A., Leon, A. and Toffano, G.
      (1979): Brit. J. Pharmacol. 67:611-616.
5.   Birdsall, N.J.M. and Hulme, E.C. (1976): J. Neurochem. 27;7-16.
6.   Breese, G.R., Mueller, R.A. and Mailman, R.B. (1979): J.
      Pharmacol. Exp. Ther. 209:262-270.
7.   Casamenti, F., Mantovani, P., Amaducci, L. and Pepeu, G.
      (1979): J. Neurochem. 32:529-533.
8.   Cohen, E.L. and Wurtman, R.J. (1975): Life Sci. 16:1095-1102.
9.   Cohen, E.L. and Wurtman, R.J. (1976): Science 191:561-562.
10.   Douglas, R.J. and Isaacson, R.L. (1966): Psychonom. Sci. 4:
      283-284.
11.   Ferrendelli, J.A., Steiner, A.L., McDougal, D.B. Jr. and
      Kipnis, D.M. (1970): Biochem. Biophys. Res. Comm. 41:
      1061-1067.
12.   Flentge, F. and Van DenBerg, C.J. (1979): J. Neurochem. 32:
      1331-1333.
13.   Gilman, A.G. (1970): Proc. Nat. Acad. Sci. 67:305-312.
14.   Goldberg, N.D., Haddox, M.K., Nicol, S.E., Glass, D.B., San-
      ford, F.A., Kuehl, F.A. and Estensen, R. (1975): IN Advances
      in Cyclic Nucleotide Research Vol. 5 (eds) G.I. Drummond,
      P. Greengard, G.A. Robison, G.A., Raven Press, New York,
      pp. 307-330.
15.   Goldberg, N.D., O'Dea, R.F. and Haddox, M.K. (1973): IN
      Advances in Cyclic Nucleotide Research Vol. 3 (eds) P. Green-
      gard and G.A. Robison, Raven Press, New York, pp. 155-223.

16. Greengard, P. (1979): TIPS $\underline{1}$:27-29.

17. Gumulka, S.W., Dinnendahl, V., Peters, H.D. and Schonhofer, P.S. (1976): N.-S. Arch. Pharmacol. $\underline{293}$:75-80.

18. Hanley, M.R. and Iversen, L.L. (1978): Molec. Pharmacol. $\underline{14}$: 246-255.

19. Ladinsky, H., Consolo, S. and Peri, G. (1974): Biochem. Pharmacol. $\underline{23}$:1187-1193.

20. Leaton, R.N. and Utell, M.J. (1969): Physiol. Behav. $\underline{5}$:331-334.

21. Lee, T.P., Kuo, J.F. and Greengard, P. (1972): Proc. Nat. Acad. Sci. $\underline{69}$:3287-3291.

22. Lowry, O.H., Rosebrough, N.J., Farr, A.L. and Randall, R.J. (1951): J. Biol. Chem. $\underline{193}$:265-275.

23. Mantovani, P., Pepeu, G. and Amaducci, L. (1976): Adv. Exp. Med. Biol. $\underline{72}$:285-292.

24. Mao, C.C. and Guidotti, A. (1974): Anal. Biochem. $\underline{59}$:63-68.

25. Nahorski, S.R., Pratt, C.N.F.W. and Rogers, K.J. (1976): Brit. J. Pharmacol. $\underline{57}$:445P.

26. Opmeer, F.A., Gumulka, S.W., Dinnendahl, V. and Schonhofer, P.S. (1976): N.-S. Arch. Pharmacol. $\underline{292}$:259-265.

27. Pepeu, G., Gori, G. and Bartolini, L. (1980): IN Aging of the Brain and Dementia: Aging, Vol. 13 (eds) L. Amaducci, A.N. Davison and P. Antuono, Raven Press, New York, 271-274.

28. Speth, R.C. and Yamamura, H.I. (1979): Eur. J. Pharmacol. $\underline{58}$: 197-201.

29. Steiner, A.L., Wehmann, R.E., Parker, C.W. and Kipnis, D.M. (1972): IN Advances in Cyclic Nucleotide Research, Vol. 2 (eds) P. Greengard, G.A. Robison and R. Paoletti, Raven Press, New York, pp. 51-61.

30. Yamamura, H.I. and Synder, S.H. (1974): Proc. Nat. Acad. Sci. $\underline{71}$:1725-1729.

REGULATION OF MUSCARINIC RECEPTOR BINDING

F.J. Ehlert, W.R. Roeske and H.I. Yamamura

Departments of Pharmacology, Psychiatry
and Internal Medicine
University of Arizona Health Sciences Center
Tucson, Arizona 85724 USA

INTRODUCTION

The specific binding of [3H] antagonists to muscarinic receptors from neuronal tissue has been demonstrated by several investigators (6,19,24,28,32,37). In each case, the binding of [3H] antagonists is consistent with the law of mass action for a single
class of independent receptors. Similarly, the results of antag-
onist/[3H] antagonist competition experiments are also consistent
with the law of mass action (23). Perhaps the most convincing
evidence demonstrating that the binding of [3H] antagonists represents a specific interaction with the muscarinic receptor is the
excellent quantitative agreement between the values of dissociation
constants determined by binding experiments and by antagonism of
agonist induced contractions of the guinea pig ileum (8,10).

In contrast to antagonists, the binding of agonists is incompatible with simple mass action behavior. This phenomenon is clearly demonstrated by agonist/[3H] antagonist competitive inhibition
curves which typically have flatter slopes than a mass action curve
(6,7,20). A simple explanation for this phenomenon is that agonists
bind to multiple classes of receptors which have equal affinity for
antagonists (7). In support of this hypothesis, Birdsall and co-
workers (9) have presented evidence demonstrating the existence of
two major populations of muscarinic receptors (high and low affinity)
having different affinities for agonists and equal affinity for
antagonists. A third minor population of superhigh affinity agonist

binding sites has also been detected (9,10). The results of binding
studies with [3H] agonists agree stochiometrically with agonist/
[3H] antagonist competition experiments and demonstrate that [3H]
agonists bind to a heterogeneous population of muscarinic receptors
(9,10). However, it is difficult to measure [3H] agonist binding
at ligand concentrations producing greater than half maximal saturation since nonspecific binding is high at large concentrations
of [3H] agonists (9,10,16).

Although the significance of the multiple forms of the muscarinic receptor remains to be firmly elucidated, the binding parameters of muscarinic agonists have been shown to be correlated
with relevant pharmacological parameters (8). In a study investigating the binding of several muscarinic agonists, it was noted
that the most efficacious agonists had very flat agonist/[3H]
antagonist competition curves while partial agonists had steeper
competition curves (8,9). Thus, there appears to be a correlation
between the efficacy of muscarinic cholinergic ligands and the ratio
of dissociation constants (K_L/K_H) of the high and low affinity
agonist binding sites. Furthermore, Birdsall et al. (8) have demonstrated that the high affinity dissociation constant (K_H) is
correlated with the ED_{50} for smooth muscle contraction while the
low affinity dissociation constant (K_L) is correlated with the
pharmacologically determined dissociation constant after elimination of spare receptors. An interesting model (8), based on the
assumption that the high and low affinity forms of the muscarinic
receptor represent uncoupled and coupled states, respectively, accounts for the correlations described above.

Modulation of Muscarinic Receptor Binding
by Guanine Nucleotides

During the past year, we have investigated the modulation of
muscarinic receptor binding by various agents in an attempt to
elucidate important regulatory mechanisms and to ascertain the
significance of muscarinic receptor heterogeneity. In studies investigating the influence of guanine nucleotides on muscarinic
receptor binding, we have observed that GTP and the nonhydrolyzable
analog, guanyl-5'-yl imidodiphosphate (Gpp(NH)p), caused an increase
in agonist IC_{50} values for competitive inhibition of [3H]-quinuc-
lidinyl benzilate ([3H]-QNB) binding to muscarinic receptors in
the rat forebrain, heart and longitudinal muscle of the ileum (17,
29,30). In each case, only minimal effects on antagonist binding
were detected. These results are consistent with previous reports
concerning the effects of guanine nucleotides on several neurotransmitter receptors (4,11,13-15,22,26,27,33-35,39).

The results of some of our experiments on the modulation of
muscarinic receptor binding by guanine nucleotides are summarized
in Table 1 and demonstrate that guanine nucleotides produce a preferential effect on the high affinity agonist state of the muscarinic receptor. In the absence of guanine nucleotides, the
agonist/[3H] antagonist competition curves were adequately described by a two-site competitive inhibition equation in both the
forebrain and ileum. In the presence of 30 μM Gpp(NH)p, there was
both an increase in the agonist IC_{50} and a steepening of the agonist
competition curve. When the competition curves were analyzed by
nonlinear regression analysis, the predominant effect of Gpp(NH)p
was a reduction in affinity of the high affinity agonist binding
site in both the forebrain and ileum (Table 1). A comparison of
the effect of Gpp(NH)p on agonist binding in the ileum and forebrain revealed that Gpp(NH)p produced a greater shift in the agonist
competition curve in the ileum; an effect which is partially due
to the greater proportion of high affinity sites in the ileum. A
further comparison of these tissues shows that Gpp(NH)p caused a
larger increase in the high affinity dissociation constants of
agonists in the ileum. No significant effects of Gpp(NH)p on
agonist binding were detected in the cerebellum and brainstem despite the 68% proportion of high affinity agonist binding sites in
these brain regions (17). Thus, the propensity of guanine nucleotides to alter the high affinity agonist muscarinic receptor varies
in different brain regions and tissues.

The preferential effect of Gpp(NH)p on the high affinity
agonist muscarinic receptor is consistent with the reported effects
of guanine nucleotides on β-adrenergic receptors (36). With regard
to β-adrenergic receptors in frog erythrocytes, Wessels et al. (36)
have demonstrated that, in the presence of guanine nucleotides, the
isoproteronol/[3H]-dihydroalprenolol competition curve shifts to
the right and becomes steeper. Analysis of control competition
curves for β-adrenergic agonists reveals the presence of high and
low affinity binding sites (36). In the presence of guanine nucleotides, the agonist competition curve assumes the shape of a single
low affinity binding site curve, demonstrating that guanine nucleotides convert high affinity β-adrenergic receptors into low affinity binding sites. Since guanine nucleotides have been shown
to enhance β-adrenergic stimulation of adenyl cyclase (26), it has
been suggested that the guanine nucleotide induced reduction in
agonist affinity represents a state of facilitated coupling between
β-adrenergic receptors and adenyl cyclase (26,36). A similar hypothesis has been proposed for muscarinic receptors. In a study investigating the influence of guanine nucleotides on cardiac muscarinic receptor binding, it was noted that GTP converted a predominantly high affinity agonist/[3H] antagonist competition curve
into a low affinity competition curve (5). Since evidence has been
obtained which suggests that the low affinity agonist state of the

TABLE 1: Effect of Gpp(NH)p on agonist binding parameters.

	Ileum		Forebrain	
	Control	Gpp(NH)p (30 μM)	Control	Gpp(NH)p (30 μM)
Oxotremorine				
K_H (μM)	0.0042 (0.0024–0.0072)	0.018** (0.0091–0.044)	0.014 (0.0071–0.028)	0.051* (0.023–0.11)
K_L (μM)	0.34 (0.22–0.51)	0.55 (0.32–0.98)	1.37 (1.04–1.80)	1.89 (1.26–2.88)
High Affinity Sites (%)	44 (34–54)	38 (30–46)	21 (13–29)	18 (12–24)
Carbamylcholine				
K_H (μM)	0.11 (0.041–0.31)	0.69* (0.25–0.91)	0.52 (0.33–0.83)	1.15* (0.69–1.90)
K_L (μM)	5.49 (3.46–8.7)	14 (8.1–24.5)	62 (42–94)	65 (48–91)
High Affinity Sites (%)	50 (42–58)	61 (49–73)	30 (18–42)	28 (20–36)

*   Significantly different from control, $p < 0.05$.
** Significantly different from control, $p < 0.01$.

Estimates of the high and low affinity dissociation constants (K_H and K_L) and percentage of high affinity sites were determined by nonlinear regression analysis from oxotremorine/[3H]-QNB and carbachol/[3H]-QNB competitive inhibition curves in the presence and absence of 30 μM Gpp(NH)p. The range of xootremorine and carbachol concentrations was 10^{-8}–10^{-3} M, each concentration being geometrically spaced every 0.5 log molar units. 95% Confidence limits are indicated in parentheses.

TABLE 2: Competitive inhibition of $[^3H]$-CD binding by
           cholinergic drugs.

Drugs	IC_{50}	95% Confidence Limits
Antagonists (nM)		
3-Quinuclidinylbenzilate	0.1	0.071–0.14
Atropine	3.0	2.3–4.0
Scopalamine	2.0	1.6–2.5
Levetimide	1000	630–1600
Dexetimide	0.10	0.047–0.21
Agonists (µM)		
Acetylcholine	0.0087	0.003–0.025
Carbamylcholine	0.017	0.0089–0.032
Cis-methyl-dioxolane ·	0.10	0.047–0.21
Trans-methyl-dioxolane	0.15	0.060–0.32
Oxotremorine	0.0010	0.00078–0.0013
Pilocarpine	0.13	0.10–0.17
1-Acetyl-β-methyl- choline	0.11	0.095–0.12
d-Acetyl-β-methyl- choline	57	51–63

The IC_{50} of each drug represents the geometric mean of IC_{50} values
from at least four experiments. The concentration of $[^3H]$-CD used
in these experiments was 5 nM. The competitive inhibition of $[^3H]$-
CD binding by ACh was determined in forebrain homogenates which
had been incubated with 0.1 mM paraoxon for 15 min at 37°C to inhibit cholinesterase activity. Following paraoxon treatment, the
homogenates were washed twice by centrifugation at 47,000 x g for
10 min followed by resuspension in fresh phosphate buffer. The
competitive inhibition of $[^3H]$-CD binding by the enantiomers of
acetyl-β-methylcholine was determined in the presence of 1 µM
physostigmine to inhibit enzymatic hydrolysis of these esters.

muscarinic receptor is coupled (8), Birdsall et al. (5) proposed
that guanine nucleotides facilitate muscarinic receptor coupling.
A strict interpretation of this hypothesis would suggest that,
at relatively low concentrations of guanine nucleotides, a partial
conversion from high to low affinity sites should occur without a
significant change in the affinity of the two sites. However, the
results of our studies investigating the influence of guanine
nucleotides on muscarinic receptor binding in the forebrain and

TABLE 3:   Effect of N-ethylmaleimide on $[^3H]$-CD binding in the rat forebrain.

$[^3H]$-CD (nM)	$[^3H]$-CD Bound (fmol/mg protein)	
	Control	NEM treated
0.5	34 ± 4.8	77 ± 5.0*
1.0	50 ± 8.3	101 ± 8.1**
2.0	77 ± 10	152 ± 18 **
4.0	112 ± 15	205 ± 12 *
8.0	162 ± 18	266 ± 21 **
16.0	263 ± 44	359 ± 30
32.0	349 ± 70	449 ± 30
64.0	466 ± 80	545 ± 39

*   Significantly different from control, p < 0.01
**   Significantly different from control, p < 0.02
Mean binding values ± S.E.M. from three experiments are shown.

ileum demonstrate that Gpp(NH)p (30 μM) causes a partial reduction
in the affinity of the high affinity binding site without s significant change in the proportion of sites (Table 1).   At the present
time, it is not clear whether low concentrations of guanine nucleotides cause a partial reduction in the proportion or a partial decrease in the affinity of high affinity agonist binding sites in
the rat myocardium.   Thus, if guanine nucleotides facilitate coupling of muscarinic receptors by conversion of high affinity sites
to low affinity, the coupling process may be complex and not a simple unitary process.

Binding of the Muscarinic Agonist
$[^3H]$-Cis methyldioxolane: Regulation by
Guanine Nucleotides and N-Ethylmaleimide

With the intention of developing an $[^3H]$ agonist ligand for
the muscarinic receptor, we have recently characterized the binding
of $[^3H]$-cis methyldioxolane ($[^3H]$-CD) to rat forebrain preparations.
Previous experimentation, concerning the pharmacological activity
of CD in the isolated guinea pig ileum, has demonstrated that CD is
a potent muscarinic agonist (12).   The results of equilibrium binding
measurements using $[^3H]$-CD concentrations between 0.5-64 nM revealed

the presence of two receptor populations in the rat forebrain having dissociation constants of 1.83 nM (95% confidence limits: 1.11-3.01 nM) and 123 nM (95% confidence limits: 32-458 nM) (16). The density of the lower affinity site (1320 $\pm$ 440 fmol/mg protein) was approximately 15 times greater than that of the higher affinity site. These two binding sites most likely represent superhigh and high affinity muscarinic receptors.

To determine the pharmacological specificity of [3H]-CD binding, we examined the competitive inhibition of [3H]-CD binding by several muscarinic agonists and antagonists. The results of these experiments are shown in Table 2 and demonstrate that [3H]-CD binding is readily displaced by a variety of muscarinic agonists and antagonists with a rank order of potency which generally agrees with pharmacological data (12,21,28,38). The stereospecificity of [3H]-CD binding was demonstrated in competition experiments using the stereoisomers of benzetimide and acetyl-β-methylcholine. Dexetimide (IC_{50} = 0.1 nM) was 10,000 times more potent than levetimide, and l-acetyl-β-methylcholine (IC_{50} = 0.1 μM) was 520 more times more potent than d-acetyl-β-methylcholine. A variety of non-muscarinic cholinergic drugs were ineffective at inhibiting [3H]-CD binding at a concentration of 10 μM.

The results of agonist/[3H]-CD and antagonist/[3H]-CD competitive inhibition experiments are in agreement with the concept of muscarinic receptor heterogeneity. In these experiments, the IC_{50} values of agonists were of higher affinity than values determined by competitive inhibition of [3H] antagonist binding (25,31,37); a finding consistent with the fact that only superhigh and high affinity muscarinic receptors were labelled with [3H]-CD. In contrast, the IC_{50} values of antagonists, determined by competitive inhibition of [3H]-CD binding, were in general agreement with the affinity of antagonists determined by saturation experiments with [3H] antagonists or by antagonists/[3H] antagonist competition experiments (25,31,37). These results are consistent with the concept that muscarinic antagonists have equal affinity for the different populations of muscarinic receptors.

We have recently investigated the influence of guanine nucleotides on [3H]-CD binding in the rat forebrain and ileum and have obtained results which are consistent with the effects of guanine nucleotides on agonist/[3H] antagonist competition curves (18). The results of these experiments showed that Gpp(NH)p (100 μM) caused 85% inhibition of [3H]-CD binding in the rat ileum. A similar reduction in [3H]-CD binding was observed in the forebrain (18), however, recent experiments in our laboratory using several concentrations of GTP and Gpp(NH)p suggest that the guanine nucleotide effect on [3H]-CD binding in the forebrain is not as great as that observed in the ileum. In the experiments described above, a

[3H]-CD concentration (10 nM) which labelled both superhigh and high affinity muscarinic receptors was used. Thus, the Gpp(NH)p induced reduction in [3H]-CD binding demonstrates that guanine nucleotides reduced the affinity of superhigh and high affinity muscarinic receptors in the forebrain and ileum or convert these higher affinity sites into low affinity receptors. The reduction in [3H]-CD binding by Gpp(NH)p was not caused by competitive inhibition of [3H]-CD binding by Gpp(NH)p since no inhibition of [3H]-QNB binding by 100 µM Gpp(NH)p was detected in the forebrain or ileum.

The experimental use of [3H] agonist ligands like [3H]-CD is a powerful method for detecting guanine nucleotide effects on muscarinic receptor binding. A relatively large guanine nucleotide induced shift in an agonist/[3H] antagonist competition curve may only result in relatively small changes in the binding values of a competition curve. In contrast, the same effect may be manifested as a relatively large reduction in [3H] agonist binding to superhigh and high affinity muscarinic receptors.

In a study investigating the effects of sulfhydryl reagents on muscarinic receptor binding, it was noted that N-ethylmaleimide (NEM) caused an increase in agonist affinity for the muscarinic receptor while producing only minimal effects on antagonist binding (1-3). Analysis of agonist/[3H] antagonist competition curves from experiments on NEM treated rat brain homogenates showed that the increase in agonist affinity was the result of a conversion of low affinity agonist binding sites into high affinity sites (1-3). Since these experiments demonstrated a specific effect on NEM on agonist binding, we investigated the effects of NEM treatment on [3H]-CD binding. The results of these studies are shown in Table 3. For our experiments, homogenates of the rat forebrain were incubated with 1 mM NEM for 20 min at 37°C. Subsequently the homogenates were washed twice by centrifugation, and [3H]-CD binding experiments were performed on the NEM treated homogenates and control homogenates which were treated in the same manner except for exposure to NEM. The results shown in Table 3 demonstrate that NEM treatment produces a 2.2 to 1.2-fold increase in [3H]-CD binding when measured at ligand concentrations between 0.5 and 64 nM. These experiments demonstrate that alkylation of sulfhydryl groups in rat forebrain preparations causes an increase in the binding agonists to superhigh and high affinity muscarinic receptors.

CONCLUSION

A variety of agents have been shown to influence the binding properties of muscarinic receptors. Of particular interest are

agents like guanine nucleotides which produce substantial changes
in agonist binding parameters without affecting antagonist binding
to a great extent. Clearly, a greater understanding of muscarinic
receptor heterogeneity and the nature of the coupling process of
muscarinic receptors is required to realize the significance of
guanine nucleotide induced changes in muscarinic receptor binding.
It is possible that investigations concerning the regulation of
agonist states of the muscarinic receptor may have practical application to other heterogeneous neurotransmitter receptors.

ACKNOWLEDGEMENTS

The authors wish to thank D.J. Triggle and K.J. Chang for
providing the CD and to acknowledge the excellent technical assistance of R. Yamada and A. Chen, the expert computer programming
assistance of S. Yamamura and the excellent secretarial assistance
of C. Thomas. Portions of this work were supported by USPHS grants
MH-27257, MH-30626, HL-21486 and Program Project grant HL-20984.
HIY is a recipient of a USPHS RSDA (MH-00095) from NIMH.

REFERENCES

1.   Aronstam, R.S., Abood, L.G. and Hoss, W. (1978): Molec.
          Pharmacol. 14:575-586.
2.   Aronstam, R.S. and Eldefrawi, M.E. (1979): Biochem. Pharmacol.
          28:701-703.
3.   Aronstam, R.S., Hoss, W. and Abood, L.G. (1977): Eur. J.
          Pharmacol. 46:274-282.
4.   Berrie, C.P., Birdsall, N.J.M., Burgen, A.S.V. and Hulme, E.C.
          (1979): Biochem. Biophys. Res. Comm. 87:1000-1005.
5.   Birdsall, N.J.M, Berrie, C.P., Burgen, A.S.V. and Hulme, E.C.
          (1980): IN Receptors for Neurotransmitter and Peptide Hor-
          mones (eds) G. Pepeu, M.J. Kuhar and S.J. Enna, Raven Press,
        ʼ New York, pp. 107-116.
6.   Birdsall, N.J.M., Burgen, A.S.V., Hiley, C.R. and Hulme, E.C.
          (1976): J. Supramolec. Struct. 4:367-371.
7.   Birdsall, N.J.M. and Hulme, E.C. (1976): J. Neurochem. 27:7-16.
8.   Birdsall, N.J.M., Burgen, A.S.V. and Hulme, E.C. (1977): IN
          Cholinergic Mechanisms and Psychopharmacology (ed) D.J.
          Jenden, Plenum Press, New York, pp. 25-33.
9.   Birdsall, N.J.M., Burgen, A.S.V. and Hulme, E.C. (1978):
          Molec. Pharmacol. 14:723-726.

10. Birdsall, N.J.M., Burgen, A.S.V. and Hulme, E.C. (1979): IN
     Recent Advances in Receptor Chemistry (eds) F. Gualtieri,
     M. Giannella and C. Melchiorre, Elsevier/North-Holland,
     New York, pp. 71-96.
11. Blume, A.J. (1978): Proc. Nat. Acad. Sci. 75:1713-1717.
12. Chang, K.J., Deth, R.C. and Triggle, D.J. (1972): J. Med.
     Chem. 15:243-247.
13. Childers, S.R. and Snyder, S.H. (1978): Life Sci. 23:759-762.
14. Creese, I., Prosser, T. and Snyder, S.H. (1978): Life Sci.
     23:495-500.
15. Creese, I. and Snyder, S.H. (1978): Eur. J. Pharmacol. 50:
     459-461.
16. Ehlert, F.J., Dumont, Y., Roeske, W.R. and Yamamura, H.I.
     (1980): Life Sci. 26:961-976.
17. Ehlert, F.J., Roeske, W.R., Rosenberger, L.B. and Yamamura,
     H.I. (1980): Life Sci. 26:245-252.
18. Ehlert, F.J., Yamamura, H.I., Triggle, D.J. and Roeske, W.R.
     (1980): Eur. J. Pharmacol. 61:317-318.
19. Fewtrell, C.M.H. and Rang, H.P. (1973): IN Drug Receptors (ed)
     H.P. Rang, MacMillan, New York, pp. 211-224.
20. Fields, J.Z., Roeske, W.R., Morkin, E. and Yamamura, H.I.
     (1978): J. Biol. Chem. 253:3251-3258.
21. Furchgott, R.F. and Bursztyn, P. (1967): Ann. NY Acad. Sci.
     144:822-889.
22. Howlett, A.C., Van Arsdale, P.M. and Gilman, A.G. (1978):
     Molec. Pharmacol. 14:531-539.
23. Hulme, E.C., Birdsall, N.J.M., Burgen, A.S.V. and Mehta, P.
     (1978): Molec. Pharmacol. 14:737-750.
24. Kloog, Y. and Sokolvosky, M. (1977): Brain Res. 134:167-172.
25. Laduron, P.M., Verwimp, M. and Leysen, J.E. (1979): J. Neuro-
     chem. 32:421-427.
26. Lefkowitz, R.J., Mullikin, D. and Caron, M.J. (1976): J. Biol.
     Chem. 251:4686-4692.
27. Maguire, M.E., Van Arsdale, P.M. and Gilman, A.G. (1976):
     Molec. Pharmacol. 12:335-339.
28. Paton, W.D.M. and Rang, H.P. (1965): Proc. R. Soc. Lond. B
     163:1-44.
29. Rosenberger, L.B., Roeske, W.R. and Yamamura, H.I. (1979):
     Eur. J. Pharmacol. 14:179-180.
30. Rosenberger, L.B., Yamamura, H.I. and Roeske, W.R. (1980):
     J. Biol. Chem. 255:820-823.
31. Snyder, S.H., Chang, K.J., Kuhar, M.J. and Yamamura, H.I.
     (1975): Fed. Proc. 34:1915-1921.
32. Soudijn, W., Van Wijngaarden, I. and Ariens, E.J. (1973):
     Eur. J. Pharmacol. 24:43-48.
33. Tsai, B.S. and Lefkowitz, R.J. (1979): Molec. Pharmacol.
     16:61-68.

34. U'Prichard, D.C. and Snyder, S.H. (1978): J. Biol. Chem.
253:3444-3452.
35. Watanabe, A.M., McConnaughey, M.M., Strawbridge, R.A., Fleming,
J.W., Jones, L.R. and Becsh, H.R. (1978): J. Biol. Chem.
253:4833-4836.
36. Wessels, M.R., Mullikin, D. and Lefkowitzs, R.J. (1979):
Molec. Pharmacol. 16:10-20.
37. Yamamura, H.I. and Snyder, S.H. (1974): Proc. Nat. Acad. Sci.
71:1725-1729.
38. Yamamura, H.I. and Snyder, S.H. (1974): Molec. Pharmacol. 10:
861-867.
39. Zahniser, N.R. and Molinoff, P.B. (1978): Nature 275:453-455.

STRUCTURAL AND STERIC ASPECTS OF COMPOUNDS RELATED TO OXOTREMORINE

R. Dahlbom

Department of Organic Pharmaceutical Chemistry
Biomedical Center, University of Uppsala
751 23 Uppsala, Sweden

INTRODUCTION

In 1956 it was reported by Everett et al. (14) that an acetylenic compound, i,4-dipyrrolidino-2-butyne (tremorine, 1), induced
tremor, hypokinesia and rigidity, accompanied by marked parasympathetic stimulation, in a variety of experimental animals. These effects, which resemble the symptoms of the parkinsonian patient,
could be antagonized by drugs used clinically in parkinsonism, and
it was suggested that tremorine induced tremor would be a convenient
model for experimental evaluation of antiparkinsonian agents. However, it was soon suggested that a metabolite of tremorine might be
the biologically active species (2), and evidence was presented
that tremorine is metabolized in the liver to a compound responsible
for the pharmacological effects (47). The structure of this metabolite was determined in 1961 by Cho, Haslett and Jenden (5), who
isolated and identified it as N-(4-pyrrolidino-2-butynyl)-2-pyrroli-
done (oxotremorine, 2). They also confirmed the structure of the
compound.

$$N-CH_2-C\equiv C-CH_2-N \qquad \overset{O}{N}-CH_2-C\equiv C-CH_2-N$$

1   2

Oxotremorine is a specific muscarinic agent equal in potency to acetylcholine (ACh)(6,15). Unlike the quaternary compound ACh it readily penetrates into the central nervous system (CNS) after system administration and is relatively selective in producing central as opposed to peripheral effects, probably due to its favorable distribution to the brain. The extraordinary potency of oxotremorine is surprising in view of its structural dissimilarity to other muscarinic agents. Whereas most powerful muscarinic agents possess a quaternary trimethylammonium group, oxotremorine is a tertiary amine with no methyl group at all, and it contains an acetylenic bond at the position in the molecule where strong muscarinic agents usually have an oxygen atom. The mechanism of action of oxotremorine is not yet fully understood. It has been suggested (3) that it is able to assume a conformation which would make it possible for oxotremorine to interact directly with the muscarinic receptor. On the other hand, the possibility that oxotremorine acts either by stimulation of the synthesis of ACh or by release of ACh from an inactive precursor has been proposed (17). However, X-ray determinations (8,9) of the crystal and molecular structure of oxotremorine sesquioxalate show that the conformation in the solid state is not incompatible with the geometry of the muscarinic receptor as delineated by Beckett et al. (4). However, it must be kept in mind that the conformation in the solid state may not reflect exactly the conformation of the drug at the receptor.

We decided, in 1961, to undertake the synthesis and testing of potential antagonists to oxotremorine in the hope of finding drugs with specific action on the CNS and less peripheral effects, which might be clinically useful in Parkinson's disease. The starting point was to vary systematically the structure of the oxotremorine molecule in order to obtain analogs with the same high ability as oxotremorine to penetrate freely into the CNS but with antagonistic rather than agonistic properties.

During the course of our investigations some similar work was reported by other research groups (3,19,20,32). It became obvious that the structural requirements for muscarinic activity in this type of compound are very specific and that oxotremorine apparently possesses an optimal structure, as even slight changes in the structure lead to loss of the muscarinic activity or to change of the type of activity from agonistic to antagonistic. The only amino compound which has appreciable agonistic activity is the dimethyl-amino analog which in different tests had 1/10-1/15 of the activity of oxotremorine (3). Derivatives of cyclic amines with greater ring size than pyrrolidine are antagonists (19). Replacement of the triple bond by a double or single bond abolishes the activity (3).

A great number of mono-, di- and tricyclic amides and imides
were substituted for the 2-pyrrolidone moiety (3,11,18,20,25-27,31,
32). In no case was any appreciable agonistic activity observed in
these cyclic amides and imides. However, a most interesting tremorogenic activity in derivatives of N-methylacetamide and N,N,N-tri-
methylcarbamide was found (3). Antagonistic activity of varying
potency was observed in several N-(4-tert-amino-2-butynyl)-substi-
tuted cyclic imides (3,20). Our research group found appreciable
antagonistic activity in derivatives of succinimide (10,11,18),
maleimide (18), 3,4-dihydrocarbostyril (27) and 2-thiazolone (31).

N-(tert-Aminoalkynyl)-Substituted Succinimides

We decided to concentrate our efforts on succinimides and
thought it should be possible to enhance the antagonistic activity
of the parent compound by the branching or lengthening of the intermediate butyne chain. A great number of compounds of general formula

$$N-Y-C\equiv C-Z-Am$$

were prepared (11,18,24,28,29,36,43,45) where Y and Z are unbranched
or branched alkylene groups and Am is an acyclic or a cyclic amine.

Pharmacological Testing: The pharmacological tests were made
mainly by Professor D.J. Jenden (UCLA) or Dr. M.R. Blair (Astra
Pharmaceuticals, Massachusetts). The compounds were tested in intact mice for central and peripheral anticholinergic activity. The
dose of test compound required to double the dose of oxotremorine
inducing a predetermined tremor intensity (tremorolytic dose, T)
was estimated either using an electronic device or by inspection
(18,28,44). Furthermore, the dose required to double the pupil size
relative to the control (mydriatic dose, M) was determined. The
ratio M/T may be regarded as a measure of the central specificity
of the drug. The pharmacological data for some representative compounds are shown in Table 1.

The structure-activity relationships can be summarized as follows. Introduction of a methyl group at C-1 in the butyne chain
enhances the tremorolytic activity about 70 times. Higher alkyl
groups (ethyl, propyl) at this position or two alkyl groups also
give very active antagonists though weaker than the methyl compound.
Lengthening of the butynyl chain with a methylene group between the
acetylenic bond and the imide nitrogen also enhances the activity

TABLE 1: Pharmacological properties of N-(<u>tert</u>-aminoalkynyl)-substituted succinimides.

$$\text{succinimide}-N-Y-C\equiv C-CH_2-N\langle(CH_2)_n$$

Y	n	Tremorolytic Dose (μmol/kg)	Mydriatic Dose (μmol/kg)
CH_2	4	80	250
$CH(CH_3)$	4	1.2	7.3
$CH(CH_3)$	5	1.0	4.9
$CH(CH_3)$	6	0.45	5.6
$CH(CH_3)$	7	4.6	23
$CH(C_2H_5)$	4	3.1	4.1
$CH(C_3H_7)$	4	6.3	2.9
$C(CH_3)_2$	4	3.8	7.7
CH_2CH_2	4	4.5	13
$CH(CH_3)CH_2$	4	4	7.3
$CH_2CH_2CH_2$	4	32	230
Atropine		2.8	0.29

of the parent compound. Introduction of 1 or 2 methyl groups at C-4
or lengthening of the chain with a methylene group at this position
has very little or no effect on the activity.

With respect to the amino moiety of the molecule the cyclic
amines pyrrolidine, piperidine and perhydroazepine give the most
active compounds. Acyclic amines having at least one methyl group
on the nitrogen appear to be partial agonists. In some cases this
is also observed with pyrrolidines.

For the most active compounds the dose producing oxotremorine
blockade is considerably lower than that producing mydriasis. This
is in marked contrast to atropine, which is less effective in blocking
oxotremorine than in producing mydriasis. Consequently, these compounds can be regarded as anticholinergic agents with selectivity
for the CNS.

N-(_tert_-Aminoalkynyl)-Substituted 2-Pyrrolidones

Thus we had found that introduction of one methyl group at C-1
of the butyne chain of the succinimide analog of oxotremorine as well
as other small modifications of the chain enhanced the tremorolytic
potency nearly 2 powers of 10. It was of interest to enquire what
would be the result if such alterations were made in oxotremorine
itself. This had not been done so far. Would we get a superactive
oxotremorine-like compound? We made these modifications and found
to our great surprise that introduction of one single methyl group
at C-1 in oxotremorine as well as other changes in the chain reversed the activity and yielded strong antagonists, some of which
were in fact the most potent antagonists we had found so far (22-24,
35,43,45). Pharmacological data for some of these compounds are
shown in Table 2. It should be noted that the structure-activity
relationships observed in the succinimide series are valid for this
series too.

The Influence of Methyl Groups on Biological Action

We have just seen the profound influence that the introduction
of a methyl group at the C-1 position exerts on the activity. Oxotremorine has seven nonequivalent carbon atoms at which a methyl
group can be introduced, and we decided to prepare and investigate
the remaining six isomers.

Table 3 shows the pharmacological activity of the seven methyl
derivatives of oxotremorine. Only one is a tremorogenic agent,
though weak in comparison with oxotremorine (1/17). The other six
isomers are antagonists to oxotremorine, the compounds having the
methyl group in the 1-position of the butynyl chain and in the 5-
position of the pyrrolidone ring being extremely potent.

TABLE 2: Pharmacological properties of N-(<u>tert</u>-aminoalkynyl)-
         substituted 2-pyrrolidones.

$$\text{pyrrolidone}-N-Y-C\equiv C-CH_2-N(CH_2)_n$$

Y	n	Tremorolytic Dose (μmol/kg)	Mydriatic Dose (μmol/kg)
CH_2 (agonist, oxotremorine)	4	–	–
$CH(CH_3)$	4	0.45	1.65
$CH(CH_3)$	5	0.9	5
$CH(CH_3)$	6	0.8	4.85
$CH(CH_3)$	7	3.45	19.5
$CH(C_2H_5)$	4	5.6	1.8
$CH(C_3H_7)$	4	6.2	2.4
$C(CH_3)_2$	4	15.4	65
CH_2CH_2	4	8.3	33
$CH(CH_3)CH_2$	4	2.6	2.4
Atropine		2.8	0.29

Some Special Modifications
of the Oxotremorine Molecule

The allenic analog: A research group in our laboratories
working on the medicinal chemistry of allenes prepared the allenic
analog of oxotremorine (<u>3</u>). This compound was found to be inactive,
and it is evident that the rigid allenic structure does not fulfill
the steric and/or electronic requirements for affinity to the muscarinic receptor (7). It should be observed that <u>3</u> has an intermediate chain containing five carbon atoms, whereas oxotremorine
has four. However, the acetylenic analog of oxotremorine having
five carbon atoms in the intermediate chain is quite a strong
antagonist (Table 2).

TABLE 3:   Pharmacological properties of methyl derivatives of oxotremorine.

Position of CH_3	Tremorogenic Dose (μmol/kg)	Tremorolytic Dose (μmol/kg)
3"		5.2
2"		5.0
4		11
1		0.45
3'	11	
4'		37
5'		0.4
Oxotremorine	0.63	

3

4

        The thiolactam analog: The carbonyl group apparently plays
a vital role in the muscarinic activity of oxotremorine. We found
it of interest to prepare the thiolactam analog (4) to see what effect substitution of sulfur for oxygen might have on the activity.
We found that 4 is inactive itself, but is transformed in vivo to
oxotremorine (21).

        The azetidino analog: The only amino component which can be
substituted for pyrrolidino in oxotremorine with retained activity
is the dimethylamino group. It appeared to us that it might be
worthwhile to use azetidine, which would be close to both dimethyl-
amine and pyrrolidine in its steric requirements, as the amino component in an oxotremorine analog (5).

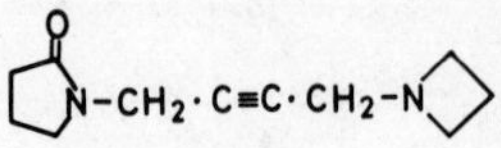

5 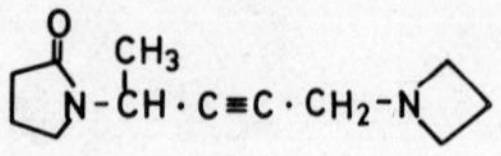   6

Compound 5 turned out to be a powerful muscarinic agonist at least
as potent as oxotremorine itself, in certain tests twice as active
(34). Introduction of a methyl group at C-1 in the butyne chain
reverses the activity and yields an antagonist (6) of medium
strength (tremorolytic dose 12 µg/kg)(45).

N-(4-tert-amino-2-butynyl)carboxamides: A most interesting
observation was made by the English research group. They prepared
and tested N-(4-pyrrolidino-2-butynyl)-N-methylacetamide and found
it to be a tremorogenic agent about half as active as oxotremorine.
The dimethylamino analog showed activity too, though quantitatively
less (3). This aroused our interest in this type of compound and
we prepared a number of analogs in order to study structure-activity
relationships (37,46). The results are summarized in Table 4. Only
four compounds are active (7-10), two agonists and two antagonists.
It is evident that only small variations in the amide moiety are
possible without complete loss of activity. The acyl group has to
be either acetyl or propionyl both in agonist and antagonists. The
alkyl substituent on the amide nitrogen determines the type of
activity. A methyl group is necessary for agonistic and an ethyl
group for antagonistic activity.

The amides in this study exist in solution as equilibrium
mixtures of E and Z conformers, as is evident from their NMR specta.
Since all atoms of the Z conformer unlike those of the E form are
superimposable on the corresponding atoms of oxotremorine, the
immediate suggestion arises that the muscarinic activity of 7 should
be associated primarily with the Z conformer.

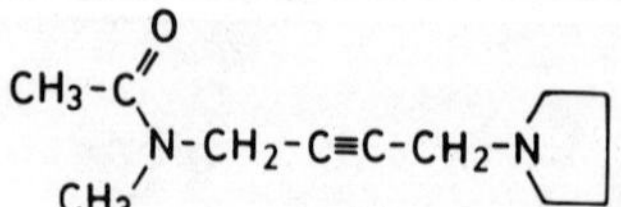

Z-7 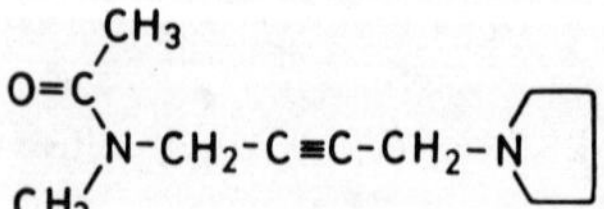   E-7

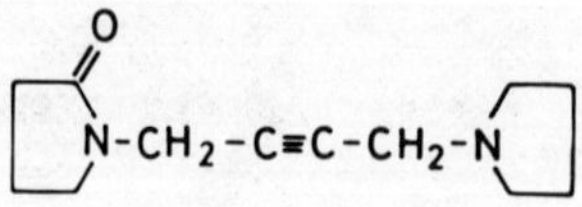

Oxotremorine

TABLE 4:   Pharmacological properties of N-(4-_tert_-amino-2-butynyl) carboxamides.

$$R-CO-N(R')-CH_2-C{\equiv}C-CH_2-N\langle\rangle$$

Compound	R	R'	Tremorogenic Dose (μmol/kg)	Tremorolytic Dose (μmol/kg)
	H	CH_3	–	–
	CH_3	H	–	–
7	CH_3	CH_3	0.24	
8	CH_3	C_2H_5		4.8
9	C_2H_5	CH_3	0.46	
10	C_2H_5	C_2H_5		8.5
	C_3H_7	H	–	–
	C_3H_7	CH_3	–	–
	C_3H_7	C_2H_5	–	–

There is a striking parallelism between oxotremorine and 7 concerning the structural requirements for biological activity. Introduction of a methyl group at the carbon next to the carbonyl group affords in both cases compounds which are still muscarinic agonists though weaker than the parent compounds. If a methyl group is introduced at the carbon next to the amide nitrogen, the 5-position in the pyrrolidine ring in oxotremorine or the N-methyl group of 7, the activity is reversed and strong antagonists are obtained (41,46).

The β-lactam analog of oxotremorine: The amides described above may be regarded as oxotremorine analogs in which the lactam ring has been opened between two carbon atoms. However, the most active agonist 7 does not correspond to an opened pyrrolidone ring as it contains only three carbon atoms in the amide part of the molecule - but to a 4-membered β-lactam ring. Consequently we prepared the β-lactam analog of oxotremorine (11) and found it to be an agonist having a tremorogenic activity of about one fourth of that of oxotremorine. Introduction of a methyl group at the C-1 position of the butyne chain afforded an antagonist (12), the tremorolytic dose being 4 μmol/kg (36).

TABLE 5: Tremorolytic activity of optically active N-(1-alkyl-4-
pyrrolidino-2-butynyl)-2-pyrrolidones.

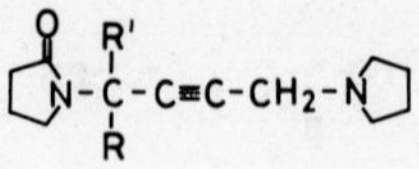

Compound	R	R'	Tremorolytic Dose (μmol/kg)	R/S Potency Ratio
(R)-13 (S)-13	H	CH_3	0.26 Inactive	
(R)-14 (S)-14	H	C_2H_5	0.52 27	52
(R)-15 (S)-15	H	C_3H_7	3.5 51	15
(R)-16 (S)-16	CH_3	C_2H_5	28 56	2

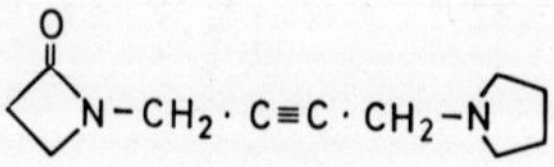

11

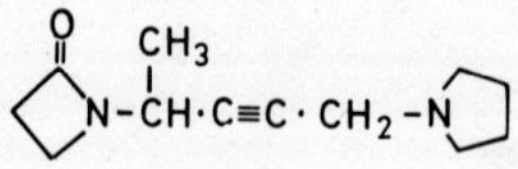

12

Stereoselectivity of Oxotremorine Antagonists
Having a Chiral Center in the Butynyl Chain

The most active oxotremorine antagonists we had prepared so far
had a methyl group at the C-1 position in the butyne chain. As this
chain contains an asymmetrically substituted carbon atom we decided
to prepare the enantiomers of these compounds in order to study
their stereoselectivity. We found that the R-isomers were about
twice as active as the corresponding racemates, whereas the S-isomers were practically inactive (12,30). This indicates that the R-
isomers have high affinity for the muscarinic receptors, whereas
the S-isomers have not.

TABLE 6:   Tremorolytic activity of optically active N-(1-alkyl-4-
           pyrrolidino-2-butynyl) succinimides.

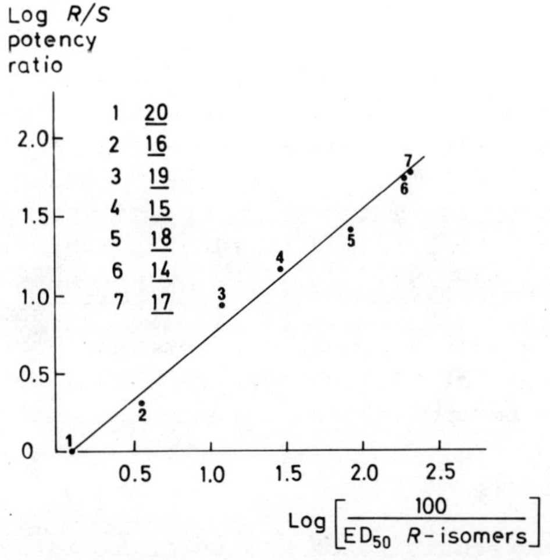

Compound	R	Tremorolytic Dose (μmol/kg)	R/S Potency Ratio
(R)-17 (S)-17	CH_3	0.48 28	58
(R)-18 (S)-18	.C_2H_5	1.2 30	25
(R)-19 (S)-19	C_3H_7	8.2 68	8

FIGURE 1: Tremorolytic activity. The logarithm for the enantiomeric
          potency ratio is plotted against the logarithm for the
          relative activity ($100/ED_{50}$) of the most active enantiomer
          (R).

In continuation of this research we also prepared the enantiomers of analogous compounds substituted with an ethyl, a propyl and
both an ethyl and a methyl group in the 1-position of the butynyl
chain. Thus, a series of closely related isomeric pairs became
available for a more thorough investigation of the stereochemical
requirements for high central activity in this class of anticholinergic compounds.

Tables 5 and 6 show that the R-isomers throughout are more
active than the S-isomers. The enantiomeric potency ratio (stereo-
selective index) increases with increasing activity of the R-isomer.
This satisfies Pfeiffer's rule (1,33), which states that the enantiomeric potency ratio of a more active compound is higher than that
of a less active one.

Figure 1 shows a plot of the enantiomeric potency ratios against
the relative tremorolytic activity of the more active member of each
enantiomeric pair (on logarithmic scales) for the compounds in
Tables 5 and 6. The achiral lower homolog of 17, N-(4-pyrrolidino-
2-butynyl)succinimide (20), is also included. The excellent correlation obtained is remarkable since the results are from in vivo
assays on whole animals.

Stereoselectivity of Oxotremorine Antagonists Containing a Pyrrolidine Group

In view of the stereoselectivity of oxotremorine antagonists
having a chiral center close to the lactam nitrogen we found it of
interest to investigate compounds with a center of chirality in the
vicinity of the other polar group in the molecule, the basic nitrogen atom. Consequently, a methyl group was introduced into the 2-
and 3-positions of the pyrrolidine ring of oxotremorine and its
succinimide analog (38,39). Only the enantiomers of the 2-substi-
tuted compounds show stereoselectivity, the S-isomers being the
most active (Table 7).   However, the stereoselectivity is less
pronounced than that of the enantiomers with a methyl group in the
1-position of the butynyl chain.

Stereoselectivity of Oxotremorine Antagonists Containing two Chiral Centers

These findings led us to investigate oxotremorine analogs with
two chiral centers, one in the 1-position of the butynyl chain and
one in the 2-position of the pyrrolidine ring. We prepared eight
optical isomers of this type (42) and the results of the tests for

TABLE 7: Tremorolytic activity of optically active N-[4-methyl-pyrrolidino)-2-butynyl]-substituted 2-pyrrolidones (21,22) and succinimides (23,24)

Compound	R	R'	Tremorolytic Dose (μmol/kg)	S/R Potency Ratio
(R)-21	CH$_3$	H	56	
(S)-21			2.6	22
(R)-22	H	CH$_3$	6.5	
(S)-22			5.8	1.1
(R)-23	CH$_3$	H	17.4	
(S)-23			8.3	2.1
(R)-24	H	CH$_3$	20	
(S)-24			17.2	1.2

tremorolytic activity are shown in Table 8. As might be expected, the isomer having methyl groups at both chiral centers and the "right" configuration at these centers [(R,S)-25] is the most active one and the most potent oxotremorine antagonist we have prepared so far.

The potency ratios of the enantiomeric pairs of 25 and 26 increase with increasing activity of the most active isomer of each pair in agreement with Pfeiffer's rule (Fig. 2).

Epimeric derivatives, which have the opposite configuration in the butyne chain but the same configuration in the pyrrolidine ring, also exhibit large differences in activity. In Fig. 3 the potency ratios of such epimeric pairs (1-4) have been plotted against the relative tremorolytic activity of the most active member of each pair. The epimeric potency ratios clearly increase with increasing activity of the more active epimer. The figure also shows a comparison between the epimeric pairs with the same configuration in the butyne chain but with the opposite configuration in the pyrrolidine ring (5-8). In this case a small and almost constant potency ratio is observed. These results indicate that only the chiral

TABLE 8:   Tremorolytic activity of optically active N[1-alkyl-4-(2-
           methylpyrrolidino)-2-butynyl]-2-pyrrolidones.

Compound*	R	Tremorolytic Dose (μmol/kg)	Enantiomeric Dose Potency Ratio
(R,S)-25 (S,R)-25	CH_3	0.10 20	200
(R,R)-25 (S,S)-25	CH_3	0.50 12	24
(R,S)-26 (S,R)-26	C_3H_7	1.5 24	16
(R,R)-26 (S,S)-26	C_3H_7	4.2 8.9	2.1

*   The first configurational symbol refers to the configuration of
    the chiral center in the butynyl chain and the second symbol to
    the configuration of the chiral center in the pyrrolidine ring.

center in the butynyl chain is critical to stereoselectivity, the
most active isomers having the R-configuration. The configuration
in the pyrrolidine ring is of less importance, the S-configuration
being somewhat more favorable.

It is of interest to compare these results with those reported
by Ellenbroek et al. (13) for anticholinergic drugs of a quite different type also containing two chiral centers, viz the stereoisomers of the β-methylcholine esters of hexahydrobenzilic acid and
α-methylatropic acid. They concluded from a study of the affinity
constants for the rat jejunum that only the chiral center in the
acyl moiety is critical to stereoselectivity, whereas that in the
β-methylcholine moiety is not.

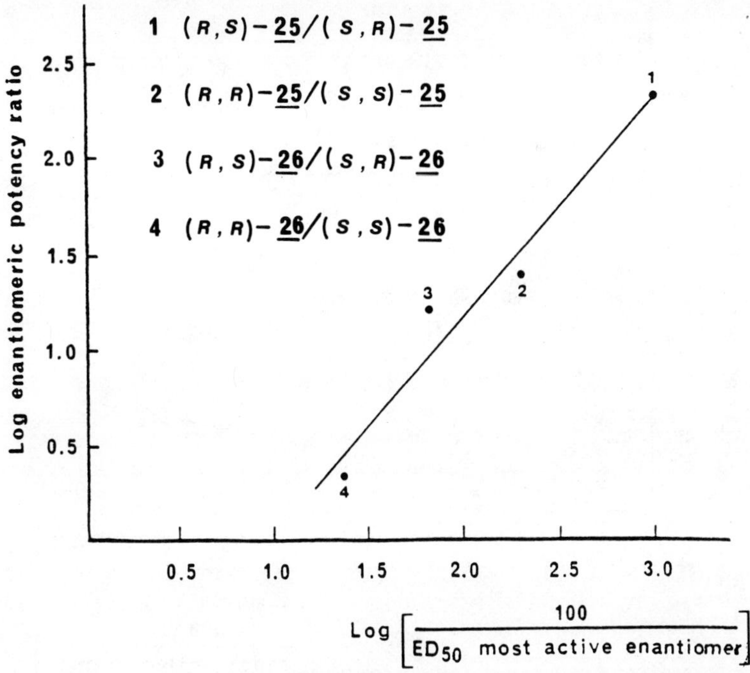

FIGURE 2: Tremorolytic activity. The logarithm for the enantiomeric potency ratio is plotted against the logarithm for the relative activity $(100/ED_{50})$ of the most active enantiomer.

CONCLUSIONS

One of the questions discussed during the first conference on cholinergic mechanisms ten years ago was "Do tremorogenic agents resemble ACh structurally?" (16). With respect to oxotremorine and related compounds the answer must be no. On the other hand there are very strict structural and steric requirements for biological activity in this class of compounds.

ACKNOWLEDGEMENTS

The extensive synthetic work has been carried out mainly by Drs. B. Karlen, B. Lindeke, S. Lindgren, A. Lindquist, B. Resul, B. Ringdahl and U. Svensson. This work was supported by grants from the Swedish Medical Research Council and Astra Pharmaceuticals, Sodertalje, Sweden.

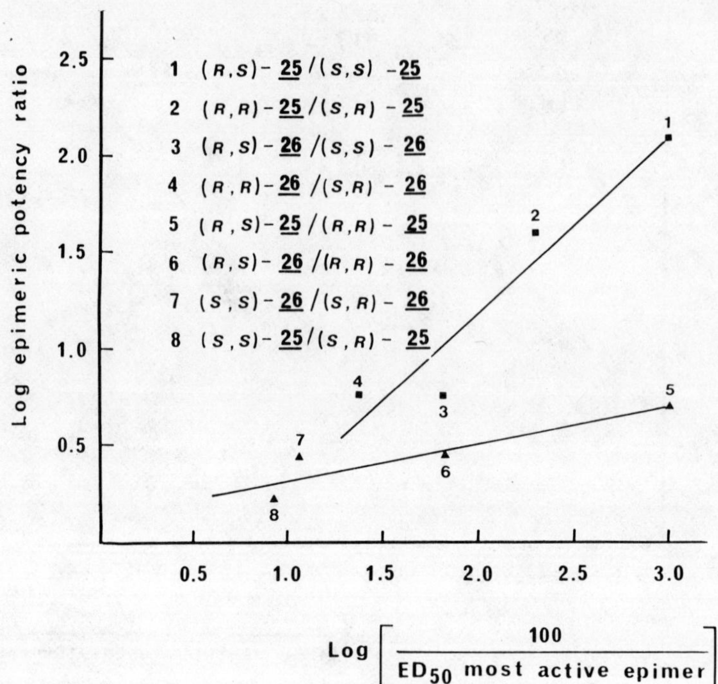

$$\text{Log}\left[\frac{100}{ED_{50}\ \text{most active epimer}}\right]$$

FIGURE 3: Tremorolytic activity. The logarithm for the epimeric
potency ratio is plotted against the logarithm for the
relative activity ($100/ED_{50}$) of the most active epimer.
■——— Compounds epimeric in the butynyl chain.
▲——— Compounds epimeric in the pyrrolidine ring.

REFERENCES

1. Ariens, E.J. (1966): Advan. Drug Res. 3:235–285.
2. Baker, W.W., Hosko, M.J., Rutt, W.J. and McGrath, J.R. (1960).
     Proc. Soc. Exp. Biol. Med. 104:214–217.
3. Bebbington, A., Brimblecombe, R.W. and Shakeshaft, D. (1966):
     Brit. J. Pharmacol. 26:56–57.
4. Beckett, A.H., Harper, N.J. and Clitherow, J.W. (1963): J.
     Pharm. Pharmacol. 15:362–371.
5. Cho, A.K., Haslett, W.L. and Jenden, D.J. (1961): Biochem.
     Biophys. Res. Commun. 5:276–279.
6. Cho, A.K., Haslett, W.L. and Jenden, D.J. (1962): J. Pharmacol.
     Exp. Ther. 138:249–257.

7.   Claesson, A., Asell, A., Bjorkman, S. and Olsson, L.-I. (1978):
     Acta Pharm. Suec, 15:105-110.
8.   Clarke, P.J., Pauling, P.J. and Petcher, T.J. (1975): J. Chem.
     Soc. Perkin Trans. 2:774-778.
9.   Dahlbom, R. and Jonsson, P.-G. (1975): J. Pharm. Pharmacol.
     27:544-546.
10. Dahlbom, R., Karlen, B., George, R. and Jenden, D.J. (1966):
     Life Sci. 5:1625-1631.
11. Dahlbom, R., Karlen, B., George, R. and Jenden, D.J. (1966):
     J. Med. Chem. 9:843-846.
12. Dahlbom, R., Lindquist, A., Lindgren, S., Svensson, U., Ringdahl,
     B. and Blair, M.R. Jr. (1974): Experientia 30:1165-1166.
13. Ellenbroek, B.W.J., Nivard, R.J.F., Van Rossum, J.M. and
     Ariens, E.J. (1965): J. Pharm. Pharmacol. 17:393-404.
14. Everett, G.M., Blockus, L.E. and Shepperd, I.M. (1956): Science
     124:79.
15. George, R., Haslett, W.L. and Jenden, D.J. (1962): Life Sci.
     1:361-363.
16. Heilbronn, E. and Winter, A. (1970): IN Drugs and Cholinergic
     Mechanisms in the CNS (eds) E. Heilbronn, E. and A. Winter,
     Res. Instit. Nat. Def., Stockholm, pp. 572-575.
17. Holmstedt, B. (1967): Ann. N.Y. Acad. Sci. 144:433-458.
18. Karlen, B., Lindeke, B., Lindgren, S., Svensson, K.-G., Dahlbom,
     R., Jenden, D.J. and Giering, J.E. (1970): J. Med. Chem. 13:
     651-657.
19. Levy, J. and Michel-Ber, E. (1966): C.R. Acad. Sci. Paris,
     Ser. D. 262:2546-2549.
20. Levy, J. and Michel-Ber, E. (1967): C.R. Acad. Sci. Paris,
     Ser. D 264:2250-2253.
21. Lindeke, B., Karlen, B., Dahlbom, R., Green, R. and Jenden, D.
     J. (1972): J. Pharm. Pharmacol. 24:25-31.
22. Lindgren, S., Lindquist, A., Lindeke, B., Svensson, U., Karlen,
     B., Dahlbom, R. and Blair, M.R. Jr. (1970): Experientia 26:
     1232-1233.
23. Lindgren, S., Lindquist, A., Lindeke, B., Svensson, U., Karlen,
     B., Dahlbom, R. and Blair, M.R. Jr. (1973): Acta Pharm. Suec.
     10:435-440.
24. Lindgren, S., Svensson, U. and Dahlbom, R. (1975): Acta Pharm.
     Suec. 12:503-506.
25. Lindquist, A. and Dahlbom, R. (1968): Acta Pharm. Suec. 5:
     227-232.
26. Lindquist, A. and Dahlbom, R. (1970): Acta Pharm. Suec. 7:
     527-530.
27. Lindquist, A., Lagerstrom, P.-O. and Dahlbom, R. (1972): Acta
     Pharm. Suec. 9:99-106.
28. Lindquist, A., Lindgren, S., Lindeke, B., Karlen, B., Dahlbom,
     R. and Blair, M.R. Jr. (1970): J. Pharm. Pharmacol. 22:707-
     708.

29. Lindquist, A., Lindgren, S., Lindeke, B., Karlen, B., Dahlbom, R. and Blair, M.R. Jr. (1972): Acta Pharm. Suec. 9:93-98.
30. Lindquist, A., Ringdahl, B., Svensson, U. and Dahlbom, R. (1976): Acta Chem. Scand. B 30:517-520.
31. Moses, P. and Dahlbom, R. (1978): Acta Pharm. Suec. 15:175-180.
32. Neumeyer, J.L., Moyer, U.V., Richman, J.A., Rosenberg, F.J. and Teiger, G.D. (1967): J. Med. Chem. 10:615-620.
33. Pfeiffer, C.C. (1956): Science 124:29-31.
34. Resul, B., Lewander, T., Zetterstrom, T., Ringdahl, B., Muhi-Eldeen, Z. and Dahlbom, R. (1980): J. Pharm. Pharmacol. (in press).
35. Resul, B., Ringdahl, B. and Dahlbom, R. (1979): Acta Pharm. Suec. 16:161-165.
36. Resul, B., Ringdahl, B. and Dahlbom, R. (1980):(in press).
37. Resul, B., Ringdahl, B., Hacksell, U., Svensson, U. and Dahlbom, R. (1979): Acta Pharm. Suec. 16:225-232.
38. Ringdahl, B. and Dahlbom, R. (1978): Experientia 34:1334-1335.
39. Ringdahl, B. and Dahlbom, R. (1978): Acta Pharm. Suec. 15:255-263.
40. Ringdahl, B. and Dahlbom, R. (1979): Acta Pharm. Suec. 16:13-20.
41. Ringdahl, B., Muhi-Eldeen, Z., Ljunggren, C., Karlen, B., Resul, B., Dahlbom, R. and Jenden, D.J. (1979): Acta Pharm. Suec. 16:89-94.
42. Ringdahl, B., Resul, B. and Dahlbom, R. (1979): J. Pharm. Pharmacol. 31:837-839.
43. Ringdahl, B., Svensson, U. and Dahlbom, R. (1975): Acta Pharm. Suec. 12:297-304.
44. Silverman, R.W. and Jenden, D.J. (1970): J. Appl. Physiol. 28:513-514.
45. Svensson, U., Dahlbom, R. and Blair, M.R. Jr. (1975): Acta Pharm. Suec. 12:209-214.
46. Svensson, U., Hacksell, U. and Dahlbom, R. (1978): Acta Pharm. Suec. 15:67-70.
47. Welch, R.M. and Kocsis, J.J. (1961): Proc. Soc. Exp. Biol. Med. 107:731-734.

SEARCH FOR NICOTINE-LIKE RECEPTOR BINDING SITES IN BRAIN

A. Nordberg and C. Larsson

Department of Pharmacology, University of Uppsala
Uppsala, Sweden

Our knowledge of nicotine-like receptors in brain, their distribution and functional importance is fragmentary. Most information has been obtained by using iontophoretic and electrophysiological techniques. The introduction of radio-receptor binding
techniques has, however, increased the possibility of characterizing
the nicotine-like binding sites in brain by using ligands with high
specific radioactivity. It is well established that α-bungarotoxin
(α-Btx) binds to the nicotine-like acetylcholine (ACh) receptor on
electric organ of electric eel and ray and on skeletal muscle and
functionally acts as an irreversible antagonist. Binding of α-Btx
to various nervous tissues such as sympathetic ganglia, retina and
brain have been reported (for review, see 5), but it has not yet
been proved that this binding is to a functionally active ACh receptor binding site (2,3,8). For nicotine receptor binding studies
in brain, α-Btx has been used by most investigators (5), but in a
few studies other ligands such as nicotine (1,7,10,17), and the
nicotinic antagonist tubocurarine (1,7) have been used.

Table 1 shows the binding constants (B_{max} and k_d) that are obtained when increasing concentrations of tritium labelled tubocurarine (^{3}H-TC; specific activity 16.48 Ci/mmol) are incubated
with P_2 fractions from the hippocampus of the mouse. Scatchard
plots of the binding data indicate two binding sites for ^{3}H-TC (a
high affinity and a low affinity binding site) and one binding site
for ^{3}H-Btx. As can been seen from Table 1 the B_{max} and the k_d
values for the high affinity site of ^{3}H-TC are rather similar to
the corresponding values for ^{3}H-Btx. Recently it has been suggested
by Carbonetto et al. (3) from their binding studies on Btx on chick

TABLE 1: ^{3}H-TC and ^{3}H-Btx binding to crude synaptosomes (P_2 frac-
          tion) from hippocampus of the mouse. Increasing concen-
          trations of ^{3}H-TC (1-10 nM) and ^{3}H-Btx (0.7-11 nM) were
          incubated for 15 and 60 min, respectively (see 7 for
          method). (Larsson and Nordberg, unpublished observations)

	B_{max} (pmol/g protein)	k_d (nM)
^{3}H-TC	275	14
	75	2
^{3}H-Btx	50	4

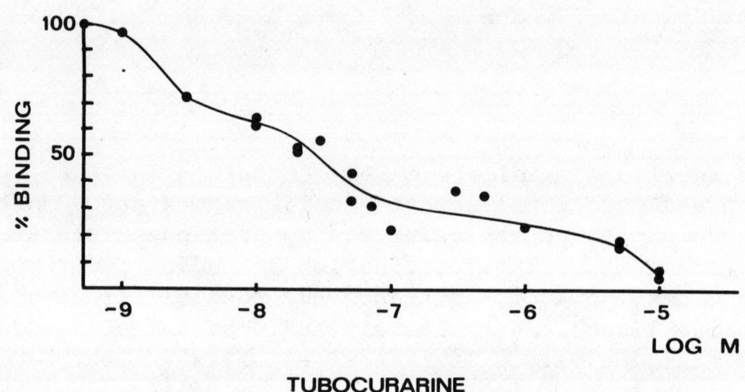

FIGURE 1: Displacement of ^{3}H-TC from its binding sites in brain
          by tubocurarine.

sympathetic ganglia that TC binds to two sites, one of which also
binds Btx.

     Figure 1 shows that increasing concentrations of unlabelled TC
displace ^{3}H-TC from its binding sites. The inhibition curve is flat
and its shape might indicate more than one binding site.

     Table 2 summarizes, from the literature, some of the IC50
values for different nicotinic agonists and antagonists that have
been obtained after incubation with different labelled ligands. In

TABLE 2:   Inhibition of binding of different labelled ligands to brain extracts.

Inhibitor	IC_{50} (µM)			
	^{3}H-TC	^{3}H-Nic	^{125}I-Btx	^{125}I-α-NNT
Nicotine	0.65[f]	1[a]; 1.0[f]	7[b]; 3.1[c] 1.3[d]; 2.0[g]	1.9[c]; 0.7[h]
Tubocurarine	0.02[f]	0.4[a]; 0.5[f]	7[b]; 1.9[c] 0.74[d]; 1.7[g]	6.3[e]; 0.5[h]
Carbachol		1[a]	90[c]; 140[d]	6[h]
ACh			60[b]; 30[c] 25[d]	21[e]; 2[h]

a = Ref. 17
b = Ref. 12
c = Ref. 11
d = Ref 4

e = Ref. 13
f = Ref. 7
g = Larsson & Nordberg: unpublished observations
h = Ref. 15

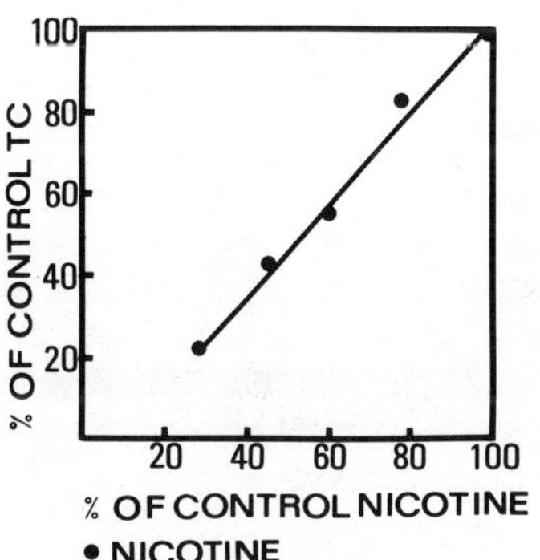

FIGURE 2:   Inhibition of ^{3}H-TC and ^{3}H-Nic binding to hippocampal synaptosomes by nicotine (10^{-9}–$5 \cdot 10^{-6}$ M). (Larsson and Nordberg, unpublished observations)

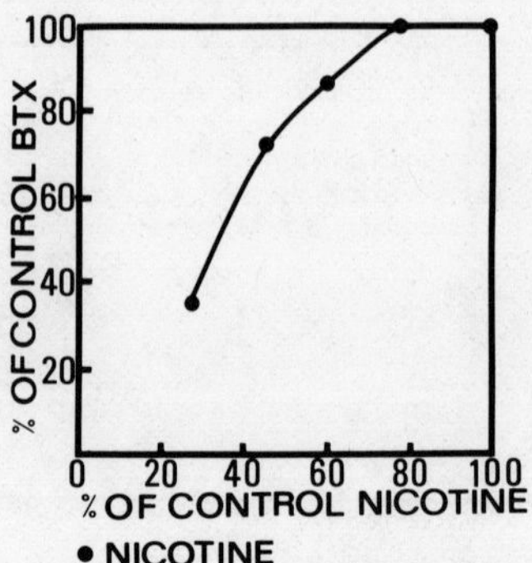

FIGURE 3: Inhibition of ^{3}H-Btx and ^{3}H-Nic binding to hippocampal synaptosomes by nicotine $(10^{-9}$–$5\cdot10^{-6}$ M). (Larsson and Nordberg, unpublished observations)

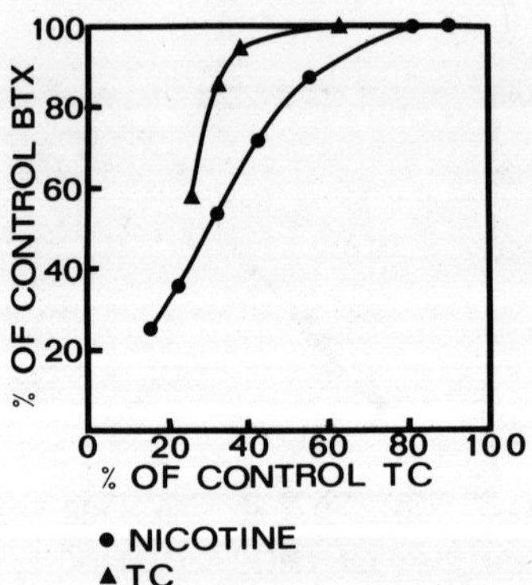

FIGURE 4: Inhibition of ^{3}H-Btx and ^{3}H-TC binding to hippocampal synaptosomes by nicotine $(10^{-8}$–10^{-5} M) and tubocurarine $(10^{-8}$–10^{-6} M). (Larsson and Nordberg, unpublished observations)

TABLE 3: Nicotine-like binding sites in different rat brain regions; comparison of published results (pmol/g protein)

| Brain Region | $^{125}I-\alpha-Btx$ | | | | $^{125}I-\alpha-NNT$ | | $^{3}H-Nic$ | $^{3}H-TC$ |
	3 nM Ref. a	1.5 nM Ref. b	15 nM Ref. c	10 nM Ref. d	1 nM Ref. e	4 nM Ref. f	Ref. g	3 nM Ref. h
Hippocampus	17	50	25	80	3.11	12.58	89	99
Cortex	21	52	14	58	1.62	6.36	25	144
Hypothalamus	--	75	26	--	2.69	--	95	135
Striatum	27	16	4	10	0.92	10.60	--	104
Cerebellum	5	8	1	10	0.23	1.26	25	99

a = Ref. 9
b = Ref. 6
c = Ref. 12
d = Ref. 16

e = Ref. 13
f = Ref. 15
g = Ref. 17
h = Ref. 7

TABLE 4: ^{3}H-TC and ^{3}H-QNB binding in mouse, rat and human hippocampus and human fetal brain. A P_2 fraction was incubated
with ^{3}H-TC (3 nM; specific activity 16.48 Ci/mmol) and
^{3}H-QNB (4 nM racemate; specific activity 16 Ci/mmol) for
15 and 60 min, respectively. Unspecific binding in the
presence of 10^{-4} M TC and 10^{-4} M atropine was subtracted.

	^{3}H-TC (pmol/g protein)	^{3}H-QNB (pmol/g protein)	$\dfrac{^3\text{H-TC}}{^3\text{H-QNB}}$
Hippocampus			
Mouse	50[b]	471[a]	0.1
Rat	99[a]	--	--
Human	130[c]	402[c]	0.3
Human fetal brain	220[d]	150[d]	1.5

a = Data taken from Ref. 7   c = Nordberg and Winblad;
b = Larsson and Nordberg;   unpublished observations
    unpublished observations   d = Nordberg; unpublished
                                observations

general the competitors seem to be more active in inhibiting labelled TC and nicotine (Nic) than labelled Btx and α-naja-naja toxin
(α-NNT) binding.

To be able to detect possible selectivity in the inhibition of
different ligands by the same competitor, displacement data obtained
in our laboratory were analyzed according to the method described
by Terenius and Wahlstrom (14). In Figs. 2-4 the inhibition of 3
labelled ligands (^{3}H-TC; ^{3}H-Nic; ^{3}H-Btx) by the same serial dilution
of competitors were compared. Figure 2 shows the inhibition of ^{3}H-
Nic and ^{3}H-TC by identical serial dilutions of unlabelled nicotine
(10^{-9}–$5\cdot10^{-6}$ M).   Each point represents percent inhibition of
binding of the competitor. As can be seen from Fig. 2 the remaining
binding of ^{2}H-TC and ^{3}H-Nic in the presence of a certain concentration of the competitor is the same, indicating that the competitor
displaces the labelled ligands equally.

Figure 3 shows the inhibition of ^{3}H-Nic and ^{3}H-Btx by unlabelled nicotine. In contrast to Fig. 2 the points do not form a straight
line. The curved line indicates that unlabelled nicotine displaces
^{3}H-Nic with a greater efficiency than ^{3}H-Btx. A similar curve is
also obtained for nicotine upon incubation with ^{3}H-TC and ^{3}H-Btx
(Fig. 4). Figure 4 also shows that unlabelled TC has a greater
preference for the ^{3}H-TC binding sites than the ^{3}H-Btx binding sites

(an even greater preference than unlabelled nicotine).

In Table 3 a comparison of published values for nicotine-like binding sites in different brain regions is made. It is somewhat difficult to compare the data since different ligands have been used and the binding values do not always represent B_{max} values. Although individual values obtained by different research groups are variable, a regional difference in binding of labelled Btx (^{125}I-Btx) can be seen. The number of ^{125}I-Btx binding sites is highest in the hippocampus and hypothalamus intermediate in the cortex and striatum, and lowest in the cerebellum. The ^{125}I-α-NNT binding follows about the same regional distribution as Btx. Yoshida and Imura (17) measured a fourfold higher binding of ^{3}H-Nic in the hippocampus and hypothalamus than in the cortex and cerebellum. Thus, a more marked difference in binding sites between cortex and hippocampus was found by using ^{3}H-Nic in comparison to using labelled Btx. Table 3 also gives the binding data for ^{3}H-TC binding in different regions. Small regional differences are found and in contrast to other ligands, a high ^{3}H-TC binding is measured in the striatum and cerebellum. Since ^{3}H-TC seems to have two binding sites (at least in the hippocampus; Table 1) an analysis of the regional distribution of the two binding sites is necessary and this is presently being investigated in this laboratory.

Binding of ^{3}H-TC to human hippocampus and to human fetal whole brain have been measured. Table 4 compares the binding of ^{3}H-TC in mouse, rat and human hippocampus, and in human fetal brain. Binding values for the labelled muscarinic antagonist quinuclidinyl benzilate (^{3}H-QNB) are also given. As seen in Table 4 the ^{3}H-TC binding is slightly higher in the human hippocampus compared with the hippocompus from rat and mouse. Human fetal brain shows a binding value which is about twice the value for rat hippocampus. The number of muscarinic binding sites is about the same in the mouse and human hippocampus, but much lower in the human fetal brain. The ratio ^{3}H-TC/^{3}H-QNB is higher in human fetal brain than in human or rat hippocampus. It is interesting to compare these findings with recent studies on ontogenesis of cholinergic binding sites in rat (2,16).

ACKNOWLEDGEMENT

These studies were supported by the Swedish Medical Research Council and the Swedish Tobacco Company.

REFERENCES

1. Abood, L.G., Tometsko, A. and MacNeil, M. (1979): Arch. Int. Pharmacodyn. 237:213-229.
2. Ben-Barak, J. and Dudai, Y. (1979): Brain Res. 166:245-257.
3. Carbonetto, S.T., Fambrough, D.M. and Muller, K.J. (1978): Proc. Nat. Acad. Sci. 75:1016-1020.
4. Eterovic, V.A. and Bennett, E.L. (1974): Biochem. Biophys. Acta 362:346-355.
5. Morley, B.J., Kemp, G.E. and Salvaterra, P. (1979): Life Sci. 859-872.
6. Morley, B.J., Lorden, J.F., Brown, G.B., Kemp, G.E. and Bradley, R.J. (1977): Brain Res. 134:161-166.
7. Nordberg, A. and Larsson, C. (1980): Acta Physiol. Scand. Suppl. 479:19-23.
8. Patrick, J. and Stallcup, W.B. (1977): Proc. Nat. Acad. Sci. 74:4689-4692.
9. Salvaterra, P.M., Mahler, H.R. and Moore, W.J. (1975): J. Biol. Chem. 250:6469-6475.
10. Schleifer, L.S. and Eldefrawi, M.E. (1974): Neuropharmacology 13:53-63.
11. Schmidt, J. (1977): Molec. Pharmacol. 13:283-290.
12. Segal, M., Dudai, Y. and Amsterdam, A. (1978): Brain Res. 148:105-119.
13. Speth, R.C., Chen, F.M., Lindstrom, J.M., Kobayshi, R.M. and Yamamura, H.I. (1977): Brain Res. 131:350-355.
14. Terenius, L. and Wahlstrom, A. (1976): Eur. J. Pharmacol. 40:241-248.
15. Tindall, R.S.A., Kent, M., Baskin, F. and Rosenberg, R.N. (1978): J. Neurochem. 30:859-863.
16. Wade, P.D. and Timiras, P.S. (1980): Brain Res. 181:381-389.
17. Yoshida, K. and Imura, H. (1979): Brain Res. 172:453-459.

IONIC EFFECTS ON ANTAGONIST BINDING TO THE MUSCARINIC RECEPTOR

FROM RAT CEREBRAL CORTEX

B. Hedlund

Department of Biochemistry, Arrhenius Laboratory
University of Stockholm
S-106 91 Stockholm, Sweden

Studies on the binding of the muscarinic antagonists ^{3}H-N-methyl-4-piperidinyl benzilate (^{3}H-4-NMPB) and ^{3}H-3-quinuclidinyl benzilate (^{3}H-3-QNB) have indicated deviations from first order kinetics of the association reaction under pseudofirst order conditions (8,9). Evidence has been provided for a mechanism of antagonist binding which involves two consecutive equilibria (6). Binding of both benzilates to the muscarinic receptor is followed by an isomerization of the receptor-antagonist complex:

$$R + A \underset{k_{-1}}{\overset{k_1}{\rightleftharpoons}} AR \underset{k_{-2}}{\overset{k_2}{\rightleftharpoons}} AR^* \qquad (1)$$

where R stands for receptor, A for antagonist, AR and AR* for the two different receptor-antagonist complexes and k_1, k_{-1}, k_2 and k_{-2} are rate constants. This mechanism, involving an isomerization reaction, has been proposed for antagonist binding to receptors both in the central nervous system and in the periphery (6). Using ^{3}H-4-NMPB further studies have been carried out for investigation of the ionic effects on the association and dissociation rates and on the equilibrium binding of the ^{3}H-antagonist to the receptor.

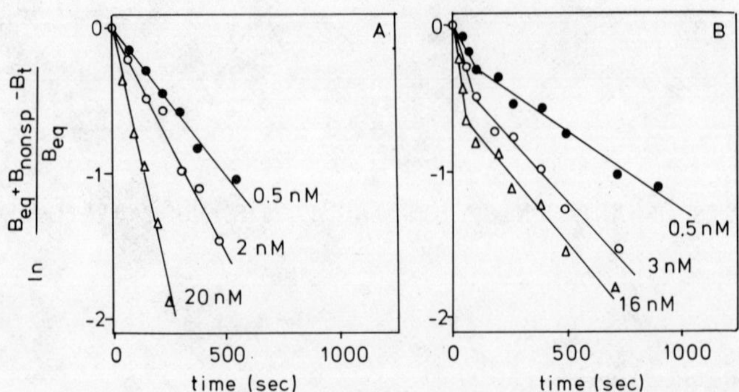

FIGURE 1: Binding of ^{3}H-4-NMPB (55.4 Ci/mmol) to muscarinic receptors in 50 mM phosphate buffer (A) and Krebs Ringer buffer
(B)(4). B_t stands for bound at time t. B_{eq} and B_{nonsp}
characterize amount bound at equilibrium.

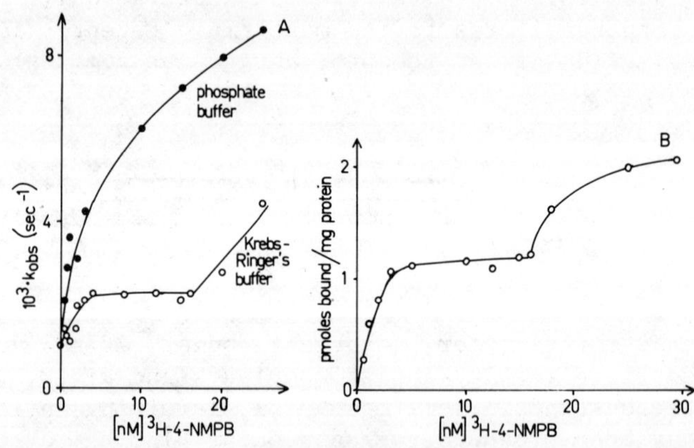

FIGURE 2: A – Dependence of the pseudofirst order rate constant for
^{3}H-4-NMPB binding on antagonist concentration.
B – Pmol bound ^{3}H-4-NMPB/mg protein under equilibrium
conditions.

TABLE 1:   Kinetic constants for cerebral cortex-muscarinic receptor
           ^{3}H-4-NMPB interactions.

Medium	$10^3 \times k_2$ (sec^{-1})	$10^3 \times k_{-2}$ (sec^{-1})	k_{-1}/k_1 (nM)	K_D (nM)
Phosphate buffer*	8.8 ± 1.1	1.2 ± 0.3	6.7 ± 2.5	1.04 ± 0.09
Krebs Ringer buffer	1.2 ± 0.1	1.0 ± 0.2	1.2 ± 0.2	1.0 ± 0.1

* Data from ref. 6.
K_D is the overall dissociation constant for the receptor-antagonist
complex and k_{-1}/k_1 the dissociation constant for the first reaction
step according to equation 1:

$$K_D = k_{-1}/k_1 \cdot k_{-2}/k_2$$

RESULTS AND DISCUSSION

The importance of the ionic composition of the medium is well
known for generation of the physiological response. Extracellular
calcium is essential for muscarinic receptor stimulated cGMP synthesis (3). Omission of calcium from the medium does also change
the properties of the receptor (4). In phosphate buffer the observed association rate under pseudofirst order conditions of ^{3}H-
4-NMPB to the receptor yields a straight semilogarithmic plot
(Fig. 1A). The kinetic constants calculated are identical with
those earlier obtained in smooth muscle using the same medium (6).
Changing phosphate buffer to Krebs Ringer buffer gave a significant
decrease in the association rate (Fig. 1B). This effect cannot be
due to change in ionic strength since the two media have almost
identical ionic strengths.

Kinetic curves at all concentrations up to 16 nM ^{3}H-4-NMPB
were biphasic: the first reaction step lasting up to 5 min. The
observed rate constant of this part decreases as the concentration
of ^{3}H-4-NMPB increases. This reaction step is followed by a second
one where the observed rate constant shows a hyperbolic dependence
on the concentration of ^{3}H-4-NMPB, indicating the same isomerization mechanism earlier shown in phosphate buffer (6,7) (Fig. 2A).
In phosphate buffer the dissociation rate under pseudofirst order

conditions yields a straight semilogarithmic plot independent of
the concentration of ^{3}H-4-NMPB. The value of the rate constant is
identical to that of k_{-2}. The dissociation in Krebs Ringer buffer
is biphasic with the second step independent of ^{3}H-4-NMPB concentration used to reach RA equilibrium. Its value corresponds k_{-2}.
The rate of dissociation during the first minutes increased as the
concentration of atropine, added to prevent rebinding of dissociated
^{3}H-4-NMPB, is increased. From these data the kinetic constants for
the isomerization reaction in Krebs Ringer buffer can be calculated.
In phosphate buffer the results were identical with those earlier
obtained (6). The appearance of biphasic on and off-rate curves
seems to be specific for the more physiological composition of the
medium indicating a different receptor conformation under these
conditions than in phosphate buffer. It is also reported that at
10°C no biphasic curves can be detected (9). These biphasic association and dissociation curves may be explained in several ways.
For example, in phosphate buffer only AR* is stable enough to be
detected by the filtration assay used. It may be that in Krebs
Ringer buffer AR dissociates more slowly and can also be detected.
In that case the first part of the biphasic curves should reflect
the first reaction step according to Equation 1.

A different explanation is a heterogenity in antagonist binding
sites similar to that found for agonists (2). This has also been
observed for two antagonists; atropine (1) and pirenzepine (5). The
same heterogenity can be detected for both 3-QNB (data not shown)
and 4-NMPB (Fig. 2B) in equilibrium binding experiments in Krebs
Ringer buffer. It may be that ^{3}H-4-NMPB binds to a second site in
a one-step association reaction

$$A + R \rightleftarrows AR$$

or that the antagonist can bind to two different subsites of the
same receptor with a negative cooperativity between the binding
sites. In equilibrium binding studies with ^{3}H-4-NMPB two sites
with K_D values of 1 nM and 18 nM, respectively, can be detected.
Whether appearance of a second binding site at high ^{3}H-4-NMPB concentrations and the biphasic kinetic curves reflect the same interaction is a topic for further investigation. The present study
provides evidence for "subclasses of antagonist binding sites" and
indicates that these are more easily detected in Krebs Ringer buffer than in phosphate buffer. Thus, since the ionic strength is
identical in the two media, ionic influence must prevail in the
behavior of the receptor.

ACKNOWLEDGEMENTS

This study was supported by grants to Dr. T. Bartfai from the Swedish Medical Research Council and USPHS/NIMH, USA. The supply of ^{3}H-4-NMPB by Drs. Y. Kloog and M. Sokolovsky (Tel Aviv University) and the experimental help by Ms. M. Grynfarb is gratefully acknowledged.

REFERENCES

1.   Alberts, P. and Bartfai, T. (1976): J. Biol. Chem. 251:1543-1547.
2.   Birdsall, N.J.M. and Hulme, E.C. (1976): J. Neurochem. 27:7-16.
3.   Bartfai, T., Study, R.E. and Greengard, P. (1977): IN Cholinergic Mechanisms and Psychopharmacology (ed) D.J. Jenden, Plenum Press, New York, pp. 285-295.
4.   Hedlund, B. and Bartfai, T. (1979): Molec. Pharmacol. 15:531-544.
5.   Hammer, R., Berrie, C.P., Birdsall, N.J.M., Burgen, A.S.V. and Hulme, E.C. (1980): Nature 283:90-92.
6.   Jarv, J., Hedlund, B. and Bartfai, T. (1979): J. Biol. Chem. 254:5595-5598.
7.   Jarv, J., Hedlund, B. and Bartfai, T. (1980): J. Biol. Chem. 255:2649-2651.
8.   Kloog, Y., Egozi, Y. and Sokolovsky, M. (1979): FEBS Lett. 97:265-268.
9.   Kloog, Y. and Sokolovsky, M. (1978): Brain Res. 144:31-48.

STUDIES ON THE STRUCTURE OF ACETYLCHOLINE RECEPTOR

V. Witzemann and J.H. Walker

Abteilung Neurochemie, Max-Planck-Institute fur
Biophysikalische Chemie
Gottingen, Federal Republic of Germany

INTRODUCTION

The subunit composition of the nicotinic acetylcholine receptor (AChR) from elasmobranch fish is still a matter of controversy.
Results obtained from affinity chromatography (6,9), immunology
(2,4), crosslinking experiments (10,12) and labelling studies using
photolabile ligands (11,13) support the notion that AChR is composed of more than one polypeptide chain and that the complex of
molecular weight (MW) approximately 250,000 Dalton (7) is composed
of four different polypeptides of apparent MW 40,000, 50,000,
60,000 and 65,000 Dalton. Nevertheless, other workers report that
the 40,000 MW polypeptide chain is the predominant polypetide (3,8)
with the other polypeptides present only in trace amounts and co-
purified with the receptor in a rather irreproducible manner. Using
crossed rocket immunoelectrophoresis with antibodies raised to
detergent solubilized, purified AChR from Torpedo californica
we present further evidence for the multisubunit structure of the
AChR.

RESULTS AND DISCUSSON

Antiserum was raised to AChR from Torpedo californica purified
by affinity chromatography as described by Raftery et al. (6) and
Claudio and Raftery (2). As shown in Fig. 1, a singly immuno-

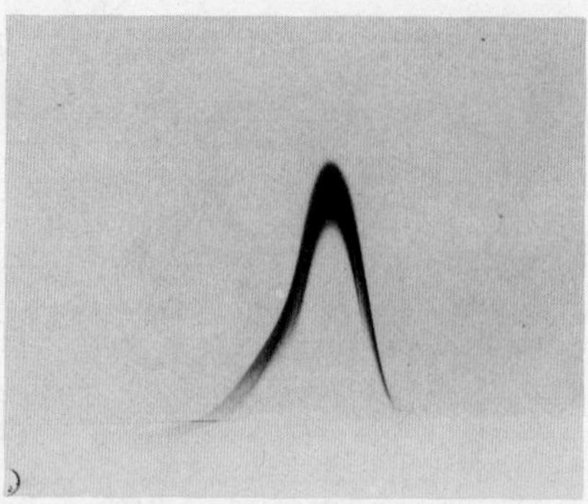

FIGURE 1: Crossed rocket immunoelectrophoresis. A triton X-100 extract of freshly prepared membrane fragments containing
AChR was separated by electrophoresis in agarose in the
first dimension followed by electrophoresis into a section
of gel containing antibodies specific for AChR.

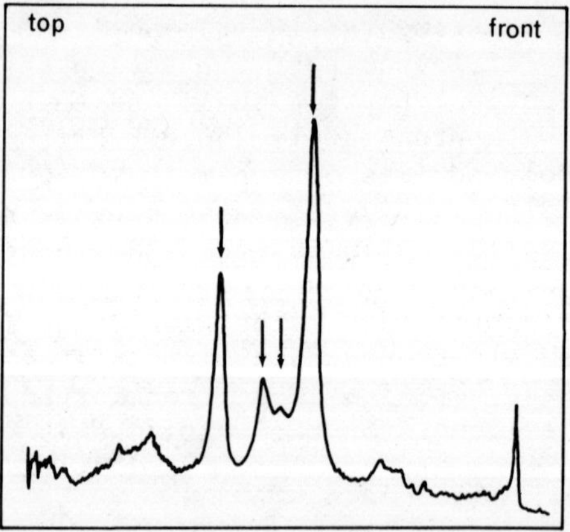

FIGURE 2: Densitometric scan of the immunoprecipitate. The immunoprecipitate containing radiolabelled antigen was analyzed
by SDS-PAGE and the polypeptides subsequently visualized
by means of fluorography. Arrows indicate polypeptides
of apparent MW 40,000, 50,000, 60,000 and 65,000.

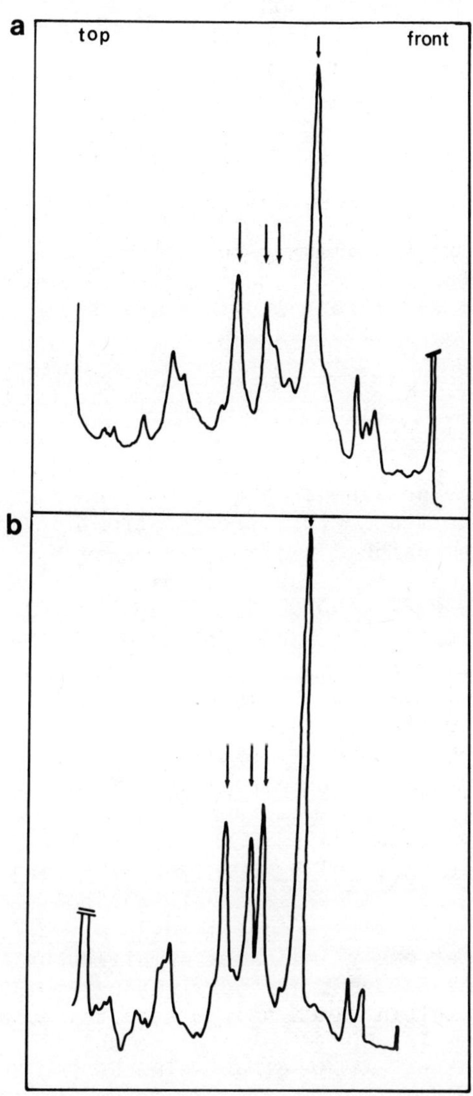

FIGURE 3: SDS–PAGE of membrane fragments containing AChR. Identical amounts of membrane fragments were loaded onto SDS polyacrylamide gels: a) 11.1:0.15 (acrylamide:bis–acrylamide); b) 8.75:0.23 (acrylamide:bis–acrylamide. Arrows indicate the positions of the receptor polypeptides.

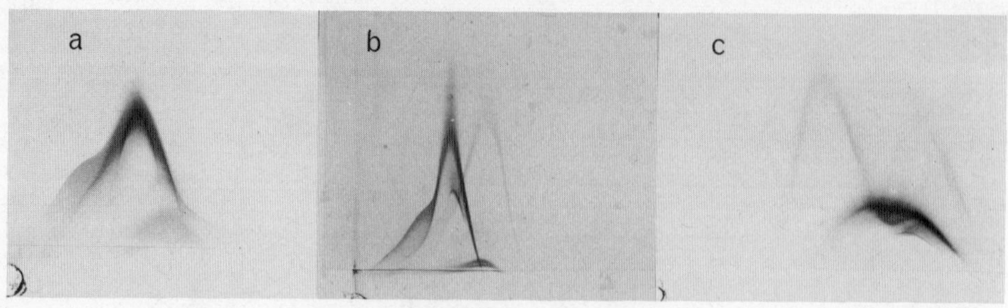

FIGURE 4: Crossed rocket immunoelectrophoresis. Effects of storage
         of membrane fragments upon the immunoprecipitate pattern
         of AChR: a) storage at 4°C for 3 days; b) storage at 4°C
         for 14 days; c) storage at -20°C.

---

precipitin line was obtained, when a Triton X-100 extract of AChR
containing membrane fragments, isolated from Torpedo marmorata
electric organ was analyzed against this serum. Incubation of the
immunoprecipitate with ^{125}I-α-bungarotoxin (^{125}I-α-Btx) containing
buffer followed by extensive washing steps showed that the immuno-
precipitated material would bind the radiolabelled toxin as evidenced by autoradiography and therefore contained the AChR. Analysis
of this immunoprecipitate by sodium dodecylsulfate polyacrylamide
gel electrophoresis (SDS-PAGE) should conclusively demonstrate the
subunit composition of AChR. A similar approach has been used to
demonstrate the multisubunit composition of cytochrome oxidase (5).
With radioactively labelled membrane fragments results identical to
those shown in Fig. 1 were obtained and when the radiolabelled immunoprecipitate was analyzed by SDS-PAGE, four major polypeptides
were observed (Fig. 2). The fact that four polypeptides of different molecular weights were contained in a single toxin binding
immunoprecipitate strongly indicates a multisubunit composition of
the AChR. If these polypeptides were not part of a stable complex
several immunoprecipitin lines should be obtained and only that
which contained the 40,000 MW polypeptide should bind ^{125}I-α-Btx.
The poor resolution of polypeptides of MW 50,000, 60,000 in Fig. 2
reflexts the anomalous migration behavior of these carbohydrate
containing polypeptides, a generally observed phenomenon of glycoproteins, and is a result of the gel system used (see legend to
Fig. 3). Complete resolution of these polypeptides occurred when
the bis-acrylamide concentration was increased about 1.6-fold (Fig.3).

As shown in Fig. 4, the immunoprecipitin line obtained with
freshly prepared membrane fragments underwent dramatic changes upon
storage. After three days at 4°C a decrease occurred in the size of

the strong immunoprecipitin line, other faint rockets became evident
(Fig. 4A), and the pattern became much more complex. Prolonged
storage at 4°C enhanced the complexity of the pattern (Fig. 4B). The
results are consistent with proteolytic modification of the AChR
complex to produce breakdown products of different electrophoretic
mobilities. Similar changes in the pattern of rocket immunoprecipitin
lines have been observed with other antigens (1). Changes in the
pattern of immunoprecipitin lines were also observed with membrane
fragments stored frozen at −20°C before analysis (Fig. 4C). In this
case proteolysis is less likely to have been responsible for the
observed change in rocket pattern and it may be that freezing and
rethawing of membrane fragments resulted in a partial dissociation
of the multisubunit complex. Multiple immunoprecipitin lines have
been observed previously for multisubunit proteins due to the presence of both holoenzyme and subunits in the extract analyzed (5).
In all cases the major rocket seemed to contain several polypeptides
similar to those obtained with freshly prepared membrane fragments
but, in addition, the number of high molecular weight polypeptides
increased. The minor immunoprecipitin lines contained a polypeptide
of apparent MW 40,000 which most probably is derived from the receptor complex. No traces of the 50,000, 60,000 and 65,000 MW
species were detected in the minor immunoprecipitin lines although
in some cases substantial amounts of a higher molecular weight species
(approximately MW 110,000) was obtained.

The results support the notion that phenomena such as proteolysis, aggregation and splitting of the AChR complex may occur
during lengthy purification procedures and give rise to variable
subunit compositions of the AChR complex.

CONCLUSIONS

With a specific antiserum to the AChR, a single immunoprecipitate is obtained. The immunoprecipitate binds ^{125}I-α-Btx and
thus contains the receptor complex.

Analysis of the immunoprecipitate obtained with radiolabelled
antigen on SDS-PAGE demonstrates the presence of several polypeptides within the receptor complex which seem to correspond to
four different polypeptides of approximately MW between 40,000 and
65,000.

The AChR can undergo rapid changes in structure upon freezing
or storage at 4°C, as demonstrated by crossed-rocket immunoelectrophoresis.

REFERENCES

1.   Bjerrum. O.J. and Bog-Hansen, T.C. (1976): IN Biochemical
     Analysis of Membranes (ed) A.H. Maddy, Chapman & Hall, London,
     pp. 378-426.
2.   Claudio, T. and Raftery, M.A. (1977): Arch. Biochem. Biophys.
     181:484-489.
3.   Heidmann, T. and Changeux, J.-P. (1978): Ann. Rev. Biochem.
     47:317-357.
4.   Lindstrom, J., Einarson, B. and Merlie, J. (1978): Proc. Nat.
     Acad. Sci. 75:769-773.
5.   Poyton, R.O. and Schatz, G. (1975):J. Biol. Chem. 250:762-766.
6.   Raftery, M.A., Vandlen, R., Michaelson, D., Bode, J., Moody,
     T., Chao, Y., Reed, K., Deutsch, J. and Duguid, J. (1974):
     J. Supramolec. Struct. 2:582-592.
7.   Reynolds, J.A. and Karlin, A. (1978): Biochemistry 17:2035-2038.
8.   Sobel, A., Weber, M. and Changeux, J.-P. (1977): Eur. J. Bio-
     chem. 80:215-224.
9.   Weill, C.L, McNamee, M.G. and Karlin, A. (1974): Biochem.
     Biophys. Res. Commun. 61:997-1003.
10. Witzemann, V., Muchmore, D. and Raftery, M.A. (1980): Bio-
     chemistry 24:5511-5518.
11. Witzemann, V. and Raftery, M.A. (1977): Biochemistry 16:5862-
     5868.
12. Witzemann, V. and Raftery, M.A. (1978): Biochem. Biophys. Res.
     Commun. 85:623-631.
13. Witzemann, V. and Raftery, M.A. (1978): Biochemistry 17:3598-
     3604.

NEWLY SYNTHESIZED ACETYLCHOLINE RECEPTOR IN DENERVATED

SKELETAL MUSCLE

M.G. Giacobini-Robecchi and M. Garelli

Istituto di Anatomia Umana Normale
Universita di Torino
Corso Massimo d'Azeglio, 10126 Torino, Italy

It is well known that the acetylcholine receptor (AChR) remains only in the neuromuscular junction, very soon after its formation in the chick embryo between the 12th and 14th incubation day (1,4), following its gradual disappearance from the entire length of the skeletal muscle fiber. Whatever the causes of this disappearance of extrajunctional AChR may be, it is also well known that when the muscle fiber is denervated (2,3,5) the extrajunctional AChR is then formed ex novo and reappears throughout the entire sarcolemma. We therefore looked to see where this resynthesis takes place in the denervated fiber.

The large pectoral muscle (a singly innervated fast muscle) was denervated in ten 8- to 10-day-old chicks). Ten days later, these and other control muscles were fixed _in situ_ for about 30 min prior to removal. During this period, bundles containing 50–100 fibers were carefully separated mechanically. The muscles were treated with [^{125}I] α-bungarotoxin (specific activity 20–200 Ci/mmol) and prepared for EM autoradiography. Prior to labelling, the fibers were treated with saponine (as suggested by Fambrough and Devreotes [5]) to make the intracellular binding sites accessible to the toxin without altering the ultrastructural characteristics.

Intracellular binding sites outside the endplate was noted in only 5% of the perinuclear areas of approximately 700 nuclei examined from the control muscles, whereas in the denervated muscles, in spite of the saponine induced damage, it was observed throughout the sarcolemma, as well as in the intracellular site in most of the approximately 600 perinuclear areas examined. The receptor was primarily located in the perinuclear cisterna and Golgi complex

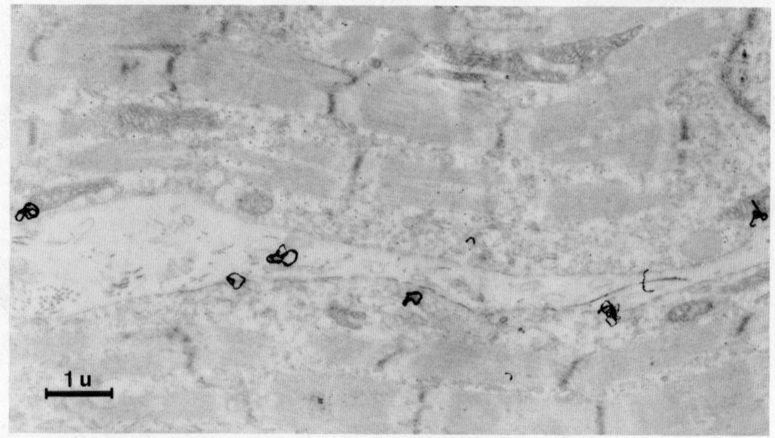

FIGURE 1: Denervated muscle – silver grains are located at the sarcolemma outside the end plate.

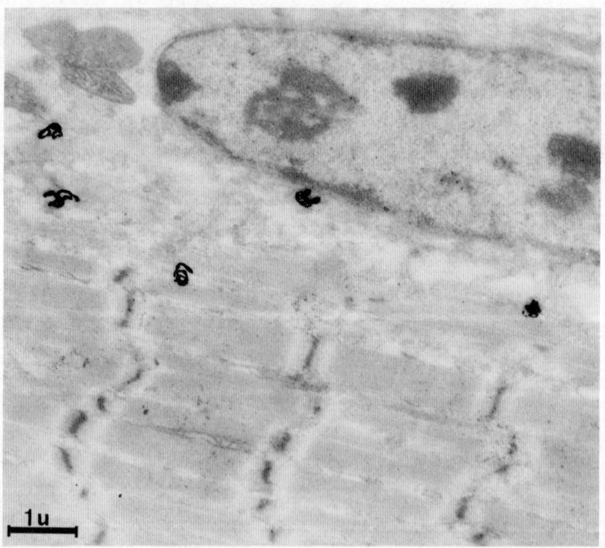

FIGURE 2: Denervated muscle – an extensive area of perinuclear cytoplasm present silver grains among the endoplasmic reticulum.

and at the internal membrane systems (Figs. 1 and 2). These areas
may be regarded as the synthesis sites (5).

Our data show that in all probability the extrajunctional AChR
appearing throughout the sarcolemma after denervation is very likely
produced by the extrajunctional machinery, and not by that of the
endplate. This machinery re-establishes the process of synthesis
that it lost when the junction was formed. This finding is surprisingly in parallel with other ultrastructural and cytochemical
signs displayed by denervated fibers (7), such as the clumps of
basophil material and subsarcolemmal clumping of ribosomes noted
along the syncytium. It thus offers further evidence for the view
that in the singly innervated fast muscle the junctional and extrajunctional machineries for AChR synthesis are separate (as suggested
by Sisto-Daneo and Filogamo [8]) but interdependent.

It is not certain how the extrajunctional activity is dis-
inhibited in the absence of neural influence. The mechanism
probably involves the T tubules that plunge from the sarcolemma
into all the sarcomeres of the striated syncytium.

ACKNOWLEDGEMENTS

This work was supported by the CNR.

REFERENCES

1.   Betz, H., Bourgeois, J.P. and Changeux, J.P. (1977): FEBS
       Lett. 77:219-224.
2.   Brockes, J.P. and Hall, Z.W. (1975): Biochemistry 14:2100-2106.
3.   Brockes, J.P. and Hall, Z.W. (1975): Proc. Nat. Acad. Sci.
       72:1368-1372.
4.   Burden, S. (1977): Develop. Biol. 57:317-329.
5.   Devreotes, P.N. and Fambrough, D.M. (1975): Proc. Nat. Acad.
       Sci. 73:161-164.
6.   Fambrough, D.M. and Devreotes, P.N. (1978): J. Cell Biol.
       76:237-244.
7.   Gauthier, G.F. and Schaeffer, S.F. (1974): J. Cell Sci. 14:
       113-137.
8.   Sisto-Daneo, L. and Filogamo, G. (1974): J. Submicro. Cytol.
       6:219-228.

ANATOMICAL ORGANIZATION OF SOME CHOLINERGIC SYSTEMS

IN THE MAMMALIAN FOREBRAIN

H.C. Fibiger and J. Lehmann

Division of Neurological Sciences
Department of Psychiatry
University of British Columbia
Vancouver, British Columbia V6T 1W5 Canada

Although acetylcholine (ACh) was one of the first neurotransmitters to be found in the central nervous system, knowledge concerning the anatomical organization of cholinergic neurons has lag-
ged far behind other, much more recently identified types of neurons.
For example, more is currently known about the anatomy of the multi-
farious peptidergic neurons than is understood about cholinergic
neurons. The reason for this has been the absence of suitable
techniques for mapping central cholinergic systems. Earlier attempts to trace central cholinergic systems using acetylcholinesterase (AChE) histochemistry met with considerable skepticism
when it became evident that some neurons that were certainly not
cholinergic contained high levels of AChE activity (3,18,24). AChE
activity therefore came to be considered an unreliable marker for
cholinergic neurons.

Another approach which has been applied to the elucidation of
central cholinergic anatomy is based on the purification of the
enzyme which is considered to be specific for cholinergic neurons,
choline acetyltransferase (CAT), and subsequent immunohistochemistry using antibodies raised against the puried CAT (9,13,29). To
date, the published results that have been obtained with this approach have been criticized on the basis that the CAT was not purified to homogeneity (31). These criticisms do not appear to be
without foundation because more recent immunohistochemical attempts
to map cholinergic systems have resulted in findings that are substantially different from the earlier reports (19).

Recent experiments in this laboratory have suggested that the introduction of certain modifications of the AChE histochemical procedure provides a potentially powerful and simple procedure for the identification of cholinergic cell bodies. Specifically, the use of the irreversible inhibitor, diisopropylphosphorofluoridate (DFP), introduced by Fukuda and Koelle (10) in 1959 and later employed by Lynch et al. (25) and extensively by Butcher and coworkers (1-3), has revealed that there are striking differences in the rate at which neurons synthesize AChE de novo. In particular, under this histochemical regimen, known cholinergic neurons of the central nervous system consistently have high levels of perikaryal AChE. On the basis of the results of this modified histochemical technique for AChE, we have proposed that intense staining for AChE 4-8 hr after DFP administration is a necessary, but not sufficient, criterion for the identification of cholinergic perikarya (21). For the purposes of the present discussion, "intense" staining can best be defined with reference to that which is seen in the strongly AChE reactive neurons of the medial septal nucleus and the nucleus of the diagonal band (26) or the large neurons of the striatum (25) 4-8 hr after DFP pretreatment. As more parts of the brain are viewed in the context of the necessary but not sufficient criterion, it becomes apparent that subjective comparison must be replaced by a quantitative technique. Electron microprobe and microphotometry may be profitably developed toward that end.

According to this hypothesis neurons that do not show the same degree of AChE staining as do these reference cells cannot be cholinergic. Because there are surprisingly few groups of neurons in the CNS that meet this criterion, application of this rule can greatly reduce the regions of the brain on which additional experiments must be focussed to confirm the presence of cholinergic perikarya. Finally, it is imperative to bear in mind that the DFP-AChE histochemical procedure and the subsequent application of the criterion of intense AChE reactivity can lead to false positives with regard to cholinergic perikarya. For example, dopamine neurons of the substantia nigra (3,20) and noradrenergic neurons of the locus coeruleus (24) appear to fall into the intensely AChE reactive category and yet these neurons are not cholinergic. This underscores the importance of using other complementary approaches to confirm or deny the cholinergic nature of neurons that have been identified as being potentially cholinergic by application of the "necessary, but not sufficient" criterion. In the following examples, the utility of this approach is demonstrated.

The Cerebral Cortex

There is presently little agreement concerning the anatomical

substrates of cholinergic activity in the neocortex. The first
experiments which suggested an extrinsic source of cholinergic innervation of the cerebral cortex were performed in 1963 by Hebb et
al. (14). They found that in the cat CAT activity decreased by 83%
1 to 5 weeks after the cortex was undercut, i.e. after all afferents
had degenerated. Green et al. (11,12) obtained a 65% decrease in
the rat following a similar operation. Collier and Mitchell (6)
found that both the·spontaneous and stimulus evoked release of ACh
from the cortex fell by 80% following undercutting. In contrast,
other workers have found either a transient or no decrease in CAT
following partial cortical isolation (33). In addition, electrophysiological experiments have indicated that a response to adjacent cortical stimulation, blocked by atropine, was unaffected by
cortical isolation (15), thus suggesting the existence of cortical
cholinergic interneurons.

In this laboratory a complete cortical isolation was achieved
by making a hemitransection at the rostral pole of the striatum.
Very close agreement with Green et al. (11) with respect to CAT and
AChE was obtained (Table 1). Some residual CAT activity was found.
These results suggest that the bulk of CAT and AChE are localized
to afferents to the cortex. The residual CAT activity, however,
suggests that some cholinergic neurons reside in the cerebral cortex. This conclusion presented difficulties for the hypothesis
that high AChE activity is a necessary but not sufficient criterion
for cholinergic neurons: in the cerebral cortex, there are no intensely AChE reactive neurons. Thus, according to the hypothesis,
there should be no cholinergic perikarya in the cortex, only cholinergic afferents. Yet, obviously, some CAT survived total de-
afferentation.

Part of the answer to this paradox came from the use of the
neurotoxin kainic acid in this laboratory. This toxin selectively
destroys perikarya while leaving axons and terminals of extrinsic
origin predominantly intact (2,7,27). Following injection of kainic
acid into the same frontal cortical region, histological analysis
showed complete and uniform perikaryal destruction over approximately 50% of the tissue analyzed biochemically. Concomitantly, no
decrease in CAT was observed, along with a small decrease in AChE
(Table 2). Extensive and uniform neuronal destruction was indicated
not only by cresyl violet histology, but also by large decreases in
glutamic acid decarboxylase (GAD) and high affinity glutamate uptake, markers for GABAergic and glutamatergic neurons, respectively.
Together, the above experiments suggest that all, or at least the
vast majority of CAT activity in the neocortex is contained in afferents that originate outside the cortex.

TABLE 1:   Neurotransmitter related enzymes in the frontal cortex
3 weeks following hemitransection at the anterior pole
of the striatum.

	Lesioned Side Percent Control	Contralateral Percent Control	Control (unoperated) (activity $\pm$ S.E.M.)
CAT	33.6 $\pm$ 5.3***	97.5 $\pm$ 4.4	30.0 $\pm$ 1.2[+]
AChE	53.7 $\pm$ 8.4***	84.6 $\pm$ 5.6*	3.98 $\pm$ 0.10[++]
GAD	128.8 $\pm$ 11.0**	109.4 $\pm$ 4.8*	210 $\pm$ 6.8[+]
GLU-up	62.3 $\pm$ 8.1**	83.4 $\pm$ 4.9*	1.51 $\pm$ 0.10[++]

n = 7.   * p < 0.05;   ** p < 0.01;   *** p < 0.001; Student's
two-tailed test
+ = nmol/mg protein/h;   ++ =   µmol/mg protein/h

TABLE 2:   Neurotransmitter related enzymes in the frontal cortex
1 week following local kainic acid injection (10 nmol/
2 µl).

	Lesioned Side Percent Control	Contralateral Percent Control	Control (unoperated) (activity $\pm$ S.E.M.)
CAT	92.2 $\pm$ 4.9	96.4 $\pm$ 5.5	29.1 $\pm$ 2.0[+]
AChE	83.4 $\pm$ 5.3*	86.6 $\pm$ 4.6	3.59 $\pm$ 0.22[++]
GAD	59.3 $\pm$ 5.0**	100.4 $\pm$ 4.1	190 $\pm$ 7.3[+]
GLU-up	56.6 $\pm$ 4.5**	84.3 $\pm$ 6.0*	1.23 $\pm$ 0.04[++]

n = 6.   * p < 0.05;   ** p < 0.001, Student's two-tailed test.
+ = nmol/mg protein/hr;   ++ =   µmol/mg protein/h

Nucleus Basalis Magnocellularis

It is only recently that the source of these cholinergic fibers
has been established. Not surprisingly, AChE played a key role in
suggesting the specific cell group. In 1967 Shute and Lewis (32)
reported that certain neurons in the globus pallidus sent AChE containing axons to the cortex. Divac (8) and Mesulam and van Hoesen
(30) performed neuroanatomical studies and noted the correspondence
between neurons which projected to the cortex and those which contained high levels of AChE, leading them to speculate that a cell

TABLE 3:   Choline acetyltransferase and acetylcholinesterase activities in the frontal cortex after lesions in the region of
nucleus basalis magnocellularis.

	Percent Control	Control Activity $\pm$ S.E.M.
Kainic acid lesions (n = 11)		
CAT	78.0 $\pm$ 2.2***	30.5 $\pm$ 0.70[+]
AChE	79.3 $\pm$ 4.0***	5.51 $\pm$ 0.10[++]
Electrolytic lesions (n = 9)		
CAT	76.2 $\pm$ 4.9*	27.8 $\pm$ 1.94[+]
AChE	74.2 $\pm$ 3.9**	5.85 $\pm$ 0.36[++]
Hemitransections caudal to nBM (n = 7)		
CAT	95.7 $\pm$ 5.3	26.9 $\pm$ 1.63[+]
AChE	96.1 $\pm$ 7.5	5.20 $\pm$ 0.45[++]

* $p < 0.02$; ** $p < 0.01$; *** $p < 0.001$, Student's two-tailed test.
+ nmol/mg protein/h;   ++ μmol/mg protein/h

TABLE 4:   Neurotransmitter related enzymes in the region of the
nucleus basalis magnocellularis 6 months after extensive
unilateral cortical lesions.

	Percent Control	Control Activity $\pm$ S.E.M.
CAT	64.1 $\pm$ 6.0**	63.9 $\pm$ 2.9[+]
AChE	80.6 $\pm$ 3.4*	19.5 $\pm$ 1.6[++]
GLU-up	107.0 $\pm$ 9.3	296 $\pm$31[+]

n = 4.   * $p < 0.02$; ** $p < 0.001$, Student's two-tailed test.
+ = nmol/mg protein/h;   ++ = μmol/mg protein/h.

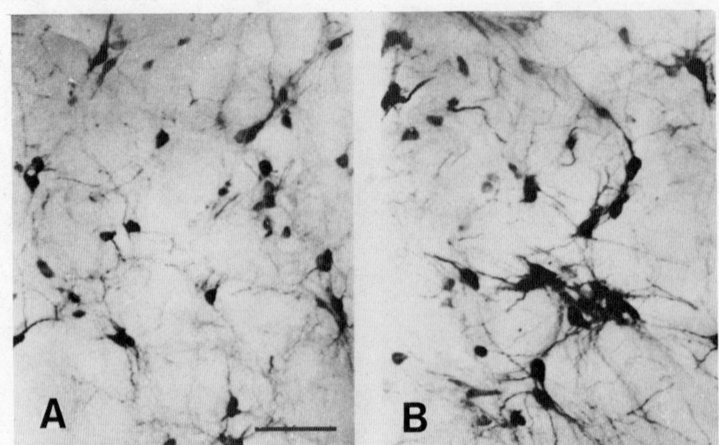

FIGURE 1: Retrograde degeneration of nucleus basalis magnocel-
          lularis following cortical lesion. Six months following
          extensive cortical ablation on the left side of the
          brain, the nBM on the left side (A) suffered a reduction
          in number of intensely AChE reactive neurons. Note also
          the decrease in diameter and dendritic arborization of
          surviving nBM neurons, compared to the contralateral
          control (B). AChE histochemical stain; 5 hr survival
          following 1.5 mg/kg DFP. Calibration bar: 50 μm.
          CP = caudate-putamen; F = fornix; GP = globus pallidus;
          IC = internal capsule; LHA = lateral hypothalamic area;
          LPO = lateral preoptic area; MaPO = magnocellular pre-
          optic nucleus; nBM = nucleus basalis magnocellularis;
          TAD = anterodorsal thalamic nucleus; TAV = anteroventral
          thalamic nucleus; TR = reticular thalamic nucleus;
          TV = ventral thalamus.

---

group known as the nucleus basalis of Meynert, in close proximity
to the globus pallidus, may be the source of the cortical cholinergic fibers. Subsequently, Kelly and Moore (16) found that large
lesions of the globus pallidus resulted in decreases in cortical
CAT activity. In this laboratory both electrolytic and kainic acid
lesions in a restricted part of the nucleus basalis resulted in decreases in cortical CAT and AChE, while slightly more caudal hemi-
transections did not result in such decreases Table 3). These experiments strongly pointed to the existence of a cell group on the
medial border of the globus pallidus which was cholinergic and
projected to the cortex. This hypothesis was further supported by
the demonstration of decreases in CAT activity in the region of
the nucleus basalis following cortical lesions (Table 4) which

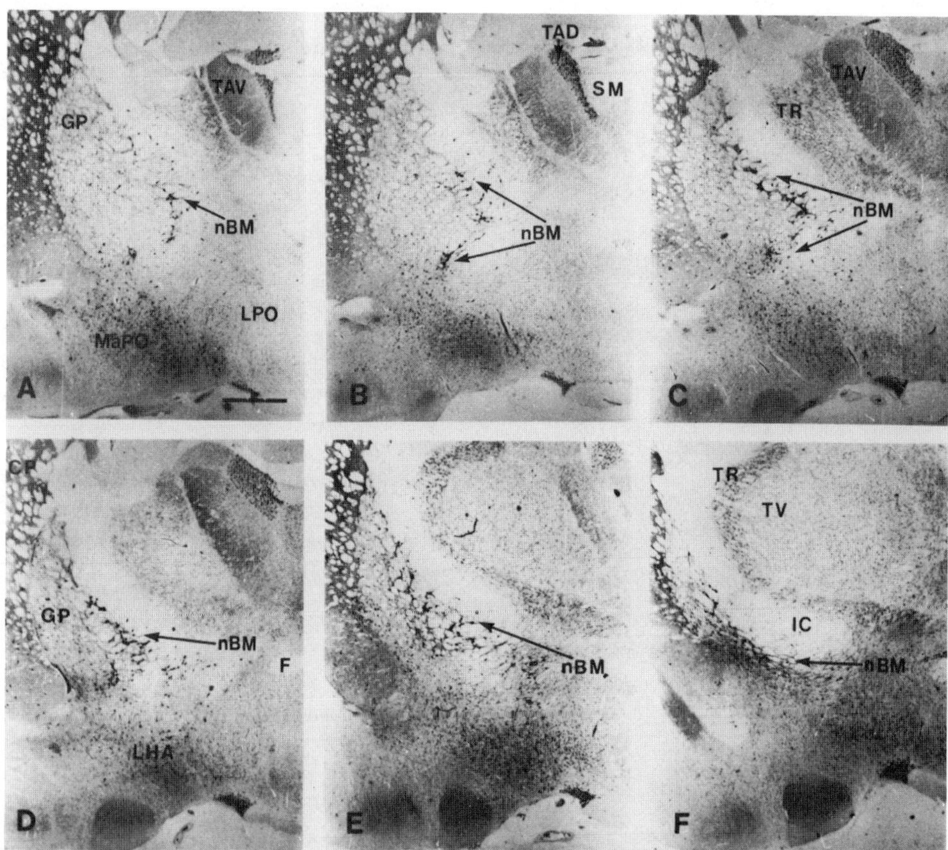

FIGURE 2: The distribution of intensely AChE reactive neurons in
        rat forebrain. Five hours after inactivation of AChE by
        intramuscular injection of 1.5 mg/kg DFP, newly syn-
        thesized enzyme has been transported as far as the proxi-
        mal dendrites of neurons which synthesize the enzyme.
        Some "background" AChE is still found, partly due to less
        than 100% complete inactivation by the sublethal DFP
        administration. The nucleus basalis magnocellularis,
        situated as it is on the border between the globus pal-
        lidus and the internal capsule, resides in a region of
        very low "background" AChE activity.

correlated with the retrograde loss of intensely AChE reactive
neurons in the same region (Fig. 1). These intensely AChE reactive
neurons could be shown to accumulate retrogradely transported horseradish peroxidase which was injected in the cortex (23). A major
source of the cortical cholinergic axons was therefore demonstrated
to be the nucleus basalis magnocellularis and could be mapped both
by AChE histochemistry (Fig. 2) and by retrograde tracing techniques
(23).

The Striatum

On the basis of lesion experiments, cholinergic neurons intrinsic to the striatum have been thought to exist for some time
(4,28). The first attempt to morphologically identify these neurons
employed the technique of immunohistochemistry (13). The neuron
type identified as the cholinergic neuron in these experiments was
the medium spiny neuron, which is thought to make up over 90% of
the total striatal neuron population (17). On the basis of neuroanatomical considerations and the observation that medium spiny
neurons contained little if any AChE, this observation was chal-
lenged (22). Instead, a large neuron comprising about 1% of the
total neuron population and thought to have few if any spines (17),
was proposed as the putative cholinergic neuron, since of all striatal neurons it alone had high AChE levels (22).

The opportunity to test this prediction presented itself in
the form of a lesion which selectively decreases CAT but not GAD
in the striatum (5). Kainic acid injected into the striatum of
10 day old rat pups produced a 25% depletion in CAT activity without
affecting GAD activity (21). When similarly lesioned litter mates
of these animals were examined histologically, a selective depletion
of the large, intensely AChE reactive neurons was observed, while
there was no decrease in the population density of two other histologically identifiable neurons, the cresyl violet impregnated population (made up almost entirely of medium spiny neurons) and smaller
weakly AChE reactive neurons (21). These results made the intensely
AChE reactive neuron an attractive candidate as the cholinergic
neuron. In addition, this large neuron appears to be an interneuron (22), consistent with biochemical indications that the cholinergic neuron of the striatum is also an interneuron. In the
striatum the most sophisticated available histological techniques
have been brought to bear on the morphological identification of
the cholinergic neuron. Surprisingly, AChE histochemical techniques
known for two decades together with a simple empirical rule led most
directly to the identification of the cholinergic neuron.

CONCLUSION

The mapping of cholinergic pathways in the brain has only begun, and without doubt the most advanced histological techniques will be called upon to accomplish this task. In achieving the primary step of identifying cholinergic perikarya, the rapid, versatile, and easily reproducible technique of AChE histochemistry used together with an irreversible AChE inhibitor has much to offer (Figs. 1 and 2). If indeed the empirical finding that high AChE levels are always found in cholinergic neurons continues to hold true, the consideration of its function in cholinergic neurons will become a pressing question.

ACKNOWLEDGEMENTS

Supported by the Medical Research Council. The excellent technical assistance of S. Atmadja is gratefully acknowledged.

REFERENCES

1.   Butcher, L.L. (1977): Life Sci. 21:1207-1226.
2.   Butcher, L.L. and Rogers, R.C. (1978): Eur. J. Pharmacol. 50:287-289.
3.   Butcher, L.L., Talbot, K. and Bilezikjian, L. (1975): J. Neural Transm. 37:127-153.
4.   Butcher, S.G. and Butcher, L.L. (1976): Brain Res. 71:167-171.
5.   Campochiaro, P. and Coyle, J.T. (1978): Proc. Nat. Acad. Sci. 75:2025-2029.
6.   Collier, B. and Mitchell, J.F. (1967): J. Physiol. 188:83-98.
7.   Coyle, J.T. and Schwarcz, R. (1976): Nature 263:244-246.
8.   Divac, I. (1976): Brain Res. 93:385-398.
9.   Eng, L.F., Uyeda, C.T., Chao, L.P. and Wolfgram, F. (1974): Nature 250:243-245.
10. Fukuda, T. and Koelle, G.B. (1959): J. Biol. Biochem. Cytol. 5:433-440.
11. Green, J.R., Halpern, L.M. and Van Niel, S. (1970): Life Sci. 9:481-488.
12. Green, J.R., Halpern, L.M. and Van Niel, S. (1970): Brain 93:57-64.
13. Hattori,T., Singh, V.K., McGeer, E.G. and McGeer, P.L. (1976): Brain Res. 102:164-173.
14. Hebb, C.O., Krnjevic, K. and Silver, A. (1963): Nature 198:692.
15. Jordan, L.M. and Phillis, J.W. (1972): Brit. J. Pharmacol. 45:584-595.

16. Kelley, P.H. and Moore, K.E. (1978): Exp. Neurol. 61:479-484.
17. Kemp, J.M. and Powell, T.P.S. (1971): Phil. Trans. R. Soc.
      Ser. B 262:383-410.
18. Koelle, G.B. (1955): J. Pharmacol. Exp. Ther. 114:167-184.
19. Kimura, H., McGeer, P.L., Peng, F. and McGeer, E.G. (1980):
      Science 208:1057-1059.
20. Lehmann, J. and Fibiger, H.C. (1978): J. Neurochem. 30:615-624.
21. Lehmann, J. and Fibiger, H.C. (1979): Life Sci. 25:1939-1947.
22. Lehmann, J., Fibiger, H.C. and Butcher, L.L. (1979): Neuro-
      science 4:217-225.
23. Lehmann, J., Nagy, J.I., Atmadja, S. and Fibiger, H.C. (1980):
      Neuroscience 5:1161-1174.
24. Lewis, P.R. and Schon, F.E.G. (1975): J. Anat. 120:373-385.
25. Lynch, G.S., Lucas, P.A. and Deadwyler, S.A. (1972): Brain
      Res. 45:617-621.
26. Lynch, G., Rose, G. and Gall, C. (1978): IN Functions of the
      Septo-Hippocampal System, Elsevier, Amsterdam, pp. 5-24.
27. McGeer, P.L. and McGeer, E.G. (1976): Nature 263:517-519.
28. McGeer, P.L., McGeer, E.G., Fibiger, H.C. and Wickson, V.
      (1971): Brain Res. 35:308-314.
29. McGeer, P.L., McGeer, E.G., Singh, V.K. and Chase, W.H. (1974):
      Brain Res. 81:373-379.
30. Mesulam, M.M. and Van Hoesen, G.W. (1976): Brain Res. 109:152-
      157.
31. Rossier, J. (1980): Brain Res. 98:619-622.
32. Shute, C.C.D. and Lewis, P.R. (1967): Brain 90:497-521.
33. Ulmar, G., Ljungdahl, A. and Hokfelt, T. (1975): Exp. Neurol.
      46:199-208.

CHOLINERGIC PATHWAYS TO THE CEREBRAL CORTEX IN RATS

H. Wenk, V. Bigl, U. Meyer and D. Biesold

Institute of Anatomy, Humboldt University Medical School
Berlin, Germany (GDR), and
Paul Flechsig Institute of Brain Research, Karl Marx
University, Leipzig, Germany (GDR)

INTRODUCTION

Recent experiments from several laboratories revealed that
magnocellular nuclei of the substantia innominata (S.I.) of the
basal forebrain of primates and rats show a direct and widespread
projection towards the ipsilateral cerebral cortex (2,6). The
large multipolar neurons of this area are usually referred to as
the nucleus basalis of Meynert (NBM). Using a combined horseradish
peroxidase (HRP) acetylcholinesterase (AChE) method those cells
which project to the cortex were found to stain heavily for AChE
(14). In addition, the basal nuclei are rich in choline acetyltransferase (CAT)(16,17) and the NBM neurons are characterized as
"cholinergic" by the pharmacohistochemical technique for AChE
(Lehmann and Fibiger, this volume; and 18). This evidence, together with recent results using different lesion techniques suggests that magnocellular neurons of the forebrain represent the
main, if not the only, origin for the cortical cholinergic projection (3,15,20,21,24).

In previous papers from our laboratory we quantitatively
evaluated the extent of this cholinergic cortical projection
originating in the NBM neurons of the basal forebrain using a
combination of the stereotaxic lesion technique with bio- and
histochemical estimation of CAT as well as AChE activity (15,21,
23,25). This paper summarizes results on the distribution of chol-

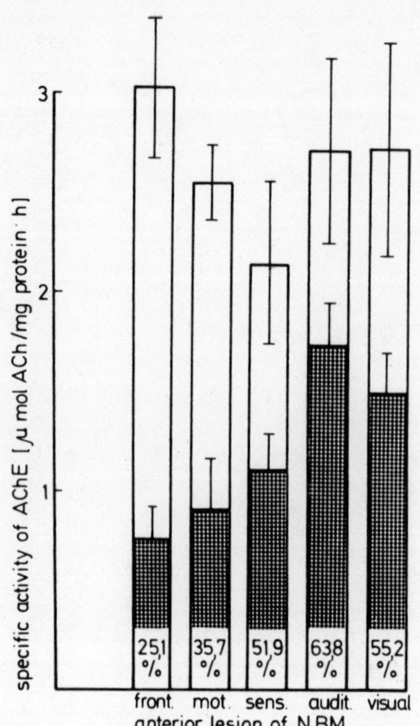
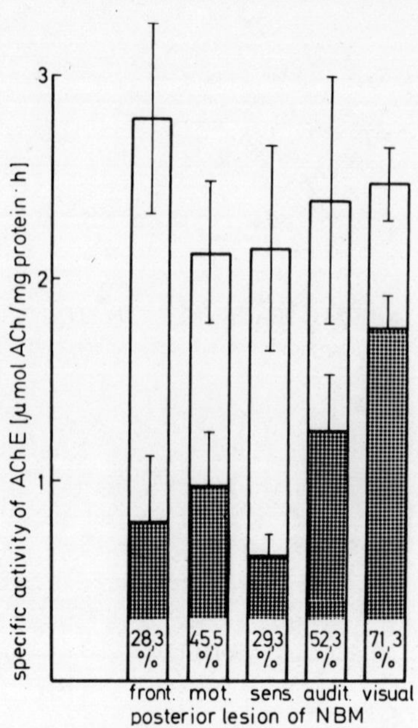

FIGURE 1: Specific activities of AChE in 5 neocortical areas of
rat brain 7 days after unilateral lesions at different
levels of the S.I. Stereotaxic coordinates: anterior
level AP 7.0, lateral 2.0, vertical +2.4; posterior
level: AP 5.8, lateral 3.2, vertical +1.5. Data are
given as mean ± S.E.M. (n = 5). Empty columns: normal
side, Dark columns: operated side. Numbers inside the
columns are residual activities of percent of contralateral side.

inergic afferents from the NBM neurons of the basal forebrain to
neo-, paleo- and archicortical structures in rats.

MATERIAL AND METHODS

Electrolytic lesions were placed in rats (strain BD III or
Wistar) anesthetized with nembutal or hexobarbital using a stereotaxic apparatus and insulated steel electrodes. Stereotaxic coordinates refer to the atlas by Koenig and Klippel (10). The size
of the lesions ranged between 500 and 2,000 μm. Extent and location of the lesions were checked histologically. Sham operated
rats served as controls.

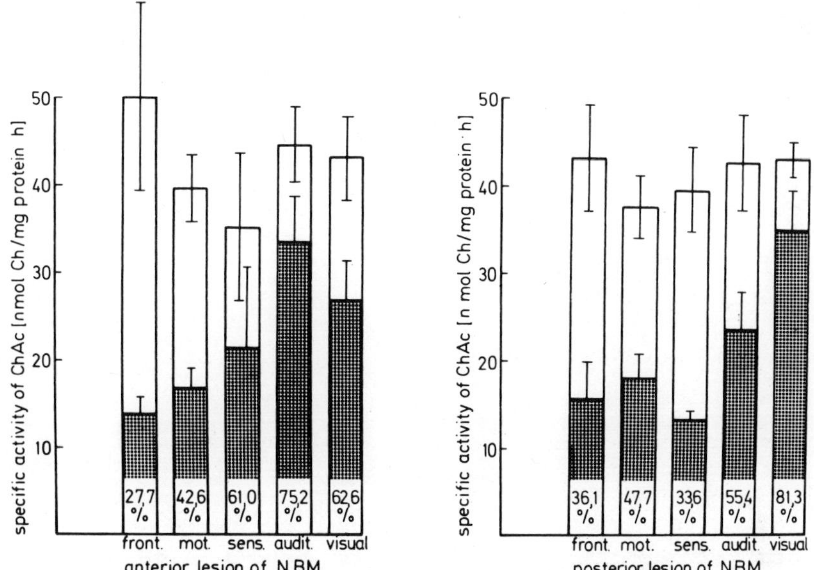

FIGURE 2: Specific activities of CAT in 5 neocortical areas of rats
after unilateral lesions in the anterior and posterior
level of S.I. Details as in Fig. 1.

Rats were killed 4 to 7 days after operation and AChE and CAT
activities in cortical areas assayed. For the histochemical evaluation of AChE activity the method of Karnovsky and Roots (9) was
used and quantified by a histophotometric scanning procedure (22).
For the biochemical estimations cortical areas were dissected using
templates according to the atlas of Krieg (11). AChE activity was
assayed by the method of Fonnum (4), CAT by a radiochemical procedure (1). Protein content was determined in aliquots of the
homogenates used for biochemical estimations by the method of Lowry
et al. (12).

RESULTS

Cholinergic afferents to neocortical areas

Seven days after unilateral lesions were placed in the S.I.
ventral to the nc lentiformis, a widespread decrease in AChE staining
was observed in the whole ipsilateral neocortex. As the medial

septum, Broca's diagonal band nucleus and the lateral preoptic
region were not affected by the lesion, the changes in AChE and
CAT activities were confined to the neocortical fields (21). Figs.
1 and 2 show quantitative estimates from 5 neocortical areas after
lesioning the S.I. at two different levels. Following more posteriorly placed lesions of the S.I., the ipsilateral frontal and
sensory motor cortex showed the highest loss of both AChE and CAT
as compared to the contralateral side. The auditory cortex and
visual cortex were less affected. Lesions placed more anteriorly
to the S.I. affected the frontal and motor cortex equally to the
posteriorly placed lesions. Residual activities of both enzymes
were considerably lower in the visual cortex after anteriorly placed
lesions but higher residual values were seen in the sensory and
auditory cortex.

Residual activities of AChE and CAT in individual animals
were often considerably lower than might be depicted from the
average data shown in Figs. 1 and 2. As changes in CAT and AChE
activity in our lesion experiments were always in excellent agreement, the maximum possible decrease in cortical AChE activity was
measured with the quantitative histochemical procedure which permits
a better localization of the effects of the lesion upon the cortex.
In this way residual values of cortical AChE activity ranging from
15 to 45% of the contralateral side were estimated (21).

Cholinergic Afferents
to Olfactory and Entorhinal Cortex

The magnocellular preoptic nucleus (10) in the lateral preoptic area exhibits high AChE activity (23). Lesions placed in
that nucleus resulted in a marked decrease in the AChE level in
several olfactory regions (Table 1). Rostral to the lesion (olfactory bulb, piriform cortex and anterior olfactory nucleus)
residual activities between 13-16% only could be detected histochemically. Enzyme activity in areas caudal to the lesion were
less markedly but still significantly affected (cortico-basal
nucleus of amygdala 40%, entorhinal cortex 30% residual AChE
activity ipsilaterally). AChE activity of olfactory tuberculum
and other parts of the amygdaloid complex was unaffected by the
lesion.

In the olfactory bulb the histochemical results were confirmed
biochemically (Fig. 3). On the side of the lesion the activity of
both AChE and CAT was decreased to about 40% of that found in control animals. Slight losses occurred contralaterally if the lesion
had encroached on that side.

TABLE 1: Relative activites of AChE ($\overline{E}$) determined by scanning histophotometry in several olfactory areas after unilateral lesion of the magnocellular preoptic nucleus.

| | AChE Activity | | Residual |
Brain Region	Lesioned Side	Contralateral	Activity
Olfact. bulb	91 ± 39	697 ± 74	13*
Ant. olfact. nc.	91 ± 72	561 ± 84	16*
Piriform cortex	81 ± 56	634 ± 246	13*
Olfact. tubercle	970 ± 126	884 ± 154	110
Amygdala:			
lat. nc.	825 ± 160	806 ± 175	102
cortico-bas. nc.	247 ± 89	612 ± 141	40*
Entorhin. cort.	117 ± 44	401 ± 74	29*

AChE activity expressed in relative extinction units x 10^3 (mean ± S.E.M. from 10 measurements each from 3-5 animals).
Residual activity expressed in percent of the activity of the contralateral side.
* Significant for P < 0.001 (Student's t-test).

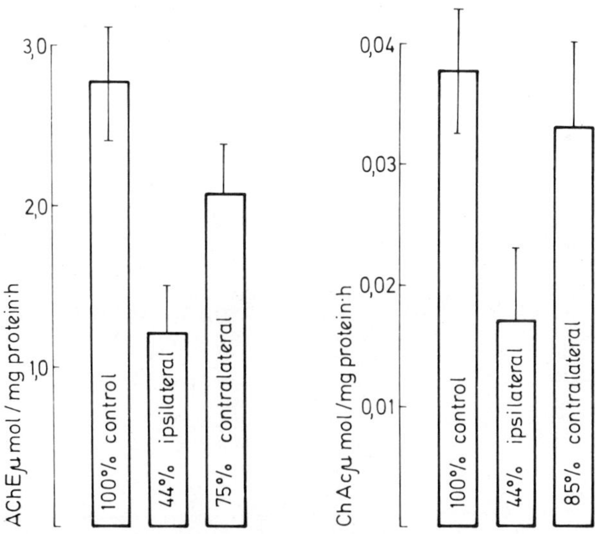

FIGURE 3: Specific activities of AChE and CAT in the olfactory bulb 4 days after unilateral lesions of the magnocellular preoptic nucleus. Columns are mean ± S.E.M. from 6 rats. Numbers inside columns give percentage of remaining activity compared to sham operated controls.

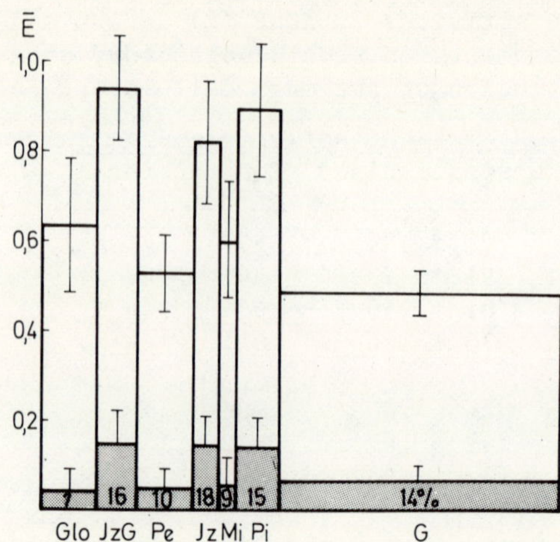

FIGURE 4: Activity of AChE in the layers of the olfactory bulb 4
days after unilateral lesions of the magnocellular preoptic nucleus as determined by scanning histophotometry.
Data are relative extinction units from rats (n = 3) arranged as in Fig. 1. G = granular layer; GLo = outer
part of the glomerular layer; IzG = inner part of the
glomerular layer; Iz = inner part of the outer plexiform
layer; Mi = mitral cell layer; Pe = outer part of the
outer plexiform layer; Pi = inner plexiform layer. Width
of the columns represent relative thickness of layers.

AChE activity in the olfactory bulb of rats is concentrated
around the olfactory glomerula (inner part of the glomerular layer),
the outer plexiform layer (inner part) and the inner plexiform
layer (Fig. 4). After optimally placed lesions of the magnocellular preoptic nucleus AChE activity is decreased in all layers
to almost the same extent (7-15%)(Fig. 4). The residual activities
compared well with those of neocortical fields after lesioning
neurons of the S.I.

DISCUSSION

Organization of Cholinergic Cortical Afferents

In primates the large multipolar neurons of the S.I. of the
basal forebrain which form the NBM and cover the full length of the
ventral aspect of the lentiform nucleus are arranged as more or less

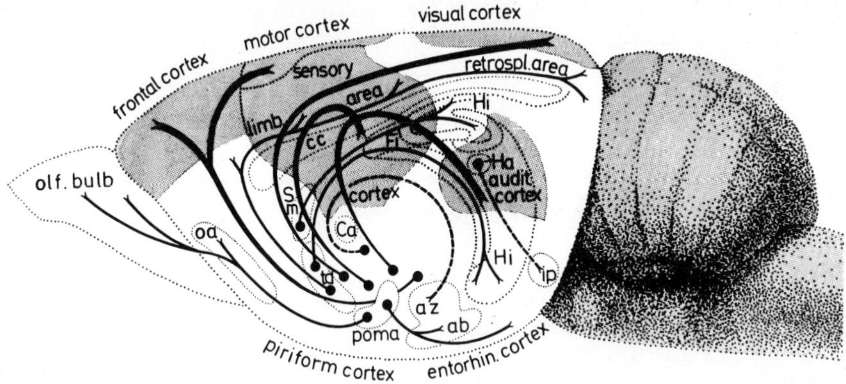

FIGURE 5: Schematic drawing of the cholinergic projections from the
basal nucleus of Meynert to cortical regions in rats.
Neurons of the NBM are indicated by black dots. Neurons
outside the indicated morphological structures refer to
the substantia innominata. For the pathways indicated by
broken lines no quantitative data are yet available.
ab = nc amygdaloideus basalis et corticalis; az = central
amygdaloid nucleus; Ca = commissura anterior; CC = corpus
callosum; Fi = fimbria hippocampi; Ha = nc habenulae
lateralis; Hi = hippocampus; ip = nc interpeduncularis;
oa = nc olfactorius anterior (from ref. 21). Sm = nucleus
septi medialis; td = nucleus tractus diagonalis Broca;
poma = nucleus preopticus magnocellularis.

---

connected islands. In addition, many single neurons invade the
medullar lamina between putamen and pallidus, infiltrate the septal
nuclei and are intermingled with neurons of the horizontal limb of
the diagonal band nucleus of Broca. Cytological as well as connectional criteria differentiate these neurons clearly from the
intrinsic neurons of these different nuclei. In the rat the arrangement of the neurons, which are homologous to the NBM neurons
in primates, is even less ordered, but on connectional grounds as
well as cytological and neurochemical criteria they must be regarded as an entity (3,6). Many discrepancies in the anatomical,
histochemical and neurophysiological literature regarding cortical
cholinergic pathways seem to be most likely due to the fact that
these neurons have been wrongly allocated to adjacent structures
(see 21).

The projections of these magnocellular neurons to several
cortical areas as demonstrated by the HRP labelling technique (2,
6,14) is widespread but not diffuse. In agreement with these neuro-
biological findings, the observed changes of AChE and CAT in

different cortical fields after lesioning several forebrain structures in which NBM neurons distribute, also points to a certain
topistic organization of the cholinergic cortical input. In spite
of the seemingly unordered arrangement of the NBM neurons in the
rat forebrain, at least three main cholinergic pathways may be
differentiated:

1.   The scattered, disaggregated neurons of the S.I. of the
basal forebrain, which must be assumed to be homologous to the
nucleus basalis in higher vertebrates (see also, 2) supply the
cholinergic afferents to all neocortical fields. The topography
of this projection is not yet established in detail, but the results
obtained with lesions, which differ in their antero-posterior position indicate a crude topographical relationship to the location
of the NBM cells along the rostro-caudal axis of the brain (21).

2.   Paleocortical fields (olfactory cortex) receive their
cholinergic input from NBM cells of the lateral preoptic region
(nucleus preopticus magnocellularis)(24). The axons of these cells
run forward to reach the cortical olfactory regions and backward
to the basal amygdaloid nuclei and the entorhinal area. Recent
experiments with the pharmacohistochemical procedure (see Lehmann
and Fibiger, this volume). support this conclusion although no HRP
labelled AChE positive cells were found in this nucleus following
injection of HRP into the olfactory bulb (14).

3.   Taking into consideration morphological and connectional
criteria (2,6) the well documented "classical" septo-hippocampal
pathway, which supplies cholinergic innervation to the hippocampal
formation, must also be confined to NBM cells which invade the
medial septal nuclei. Those basal nuclei neurons situated at the
border or within the nucleus of the diagonal band of Broca (between
the vertical and the horizontal limb) provide the cholinergic innervation to the medial limbic cortex. These data are supported
by a great number of experimental results involving lesions of the
fimbria, the septum as well as HRP labelling (see 21).

Thus it seems that the cholinergic innervation of the whole
cerebral cortex originates from a uniform, continuous population
of large, multipolar neurons in the basal forebrain. Their axons
are probably the terminal corticopetal fibers of the "ascending
cholinergic reticular system" (19). Due to an enormous preterminal
ramification, the surprisingly small number of neurons provides the
monosynaptic cholinergic projection to all ipsilateral cortical
structures. Available experimental evidence support neither the
idea of a direct cholinergic ascending mesencephalic reticular
projection (19; for discussion see 3), nor a direct cholinergic
projection from the globus pallidus and the nc entopeduncularis to
neocortical fields (19). Most probably small changes in cholinergic

enzyme activities after discrete lesions in the globus pallidus
are attributed to NBM neurons situated along the ventral border
of the nucleus and spread into the intramedullary laminae (21).

Quantitative Evaluation of Cholinergic Projections
to the Cerebral Cortex

Because of its irregular spatial extension it is not possible
to completely destroy the different parts of the NBM in the rat.
Adjacent structures are often involved in the lesions. Depending
on the size and the exact location of the lesions, the changes of
cholinergic enzyme activities, especially in the neocortical fields,
were often considerably lower in individual animals as can be
depicted from the averaged values. In none of the cortical areas
studied could be achieved a complete loss of AChE or CAT activities
by any of the various lesions of the NBM. Maximum loss of AChE
amounts to about 80 to 90% of the control values.

The residual values of 10-20% are in excellent agreement with
the reduction of CAT in cortical tissue following undercutting (5).
Thus the remaining enzyme activity might most probably indicate
some cholinergic interneurons in the cortex (13).

Functional Relevance
of the Cholinergic Cortical Pathways

In its morphological arrangements the cholinergic neurons of
the NBM show a striking parallelism to the noradrenergic cortical
pathway originating in the locus coeruleus. In addition, they are
well supplied by afferents from almost all sensory systems and the
hypothalamus (21). Both pathways are widespread and project ipsilaterally to the cortex without being relayed via thalamic nuclei.
However, little is known about the terminal intracortical synaptic
connection of the cholinergic and aminergic afferents. The precise
functional significance of the cholinergic pathways which arise
from the NBM neurons of the forebrain is therefore by no means
clear, but obviously they are involved in a number of behavioral
situations (2).

On the basis of pharmacological as well as electrophysiological
data, it is commonly accepted that the cholinergic afferents trans-
mit cortical arousal. ACh mainly exerts an excitatory action on
cortical cells. On the other hand, it was concluded that stimulation of some cholinergic pathways might ultimately lead to behavioral

inhibition (8). In previous papers (23,24) we discussed suggestive
evidence that cholinergic afferents from the NBM to the olfactory
bulb terminate on local neurons which are most probably inhibitory
GABAergic interneurons. In addition, for a second pathway from
the basal forebrain nuclei, the septo-hippocampal projection, an
inhibitory action was assumed (7). This might, therefore, apply
to the neocortical projections as well, but far more data on the
intracortical arrangement of the cholinergic network are needed to
answer these important questions. It is, however, tempting to
speculate that in a number of behavioral reactions the cholinergic
projections from the NBM represent the final link in a complex chain
of neurophysiological events which finally modulates the activity
pattern of cortical neurons. By their interaction the cholinergic
and noradrenergic afferents might control cortical information
processing in a way roughly comparable to the control of peripheral
activity by the parasympathetic and sympathetic system.

ACKNOWLEDGEMENTS

This study was supported by a grant from the Ministry of Science
and Technology of the GDR. The skilful technical assistance of Mrs.
B. Bigl, Mrs. H. Daniel and Mrs. H. Kaminski is gratefully acknowledged.

REFERENCES

1.   Bigl, V. (1975): Acta Biol. Med. Germ. 34:1437-1440.
2.   Divac, I. (1975): Brain Res. 93:385-398.
3.   Emson, P.C. and Lindvall, O. (1979): Neuroscience 4:1-30.
4.   Fonnum, F. (1969): Biochem. J. 115:465-472.
5.   Hebb, C.O., Krnjevie, K. and Silver, A. (1963): Nature 198:
       692.
6.   Jones, E.G., Burton, H., Saper, C.B. and Swanson, L.W. (1976):
       J. Comp. Neurol. 167:385-420.
7.   Karczmar, A.G. (1976): IN Biology of Cholinergic Function
       (eds) A.M. Goldberg and I. Hanin, Raven Press, New York,
       pp. 395-449.
8.   Karczmar, A.G. and Dun, N.J. (1978): IN Psychopharmacology:
       A Generation of Progess (eds) M.A. Lipton, A. DiMascio and
       K.F. Killam, Raven Press, New York, pp. 293-305.
9.   Karnovsky, M.J. and Roots, L. (1964): J. Histochem. Cytochem.
       12:219-221.

10. Koenig, J.F.R. and Klippel, R.A. (1963): The Rat Brain: A
     Stereotaxic Atlas of the Forebrain and Lower Parts of the
     Brainstem, The Williams and Wilkins Company, Baltimore.
11. Krieg, W.J.S. (1946): J. Comp. Neurol. 84:221-275.
12. Lowry, O.H., Rosebrough, N.J., Farr, A.L. and Randall, R.J.
     (1951): J. Biol. Chem. 193:265-275.
13. McGeer, P.L., McGeer, E.G., Singh, V.K. and Chase, W.H. (1974):
     Brain Res. 81:373-379.
14. Mesulam, M.M. and Van Hoesen, G.W. (1976): Brain Res. 109:
     152-157.
15. Meyer, U., Wenk, H. and Bigl, V. (1978): Abh. Akad. Wiss, DDR,
     Abt. Mathematik, Naturwissenschaft, Tech. 5:479-481.
16. Packman, P.M., Godfrey, D.A., Williams, A.D. and Matschinksky,
     F.M. (1977): J. Histochem. Cytochem. 25:1237-1246.
17. Palkovits, M., Saavedra, J.M., Kobayashi, R.M. and Brownstein,
     M. (1974): Brain Res. 79:443-450.
18. Parent, A., Poirier, L.J., Boucher, R. and Butcher, L.L. (1977):
     J. Neurol. Sci. 32:9-28.
19. Shute, C.C.D. and Lewis, P.R. (1967): Brain 90:497-520.
20. Wenk, H., Bigl, V. and Meyer, U. (1979): Prog. Brain Res.
     49:497.
21. Wenk, H., Bigl, V. and Meyer,U. (1980):Brain Res. Rev. 2:295-376.
22. Wenk, H., Krug, H. and Fletcher, A.M. (1973): Acta Histochem.
     (Jena) 45:37-60.
23. Wenk, H., Meyer, U. and Bigl, V. (1976): Z. Mikroskop. Anat.
     Forsch. 90:940-958.
24. Wenk, H., Meyer, U. and Bigl, V. (1977): Neuroscience 2:797-800.
25. Wenk, H., Ritter, J. and Meyer, U. (1975): Acta Histochem.
     (Jena) 53:77-92.

LESIONS OF THE GLOBUS PALLIDUS: CHANGES IN CORTICAL CHOLINE

ACETYLTRANSFERASE, CHOLINE UPTAKE AND ACETYLCHOLINE OUTPUT

IN THE RAT

F. Casamenti, F. Pedata, S. Sorbi, G. Lo Conte and
G. Pepeu

Department of Pharmacology, University of Florence
School of Medicine
Viale Morgagni 65, 50134 Florence, Italy

INTRODUCTION

The presence of acetylcholine (ACh) (16), choline acetyltransferase (CAT) (6), the histochemical visualization of acetyl-
cholinesterase-rich fibers (17), and the output of ACh from the
cortical surface, demonstrate the existence of a dense cholinergic
network in the cerebral cortex of all animal species so far
investigated. However, the origin of this cortical cholinergic
network is still mostly unknown. According to McGeer (10), it could
be partly formed by intrinsic cholinergic neurons. The destruction
of the septum in the rat is followed by approximately 40% decrease
in ACh content of (15) and by a similar decrease in spontaneous ACh
output from the cerebral cortex (1). These findings suggest that
part of the cortical cholinergic network is formed by fibers
originating from or passing through the septum.

Kelly and Moore (4) demonstrated that a pathway, either cholinergic or exerting a trophic function on cholinergic neurons,
distributed preferentially to frontal sensorimotor cortical regions,
is destroyed by lesions of the globus pallidus. This pathway, however, could be considered part of a wider cholinergic projection to
different neocortical areas originating from the magnocellular forebrain nuclei in the rat (2) or from the nucleus basalis of the
substantia innominate in primates (11).

In this study we attempt to evaluate the functional impairment of the cortical cholinergic network in rats with a lesion of
the globus pallidus by measuring CAT activity and high affinity
choline (Ch) uptake (HACU) of, and ACh output from, the cerebral
cortex. Not only can HACU be considered a marker of cholinergic
neurons (7), but it also appears to be coupled to neuronal activity
(7,19). Changes in ACh output are considered to indicate changes
in the activity of the cholinergic neurons (14).

MATERIALS AND METHODS

The experiments were carried out on adult male Wistar rats.
Two unilateral lesions of the globus pallidus were made by electro-
coagulation under ketamine anesthesia (100 mg/kg i.p.) according
to the following stereotaxic coordinates: A, 0.2; H, 5.4 and 6.3,
and L, 3.1 (5). At the end of the experiments a histological examination was carried out on each animal in order to assess the
placement and size of the lesion. In the sham-operated rats the
electrode was only lowered into the cortex without passing the
current.

In order to measure CAT activity and HACU the rats were sacrificed by decapitation 20 days after the lesion. The brain was
quickly removed, dissected out and the samples homogenized in
20 vol of 10 mM EDTA sodium salt and 0.2 Triton X100 and ice-cold
0.32 M sucrose, respectively.

CAT was assayed by the micromethod of Fonnum (3). Aliquots of
the homogenate were added to an incubation medium containing ^{14}C
acetylCoA (59 mCi/mmol: Amersham) at a final concentration of
16.9 mCi/mmol. The incubation lasted 20 min at 37°C. At the end
of the incubation the reaction tubes were transferred directly to
scintillation vials and ^{14}C ACh was extracted into 10 ml toluene
scintillation mixture plus 2 ml acetonitrile containing 10 mg of
tetraphenylboron. When the two layers were separated ^{14}C was counted
in a liquid scintillation counter.

HACU determination (18) was performed on a crude mitochondrial
preparation. Aliquots of the preparation (100 µl; 25 mg of protein)
were added to 900 µl of a Krebs-Ringer phosphate medium containing
0.062 µM final concentration of ^{3}H methyl Ch chloride (6.4 Ci/mmol:
Amersham). The samples were incubated for 4 min at 30°C. At the
end of the incubation time the samples were poured over Millipore
filters (pore size 0.65 µm) on a Millipore 12 place manifold. The
filters were allowed to dry and then transferred to scintillation
vials. After addition of 1 ml of ethylen–glycolmonomethylether
and 10 ml of Instagel (Packard), radioactivity was counted in a

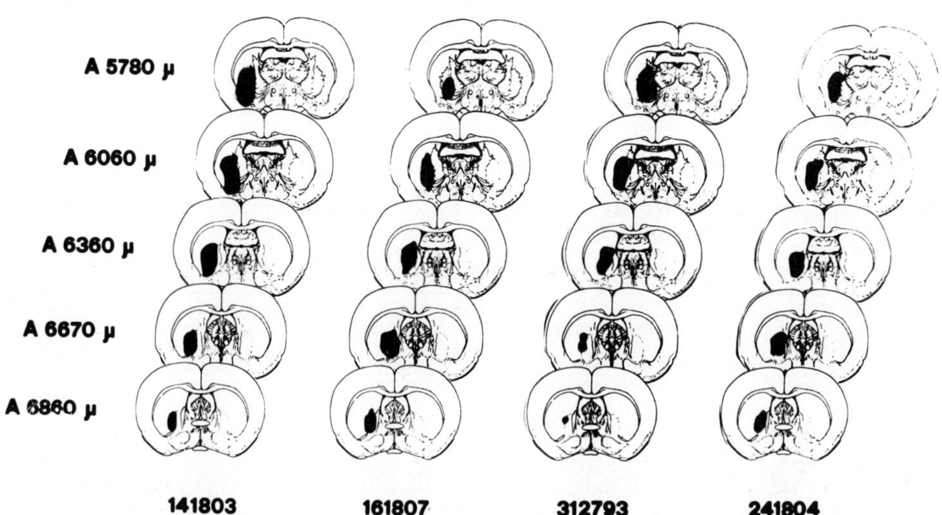

FIGURE 1: Four examples of typical lesions of the globus pallidus in the rat.

TABLE 1:   CAT activity (µnol/hr/100 mg protein ± S.E.) in the cerebral cortex of sham-operated rats.

Cortical Area	Hemispheres	
	Right	Left
Frontal	2.03 ± 0.04	2.04 ± 0.03
Parietal	1.97 ± 0.04	1.98 ± 0.02
Occipital	1.83 ± 0.07	1.90 ± 0.02
Number of rats = 7		

TABLE 2: Regional changes in CAT activity (μmol/hr/100 mg protein
± S.E.) in rats with unilateral destruction of the pallidum.

Brain Region	Hemispheres		Change	
	Controlateral	Ipsilateral	(percent)	P
Frontal cortex	0.43 ± 0.18	1.26 ± 0.08	−48	0.001
Parietal cortex	2.12 ± 0.14	1.13 ± 0.09	−47	0.001
Occipital cortex	1.93 ± 0.21	1.15 ± 0.01	−40	0.001
Head of caudate N	11.74 ± 0.77	9.52 ± 1.81	−18	n.s.
Thalamus ventralis	6.78 ± 0.41	5.48 ± 1.11	−19	n.s.

Number of rats = 7; n.s. = not significant

Packard Tricarb (model 577) scintillation spectrometer. The rates
of Ch uptake were expressed as pmol of Ch per 4 min incubation time
per mg of protein. Protein content of the homogenates was
measured by the method of Lowry (9).·

ACh output from the cerebral cortex was estimated 20 days after
lesion in freely moving rats (1). Under ketamine anesthesia a
Perspex cylinder was screwed into the left or right parietal bone
so as to exert a slight compression on the dura mater, and secured
with dentist cement. ACh output was measured three days after
surgery. During this time the dura was washed with a terramycin
solution (300 μg/ml). Ringer solution containing eserine sulphate (100 μg/ml) was placed in the cylinder. Every 20 min the
solution was removed and bioassayed for ACh on the dorsal muscle
of the leech.

RESULTS

Figure 1 illustrates 4 examples of typical electrolytic lesions
found in the pallidum of the operated rats at histological examination. The lesions destroyed mostly the rostral part of the pallidum.
In some cases the head of the caudate nucleus and/or the capsula
interna were also damaged.

Table 1 shows that symmetrical areas from the cerebral cortex
in sham-operated rats had the same CAT activity. However, there
was a marked difference between the two hemispheres, as found in
the cerebral cortex of rats with a lesion of the globus pallidus.

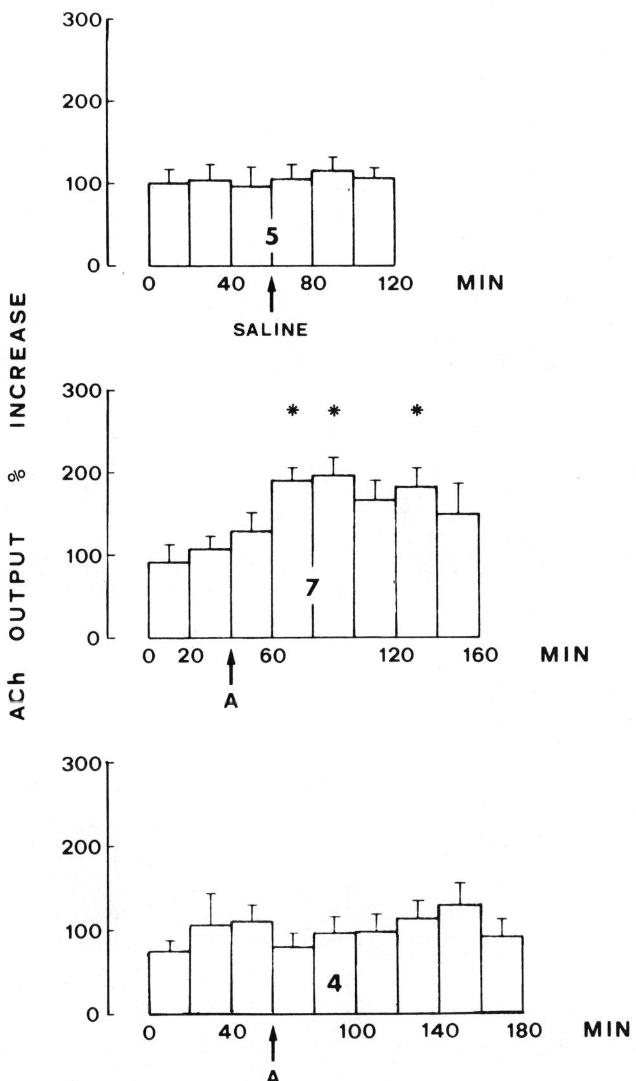

FIGURE 2: Effect of amphetamine 1 mg/kg i.p. (A) on ACh output
          from the cerebral cortex in freely moving rats. The
          numbers indicate the number of experiments. * P < 0.01.

TABLE 3: High affinity choline uptake (pmol/4 min/mg protein $\pm$ S.E.) in the cerebral cortex of rats with unilateral lesion of the pallidum.

Conditions	Hemispheres	Frontal Cortex	Change (percent)	P	Parietal Cortex	Change (percent)	P
Sham operated	Controlateral	4.20 ± 0.14	–	–	4.94 ± 0.27	–	–
	Ipsilateral	4.21 ± 0.15	–	–	4.95 ± 0.25	–	–
Operated	Controlateral	5.21 ± 0.28	+24	< 0.02	5.82 ± 0.67	+18	n.s.
	Ipsilateral	3.37 ± 0.30	−20	< 0.05	4.87 ± 0.27	− 2	n.s.

Number of rats = 5; n.s. = not significant

TABLE 4:   Spontaneous ACh output from the cerebral cortex in freely
           moving rats with a pallidum lesion.

Conditions	Hemispheres		ACh Output $(ng/min/cm^2 \pm S.E.)$	Change (percent)
Sham operated	Ipsilateral	(6)	0.96 ± 0.06	–
	Controlateral	(4)	0.94 ± 0.09	–
Pallidum lesion	Ipsilateral	(6)	0.67 ± 0.06	–31
	Controlateral	(4)	1.09 ± 0.05	+13

Number of rats in parentheses. Difference statistically significant from ipsilateral sham-operated hemisphere ($P < 0.01$).

As shown in Table 2, CAT activity in the frontal, parietal and
occipital cortex on the lesioned side was at least 40% lower than
the controlateral side. A small but not statistically significant
decrease in CAT activity was also found in the head of the caudate
nucleus and in the thalamus ventralis.   However, if CAT activity
in the lesioned rats was compared with that in the sham-operated
rats, the decrease in the frontal cortex on the lesioned side was
smaller than when compared with the controlateral cortex in the
lesioned rats. In these rats the controlateral frontal cortex
showed a 19% increase in CAT activity over that in the sham-
operated rats.

The cerebral cortex of both hemispheres in the sham-operated
rats had the same HACU. In the lesioned rats a 35% difference
between the frontal cortices of the two hemispheres was found.
However, as demonstrated in Table 3, if HACU in the frontal cortex
of the lesioned rats was compared with that in the sham-operated
animals, it showed a decrease on the side of the lesion and an
increase on the opposite side.

The spontaneous ACh output from the cerebral cortex was also
measured in unanesthetized freely moving rats 20 days after placing
a lesion in the pallidum. Three samples were collected for each
rat. It can be seen in Table 4 that there were no differences in
ACh output from the two hemispheres in the sham-operated rats. On
the contrary, when the lesioned rats were compared with the sham-
operated rats there was a 31% decrease in the spontaneous ACh
output on the side of the lesion and a small, not statistically
significant increase on the controlateral side.

Figure 2 shows that the administration of amphetamine (1 mg/ kg i.p.) was followed in the unoperated rats by an increase in ACh output from the cerebral cortex lasting more than 1 hr. This amphetamine effect was not present in the rats with a lesion of the pallidum. It was, however, still present in a rat (not shown) with a lesion of the capsula interna.

DISCUSSION

The lesion of the globus pallidus was followed by a marked decrease in CAT activity in all cortical areas investigated. Since CAT activity in the cholinergic neurons is located mostly at the nerve endings (20), its decrease would indicate that the lesion of the pallidum brings about the destruction of a large number of cholinergic corticopetal fibers originating from or passing through this region. The extent of the decrease in CAT activity found in our experiments was the same as that found by Kelly and Moore (4) after a similar lesion of the pallidum.

It is interesting to note that when a comparison is made between the lesioned and sham-operated rats, the decrease in CAT activity in the frontal cortex is smaller than when the comparison is made between the two hemispheres in the lesioned rats, and a small increase on the side opposite to the lesion can also be detected. It appears that a controlateral compensatory increase in CAT activity might take place.

The marked decrease in CAT activity following the lesion of the pallidum was not associated with a comparable decrease in HACU. The finding was unexpected since it has been shown (8) that after a septal lesion there is a large and permanent decrease in hippo- campal HACU. It has also been demonstrated (19) that an increase in nerve impulse flow is associated with HACU stimulation. There- fore an enhanced HACU in the remaining cholinergic fibers could mask its disappearance in the fibers destroyed by the lesion. The observation that HACU also was increased in the frontal cortex op- posite to the lesion supports the possibility of a compensatory HACU stimulation. Further confirmation will derive from HACU determinations at different times after the lesion. In spite of the relatively small changes in HACU, the decrease in CAT activity affects ACh synthesis as shown by the significant reduction in spontaneous ACh output from the frontoparietal cortex on the side of the lesion.

If changes in cortical ACh output indicate changes in the activity of cortical cholinergic nerve endings (14), the decrease in ACh output demonstrates the functional impairment caused by the

lesion of the globus pallidus. The disappearance of the stimulatory
effect of amphetamine on ACh output is a further evidence of the
functional impairment of the corticopetal cholinergic fibers.
Amphetamine stimulates cortical ACh output through an indirect
mechanism involving catecholamine neurotransmission (13). This effect of amphetamine is also abolished by a septal lesion (12). The
finding that both septal and pallidum lesions prevent amphetamine
stimulation on ACh output suggests that the same cholinergic pathway is destroyed by both lesions. The precise origin and course
of this pathway are still a matter of investigation.

ACKNOWLEDGEMENT

This work was supported by CNR grant No. 7901950.

REFERENCES

1.   Casamenti, F., Pedata, F., Corradetti, R. and Pepeu, G.
        (1980): Neuropharmacology 19:597-605.
2.   Emson, P.C. and Lindvall, O. (1979): Neuroscience 4:1-30.
3.   Fonnum, F. (1975): J. Neurochem. 24:407-409.
4.   Kelly, P.H. and Moore, K.E. (1978): Exp. Neurol. 61:479-484.
5.   Koenig, J.F. and Klippel, R. (1963): The Rat Brain: A Stereo-
        taxic Atlas, William and Wilkins, Baltimore.
6.   Kuhar, M.J. (1976): In Biology of Cholinergic Function (eds)
        A.M. Goldberg and I. Hanin, Raven Press, New York, pp. 3-27.
7.   Kuhar, M.J. (1977): In Cholinergic Mechanisms and Psycho-
        pharmacology (ed) D.J. Jenden, Plenum Press, New York,
        pp. 447-456.
8.   Kuhar, M.J., Sethy, V.H., Roth, R.H. and Aghajanian, G.K.
        (1973): J. Neurochem. 20:581-593.
9.   Lowry, O.M., Rosebrough, N.J., Farr, A.L. and Randall, R.J.
        (1951): Biol. Chem. 193:265-275.
10. McGeer, P.L. and McGeer, E.G. (1979): In Nutrition and the
        Brain, Vol. 5 (eds) A. Barbeau, J.H. Growden and R.J.
        Wurtman, Raven Press, New York, pp. 177-199.
11. Mesulam, M.M. and Van Hoesen, G.W. (1976): Brain Res. 109:
        152-157.
12. Mulas, A., Mulas, M.L. and Pepeu, G. (1974): Psychopharmacology
        (Berl.) 30:223-230.
13. Nistri, A., Bartolini, A., Deffenu, G. and Pepeu, G. (1972):
        Neuropharmacology 11:665-679.

14. Pepeu, G. (1973): Prog. Neurobiol. 2:257-288.
15. Pepeu, G., Mulas, A., Mulas, M.L. (1973): Brain Res. 57:153-164.
16. Saèlens, J.K. and Simke, J.P. (1976): In Biology of Cholinergic
       Function (eds) A.M. Goldberg and I. Hanin, Raven Press, New
       York, pp. 661-706.
17. Silver, A. (1974): The Biology of Cholinesterases, North
       Holland Publishing Company, Amsterdam.
18. Simon, J.R., Atweh, S. and Kuhar, M.J. (1976): J. Neurochem.
       26:909-922.
19. Simon, J.R. and Kuhar, M.J. (1976): J. Neurochem. 27:93-99.
20. Tucek, S. (1978): Acetylcholine Synthesis in Neurons, Chapman
       and Hall, London.

MAPPING OF CHOLINERGIC SYSTEMS IN ROSTRAL FOREBRAIN OF THE RODENT

H. Kimura, P.L. McGeer, J.H. Peng and E.G. McGeer

Kinsman Laboratory, Department of Psychiatry
University of British Columbia
Vancouver, British Columbia, Canada V6T 1W5

Visualization of central cholinergic system has been hampered
by technical difficulties. No suitable method is yet available
for the visualization of acetylcholine (ACh). The conventional
histochemistry for acetylcholinesterase (AChE), as discussed in
another chapter (9), is not a reliable marker for cholinergic
neurons. Although the modified histochemistry for AChE has provided useful information (10) it is still not definitive for cholinergic perikarya nor applicable to the visualization of fibers and
terminals. Several attempts have been made to visualize choline
acetyltransferase (CAT), a specific marker enzyme for cholinergic
systems in the CNS. A proposed histochemical method is not completely specific, since it is based on the precipitation of CaA
which is produced not only by CAT but by several enzymes (2).

A substantial effort in several laboratories has been devoted
to develop an immunohistochemical technique for CAT since similar
techniques for other specific synthetic enzymes have been very
valuable in mapping various neurotransmitter systems. Previous
reports on CAT immunohistochemistry, however, have been questioned
because of insufficient evidence as to the specificity of the
staining (3,16). To answer this question, it is desirable to
demonstrate not only immunoreactive staining in well established
cholinergic pathways in the CNS as well as the peripheral nervous
system, but also the disappearance of immunoreactivity following
lesions to appropriate pathways.

We have previously shown that our present CAT immunohistochemical technique stains such known cholinergic structures as the
ventral horn motor neurons in the rat spinal cord and the neuromuscular junction of the guinea pig diaphragm (7). In the present

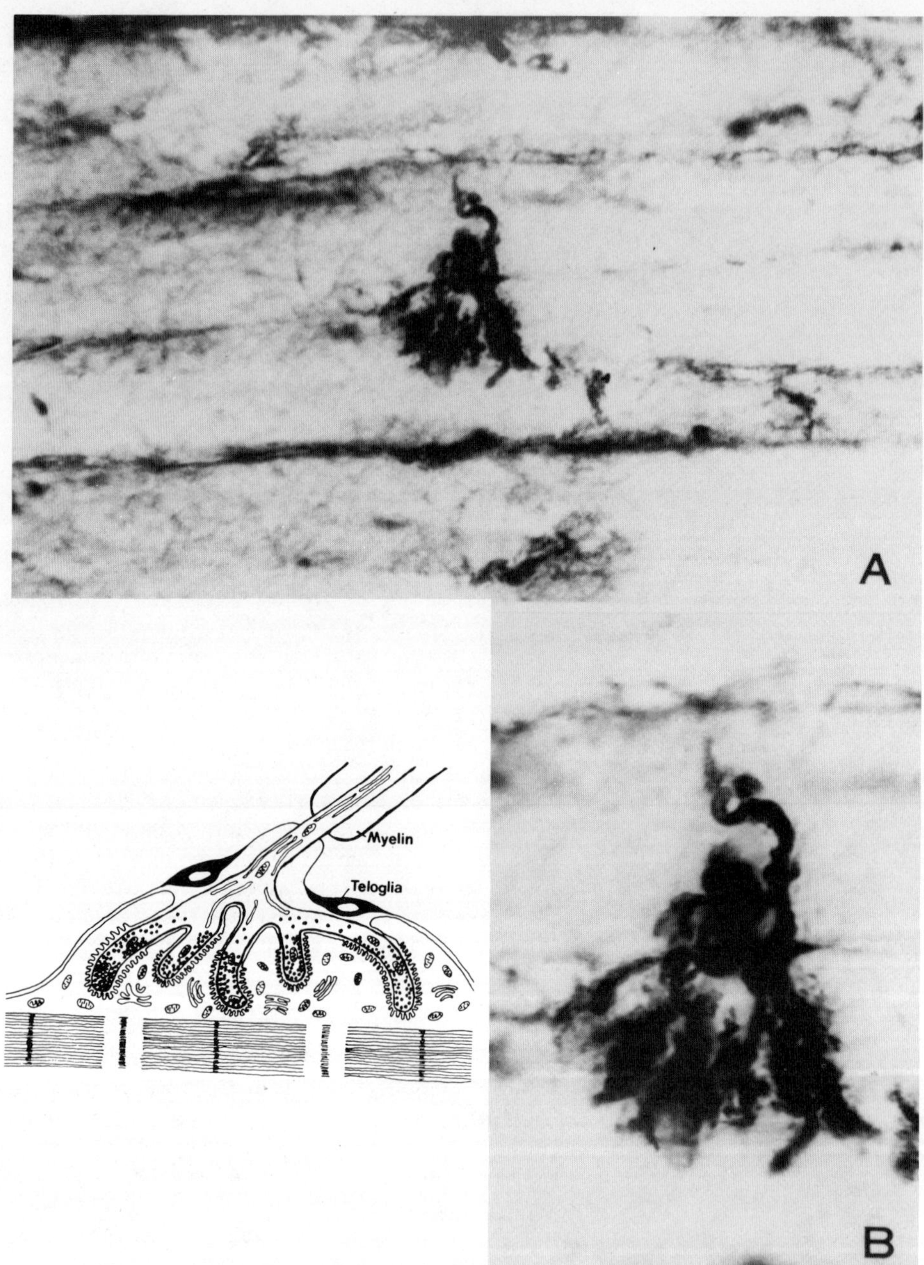
Myelin
Teloglia
A
B

FIGURE 1:   Immunohistochemical micrograph of CAT stained by the
 ⟵   PAP A: low power magnification of CAT-positive stain-
             ing seen at the neuromuscular junction (arrowhead) of
             guinea pig diaphragm. mf: muscle fibers. Bar indicates
             25 μm. B: higher power magnification of neuromuscular
             junction of A1. The axon (double arrowhead) just be-
             fore forming "en plaque" nerve ending is seen. Bar
             indicates 10 μm.

---

study it is also shown that CAT immunoreactive perikarya in the
nucleus of the diagonal band disappeared following lesions removing
the olfactory bulb. Furthermore, as reported previously for human
(8), and here for rodent forebrain, the specific localization of
CAT in certain neuronal cell bodies, fibers and possibly terminals
indicated by our present immunohistochemical technique is in concordance with other types of available evidence on cholinergic
pathways.

Well known pathways which are believed to use ACh include:
a) ventral horn cells to all voluntary muscles; b) neostriatal
interneurons; c) the habenulo-interpenduncular tract (which may
also be derived from the nucleus diagonal band of Broca); and d)
the septo-hippocampal pathway. In addition, it has been also
strongly suggested that the cholinergic system in the olfactory
bulb is exclusively extrinsic and may be derived from the basal
part of the forebrain (15,17). Therefore prior to mapping the
whole brain we attempted to stain CAT immunohistochemically in the
rostral part of the forebrain and in the spinal cord, areas which
are concerned with all of the above pathways.

The PAP immunohistochemical visualization of CAT was carried
out on cryostat sections of paraformaldehyde-glutaraldehyde fixed
tissue using techniques which will be described in detail elsewhere
(manuscript in preparation). The preparation of anti-serum in rabbits against human CAT has been described previously (14). The
antibodies cross-react with CAT from several other mammalian species
Fab fragments of the CAT antibodies were employed in the present
study. Anti-rabbit IgG raised in goat and rabbit peroxidase-anti-
peroxidase (PAP) were from commercial sources (Cappel Laboratories,
Cochranville, PA). In control experiments, rabbit serum collected
before immunization and Fab fragments of anti-serum to CAT that had
been treated with purified human CAT were used instead of the Fab
fragments of antibodies to CAT. No positive staining was obtained
in these control experiments.

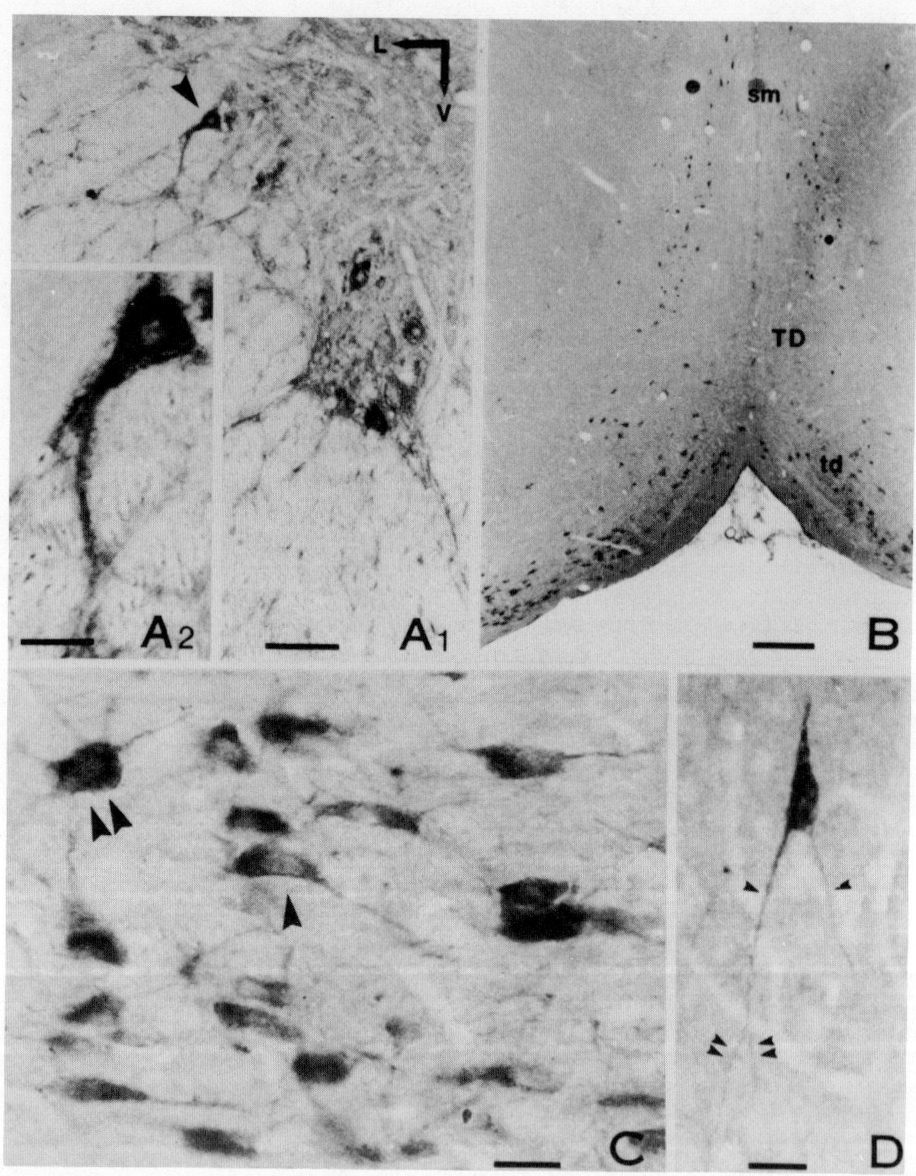

FIGURE 2: Low power magnification of the ventral horn
           of the rat spinal cord. Motor neurons are positively
           stained. L: lateral; V: ventral. Bar indicates 100 µm.
           A2: higher power magnification of the same cell seen in
           A1 (arrowhead). Bar indicates 25 µm. B: the basal mid-
           line part of a coronal section through the septal nuclei
           complex of the rat brain. SM: the medial septal nucleus.
           td: diagonal band of Broca. TD: the tractus diagonalis.
           Bar indicates 250 µm. C: higher power magnification of
           the CAT-staining cells in the TD. No staining is seen in
           nuclei of these neurons (single arrowhead). The majority
           of these neurons have large cell bodies with few processes
           but large multipolar cells are also occasionally found
           (double arrowheads). Bar indicates 25 µm. D: a typical
           CAT-positive neuron which is found in the guinea pig neo-
           striatum. Few long dendrites (single arrowhead) with few
           branches (double arrowheads), are seen tapering off rapid-
           ly from the large cell body. Bar indicates 25 µm.

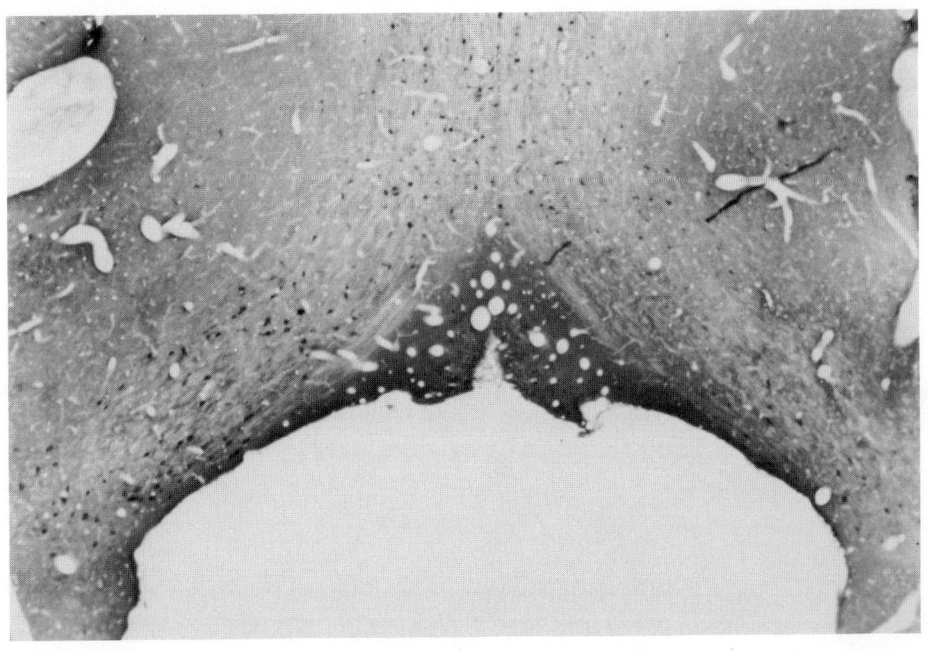

FIGURE 3: Disappearance of neurons (right side) in the horizontal
          limb of the diagonal band following ipsilateral lesion of
          the olfactory bulb by electrocoagulation. The CAT-posi-
          tive cells in the septal area show symmetrical distribution
          pattern.

Specificity of the CAT Immunoreactive Staining

In the neuromuscular junction of the guinea pig diaphragm
strong staining was seen in the terminals of an axon leaving the
myelin sheath (Fig. 1A, 1B). In the rat spinal cord intense positive reactions were localized in cells of the cells of the ventral
horn motor neurons (Fig. 2A1, 2A2), while no cells were found in
the apex and head of the dorsal gray column. The nucleus of CAT
staining perikarya in the anterior horn did not stain positively,
but there was positive staining of the axon bundles of these cells
through the anterior white matter on their way to form the ventral
roots. These CAT immunoreactive motor neurons, which have been
shown to stain strongly for AChE histochemistry and have been
visualized in CAT histochemical (2) and previous immunohistochemical
studies (6,8,13), are known to be cholinergic.

Furthermore, Fig. 3 shows the unilateral disappearance of the
CAT immunoreactive neurons in the nucleus of the diagonal band
after ipsilateral lesioning of the olfactory bulb by electrocoagula-
tion. Disappearance of neurons was also observed in the medial
forebrain bundle region. From these data, we suggest that neurons
in the horizontal limb of the diagonal band as well as neurons in
the area of the medial forebrain bundle are the sources of cholinergic afferents to the olfactory bulb. ·These results indicate
that, with this immunohistochemical technique, we are able to stain
specifically cell bodies, dendrites, axons and nerve terminals of
cholinergic neurons.

Mapping of Cat Containing Neurons
in the Rostral Forebrain of the Rodent

Although the results presented here were obtained in rat brain,
our preliminary experiments with human and guinea pig brain have
also shown specific staining for CAT. The morphological features
and distribution pattern of CAT positive neurons of the guinea pig
closely resemble those of the rat in areas so far examined.

Results using the immunohistochemical method for CAT in a
frontal plane of the rostral forebrain are represented in Fig. 4.
CAT positive cells show a heterogeneous distribution pattern and
can be conveniently divided into three major groups on the basis
of cell continuity and density.

The most distinctive group lies in the area of the medial septum
and the diagonal band. Within this group, the highest accumulation
of CAT positive cells exists in the so-called horizontal limb of the
diagonal band which is situated in the area adjacent to the ventral

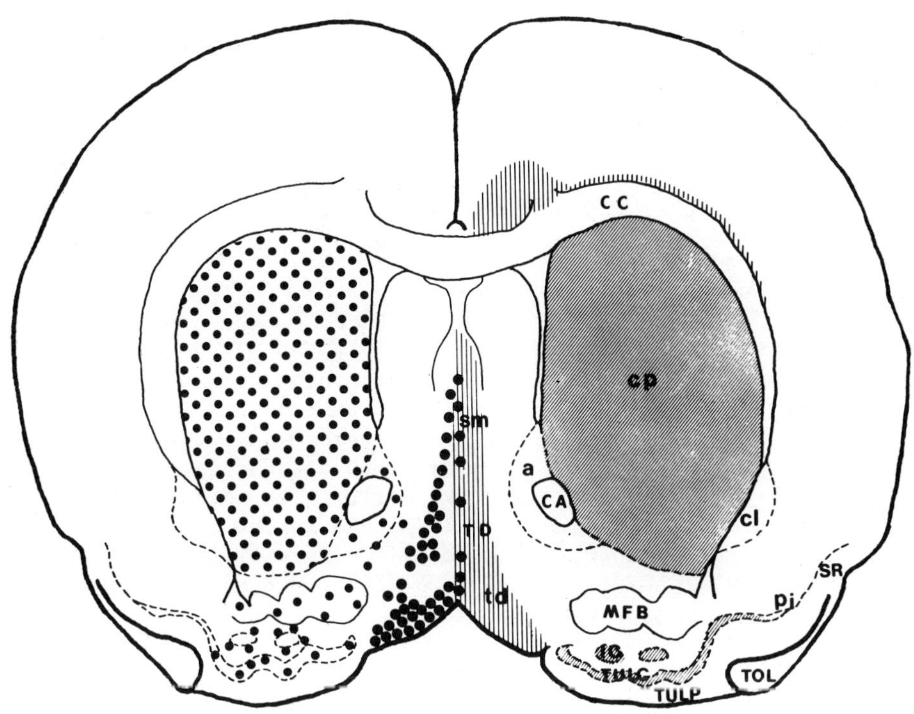

FIGURE 4: Mapping of CAT-containing neurons in the rostral forebrain
of rodent. CAT-positive cell bodies (dots), fibers (lines)
and probably terminals (shading) are represented. The
relative densities of the cells are designated by the
abundance of dots. Abbreviations: a: nucleus accumbens;
CA: commisura anterior; CC: corpus callosum; cl: claus-
trum; cp: nucleus caudate – putamen; IC: insulae Calleja;
MFB: medial forebrain bundle; pi: cortex piriformis;
sm: nucleus medialis septi; SR: sulcus rhinalis; tdB:
nucleus tractus diagonalis Broca; TD: tractus diagonalis
Broca; TOL: tractuc olfactorius lateralis; TULC: tuberculum olfactorium, pars corticalis, lamina pyramidalis;
TULP: tuberculum olfactorium, pars corticalis, lamina
plexiformis.

surface of the brain. Their packing density decreases in ventro-
lateral direction forming the ventral limb of the diagonal band.
The cells which lie in the midline can be further divided into two
divisions. The dorsal division corresponds to the medial septal
nucleus which fuses with the ventral part of the ventral limb of
the diagonal band, and the ventral division is equivalent to the
septal part of the diagonal band which fuses with the medial horizontal limb of the diagonal band. In general, these boundaries are
not clear on the basis of cell continuity. However, the features
of the cells in the diagonal band and the septal (midline) area can
be easily distinguished. The CAT reactive neurons in the diagonal
band region are characterized by their large perikarya with few
dendritic branches. These cells are usually intensely stained and
their axonal processes are occasionally visible (Fig. 2C). In the
central portion of the medial septal nucleus, CAT staining neurons
are less densely packed and are distributed in the area which is
bounded by the cells of the septal part of the diagonal band (Fig.
2B). The processes of these neurons and the fibers passing through
this cell group are always aligned dorsoventrally. CAT neurons of
the medial septum extend ventrally and laterally and seem to form
a group of cells continuous with those in the nucleus of the diagonal band (Fig. 2B). CAT reactive cells in the medial septum and
a part of the diagonal band seem to be the source of the cholinergic
septohippocampal pathway. Some of the positive neurons in the diagonal band may be the origin of part of the presumed cholinergic
pathway to the interpeduncular nucleus via the fasciculus retro-
flexus of Meynert. Moreover, our lesioning experiments on the olfactory bulb indicate that some of the neurons of the horizontal
limb of the diagonal band send cholinergic projections to the olfactory bulb. The morphological basis of these pathways has been
suggested by histochemistry for AChE, but that method is not a
sufficient criterion for the identification of cholinergic neurons.
Our results strongly indicate that these pathways are truly
cholinergic.

    The second prominent group occurs in the neostriatum and nucleus
accumbens. The CAT reactive neurons in these structures have similar if not identical morphological features with large (> 25 μm)
cell bodies which possess few long dendrites with few spines and
few branches (Fig. 2D). Positive cells are scattered evenly in the
gray matter of the striatum and nucleus accumbens. The cells of
the nucleus accumbens are fewer in number than those of the striatum.
The striatum also appears to contain dense, fine terminals which
give the area a diffuse background stain. As discussed in another
chapter on the morphology of striatal neurons, recent evidence
indicates that striatal efferents may be derived from one population
of medium sized spiny cells (1,4,5). In contrast the large neurons,
which were at one time thought to be striatal efferent neurons, are
probaby interneurons (10). Inasmuch as the cholinergic neurons in

the neostriatum are believed to be intrinsic (12), it is very
reasonable to consider that large CAT containing aspiny neurons are
the intrinsic cholinergic neurons of the striatum.

A third smaller group occurs in the olfactory tubercle region
and the area of the medial forebrain bundle. In the former region
the positive cells are localized mainly in the lamina pyramidalis
(pars corticalis) of the olfactory tubercle and distribute laterally
to the olfactory tract. In the latter area a few CAT reactive
neurons can be seen in the gaps between these fiber bundles. The
disappearance of some of the CAT positive cells in this region was
observed following lesions of the olfactory bulb, indicating not
only the specificity of the immuno-staining but also that the olfactory bulb receives some projections from these neurons. Positively stained cells are very rarely observed in the lamina plexiformis (pars corticalis) of the olfactory tubercle.   Although it
is seldom possible to find positive cells in the islands of Calleja,
numerous CAT positive dots, assumed to be nerve terminals, are
localized in this structure as well as in the olfactory tubercle.

In other areas such as the neo- and piriform-cortex in this
A-P plane, we could not detect any positive cell staining, but a
few fibers appear in the indusium griseum which covers the superior
surface of the corpus callosum. Some of these fibers pass parallel
with the dorsal corpus callosum and others radiate, rapidly taper
off and fade away into the cingulate cortex.

CONCLUSION

Our results indicate that CAT, a reliable marker for cholinergic neurons, occurs in certain neurons of the rostral forebrain
as well as the spinal cord. Cell bodies, fibers and probably nerve
terminals were stained. These findings in the rostral forebrain
provide a firm morphological basis for the view that ACh may be the
transmitter substance in some of the neuronal pathways postulated
as a result of biochemical, neurophysiological and histochemical
studies (11).

ACKNOWLEDGEMENTS

Supported by the MRC. We thank S.J. Suzuki for excellent
technical assistance.

REFERENCES

1.   Bunny, B.S. and Aghajanian, G.K. (1976): Brain Res. <u>117</u>:
     423-425.
2.   Burt, A.M. and Silver, A. (1973): Brain Res. <u>62</u>:509-516.
3.   Emson, P.C. and Lindvall, O. (1979): Neuroscience <u>4</u>:1-30.
4.   Gabriel, A.M., Ragsdale, C.W. and Edley, S.M. (1979): Exp.
     Brain Res. <u>34</u>:189.
5.   Kanazawa, I., Marshall, G.R. and Kelly, J.S. (1976): Brain
     Res. <u>115</u>:485-491.
6.   Kan, S.S.K., Chao, L.P. and Eng. L.F. (1978): Brain Res.
     <u>146</u>:221-230.
7.   Kimura, H., McGeer, P.L., Peng, J.H. and McGeer, E.G. (1980):
     Science <u>208</u>:1057-1059.
8.   Kimura, H., McGeer, E.G., McGeer, P.L. and Peng, J.H. (1979):
     Proc. Sco. Neurosci. 5:240.
9.   Lehmann, J. and Fiberger, H.C. (1980): This Volume.
10. Lehmann, J., Fibiger, H.C. and Butcher, L.L. (1979): Neuro-
     science <u>4</u>:217-225.
11. McGeer, P.L. and McGeer, E.G. (1979): In <u>Nutrition and the
     Brain</u> (eds) A. Barbeau, Growdon and R.J. Wurtman, Raven
     Press, New York, pp.177-195.

12. McGeer, P.L., McGeer, E.G., Fibiger, H.C. and Wickson, V.
     (1971): Brain Res. <u>35</u>:308-314.
13. McGeer, P.L., McGeer, E.G., Singh, V.K. and Chase, W.H. (1974):
     Brain Res. <u>81</u>:373-379.
14. Peng, J.H., McGeer, P.L., Kimura, H., Sung, S.C. and McGeer,
     E.G. (1980): Neurochem. Res. <u>5</u>:943-961.
15. Ross, C.D., Godfrey, D.A., Williams, A.D. and Matschinsky,
     F.M. (1978): Proc. Soc. Neurosci. <u>4</u>:264.
16. Rossier, J. (1975): Brain Res. <u>98</u>:619-622.
17. William, M., Youngs, N., Nadi, S., Davis, B.J., Margolis, F.L.
     and Macrides, F. (1979): Proc. Soc. Neurosci. <u>5</u>:436.

REGULATION OF SEPTAL-HIPPOCAMPAL CHOLINERGIC NEURONS

BY CATECHOLAMINES

S.E. Robinson*, D.L. Cheney and E. Costa

Laboratory of Preclinical Pharmacology, NIMH
St. Elizabeths Hospital, Washington, D.C. USA and
*Department of Medical Pharmacology and Toxicology
College of Medicine, Texas A&M University
College Station, Texas, USA

INTRODUCTION

The septum occupies a pivotal site in the brain, connecting
the lower centers of the central nervous system with higher centers including the limbic areas. It is, therefore, in a position
to be involved in such important processes as the organization of
memory and emotional responsiveness. Of the neuronal projections
from the septum a cholinergic projection from the septum to the
hippocampus has been described in particular detail.

The septal-hippocampal cholinergic pathway has cell bodies
located within the medial septum and the nucleus of the diagonal
band and projects through the fimbria to synapse with the dendrites
of the pyramidal and granular cells of the hippocampus and dentate
gyrus (21,43). In return, the hippocampus projects back to the
lateral septum (31). The septal-hippocampal pathway has been well
described histologically in the rat by the use of degeneration
techniques staining for acetylcholinesterase (AChE) (17,24,43) and
techniques for choline acetyltransferase (CAT)(18). Furthermore,
using autoradiographic binding studies with 3-quinuclidinyl benzilate (QNB), Kuhar and Yamamura (15) have demonstrated muscarinic
receptors in many of the same areas of the hippocampus and dentate
gyrus that contain AChE. Stimulation of the septal area activates
neurons in the hippocampus (2,4,12), increases the release of
acetylcholine (ACh) from the hippocampus in the rabbit (9,39), and

decreases the level of hippocampal ACh in the rat (35). Furthermore, in the cat, microelectrophoretic application of ACh results
in the activation of cells in the hippocampus (41). Acute and
chronic lesions of the medial septum in the rat have been found to
decrease ACh in the hippocampus, as well as to decrease the activity
of CAT and choline (Ch) uptake in this area (16,22). Section of the
fimbria, through which the cholinergic neurons pass, has a similar
effect in the hippocampus (28).

Significant amounts of dopamine (DA) have been measured in the
lateral septal nucleus (3). Histochemical mapping experiments
have demonstrated a high density of DA terminals in the medial aspect of the lateral septal nucleus which remain after bilateral
transection of noradrenergic projections (19). Experiments have
shown the dopaminergic projection to the lateral septum to arise
in the cell body group known as A10 (8) located in the ventral
medial tegmentum: lesion of the area A10 or of the medial forebrain
bundle results in a complete loss of DA in the lateral septum (19).
Retrograde uptake experiments with horseradish peroxidase confirm
the existence of this pathway (1).

Furthermore, Assaf and Miller (1) have demonstrated that electrical stimulation of the A10 cell group excites neurons in the
medial aspect of the lateral septum. Since the direct DA agonist
apomorphine decreases the activity of septal-hippocampal cholinergic neurons (32), it is unlikely that the neurons excited by the
stimulation of the area A10 are cholinergic. It is more likely
that the dopaminergic neurons excite an inhibitory interneuron that,
in turn, inhibits the cholinergic neurons, as discussed below.

Norepinephrine (NE) has also been detected in the septum, in
both the medial and lateral septal nuclei (3). Innervation of the
septum by noradrenergic neurons is more complex than the innervation
by dopaminergic neurons. Noradrenergic neurons arise from cell
bodies located in several areas of the brainstem and these cell
groups project directly to the hippocampus as well as to the septum.
The locus coeruleus (cell group A6) sends fibers to the forebrain
through the dorsal noradrenergic bundle. Both anterograde and
retrograde autoradiographic techniques have confirmed the existence
of noradrenergic projections from the locus coeruleus to the medial
septum in the rat (13,14,29,38). Fluorescence histochemistry has
demonstrated projections from the locus coeruleus to the medial
septal nucleus and nucleus of the diagonal band, with a sparse innervation of the lateral septal nucleus (20). Furthermore, Segal
(36,37) has demonstrated that stimulation of the locus coeruleus
as well as the iontophoretic application of NE exerts an inhibitory
action on neurons in the septum. The locus coeruleus appears to be
the source of approximately half of the noradrenergic innervation
of the septal area, as lesions of the locus coeruleus have been

found to reduce septal NE by 48% (25).

Since lesions of the locus coeruleus result in a loss of only half of the NE present in the septum, it is clear that other groups of neurons must also supply the septal area. Ungerstedt (44) found that lesions of the ventral noradrenergic bundle, which is supplied by the A1, A2, A5 and A7 noradrenergic cell groups of the pons and medulla, cause a decrease in NE in the septum. Furthermore, transsections of the ventral noradrenergic bundle caudal to input from the A5 and A7 cell groups cause a significant decrease in the level of NE in the lateral septal nucleus (27). However, lesions of the area A5 (40) do not affect NE in the lateral septal nucleus so this area does not appear to be a source of noradrenergic neurons. Lindvall and Stenevi (20) report that a medulla oblongata noradrenergic system, originating in the A1, A2 or A3 cell groups, projects more heavily to the nucleus of the diagonal band and the lateral septal nucleus, with a sparse innervation of the medial septum. Thus, the locus coeruleus appears to project mainly to the medial septal nucleus and nucleus of the diagonal band, whereas neurons from the remaining brainstem noradrenergic cell groups project to the lateral septum and nucleus of the diagonal band.

METHOD

The index used to measure the activity of cholinergic neurons is the turnover rate of ACh (TR_{ACh}). The present consensus is that the turnover rate of a given neurotransmitter reflects the activity of the neurons which contains that neurotransmitter. It has been shown that TR_{ACh} is correlated with neuronal activity in the septal-hippocampal pathway (26). TR_{ACh} is monitored mass fragmentographically in the rat by measuring the incorporation of deuterium into ACh and Ch after constant rate infusion with deuterated phosphoryl-choline. The efflux rate for ACh, and hence the TR_{ACh}, are determined under steady state conditions from the relative incorporation of label into ACh and Ch by equations derived by Racagni et al. (30) and Cheney et al. (5).

Dopaminergic Control of Septal-Hippocampal
Cholinergic Neurons

The direct DA agonist apomorphine has been found to decrease TR_{ACh} in the hippocampus without affecting the level of ACh (Fig. 1). Thus, it appears that DA neurons exert an inhibitory influence on cholinergic neurons in the hippocampus. In order to determine if

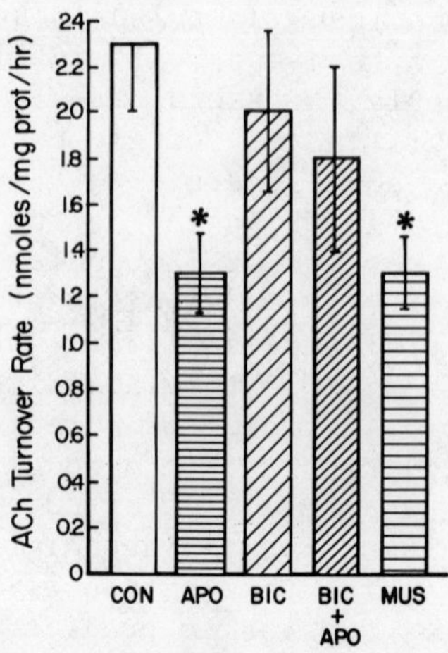

FIGURE 1: TR$_{ACh}$ in hippocampus in control animals (CON) and animals
           treated with subcutaneous apomorphine (2 mg/kg, 27 min,
           APO), intraseptal bicuculline (35 μg, 30 min, BIC),
           intraseptal bicuculline + subcutaneous apomorphine, or
           intraseptal muscimol (0.1 μg, 30 min.

TABLE 1:   Hippocampal TR$_{ACh}$ and septal catecholamines after local
           injection of haloperidol or 6-hydroxydopamine into sep-
           tum or 6-hydroxydopamine into area A10.

Treatment	TR$_{ACh}$ (%)	DA (% control)	NE (%)
Intraseptal Haloperidol (5 μg, 45 min)	250*	–	–
Intraseptal 6-Hydroxydopamine (8 μg base, 11 days)	188*	54*	45*
6-Hydroxydopamine into area A10 (8 μg base, 11 days)	247*	28*	88

* = p < 0.001 by Student's t-test.

the septum is the site of the inhibitory influence of DA, a DA
receptor blocker, haloperidol, was injected into the septum. Intraseptal haloperidol was found to increase TR_{ACh} in the hippocampus
(Table 1). To further study the control of septal cholinergic
neurons by mesolimbic dopaminergic neurons, DA in the septal area
was depleted by local injection of the neurotoxin 6-hydroxydopamine.
As can be seen in Table 1, there was again a considerable increase
in TR_{ACh} in the hippocampus. However, since haloperidol blocks NE
receptors in addition to DA receptors and intraseptal injection of
6-hydroxydopamine depletes NE as well as DA (Table 1), additional
experiments had to be performed to prove that stimulation of DA
receptors in the septum inhibited the ACh metabolism and presumably
the firing rate of cholinergic neurons in the hippocampus. Since
the cell group A10 appears to be the sole source of dopaminergic
terminals in the septum (19), specific lesions of that area would
disrupt only the DA innervation of the septum. The specific depletion of DA from the septum by the injection of 6-hydroxydopamine
into the A10 mesencephalic cell group increases the TR_{ACh} in the
hippocampus to more than double (Table 1). Therefore, since apomorphine decreases TR_{ACh} in the hippocampus and the selective depletion of DA in the septum increases TR_{ACh} in the hippocampus, it
appears that dopaminergic neurons in the septum exert an inhibitory
influence on the septal hippocampal cholinergic pathway (34).

However, since the DA terminals are restricted to the lateral
septum and the cholinergic cell bodies are located in the medial
septal nucleus and nucleus of the diagonal band, an interneuron
must exist between the dopaminergic neurons and the cholinergic
neurons. Furthermore, since the stimulation experiments of Assaf
and Miller (1) indicate that the A10 DA neurons activate neurons
in the septum, it is likely that this interneuron is inhibitory in
nature. Since γ-aminobutyric acid (GABA) has been implicated as
an inhibitory neurotransmitter in the septum (23), it is possible
that Gabaergic interneurons mediate the link between the dopaminergic terminals and the cholinergic neurons in the septum. Local
injection of the GABA agonist muscimol into the septum decreases
TR_{ACh} in the hippocampus in a dose dependent fashion. Therefore,
it appears that the cholinergic neurons projecting from the septum
to the hippocampus are controlled by inhibitory GABA receptors (6).
To determine if the dopaminergic neurons act through an inhibitory
GABA interneuron, the effect of the GABA blocker, bicuculline, was
studied on the action of apomorphine. Intraseptal injection of
bicuculline prevents the decrease in hippocampal TR_{ACh} seen after
systemic administration of apomorphine (Fig. 1). This finding,
combined with the action of muscimol, suggests that the A10 cell
group of DA neurons inhibit septal-hippocampal cholinergic neurons
by exciting an inhibitory GABA interneuron.

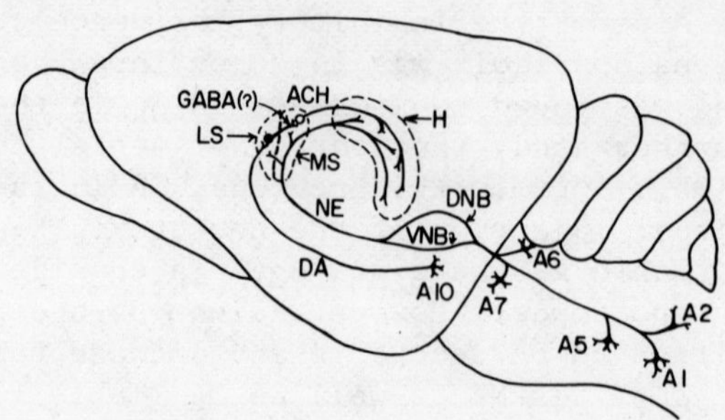

FIGURE 2: Catecholaminergic innervation of the septum. DNB = dorsal
          noradrenergic bundle; VNB = ventral noradrenergic bundle;
          MS = medial septum; LS = lateral septum; H = hippocampus.

TABLE 2: Effect of intraseptal phenoxybenzamine on increase of
          TR_{ACh} in the hippocampus by amphetamine or nomifensine.

Treatment	TR_{ACh} (nmol/mg protein/hr)
Vehicle	2.2 ± 0.3
Phenoxybenzamine (5 μg, 90 min)	2.1 ± 0.3
Amphetamine (11 mg/kg, 60 min)	5.3 ± 1.4*
Phenoxybenzamine (5 μg, 90 min) + Amphetamine (11 mg/kg, 60 min)	3.2 ± 0.6
Nomifensine (25 mg/kg, 45 min)	3.4 ± 0.6*
Phenoxybenzamine (5 μg, 90 min) + Nomifensine (25 mg/kg, 45 min)	2.2 ± 0.6

* = $p < 0.05$ by Dunnett's t-test.

Noradrenergic Control of Septal-Hippocampal Cholinergic Neurons

Nomifensine and d-amphetamine both increase the turnover rate
of DA and NE and the release of these two neurotransmitters (7,11).
Therefore, it was interesting to find that the two drugs increased
TR_{ACh} in the hippocampus (Table 2). This action is not likely due
to the DA agonist properties of these drugs, because DA stimulation

actually decreases TR$_{ACh}$ in the hippocampus. Thus, the activation of the septal-hippocampal cholinergic neurons is most likely due to the increase in noradrenergic stimulation.

Since noradrenergic neurons project direction to the hippocampus as well as to the septum, two sets of experiments were performed to determine if the two drugs were acting in the septum or through some other pathway (32,33). First, the fimbria were transected to isolate hippocampal cholinergic terminals from septal influences. Amphetamine (7.5 mg/kg) was then administered after acute fimbrial transection (2 hr). Since the amphetamine induced increase in hippocampal TR$_{ACh}$ is blocked by acute fimbrial transection, it appears that the amphetamine effect is exerted on neurons passing through the fimbria (33). In order to ascertain that the effects of nomifensine and amphetamine represent an increase in noradrenergic stimulation in the septum, the irreversible α-adrenergic blocker, phenoxybenzamine, was injected intraseptally prior to systemic administration of nomifensine or amphetamine. α-Adrenergic receptors have been identified in the septal area by autoradiographic binding studies (45). Intraseptal phenoxybenzamine was found to block the increase in TR$_{ACh}$ in the hippocampus caused by amphetamine or nomifensine (Table 2). Since phenoxybenzamine alone does not affect TR$_{ACh}$ in the hippocampus, it would appear that the noradrenergic neurons are not tonically active. Thus, noradrenergic neurons appear to excite septal-hippocampal cholinergic neurons by acting on α-adrenergic receptors in the septum.

It is not yet known whether excitatory noradrenergic receptors are located on the cholinergic cell bodies in the septum or if noradrenergic neurons are activating the cholinergic septal-hippocampal neurons by, somehow, inhibiting an inhibitory interneuron. Two findings are of interest in this respect. One is the finding that iontophoretic application of NE exerts an inhibitory action in the septum (36,37). The other is the suggestion by O'Donohue et al. (27) that lesions of the ventral noradrenergic bundle may excite the dopaminergic neurons in the nucleus of the diagonal band.

CONCLUSIONS

The sources of catecholaminergic innervation to the septum and hippocampus are illustrated schematically in Fig. 2. The dopaminergic projection originating in the cell group A10 and terminating in the lateral septum appears to exert an inhibitory influence on the septal-hippocampal cholinergic pathway. It is likely that these neurons excite an inhibitory GABA interneuron which then may act on the cholinergic neurons (Fig. 2). In contrast, noradrenergic neurons appear to exert an excitatory effect on the cholinergic septal-hippocampal neurons. However, the source of these noradrenergic neurons has not yet been identified, nor has the nature of the interaction between dopaminergic and noradrenergic neurons in the septum been clarified. Considering the proposed involvement of these noradrenergic and dopaminergic neurons in such disease states as schizophrenia (10,42), this remains a very exciting area for investigation. We are presently carrying out experiments to investigate these areas.

REFERENCES

1.   Assaf, S.Y. and Miller, J.J. (1977): Brain Res. 129:353-360.
2.   Biscoe, T.J. and Straughan, D.A. (1966): J. Physiol. 183: 341-359.
3.   Brownstein, M., Saavedra, J.M. and Palkovits, M. (1974): Brain Res. 79:431-436.
4.   Brucke, F., Gogolak, G. and Stumpf, C. (1963): Pflug. Arch. Ges. Physiol. 276:456-470.
5.   Cheney, D.L., Trabucchi, M., Racagni, G., Wang, C. and Costa, E. (1974): Life Sci. 15:1977-1990.
6.   Cheney, D.L., Robinson, S.E., Malthe-Sørenssen, D., Wood, P. L., Commissiong, J.W. and Costa, E. (1978): IN Advances in Pharmacological Therapeutics, Vol. 5, Neuropsychopharmacology (ed) C. Dumont, Pergamon Press, New York, pp. 241-250.
7.   Costa, E. and Groppetti, A. (1970): IN Amphetamines and Their Related Compounds (eds) E. Costa and S. Garattini, Raven Press, New York, pp. 231-255.
8.   Dahlstrom, A. and Fuxe, K. (1964): Acta Physiol. Scand, Suppl. 232, 62:1-55.
9.   Dudar, J.D. (1975): Brain Res. 83:123-133.
10. Farley, I.J., Price, K.S., McCullough, E., Deck, J.H., Hordynski, W. and Hornykiewicz, O. (1978): Science 200:456-458.
11. Gerhards, H.J., Carenzi, A. and Costa, E. (1974): N.-S. Arch. Pharmacol. 286:49-63.
12. Gogolak, G., Stumpf, C.H., Petshe, H. and Sterc, J. (1968): Brain Res. 7:201-207.
13. Jones, B.E. and Moore, R.Y. (1977): Brain Res. 127:23-53.
14. Jones, B.E., Halaris, A.E., McIlhany, M. and Moore, R.Y. (1977): Brain Res. 127:1-21.
15. Kuhar, M.J. and Yamamura, H.I. (1976): Brain Res. 110:229-243.
16. Kuhar, M.J., Sethy, V.H., Roth, R.H. and Aghajanian, G.K. (1973): J. Neurochem. 20:581-593.
17. Lewis, P.R. and Shute, C.D. (1967): Brain 90:521-539.

18. Lewis, P.R., Shute, C.C.D. and Silver, A. (1967): J. Physiol. 191:215-224.

19. Lindvall, O. (1975): Brain Res. 87:89-95.

20. Lindvall, O and Stenevi, U. (1978): Cell. Tiss. Res. 190: 383-407.

21. Lowry, O.H., Roberts, N.R., Leiner, K.Y., Wu, M.L., Farr, A. L. and Albers, R.W. (1954): J Biol. Chem. 207:30-49.

22. McGeer, E.G., Wada, J.A., Terao, A. and Jung, E. (1969): Exp. Neurol. 24:277-284.

23. McLennan, H. and Miller, J.J. (1974): J. Physiol. 237:625-633.

24. Mellgren, S.I. and Srebro, B. (1973): Brain Res. 52:19-36.

25. Moore, R.Y. (1978): J. Comp. Neurol. 177:665-684.

26. Moroni, F., Malthe-Sørenssen, D., Cheney, D.L. and Costa, E. (1978): Brain Res. 150:333-341.

27. O'Donohue, T.L., Crowley, W.R. and Jacobowitz, D.M. (1979): Brain Res. 172:87-100.

28. Pepeu, G., Mulas, A. and Mulas, M.L. (1973): Brain Res. 57: 153-164.

29. Pickel, V.M., Segal, M. and Bloom, F.E. (1974): J. Comp. Neurol. 155:15-42.

30. Racagni, G., Cheney, D.L., Trabucchi, M., Wang, C. and Costa, E. (1974): Life Sci. 15:1961-1975.

31. Raisman, G. (1966): Brain 89:317-348.

32. Robinson, S.E., Cheney, D.L. and Costa, E. (1978): N.-S. Arch. Pharmacol. 304:263-269.

33. Robinson, S.E., Cheney, D.L. and Costa, E. (1979): IN Catecholamines: Basic and Clinical Frontiers, Vol. 2 (eds) E. Usdin, I. Kopin and J. Barchas, Pergamon Press, New York, pp. 1077-1079.

34. Robinson, S.E., Malthe-Sørenssen, D., Wood, P.L. and Commissiong, J. (1979): J. Pharmacol. Exp. Ther. 208:476-479.

35. Rommelspacher, H. and Kuhar, M.J. (1974): Brain Res. 81:243-251.

36. Segal, M. (1974): Life Sci. 14:1345-1351.

37. Segal, M. (1976): J. Physiol. 261:617-631.

38. Segal, M. and Landis, S.C. (1974): Brain Res. 82:263-268.

39. Smith, C.M. (1974): Life Sci. 14:2159-2166.

40. Speciale, S.G., Crowley, W.R., O'Donohue, T.L. and Jacobowitz, D.M. (1978): Brain Res. 154:128-133.

41. Steiner, F. (1968): Pflug. Arch. 303:173-180.

42. Stevens, J.R. (1973): Arch. Gen. Psychiat. 29:177-189.

43. Storm-Mathisen, J. (1970): J. Neurochem. 17:739-750.

44. Ungerstedt, U. (1971): Acta Physiol. Scand. Suppl. 367.

45. Young, W.S. and Kuhar, M.J. (1979): Eur. J. Pharmacol. 59: 317-319.

INTERACTIONS OF NEUROPEPTIDES WITH CHOLINERGIC SEPTAL-HIPPOCAMPAL

PATHWAY: INDICATION FOR A POSSIBLE TRANS-SYNAPTIC REGULATION

P.L. Wood*, D.L. Cheney and E. Costa

Laboratory of Preclinical Pharmacology
National Institute of Mental Health
St. Elizabeths Hospital, Washington D.C. 20032 USA, and
*Department of Pharmacology, Merck Frosst Laboratory
Pointe Claire, P.O. Box 1005,
Dorval 700, Quebec H9R 4P8 Canada

The peculiar way in which the brain processes, stores and
retrieves information is determined by an interplay of biochemical
and electrical mechanisms which are contained and function within
complex neuronal loops. The nature of these interactions and the
way in which they relate to information processing is not understood. These circuits possess intrinsic regulatory features presumably related to memory which are dependent upon feedback and
extrinsic modulatory influences for their function. Modulatory
processes by definition include local interneurons, afferent fibers
from other brain nuclei and neuroendocrine factors. The coexistence
of neuropeptides with other primary transmitters in neurons adds
further flexibility to the system but reduces our ability to comprehend the neurochemical processes in defined anatomical pathways.
Thus, more sensitive and specific techniques are required to study
these processes. To this end we have used gas chromatography mass
spectrometry (GCMS) to examine the actions of peptides on the activity of the septal hippocampal cholinergic pathway (SHCP). The
assay of acetylcholine turnover (TR_{ACh}) by GCMS has been shown to
be an accurate index for estimating biochemically the activity of
septal cholinergic neurons (13).

The cholinergic cell bodies of the SHCP are located in the
medial septal nucleus and the nucleus of the diagonal band (8) and
project via the fimbria/fornix to innervate the pyramidal cells of

715

TABLE 1:   Decreased hippocampal TR_{ACh} after intraseptal substance
           P (40 min).

DRUG (nmol)	ACh (nmol/mg prot)	TR_{ACh} (nmol/mg prot/hr)
Control	0.27 ± 0.009	1.4 ± 0.17
Substance P (20)	0.27 ± 0.014	0.63 ± 0.058*
Bicuculline (70) + Substance P (20)	0.29 ± 0.014	0.75 ± 0.033*

* p < 0.01

TABLE 2:   Reversal by intraseptal bicuculline of the actions of
           intraseptal β-endorphin on hippocampal TR_{ACh} (30 min).

DRUG	ACh (nmol/mg prot)	TR_{ACh} (nmol/mg prot/hr)
Control	0.26 ± 0.023	2.3 ± 0.31
β-Endorphin (0.9)	0.22 ± 0.0084	0.71 ± 0.079*
Bicuculline (70) + β-endorphin (0.9)	0.22 ± 0.0099	2.2 ± 0.19

* p < 0.01

the hippocampus. This excitatory input is expressed via an activation of muscarinic recognition sites which may be coupled to
guanylate cyclase (2). In this paper we present the regulation of
this pathway at the level of the septum, presumably via axodendritic
and axosomatic synapses, and at the level of the hippocampus via
axoaxonic synapses.

Septal Modulation

     Peptidergic septal afferents which appear to regulate the
metabolism of the primary transmitter within the SHCP, include an
inhibitory β-endorphin (12) tract from the hypothalamus (20) and an
inhibitory substance P (10) input from an area ventral to the septum
(5,14). Intraventricular (i.vt.) or intraseptal (i.st.) administration of these peptides produces a dose dependent depression of the

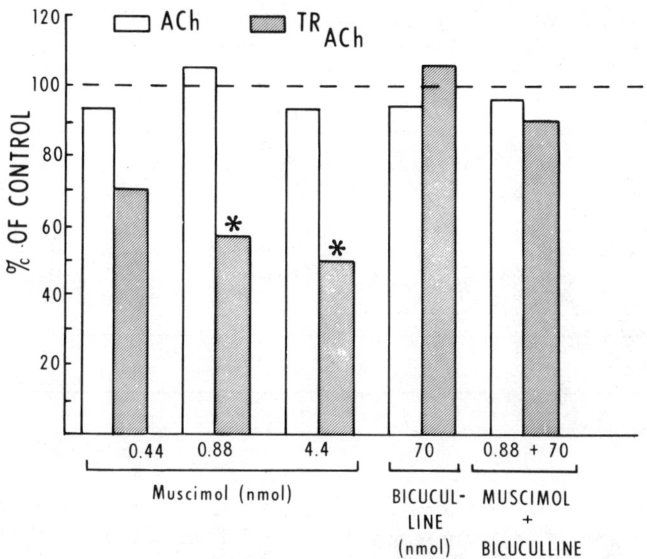

FIGURE 1: Decreased hippocampal TR_{ACh} 30 min after intraseptal
muscimol and its reversal by intraseptal bicuculline
* p < 0.01.

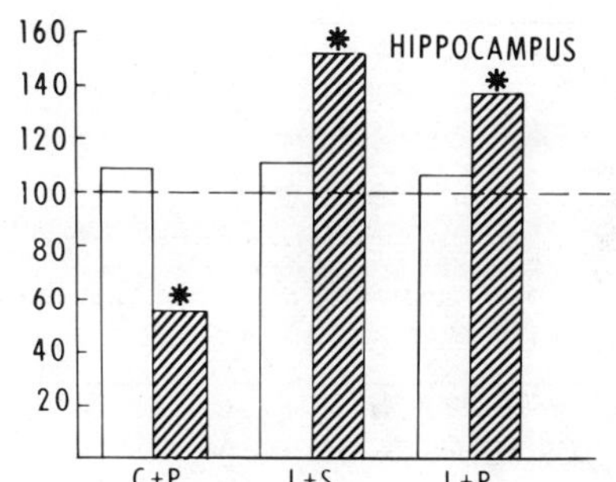

FIGURE 2: Actions of intraventricular prolactin (0.2 nmol) on
hippocampal TR_{ACh} 14 days after 6-OHDA lesions of
area A_{10}. C = control; P = prolactin; L = 6-OHDA-
lesioned; S = saline; ACh = open bars; TR_{ACh} = hatched
bars. * p < 0.05.

TABLE 3: Decreased hippocampal TR_{ACh} after intraventricular but not intraseptal α-MSH and SRIH (30 min).

DRUG (nmol)	ACh (nmol/mg prot)	TR_{ACh} (nmol/mg prot/hr)
Control	0.29 ± 0.031	1.7 ± 0.21
α-MSH (6; i.vt.)	0.21 ± 0.021	2.5 ± 0.33*
α-MSH (1.2; i.st.)	0.22 ± 0.008	1.4 ± 0.10
SRIH (21; i.vt.)	0.29 ± 0.026	2.9 ± 0.29*
SRIH (3; i.st.)	0.28 ± 0.010	1.7 ± 0.12
[Ala]-SRIH (21; i.vt.)	0.29 ± 0.031	1.7 ± 0.15

*   p   < 0.01

TABLE 4: Effects of intraventricular SRIH (21 nmol) and α-MSH (6 nmol) on hippocampal TR_{ACh} after fimbria transection.

GROUP	ACh (nmol/mg prot)	TR_{ACh} (nmol/mg prot/hr)
Control side	0.23 ± 0.021	1.4 ± 0.11
Transected side	0.63 ± 0.093*	0.46 ± 0.093*
Control + α-MSH	0.27 ± 0.024	4.0 ± 0.34*
Transected + α-MSH	0.29 ± 0.041	2.0 ± 0.33**
Control + SRIH	0.29 ± 0.033	3.4 ± 0.041*
Transected + SRIH	0.41 ± 0.035**	3.4 ± 0.55**

*   p   < 0.01 vs control;   **   p   < 0.01 vs transected

TABLE 5: Actions of intraventricular SRIH (21 nmol) and α-MSH (6 nmol) on hippocampal TR_{ACh} after bilateral section of the cingulum.

GROUP	ACh (nmol/mg prot)	TR_{ACh} (nmol/mg prot/hr)
Control	0.23 ± 0.014	1.6 ± 0.082
Sectioned	0.20 ± 0.010	1.1 ± 0.061
Sectioned + SRIH	0.20 ± 0.012	2.3 ± 0.19*
Sectioned + α-MSH	0.16 ± 0.007	1.7 ± 0.14*

*   p   < 0.05 vs sectioned.

TABLE 6:   Actions of intraventricular SRIH (21 nmol) and α-MSH
           (6 nmol) on hippocampal TR_{ACh} after lesion of the
           entorhinal cortex.

GROUP	ACh (nmol/mg prot)	TR_{ACh} (nmol/mg prot/hr)
Control	0.27 ± 0.023	1.6 ± 0.25
Lesioned	0.33 ± 0.013	1.6 ± 0.19
Lesioned + SRIH	0.28 ± 0.034	3.4 ± 0.53*
Lesioned + α-SRIH	0.27 ± 0.023	3.3 ± 0.33*

* p < 0.01 vs lesioned

TABLE 7:   Depression of hippocampal TR_{ACh} by intraventricular and
           intraseptal prolactin (30 min)

GROUP	ACh (nmol/mg prot)	TR_{ACh} (nmol/mg prot/hr)
Control	0.31 ± 0.028	2.0 ± 0.18
Prolactin (0.20 nmol; i.vt.)	0.31 ± 0.026	1.3 ± 0.18*
Prolactin (0.04 nmol; i.st.)	0.26 ± 0.014	1.1 ± 0.059*

* p < 0.05;

hippocampal TR_{ACh}, indicating an action within the septum (10,12).

We have studied these inhibitory peptidergic actions in great
detail to determine whether they are regulated by, or function via,
the abundant GABAergic interneurons which are located in the septum
(18,19). An activation of these GABAergic interneurons decreases
the TR_{ACh} in the hippocampus (Fig. 1; also Ref. 3). To examine the
possible role of GABAergic interneurons in the actions of β-endor-
phin or substance P, bicuculline was injected i.st. 6 min prior to
the i.vt. administration of the peptide. As shown in Table 1, this
pretreatment did not modify the decrease in hippocampal TR_{ACh} elicited by i.st. administered substance P. However, the actions of
i.st. β-endorphin were effectively antagonized by bucuculline pretreatment (Table 2). These data suggest that the afferent axons to

the septum that contain β-endorphin, modulate the activity within
the SHCP via GABAergic interneurons while the axons that store
Substance P inhibit SHCP independently from this interneuronal pool.
It remains, however, to be established whether a direct β-endorphin/
GABAergic linkage exists within the septum.

Hippocampal Modulation

The peptides, α-melanocyte stimulating hormone (α-MSH)(21,22)
and somatostatin (SRIH) (11) augmented hippocampal TR_{ACh} after i.vt.
administration. The septum, however, may not be the site for this
modulation since i.st. injections (Table 3) of α-MSH or SRIH fail
to modify hippocampal TR_{ACh}. We then investigated whether an action
of these peptides on the hippocampal cholinergic terminals may be
operative. To examine this possibility, we lesioned the fimbria/
fornix thereby isolating the cholinergic terminals of the hippocampus from their cell bodies in the septum and measured hippocampal TR_{ACh} immediately thereafter. As shown in Table 4, this
lesion could not inhibit the increase in hippocampal TR_{ACh} caused
by either peptide.

In addition to the fimbria/fornix, the hippocampus also receives
innervation via the cingulum and perforant pathway. However, chronic
lesions of these pathways (Tables 5 and 6) failed to modify the
response to i.vt. α-MSH or SRIH. These data suggest that the actions
of SRIH and α-MSH are not dependent upon hippocampal afferents whose
cell bodies might be under peptidergic control. Moreover, we can
exclude that peptidergic receptors may be located on the axon terminals of these afferent pathways. Since the two peptides caused inhibition after fimbria transection, it may be inferred that the
hippocampal afferents which provide a feedback to the septum are not
involved and it may suggest that a population(s) of hippocampal
cholinergic terminals contains receptors for the two peptides. Perhaps these receptors are innervated by axons of peptidergic interneurons. Immunocytochemical studies (16) have indicated the presence of SRIH containing interneurons in fields CA_1 and CA_2 of the
hippocampus. These SRIH receptors possess a certain degree of specificity; in fact, [Ala^8]-SRIH, a biologically inactive analogue of
SRIH, failed to change hippocampal TR_{ACh} (Table 3).

Neuroendocrine Modulation

Dopaminergic innervation of the septum arises from the ventral
tegmental area (A_{10}) and exerts a tonic inhibitory influence on the
cholinergic cell bodies of the SHCP (17). Since prolactin has been
shown to increase dopamine release and turnover (1,7,15) in the CNS,
we have investigated the actions of prolactin on the TR_{ACh} in the

SHCP. As shown in Table 7, both i.vt. and i.st. prolactin depressed
the TR_{ACh} in the hippocampus. To test whether this action of prolactin was dependent upon dopaminergic innervation of the septum,
6-hydroxydopamine lesions of area A_{10} were performed and the actions
of i.vt. prolactin tested when the septal dopaminergic content was
reduced to 17% of control values. In these rats the inhibition by
prolactin of the TR_{ACh} in the SHCP was antagonized (Fig. 2). These
results suggest that prolactin may modulate neuronal function in the
SHCP by increasing the release of dopamine from dopaminergic axons
afferent to the septum. However, the role of septal interneurons
and other septal afferents in this action remains to be determined.
Since extrahypothalamic prolactin containing nerve terminals have
not been found in the CNS (6), we may infer that if prolactin plays
a physiological role, it acts as a neuroendocrine modulator of
activity within the SHCP and that this modulation is dependent upon
the dopaminergic innervation of the septum. This hypothesis is
further supported by observations of parallel fluctuations in CSF
and serum prolactin levels (4,9).

In summary, we have demonstrated that the activity of cholinergic axon terminals within the SHCP is regulated by peptides in
many ways. Complex Substance P inputs reach the septum where they
modulate the ACh metabolism in the hippocampus by acting on some
inhibitory pathway; an intermediation by GABA or endorphin secreting
neurons does not appear to be operative. In contrast, SRIH and
α-MSH act on specific receptors located on cholinergic terminals in
the hippocampus. The presence of such receptors together with
histochemical evidence suggests the existence of peptidergic interneurons within the pyramidal fields of the hippocampus. Finally,
there is evidence for a neuroendocrine modulation of central neuro-
synaptic transmission. Specifically, the ability of prolactin to
regulate ACh septal neurons has been demonstrated and the possibility
must be considered that prolactin may also facilitate release of
dopamine and the release of unknown inhibitory transmitters from
interneurons regulated by dopaminergic neurons.

REFERENCES

1.   Annunziato, L. and Moore, K.E. (1978): Life Sci. 22:2037-2042.
2.   Biscoe, T.J. and Straughan, D.W. (1966): J. Physiol. 183:
         341-359.
3.   Cheney, D.L., Robinson, S.E., Malthe-Sorenssen, D., Wood, P.L.,
         Commissiong, J.W. and Costa, E. (1979): Adv. Pharmacol. Ther.
         5:241-250.
4.   Clemens, J.A. and Sawyer, B.D. (1974): Exp. Brain Res. 21:
         399-402.

5. Cuello, A.C. and Kanazawa, I.   1978): J. Comp. Neurol. 178: 129-156.
6. Fixe, K., Hokfelt, T., Eneroth, P., Gustafsson, J.-A. and Skett, P. (1977): Science 196:899-900.
7. Hokfelt, T. and Fuxe, K. (1972): Neuroendocrinology 9:100-122.
8. Lewis, P.R. and Shute, C.C.D. (1967): Brain 90:521-540.
9. Login, I.S. and MacLeod, R.M. (1977): Brain Res. 132:477-483.
10. Malthe-Sorenssen, D., Cheney, D.L. and Costa, E. (1978): J. Pharmacol. Exp. Ther. 206:21-28.
11. Malthe-Sorenssen, D., Wood, P.L., Cheney, D.L. and Costa, E. (1978): J. Neurochem. 31:685-691.
12. Moroni, F., Cheney, D.L. and Costa, E. (1977): Naun.-Schmied. Arch. Pharmacol. 299:149-153.
13. Moroni, F., Malthe-Sorenssen, D., Cheney, D.L. and Costa, E. (1978): Brain Res. 150:333-341.
14. Paxinos, G., Emson, P.C. and Cuello, A.C. (1978): Neurosci. Lett. 7:133-136.
15. Perkins, N.A. and Westfall, T.C. (1978): Neuroscience 3:59-63.
16. Petrusz, P., Sar, M., Grossman, G.H. and Kizer, J.S. (1977): Brain Res. 137:181-187.
17. Robinson, S.E., Malthe-Sorenssen, D., Wood, P.L. and Commissiong, J. (1979): J. Pharmacol. Exp. Ther. 208:476-479.
18. Storm-Mathisen, J. and Opsahl, M.W. (1978): Neurosci. Lett. 9:65-70.
19. Tappaz, M.L., Brownstein, M.J. and Palkovits, M. (1976): Brain Res. 108:371-379.
20. Watson, S.J. and Barchas, J.D. (1979): In Mechanisms of Pain and Analgesic Compounds (eds) R.F. Beers Jr. and E.G. Bassett, Pergamon Press, New York, pp. 227-237.
21. Wood, P.L., Malthe-Sorenssen, D., Cheney, D.L. and Costa, E. (1978): Life Sci. 22:673-678.
22. Wood, P.L., Cheney, D.L. and Costa , E. (1978): J. Pharmacol. Exp. Ther. 209:97-103.

CHOLINERGIC-MONOAMINERGIC INTERACTIONS IN SELECTED REGIONS OF THE

BRAIN: HISTOCHEMICAL AND PHARMACOLOGIC ANALYSES

L.L. Butcher*,**, N.J. Woolf**, A. Albanese* and
S.H. Butcher***

*Department of Psychology
**Brain Research Institute, and
***Mental Retardation Research Center
University of California, Los Angeles, California 90024

INTRODUCTION

The elucidation of neuronal circuitry has been a major goal in
the neurosciences. Anatomic studies with radioactive amino acids
and other neuronally transported substances such as horseradish
peroxidase have focussed on possible relationships among different
cell body groups in the central nervous system. In most of these
experiments, however, neurochemical features of the relevant neural
regions are usually not considered. Yet immense interest in certain of these neurochemicals exists, especially transmitters and
related substances involved in neuronal communication processes,
because of their obvious or apparent significance for functional
neurobiology. In the present report we attempt to address the
problem of transmitter neuroanatomy in three regions of the brain
thought to be involved in cholinergic-monoaminergic interactions:
the substantia nigra, corpus striatum and locus ceruleus.

Distribution of Cholinergic Substances in the Brain: Histochemical Studies

Histochemical studies of cholinergic substances in the brain
have been restricted almost entirely to four neurochemical entities
(6): choline-O-acetyltransferase (CAT, EC 2.3.1.6), acetylcholin-

esterase (AChE, EC 3.1.1.7) and muscarinic and nicotinic receptors.
Of these, morphologic detail has been best elucidated for the synthetic and degradative enzymes and, accordingly, it is to these two
substances that the remainder of this discussion is directed.

CAT: Controversy exists concerning the reliability and validity
of currently available histochemical methods for CAT (6). Although
the lead-trapping protocol for the enzyme has long ceased to be a
viable histochemical technique (see 6), immunohistochemical methods
are being pursued vigorously but not without their difficulties (6).
As pointed out by Rossier (38), a major concern has been the purification of CAT and the specificity of the antibodies directed
against the enzyme. Hattori and his associates (21,22), for example,
initially described the association of CAT with medium sized spiny
neurons in the caudate-putamen complex. In a more recent report
from the same laboratory, however, Kimura et al. (23), using a newly
purified enzyme and new antibodies, have reported that CAT is
localized within large aspiny or small multipolar cells in the
striatum.

AChE: Whereas controversy clouds histochemical studies of CAT,
AChE remains one cholinergic substance that apparently has been
demonstrated reliably and validly with current histochemical procedures (6). Indeed, recent developments in histochemical protocols
for the enzyme have permitted improved morphologic detail to be
obtained (6). The pharmacohistochemical regimen of Butcher et al.
(12), for example, derived essentially from the observations of
Nichols and Koelle (34,35) that AChE reappears first in the somata
of retinal amacrine cells following irreversible enzyme inhibition
by bis(1-methylethyl)phosphorofluoridate, has been used by many different laboratories to "rewrite" the AChE neuroanatomy of the brain
(e.g. 19,36,37).

Acetylcholinesterase and Cholinergic Neurons

Beginning with the development by Koelle and Friedenwald (26)
of a sensitive histochemical method for AChE, there has been speculation concerning to what extent the presence of the enzyme in
neural tissue could be used as a marker for cholinergic neurons
(i.e. those containing and utilizing acetylcholine [ACh] as a
neurotransmitter; see 10). Although most researchers investigating
cholinergic mechanisms realize that the mere presence of AChE within
and/or on a particular neuron is not a sufficient condition to label
that neuron cholinergic (see 10), hypotheses have been formulated
attempting to relate the intensity of AChE staining to the simultaneous presence of ACh and, by implication, its synthetic enzyme,
CAT. One such proposal, which for want of a better appellation

will be termed the Koelle formulation, states that "Neurons...in...
intensely and moderately stained categories are...cholinergic (25,
p.224)." At least two exceptions to this rule appear to exist currently: (A) the neuronal somata of pars compacta of the substantia
nigra containing dopamine (5,6,8,9,12); (B) the cell bodies of locus
ceruleus that possess norepinephrine in addition to AChE (2,3). The
utility of the Koelle hypothesis must therefore be questioned. Nonetheless, a weak corrollary of the Koelle proposal, hereafter termed
the AChE corollary, has been proposed recently (29; cf. 39, p.3).
The AChE corollary states that "High AChE activity is a necessary
but not sufficient characteristic for identifying cholinergic
neurons (29, p.1940)." Despite claims that the validity of the
AChE corollary will "...accelerate the elucidation of cholinergic
neuroanatomy (29, p.1940)," it is unlikely that by itself it will
actually do so. Stated another way, the AChE corollary becomes:
both non-cholinergic and cholinergic neurons may stain intensely
for AChE, but cholinergic neurons stain intensely for AChE. Although all cholinergic neurons may indeed stain heavily for AChE
(see 39), there is no way to determine on the basis of staining
intensity alone whether or not a given AChE containing neuron is
cholinergic. Independent methods are needed to assess the presence
or absence of ACh and CAT, and their functions in neurons. Only
when a transmitter role can be demonstrated for ACh so localized
will it be possible to call a neuron cholinergic. Unfortunately,
no reliable or valid histochemical procedures currently exist to
assess ACh or CAT (6). Precise correlations among the distributions
of AChE, ACh and CAT at the neuronal level are currently impossible.
As an empirical rule, therefore, the AChE corollary is considerably
less powerful than the Koelle hypothesis, and its value as an eluci-
dative device is not yet clear. The fact that the AChE corollary
has been put forth recently (29, cf. 39, p.3) should not diminish
efforts to develop viable histochemical procedures for CAT, ACh,
and other indices of cholinergic function. The insightful comments
of Silver (39) can be recalled profitably in this regard: "Because
proven cholinergic cells give a strong histochemical reaction for
AChE, authors have come to use 'cholinergic' as an epithet for any
AChE containing cell, regardless of whether or not it is known to
release ACh. To use 'cholinergic' as a synonym for 'acetylcholin-
esterase-containing' may be convenient but it can be misleading:
although all cells known to use ACh as their transmitter contain
AChE, not all AChE containing cells are functionally cholinergic."

AChE-Dopaminergic Relationships in the Substantia Nigra

    That the somata, dendrites and axons of most, if not all,
dopamine neurons of pars compacta of the substantia nigra (group A9
of Dahlstrom and Fuxe [16]) contain AChE is suggested by the following

experimental evidence. (A) The organization, morphologies, and dimensions of dopamine containing cell bodies in pars compacta are similar, if not identical, to the distribution, geometry, and sizes of many AChE containing somata in that nigral subdivision (Fig. 1 and ref. 5). (B) Radio-frequency ablations at various locations in the nigro-striatal pathway result, after 1-4 weeks depending on location of the lesion, in retrograde degeneration and consequent loss of AChE containing somata in pars compacta of the substantia nigra (5). Similarly placed lesions produce retrograde degeneration of pars compacta dopamine somata, as well as a decrease in histochemically assessed dopamine in the caudate-putamen complex (unpublished observations, this laboratory). (C) Colchicine infusion into brain regions through which the nigro-striatal pathway projects results in AChE accumulation within axons of nigro-striatal neurons caudal to the injection site. That these fibers originate, in major part at least, from cell bodies in pars compacta is supported by the findings that intra-axonal accumulation of AChE does not occur to any great extent if a lesion is made in the substantia nigra prior to colchicine infusion into the globus pallidus (5).

Although the preceding arguments are cogent, direct and compelling histochemical evidence for the localization of AChE within and/or on pars compacta dopamine neurons has been obtained only recently. We have developed a histochemical procedure for visualizing catecholamines and AChE on the same brain section (2,3,9). Catecholamines are demonstrated with the glyoxylic acid procedure of de la Torre and Surgeon (17), and AChE is demonstrated according to the pharmacohistochemical regimen of Butcher et al. (12).

Virtually all neuronal somata in the substantia nigra stained for AChE. As a function of the animal and level of substantia nigra examined, 70-98% of these AChE positive cells also contained dopamine (see Fig. 1). The converse was true in only one neuron in one rat and in six somata in another; in all other animals studied each dopamine containing neuron examined also possessed AChE. Caudal substantia nigra demonstrated a greater proportion of neurons possessing both AChE and dopamine than did rostral regions of the nucleus. In addition, many processes in the substantia nigra, probably dendrites primarily, contained both AChE and dopamine (Fig. 1, compare A with C and B with D).

It is open to speculation why dopamine containing neurons originating in the substantia nigra and other areas of the mesencephalon (11) should contain AChE, but some hypotheses are: (A) AChE in dopamine somata and dendrites could be so situated to inactivate ACh released from cholinergic afferents to those neuronal elements (5), (B) AChE in striatal dopamine terminals could inactivate ACh released from dendrites of striatal intraneurons (11) and (C) both dopamine and ACh could be neurotransmitters and/or

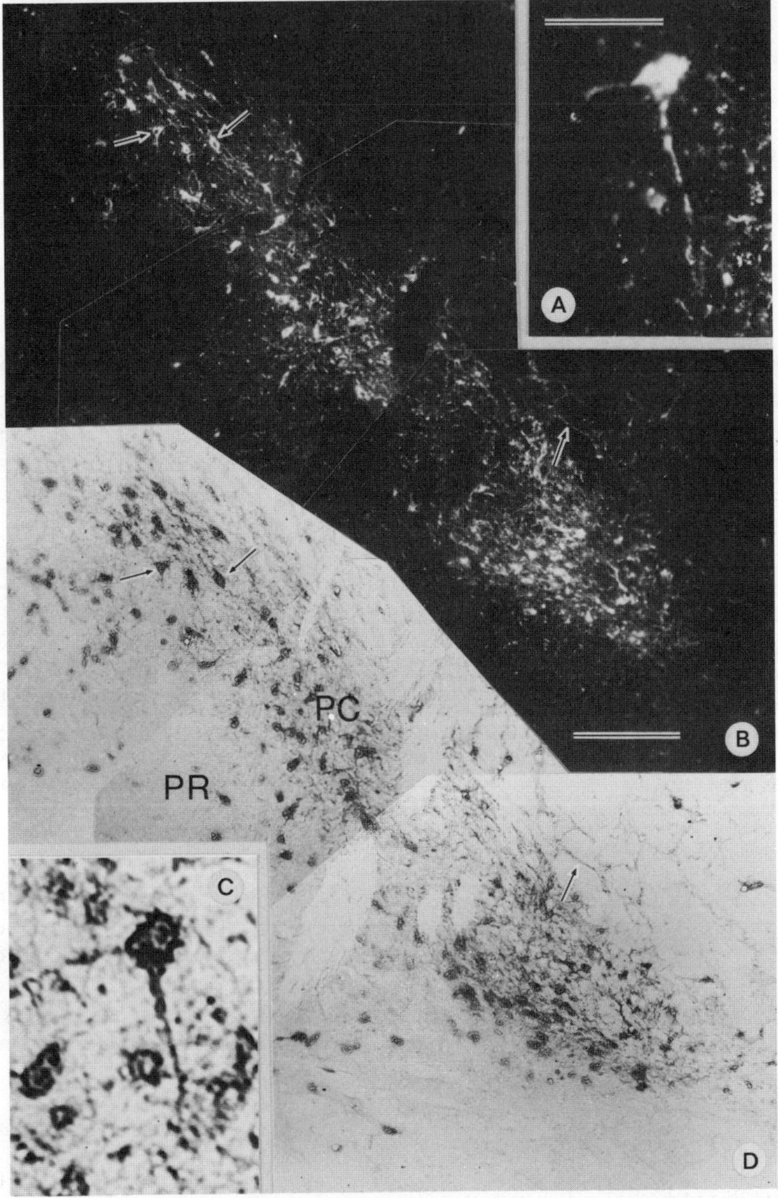

FIGURE 1: Dopamine (A,B) demonstrated on the same tissue section of
the substantia nigra stained for AChE (C,D). Pharmacohistochemical regimen for AChE. Arrows in B and D point
to corresponding dopamine and AChE containing elements.
Scale in A is 50 µm and applies also to C; scale in B is
200 µm and applies also to D. PC = pars compacta, substantia nigra. PR = pars reticulata, substantia nigra.

neuromodulators in the same neuron, either at the same or different
synapses (11). Relevant to this last possibility are the findings
of Agid et al. (1) that neostriatal ACh is reduced by 26-28% at
short intervals following nigral ablations. Similarly, Nagy et al.
(33) observed a permanent 18% reduction of CAT in the caudal caudate-
putamen complex after lesions in the nigro-striatal bundle.

Cholinergic-Monoaminergic Relationships in the Striatum:
Histochemical Studies

Application of the pharmacohistochemical procedure of Butcher
et al. (12) to the study of the rat caudate-putamen complex has revealed the existence of two major categories of AChE neurons (40):
(A) Type A neurons have maximum cell body extents within the range
7-20 μm. They stain lightly for AChE (Fig. 2, arrows in frames A,
B,D,E,F and G), the reaction product being homogeneously distributed
throughout the soma and proximal processes. Although somata can be
ovoid, fusiform, triangular, or spheroid, the most prevalent shape
in this category of striatal AChE neurons is oval. Type A caudate-
putamen cells have an average of one AChE containing process per
soma, and they comprise approximately 2.2% of the total population
of striatal neurons. (B) Type B neurons have maximum cell body
extents of 20-44 μm; somata shapes are fusiform (Fig. 2; D,E.F.),
oval (Fig. 2; F,G), triangular (Fig. 2; C,D), or complex (no readily
identifiable geometric shape; Fig. 2, H). These neurons stain intensely for AChE (Fig. 2, C-H), possess an average of 2.4 AChE containing processes per cell body, and comprise roughly 1.2% of the
total number of neurons in the rat caudate-putamen complex.

If the Koelle formulation is valid when applied to AChE containing neurons in the striatum, then the Type B cells we observed
would be cholinergic. Furthermore, they should be the local cir-
cuit, intrinsically organized neurons whose existence has been deduced from previous chemical and histochemical studies (13,14,32).
It is unknown what functions the Type A neurons might subserve, but
one possibility is that they are cholinoceptive.

If the Type B neurons are indeed cholinergic intraneurons,
then a unique opportunity presents itself to determine the possibility of direct anatomic relationships between afferent dopamine
terminals deriving from the mesencephalon and AChE containing,
possibly cholinergic, neurons in the striatum. Using the combined
catecholamine-AChE histochemical protocol developed by Butcher and
Marchand (9) and Albanese and Butcher (2,3), we have attempted to
shed light on that possibility.

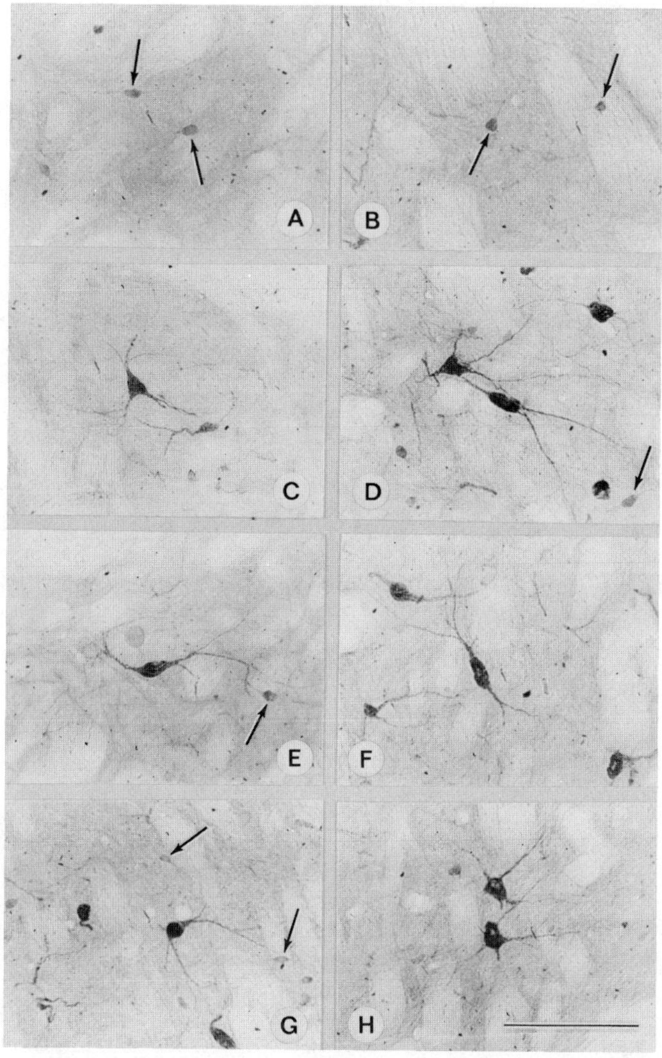

FIGURE 2: Various AChE containing neurons in the rat caudate-putamen complex demonstrated with the pharmacohistochemical
procedure. Arrows point to Type A neurons (see text).
Scale = 100 μm.

There are marked differences between the distribution of
dopamine terminals and the organization of striatal AChE somata and
their proximal processes (Fig. 3). Whereas the number of striatal
AChE neurons constitutes only about 3.6% of the total number of
cell bodies in that neural region, dopamine containing terminals
are distributed profusely throughout the structure (Fig. 3). This
disparity in topography possibly reflects, in part, the participation
of dopamine in striatal interactions in addition to those involving
cholinergic mechanisms. Notwithstanding this possibility, however,
the pattern of AChE and dopamine staining most characteristic of
the striatum revealed that no obvious relationship existed between
dopamine terminals and intensely staining AChE, probably cholinergic,
cell bodies and proximal processes. In some regions of the caudate-
putamen complex there were no AChE containing somata or proximal
processes yet the same field was densely innervated by dopamine
terminals (Fig. 4). In other regions no obvious relationship e-
merged between AChE and dopamine neuronal elements. Infrequently,
however, dopamine terminals were observed to follow in apparent
association with AChE containing processes, probably dendrites
(Fig. 5; compare with Fig. 20 in ref. 5).

Although the implications of these data are that few, if any,
direct relationships exist between dopamine afferents to the striatum
and intrinsically organized cholinergic neurons, final evaluation
of such a speculation must await electron microscopic data.

Pharmacologic Studies

Much of the initial data relevant to cholinergic-monoaminergic
interactions in the brain was derived from pharmacologic studies in
which it was assumed that the drugs used selectively affected one
or another neurotransmitter or neuromodulator system (for overview
see 7,20). An early favorite in such investigations was the nigro-
striatal pathway, and many investigators assumed a direct synaptic
relationship between afferent dopamine terminals and cholinergic
intraneurons in the caudate-putamen complex (e.g. 27). Subsequent
drug effects were interpreted in terms of this neuroanatomic model
(e.g. 27). As indicated previously in this paper, however, such
an assumption of direct connectivity may be tenuous. Nonetheless,
predicted on the validity of direct monoamine-ACh linkages in the
striatum, drugs increasing or decreasing levels and/or the efficacy
of dopamine and serotonin have been examined for their effects on
striatal cholinergic mechanisms (e.g. 27).

Most of these drug studies have indicated that dopamine, as
well as serotonin, is inhibitory on striatal cholinergic neurons,
and, indeed, if a direct anatomic relationship exists between neurons

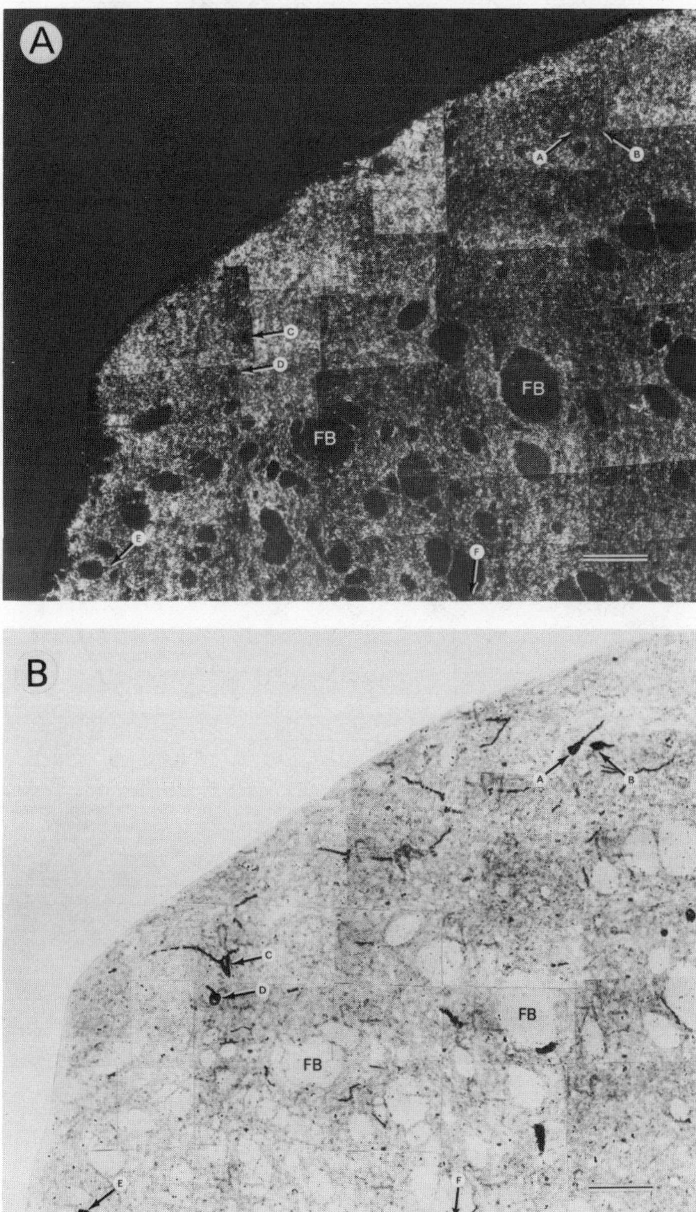

FIGURE 3: Photomontage showing the same segment of the caudate-putamen complex stained for dopamine (frame A) and AChE (frame B, pharmacohistochemical regimen). Small letters with attached arrows point to corresponding portions of the two photomontages. FB = fiber bundle perforating striatum. Scales = 100 µm.

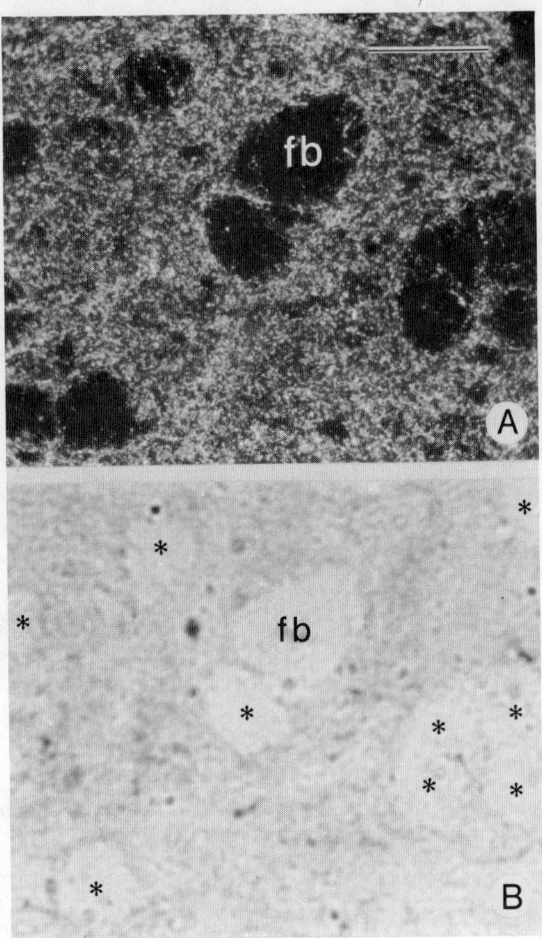

FIGURE 4: The same striatal brain section stained for dopamine (A)
and AChE (B, pharmacohistochemical regimen). No AChE
staining elements are observed in B. fb = fiber bundle
perforating striatum. Asterisks in B denote fiber bundles shown in black in A. Scale = 50 μm.

containing the two neurochemicals, such a conclusion seems warranted.
Ladinsky et al. (27) have shown that dopamine agonists increase
striatal ACh levels whereas dopamine antagonists decrease those
concentrations. Presumably, dopamine agonists inhibit firing rates
of cholinergic intraneurons, resulting in decreased release and an
intraneuronal accumulation of ACh. The converse is argued for
dopamine antagonists. We have corroborated the drug effects seen

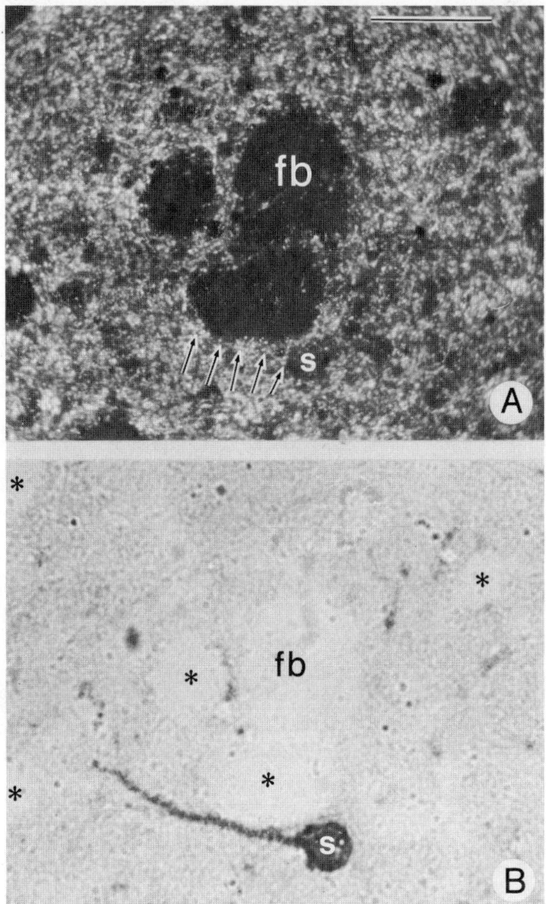

FIGURE 5: The same striatal brain section stained for dopamine (A)
          and AChE (B, pharmacohistochemical regimen). Arrows in A
          point to dopamine terminals that appear to be associated
          with an AChE staining process in B, probably a dendrite.
          fb = fiber bundle perforating striatum. Asterisks in B
          denote fiber bundles shown in black in A. s = neuron soma,
          scale = 50 μm.

by Ladinsky et al. (27) but, in addition, have examined the synthesis of ACh (Table 1) and cannot agree entirely with their viewpoint that dopamine and serotonin are inhibitory upon striatal cholinergic neurons.

One difficulty in drug studies on striatal dopaminergic-chol-
inergic and serotonergic-cholinergic interactions is that the

TABLE 1: Dopaminergic, serotonergic and cholinergic drugs and their effects on levels and synthesis of striatal ACh and Ch.

TREATMENT	Cholinergic Substances in the Striatum (mean nmol/g tissue, n = 10)			
	Endogenous ACh	Endogenous Ch	$^2H_9$–ACh	$^2H_9$–Ch
Control	92.18	20.18	0.86	1.86
Ro + L–DOPA	97.76*	20.21	0.68*	1.65*
Ro + 5–Hydroxytryptophan	80.37*	21.10	0.85	1.68*
Control	85.65	24.98	1.11	1.86
Haloperidol	51.10*	26.27	0.50*	1.74
Pimozide	58.62*	26.12	0.60*	1.67*
Control	85.63	24.90	0.97	2.83
p–Cl–amphetamine	94.16*	22.56	0.47*	1.49*
Control	90.90	22.17	0.93	2.06
Amphetamine	98.21*	23.11	1.28	2.27
Control	83.80	20.39	0.78	2.15
Oxotremorine	90.75	20.36	0.36*	2.76
Atropine	65.80*	17.57	0.64	2.00
Oxotremorine + Atropine	94.24	24.16	0.83	1.88

Treatment: Abbreviations & i.p. doses: Ro-4-4602, 50 mg/kg; L–DOPA, 100 mg/kg; 5-hydroxytryptophan, 100 mg/kg; haloperidol, 1 mg/kg; pimozide, 1 mg/kg; p–Cl–amphetamine, 1 mg/kg; oxotremorine, 0.3 mg/kg; atropine, 1 mg/kg.

Cholinergic substances in the striatum: A pulse intravenous injection of 20 μmol/kg $^2H_9$–Ch in isotonic NH_4Cl was given 1 min before rat was sacrificed by microwave irradiation (15) during peak effect of intraperitoneally injected drug.

* p < 0.05 compared to control for that drug series.

conclusions of most investigators are at variance with a large body
of electrophysiologic data indicating that both dopamine and serotonin are excitatory on monosynaptically innervated target cells
in the caudate and putamen nuclei (30,41).

Reconciliation of electrophysiologic with pharmacologic data
is complicated by many factors including ignorance of the precise
neuronal circuitry involved and the fact that most drugs have other
central effects in addition to those given prominence in published
works. Central blood flow, for example, should be considered.
Amphetamine produces a marked increase in blood flow through the
caudate nucleus and cortex (28), whereas haloperidol substantially
reduces that flow. McCulloch and Harper (31) have demonstrated
similar effects for apomorphine and pimozide. Dross and Kewitz
(18) have shown that the brain does not appreciably synthesize
choline (Ch) de novo and primarily depends on the blood for supplies of that substance; a major means for removal of Ch from the
brain is also via the blood. This information, taken together with
the data of Atweh and Kuhar (4) and Butcher et al. (15) suggests
that measurement of the specific activity of Ch in calculating turnover can be misleading, but assessment of turnover on the basis of
endogenous levels and synthesis of ACh can provide a reasonable
indication of the activity of the cholinergic neuron.

AChE-Noradrenergic Relationships in Locus Ceruleus

The localization of AChE within and/or on norepinephrine containing neurons in locus ceruleus (group A6 of Dahlstrom and Fuxe,
16) has been deduced from several experimental observations (5,24).
(A) The morphologies and cytoarchitectural organization of norepinephrine and AChE somata in locus ceruleus are similar, if not
identical, to one another (5,24). (B) Retrograde degeneration of
AChE containing cell bodies in locus ceruleus is produced in rats
having unilateral radio-frequency lesions in both the cerebellum
and dorsal tegmental bundle (5), two areas possessing norepinephrine
containing afferents from locus ceruleus. And (C), injection of
colchicine into the dorsal tegmental bundle and cerebellum produces
accumulation of AChE within processes, probably axons, deriving
from locus ceruleus neuronal somata (5).

Comparison of fluorescence patterns and AChE staining on the
same brain sections provided direct evidence that norepinephrine
somata in locus ceruleus contained AChE (compare Fig. 6A with 6B).
Such cell bodies were round, oval, fusiform or pyramidal. The deposits of russet colored reaction product representing AChE loci
were organized in a peri-nuclear array; norepinephrine staining
was similarly organized and followed patterns of AChE staining

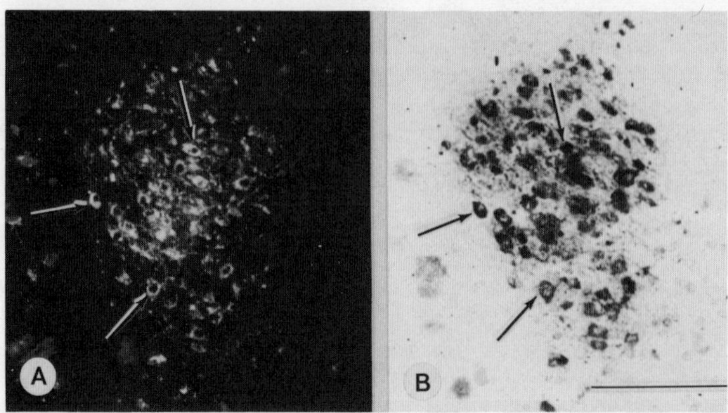

FIGURE 6: The same brain section of locus ceruleus stained for both
          norepinephrine (A) and AChE (B, pharmacohistochemical
          regimen). Arrows point to corresponding somata.
          Scale = 200 µm.

(compare Fig. 6A with 6B). All norepinephrine somata in locus
ceruleus appeared to contain AChE.   Why norepinephrine somata in
locus ceruleus should also contain AChE is, like the situation in
pars compacta of the substantia nigra, not yet known, but perhaps
the same interactive mechanisms operating in the nigra (<u>vide</u> <u>supra</u>)
also operate in locus ceruleus.

CONCLUSIONS

Perhaps none of the interactive schema proposed in this report
will prove valid, or even heuristically useful, when additional
data are collected. Nonetheless, it seems reasonable to suggest
that the nature of cholinergic–monoaminergic interactions in the
brain is a topic worthy of further investigation. Major impetus
to such an effort will be provided when histochemical methods for
the entire gamut of cholinergic substances and mechanisms are
developed that are comparable to currently available procedures
for AChE.

ACKNOWLEDGEMENTS

This research was supported by USPHS grant NS-10928 to LLB.
AA is a Fulbright-Hays scholar from the Istituto di Neurologia;
Universita Cattolica, Rome.

REFERENCES

1. Agid, Y., Guyenet, P., Glowinski, J., Beaujouan, J.C. and
    Javoy, F. (1975): Brain Res. 86:488-492.
2. Albanese, A. and Butcher, L.L. (1979): Neurosci. Lett. 14:101-
    104.
3. Albanese, A. and Butcher, L.L. (1980): Brain Res. Bull. 5:
    127-134.
4. Atweh, S.F. and Kuhar, M.J. (1976): Eur. J. Pharmacol. 37:
    311-319.
5. Butcher, L.L. (1977): Life Sci. 21:1207-1226.
6. Butcher, L.L. (1977): IN Cholinergic Mechanisms and Psycho-
    pharmacology (ed) D.J. Jenden, Plenum Press, New York,
    pp. 93-124.
7. Butcher, L.L., ed (1978): Cholinergic-monoaminergic Interac-
    tions in the Brain. Academic Press, New York.
8. Butcher, L.L. and Bilezikjian, L. (1975): Eur. J. Pharmacol.
    34:115-125.
9. Butcher, L.L. and Marchand, R. (1978): Eur. J. Pharmacol. 52:
    415-417.
10. Butcher, L.L. and Talbot, K. (1978): IN Cholinergic-monoaminer-
    gic Interactions in the Brain (ed) L.L. Butcher, Academic
    Press, New York, pp. 3-22.
11. Butcher, L.L. and Talbot, K. (1978): IN Cholinergic-monoaminer-
    gic Interactions in the Brain (ed) L.L. Butcher, Academic
    Press, New York, pp. 25-95.
12. Butcher, L.L., Talbot, K. and Bilezikjian, L. (1975): J. Neural
    Transm. 37:127-153.
13. Butcher, S.H. and Butcher, L.L. (1974): Brain Res. 71:167-171.
14. Butcher, S.H. and Butcher, L.L. (1978): IN Cholinergic-Mono-
    aminergic Interactions in the Brain (ed) L.L. Butcher,
    Academic Press, New York, pp. 109-122.
15. Butcher, S.H., Butcher, L.L. and Cho, A.K. (1976): Life Sci.
    18:733-744.
16. Dahlstrom, A. and Fuxe, K. (1965): Acta Physiol. Scand. Suppl.
    232:1-55.
17. de la Torre, J.C. and Surgeon, J.W. (1976): Histochemistry
    49:81-93.
18. Dross, K. and Kewitz, H. (1972): N.-S. Arch. Pharmacol. 274:
    91-106.
19. Emson, P.C. and Lindvall, O. (1979): Neuroscience 4:1-30.
20. Garattini, S., Pujol, J.F. and Samanin, R., eds (1978): Inter-
    actions Between Putative Neurotransmitters in the Brain,
    Raven Press, New York.
21. Hattori, T. (1978): IN Cholinergic-Monoaminergic Interactions
    in the Brain (ed) L.L. Butcher, Academic Press, New York,
    pp. 97-108.
22. Hattori, T., Singh, V.K., McGeer, E.G. and McGeer, P.L. (1976):
    Brain Res. 102:164-173.

23. Kimura, H., McGeer, E.G., McGeer, P.L. and Peng, F. (1979):
     Soc. Neurosci. Abst. 5:74.
24. Knight, D.P. (1970): Proc. Roy. Microscop. Soc. 6:26-27.
25. Koelle, G.B. (1954): J. Comp. Neurol. 100:211-235.
26. Koelle, G.B. and Friedenwald, J.S. (1949): Proc. Soc. Exp.
     Biol. Med. 70:617-622.
27. Ladinsky, H., Consolo, S., Bianchi, S., Ghezzi, D. and
     Samanin, R. (1978): IN Interactions Between Putative Neuro-
     transmitters in the Brain (eds) S. Garattini, J.F. Pujol and
     R. Samanin, Raven Press, New York, pp. 3-21.
28. Lavyne, M.H., Koltun, W.A., Clement, J.A., Rosene, D.J.,
     Pickren, K.S., Zervas, N.T. and Wurtman, R.J. (1978): Brain
     Res. 135:77-86.
29. Lehmann, J. and Fibiger, H.C. (1979): Life Sci. 25:1939-1947.
30. Levine, M.S., Hull, C.D., Garcia-Rill, E., Fuller, D.R.G. and
     Buchwald, N.A. (1978): IN Cholinergic-Monoaminergic Inter-
     actions in the Brain (ed) L.L. Butcher, Academic Press, New
     York, pp. 125-139.
31. McCulloch, J. and Harper, A.M. (1977): Amer. J. Physiol. 233:
     222-227.
32. McGeer, P.L., McGeer, E.G., Fibiger, H.C. and Wickson, V.
     (1971): Brain Res. 35:373-379.
33. Nagy, J.I., Vincent, S.R. and Fibiger, H.C. (1978): Life Sci.
     22:1777-1782.
34. Nichols, C.W. and Koelle, G.B. (1967): Science 155:477-478.
35. Nichols, C.W. and Koelle, G.B. (1968): J. Comp. Neurol. 133:
     1-16.
36. Parent, A. and Butcher, L.L. (1976): J. Comp. Neurol. 170:
     205-226.
37. Parent, A., Gravel, S. and Olivier, A. (1979): IN Advances in
     Neurology, Vol. 24: The Extrapyramidal System and its Dis-
     orders (eds) L.J. Poirier, T.L. Sourkes and P.J. Bedard,
     Raven Press, New York, pp. 1-11.
38. Rossier, J. (1975): Brain Res. 98:619-622.
39. Silver, A. (1974): The Biology of Cholinesterases, North
     Holland/American Elsevier, New York.
40. Talbot, K., Woolf, N.J. and Butcher, L.L. (1981): Brain Res.
     Bull. (in press).
41. Vandermaelen, C.P., Bonduki, A.C. and Kitai, S.T. (1979): Brain
     Res. 175:356-361.

ON MODULATION OF CEREBRAL CHOLINERGIC MECHANISMS BY

ENDOGENOUS INDOLEAMINES AND THEIR DERIVATIVES

B. Oderfeld-Nowak* and M.H. Aprison

The Institute of Psychiatric Research
Departments of Psychiatry and Biochemistry
Indiana University School of Medicine
Indianapolis, Indiana 46223 USA, and
*Nencki Institute of Experimental Biology
Polish Academy of Sciences
Pasteur 3, 02093 Warsaw, Poland

INTRODUCTION

Since the early 1960s, a few investigators, including one of
the present authors, began to point out in the literature that due
to the complexity of the CNS, the probability is extremely great
that several neurotransmitter systems or neurotransmitter-modulator
systems are involved in the various functional illnesses rather than
a single system (1,2).   Further, one could envisage the possibility
that changes in excitation or inhibition at one or more synapses
could affect neurons further down a neuronal pathway utilized by a
living organism in emitting specific types of responses (3). The
malfunction of several of these synapses could result in directly
imbalancing the neuronal system in which the pathway normally plays
an important role. Thus, the "problem" at the end of an important
common pathway could be due to one or more different transmitter or
modulator defects in the input pathways (5,28). Therefore, it is
not surprising that recently the concept was again enunciated, that
a number of important functions of the brain result from a balance
between two or more neuroregulators (7,13). The full impact of
these concepts a) provides a basis to explain why the 1-drug therapy
often used does not provide a universal cure and hence may not be
the method of choice in the future (3,7,13), and b) forms the basis
of several newer hypotheses to explain psychiatric disorders. Certainly clarifying as well as identifying the specific malfunctioning

mechanisms or "lesions" within the CNS should provide a better method
to facilitate the treatment (5,6,28).

One additional point should be made concerning transmitter or
modulator actions at synapses. Both types of compounds, or multiple
transmitters, may exist within the same nerve endings (45); evidence
is increasing to support those ideas. If a transmitter and a modulator are released together at a single synapse, it is useful to
make a distinction between transmitters and modulators on the basis
of theoretical concepts of physical chemistry (1). Thus a modulator
can be defined as any compound which, once released from specific
nerve endings into the synaptic cleft, may influence the effective
concentration of the transmitter normally acting at its receptor
sites in the postsynaptic membrane by competing with it and thereby
reducing or modulating the action of the transmitter (5). The postsynaptic effect of the modulator should vary as a function of the
concentration of the transmitter and the modulator in the cleft. A
modulator could also produce its effect by influencing the release
or uptake of the true transmitter. Hingtgen and Aprison (28) used this
broad definition to suggest that a) a modulator can also function as
a transmitter at other synapses within the same organism or within
other animals, and b) effects of modulators in the CNS are due to
their ability to influence the activity of transmitters at receptor
sites or sensitive portions of the postsynaptic membrane located in
the synapses of various neuronal pathways being activated or utilized
by an organism to generate a specific behavior. Thus it is conceivable that several compounds, including the normal transmitter, may
be acting in the region of the receptor during the normal act of
neurotransmission. Although the distinction suggested by Hingtgen
and Aprison (28) between transmitters and modulators can be considered a hypothetical concept, subsequent work in their laboratories
based on this concept led to an investigation of the role of acetylcholine (ACh) in the brain of rats injected with iproniazid and
tetrabenazine, and exhibiting increased behavioral (excitatory)
responses. Without this concept it would not have occurred to
Aprison, Hingtgen and co-workers to measure ACh after injecting
iproniazid and tetrabenazine, two drugs known to affect both the
serotonergic and catecholaminergic systems. That they found a good
correlation between the changes in content of ACh in the telencephalon and the changes in behavioral excitation of the animal in
these experiments is reinforcing support for these concepts (4, see
28 for other references).

Although there has been an increasing interest in the interactions of cholinergic neurons with other chemically defined neuronal
systems in the brain (11,55), comparatively little is known about
the interactions of the cholinergic and serotonergic systems at the
molecular and neuronal levels. However, a number of different types
of experiments employing various experimental approaches are reported

in the literature which provide some data for consideration in planning future studies to investigate cholinergic and serotonergic system interactions.

With invertebrate preparations it has been reported that ACh and choline acetyltransferase (CAT) activity are present in the giant serotonin cell of _Helix pomatia_ (25,26). Whether the functional role of ACh in this 5-HT cell is hormonal, regulatory, modulatory, or transmitter-like is not clear at present. In another study (23) using both serotonergic and cholinergic neurons from _Aplysia californica_, it was noted that [^{3}H]-5-HTP can be converted to [^{3}H]-5-HT in each type of cell. The former neurons possess specific mechanisms for both storage and axonal transport of 5-HT, whereas these properties are absent in the latter cells. The significance of this observation awaits additional data.

In the area of neuropharmacological studies, Haubrich and Reid (27) demonstrated that arecoline and pilocarpine, two cholinergic agonists, increase the turnover of 5-HT in rat brain, whereas Pepeu et al. (46) reported that depletion of 5-HT by pretreatment with p-chlorophenylalanine or by placement of a lesion in the midbrain raphe is without significant effect on the release of endogenous ACh from the cerebral cortex of the rat. Samanin et al. (56), using quipazine and D-fenfluramine concluded that these drugs probably activated 5-HT transmission which in turn reduced the release of ACh and increased its level in the striatum and hippocampus. These data support the concept that the serotonergic neurons which are located in the raphe nuclei may serve to regulate cholinergic neurons both in the corpus striatum and the hippocampus, thus confirming an earlier suggestion by Butcher et al. (12). However, further data are necessary to determine conclusively whether the regulation of the cholinergic neurons is inhibitory (15,24,56) or excitatory (12).

Finally, there is a large body of enzymatic data that is suggestive. Many papers have been published since the mid-1950's which describe the inhibitory action of 5-HT on acetylcholinesterase (AChE) activity from brain and red blood cells (RBC)(1,4,63,64,69; see also 42 for additional references). Investigators have wondered about the possible physiological significance, if any, of these observations.

In the early 1960's Aprison (1) suggested that 5-HT might function in the nervous system both as a modulator of the cholinergic system as well as a neurotransmitter in its own right at specific synapses (see above). The modulator effect was seen simply as an interaction of 5-HT and ACh at two receptors in the postsynaptic membrane: a) at the transmitter site (ACh) and b) at the enzyme site (AChE). This concept was used to explain how "fine" behavioral control could be produced if "coarse" control resulted from the antagonistic action of the transmitters for the trophotropic and

ergotropic divisions in brain. Further, an additional point was
made to underscore the importance of the inhibition of AChE by 5-HT.
Although there are many compounds that can inhibit AChE activity _in
vitro_ (35,59), few are as unique as 5-HT. Thus, in addition to possessing the property of being able to inhibit AChE activity _in vitro_
at extremely low concentrations (42), this biologically active indole compound which is normally present in varying amounts in different areas of the CNS, also can inhibit or excite various neurons
when iontophoretically applied to their specific receptors. Very
few compounds possess these four properties, i.e. a) are inhibitors
of AChE activity at very low concentrations, i.e. 10-50 µM; b) are
normally present as natural compounds in the mammalian CNS, c) present at different levels in specific parts of the brain, and d) can
inhibit or excite neurons when applied in minute amounts to receptors
of such cells by the iontophoretic method.

While testing this hypothesis in an _in vitro_ system, Aprison
(1) found that when 5-HT and ACh are present in equimolar concentrations, the inhibition of AChE activity at maximum substrate concentrations (i.e. 10^{-3} M) varied up to 50% depending on the source of
the enzyme. Although realizing that at 10^{-3} M the 5-HT effect may
be non-physiological, the data suggested an explanation for the biphasic action of 5-HT reported to occur in the CNS (see 1,42). As
noted above, Aprison, Hingtgen and co-workers used this idea to test
whether cerebral ACh was involved in animals made to exhibit excitatory behavioral responses following administration of iproniazid
16 hr before tetrabenazine (3,4,28). From the correlation of changes
in cerebral ACh and behavioral excitation, it was assumed that _in
vivo_ an interaction between the 5-HT and ACh systems occurred during
the period following the administration of these drugs (3,4,28).
Recently Oderfeld-Nowak et al. (42) re-evaluated the conditions for
inhibition of AChE by 5-HT to see if this relationship still occurred when both 5-HT and ACh were present at normal physiological
levels (5 x 10^{-5} to 5 x 10^{-6} M). In addition, to investigaté specificity of action, other parameters were studied which included muscarinic receptor binding, high affinity uptake of choline (Ch),
choline acetyltransferase (CAT) activity, as well as the effect of
other endogenous indoleamines and their derivatives. These studies
are discussed in the following sections.

This review indicates that there are two groups of studies in
the area of serotonergic-cholinergic interactions. In one, the
data support the position that these interactions are due to the
action of a transmitter from one neuron inhibiting or exciting a
second neuron with a different transmitter in a specific pathway
in the brain. In this case, one is not dealing with the concept of
modulation as defined above. In the other group, the data suggest
that 5-HT, or one of its metabolites, may act as a modulator of the
cholinergic system at specific sites in the postsynaptic membrane

of such synapses. Should this phenomenon be proven to occur in vivo, it is definitely an example of modulation.

Effect of 5-Hydroxytryptamine, other Indoleamines and other Indole Containing Derivatives on Cholinergic Parameters

Using the radiometric method of Fonnum (16) with slight modifications, Oderfeld-Nowak et al. (42) measured the effect of low concentrations of 5-HT (10-100 μM) on the activity of AChE from the caudate nucleus (CN) of the rat brain and of purified AChE from bovine erythrocytes (RBC). The degree of inhibition of AChE from these two sources in the presence of 50 μM ACh was found to be dependent on the ratio of the 5-HT concentration to the amount of enzyme. The inhibition was concentration dependent over the range studied (10-50 μM), whereas the degree of inhibition was essentially unaltered when the concentration of ACh was raised from 10 μM to 50 μM. The inhibition of 5-HT was found to be non-competitive regardless of whether the source of AChE was CN or RBC (see Fig. 1). The mean K_i was 2×10^{-4} M for CN and 2.5×10^{-4} M for RBC.

Oderfeld-Nowak et al. (42) also investigated the specificity of the observed inhibitory effects of 5-HT by checking its interaction with three additional cholinergic parameters. The data in Table 1 show that 50 μM 5-HT had no effect on high affinity uptake of Ch, CAT activity, and [3H]-QNB binding.

In addition to 5-HT, other indoleamines such as tryptamine, bufotenin and N-methyl-serotonin can inhibit brain AChE activity when the concentration of the substrate, ACh, is present at physiological levels. The type of inhibition and the kinetics is the same as for 5-HT. Even the K_i for each compound was found to be in the same range as for 5-HT. Further, when other indole containing derivatives such as 5-hydroxytryptophol, tryptophol, and 5-hydroxyindoleacetic acid were tested, all inhibit AChE activity to some extent.

The results reported above by Oderfeld-Nowak et al. (42) which were obtained using ACh and 5-HT at physiological levels, confirm the earlier in vitro data reported by Aprison (1), Wiggs and Aprison (67), Tonini (63) and Zsigmond et al. (69), as well as the more recent data of Mohammed et al. (38) and Gilboa-Garber et al. (22). All these investigators used substrate and inhibitor levels 60-100 times higher than those used by Oderfeld-Nowak (42).

At present it is difficult to understand fully the nature of the inhibition of AChE activity by 5-HT and related indole metabolites.

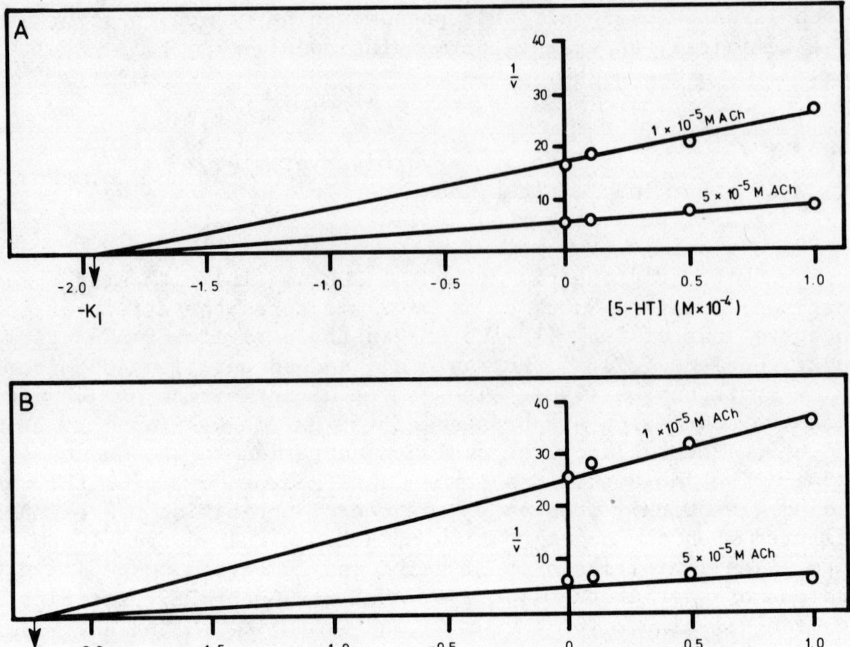

FIGURE 1: Determination of K_i values for 5-HT inhibition of AChE by
the method of Dixon. A) AChE from rat caudate nucleus
homogenate (CN). B) purified AChE from bovine red blood
cells (RBC). On the ordinate axes, 1/V is expressed as
(nmol ACh hydrolyzed/15 min/1 g wet weight tissue)$^{-1}$ for
CN or (nmol ACh hydrolyzed/15 min/60 ng solid material)$^{-1}$
for RBC. The data are representative of 4 determinations:
K_i is 2 x 10^{-4} M for CN and 2.5 x 10^{-4} M for RBC.

The role of the indole group in the inhibition of AChE by all the
compounds referred to above is intriguing since it, or some portion
of it (35), appears to be important in this mechanism. Although it
is not apparent from an examination of the ACh molecule, there are
suggestions from a number of enzyme inhibitor studies that the
enzymatic receptor site on the AChE molecule might contain a subsite
which is complimentary to aromatic groups (see 35). The observation
that aromatic esters such as indoxylacetate and phenyl acetate are
hydrolyzed by AChE supports the suggestion that such a site is
active. In addition, Oommen and Balasubramanian (44) and Fujimoto
(17) reported the non-competitive inhibition of aryl acylamidase by
5-HT and tryptamine and that this enzyme is associated with AChE,
representing a small fraction of this activity. They suggest that

TABLE 1:   Lack of effect of 5-HT on high affinity uptake of Ch, CAT
           and muscarinic cholinergic receptor binding (QNB) in rat
           caudate nucleus.

Condition	High Affinity Ch uptake	CAT	[³H]-QNB Binding
Control	1.1 ± 0.14	12.8 ± 2.39	961 ± 83.1
5-HT (5×10^{-5} M)	1.1 ± 0.21	12.1 ± 1.90	1099 ± 48.6

High affinity uptake of Ch - pmol/mg protein/4 min;CAT activity -
nmol ACh formed/mg protein/30 min;[³H]-QNB binding - fmol/mg protein.
From Oderfeld-Nowak et al. (42).

the site at which 5-HT acts to inhibit aryl acylamidase activity,
could be an active site for AChE. The non-competitive nature of
the inhibition suggests that some allosteric mechanisms are involved
and the subsite for aromatic groups may be outside of the active
site or center of the enzyme.

Since 5-HT and a number of its metabolites can inhibit brain
AChE activity, Aprison (1) has suggested one possible physiological
role for this mechanism, as noted above. Further discussion of the
physiological significance of this interaction and other factors
will be discussed in the last section.

Effect of β-Carbolines
on Cholinergic Mechanisms

It has been known since 1934 that a class of tricyclic compounds
called β-carbolines or tryptolines are formed after a condensation
reaction occurs between tryptamine (and its derivatives) and alde-
hydes at physiological pH and temperature. Now research with these
β-carbolines using tissues from brain is regaining interest among
neurobiologists. The reason for this renewed interest stems from
at least three observations: a) some β-carbolines were shown to be
reversible inhibitors of monoamine oxidase (MAO) activity as well
as uptake blockers of 5-HT, norepinephrine and dopamine; b) investigators showed that tryptamine and some of its derivatives occur
naturally in the CNS and c) alcohol is readily converted to alde-
hydes in similar tissues. Thus, investigators interested in the
nervous system wondered whether such compounds, if and when formed

in the brain, could provide abnormal physiological and behavioral
effects in man.

Although many research groups have demonstrated that enzymatic
activity from brain tissue can convert indoleamines to tetrahydro-
β-carbolines in the presence of 5-methyltetrahydrofolic acid (10,
32,36,53,68), it is generally believed that this derivative of folic
acid is enzymatically converted to formaldehyde and tetrahydrofolic
acid followed by spontaneous condensation of HCHO with tryptamine.
Thus, when tryptamine condenses with formaldehyde, tetrahydronorharmane (THN) is formed (53); if the condensation occurs with
acetyldehyde instead of formaldehyde, 1,2,3,4-tetrahydroharmane (THH)
is formed. Apparently serotonin also will react with formaldehyde
to yield 5-hydroxytryptoline (6-hydroxy-1,2,3,4-tetrahydro-β-carbo-
line). The question whether these compounds were synthesized <u>in</u>
<u>vivo</u> in brain remained unanswered.

In 1979, Barker et al. (8) published gas chromatographic mass
spectrometric evidence for the identification of 1,2,3,4-tetrahydro-
β-carboline in rat brain. Other groups as well have reported identification of β-carbolines in brain (29,51,57) and specific brain areas
such as the arcuate nucleus of the rat brain (58). With more data
now reported in the literature on β-carboline derivatives, it is
clear that a) these compounds are excreted with the urine in rats
and man (29); b) their levels in tissues can be increased by administration of precursors; c) THN and 6-OH-THN exert hypothermia and
decrease motility (51); d) THN elicits catalepsy; e) harmane elicits
excitation, euphoria and colorful hallunications (53,62); f) tryptolines also occur naturally in certain plants; g) β-carbolines inhibit high affinity uptake of 5-HT (34,53,65), and to a small extent NE and dopamine (34,53); some stimulate release of 5-HT (54);
i) some may act as false transmitters (54,57); some possibly may
modify the action of endorphins and ACTH in the arcuate nuclei (59).

Oderfeld-Nowak, Rommelspacher and co-workers have turned their
attention to a number of these β-carbolines to see if the compounds
can influence the cholinergic system. Since an indole group constitutes part of the tricyclic structure of β-carbolines, it was of
interest to question whether some of these compounds can affect any
of the parameters of the cholinergic system. These investigators
tested tetrahydronorharmane (THN), tetrahydroharmane (THH) and
harmane on AChE activity, CAT activity and on muscarinic receptor
binding. The three methods were described by Oderfeld-Nowak et al.
(42). Harmine was purchased from Sigma, THN and THH were synthesized
in Institute of Pharmacy, Free University of Berlin (Federal Republic
of Germany). For the AChE activity studies, several sources of
enzyme were used: caudate nuclei, hypothalamus and hippocampus from
the rat and purified bovine red blood cells. For the CAT activity
and QNB binding studies, the caudate nuclei and hippocampus from rat
brain were used.

The three β-carbolines, THN, THH and harmane, inhibit the AChE activity from all the sources tested (Fig.2, A,B,C). The inhibition was concentration dependent over the range used (5 x 10^{-7} M to 1 x 10^{-4} M) when employing ACh as substrate at biological concentrations (10 μM to 50 μM). The data in Fig. 2C show that harmane had the greatest inhibitory effect on AChE activity (approximately 55% inhibition at 50 μM), with THH and THN inhibiting AChE activity at this concentration by 50% and 34%, respectively. These β-carbolines were better inhibitors than the indole derivative, as discussed in the previous section.

No effect of THN, THH or harmane was observed on CAT activity (see Fig.2, A,B,C). As to QNB binding, the three drugs were different: a) THN had only a slight effect; b) THH had a very small effect (6-7%) until the dose was raised to 10^{-4} M when the inhibition reached 23%; and c) harmane showed a dose dependent effect, increasing linearly throughout the range used (5 x 10^{-7} M to 10^{-4} M), passing 35% inhibition at 50 μM. The latter observation with harmane is very interesting since the presence of the aromatic C-ring in the β-carboline structure was essential for its simultaneous effect on AChE activity and receptor binding. Perhaps similar aromatic complimentary sites exist for both the receptor and enzyme molecules.

When the type of inhibition was determined for harmane and THH, the data (Fig. 3A, 3B) show that the inhibition by these compounds was non-competitive. Further, Oderfeld-Nowak and co-workers found that in no case was the inhibition of AChE activity influenced by the substrate concentration. The K_i for THH (Fig. 3A) was approximately 5 x 10^{-5} M, whereas for harmane, it was 3 x 10^{-5} M. When THN was used, the K_i was about 1 x 10^{-4} M. Similar enzymatic data were obtained when the enzyme source of AChE was obtained from the hypothalamus or hippocampus of the rat brain (data now shown).

Employing pharmacological tests to estimate the effect of β-carbolines on AChE and ChE activity, Ghosal et al (21) noted that some aromatic quaternary β-carbolines exerted inhibitory effects on both enzymes, whereas tetrahydro-β-carbolines did not. Oderfeld-Nowak and co-workers also noted greater inhibition by quaternary aromatic β-carbolines, suggesting that this stereochemistry is more favorable for binding of the inhibitor to the enzyme. It was recently shown (53) that β-carbolines, like harmane and norharmane, compounds with oxidized C-rings, are more potent inhibitors of the high affinity uptake of Ch into a synaptosome-rich fraction from the hippocampus than are the compounds with reduced C-rings such as THN. This finding is the reverse situation as reported for the inhibitory potency of these β-carbolines toward the high affinity uptake of 5-HT into a similar preparation (53).

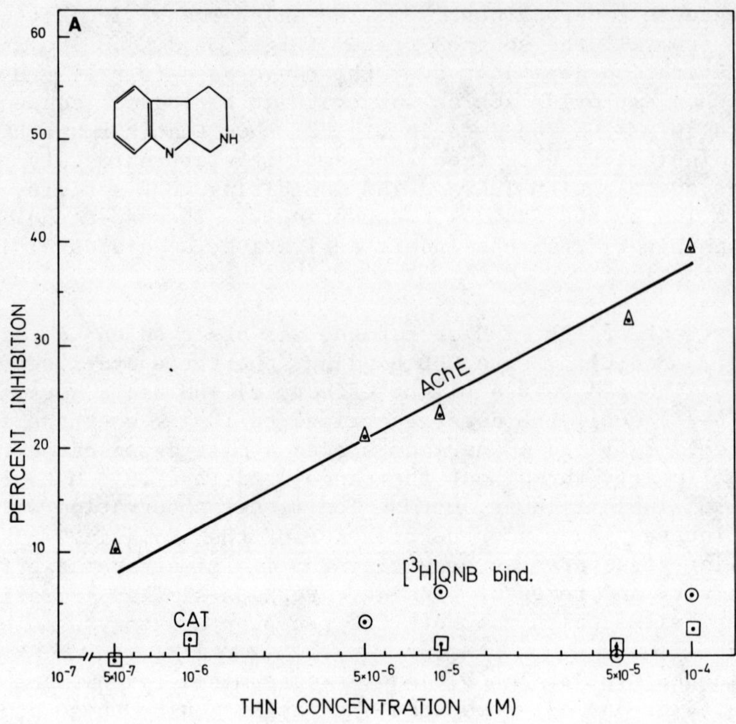

FIGURE 2: Effect of three β-carbolines (A-tetrahydronorharmane [THN],
B-tetrahydroharmane [THH], C-Harmane) on AChE activity,
muscarinic cholinergic receptor binding and CAT activity.
The source of AChE is purified bovine red blood cells
(RBC).   [3H]-QNB binding and CAT activity were determined
using rat hippocampal homogenates.   The data are shown as
means of 3 experiments.   Panel A (THN) is shown on
page and panels B (THH) and C on the following page.
From Oderfeld-Nowak et al. (43), Skup et al (60) and
Oderfeld-Nowak (in preparation).

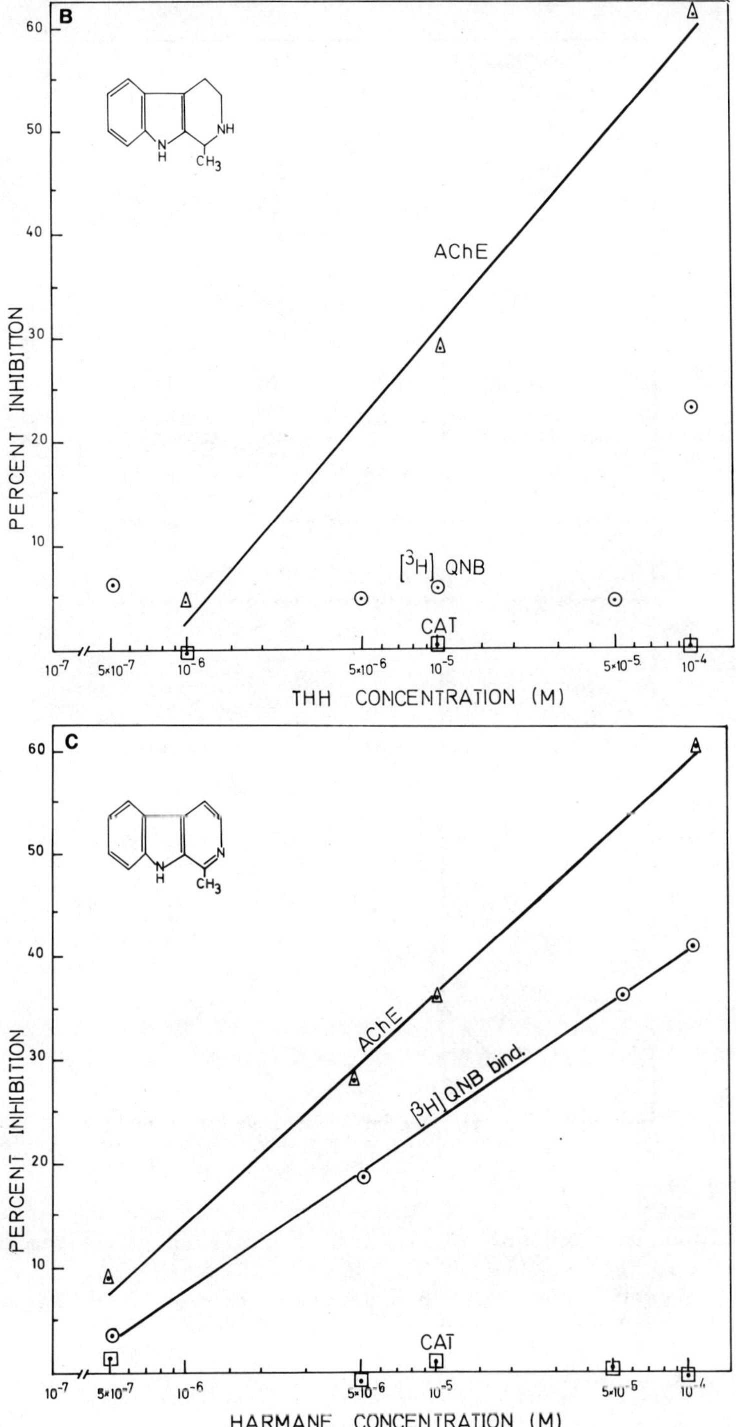
B
AChE
[3H] QNB
CAT
PERCENT INHIBITION
THH CONCENTRATION (M)
60
50
40
30
20
10
10⁻⁷
5×10⁻⁷
10⁻⁶
5×10⁻⁶
10⁻⁵
5×10⁻⁵
10⁻⁴
C
AChE
[3H] QNB bind.
CAT
PERCENT INHIBITION
HARMANE CONCENTRATION (M)
60
50
40
30
20
10
10⁻⁷
5×10⁻⁷
10⁻⁶
5×10⁻⁶
10⁻⁵
5×10⁻⁵
10⁻⁴

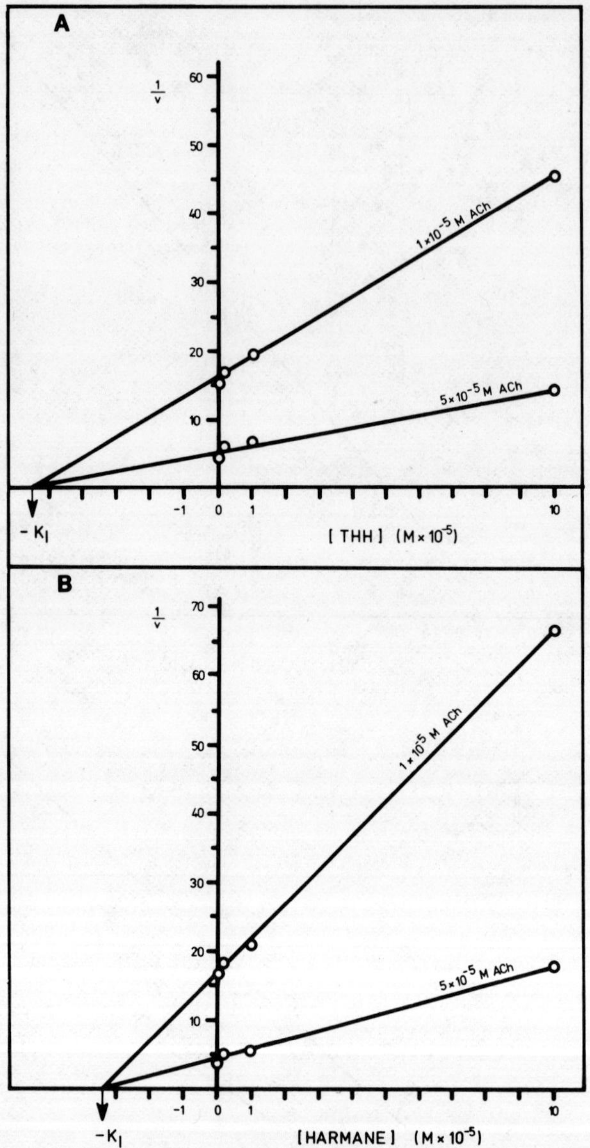

FIGURE 3: Determination of K_i values for tetrahydroharmane (THH)(A) and harmane (B) inhibition of purified AChE from bovine red blood cells (RBC). On the ordinate axes, 1/V is expressed as [nmol ACh hydrolyzed/15 min/60 ng solid material]$^{-1}$. The K_i value for THH is 5.5×10^{-5} M and for harmane is 3.5×10^{-5} M. From: A) Skup et al. (60); B) Oderfeld-Nowak et al. (43).

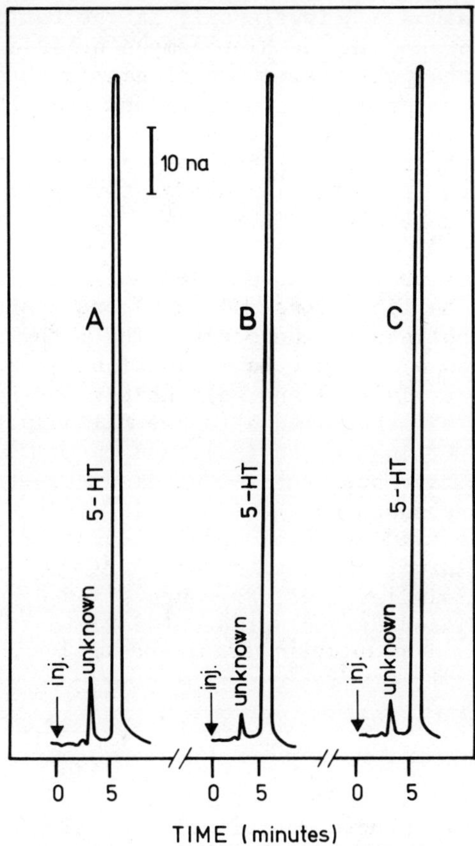

FIGURE 4: HPLC chromatograms of three 50 μM 5-HT solutions which were maintained under different conditions: A = 17 hr of light (40 cm from a 250W Westinghouse Krypton flood lamp) with the first 3 hr being under a constant stream of oxygen. B = 3 hr under a constant stream of oxygen in the dark. C = 17 hr of light (40 cm from a 250W Westinghouse Krypton flood lamp). From Oderfeld-Nowak et al. (42).

Javoy et al. (33) provide an indication of a modulatory effect of β-carbolines on cholinergic mechanisms in the brain when they found that harmaline, a β-carboline with hallucinogenic properties induced an increase in striatal ACh content. These authors suggest that an increase in ACh levels probably reflects a decrease in the rate of utilization of this transmitter, but not through an effect on dopaminergic transmission or GABA mechanisms. They found this

effect of harmaline on ACh levels only in the rat striatum and not
in the parietal cortex, in the hippocampus or in certain limbic nuclei, and suggest that this selective effect is perhaps mediated by
interneuronal processes. It is of interest that their results fit
well with the findings of Oderfeld-Nowak, Rommelspacher and co-
workers, who show that β-carbolines have an inhibitory effect on
AChE. Certainly the reduction in AChE activity could result in an
increase in the ACh levels.

It is instructive to point out some of the similarities of
action of several hallucinogens and antidepressants on the one hand,
and several β-carbolines on the other. The methylated indoles, bufotenine and LSD, can a) produce hallucinations in man, b) inhibit the
uptake of 5-HT in brain (30) and, c) inhibit ChE activity (69). Some
β-carbolines like harmane, also a) cause hallucinations (62), b) inhibit uptake of 5-HT in the CNS (53), and c) inhibit AChE activity
(Fig. 2C). There have been several studies reporting the effect of
tricyclic antidepressants on the inhibition of the uptake of 5-HT
into structures containing serotonergic neurons (see 31). Imipramine,
an antidepressant drug, has been shown to inhibit AChE activity (37).
In light of these similarities, the suggestion can be made that both
properties, inhibition of the uptake of 5-HT and inhibit AChE activity,
may play a role in the antidepressant and hallucinogenic actions of
these compounds.

New Potent Unknown Inhibitor(s) of AChE activity
Derived from Serotonin During Photosensitized Oxygenation:
Some Considerations
Concerning its Possible Structure

An unexpected finding was made recently by Oderfeld-Nowak et
al. (41,42) when these investigators observed that solutions of
50 μM 5-HT, exposed to light and oxygen for short periods of time,
under normal laboratory conditions, caused a greater inhibition of
AChE activity than freshly prepared 5-HT solutions. Oderfeld-Nowak
and Aprison were led to the assumption that a new derivative was
formed under these conditions which was a more potent inhibitor of
AChE activity than 5-HT itself. Consequently the effect of sub-
jecting 5-HT solutions to controlled conditions of light and oxygen
were carried out. Freshly prepared solutions of 5-HT were placed
40 cm from a 250W Westinghouse Krypton flood lamp either for 17 hr
of light, with the first 3 hr under a constant stream of oxygen, or
just exposed for 17 hr of light under room conditions. In a third
condition, the 5-HT solution was exposed for 3 hr to a constant
stream of oxygen in the dark. Samples from these 3 solutions were
subjected to high performance liquid chromatography (HPLC). The

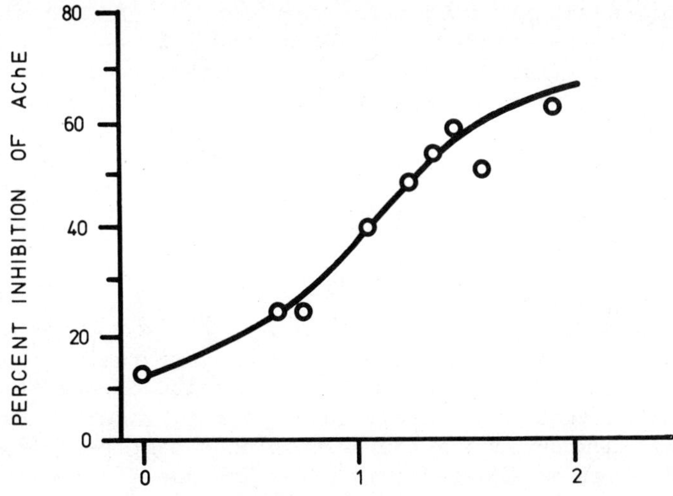

FIGURE 5: The relationship between the percent of inhibition of AChE
activity in homogenates of rat CN by 50 μM solutions of
5-HT exposed to the atmosphere and light of the laboratory
for different lengths of time (up to 14 days) and the
relative height of the unknown peak (compared to height on
day 2) as determined by HPLC. From Oderfeld-Nowak et al.
(42).

data which are shown in Fig. 4A,B,C, reveal the presence of an additional peak to the one identified as 5-HT. The unknown peak had
a shorter retention time than 5-HT. Note that the effects of light
and oxygen were synergistic (Fig. 4A).

The unknown was separated from 5-HT by running the 50 μM 5-HT
solutions exposed to light and oxgen on thin layer chromatography
plates under specific conditions (42). Each spot was eluted separately from the silica gel of the plate, and individually applied to
the HPLC. The data for the unknown revealed a peak with retention
time less than 5-HT and identical to that obtained when the exposed
5-HT solutions were tested. This compound inhibited AChE activity
from rat CN and purified RBC to a greater degree than 5-HT itself,
indicating that an inhibitory substance of high potency was formed
under the conditions noted above.

In an additional experiment, solutions of 50 µM 5-HT were kept exposed to the atmosphere and light of the laboratory during a 14 day period. When several of these solutions (exposed for different lengths of time) were tested for their anti-AChE activity and analyzed by HPLC, it was found that inhibition of AChE activity was directly related to exposure time (42). Further, with the HPLC data showing an increase in the peak height of the unknown compound, a monotonic relationship was evident between the percent of inhibition of AChE and the peak height of the unknown. Thus the percent inhibition of AChE was found to increase with increasing peak height of the unknown (see Fig. 5).

Although the data presented above support the concept that this unknown peak is responsible for the inhibition of AChE (see also 42), it is possible that it may represent a final product. Evidence for intermediate compounds was not seen in most samples assayed by HPLC, as shown in Fig. 4. However, in a few instances and under certain conditions, several smaller peaks of longer retention time were seen in the HPLC analyses in addition to finding the unknown and 5-HT peaks (Fig. 6). Further work is necessary to identify all of these peaks in this type of HPLC separation. Perhaps then investigators will have a better insight into the mechanism of production of the peaks from 5-HT.

One can only speculate both on the mechanism of formation of this unknown "metabolite" of 5-HT seen in Fig. 4, and its possible physiological significance. The fact that the unknown compound is formed from 5-HT in the presence of light and oxygen permits the formulation of several hypotheses. Some speculations can be based on information obtained from _in vitro_ experiments on photosensitized oxygenation of N_b-methyltryptamine and N_b-methyloxycarbonyltrypt-amines (39,40) to gain an insight to the _in vivo_ oxidation pathways leading to kynurenine. Perhaps during photosensitized oxygenation, there is also a re-arrangement of the 5-HT molecule which leads to some kynurenine-like derivatives. In the metabolic pathway from tryptophan to kynurenine, N-formyl-kynurenine is formed by the action of the enzyme tryptophan 2,3-dioxygenase on the amine acid. Recently, Gal and Sherman (19) reported on such a pathway in brain. Nakagawa et al. (39,40) postulates a hypothetical dioxetane intermediate before a kynurenine derivative is formed. In addition, a tricyclic intermediate is also postulated (see Fig. 7), which has a striking resemblence to the physostigmine structure (see Fig. 8). Thus, if similar re-arrangements can take place under our experimental conditions, then it is possible that some tricyclic derivatives of physo-stigmine-like ring structures might occur which would account for the high anti-AChE activity of our unknown (see also 90). In this respect, it is interesting to note that a) in 1937, Sobotka and Antopol (61) first noticed the structural similarity between physostigmine and the indole derivative bufotenine and, b) in 1960, a scheme was

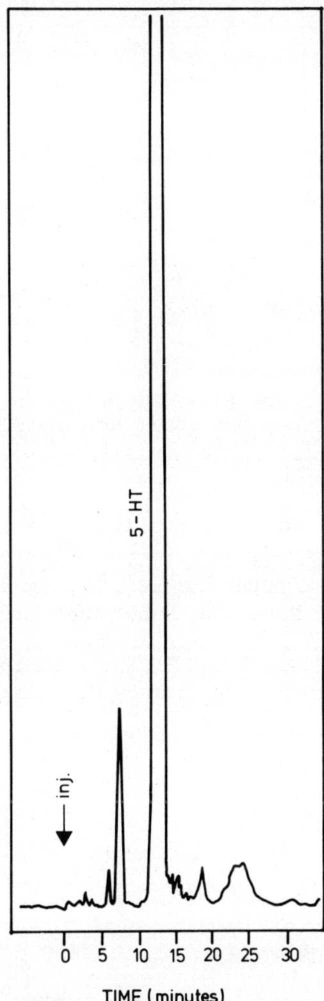

FIGURE 6: HPLC chromatogram of a 50 µM 5-HT solution exposed to light
and oxygen. Note that in addition to the unknown peak with
a shorter retention time than 5-HT, several peaks are
present with longer retention times (Shea, Simon, Oderfeld-
Nowak and Aprison: unpublished observations).

published in the "Handbuch of Alkaloids" (9), which illustrates a
pathway from 5-hydroxytryptamine to physostigmine!

Perhaps in the final comments for this section it is useful to
speculate further in order to arrive at additional ideas or a

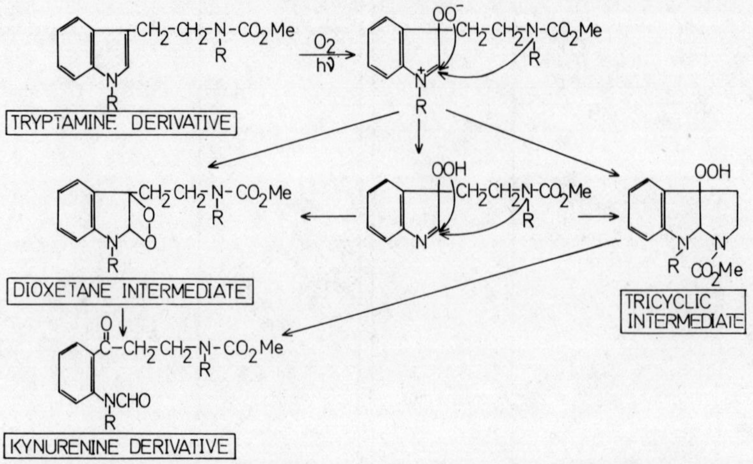

FIGURE 7: Hypothetical scheme simplified from Nakagawa et al. (39)
showing intermediates in photosensitized oxygenation reaction of a tryptamine derivative leading to a kynurenine
derivative. Note the suggested tricyclic intermediate
and its similarity to the structure of physostigmine.

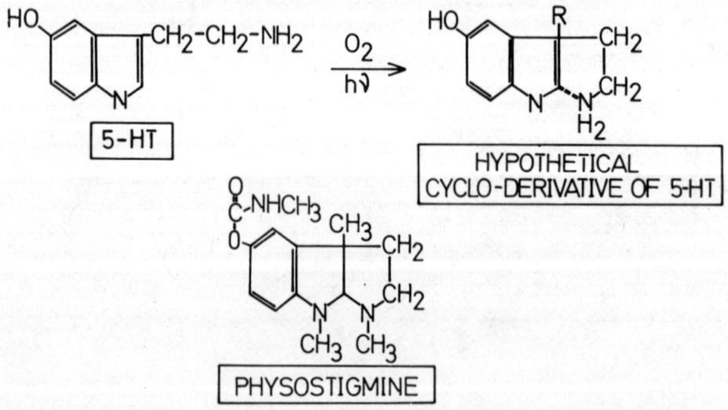

FIGURE 8: Hypothetical formula of a cyclo-derivative of 5-HT formed
in the presence of oxygen and light. The formula of the
anticholinesterase drug physostigmine is shown for
comparison.

hypothesis which can be tested in the laboratory. Under our experimental conditions, it is possible that if tricyclic intermediates are formed, these compounds would correspond to the small peaks with higher retention times than 5-HT (referred to in the HPLC shown in Fig. 6), whereas the final product, perhaps a kynurenine derivative, could correspond to the unknown possessing a shorter retention time (referred to in Figs. 4-6). As Gal and Sherman recently pointed out (19), the mechanisms involved in the cerebral pathways by which formylkynurenine and kynurenine are synthesized by indoleamine 2,3-dioxygenase and kynurenine formamidase, respectively, are not fully understood, but it is likely that this pathway has an important regulatory function. The in vivo appearance of several products of mono- and dioxygenase activites was reported (14). It appears that sufficient information and instrumentation is at hand to a) identify the unknown compounds, b) clarify their properties, and c) determine their physiological significance. This effort would be important both for basic and clinical points of view since many investigators have postulated over the years that there is a correlation between the appearnace of normal and abnormal indole derivatives and some mental disorders.

CONCLUSIONS

In recent years, it has become increasingly clear that post-synaptic receptors and presynaptic receptors on neurons are extremely important, and when these biological entities cease to function properly, the organism suffers. Many diseases have now been identified as having cell receptor disorders. In psychiatry, there is also a strong interest in receptors and just how the malfunction of these receptors might be implicated in mental illness (6). Thus, the idea that the behavior of an organism can be changed or modified by biochemical or neurophysiological action at the receptor is not difficult to understand.

Neurotransmitters and modulators function at these receptors. It is as important to identify the latter compounds as it is to identify new transmitters. This whole area of research as well as the study of the interaction of transmitter systems is gaining interest among neurobiologists. Certainly the balance between different transmitter systems, including their modulators, when present, has an important role in the regulation of brain function. This concept has been stressed in the research approach from our laboratories.

Rapid progress is now being made in identifying the detailed neuroanatomical organization of several of these interacting transmitter systems. At present the new information being reported on

many systems (47-50,66) is not concerned directly with the modulation
of cerebral cholinergic mechanisms. More investigators should turn
their attention to this important area of research.

ACKNOWLEDGEMENTS

Since a portion of the data presented hwere was obtained in
M.H. Aprison's laboratory, while B. Oderfeld-Nowak was a 1978-79
recipient of a Fogarty International Research Fellowship, we both
express our sincere thanks and appreciation to the Fogarty International Center of USPHS.

REFERENCES

1.   Aprison, M.H. (1962): Recent Adv. Biol. Psychiat. $\underline{4}$:133-146.
2.   Aprison, M.H. (1965): IN Horizons in Neuropsychopharmacology
       (eds) W. Himwich and J. Schade, Elsevier, Amsterdam, pp. 48-80.
3.   Aprison, M.H. (1977): IN Neuroregulators and Psychiatric Dis-
       orders (eds) E. Usdin, D. Hamberg and J. Barchas, Oxford
       University Press, New York, pp. 381-390.
4.   Aprison, M.H., Kariya, T., Hingten, J.N. and Toru, M. (1968):
       J. Neurochem. $\underline{15}$:1131-1139.
5.   Aprison, M.H. and Hingtgen, J.N. (1970): Int. Rev. Neurobiol.
       $\underline{13}$:325-341.
6.   Aprison, M.H. and Hingtgen, J.N. (1980): IN Serotonin- Current
       Current Aspects of Neurochemistry and Function (eds) B. Haber,
       S. Gabay, S. Alivisatos and M. Issidorides, New York, Plenum
       Press, New York (in press).
7.   Barchas, J.D., Akil, H., Elliott, G.R., Holman, B. and Watson,
       S.J. (1978): Science $\underline{200}$:964-973.
8.   Barker, S.A., Harrison, R.E., Brown, G.B. and Christian, S.T.
       (1979): Biochem. Biophys. Res. Commun. $\underline{87}$:146-154.
9.   Boit, H.A. (1960): IN Ergebnisse der Alkaloid Chemis, Akademic
       Verlag, Berlin, p. 487.
10. Buckholtz, N.S. and Boggan, W.O. (1977): Life Sci. $\underline{20}$:2093-2100.
11. Butcher, L.L.,ed (1978): Cholinergic-Monoaminergic Interactions
       in the Brain, Academic Press, New York.
12. Butcher, S.H., Butcher, L.L. and Cho, A.K. (1976): Life Sci.
       $\underline{18}$:733-744.
13. Butcher, L.L. and Talbot, K. (1978): IN Cholinergic-Monoaminergic
       Interactions in the Brain (ed) L.L. Butcher, Academic Press,
       New York, pp. 3-22.

14. Daly, J.W., Jerina, D.M. and Witkop, B. (1972): Experientia
         28:1129-1149.
15. Euvrard, C., Javoy, F., Herbet, A. and Glowinski, J. (1977):
         Eur. J. Pharmacol. 41:281-289.
16. Fonnum, F. (1969): Biochem. J. 115:465-472.
17. Fujimoto, D. (1976): FEBS Lett. 71:121-123.
18. Fuller, R.W. (1976): Life Sci. 19:625-628.
19. Gal, E.M. and Sherman, A.D. (1978): J. Neurochem. 30:607-613.
20. Galli, A., Renzi, G., Bartolini, A., Bartolini, R. and Malmberg-
         Aiello, P. (1980): Abst. Int. Symp. Chol. Mech. p.113.
21. Ghosal, S., Bhattacharya, S.K. and Mehta, R. (1972): J. Pharm.
         Sci. 61:808.
22. Gilboa-Barber, N., Katz-Bergman, Y. and Pinsky, A. (1978):
         Experientia 34:992-993.
23. Goldman, J.E. and Schwartz, J.H. (1974): J. Physiol. 242:61-76.
24. Guyenet, P., Euvrard, C., Javoy, F., Herbert, A. and Glowinski,
         J. (1977): Brain Res. 136:487-500.
25. Hanley, M.R. and Cottrell, G.A. (1974): J. Pharm. Pharmacol.
         26:980.
26. Hanley, M.R., Cottrell, G.A., Emson, P.C. and Fonnum, F. (1974):
         Nature 251:631-633.
27. Haubrich, D.R. and Reid, W.D. (1972): J. Pharmacol. Exp. Ther.
         181:19-27.
28. Hingtgen, J.N. and Aprison, M.H. (1976): IN Biology of Chol-
         inergic Function (eds) A.M. Goldberg and I. Hanin, Raven
         Press, New York, pp. 515-566.
29. Honecker, H. and Rommelspacher, H. (1978): N.-S. Arch. Exp.
         Path. Pharmakol. 305:135-141.
30. Horn, A.S. (1973): J. Neurochem. 21:883-888.
31. Horn, A.S. and Trace, R.C.A.M. (1974): Brit. J. Pharmacol.
         51:399-403.
32. Hsu, L.L. and Mandell, A.J. (1975): J. Neurochem. 24:631-636.
33. Javoy, F., Euvrard, C., Herbet, A., Bockaert, J. Enjalbert, A.,
         Agid, Y. and Glowinski, J. (1977): N.-S. Arch. Exp. Path.
         Pharmakol. 297:233-239.
34. Kellar, K.J., Elliott, G.R., Holman, R.B., Barchas, J.D. and
         Vernikos-Danellis, J. (1976): J. Pharmacol. Exp. Ther. 198:
         619-625.
35. Main, A.R. (1976): In Biology of Cholinergic Function (eds)
         A. Goldberg, and I. Hanin, Raven Press, New York, pp. 269-353.
36. Mandel, L.R., Rosegay, A., Walker, R.W., Vanden-Heuvel, W.J.A.
         and Rokach, J. (1974): Science 136:741-743.
37. Michatek, H. (1973): Biochem. Pharmacol. 22:1067-1074.
38. Mohammed, Y.S., Osman, M.Y. and Gabr, Y. (1975): Arz. Forsch.
         Drug Res. 25:1714-1715.
39. Nakagawa, M., Okajima, H. and Hino, T. (1977): J. Amer. Chem.
         Soc. 99:4424-4429.
40. Nakagawa, M., Yoshikawa, K. and Hino, T. (1975): J. Amer. Chem.
         Soc. 97:6496-6501.

41. Oderfeld-Nowak, B., Simon, J.R., Chang, L. and Aprison, M.H. (1979): Soc. Neurosci. Abst. 5:412.

42. Oderfeld-Nowak, B., Simon, J.R., Chang, L. and Aprison, M.H. (1980): Gen. Pharmacol. 11:37-45.

43. Oderfeld-Nowak, B., Skup, M., Ulas, J. and Rommelspacher, H. (1980):Abst. 3rd Meet. ESN (in press).

44. Oommen, A., Balasubramanian, A.S. (1978): Biochem. Pharmacol. 27:891-895.

45. Osborne, N.N. (1979): Trends Neurosci. 2:73-75.

46. Pepeu, G., Garau, L. and Mulas, M.L. (1974): Adv. Biochem. Psychopharmacol. 10:247-252.

47. Reader, T.A., de Champlain, J. and Jasper, H. (1976): Brain Res. 111:95-108.

48. Reader, T.A. (1978): Experientia 34:1586-1587.

49. Reader, T.A., Ferron, A., Descarries, L. and Jasper, H.H. (1979): Brain Res. 160:217-229.

50. Reader, T.A., Masse, P. and de Champlain, J. (1979): Brain Res. 177:499-513.

51. Rommenspacher, H., Honecker, H., Barbey, C. and Meinke, B. (1980): N.S. Arch. Exp. Path. Pharmakol. (in press).

52. Rommelspacher, H., Strauss, S. and Lindemann, J. (1980): FEBS Lett. 109:209-212.

53. Rommelspacher, H., Strauss, M. and Rehse, K. (1978): J. Neurochem. 30:1573-1578.

54. Rommelspacher, H. and Subramanian, N. (1979): Eur. J. Pharmacol. 56:81-86.

55. Roth, R.H. and Bunney, B.S. (1976): IN Biology of Cholinergic Function (eds) A. Goldberg and I. Hanin, Raven Press, New York, pp. 379-394.

56. Samanin, R., Quattrone, A., Peri, G., Ladinsky H. and Consolo, S. (1978): Brain Res. 151:73-82.

57. Shoemaker, D.W., Cummins, J.T. and Bidder, T.G. (1978): Neuroscience 3:233-239.

58. Shoemaker, D.W., Cummins, J.T., Bidder, T.G., Boettger, H.G. and Evans, M. (1980): N.-S. Arch. Pharmacol. 319:227-230.

59. Silver, A. (1974): The Biology of Cholinesterases. North-Holland Publ., Amsterdam, pp. 449-487.

60. Skup, M., Mouleshlova, N.S. and Oderfeld-Nowak, B. (1980): VII Meeting Polish Soc. Pharmacol. Abst. (in press).

61. Sobotka, H. and Antopol, W. (1937): Enzymologia 4:189-191.

62. Thomas, T.N., Buckholtz, N.S. and Zemp, J.W. (1979): Life Sci. 25:1435-1442.

63. Tonini, G. (1955): Soc. Ital. Biol. Sper. 31:766-771.

64. Toru, M., Hingtgen, J.N. and Aprison, M.H. (1966): Life Sci. 5:181-189.

65. Tuomisto, L. and Tuomisto, J. (1973): Arch. Pharmacol. 279: 371-380.

66. Vizi, E.S. (1979): Prog. Neurobiol. 12:181-290.
67. Wiggs, J.W. and Aprison, M.H. (1961): Physiologist 4:134.
68. Wyatt, R.J., Erdelyi, E., DoAmaral, J.R., Elliot, G.R., Rensor, J. and Barchas, J.D. (1975): Science 187:853-855.
69. Zsigmond, E.K., Foldes, F.F. and Foldes, V.M. (1961): J. Neurochem. 8:72-80.

INFLUENCE OF GABA ON ACETYLCHOLINE RELEASE FROM THE GUINEA PIG BRAIN

L. Beani, C. Bianchi and S. Tanganelli

Department of Pharmacology, University of Ferrara
Via Fossato di Mortara 64/B
44100 Ferrara, Italy

INTRODUCTION

Gamma-aminobutyric acid (GABA) is generally considered as an
inhibitory transmitter in the CNS at both pre- and postsynaptic
sites (12). Since GABA distribution is widespread, multiple interactions with other neurotransmitters have been postulated.

Many reports described the influence of GABA on catecholamine
release and turnover (1,2,7), but the investigations concerning
GABAergic control of cholinergic neurons in the telencephalon are
few and not unequivocal. Some authors reported that GABA and GABA-
like compounds, administered _in vivo_, increased acetylcholine (ACh)
tissue levels and decreased its turnover and release in various
areas (3,14-16). Others found an evident GABAergic inhibition of
ACh neurones in the neocortex and mesencephalon, but not in the
striatum (18), where a dopamine mediated disinhibition seemed to
be operative (9,10). Data concerning GABA effects _in vitro_ are
still more scanty and preliminary. Recently, Stoof and Mulder (17)
found that GABA reduced ACh release (and decreased dopamine outflow)
from KCl stimulated rat caudate slices. However, this effect did
not seem specific, since it was bicuculline insensitive. The lack
of further work in this field prompted us to investigate the nature
of GABA control of central ACh neurones by comparing the results obtained from _in vitro_ and _in vivo_ models.

763

TABLE 1: Effect of GABA, ethanolamine-O-sulphate (EOS) and picrotoxin on ACh release (ng/g/min $\pm$ S.E.M.) from slices of cerebral cortex and caudate nucleus, kept at rest and stimulated for 5 min every 20 min at 1 Hz and 5 Hz.

Stimulation Frequency	Cerebral Cortex			Caudate Nucleus		
	No Stimulation	1 Hz	5 Hz	No Stimulation	1 Hz	5 Hz
Control(6)	8.1 ± 0.7	45.2 ± 5.3	147 ± 17	130 ± 11	358 ± 37	1040 ± 118
GABA(6) (3 x 10⁻³ M)	11.2 ± 1.1*	36.2 ± 6.1**	128 ± 10.9**	173 ± 15*	265 ± 41*	740 ± 75*
GABA(6) (6 x 10⁻³ M)	15.3 ± 2.2**	20.9 ± 3.6*	94 ± 11.8**	193 ± 13**	179 ± 21**	502 ± 99**
EOS(5) (2 x 10⁻³ M)	11.0 ± 0.8*	35.7 ± 2.2*	110 ± 5**	---	---	---
Picrotoxin(6) (8 x 10⁻⁵ M)	7.3 ± 0.9	68.0 ± 2.8*	175 ± 13*	95 ± 16*	407 ± 34**	1157 ± 180*
Picrotoxin + GABA(5) (6 x 10⁻³ M)	9.5 ± 0.34	43.0 ± 1.8	149 ± 11	120 ± 11	347 ± 72	997 ± 94
Picrotoxin + EOS(5)	9.1 ± 0.82	54.0 ± 10	165 ± 15	---	---	---

Number of experiments in parentheses. The stimulus triggered release is given as net extra-release. Statistically different from control: * p < 0.05; ** p < 0.01.

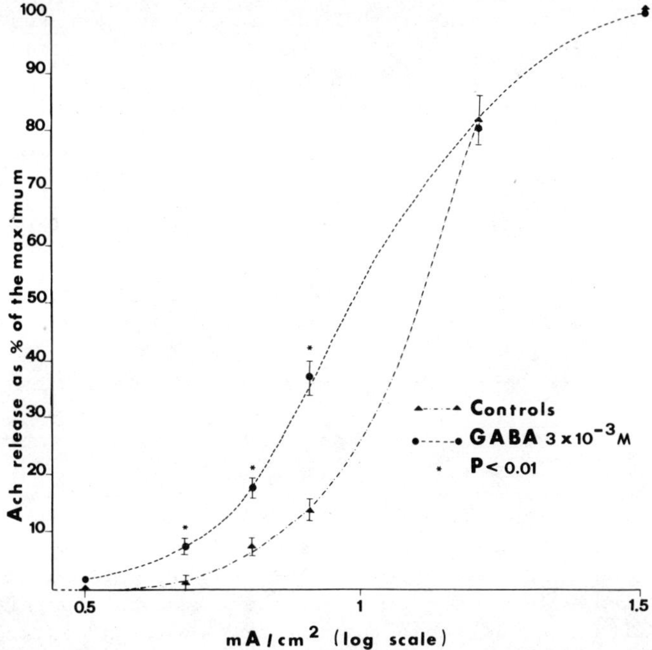

FIGURE 1: Relationship between stimulus intensity and ACh release
as percent of the outflow caused by pulses of maximal
strength (30 mA/cm^2) in normal (–·–·–·–) and GABA treated
(– – – –) slices of guinea pig cerebral cortex. Note that
at the lower intensities GABA shifts the position of the
curve to the left. Vertical bars = S.E.M.; n = 5.

MATERIALS AND METHODS

Guinea pig brain slices, prepared, set up and superfused as
recently described (4,5) were used to investigate GABA effects on
ACh release from different brain areas. Freely moving, unanesthetized guinea pigs, equipped with epidural parietal cups and intraventricular guide cannulas were used to study the cortical ACh outflow. The tetrodotoxin pretreated guinea pig ileum was employed
for the bioassay of ACh (4).

RESULTS

In vitro: The influence exerted by GABA (3–6 x 10^{-3} M) on
ACh release from slices of cerebral cortex (CC) and caudate nucleus

TABLE 2: Effect of GABA (6×10^{-3} M) on ACh release from slices of
          cerebral cortex and caudate nucleus in the absence and
          in the presence of tetrodotoxin (TD) (5×10^{-7} M).
          Number of experiments in parentheses. Statistically
          different from control: * = $p < 0.01$.

Experimental Conditions	Cerebral Cortex		Caudate Nucleus	
	No Stimulation	5 Hz	No Stimulation	5 Hz
Normal(6)	9.6 ± 1.6	154 ± 16.7	128 ± 15	798 ± 86
GABA(8)	15.3 ± 2.2*	94 ± 11.8*	193 ± 13*	502 ± 99*
TD(5)	8.4 ± 0.4	---	33 ± 3.7*	---
TD + GABA(5)	8.8 ± 0.8	---	29 ± 2*	---

(CN) is shown in Table 1. Clearly, GABA caused an opposite and
dose dependent influence on resting and stimulus triggered release,
both in CC and CN. Indeed, while the resting release was increased,
the stimulated release was reduced. To assay the specificity of
these facilitating and inhibitory effects, various antagonists were
tested. Picrotoxin (8×10^{-5} M) added to CC and CN untreated
slices, enhanced the evoked ACh release and prevented any GABA
influence on neurotransmitter efflux from resting and stimulated
tissue (Table 1). Bicuculline (1×10^{-4} M) was less effective than
picrotoxin, while phentolamine (3×10^{-6} M) and spiroperidol (2.5×10^{-7} M) added to CC and CN, respectively, did not influence GABA
effects at all (data not shown).

To ascertain whether GABA modulation on ACh release could be
reproduced by enhancing its tissue levels, slices of CC were preincubated and perfused with EOS (2×10^{-3} M), so as to inhibit
GABA-T. As summarized in Table 1, EOS enhanced the ACh resting
release and reduced the stimulus triggered ACh outflow, as GABA
did. Both EOS effects were prevented by picrotoxin.

Some experiments were performed to ascertain whether GABA
modified the electrical excitability of the cholinergic nerve elements. For this purpose, slices of CC were stimulated at 5 Hz for
5 min every 20 min with stepwise increasing pulse intensities. In
the presence of GABA the actual extra-release, given as percentages
of the maximum attainable with fully effective current (30 mA/cm^2),
was higher at the lower pulse intensities, suggesting enhanced
excitability of the cholinergic nerve endings (Fig. 1). Other
attempts, carried out to test a possible influence of GABA on KCl
depolarization, are in keeping with these findings. When KCl concentration was raised to 25 mM the CC slices released up to 15
times more ACh than in normal conditions (from 9.5 ± 1 to 150 ± 6;

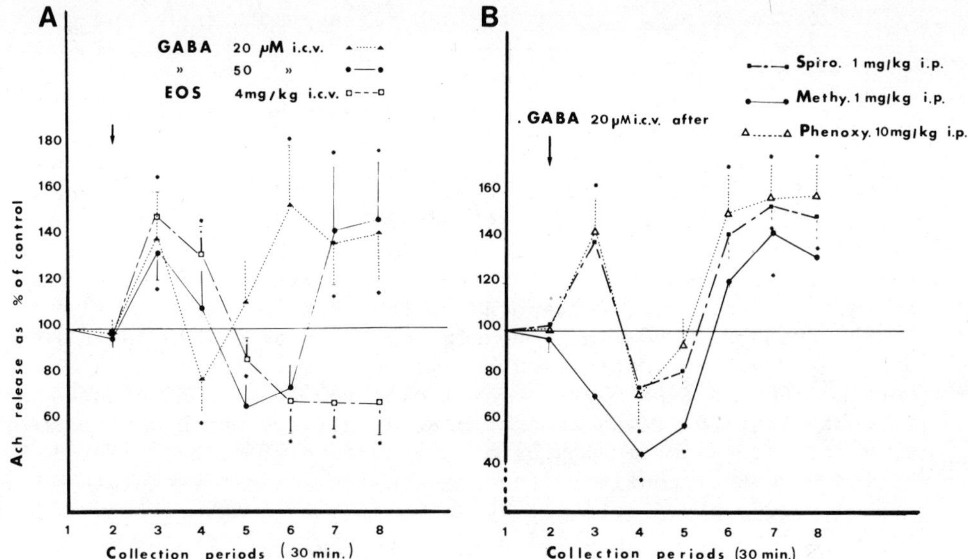

FIGURE 2: ACh release as percent of basal values, from guinea pig
cerebral cortex after GABA, EOS and various antagonists.
A: effect of GABA 20 µMol (•••• ; 8 experiments) and
50 µMol (______; 6 experiments) and EOS 4 mg/kg (——•——;
5 experiments) injected i.c.v..
B: effect of GABA 20 µMol in guinea pigs pretreated with
spiroperidol 1 mg/kg (____•____; 4 experiments, phenoxybenzamine 10 mg/kg i.p. (•••••; 4 experiments) and methy-
sergide 1 mg/kg i.p. (______; 4 experiments). Vertical
bars: S.E.M. * = significantly different (p < 0.05 or
lower) from pretreatment values. Student's t-test for
paired data.

ng/g/min). GABA (6 x 10^{-3} M) added áfterwards, further enhanced
KCl evoked release (from 148 ± 10 to 221 ± 18; ng/g/min; p < 0.05;
n = 6). On the other hand, GABA was ineffective in the presence
of complete depolarization caused by KCl (60 mM). Similar results
were obtained in CN slices.

Finally, in order to establish whether the effects of GABA on
resting and electrically evoked release were linked to sodium
dependent membrane processes, slices of CC were pretreated for
15 min with tetrodotoxin (5 x 10^{-7} M). Tetrodotoxin abolished the
effect of electrical stimulation and reduced the ACh resting release in CN, but not in CC. GABA (6 x 10^{-3} M), added thereafter,

no longer influenced the resting ACh release both in CC and CN
(Table 2).

Ex vivo: A few minutes after injecting GABA 20-50 µmol i.c.v.
(10-20 µl) the animals showed increased locomotor activity and enhanced exploratory behavior. The intensity and duration of the
stimulation was greater after the lower GABA dose; then neurodepres-
stion followed, the guinea pigs showed reduced muscle, loss of the
righting reflex, slowed and superficial breathing and fall of core
temperature by 1-2°C. However, the reactivity to noxious stimuli
was maintained, even at the peak of the neurodepression, which
lasted from 30 to 90 min, depending on the dose. Then, a third
phase of "rebound" activation became evident. The pattern of
cortical ACh release clearly paralleled the behavioral changes,
i.e. at first there was an increase in outflow, which was followed
by a reduction and finally by a late rebound (Fig. 2A). The inhibition of GABA-T confirmed the presence of a first "excitatory"
period. EOS, given both i.c.v. and i.p. raised ACh outflow and then
caused its progressive decline until the animal's death, which occurred many hours later (Fig. 2A). Since these results, characterized by activation and inhibition of ACh release, did not fit well
with those obtained in vitro, possible influences of GABA on other
neuronal systems functionally linked to cholinergic neurons were
considered. Therefore, the problem was re-examined in guinea pigs
pretreated with various antagonists. The convulsant properties of
picrotoxin and bicuculline did not allow us to employ them at high
doses in the unanesthetized animals. Bicuculline 2 µg i.c.v.
(maximum tolerated dose) solely dampened the entire response of
ACh output to GABA injection, but did not prevent it.

Phenoxybenzamine and spiroperidol were unable to change the
polyphasic pattern of GABA effects (Fig. 2B). Conversely, methy-
sergide cancelled the initial stimulatory phase, potentiated the
subsequent inhibition by GABA (Fig. 2B) and increased the basal
ACh output from 30 $\pm$ 3 to 45 $\pm$ 6 ng/cm^2/30 min.

DISCUSSION

The data, gathered from the electrically stimulated slices,
support the inference that GABA actually modulates the cholinergic
nervous elements of CC and CN, according to the classical concept
of "presynaptic" inhibition (12). In fact, GABA increased both
the resting and KCl evoked release, reduced the pulse triggered
transmitter output and enhanced nerve ending excitability. The
effects on resting and stimulated release were prevented by picrotoxin and, to a lesser extent, by bicuculline, thus demonstrating

the specificity of the effect. The facilitating influence of
picrotoxin on the evoked output, together with the GABA-like action
of EOS, are suggestive for an intrinsic control on the cholinergic
structures, exerted by GABA neurones present not only in CN, but
also in CC (6). Neither phentolamine nor spiroperidol affected
GABA modulation on ACh release. Thus any noradrenaline or dopamine
mediated effect could be ruled out (4,5). Obviously, the possible
involvement of other neurotransmitters needs testing. The experiments performed with tetrodotoxin suggest that a large part of the
"resting" release in CN depended on the spontaneous firing of ACh
interneurones, possibly still intact in 400 µ thick slices. In
fact, the toxin reduced the outflow to one third in CN, but not in
CC, where the cholinergic elements are known to be represented
mostly by nerve endings.

    More pertinent to the present discussion is the finding that
tetrodotoxin prevented any facilitation by GABA of the "actual"
rest-release in both areas (7). Therefore, a sodium dependent
membrane process seems to be involved in the effect of the amino
acid, even in unstimulated slices. This dispute does not entitle
us to postulate the existence of an intermediate neuronal link,
subserving GABA modulation on the cholinergic structures, because
changes in Na^+ conductance may participate in the mechanism of
GABAergic presynaptic inhibition (12).

    As far as the results in the unanesthetized animals are con-
cerned, the initial enhancement of ACh output seen after GABA or
EOS does not comply with the conclusion reached from the in vitro
experiments, showing a (probable) direct inhibition of GABA on the
cholinergic elements. Thus, an indirect interference on the corticopetal ACh pathways can be considered. GABA could, at first, inhibit
certain neurones involved in the tonic control of the ascending
cholinergic system (whose distribution would be responsible for the
ACh output).

    Subsequently, GABA could directly inhibit the corticopetal
cholinergic input, thus reducing the neurotransmitter efflux. Our
results suggest that the neuronal pool primarily affected by GABA
is serotonergic in nature (13), particularly since methy-sergide
per se enhanced ACh outflow (as if removing an inhibitor 5-HT tone),
prevented the initial increase of ACh outflow by GABA and rendered
tee secondary inhibition more evident. A higher sensitivity to
microiontophoretically applied GABA has been recently demonstrated
for pars reticulata interneurones in comparision with pars compacta
dopaminergic cells (8). A situation similar to the 5-HT and ACh
neurones present in the brain stem, which seem to receive a GABA-
ergic inhibitory input (11) could exist.

ACKNOWLEDGMENT

The research presented in this paper was supported by Grant No. 79-01837-04 from Consiglio Nazionale delle Ricerche, Rome.

REFERENCES

1. Anden, N.E., Grabowska-Anden, M. and Wachtel, H. (1979): Acta Pharmacol. Toxicol. 44:191-196.
2. Arbilla, S. and Langer, S.Z. (1979): Arch. Pharmacol. 306: 161-168.
3. Beani, L., Bianchi, C. and Bolletti, A. (1964): Acta Anesth. 15:89-93.
4. Beani, L., Bianchi, C., Giacomelli, A. and Tamberi, F. (1978): Eur. J. Pharmacol. 48:179-193.
5. Bianchi, C., Beani, L. and Tanganelli, S. (1979): Eur. J. Pharmacol. 58:235-246.
6. Emson, P.C. and Lindwall, O. (1979): Neuroscience 4:1-30.
7. Giorguieff-Chasselet, M.F., Kemel, M.L., Wandscheer, D. and Glowinski, J. (1979): Brain Res. 175:383-386.
8. Grace, A.A. and Bunney, B.S. (1979): Eur. J. Pharmacol. 59: 218-221.
9. Javoy, F., Envrad, C., Herbert, A. and Glowinski, J. (1977): Brain Res. 126:982-986.
10. Ladinsky, H., Consolo, S., Bianchi, S. and Jori, A. (1976): Brain Res. 108:351-361.
11. Lewis, P.R. and Shute, C.C.D. (1978): Handbook Pharmacol. 9: 315-355.
12. Nistri, A. and Constanti, A. (1979): Prog. Neurobiol. 13:117-235.
13. Samanin, R., Quattrone, A., Consolo, S., Ladinsky, H. and Algeri, S. (1978): IN Interactions Between Putative Neuro-transmitters in the Brain (eds) S. Garattini, F. Pujol and R. Samanin, Raven Press, New York, pp. 383-400.
14. Scatton, B. and Bartholini, G. (1979): Eur. J. Pharmacol. 56: 181-182.
15. Scherber, A., Staib, A.H. and Oelrzner, W. (1975): Acta Biol. Med. Germ. 3:1517-1524.
16. Sethy, V.H. (1978): Res. Comm. Chem. Pathol. Pharamcol. 21: 359-362.
17. Stoof, J.C. and Mulder, A.A. (1977): Eur. J. Pharmacol. 46: 177-180.
18. Zsilla, G., Cheney, D.C. and Costa, E. (1976): Arch. Pharmacol. 234:251-256.

GABA-ACETYLCHOLINE INTERACTION IN THE RAT STRIATUM

B. Scatton and G. Bartholini

Synthelabo-L.E.R.S., Research Department
58 rue de la Glaciere
75013 Paris, France

INTRODUCTION

The corpus striatum contains a massive cholinergic innervation
which is mainly of intrinsic origin. Evidence has been provided that
striatal acetylcholine (ACh) (inter)neurons may be regulated by the
afferent nigrostriatal dopaminergic (13,25) and raphe-striatal serotonergic pathways (5). Little is known, however, of a possible regulation of cholinergic neurons by GABA. Previous studies with picrotoxin have suggested that GABA may affect the activity of cholinergic
neurons indirectly via changes in the activity of the nigrostriatal
dopaminergic pathway (10,12). Recent data obtained in our laboratory have also indicated the existence of a GABA-mediated inhibitory
influence on cholinergic cells which is intrinsic to the striatum (17,
18,20). The present report reviews the available evidence for, and
analyzes the possible functional implications of, this intrastriatal
GABA-ACh link.

Effect of GABAergic Agents
on Striatal Acetylcholine Turnover

GABAergic compounds increase rat striatal ACh concentrations
(Table 1)(17,18,20). This effect is induced, not only by GABA receptor agonists, e.g. muscimol. SL 76,002 (4[2'hydroxy-4'-fluoro-α-
(4"-chorophenyl) benzenemethaniminol] butyramide)(2,11), its acidic
metabolite SL 75,102 or δ-aminovaleric acid, but also by GABA transaminase inhibitors, GABA uptake inhibitors and drugs chemically related to GABA. Striatal choline (Ch) levels are unaffected by

the drugs (Table 1). The effect of GABA receptor agonists on
striatal ACh concentration does not result from a direct action on
ACh metabolism as: 1) neither drug affected _in vitro_ (up to 10^{-4} M),
or _in vivo_, choline acetyltransferase (CAT) and acetylcholinesterase
(AChE) activity in striatal homogenates (17,20); 2) GABA, muscimol
and SL 75,102 failed to affect the high affinity Ch uptake in
striatal synaptosomes ($IC_{50} > 10^{-4}$ M). An interaction of GABA
mimetics with ACh receptors has also to be excluded as SL 76,002,
SL 75,102 and muscimol (5 x 10^{-4} M) failed to alter the binding of
^{3}H-quinuclidinylbenzylate (QNB) to rat brain membrane preparations
(Briley, personal communication). The enhancement of striatal ACh
levels induced by GABA receptor agonists rather reflects a decrease
in the activity of striatal cholinergic cells. In fact, previous
studies have shown that changes in ACh concentrations are in general inversely related to the rate of impulse flow (7). Moreover,
systemic administration of muscimol or SL 75,102 retards the ACh
disappearance induced by intrastriatal hemicholinium-3 (HC-3) infusion in the rat (20). These data indicate, therefore, that GABA
receptor agonists reduce striatal ACh turnover. This suggests that
a GABA input is involved in the regulation of the activity of
striatal cholinergic cells.

Evidence of an Intrastriatal GABA-ACh Link

The effect of GABA receptor agonists on striatal ACh turnover
does not involve dopaminergic mechanisms. Evidence for this view
is the following: 1) striatal and/or pallidal GABAergic neurons
are known to innervate the substantia nigra (6) and inhibit dopaminergic neurons (15). As dopaminergic neurons tonically inhibit
striatal ACh cells (13,25), GABA receptor agonists would be expected to reduce dopaminergic activity resulting in an enhanced
ACh turnover. In contrast, these compounds reduce ACh utilization
(see above) and therefore a possible mediation by dopaminergic
neurons does not become apparent; 2) SL 76,002 induced a similar
percentage increase in rat striatal ACh levels when given alone or
in combination with apomorphine or haloperidol as compared with the
effect of either of the latter drugs (17,20); 3) muscimol or
SL 76,002 retain their ability to increase striatal ACh concentration after a unilateral 6-hydroxydopamine induced destruction of
the nigrostriatal dopaminergic pathway or cerebral hemitransection
between striatum and substantia nigra (17,18,20). The transection
experiments also rule out the possibility that the GABA receptor
agonist induced alteration of striatal cholinergic neuron activity
is mediated via non-dopaminergic ascending striatal afferents, e.g.
the serotonergic pathway originating from the raphe nuclei.

TABLE 1:   Effect of GABAergic agents on striatal ACh levels.

Treatment	Dose (mg/kg i.p.)	ACh (nmol/g)	Ch (nmol/g)
Saline	--	40.5 ± 1.3	133.9 ± 5.3
Muscimol	2.5	70.1 ± 4.0*	131.8 ± 8.0
SL 76,002	400	68.8 ± 2.4*	127.5 ± 10.0
SL 75,102	400	67.2 ± 2.4*	
δ-aminovaleric acid	300	55.5 ± 0.8*	
Amino-oxyacetic acid	50	54.3 ± 2.8*	
γ-acetylenic GABA	100	53.8 ± 2.4*	
Dipropylacetamide	200	67.6 ± 2.0*	
Depakine	200	52.7 ± 2.4*	
Diaminobutyric acid	300	52.2 ± 2.4*	
Nipecotic acid	1000	61.1 ± 1.6*	
Nipecotamide	300	64.0 ± 1.0*	
2-Pyrrolidone	1000	62.0 ± 2.4*	
Piracetam	4000	70.9 ± 2.8*	135.2 ± 3.2

Rats were decapitated 2 hr after administration of γ-acetylenic
GABA or amino-oxyacetic acid, or 30 min after administration of
other agents.   ACh concentrations were measured by the modified
(19) method of Guyenet et al. (7).   Results are means with S.E.M.
of data from 12 animals.   * p <   0.001 as compared to controls
(Student's t-test).

TABLE 2:   Effects of GABAergic compounds and apomorphine on rat
striatal ACh levels after lesion of the corticostriatal
projections.

Treatment	Dose (mg/kg)	Striatal ACh Levels (nmol/g) Sham Operated	Lesioned
Saline		37.9 ± 1.1	38.0 ± 1.9
Muscimol	5	62.9 ± 3.0*	44.8 ± 1.9**
	7.5	72.7 ± 2.2*	50.5 ± 2.2*,**
SL 76,002	400	51.5 ± 1.8*	44.0 ± 1.8**
Dipropylacetamide	200	69.0 ± 3.0*	46.0 ± 3.4**
Apomorpine	0.3	51.2 ± 2.2*	56.2 ± 3.0*
	1	54.2 ± 2.3*	52.1 ± 2.3*

The corticostriatal projections were lesioned according to McGeer
et al. (14) three weeks before drug administration.   GABAergic
compounds were administered intraperitoneally and apomorphine subcutaneously.   Rats were sacrificed 30 min after drug injection.
Results are means ± S.E.M of data from 16 animals (2 experiments).
* p < 0.001 vs controls;   ** p < 0.01 vs sham operated rats (t-test).

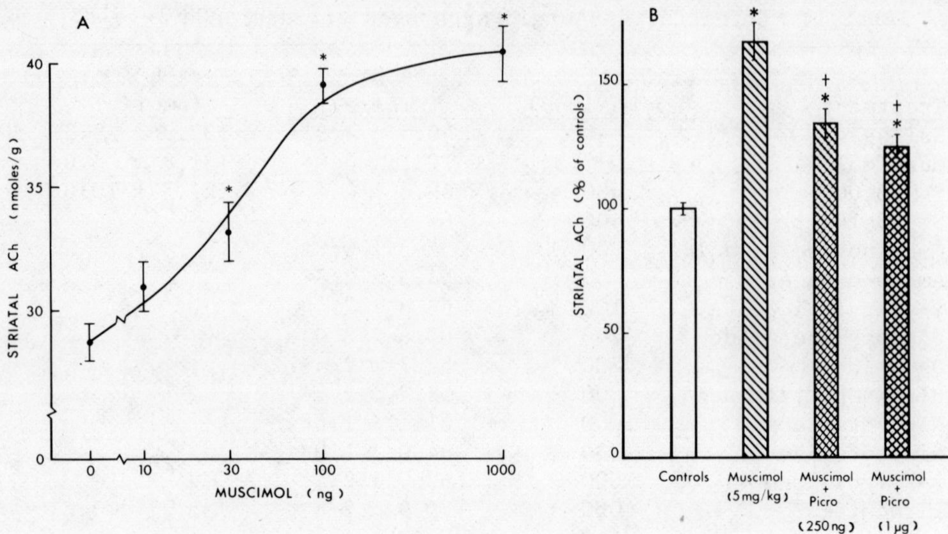

FIGURE 1: Effect of intrastriatal infusion of muscimol or picrotoxin in normal or muscimol treated rats. A) rats
received an intrastriatal microinjection of muscimol dissolved in Ringer medium at a rate of 1 µl/120 sec through
in-dwelling cannulae implanted stereotaxically 7 days
before the experiment. Animals were sacrificed 30 min
later. Results are means $\pm$ S.E.M. of 4-15 determinations.
* p < 0.01 vs controls (Ringer infusion). B) rats received an i.p. injection of muscimol (5 mg/kg) immediately following termination of an intrastriatal infusion of
picrotoxin (0.25 or 1 µg) or Ringer medium and were sacrificed 30 min later. Columns represent means $\pm$ S.E.M.
of 4-15 determinations. * p < 0.001 vs controls;
+ p < 0.001 vs muscimol treatment.

In addition, extrastriatal GABAergic mechanisms on thalamic
cholinergic afferents to the striatum do not seem to play a role in
the GABA mediated alterations of striatal ACh. Thus, after systemic administration of muscimol. a similar percentage increase in
ACh levels was observed in the head (which receives cholinergic afferents from the parafascicular nucleus of the thalamus [23]), as
well as in the remaining part of the corpus striatum and in globus
pallidus (20).

The reduction of ACh turnover in the striatum induced by GABA-
ergic compounds and possibly by GABA, might occur via an intra-

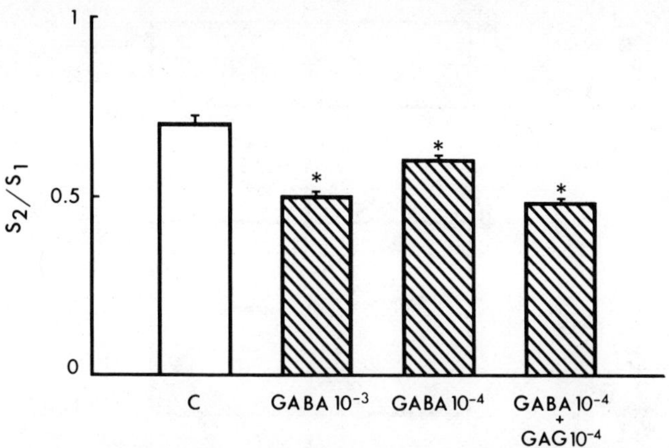

FIGURE 2: Effects of GABA on potassium evoked ^{3}H–ACh overflow in
rat striatal slices. Striatal slices were incubated with
^{3}H–Ch (10^{-7} M) for 30 min at 37°C, rinsed, transferred
to a chamber and continuously superfused with Krebs solution (0.5 ml/min) containing HC–3 (10 µM). ^{3}H–ACh release
was induced by a 2 min exposure to 20 mM potassium. The
first stimulation (S_1) was carried out 45 min and the
second one (S_2) 60 min after starting sample collection.
GABA and γ-acetylenic-GABA (GAG) were added to the medium
10 min before S_2. The potassium evoked overflow is expressed as a percentage of the total tritium present in
the tissue at the onset of stimulation (fractional release).
Columns are means $\pm$ S.E.M. (5-11 determinations).
*   p < 0.001 vs controls.

striatal mechanism. Several pieces of evidence support this view.
Thus 1) intrastriatal infusion of muscimol (10 to 1000 ng) induces
a concentration dependent rise in rat striatal ACh levels (Fig. 1A)
(17,20) as measured 30 min later. A similar change was induced by
GABA (750 ng) infusion (data not shown). In contrast, intrastriatal
infusion of picrotoxin (0.25-1 µg) reduces striatal ACh levels (17)
and antagonizes the increase of striatal ACh concentrations caused
by a systemic injection (Fig. 1B) (17,20) or an intrastriatal infusion (data not shown) of muscimol; 2) GABA (10^{-4}-10^{-3} M) reduces
the potassium (20 mM) evoked ^{3}H–ACh overflow in rat striatal slices
preloaded with ^{3}H–Ch (Fig. 2). The effect of GABA was potentiated
by addition of a GABA transaminase inhibitor into the incubation
medium. A similar effect of GABA on the release of endogenous ACh
was recently observed in guinea pig striatal slices stimulated with
electrical pulses (3). Altogether, these data therefore strongly

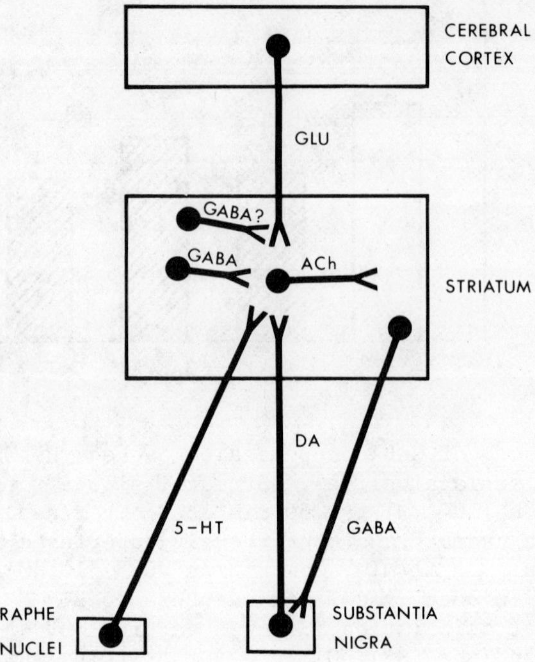

FIGURE 3: Schematic representation of the possible connections
between GABAergic (GABA), dopaminergic (DA), serotonergic
(5-HT), glutamatergic (GLU) and cholinergic (ACh) neurons
in basal ganglia and related brain structures.

suggest the existence of an intrastriatal GABAergic control of
cholinergic neurons which is tonic in nature.

This influence is probably exerted by GABAergic neurons. Thus,
GABA uptake has been shown in striatal synaptosomal preparations (9)
as well as by means of autoradiographic experiments (8). Therefore "extrajunctional" GABA receptors (not apposed to GABA neurons)
on striatal cholinergic cells are unlikely.

Involvement of the Corticostriatal Projections
in the Intrastriatal GABA-ACh Link

The GABA influence on cholinergic neurons may depend on the
activity of other afferents to cholinergic neurons. In this respect,

in addition to the nigrostriatal dopaminergic, the raphe-striatal
serotonergic and the thalamostriatal cholinergic projections (which
do not appear to participate in the GABAergic influence on striatal
cholinergic activity; see above) the striatum receives a massive
excitatory input from the cerebral cortex (24). The corticostriatal
projection appears to be glutamatergic in nature as 1) lesion of
the corticostriatal tract reduces glutamate levels, uptake (4,14),
release and synthesis (16) in the rat striatum, and 2) iontophoretic
application of the putative glutamate antagonist,glutamic acid diethylester, suppresses the excitation of striatal cells induced by
cortical stimulation (24). These cortical glutamatergic afferents
to the striatum may terminate <u>(inter alia)</u> on cholinergic neurons.
Thus, 1) injection of the glutamate analogue, kainic acid, leads
to a destruction of striatal cholinergic cells (22); 2) cortical
ablation reduces striatal ACh turnover (26); and 3) infusion of
glutamic acid diethylester increases striatal ACh levels (Scatton,
unpublished data). The cortical glutamatergic projection to striatal cholinergic cells may play a role in the inhibitory action of
GABA on ACh cells. Indeed, lesion of the corticostriatal projections reduced the ability of muscimol, SL 76,002 and dipropylacet-
amide to enhance striatal ACh levels (Table 2) (21). In these
experiments, the surgical lesion induced a 44% reduction of glutamate uptake (nmol/5 min/mg protein, unoperated controls: $1.35 \pm$
0.03; sham operated controls: 1.40 ± 0.05; lesioned rats: $0.79 \pm$
0.1, $p < 0.001$), whereas striatal ACh levels and choline acetyltransferase activity were unaffected. In addition, in preliminary
experiments, a reduction of the ability of GABA to inhibit the
potassium (20 mM) evoked ^{3}H-ACh release from rat striatal slices
was observed after lesion of the corticostriatal projections (unpublished observations). These data suggest that the inhibitory
GABAergic influence on striatal cholinergic transmission depends,
at least in part, upon the corticostriatal (possibly glutamatergic)
tract. The cortical projections may participate in the GABAergic
inhibition of striatal ACh cells in various ways; 1) GABA receptors
may be localized on excitatory cortical (glutamatergic) afferents
to striatal cholinergic neurons; increase in GABAergic transmission
may lead to reduction of the excitatory input on cholinergic cells
resulting in decreased turnover of ACh. However, we failed to observe a reduction of ^{3}H-muscimol binding in rat striatal membrane
preparations after lesion of the corticostriatal projection (fmol/
mg protein, sham operated controls: 22.0 ± 5.0; lesioned rats,
25.7 ± 9.4) (Beaumont and Scatton, in preparation). Nevertheless,
as the bulk of GABA or muscimol binding sites are located on striatal interneurons and efferent neurons, a limited reduction may not
be detectable; 2) GABAergic neurons may terminate on ACh cells,
the GABA mediated inhibition may be exerted only in the presence
of a parallel excitatory drive on these cells exerted by the corticostriatal glutamatergic neurons. Alternatively, adaptive mechanisms
occurring in cholinergic neurons as a consequence of corticostriatal

tract degeneration may render them refractory to GABAergic inhibition.

It is possible that the action of other inhibitory transmitters (e.g. dopamine) on ACh neurons is mediated by the corticostriatal (glutamatergic) projections. This, however, appears to be unlikely, as lesion of the corticostriatal projections fails to affect the ability of systemically administered apomorphine to increase striatal ACh concentrations (Table 2) (21) and to inhibit the in vitro potassium evoked release of ^{3}H-ACh from rat striatal slices (Scatton and Bartholini, in preparation). These data indicate, therefore, that the cortical lesion does not modify the action of dopaminergic on cholinergic cells, excluding, in addition, a non-specific effect on the lesion on cholinergic neuron excitability. The above data also rule out the possibility that the reduction of the effect of GABA receptor agonists induced by the lesion of corticostriatal projections is connected with a reduced blood supply and a consequent reduction of drug concentration in the striatum.

CONCLUSION

From previous and present studies, it appears that GABA may modulate the activity of striatal cholinergic interneurons via an inhibitory action on the nigrostriatal dopaminergic neurons (10, 12) and, in addition, a mechanism which is intrinsic to the striatum and mediated by the presence of GABA receptors on cholinergic cells and on the corticostriatal glutamatergic nerve terminals (Fig. 3). The intrastriatal GABA influence on ACh cells appears to predominate as compared to the dopamine mediated mechanisms as GABA mimetic agents decrease the activity of striatal cholinergic neurons whereas an increased ACh turnover should be expected from an indirect action on the nigrostriatal dopaminergic pathway. As previously discussed (20), this predominance is possibly related to the basal rate of firing of dopamine vs ACh neurons; dopaminergic cells, being likely in a state of tonic inhibition, possibly have a lower susceptibility to further inhibition by GABA agonists than the ACh cells which, in contrast, may receive an excitatory drive by the corticostriatal glutamatergic pathway. Supportive evidence for this view is provided by the observation that the threshold dose of GABA mimetics for inhibiting cholinergic neurons is lower than that affecting the nigrostriatal dopaminergic pathway (2). However, this conclusion is based on results obtained with drugs; this does not exclude that under physiological and/or pathological conditions, the dopamine mediated GABA influence on striatal ACh cells prevails on the direct action of the amino acid.

The intrastriatal GABA-ACh link may have functional implications. Thus, while GABA receptor agonists (Sl 76,002, muscimol), given alone, do not cause stereotypies or catalepsy, they potentiate in low doses (SL 76,002, 12.5-25 mg/kg i.p.; muscimol, 0.2-0.5 mg/kg i.p.) the apomorphine induced stereotypies (1) and antagonize the catalepsy caused by haloperidol in the rat (27). The inhibitory influence of low doses of GABAergic compounds on striatal cholinergic neurons may account for these effects. In fact, previous studies have shown that these doses of SL 76,002 potentiate the apomorphine induced increase of striatal ACh concentrations and prevent the haloperidol induced reduction of striatal ACh levels (17,20). Finally, GABA appears to modulate cholinergic transmission also in various extrastriatal ACh-rich brain areas. Thus, muscimol induces an increase in ACh levels (which is antagonized by picrotoxin) in limbic areas (frontal cortex, nucleus accumbens, olfactory tubercle, hippocampus), interpeduncular nucleus, hypothalamus and brainstem (20). The effect is, however, smaller in these areas than in striatum, suggesting a less effective GABA influence on cholinergic transmission in non-striatal areas.

ACKNOWLEDGEMENTS

The authors wish to thank D. Fage for skilful technical assistance. SL 76,002 was synthesized by Dr. J.P. Kaplan of the Chemistry Department, Synthelabo.

REFERENCES

1.   Bartholini, G., Scatton, B. and Zivkovic, B. (1980): In Long
      Term Effects of Neuroleptics, (ed) F. Cattabeni, Raven
      Press, New York, pp. 207-214.
2.   Bartholini, G., Scatton, B., Zivkovic, B. and Lloyd, K.G.
      (1978): In GABA Neurotransmitters,(eds) P. Krogsgaard-Larsen,
      J. Scheel-Krueger and H. Kofod, Munksgaard, Copenhagen,
      pp. 326-339.
3.   Beani, L., Bianchi, C., Siniscalchi, A. and Tanganelli, S.
      (1978): Proc. 7th Intl. Congr. Pharmacol. Paris, Abst.2360.
4.   Divac, I., Fonnum. F. and Storm-Mathisen, J. (1977): Nature
      266:377-378.
5.   Euvrard, C., Javoy, F., Herbet, A. and Glowinski, J. (1977):
      Eur. J. Pharmacol. 41:281-283.
6.   Fonnum, F., Grofova, I., Rinvik, K., Storm-Mathisen, J. and
      Walberg, F. (1974): Brain Res. 71:77-87.

7. Guyenet, P.G., Agid, Y., Javoy, F., Beaujouan, J.C., Rossier, J. and Glowinski, J. (1975): Brain Res. 84:227-244.

8. Iversen, L.L. and Bloom, F.E. (1971): Brain Res. 41:131-143.

9. Iversen, L.L. and Johnston, G.A.R. (1971): J. Neurochem. 18: 1939-1950.

10. Javoy, F., Euvrard, C., Herbet, A. and Glowinski, J. (1977): Brain Res. 126:382-386.

11. Kaplan, J.P., Raison, B., Desarmenien, M., Feltz, P.,Headley, P.M., Worms, P., Lloyd, K.G. and Bartholini, G. (1980): J. Med. Chem. 23:702-704.

12. Ladinsky, H., Consolo, S., Bianchi, S. and Jori, A. (1976); Brain Res. 108:351-361.

13. Lloyd, K.G. (1978): In Essays in Neurochemistry and Neuropharmacology,(eds) M.B.M. Youdim, W. Lovenberg, D.F. Sharman and J.R. Lagnado, Wiley, New York, pp. 131-207.

14. McGeer, P.L., McGeer, E.G., Scherer, U. and Singh, K. (1977): Brain Res. 369-373.

15. Precht, W. and Yoshida, M. (1971): Brain Res. 32:229-232.

16. Reubi, J.C. and Cuenod, M. (1979): Brain Res. 176:185-189.

17. Scatton, B. and Bartholini, G. (1980): Brain Res. 183:211-216.

18. Scatton, B. and Bartholini, G. (1979): Eur. J. Pharmacol. 56:181-182.

19. Scatton, B. and Worms, P. (1979): J. Pharm. Pharmacol. 31: 861-863.

20. Scatton, B. and Bartholini, G. (1980): In GABA Neurotransmission (ed) H. Lal, Ankho International, Fayetteville, pp. 223-229.

21. Scatton, B. and Bartholini, G. (1980): Brain Res. 200:174-178.

22. Schwarcz, R. and Coyle, J.T. (1977): Brain Res. 127:235-239.

23. Simke, J.P. and Saelens, J.K. (1977): Brain Res. 126:487-495.

24. Spencer, H.J. (1976): Brain Res. 102:91-101.

25. Stadler, H., Lloyd, K.G., Gadea-Ciria, M. and Bartholini, G. (1973): Brain Res. 55:476-480.

26. Wood, P.L., Moroni, F., Cheney, D.L. and Costa, E. (1979): Neurosci. Lett. 12:349-354.

27. Worms, P. and Lloyd, K.G. (1980): N.-S. Arch. Pharmacol. 311:179-184.

REGULATION OF CHOLINERGIC ACTIVITY IN THE RAT HIPPOCAMPUS:

IN VIVO EFFECTS OF OXOTREMORINE AND FENFLURAMINE

H. Ladinsky, S. Consolo, A.S. Tirelli, G.L. Forloni and
M. Segal*

Istituto di Ricerche Farmacologiche Mario Negri,
20157, Milan, Italy, and
*Isotope Department, The Weizmann Institute of Science,
P.O. Box 26, Rehovot, Israel

Oxotremorine (OTMN) is a powerful muscarinic receptor agonist
whose profound effects on cholinergic neurons of the hippocampus
have been particularly well characterized. The drug increases the
content and decreases the turnover (TR_{ACh}) and release of acetylcholine (ACh). In addition, it depresses the sodium dependent high
affinity uptake of choline (SDHACU) into the cholinergic terminals
(see Table 1). The biochemical effects are indicative of a depressant action of OTMN on cholinergic neurons. It is implicit
that when the muscarinic receptor sites are occupied by this agonist
the regulatory mechanism of ACh synthesis receives a signal to slow
down its activity. The location of the sites at which the signal
starts is a matter of controversy at the moment. OTMN acts directly on muscarinic receptors which may be located either on the presynaptic cholinergic terminals as suggested by Szerb et al. (23)
or postsynaptically on non-cholinergic neurons, and thereby operates through a neuronal feedback loop. The latter hypothesis appears more likely since it has been found that reserpine, as well
as atropine blocked the effect of OTMN on whole brain (3) and
striatal (24) ACh content.

The present study investigated the action of OTMN on cholinergic activity in the hippocampus and particular attention was
directed to the role that the putative neurotransmitters present in
this area, norepinephrine (NA) and serotonin (5-HT), play in the
regulation of the cholinergic system.

781

TABLE 1: Some cholinergic effects of OTMN in rat hippocampus.

Cholinergic Marker	Control	OTMN	Ref.No.
ACh (nmol/g)	18.4 ± 0.8	24.9 ± 0.6 *	13
TR_{ACh} (nmol/g/min)	4.3 ± 0.2	3.5 ± 0.2 *	**
SDHACU (pmol/mg prot/min)	0.71 ± 0.04	0.45 ± 0.02*	**
ACh release (min^{-1})	0.06	0.03*	7

* = p < 0.01 vs control.   ** = unpublished data from this laboratory.

Female CD_1 rats (210–220 g: Charles River, Italy) were used. The animals were killed by decapitation. ACh and choline (Ch) were determined by the radioenzymatic method of Saelens et al. (20) with modifications (12). Sodium dependent high affinity uptake of Ch was determined by the method of Atweh et al. (1), and ACh turnover was measured as described by Racagni et al. (17). NA and 5-HT were measured by electrochemical detection coupled with HPLC (10,16).

Electrolytic lesions of the locus coeruleus and median raphe nucleus were made as described elsewhere (11). Kainic acid (0.5 µg in 0.5 µl) was injected bilaterally into the dorsal and ventral hippocampus (coordinates from bregma: -3.5, 2.5, 4.0 mm and -6.5, 5.5, 7.5 mm, respectively) of adult male anesthetized rats of the Weizmann Institute breeding colony. Experiments were performed 1-2 weeks after injection. Selected rats were used for histological verification of the injection sites and the magnitude of the damage produced by kainic acid. In all rats tested there was a complete or near complete disappearance of neurons in the injected hippocampus. There was no noticeable damage in sham-injected controls. The data were statistically analyzed by ANOVA (2 x 2) factorial analysis and Tukey's test for unconfounded means.

RESULTS AND DISCUSSION

At a dose of 1.4 µmol/kg (i.p.) OTMN maximally increased the ACh content of the hippocampus by 30% and this effect was blocked by pretreatment with the antimuscarinic agent, atropine sulfate (5 mg/kg i.p.) and by the monoamine depletor, reserpine (5 mg/kg i.p., 16 hr)(Table 2). These experiments indicate that the action of OTMN is initiated at muscarinic receptors and then a feedback

TABLE 2: The effect of several classes of drugs on the oxotremorine induced increase in hippocampal ACh.

Drug in Columns C & D	Hippocampal ACh (nmol/g)				
	A Saline	B OTMN	C Drug	D Drug + OTMN	Interaction
Atropine	18.1±1.0(8)	24.2±1.8(8)*	18.8±1.0(8)	18.2±1.0(8)	F 1,28= 7.4 p < 0.02
Reserpine	17.9±1.0(8)	25.1±0.9(8)*	19.1±1.0(8)	20.9±0.7(8)	F 1,28= 8.9 p < 0.01
α–MpT	17.5±1.1(8)	26.7±1.8(8)*	19.8±0.5(8)	22.9±1.3(8)**	F 1,28= 5.7 p < 0.05
FLA–63	17.8±0.5(14)	23.4±0.5(14)*	17.6±0.8(14)	18.6±0.6(14)	F 1,52=12.8 p < 0.01
Phenoxybenzamine	17.9±0.6(12)	25.6±1.4(12)*	18.5±0.7(12)	20.4±1.1(12)	F 1,44= 8.0 p < 0.01
Propranolol	18.0±1.0(8)	25.0±0.9(8)*	19.0±0.5(8)	27.3±1.0(7)*	NS
Pimozide	17.6±0.8(6)	23.5±0.7(6)*	19.3±0.9(6)	25.3±0.9(6)*	NS
Penfluridol	17.1±0.8(8)	23.5±0.8(8)*	18.7±0.8(8)	23.9±0.7(8)*	NS

Data are means ± S.E.; n = number in parentheses. * = p < 0.01; ** = p < 0.05 vs saline. Rats were killed 20 min after administration of oxotremorine sesquifumarate (1.4 μmol/kg i.p.). See text and Ladinsky et al. (13) for solvents and treatment schedules of the drugs. NS = not significant.

mechanism is triggered in which reserpine sensitive monoamines are
involved. The implication of a monoamine fits with the known ability
of OTMN to produce a widespread activation of catecholaminergic
neurons in the brain (2). However, among the reserpine sensitive
monoamines, dopamine does not appear to be involved since the blockade of its receptors by penfluridol or pimozide   not prevent
the effect of OTMN (Table 2). Instead, NA is a likely candidate
to mediate the action of OTMN as seen from the data in Table 2.
When noradrenergic transmission was interrupted in various ways,
such as by inhibiting NA synthesis with α-methyl-para-tyrosine
(α-MpT) or FLA-63, and by blocking α-adrenoceptors with phenoxybenzamine, the effect of OTMN was prevented. Propranolol, a blocker
of β-adrenoceptors, was ineffective.

In addition, the effect of OTMN was antagonized both by bilateral lesions of the locus coeruleus, from which arises the major
noradrenergic projection to the hippocampus (14), and by selective
destruction of hippocampal noradrenergic terminals with the synthetic presynaptic neurotoxin, DSP-4 (8,9,13,19)(Table 3). None
of the pretreatments or lesions shown in Tables 2 and 3 significantly altered hippocampal ACh content by themselves.

The above pharmacological and lesion data suggest that OTMN,
by activating noradrenergic neurons, provokes a release of NA which
then inhibits the cholinergic system. These findings are in line
with the results of other workers, who showed that cholinergic
agonists can affect brain NA levels _in_ _vivo_ (6) and NA release _in_
vitro (25).

The indirect consequence of the effect of OTMN on ACh via NA
argues against its acting on presynaptic muscarinic receptors
located on cholinergic nerve terminals of the septal-hippocampal
pathway. Indeed, receptor binding studies (26) indicate that the
muscarinic receptors in the hippocampus are located post-synaptical-
ly to the cholinergic neurons since the total number of cholinergic
muscarinic binding sites was not diminished upon destruction of the
septal-hippocampal pathway. However, the possibility that such
receptors exist on terminals or axons of non-cholinergic neurons
as well as on cell bodies could not be excluded in the experiments
of Yamamura and Snyder (26). Recently, Schwarcz et al. (22) found
that when virtually all of the perikarya in the hippocampus were
destroyed by the action of the neurotoxin, kainic acid, placed
locally into this area the muscarinic receptor binding decreased
only modestly, amounting to less than 15% of the total receptor
number. This result suggests that a predominant proportion of the
muscarinic receptors of the hippocampus may reside on non-cholinergic nerve terminals.

Direct localization of the important interaction between central noradrenergic and cholinergic neurons is lacking at present.

TABLE 3: Effect of electrolytic or neurotoxin induced destruction of the noradrenergic pathway on oxotremorine's increase in hippocampal ACh.

Treatment in Columns C & D	Hippocampal ACh (nmol/g)			
	A Saline	B OTMN	C Treatment	D Treatment + OTMN
Locus coeruleus lesion	18.1 ± 0.3	24.6 ± 0.6*	18.9 ± 0.6	20.8 ± 1.7
DSP-4	18.4 ± 0.8	24.9 ± 0.6*	15.7 ± 1.0	16.9 ± 1.0

* = $p < 0.01$ vs saline group. Rats were killed two weeks after bilateral electrolytic lesion of the locus coeruleus and three weeks after DSP-4 (300 µg/10 µl i.c.v). The hippocampal noradrenaline contents were reduced by 55% and 80%, respectively.
Interactions with oxotremorine: locus coeruleus lesion F 1,16=58 $p < 0.05$;
DSP-4 F 1,44=8.8 $p < 0.01$

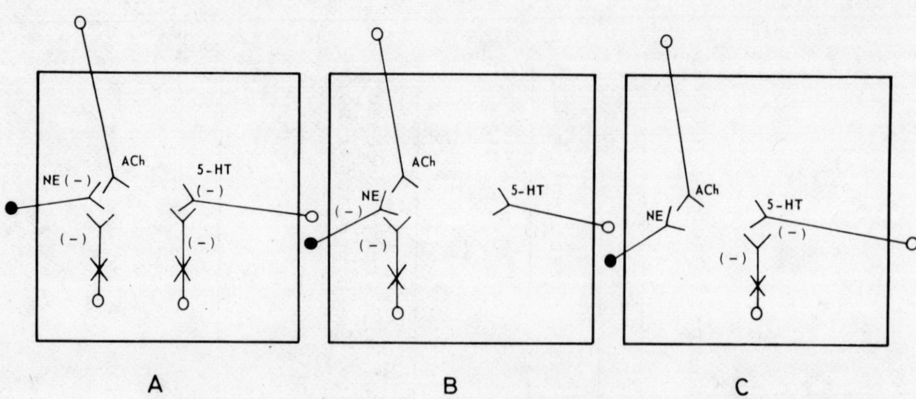

FIGURE 1: Schematic diagrams of the organization of cholinergic, noradrenergic and serotonergic neurons within the hippocampus and possible arrangements of inhibitory interneuron targets of kainic acid (denoted by crosses). In (A) the inhibitory neurons intrinsic to the hippocampus are shown to interact both with the noradrenergic and serotonergic terminals and destruction of these interneurons by kainic acid leads to simultaneous disinhibition of the monoaminergic neurons and subsequent loss of their transmitter with time. In (B) only the noradrenergic terminals are primarily affected by the loss of the inhibitory interneuron with the serotonergic terminals being affected secondarily to the depletion of NA. In (C) only the serotonergic nerve endings are affected primarily and the noradrenergic neurons secondarily to the loss of 5-HT. In this regard it is known that inhibitory GABAergic neurons intrinsic to the hippocampus are highly vulnerable to kainic acid (22).

In vivo (18) and in vitro (7) studies showed that a local receptor mediated response to muscarinic agonists and antagonists regulates ACh synthesis and release in the hippocampus. An attractive hypothesis which is consistent with the above events is that the noradrenergic afferents with cell bodies in the locus coeruleus and the cholinergic afferents with cell bodies in the medial septum and diagonal band nuclei converge in the hippocampus to form noradrenergic axo-axonic synapses (see Fig. 1A). One functional possibility of such an arrangement is that OTMN acts on heterologous muscarinic receptors located on the noradrenergic boutons to release inhibitory NA onto the cholinergic terminals. The prospect of such a simple crosstalk intercommunication between the two

terminals was investigated with the use of kainic acid, the neuroexcitant drug that induces the virtual complete disappearance of
interneurons while leaving the terminals of extrinsic neurons intact. We reasoned that under these conditions the action of OTMN
should not be affected. Instead, two weeks after kainic acid
injection bilaterally into the dorsal and ventral hippocampus, a
blockage of the action of OTMN on ACh was observed, but this was
accompanied by major falls in both NA and 5-HT contents (Table 4).
Therefore the results do not permit us to distinguish whether it
is the loss in local receptors or the loss in NA which is responsible for the antagonism of OTMN.

A non-specific destruction of nerve terminals by kainic acid
in this experiment can be largely excluded since other terminals,
such as the cholinergic nerve endings, were unimpaired as shown by
the fact that the ACh content was not significantly altered. Furthermore, under essentially similar experimental conditions,
Schwarcz et al. (22) found that choline acetyltransferase (CAT)
activity was unaltered and tyrosine hydroxylase, a marker for the
afferent noradrenergic terminals, was actually increased by kainic
acid. The fall in the steady state level of NA is compatible with
a massive increase in the firing rate of the noradrenergic neurons
subsequent to the loss of inhibitory control. Therefore, more
likely than an unspecific destruction of these terminals by kainic
acid is that the neurotransmitters fall secondarily to the degeneration of inhibitory neurons more vulnerable to the neurotoxin.
A positive result of this experiment is that the data do support
the possibility that a local release of NA is involved in the
action of OTMN.

The large increase in Ch content (Table 4) found in the hippocampus of the kainic acid treated animals is possibly related to
the extensive proliferation of glial cells stimulated by the severe
destruction of the nerve cell bodies.

The marked fall in 5-HT after kainic acid injection may occur
independently of the decrease in NA, as shown in Fig. 1A, or as
indicated from the schemes in Fig. 1B & 1C, there may be reciprocal
modulatory influences between the noradrenergic and serotonergic
neurons (see legend to Fig. 1). To clarify this aspect we investigated the role that the massive serotonergic projection to the
hippocampus played in the complexity of converging neurons in the
hippocampus. An inhibitory influence of the serotonergic system
on cholinergic neurons of the striatum and hippocampus has been
previously reported (4,21) and was further sustained in the following experiments (Table 5). Three selective releasers of 5-HT,
the dextro- and levostereoisomers of fenfluramine and quipazine
increased the content of hippocampal ACh by about 25%. Electrolytic lesion of the median raphe nucleus reduced both hippocampal

TABLE 4: Effect of the neuroexcitant, kainic acid, on the increase in hippocampal ACh induced by oxotremorine and on other parameters.

Treatment	ACh (nmol/g)	Ch (nmol/g)	Hippocampus NA (ng/g)	5-HT (ng/g)	Tissue Weight (mg)
Saline	18.8 ± 0.5(14)	49.6 ± 2.4(6)	340 ± 48(3)	280 ± 40(3)	142.0 ± 4.1(8)
Kainic acid	16.5 ± 0.7(14)	76.0 ± 5.2(6)*	137 ± 23(7)*	ND (7)	107.3 ± 3.1(7)*
OTMN	26.6 ± 0.7(14)*	49.0 ± 3.1(9)			
Kainic acid + OTMN	18.8 ± 0.9(14)	88.7 ± 2.6(8)*			
Interaction	F 1,52=13.4 p<0.01	NS			

* = p < 0.01 vs saline group.
Kainic acid, 1 µg/hippocampus, 2 weeks
OTMN, 1.4 µmol/kg i.p., 20 min

ND = not detectable
NS = not significant

TABLE 5: Effect of serotonin depletion by lesion of the median raphe nucleus on the increase in ACh content induced by three serotonin releasers in the hippocampus.

Drug in Columns C & D	Hippocampal ACh (nmol/g)			
	A Saline	B MRN Lesion	C Drug	D MRN Lesion + Drug
D-Fenfluramine	17.9 ± 0.5	19.1 = 1.0	22.8 ± 0.5*	20.6 ± 0.8
L-Fenfluramine	17.7 ± 0.6	17.7 = 0.6	23.4 ± 1.3*	18.3 ± 0.3
Quipazine	17.6 ± 0.8	17.3 ± 0.4	20.3 ± 0.6*	19.6 ± 0.4

* = $p < 0.01$ vs saline group.
Rats were killed 30 min after D-fenfluramine (7.5 mg/kg i.p.) and L-fenfluramine (15 mg/kg i.p.) and 60 min after quipazine (10 mg/kg i.p.). The 5-HT (380 ± 30 ng/g) and 5-HIAA (280 ± 20 ng/g) contents of the hippocampus were reduced by more than 90% by the MRN lesion while the NA content was unaffected. Significance of interactions with MRN lesion: D-fenfluramine F 1 36 = 5.0 $p < 0.05$; L-fenfluramine F 1,20 = 11.3 $p < 0.01$; quipazine F 1,52 = 7.3 $p < 0.01$.

TABLE 6: Effect of interference with serotonergic transmission on the increase in hippocampal ACh by OTMN.

Treatment in Columns C & D	Hippocampal ACh (nmol/g)			
	A Saline	B OTMN	C Treatment	D Treatment + OTMN
Methergoline	$17.5 \pm 0.9(7)$	$24.4 \pm 1.2(8)$*	$17.2 \pm 0.7(6)$	$24.3 \pm 0.6(6)$*
D-Fenfluramine	$17.5 \pm 0.5(15)$	$24.9 \pm 0.7(14)$*,**	$22.7 \pm 0.6(14)$*	$29.5 \pm 0.7(13)$*

Methergoline maleate (6 mg/kg i.p., 3 hr); D-fenfluranime (supramaximal dose of 7.5 mg/kg i.p., 30 min; OTMN (1.4 µmol/kg i.p., 20 min).

* = $p < 0.01$ vs saline group. Interactions with oxotremorine; methergoline, NS; D-fenfluramine, NS.
** = OTMN, supramaximal dose of 2.8 µmol/kg i.p., 20 min.   NS = not significant.

5-HT and 5-HIAA contents by more than 90% after two weeks and blocked the action of all three releasers. Inhibition of 5-HT synthesis by parachlorophenylalanine produced similarly marked reductions in hippocampal 5-HT and 5-HIAA contents and also prevented D-fenfluramine's increase in ACh content (data not shown).

These data reveal the existence of two (serotonergic and noradrenergic) inhibitory inputs on the cholinergic neurons of the hippocampus and, as shown in the following experiments, these inputs appear to operate independently of each other (Table 6). When both D-fenfluramine and OTMN were administered at supramaximal doses (i.e. doses equivalent to about twice the amount normally required to give a peak increase in hippocampal ACh), there was an evident summation of the response to the drugs. In addition, blockade of 5-HT receptors with methergoline did not suppress the effect of OTMN. These results thus present evidence that OTMN does not utilize the serotonergic system in exerting its actions on hippocampal cholinergic neurons and is in accord with the overall results from several laboratories indicating that OTMN affects the serotonergic system minimally (15) or not at all (2).

CONCLUSIONS

The results demonstrate that following the activation of heterologous muscarinic receptors, OTMN increases the content and decreases the turnover of ACh in the hippocampus. The effect has been shown to involve activation of noradrenergic neurons resulting in the local release of NA in the hippocampus which then acts on "alpha" receptors. An axo-axonic synapse between noradrenergic and cholinergic terminals is feasible. Serotonergic neurons also regulate the cholinergic neurons of the septal-hippocampal pathway and these appear to operate independently of the noradrenergic system although this aspect requires further investigation because of the apparent interdependence of the noradrenergic and serotonergic systems in brain (5,11).

ACKNOWLEDGEMENTS

The authors appreciate the contributions of Drs. V. Greenberger (Weizmann Institute of Science, Rehovot, Israel) and F. Ponzio to these experiments. We also thank G. Achilli and C. Perego for measuring tissue noradrenaline and serotonin. Part of this work was performed at the Weizmann Institute (HL and SC) and part at the Mario Negri Institute (MS) under the auspices of an

exchange program between the two institutes, supported by the "Roth
Fund" of the Mario Negri Institute Foundation, New York, USA.
Part of this work was supported by the Italian National Research
Council, Contract No. 78-02143-04.

REFERENCES

1. Atweh, S., Simon, J.R. and Kuhar, M.J. (1975): Life Sci.
       17:1535-1544.
2. Corrodi, H., Fuxe, K., Hammer, W., Sjoqvist, F. and Ungerstedt,
       U. (1967): Life Sci. 6:2557-2566.
3. Cox, B. and Potkonjak, D. (1970): Brit. J. Pharmacol. 38:171-180.
4. Euvrard, C., Javoy, F., Herbert, A. and Glowinski, J. (1977):
       Eur. J. Pharmacol. 41:281-289.
5. Frankhuyzen, A.L. and Mulder, A.H. (1980): Eur. J. Pharmacol.
       (1980): Eur. J. Pharmacol. 63:179-182.
6. Glisson, S.N., Karczmar, A.G. and Barnes, L. (1972): Neuro-
       pharmacology 11:465-477.
7. Hadhazy, P. and Szerb, J.C. (1977): Brain Res. 123:311-322.
8. Jaim-Etcheverry, G. and Zieher, L.M. (1980): Brain Res. 188:
       513-523.
9. Jonsson, G., Ponzio, F. and Ross, S. (1978): IN Abstracts
       Seventh International Congress of Pharmacology (IUPHAR)
       (eds) J.R. Boissier and P. Lechat, Pergamon Press, Oxford,
       p. 926.
10. Keller, R., Oke, A., Mefford, I. and Adams, R.N. (1976): Life
       Sci. 19:995-1004.
11. Kostowski, W., Samanin, R., Bareggi, S.R., Marc, V. Garattini,
       S. and Valzelli, L. (1974): Brain Res. 82:178-182.
12. Ladinsky, H., Consolo, S., Bianchi, S. and Jori, A. (1976):
       Brain Res. 108:351-361.
13. Ladinsky, H., Consolo, S., Tirelli, A.S. and Forloni, G.L.
       (1980): Brain Res. 187:494-498.
14. Lindvall, O. and Bjorklund, A. (1974): Acta Physiol. Scand.
       Suppl. 412:1-48.
15. Paalzow, G. and Paalzow, L. (1975): Eur. J. Pharmacol. 31:
       261-272.
16. Ponzio, F. and Jonsson, G. (1979): J. Neurochem. 32:129-132.
17. Racagni, G., Cheney, D.L., Trabucchi, M., Wang, C. and Costa,
       E. (1974): Life Sci. 15:1961-1975.
18. Rommelspacher, H. and Kuhar, M.J. (1975): N.-S. Arch. Pharma-
       col. 291:17-21.
19. Ross, S.B. (1976): Brit. J. Pharmacol. 58:521-527.
20. Saelens, J.K., Allen, M.P. and Simke, J.P. (1970): Arch. Int.
       Pharmacodyn. 186:279-286.

21. Samanin, R., Quattrone, A., Peri, G., Ladinsky, H. and Consolo,
     S. (1978): Brain Res. 151:73-82.
22. Schwarcz, R., Zaczek, R. and Coyle, J.T. (1978): Eur. J.
     Pharmacol. 50:209-220.
23. Szerb, J.C., Hadhazy, P. and Dudar, J.D. (1977): Brain Res.
     128:285-291.
24. Tirelli, A., Forloni, G., Consolo, S. and Ladinsky, H. (1978):
     IN Abstracts of the XIX Congress Italian Pharmacological
     Society Meeting, Abst. No.92, p. 415.
25. Westfall, T.C. (1974): Life Sci. 14:1641-1652.
26. Yamamura, H.I. and Snyder, S.H. (1974): Brain Res. 78:320-326.

ACETYLCHOLINE INDUCED ANTINOCICEPTION: COMPARISONS TO

OPIATE ANALGESIA

N.W. Pedigo and W.L. Dewey*

Department of Pharmacology, Louisiana State University
Medical School, New Orleans, Louisiana 70119 USA, and
*Department of Pharmacology, Medical College of Virginia,
Virginia Commonwealth University, Richmond, Virginia 23298

INTRODUCTION

The ability of a number of cholinomimetic drugs to inhibit
nociceptive responses in several animal species is well documented.
Cholinomimetic agents which have shown antinociceptive activity in a
variety of tests include the following: oxotremorine (16,19,26,32,
37,38,42), tremorine (7), arecoline (23,25,29,37,44), pilocarpine
(10,25,29) and physostigmine (10,20,22,28,34,42,44,49). The specific central muscarinic mediation of these antinociceptive effects
is agreed upon by most authors (10,25,28,37). Cholinomimetic drugs
also potentiate narcotic analgesia in laboratory animals (1,26,28,
29,42,46,56) and in man (8,15,40,50,54). However, the relationship
of cholinomimetic to narcotic analgesia remains unclear. Both
classes of drugs demonstrate antinociceptive activity in similar
tests and tolerance to each has been noted (see 48,55; also 10,26,
34). Significant cross-tolerance between narcotics and cholinomimetics has not been observed. Investigations into the possible
inhibition of cholinergic antinociception by narcotic antagonists
(7,18,20,26-28,42) and, conversely, the inhibition of narcotic
analgesia by muscarinic receptor blockers (3,7,8,10,27-29,56) have
yielded mixed results.

These various reports prompted a more thorough investigation
of cholinomimetic induced antinociception, i.e. the nature of the
antinociceptive activity of intraventricularly (i.vt.) administered
acetylcholine (ACh) in mice. Previous studies in our laboratories

(11,12,14,39,59) demonstrated ACh induced antinociception in the
tail-flick and writhing tests. These antinociceptive effects were
shown to be mediated by central muscarinic systems (11,39). Similarities between ACh and morphine induced antinociception were also
noted, especially their inhibition by various narcotic antagonists
(39). However, the stereoselectivity of three narcotic antagonists
of the benzomorphan class was opposite in blocking the antinociceptive activity of i.vt. ACh vs morphine (11,39). The effects of
divalent cations and chelating agents on ACh induced antinociception were almost identical to their effects on morphine analgesia
(59).

The results described in this chapter represent a continuation
of these studies and further differentiate ACh and morphine induced
antinociception. The antinociceptive effects of i.vt. ACh are shown
to be markedly reduced by pretreating mice with reserpine or tetrabenazine, as has been reported for morphine analgesia (31,52,58).
In contrast to morphine analgesia (see 35), specific catecholamine
and serotonin depletors have relatively minor effects on ACh induced antinociception. The studies described here also demonstrate
tolerance development to the antinociceptive effects of i.vt. ACh,
although minimal cross-tolerance to morphine is observed. Finally,
pA2 experiments using various doses of naloxone indicate that the
inhibition of ACh induced antinociception by narcotic antagonists
is probably a non-competitive event.

METHODS

The animals used throughout these studies were male Swiss-
Webster mice, weighing from 15-30 g. All drugs administered i.vt.
were dissolved in sterile, non-pyrogenic saline at a constant
volume of 0.005 ml/mouse. All other drugs were administered either
intraperitoneally (i.p.) or subcutaneously (s.c.) in saline, distilled water or dilute lactic acid at a volume of 0.1 ml/10 g body
weight.

Intraventricular Injection

The technique used for the i.vt. administration of drugs has
been previously described (39). Mice were lightly anesthetized
with ether and a small incision made in the scalp exposing the
reference point Bregma. The i.vt. injection was made through the
skull 2 mm lateral to Bregma using a 0.05 ml Hamilton syringe attached to a modified 27 gauge hypodermic needle. The injection

needle was modified so as to penetrate only 3.0 mm from the top of
the skull. The mouse was then placed in a well ventilated plastic
cage and allowed to recover from the anesthesia. Sham operated
animals and animals given i.vt. saline showed only minor effects
from this procedure. Some hyperventilation, possibly in response
to the ether anesthesia, was evident up to 5 min after the i.vt.
injection. The mice remained relatively inactive for a short period,
but were responsive to external stimuli and recovered normal activity
within 10 min. Control animals tested in the tail-flick assay prior
to and 10 min after the i.vt. injection of saline showed no antinociceptive effects.

Mouse Tail-Flick Procedure

The procedure used was similar to that of D'Amour and Smith
(9), as modified by Pedigo et al. (39). Only mice with a control
latency of 2-4 sec were used and a 10 sec cutoff time was employed.
Groups of at least 6 mice were tested and a control reaction time
established prior to any medication. The animals were retested on
the tail-flick apparatus 10 min after the i.vt. injection of ACh.
This 10 min time period ensured the animal's full recovery from
the anesthesia prior to complete degradation of ACh by brain cholinesterase. The antinociceptive effect was calculated as the percentage of the maximum possible response time using the following
formula:

$$\frac{(\text{Test-Control})}{(10 \ -\text{Control})} \ \times 100 = \text{percent maximum possible effect}$$
$$(\text{percent MPE})$$

Dose-response curves were established with at least 3 doses per
curve by plotting the percent maximum possible effect (percent MPE)
vs log-dose on probit paper. The ED_{50}, slope and potency ratios
with their 95% confidence limits (CL) were determined by the method
of Litchfield and Wilcoxin (33).

Drug Interaction Studies

A variety of drugs which have been reported to alter central
neurotransmitter systems were tested for their effects on ACh induced antinociception. At least four logarithmically spaced doses
of ACh were administered i.vt. to mice following pretreatment with
a neurochemical modulator. Pretreatment regimens were designed to
correspond with the reported peak pharmacological effect of these
drugs on specific brain systems. The following drugs were administered i.p. to mice prior to i.vt. ACh: reserpine (5 mg/kg, 16-

20 hr), α-methyl-p-tyrosine (250 mg/kg, 1 hr), diethyldithiocarbamate (500 mg/kg, 1 hr), tetrabenazine (50 mg/kg, 1 hr) and p-chlorophenylalanine and its methyl ester (300 mg/kg, daily, 3 days).

Tolerance Experiments

The time course of tolerance development to the antinociceptive
effects of i.vt. ACh was first determined in mice which were pretreated with i.vt. injections of saline or an ED_{50} dose of ACh
(20 µg) administered at hourly intervals for 1, 2, 4 or 8 hr. One
hour after the last injection, the antinociceptive activity of ACh
was determined in the tail-flick assay. Tolerance developwed following pretreatment with i.vt. ACh, but not after i.vt. saline, and
seemed to plateau after a 4 hr treatment regimen (see Results). This
study was followed by a more thorough investigation of the extent of
tolerance to the antinociceptive activity of i.vt. ACh and possible
cross-tolerance to peripherally administered morphine. Mice were
rendered tolerant to ACh as indicated above, using a 4 hr pretreatment regimen. A dose-response curve in tolerant mice was established by administering 5 logarithmically spaced doses of ACh i.vt.
10 min prior to the tail-flick test. Cross-tolerance to morphine
(4 or 10 mg/kg, s.c., 30 min prior) was also assessed in mice pretreated with i.vt. ACh or saline.

pA2 Determinations

A study to determine the apparent pA2 values for naloxine
antagonism of ACh and morphine induced antinociception was done in
the manner suggested by Schild (47) and Smits and Takemori (51).
Naloxone (0.3, 1.0 and 3.0 mg/kg, s.c.) was administered 10 min
before the i.vt. injection of ACh and 20 min prior to testing tail-
flick latency. Each dose-response curve was plotted using 3 doses
of i.vt. ACh with an n of 12 mice per dose. Similar curves were
established for morphine induced antinociception using 4 doses of
morphine per curve. Mice were tested in the tail-flick assay 30
min after the s.c. administration of morphine. Naloxone (0.064,
0.16 or 0.64 mg/kg, s.c.) was injected at a different site 10 min
prior to the morphine.

RESULTS

Neurochemical Specificity

The ED_{50} for i.vt. ACh in the tail-flick test, measured 10 min after injection,was about 20 µg. There was no increase in tail-flick latency in mice injected with an equal volume (5 µl) of saline. Antinociceptive activity seemed to peak at 10 min and decreased sharply at 20 and 40 min after injection (data not shown). The central muscarinic nature of this antinociceptive response to i.vt. ACh has been shown previously (39). Pretreating mice with reserpine drmatically inhibited ACh induced antinociception, increasing the ED_{50} for i.vt. ACh in the tail-flick test over 40-fold (Table 1). The pretreatment regimen used has been shown to extensively deplete brain dopamine, norepinephrine and serotonin (24,41). Mice previously administered reserpine showed marked debilitation, ataxia, diarrhea and hypoactivity, but their tail-flick latencies were within control range (2-4 sec). Tetrabenazine, a drug which acts similarly to reserpine but with a shorter time course (43), also reduced the antinociceptive activity of ACh (Table 1). However, these effects were not reporduced by other, more specific catecholamine or serotonin depletors (Table 1). Neither α-methyl-p-tyrosine, an inhibitor of tyrosine hydroxylase (53), nor diethyldithiocarbamate, a dopa-decarboxylase inhibitor (5), significantly shifted the dose-response curve for i.vt. ACh. Pretreatment with p-chlorophenylalanine, which inhibits the enzyme tryptophan hydroxylase (36),also failed to alter the tail-flick activity of i.vt. ACh, although pretreatment with its more soluble methyl ester did increase the ED_{50} about 2.5 times.

Tolerance Experiments

As noted in Fig. 1, there was a significant difference in the tail-flick activity of i.vt. ACh following 2, 4 and 8 hr pretreatment with ACh, but not after chronic saline administration. In further studies using a 4 hr pretreatment regimen (20 µg/hr, i.vt.) the antinociceptive activity was determined for i.vt. ACh revealing nearly a 20-fold shift to the right in the dose-response curve (from 18.8 µg [15.3-23.2] to 357.9 µg [127-1013]). Cross-tolerance to morphine was also evaluated. The analgesic effect of morphine was partially reduced after chronic ACh administration, but the degree of cross-tolerance was not marked.

TABLE 1: Effects of biogenic amine depletors on ACh induced antinociception.

DRUG	PRETREATMENT (mg/kg)		ED_{50} with 95% CL (μg)	
ACh (i.vt.)	---		19.8	(11.8-30.8)
Reserpine	5	(16-20 hr)	839	(723-924)*
Tetrabenazine	50	(1 hr)	412	(144-1182)*
α-methyl-p-tyrosine	250	(1 hr)	12.8	(2.7-61.0)
Diethyldithiocarbamate	500	(1 hr)	15.2	(1.0-26.4)
p-Chlorophenylalanine	300	(x 3 days)	18.0	(6.6-49.0)
p-Chlorophenylalaline methyl ester	300	(x 3 days)	46.9	(22.1-99.4)*

*   Significantly different from ACh (i.vt.) controls at $p < 0.05$ by the method of Litchfield and Wilcoxin (33).

TABLE 2A: Naloxone inhibition of the tail flick activity of morphine.

PRETREATMENT (mg/kg)	ED_{50} for Morphine (mg/kg, s.c.)		SLOPE	
None	4.0	(1.8-8.5)	2.0	(0.9-4.1)
Naloxone (0.064)	9.0	(5.2-15.6)	2.0	(1.2-3.4)
Naloxone (0.16)	18.2	(9.4-35.2)*	2.8	(1.1-7.2)
Naloxone (0.64)	62.0	(38.1-100.9)*	2.1	(1.3-4.7)

TABLE 2B: Naloxone inhibition of the tail flick activity of ACh (i.vt.)

PRETREATMENT (mg/kg)	ED_{50} for ACh (i.vt.) (μg)		SLOPE	
None	21.4	(14.4-31.9)	2.4	(1.2-4.6)
Naloxone (0.3)	43.0	(28.7-64.4)*	2.4	(1.2-4.7)
Naloxone (1.0)	72.0	(47.7-108.9)*	2.1	(1.2-3.7)
Naloxone (3.0)	125.9	(99.4-159.6)*	1.7	(1.1-2.7)

*   Significantly different from untreated controls at $p < 0.05$ by the method of Litchfield and Wilcoxin (33).
Naloxone was administered 10 min (s.c.) prior to morphine or ACh (i.vt.). Morphine was tested 30 min and ACh 10 min after injection. The numbers in parentheses are 95% CL.

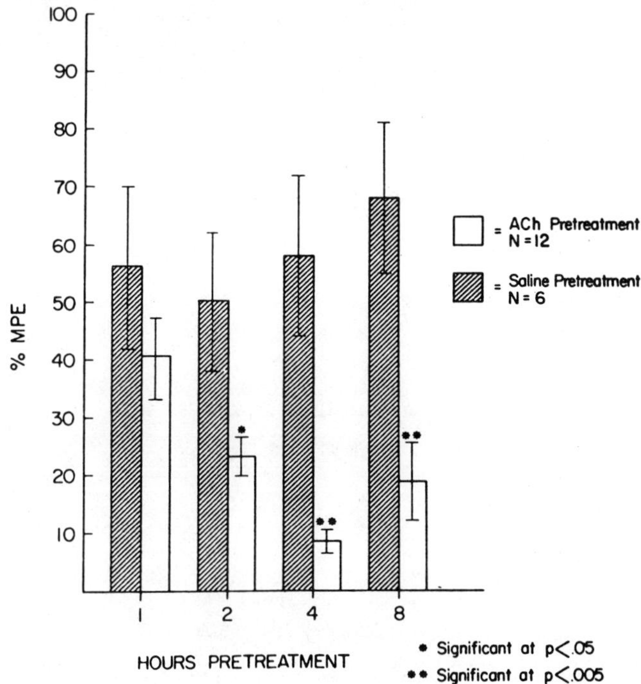

FIGURE 1: Tolerance development to ACh (20 μg i.vt.) after chronic
i.vt. saline or ACh.

pA2 Experiments

The data presented in Table 2 show ED_{50}s and slopes for the
antinociceptive activity of morphine and i.vt. ACh, alone and following pretreatment with various doses of the narcotic antagonist
naloxone. From these data a modified Schild plot was constructed
(see 51) for naloxone inhibition of ACh and morphine induced antinociception (Fig. 2A and B). The apparent pA2 for morphine was
quite similar to previously reported values (57,60) and the slope
was close to unity. In contrast, the apparent pA2 for i.vt. ACh
was much lower and the slope of the line was significantly greater
than 1.0.. Antinociceptive activity was measured in the tail-flick
assay. The slope of the line and its intersection with the abscissa
(apparent pA2) were calculated by linear regression (r > 0.99).
Data is from Table 2.

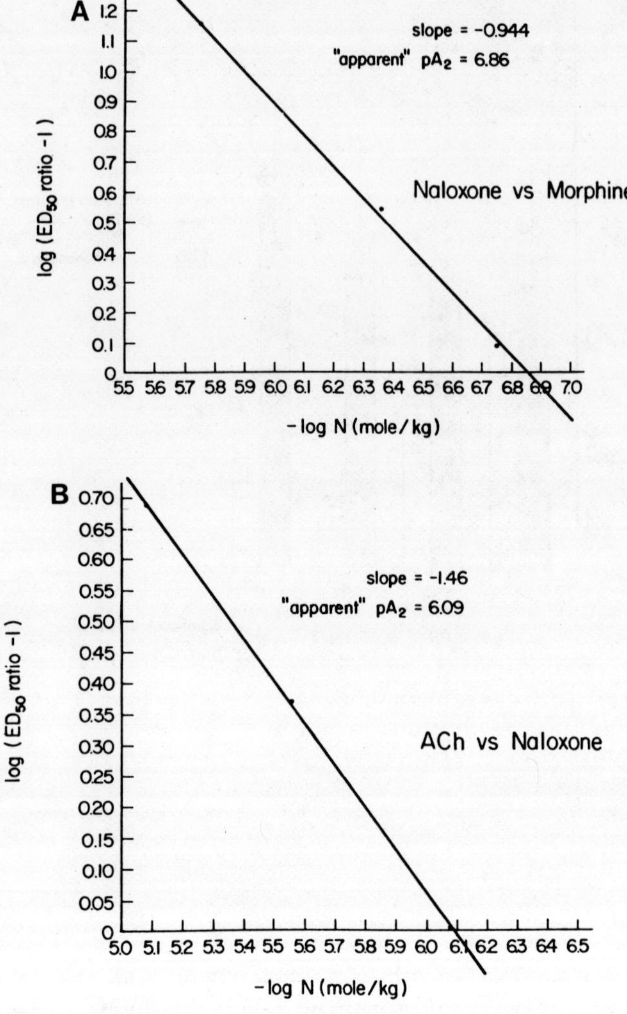

FIGURE 2: Modified Schild plots for naloxone inhibition of morphine
and ACh induced antinociception.

DISCUSSION

The results described in this chapter and previous reports
from our laboratory (11,12,14,39,59) demonstrate the antinociceptive activity of i.vt. ACh and define its relationship to morphine
induced antinociception. Three general conclusions may be derived
from these studies. First, the antinociceptive activity of i.vt.
ACh is a specific central muscarinic effect which does not require
intact catecholamine or serotonin systems. Second, the antinocicep-

tive mechanisms of both ACh and morphine may involve alterations in
the intracellular levels of calcium. Third, although there are
many similarities between morphine and ACh induced antinociception,
the latter is not a specific opiate effect.

   The antinociceptive activity of i.vt. ACh has been shown to be
blocked by atropine, but not by atropine methyl nitrate or mecamyl-
amine (39) and to be mimicked by β-methylcholine (11). The S and
R isomers of β-methylcholine exhibit a stereoselectivity in the
phenylquinone writhing test which is identical to their preferred
conformation at the muscarinic cholinergic receptor (11,12). In
contrast to previous reports (3,17), similar antinociceptive effects were not observed following the i.vt. administration of dopamine, norepinephrine, serotonin or epinephrine (unpublished observations). Furthermore, depletion of central catecholamines or
serotonin by specific drugs which inhibit their synthesis, and have
been shown to affect opiate analgesia (see 35), did not alter ACh
induced antinociception. The marked inhibition of the antinociceptive activity of i.vt. ACh by reserpine and tetrabenazine may be
more related to their effects on brain calcium (see below) than to
a depletion of biogenic amines. These data indicate that the antinociceptive activity of i.vt. ACh is a specific effect, which is
most probably mediated by central muscarinic systems and, unlike
opiate analgesia, does not involve brain catecholamines or serotonin.

   Evidence relating the effects of narcotic drugs to changes in
calcium levels in neuronal tissues has been recently reviewed (45).
Kakunga et al. (30) first reported that intracisternal injections
of calcium antagonized narcotic analgesia, and that this action of
calcium could be blocked by chelating agents. These results were
confirmed by Harris et al. (21) who noted that the effects of calcium were reversed by the chelator EGTA but not by EDTA. Nearly
identical observations were made in our laboratories where i.vt.
calcium blocked the antinociceptive activity of i.vt. ACh, an effect which was reduced by EGTA but not by EDTA (59). Additionally,
EGTA and lanthanum each potentiated the effects of i.vt. ACh (59).
These data are of interest when considering the inhibition of morphine and ACh induced antinociception by reserpine. Reserpine has
been shown to displace a specific pool of calcium in brain tissues
(4) and it has been proposed that this mobilization of calcium is
the mechanism by which reserpine blocks morphine analgesia (6). It
is tempting to speculate that the marked inhibition of ACh induced
antinociception by reserpine and tetrabenazine could also be due
to alterations in brain calcium. Changes in intracellular calcium
levels may therefore play an important role in the antinociceptive
effects of both i.vt. ACh and narcotic drugs.

   In addition to the influence of divalent cations, i.vt. ACh
and morphine are similar in that they each show antinociceptive

activity in the tail-flick assay and writhing tests (11,14,39) and
tolerance develops to the antinociceptive effects of each drug.
Perhaps most striking, the antinociceptive activity of both morphine and i.vt. ACh is blocked by several narcotic antagonists with
an identical rank order of potency (39). However, two critical observations do not support a specific narcotic mechanism for i.vt.
ACh, i.e. the different pA2 values for the antagonist naloxone and
the reversed stereoselectivity of narcotic antagonists. In the pA2
studies, naloxone blockade of morphine analgesia gave a slope of
unity for the modified Schild plot and an apparent pA2 value very
similar to previously published data (57,60). However, the modified
Schild plot for naloxone inhibition of the antinociceptive effects
of i.vt. ACh showed a slope greater than unity and a markedly lower
pA2 value. These data suggest that naloxone was acting in a non-
competitive manner vs i.vt. ACh and probably at a different site
than that at which it inhibits morphine analgesia.

The reversed stereoselectivity of narcotic antagonists in
blocking morphine vs ACh induced antinociception was first observed
with benzomorphan antagonists (11,39) and has recently been confirmed with the (+) and (−) isomers of naloxone (unpublished observations). These data differ from the results of Ramabadran and
Jacob (44), who showed a similar stereoselectivity for inhibition
of morphine analgesia and antinociception produced by peripherally
administered cholinomimetics. This discrepancy may be due to the
different routes of administration used rather than to differences
in the antinociceptive tests. It should be noted that an opposite
stereoselectivity (i.e. d-isomers more potent tha l-isomers) was
also shown for the antagonism of ACh induced decreases in specific
^{3}H-dihydromorphine binding (11). In these studies, opiate receptors
in mouse brain were measured after the mice were pretreated with
i.vt. ACh. Preliminary results from our laboratory also suggest
an opposite stereoselectivity of narcotic antagonists vs stress
induced analgesia (A. Snow, personal communication). The reversed
stereoselectivity of narcotic antagonists and the non-competitive
inhibition by naloxone of ACh induced antinociception provide strong
evidence for the hypothesis that i.vt. ACh does not act through
central opiate mechanisms.

SUMMARY

Intraventricularly administered ACh produces an antinociceptive
effect in mice which is mediated by central muscarinic systems. Other neurotransmitter systems do not appear to be involved, although
reserpine and tetrabenazine markedly inhibit this antinociceptive
effect. Changes in the levels of intracellular calcium may play a
role in the antinociceptive mechanisms of i.vt. ACh. Tolerance

develops to ACh induced antinociception, but minimal cross-tolerance
to morphine is seen. Although narcotic antagonists inhibit the
antinociceptive activity of i.vt. ACh, such inhibition is a non-
competitive event and exhibits a reversed stereoselectivity (i.e.
opposite to morphine analgesia). It is therefore probable that ACh
and morphine produce their antinociceptive effects through different, if not entirely independent, mechanisms.

ACKNOWLEDGEMENTS

Thanks are due to R. Hobson for his valuable technical assistance and to C. Thomas and N. Pedigo for their help in the preparation of this manuscript. This work was supported by a USPHS
NIH Fellowship (DA-00326).

REFERENCES

1.   Bhargava, H.N. and Way, E.L. (1972): J. Pharmacol. Exp. Ther.
     183:31-40.
2.   Calcutt, C.R., Handley, S.L., Sparkes, C.G. and Spencer, P.S.
     J. (1972): IN Agonist and Antagonist Actions of Narcotic
     Analgesic Drugs. (eds) H.W. Kosterlitz, H.O.J. Collier and
     J.E. Villarreal, MacMillan Press, London, pp. 176-191.
3.   Calcutt, C.R. and Spencer, P.S.J. (1972): Brit. J. Pharmacol.
     41:401P-402P.
4.   Cardenas, H.L. and Ross, D.H. (1975): J. Neurochem. 24:487-493.
5.   Carlsson, A., Lindqvist, M., Fuxe, K. and Hokfelt, T. (1966):
     J. Pharm. Pharmacol. 18:60-62.
6.   Cerreta, K.V., Loh, H.H. and Way, E.L. (1977): Fed. Proc.36:943.
7.   Chen, G. (1958): J. Pharmacol. Exp. Ther. 124:73-76.
8.   Christensen, E.M. and Gross, E.G. (1948): J. Amer. Med. Assoc.
     137:594-598.
9.   D'Amour, F.E. and Smith, D.L. (1941): J. Pharmacol. Exp. Ther.
     72:74-79.
10. Dayton, H.E. and Garrett, R.L. (1973): Proc. Soc. Exp. Biol.
     Med. 142:1011-1013.
11. Dewey, W.L., Chau-Pham, T.T. and Cocolas, G.C. (1976): IN
     Opiates and Endogenous Opiate Peptides (ed). H.W. Kosterlitz,
     Elseview/North-Holland, Amsterdam, pp. 185-190.
12. Dewey, W.L., Cocolas, G., Daves, E. and Harris, L.S. (1975):
     Life Sci. 17:9-10.
13. Dewey, W.L., Harris, L.S., Howes, J.F. and Nuite, J.A. (1970):
     J. Pharmacol. Exp. Ther. 175:435-442.

14. Dewey, W.L., Pedigo, N.W., Karbowski, M., Forbes, J. and
    Harris, L.S. (1976): IN Tissue Responses to Addictive Drugs,
    (eds) D.H. Ford and D.H. Clouet, Spectrum, New York, pp. 255-
    271.
15. Floodmark, S. and Wramner, T. (1945): Acta Physiol. Scand.
    9:88-96.
16. George, R., Haslett, W.L. and Jenden, D.J. (1962): Life Sci.
    1:361-363.
17. Handley, S.L. and Spencer, P.S.J. (1969): Brit. J. Pharmacol.
    35:361P-362P.
18. Harris, L.S. and Dewey, W.L. (1972): IN Agonist and Antagonist
    Actions of Narcotic Analgesic Drugs (eds) H.W. Kosterlitz,
    H.O.J. Collier and J.E. Villarreal, MacMillan Press, London,
    pp. 198-206.
19. Harris, L.S., Dewey, W.L. and Howes, J.F. (1968): Fed. Proc.
    27:753.
20. Harris, L.S., Dewey, W.L., Howes, J.F., Kennedy, J.S. and
    Pars, H. (1969): J. Pharmacol. Exp. Ther. 169:17-22.
21. Harris, R.A., Loh, H.H. and Way, E.L. (1975): J. Pharmacol.
    Exp. Ther. 195:488-498.
22. Hendershot, L.C. and Forsaith, J. (1959): J. Pharmacol. Exp.
    Ther. 125:237-240.
23. Herz, A. (1962): Arch. Exp. Pathol. Pharmakol. 242:414-429.
24. Holzbauer, M. and Vogt, M. (1956): J. Neurochem. 1:8-11.
25. Houser, V.P. (1976): Pharmacol. Biochem. Behav. 4:561-568.
26. Howes, J.F., Harris, L.S., Dewey, W.L. and Voyda, C.A. (1969):
    J. Pharmacol. Exp. Ther. 169:23-28.
27. Ireson, J.D. (1969): Brit. J. Pharmacol. 37:504P-505P.
28. Ireson, J.D. (1970): Brit. J. Pharmacol. 40:92-101.
29. Kaakkola, S. and Ahtee, L. (1977): Psychopharmacology 52:7-15.
30. Kakunga, T., Kaneto, H. and Hano, K. (1966): J. Pharmacol.
    Exp. Ther. 153:134-141.
31. Lee, R.L., Sewell, R.D.E. and Spencer, P.S.J. (1979): Neuro-
    pharmacology 18:711-717.
32. Leslie, G.B. (1969): J. Pharm. Pharmacol. 21:248-250.
33. Litchfield, S.T. and Wilcoxin, F. (1949): J. Pharmacol. Exp.
    Ther. 96:99-113.
34. Little, H.J. and Rees, J.M.H. (1974): Experientia 30:930-932.
35. Mayer, D.J. and Price, D.D. (1976): Pain 2:379-404.
36. McBride, W.J., Penn, P.E., Hyde, T.P., Lane, J.D. and Smith,
    J.E. (1976): Neurochem. Res. 1:437-449.
37. Metys, J., Wagner, N., Metysova, J. and Herz, A. (1969): Int.
    J. Neuropharmacol. 8:413-425.
38. Paalzow, G. and Paalzow, L. (1975): Eur. J. Pharmacol. 31:
    261-272.
39. Pedigo, N.W., Dewey, W.L. and Harris, L.S. (1975): J. Pharma-
    col. Exp. Ther. 193:845-852.
40. Pellanda, C.L. (1933): Lyon Med. 151:653.

41. Pletscher, A., Shore, P.A. and Brody, B.B. (1956): J. Pharmacol. Exp. Ther. 116:84-89.

42. Pleuvry, B.J. and Tobias, M.A. (1971): Brit. J. Pharmacol. 43:706-714.

43. Quinn, G.P., Shore, P.A. and Brodie, B.B. (1959): J. Pharmacol. Exp. Ther. 127:103-109.

44. Ramabadran, K. and Jacob, J.J.C. (1978): Arch. Int. Pharmacodyn. Ther. 236:27-42.

45. Ross, D.H., Lynn, S.C. and Cardenas, H.L. (1977): IN Alcohol and Opiates, Neurochemical and Behavioral Mechanisms, (ed) K. Blum, Academic Press, New York, pp. 265-281.

46. Saxena, P.N. and Gupta, G.P. (1957): Indian J. Med. Res. 45: 319-325.

47. Schild, H.O. (1957): Pharmacol. Rev. 9:242-246.

48. Shuster, L. (1971): IN Narcotic Drugs: Biochemical Pharmacology (ed) D.H. Clouet, Plenum Press, New York, pp. 408-423.

49. Sitaram, N., Buchsbaum, M.S. and Gillin, J.C. (1977): Eur. J. Pharmacol. 42:285-290.

50. Slaughter, D. and Munsell, D. (1940): J. Pharmacol. Exp. Ther. 68:104-112.

51. Smits, S.E. and Takemori, A.E. (1970): Brit. J. Pharmacol. 39:627-638.

52. Sparkes, C.G. and Spencer, P.S.J. (1971): Brit. J. Pharmacol. 42:230-241.

53. Spector, S., Sjoerdsma, A. and Udenfriend, S. (1965): J. Pharmacol. Exp. Ther. 147:86-95.

54. Szerb, J.C. (1957): Arch. Int. Pharmacodyn. Ther. 111:314-321.

55. Takemori, A.E., (1975): Biochem. Pharmacol. 24:2121-2126.

56. Takemori, A.E., Tulunay, F.C. and Yano, I. (1975): IN The Opiate Narcotics, Intl. Narcotic Res. Club Conf., Pergamon Press, New York, pp. 175-181.

57. Tulunay, F.C. and Takemori, A.E. (1974): J. Pharmacol. Exp. Ther. 190:401-407.

58. Way, E.L. and Shen, F.H. (1971): IN Narcotic Drugs: Biochemical Pharmacology (ed) D.H. Clouet, Plenum Press, New York, pp. 229-253.

59. Widman, M., Rosin, D. and Dewey, W.L.(1978): J. Pharmacol. Exp. Ther. 205:311-318.

60. Yaksh, T.I., Frederickson, R.C.A., Huang, S.P. and Rudy, T.A. (1978): Brain Res. 148:516-520.

CHRONIC BARBITAL TREATMENT AND CHOLINERGIC MECHANISMS IN BRAIN

A. Nordberg and G. Wahlstrom

Department of Pharmacology, University of Uppsala, Sweden
and Department of Pharmacology, University of Umea
Sweden

Forced long term treatment (30-40 weeks) with barbital to rats
in their drinking fluid has proved to be a useful method for inducing changes in cholinergic mechanisms in the brain which are
related to tolerance and physical dependence and which could be
regarded as adaptive changes in the cholinergic system revealed
during abstinence (18). When rats are receiving barbital in their
drinking fluid (3.33 mg/ml) the daily intake of barbital is about
200 mg/kg (14). Animals drinking barbital in this manner are
initially sedated but this sedation disappears as the treatment
continues. In the experiments reported here the treatments usually
lasted for more than 30 weeks. At the end of the treatment the
barbital solution is replaced by water (day 0); the days in abstinence are then counted consecutively. Changes in tolerance in
the brain can be followed during abstinence by a hexobarbital threshold and such a tolerance toward barbiturates has been measured up to
day 60 in the abstinence (Wahlstrom and Nordberg, this volume). That
cholinergic mechanisms could be involved in the sensitivity to barbiturates measured with the hexobarbital threshold was indicated
by the finding that an acute injection of pilocarpine or physostigmine increased the hexobarbital threshold while atropine decreased it (15-17). This idea has been strongly supported by recent results which show that that effect of atropine is restricted
to only one of the isomers of hexobarbital (10).

Physical dependence can be recorded as spontaneous convulsions.
Figure 1 shows the number of spontaneous convulsions per rat and
day during the early part of the abstinence after barbital treatment

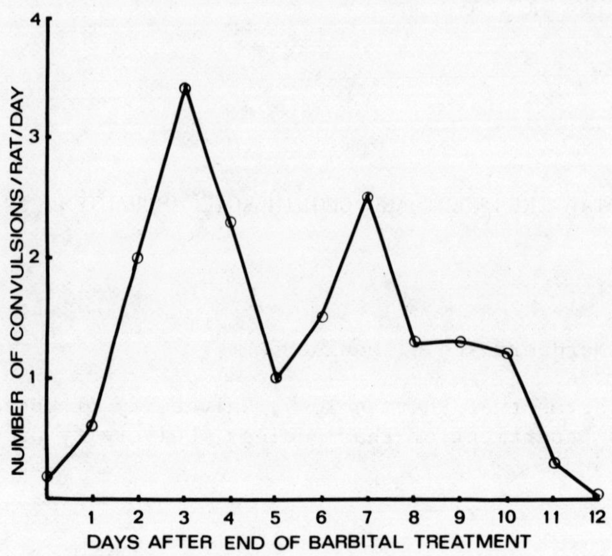

FIGURE 1: Spontaneous convulsions in the abstinence period after the end of barbital treatment. Data taken from Nordberg et al. (13).

TABLE 1:   Effect of forced oral chronic treatment of barbital on the endogenous content of ACh in the striatum.

	Day after withdrawal of barbital				
	C	0	A3	A12	A30	A80
ACh (nmol/g)	28	26	18***	19***	20**	20*

* = p < 0.05;   ** p < 0.01;   *** p < 0.001.

Data taken from Nordberg and Wahlstrom (11,13,23).
C = control; 0 = barbital until death; A3 = abstinent for 3 days; A12 = abstinent for 12 days; A30 abstinent for 30 days; A80 = abstinent for 80 days.

---

lasting 42 weeks. Convulsions were recorded by jiggle cages. A
first maximum is seen on the third day of abstinence followed by a
second maximum on the seventh day of abstinence. Very few convulsions are seen around day 12.

A decrease in brain weight after long term treatment has been
found (21). Such a reduction in brain weight is still present after
80 days of abstinence (Wahlstrom and Nordberg , this volume).

A supersensitivity to the temperature reducing effect of pilocarpine on day 3 of abstinence (20), together with the other changes
as mentioned above (15-17) focussed our interest on the biochemical
mechanisms involved in the effect of chronic barbital treatment on
the cholinergic system. In the present report the effect on several different cholinergic parameters such as the steady state concentration and biosynthesis of acetylcholine (ACh), choline (Ch)
uptake, enzyme activity and number of muscarinic receptors will be
presented.

Effect of Chronic Barbital Treatment
on the Steady State Concentration
and Biosynthesis of Acetylcholine

As shown in Table 1 the content of endogenous ACh in the striatum has been measured on several different days after the end of
chronic barbital treatment. The animals were killed by decapitation
and the ACh content analyzed by a biological method (leech muscle)
(9) or by mass fragmentography (5). On the day when the barbital
solution was withdrawn (day 0) no change in the content of endogenous ACh was found in comparison with control. On the third day
of abstinence (day A3) when the maximal number of abstinence convulsions were observed, a 33% decrease in ACh content was found.
When the amount of endogenous ACh was analyzed on the following
12th, 30th and 80th day of abstinence, a similar decrease in content of ACh was still measured. This occurred despite the fact
that the weights of the brains were decreased during the corresponding part of the abstinence (21). The endogenous ACh was
also measured in a cortex preparation (cortex + hippocampus) and
a midbrain preparation (midbrain + medulla oblongata + cerebellum)
at withdrawal of barbital, and after 3 and 12 days of abstinence.
The content of endogenous ACh was found to be unchanged in these
brain regions except for a 20% decrease of endogenous ACh content
in the midbrain preparation on day 12 of abstinence.

In order to study the effect of chronic barbital treatment on
the biosynthesis of ACh, a tracer dose of tritium labelled Ch (^{3}H-
Ch; 30 nmol Spec. Act. 10 Ci/mmol) was injected (i.v.) 1 min before

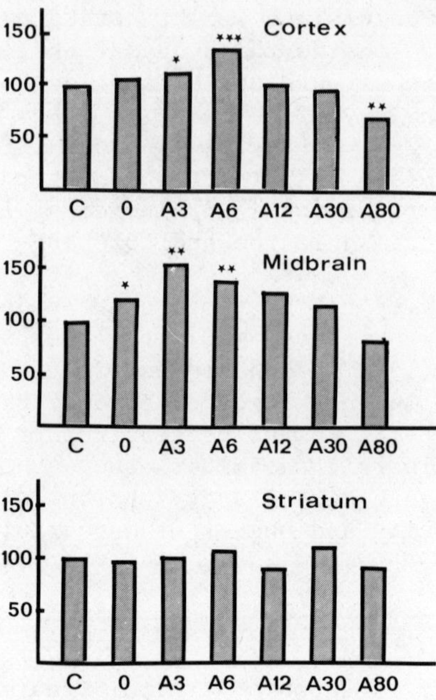

FIGURE 2: Biosynthesis of ^{3}H–ACh (in percent of control). Groups designated as in Table 1. Data taken from Nordberg and Wahlstrom (12, 23).   Animals in each group = 4–7. * = p < 0.05; ** = p < 0.01; *** = p < 0.001.

TABLE 2: Specific radioactivity of ACh on different days after chronic barbital treatment in rat. Specific radioactivity = ^{3}H–ACh in percent of ^{3}H–total/nmol endogenous ACh.

			Day			
	C	0	A3	A12	A30	A80
Striatum	0.55	0.57	0.87	0.74	0.86	0.77
Cortex preparation	0.84	0.98	1.22	1.04	–	–
Midbrain preparation	0.25	0.32	0.41	0.40	–	–

Data taken from Nordberg and Wahlstrom (11, 12, 23).
Groups designated as in Table 1.

the animals were killed (for description of method, see 6,12). As
can be seen in Fig. 2 a significant increase in the synthesis of
^{3}H-ACh was found on the day of withdrawal of barbital (day 0) in
the midbrain preparation and on the third day of abstinence (day A3)
both in the midbrain and cortex preparations. On the 12th and 30th
day of abstinence the ^{3}H-ACh formation was not significantly changed
in comparison with control. On the 80th day of abstinence a significant decrease in the ^{3}H-ACh formation was observed in the cortex
preparation. In the striatum no change in the ^{3}H-ACh formation was
found during the time interval 0 to 80 days of abstinence.

Table 2 gives the values for the specific radioactivity of
ACh in the striatum at different times during abstinence. As can be
seen, the specific radioactivity of ACh was increased in the abstinence at all observed time points in comparison with control. For the
cortex and midbrain preparations the specific radioactivity was calculated only for the first 12 days of abstinence, where an increase
in the specific activity of ACh was found.

The enzyme activity of brain choline acetyltransferase (CAT)
and acetylcholinesterase (AChE) was also measured up to day 30 in
the abstinence but no significant change was found (12).

Effect of Chronic Barbital Treatment
on the High Affinity Choline Uptake in Brain

In 1975 Atweh et al. (1) suggested that the sodium dependent
high affinity uptake of Ch by synaptosomes _in vitro_ could be used
as a simple method for measuring changes in cholinergic activity
after drug manipulation. This idea is supported by recent studies
which have shown that a correlation between _in vivo_ turnover of ACh
and _in vitro_ high affinity Ch uptake can be found between different
regions of the brain (7). Thus a study of the uptake of Ch in
chronic barbital treated rats could give valuable information. When
crude synaptosomes prepared from different areas of the brain were
incubated with different concentrations of Ch (0.05-0.50 µM) an
increased uptake of Ch by the sodium dependent high affinity system
was obtained in the striatum on days 12 and 21 of abstinence. As
shown in Table 3, the V_{max} for Ch uptake was significantly increased.
On day 12 a significant increase was also seen in the KM. No significant change in Ch uptake was found in the cortex and midbrain
preparations.

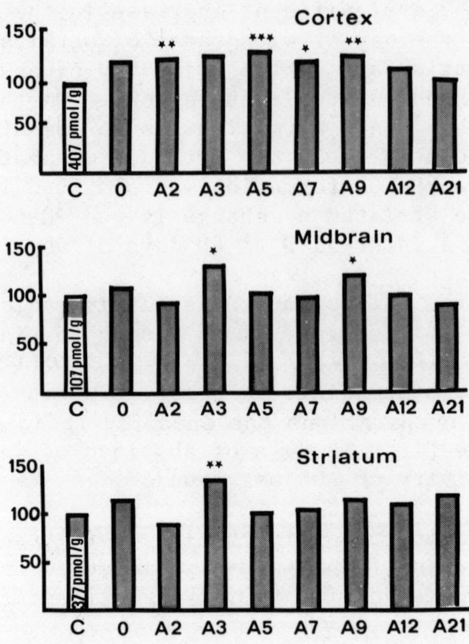

FIGURE 3:   Number of muscarinic binding sites (in percent of con-
            trol). Groups designated in same manner as in Table 1.
            Data taken from Nordberg et al. (13,23). Number of
            animals in each group = 3-17.
            * = p < 0.05;   ** = p < 0.01;   *** = p < 0.001.

TABLE 3:   Effect of forced oral chronic treatment of barbital on the
           high affinity uptake of Ch in the striatum.

	Day after withdrawal of barbital						
	C	0	A3	A5	A7	A9	A12	A21
V_{max} (pmol/mg protein/4 min)	23	21	27	25	23	22	35**	32*
K_M (μM)	0.26	0.27	0.27	0.22	0.26	0.25	0.45**	0.38

Data from Wahlstrom and Nordberg (22, 23).
Groups designated as in Table 1. Number of animals in each group =
5-6. * = p < 0.05; ** = p < 0.01.

Effect of Chronic Barbital Treatment
on Muscarinic Receptor Binding in Brain

To study whether any change in the postsynaptic activity in
the cholinergic neuron could be seen during abstinence the number
of muscarinic binding sites were measured. The muscarinic antagonist quinuclidinyl benzilate (QNB) was used as a labelled ligand
in the binding studies. Tritium labelled QNB (^{3}H-QNB; specific
activity 16 Ci/mmol) was incubated with crude synaptosomes from
different brain regions during abstinence up to day 21 (for details
of method see 13). As seen in Fig. 3, an increased number of ^{3}H-
QNB binding sites were found in the brain when the first 21 days
of abstinence was studied. A marked difference in the time course
of the increase of number of muscarinic binding sites between the
brain regions could be observed. A 40% increase in the number of
^{3}H-QNB binding sites was found in the striatum only on the third
day of abstinence. In the midbrain preparation two peaks of increased ^{3}H-QNB binding were observed with maxima on the third and
ninth day of abstinence, respectively. A somewhat different time
course in the increase of ^{3}H-QNB binding was measured in the cortex
preparation. As seen in Fig. 3, a significant increase in ^{3}H-QNB
binding was found from day 2 to day 9 of abstinence. Saturation
studies in the midbrain preparation on the third day of abstinence
indicated changes in number of receptors with no difference in receptor affinity (13).

DISCUSSION

As seen from the results, several different cholinergic parameters are affected in the abstinence after long term chronic barbital treatment. The changes are quite different from those measured after a single anesthetic dose of barbital (8-10). In the
present study the rats were killed by decapitation. Microwave irradiation would have been preferable for the measurement of endogenous ACh and biosynthesis of ACh (6) but it was not possible due
to the size of the animals (about 600 g). However, studies on the
acute effect of barbiturates on ACh have shown the same relative
effect following decapitation and microwave irradiation (10). At
the beginning of our study the brains were dissected into the
following three parts: striatum, the midbrain preparation (including medulla oblongata, cerebellum) and the cortex preparation
(including hippocampus). This dissection technique was then kept
throughout the investigation, although a more extensive microdis-
section might have revealed even more selective results. Surprisingly, a very small effect on the cholinergic system was found
at the withdrawal of barbital with only an increase of ^{3}H-ACh in the

TABLE 4: High and low affinity muscarinic binding sites and affinity constants (K_D) estimated from agonist/^{3}H–QNB competition curves. The high and low affinity binding sites are given in percent of the total specific binding and the K_D in μM.

Agonist	High Affinity Site		Low Affinity Site	
	Percent	K_D	Percent	K_D
Oxotremorine				
Control	28 ± 4	0.08 ± 0.020	72 ± 6	3 ± 0.7
A2	22 ± 3	0.02 ± 0.007	78 ± 4	2 ± 0.4
A7	19 ± 3	0.02 ± 0.009	79 ± 5	2 ± 0.6
Carbachol				
Control	39 ± 3	0.15 ± 0.08	56 ± 7	285 ± 129
A5	33 ± 2	0.16 ± 0.05	61 ± 5	266 ± 62

Groups designated in same manner as Table 1 (unpublished observations). Number of animals in each group = 5. Mean ± S.D.

midbrain preparation (+ 22%). It is remarkable that this increase in ^{3}H–ACh was found in a part of the brain which normally has a rather low cholinergic activity (6).

On the third day of abstinence, when the maximum number of spontaneous convulsions is recorded, the largest changes in cholinergic activity are found. These changes consist of a decreased endogenous ACh content, increased formation of ^{3}H–ACh and an increased number of muscarinic receptor binding sites. Thus the general pattern indicates an increased ACh turnover and post-synaptic activity. The uptake of Ch by the cholinergic nerve termonal (measured as high affinity synaptosomal uptake) does not seem to be strictly coupled to the biosynthesis of ACh since an increase in Ch uptake is not found before the 12th day of abstinence at a time when ^{3}H–ACh biosynthesis has normalized.

The different brain regions studied are affected differently by the chronic barbital treatment. In the striatum the content of endogenous ACh is decreased on the third day of abstinence and continued to decrease at least up to the 80th day of abstinence. The biosynthesis of ^{3}H–ACh in the striatum was unchanged during abstinence, which was clearly different from the results found in

the other brain parts. Since ^{3}H-ACh biosynthesis in the striatum
is normally high (6), this could be the limiting factor responsible
for such a difference. Another alternative explanation is that a
more rapid method of killing the animals might have made it possible
to detect a further increase in ^{3}H-ACh biosynthesis. A difference
in time course of the change in number of muscarinic receptor binding
sites is found among the brain regions. A correlation has been found
between the number of binding sites in the cortex of rats abstinent
for 7 and 9 days and the latency to last recorded convulsion (r =
-0.80; b = -2.9; degrees of freedom, 7; p < 0.01). Rats with a short
latency to last recorded convulsion show a higher number of muscarinic binding sites in the cortex than rats with a longer latency.
Through displacement experiments with different muscarinic agonists
it is possible to calculate the ratio of high and low affinity binding sites in the brain (2,4). Table 4 shows the percent of high and
low affinity sites in the cortex of abstinent (A2, A5, A7) and control rats observed after displacement of ^{3}H-QNB binding by the
agonists, oxotremorine and carbachol. As seen from Table 4 the
percent of high affinity muscarinic binding sites is lower in the
abstinent rats than in controls, while a higher percent of low affinity sites is found in abstinent rats. This shift towards a
higher percent of low affinity sites in abstinent rats is interesting
since the low affinity muscarinic binding sites has been suggested
to be the physiological active sites for the muscarinic receptor (3).

If the total pattern of complex changes in different cholinergic
parameters seen in this study are evaluated they indicate an increased activity in the cholinergic parts of the CNS. It is also
clear that this increased activity in such a dynamic state as abstinence cannot be explained by one simple biochemical mechanism.

ACKNOWLEDGEMENT

This study was supported by the Swedish Medical Research
Council (Projects 3771, 5641, 5817) and by the Swedish Tobacco
Company.

REFERENCES

1.   Atweh, S., Simon, J.R. and Kuhar, M.J. (1975): Life Sci.
        17:1535-1544.
2.   Birdsall, N.J.M., Burgen, A.S.V. and Hulme, E.C. (1978):
        Molec. Pharmacol. 14:723-736.

3. Birdsall, N.J.M., Burgen, A.S.V. and Hulme, E.C. (1978): IN
    Advances in Pharmacology and Therapeutics, Vol. 1 (ed) J.
    Jacob, Pergamon Press, New York, pp. 983-980.

4. Hedlund, B. and Bartfai, T. (1979): Molec. Pharmacol. 15:
    531-544.

5. Karlen, B., Lundgren, B., Nordgren, I. and Holmstedt, B. (1974):
    IN Choline and Acetylcholine: Handbook of Chemical Assay
    Methods (ed) I. Hanin, Raven Press, New York, pp.163-179.

6. Nordberg, A. (1977): Acta Physiol. Scand. Suppl. 445:1-51.

7. Nordberg, A. (1978): Life Sci. 23:937-944.

8. Nordberg, A. and Sundwall, A. (1973): Acta Physiol. Scand.
    Suppl. 396:69.

9. Nordberg, A. and Sundwall, A. (1975): IN Cholinergic Mechanisms
    (ed) P.G. Waser, Raven Press, New York, pp. 229-239.

10. Nordberg, A. and Sundwall, A. (1977): Acta Physiol. Scand.
    99:336-344.

11. Nordberg, A. and Wahlstrom, G. (1977): Eur. J. Pharmacol. 43:
    237-242.

12. Nordberg, A. and Wahlstrom, G. (1979): J. Neurochem. 32:371-378.

13. Nordberg, A., Wahlstrom, A. and Larsson, C. (1980): Life Sci.
    26:231-237.

14. Wahlstrom, G. (1974): Acta Pharmacol. Toxicol. 35:131-144.

15. Wahlstrom, G. (1976): Psychopharmacology 49:159-166.

16. Wahlstrom, G. (1976): Acta Pharmacol. Toxicol. 38:72-80.

17. Wahlstrom, G. (1977): Acta Pharmacol. Toxicol. 41:287-289.

18. Wahlstrom, G. (1979): Drug Alcoh. Depend. 4:221-234.

19. Wahlstrom, G. (1979): Eur. J. Pharmacol. 59:219-225.

20. Wahlstrom, G. and Ekwall, T. (1976): Eur. J. Pharmacol. 38:
    123-129.

21. Wahlstrom, G. and Nordberg, A. (1978): Life Sci. 23:1583-1590.

22. Wahlstrom, G. and Nordberg, A. (1979): Acta Physiol. Scand.
    473:65.

ADDED IN PROOF:

23. Nordberg, A. and Wahlstrom, G. (1981): Drug Alcoh. Depend.
    7:53-63.

INTERACTION BETWEEN ATROPINE AND HEXOBARBITAL IN THE ABSTINENCE

AFTER CHRONIC BARBITAL TREATMENTS IN THE RAT

G. Wahlstrom and A. Nordberg*

Department of Pharmacology, University of Umea
S-901 87 Umea, Sweden, and
*Department of Pharmacology, University of Uppsala
S-751 23 Uppsala, Sweden

INTRODUCTION

After chronic treatments with barbital a decreased sensitivity
to hexobarbital can be observed in rats. Such a decreased sensitivity is usually called tolerance and it can be detected for at
least 60 days after the end of a 50 week long barbital treatment
(8). There is a positive relation between the duration of the
treatment and the duration of the tolerance found in the abstinence
(2). Spontaneous convulsions during abstinence are usually regarded as a sign that physical dependence has been induced by the
treatment. The frequency of the spontaneous convulsions also seems
to depend on the duration of barbital treatment and can be recorded
at least for the first five weeks of abstinence after 50 weeks of
barbital treatment (8). Since convulsions can decrease tolerance
to hexobarbital a relation seems to exist between these two variables (3). Thus, long term treatments with barbital can induce
changes in the CNS indicating increased escitation which lasts for
a long time during the abstinence. Several changes in brain cholinergic mechanisms, indicating increased activity, have also been
found during abstinence (Nordberg and Wahlstrom, this volume) and
these changes might be related to the tolerance and physical dependence described above (6).

Atropine can increase the sensitivity to hexobarbital (5).
There is, furthermore, in normal animals a stereospecificity in the
interaction between atropine and the isomers of hexobarbital (7).
This means that treatments with atropine should be tried during the
abstinence and that hexobarbital could be used to detect changes

TABLE 1: The effects on hexobarbital threshold of chronic barbital treatment followed by atropine treatment.

		Weeks After Last Atropine Dose								
		1		2		3		4		9
EXP	DAY		DAY		DAY		DAY		DAY	
1A	51	95.9±2.7(22)	56	96.7±3.0(19)	63	94.6±2.3(21)	79	89.6±2.0(22)		
B	51	88.3±2.0(29)	56	85.3±1.6(26)	63	89.4±1.9(29)	79	86.3±1.8(26)		
DIF		7.6*		11.4***		5.2		3.3		
1C	51	89.5±2.4(15)	56	85.3±2.8(18)	63	88.2±1.4(18)	79	83.9±3.1(17)		
D	51	90.1±1.8(25)	56	86.5±2.8(22)	63	88.3±2.1(24)	79	84.3±2.1(25)		
DIF		−0.6		−1.2		−0.1		−0.4		
2A	20	94.7±3.0(9)	27	105.0±3.9(9)	38	106.4±7.2(7)	45	90.8±4.8(9)	80	91.5±3.5(8)
B	20	84.1±2.5(12)	27	88.4±3.3(11)	38	85.6±2.1(10)	45	85.0±3.9(9)	80	77.3±2.2(10)
DIF		10.6*		16.6**		20.8**		5.8		14.2**
2C	20	96.5±6.0(9)	27	100.3±2.6(8)	38	92.1±3.4(8)	45	85.1±2.7(9)	80	78.7±3.3(8)
D	20	82.2±2.2(11)	27	87.1±1.5(11)	38	86.9±3.2(10)	45	85.3±2.2(12)	80	79.6±2.3(12)
DIF		14.3*		13.2***		5.2		−0.2		−0.9

Hexobarbital thresholds are given as percent of pre-experimental value. Barbital treatments lasted 33 weeks. Day 0 is day when treatment was stopped. In both experiments group A was treated with barbital and atropine (4 mg/kg/day), group B only with atropine, group C only with barbital and group D had a control treatment. Experiment 1 was treated with atropine on day 29–44 and experiment 2 on day 3–12. In experiment 1 an atropine dose (2, 4 or 8 mg/kg) was given on day 3 to some animals in all groups (see text). EXP = experiment; DIF = difference. All values are mean ± S.E. Number of animals given in parentheses. * = p < 0.05; ** = p < 0.01; *** = p < 0.001.

TABLE 2:   Effects of Ch thresholds and brain weights of a barbital
           treatment followed by an atropine dose.

		Weeks After Atropine Dose		
		(Ch threshold mg/g)		(Brain weight g)
EXP	DAY	Week 7	DAY	Week 11
3A	53	81.9 ± 4.6(10)	81–84	2.122 ± 0.031(7)
B	53	81.4 ± 3.5(6)	81–84	2.167 ± 0.043(3)
DIF		−2.2		−0.045
3C	53	68.3 ± 4.4(6)	81–84	2.092 ± 0.015(6)
D	53	83.3 ± 3.1(10)	81–84	2.175 ± 0.021(8)
DIF		−15.0*		−0.083*

Barbital treatment lasted 33 weeks. Day 0 is day when treatment
stopped. Group A was treated with barbital and atropine (8 mg/kg
i.p. on day 3), group B only with atropine, group C only with
barbital and group D had a control treatment. EXP = experiment;
DIF = difference. All values are mean ± S.E. Number of animals
given in parentheses. * = $p < 0.02$.

induced by atropine treatments. Some experiments dealing with the
immediate effects of single doses of atropine in abstinence have
been reported earlier (4,7). This report will discuss preliminary
results of longer atropine treatments, together with some remarkably long lasting effects which have been induced by single atropine doses given to chronic barbital treated animals.

METHODS

The sensitivity to hexobarbital has been tested with a threshold method (1). Essentially, hexobarbital as sodium salt is infused in a tail vein of male rats (Sprague-Dawley). The electro-
encephalogram is recorded during the infusion and the first burst
suppression of 1 sec or more is used as a sign of depression in the
central nervous system. The dose of hexobarbital needed to induce
this "silent second" is used as the threshold dose. Since rats
have an individual sensitivity to this hexobarbital threshold, the
dose is usually calculated as percent of a pre-experimental value
obtained in the same animal.

The sensitivity to choline (Ch) has also been determined by
intravenous infusion essentially in the same manner as in the hexobarbital threshold with overt convulsions as the endpoint (5).
Barbital was given in the drinking water. A solution of sodium
barbital (3.33 mg/ml) replaced the ordinary drinking water. Animals
on these treatments usually take around 200 mg/kg/day and the treatments lasted for 32-33 weeks. Taken in this manner, such a dose
of barbital only indices a slight sedation during the first part
of the treatment. The day when the barbital solution was replaced
by water was called day 0 of the abstinence.

The barbital treatment can induce a decrease in brain weight
(9) and for this reason brain weights were usually recorded at the
end of the experiments.

Atropine sulphate has been given intraperitoneally once daily
during the treatments. Since all animals were on reversed light
dark schedules (12 hr light and 12 hr dark), all treatments and
test procedures have been performed during the early hours of dark.

RESULTS AND DISCUSSION

The results of two experiments with 10-15 days of treatment
with atropine (4 mg/kg/day i.p.) during two parts of the abstinence
after chronic barbital treatments are given in Table 1. Four
groups of animals with treatments as described in Table 1 are included in both experiments. Repeated testing with the hexobarbital
threshold was used to follow the changes induced by the treatments.

In experiment 1 (Table 1) the atropine treatment was given
during the latter part of the abstinence (day 29-44). It is evident that the controls (group D) had hexobarbital thresholds below
the pre-experimental value of 100%. The pre-experimental value
was obtained before the barbital treatment, which means that this
decrease is probably due to aging. It is also clear that there was
no remaining tolerance to hexobarbital in the animals which had
been given only the barbital treatment (group C and difference
between groups C and D). The most important result of this experiment is that the atropine treatment can induce tolerance in the
previously barbital treated rats (group A), but not in the corresponding controls (group B) which were only given atropine. The
magnitude of this difference between the two groups is clearly much
larger compared with the difference when no atropine treatment was
given (group C and D). Furthermore, no detectable change was observed in group B, which means that the difference between groups
A and B only is due to an increased sensitivity to atropine in
group A which had received barbital treatment.

In experiment 2 (Table 1) atropine treatment was given much earlier during abstinence (day 3-12). The controls (group D) showed hexobarbital thresholds in the same range as in experiment 1. In this experiment there was a clear remaining tolerance in the animals treated only with barbital (group C) which lasted at least until day 27. The comparable tolerance in group A which also had the atropine treatment, was, during this abstinence stage, of the same magnitude. This meant that no increase in tolerance had been induced by atropine treatment. However, it was evident that when tolerance induced by barbital was reduced (day 38) a tolerance was still found in the animals which had been treated with both barbital and atropine (difference between groups A and B). On day 45 no significant tolerance was recorded in any of the two barbital treated groups. On day 80 a curious renewal of tolerance was recorded in the barbital treated animals which also had had the atropine treatment. A week later no significant tolerance was recorded.

The magnitude of the recorded tolerance seemed to be somewhat larger in experiment 1 than in experiment 2. This could have been due to a single atropine dose (2, 4 or 8 mg/kg) given on day 3 in some of the animals in experiment 1. Animals not given atropine on day 3 showed the largest tolerance on day 56, while those given 8 mg/kg on day 3 showed very little tolerance on this day. Immediate results was given by 2 and 4 mg/kg. Thus in experiment 1 there were also indications that atropine given early in the abstinence could be responsible for the way in which the animals reacted several weeks later.

Further evidence that a single dose of atropine early in the abstinence could have long lasting effects is given in Table 2. In this case the animals were treated with barbital and then given atropine on day 3 (8 mg/kg). Much later (day 53) they were tested with Ch and then sacrificed (day 81-84). At sacrifice the brain weights were determined. Table 2 shows that animals treated only with barbital (group C) have a lower Ch threshold and brain weight than the corresponding controls (group D). No difference was found if the animals were also given atropine once during the abstinence (groups A and B). As seen from the difference between groups A and B, it looks as if atropine reduced the effect found in the animals which only had the barbital treatment. Thus no significant difference was observed between barbital treated animals given either atropine or saline on day 3 (group A compared to group C, Table 2).

The present experiments have shown that chronic barbital treatments can induce changes which make the animals sensitive to atropine treatments during different parts of the abstinence. These changes can be measured as a tolerance to hexobarbital and they probably involve some of the changes in cholinergic activity found

during the abstinence as discussed by Nordberg and Wahlstrom (this
volume). Since all time intervals used for the atropine treatments in the present experiments were not covered by the biochemical studies, it is at present futile to try to speculate on which
changes made the earlier barbital treated animals more sensitive
to atropine.

The possibility that a single dose of atropine could have long
lasting effects in chronic barbital treated animals was supported
by all the experiments presented here. The used variables have
been different as well as the time between the atropine dose and
testing. At present it can only be stated that it looks as if a
treatment with atropine on day 3 of the abstinence could modify
the system in such a way that far-reaching effects are obtained.
Day 3 clearly occurs during a very dynamic part of the abstinence
with complex changes in cholinergic mechanisms in the brain indicating an increased cholinergic activity (Nordberg and Wahlstrom,
this volume). Further studies on these phenomena are clearly warranted.

ACKNOWLEDGEMENT

This study was supported by the Swedish Medical Research
Council (project 3771).

REFERENCES

1.   Wahlstrom, G. (1966): Acta Pharmacol. Toxicol. $\underline{24}$:404-418.
2.   Wahlstrom, G. (1974): Acta Pharmacol. Toxicol. $\underline{35}$:131-144.
3.   Wahlstrom, G. (1976): Life Sci. $\underline{19}$:1817-1826.
4.   Wahlstrom, G. (1978): Psychopharmacology $\underline{59}$:123-128.
5.   Wahlstrom, G. (1978): Eur. J. Pharmacol. $\underline{51}$:219-227.
6.   Wahlstrom, G. (1979): Drug Alcohol Depend. $\underline{4}$:221-230.
7.   Wahlstrom, G. (1979): Eur. J. Pharmacol. $\underline{59}$:219-229.
8.   Wahlstrom, G. and Larsson, R. (1977): Pharmacol. Biochem.
       Behav. $\underline{6}$:187-189.
9.   Wahlstrom, G. and Nordberg, A. (1978): Life Sci. $\underline{23}$:1583-1587.

D.L. Cheney, A.V. Revuelta and E. Costa

Laboratory of Preclinical Pharmacology
National Institute of Mental Health
St. Elizabeths Hospital
Washington, D.C.   20032

INTRODUCTION

Modification of central cholinergic activity has been suggested to be one of the major actions of tetrahydrocannabinols (THC)(4,14). In fact there have been numerous reports implicating an anticholinergic mechanism in the action of these compounds. Moderate doses of Δ^9-THC reduce acetylcholine (ACh) release from cat cortex (7) and reduce the rate of ACh depletion after administration of hemicholinium (5). Furthermore, Δ^9-THC inhibits ACh biosynthesis in hypothalamic and striatal slices (9). In non-microwaved preparations the effects of Δ^9-THC on total brain ACh content that have been reported are inconsistent. Domino (5,6) reports that Δ^9-THC elevates total brain ACh content in mice (50 mg/kg i.p.) and rats (10-32 mg/kg), while Ashew et al. (2) report that Δ^8 or Δ^9-THC reduced rat brain ACh content and rat brain synaptosomal ACh content. The preponderance of evidence, however, is consistent with the hypothesis that THC reduces the utilization of brain ACh.

Attempts have been made to ascertain the possible neural and chemical mechanisms which mediate the behavioral effects of THC. Surprisingly, treatment with THC, treatment with anticholinergics, and lesions of the hippocampus produce strikingly similar behavioral effects (14). These three treatments have been shown to disrupt acquisition and performance of maze behaviors, impair sequential integration behavior, affect memory formation, produce inattentiveness, inhibit the acquisition of conditioned fear response, suppress trained avoidance response, increase resistance to extinction, disrupt differential reinforcement of low rate schedule, and retard habituation (for references see 14). Thus it appears that both the hippocampus and the cholinergic system may be involved in the effects of THC. One of the three major afferent pathways to the hippocampus is cholinergic. Neurons from this pathway enter the hippocampus at the level of the stratum oriens of field CA_3 and immediately divide to innervate the stratum oriens of fields CA_1, CA_2, CA_3 and the supragranular region of the dentate gyrus (16). Their axons project to the hippocampus via the fornix/fimbria (15) from cell bodies located in the medial septal nucleus and in the nucleus of the diagonal band (11).

The functional participation of the particular group of cholinergic neurons during behavioral activities or during drug action may be understood by measuring the turnover rate of ACh (TR_{ACh}). The TR_{ACh} in hippocampus increases or decreases proportionally to the stimulation of the cell body region in the septum (15). When cholinergic hippocampus afferents are activated by electrical stimulation of the septum the hippocampal TR_{ACh} is significantly increased without altering the ACh content. On the other hand, when the cholinergic regulation of hippocampal cell activity is blocked by cutting the fimbria the TR_{ACh} in hippocampus is reduced by 80% while the ACh content is slightly increased.

Modulation of the Hippocampal
TR_{ACh} by Δ^9-THC

Thus it is not surprising that Δ^9-THC, whose behavioral effects mimic those caused by administration of anticholinergic drugs and destruction of the hippocampus, reduces the TR_{ACh} in the hippocampus. Figure 1 compares the effects in hippocampus of the psychoactive Δ^9-THC with another component of Cannabis sativa, cannabidiol (CBD), which is apparently devoid of psychotomimetic activity (3, 10), although it does possess anticonvulsant activity (13). In the hippocampus Δ^9-THC causes a potent dose related inhibition of TR_{ACh} but CBD has no effect at 100 times the minimal active dose of Δ^9-THC. Both Δ^9-THC and CBD at high doses reduce the TR_{ACh} in striatum but not in parietal cortex (19). The observation that CBD is not active on hippocampal TR_{ACh} suggests the possibility that hippocampal cholinergic mechanisms may play a role in the psychotomimetic activity of Δ^9-THC.

Since the reduction in TR_{ACh} is observed only in the hippocampus and not in other brain areas, and since the effect can be seen at such a small dose of Δ^9-THC (0.2 mg/kg i.v.) it may be that Δ^9-THC is triggering its potent pharmacological response by modifying

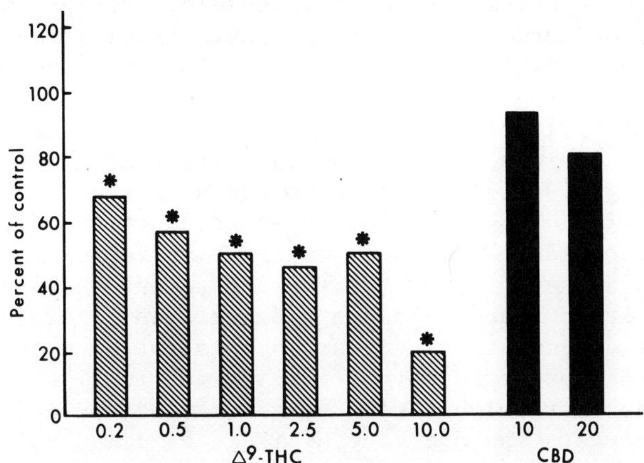

FIGURE 1: TRACh in hippocampus after various doses of (-) Δ^9-THC and cannabidiol (CBD), * = P < 0.05. Phosphoryl[$(C^2H_3)3$]-Ch (135 μmol/kg) was infused into the lateral vein of unanesthetized rats over a 9 min period. Rats were killed by focussed microwave irradiation and the deuterated fragments quantitated by GCMS. Turnover rate was calculated as described previously (21).

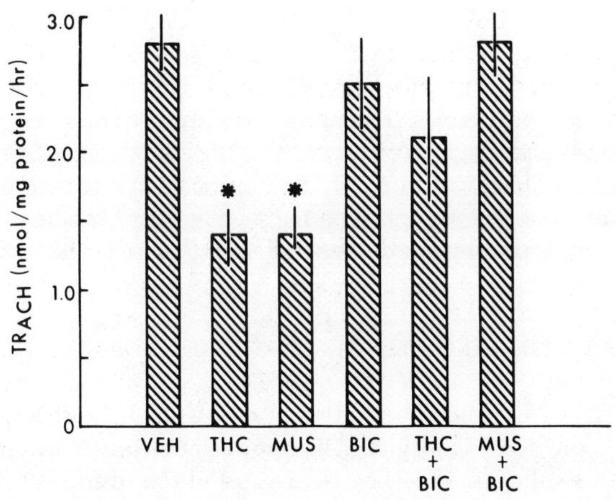

FIGURE 2: Antagonism by bicuculline (30 μg; i. sp.; 36 min) of the inhibition of hippocampal TRACh elicited by Δ^9-THC (5 mg/kg; i.v.; 30 min) or muscimol (1 μg; i. sp.; 30 min). * = P < 0.05.

the action of a neurotransmitter system which regulates the activity
of the septal-hippocampal pathway via specific receptors. Stimulation of septal GABAergic receptors by intraseptal administration of
muscimol reduces the hippocampal TR_{ACh} as does intravenously administered Δ^9-THC (Fig. 2). The reduction of hippocampal TR_{ACh} by
Δ^9-THC or muscimol is blocked by intraseptally administered bicuculline. Fimbrial lesions further substantiate that the reduction in
hippocampal TR_{ACh} is mediated via the septal-hippocampal pathway.
The administration of Δ^9-THC 2 hr after transection of the fimbria
does not further reduce the TR_{ACh} in the hippocampus (18). Moreover, THC does not act indirectly through receptors on endorphinergic or dopaminergic neurons in the hypothalamus or brain stem projecting to the septum since neither administration of naltrexone
to block endorphinergic receptors nor the destruction of septal
dopaminergic nerve terminals with 6-hydroxydopamine injected intra-
tegmentally prevents the decrease of TR_{ACh} elicited by Δ^9-THC (18).
These studies, however, do not answer the question whether GABA
receptors located in the septum are activated directly by Δ^9-THC
or indirectly through GABAergic interneurons or other septal
afferents.

Action of THC Derivatives
on the Hippocampal TR_{ACh}

     Animal behavioral studies have shown that modification of the
cannabinoid molecule alters the effectiveness of these compounds.
A shift in the double bond of Δ^9-THC to produce Δ^8-THC decreases
its psychotomimetic activity but increases its stability in solution (8). Replacement of the pentyl side chain with a 1,2-dimethyl-
heptyl (DMH) moiety enhances the potency in animal experiments (1,
12). Such a replacement in Δ^{6a}, Δ^{10a}-THC is very effective in
producing ataxia in dogs (1,12). The administration of Δ^{6a}, Δ^{10a}-
THC-DMH to humans causes psychotomimetic effects when given in
doses smaller than threshold doses of Δ^{6a}, Δ^{10a}-THC (20). In the
monkey the threshold dose for Δ^8-THC is 250 µg/kg, whereas the
1,2-dimethyl-heptyl homolog is active at one-fifth of this dose
(8). Furthermore, the effects of Δ^8-THC last about 4 hr; those of
Δ^8-THC-DMH last about 48 hr (8). In Fig. 3 the effects of Δ^9-THC,
(-) Δ^8-THC, (+) Δ^8-THC and the dimethyl-heptyl homolog of (-) Δ^8-
THC (Δ^8-THC-DMH) on the TR_{ACh} in hippocampus have been compared.
The structures are shown in Fig. 4. The data demonstrate that (-)
Δ^8-THC-DMH, (-) Δ^9-THC and (-) Δ^8-THC all reduce the TR_{ACh} in the
hippocampus in a dose dependent manner. The ED_{50} of (-) Δ^8-THC-DMH
to lower the TR_{ACh} is about 70 times smaller than that of (-) Δ^8-
THC. Moreover, the content of ACh or choline is unaltered except
for the highest dose of (-) Δ^8-THC-DMH (17). The increased

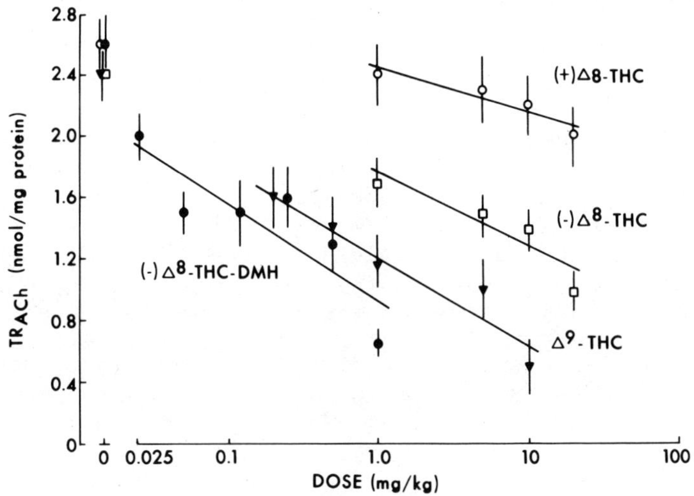

FIGURE 3: Dose-response relationship of various derivatives of THC.

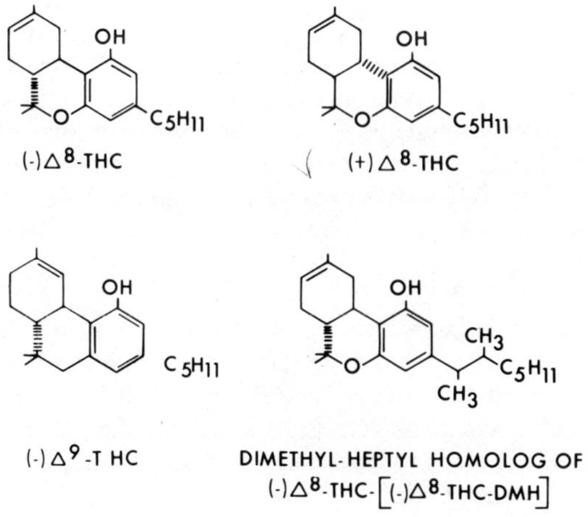

FIGURE 4: Structures of various THC derivatives.

effectiveness induced by the side chain homologs of the tetrahydrocannabinoids is consistent with a more favorable orientation of the
molecule within the vicinity of the recognition site.

The steric characteristics of the cannabinoid molecule can
contribute to the effectiveness of the compounds in eliciting psychotomimetic responses. Such studies have shown that the (-) isomers of Δ^9-THC and Δ^8-THC, with the natural configurations, are
active at 0.05 mg/kg and 0.25 mg/kg, respectively, whereas the (+)
isomers are inactive at 1.0 mg/kg (8). In Fig. 3 (-) Δ^8-THC is
active at doses lower than 5.0 mg/kg, whereas (+) Δ^8-THC is inactive at 20 mg/kg. The inactivity of the (+) Δ^8-THC supports the
hypothesis that the action of the cannabinoids on TR_{ACh} and on behavior may depend upon their interaction with specific recognition
sites.

CONCLUSIONS

The central cholinergic system must certainly be involved in
a number of the actions of the tetrahydrocannabinoids (4,14). Intravenously administered Δ^9-THC reduces the TR_{ACh} in hippocampus
in a dose related manner. This reduction in TR_{ACh} occurs specifically in the hippocampus and is not present in other brain areas
(19). Moreover, intraseptal administration of the GABA agonist,
muscimol, reduces the hippocampal TR_{ACh} as does intravenously administered Δ^9-THC. The reduction in hippocampal TR_{ACh} may be inhibited either by intraseptal injection of bicuculline or by
lesioning the fimbria (18). Thus it appears that Δ^9-THC may reduce the TR_{ACh} in the nerve terminals of the cholinergic septal-
hippocampal pathway by modifying the activity of GABAergic receptors. Whether this effect is a direct action on septal GABAergic
receptors or an indirect action on another neurotransmitter system
mediated by the septal GABAergic interneurons is not yet known.

Cannabidiol, a non-psychoactive THC analog has no effect on
the hippocampal TR_{ACh} at 100 times the minimal active dose of Δ^9-
THC (19). It is possible that the inhibition of hippocampal cholinergic mechanisms plays a role in the psychotomimetic activity of
Δ^9-THC. Such a role for the hippocampal cholinergic system would
be consistent with studies that have implicated both the hippocampus and the cholinergic system in many complex behavioral functions (see 14 for references).

Structural analogs of Δ^9-THC reduce the hippocampal TR_{ACh} in
a dose dependent manner.The slopes of the dose-response figure
(Fig. 3) do not differ significantly, suggesting a common mechanism

of action. Furthermore, the enhanced effectiveness of the dimethyl
heptyl homologs and the inactivity of the (+) isomers of the THC
supports the hypothesis that these compounds may depend upon their
interaction with specific recognition sites for their biochemical
and behavioral effectiveness. Although it may be concluded that
the tetrahydrocannabinoids activate specific transmitter receptors,
the localization of this action to the GABAergic-cholinergic septal-
hippocampal pathway seems to indicate that the endogenous ligand
for this receptor is not a transsynaptic regulator for all GABA-
ergic or cholinergic neurons.

REFERENCES

1.   Adams, R., Harfenist, M. and Loewe, S. (1949): J. Amer. Chem.
        Soc. 71:1624-1628.
2.   Ashew, W.E., Kimball, A.P. and Ho, B.T. (1974): Brain Res.
        69:375-378.
3.   Brimblecombe, R.W. (1973): Adv. Drug Res. 7:165-206.
4.   Brown, H. (1971): Psychopharmacologia (Berl.) 21:294-301.
5.   Domino, E.F. (1971): Ann. NY Acad. Sci. 191:166-191.
6.   Domino, E.F. (1976): IN Marihuana: Chemistry, Biochemistry and
        Cellular Effects (ed) G.G. Nahas, Springer-Verlag, New York,
        pp. 407-413.
7.   Domino, E.F. and Bartolini, A. (1972): Neuropharmacology 11:
        703-713.
8.   Edery, H., Grunfeld, Y., Ben-Zvi, Z. and Mechoulam, R. (1971):
        Ann. NY Acad. Sci. 191:40-53.
9.   Friedman, E., Hanin, I. and Gershon, S. (1976): J. Pharmacol.
        Exp. Ther. 196:339-345.
10. Hollister, L.E. (1973): Experientia 29:825-826.
11. Lewis, P.R. and Shute, C.C.D. (1967): Brain 90:521-540.
12. Loewe, S. (1950): N.-S. Arch. Exp. Pathol. Pharmakol. 211:
        175-193.
13. Mechoulam, R., McCallum, N.K., Lander, N., Yagen, B., Ben-Zvi,
        Z. and Levy, S. (1976): IN The Pharmacology of Marijuana
        (eds) M.C. Braude and S. Szara, Raven Press, New York,
        pp. 39-48.
14. Miller, L.L. and Drew, W.G. (1974): IN Marijuana, Effects on
        Human Behavior (ed) L.L. Iversen, Academic Press, New York,
        pp. 157-188.
15. Moroni, F., Malthe-Sørenssen, D., Cheney, D.L. and Costa, E.
        (1978): Brain Res. 150:333-341.
16. Mosko, S., Lynch, G. and Cotman, C.W. (1973): J. Comp. Neurol.
        152:163-174.

17. Revuelta, A.V., Cheney, D.L., Costa, E., Lander, N. and
        Mechoulam, R. (1980): Brain Res. $\underline{195}$:445-452.
18. Revuelta, A.V., Cheney, D.L., Wood, P.L. and Costa, E. (1979):
        Neuropharmacology $\underline{18}$:525-530.
19. Revuelta, A.V., Moroni, F., Cheney, D.L. and Costa, E. (1978):
        N.S. Arch. Pharmacol. $\underline{304}$:107-110.
20. Sim, V. (1970): IN Psychotomimetic Drugs (ed) D.H. Efron,
        Raven Press, New York, pp.332-338.
21. Wood, P.L. and Cheney, D.L. (1979): Canad. J. Physiol.
        Pharmacol. $\underline{57}$:404-411.

ANTINOCICEPTIVE ACTION OF CHOLINOMIMETICS EVALUATED WITH THE

METHOD OF THE RETURN OF CORNEAL ANESTHESIA INDUCED WITH PROCAINE

F. Aloisi, A. Scotti de Carolis and V. G. Longo

Laboratorio di Farmacologia
Istituto Superiore di Sanita, Rome, Italy

The antinociceptive action of cholinomimetic agents has been
extensively described in the literature using a variety of methods
involving mechanical, termal or chemical stimulation. Tremorine,
a cholinomimetic drug with tremorogenic activity, suppresses the
responsiveness of mice to tail pinching (3). Arecoline shows an
antinociceptive action in the tooth pulp test in the rabbit and in
several tests in mice and rats (vocalization response, tail flick,
tail pinching, hot plate, abdominal writhing)(8). Eserine and
oxotremorine are active in the tail flick test in mice (4). Cholinomimetic show antinociceptive activity also when injected into
the cerebral ventricles, or into various cerebral areas. This has
been demonstrated by Metys et al. (8) for carbachol and oxotremorine
and by Pedigo et al. (9) for acetylcholine (ACh). The antinociceptive action of cholinomimetics is abolished by atropine or by other
anticholinergic drugs but is not influenced by methyl-atropine. According to some authors, narcotic antagonists also are able to op-
pose this effect of cholinergics (4,9).

In the paper we report on the results obtained in studying
the effect of various cholinomimetics on another test for antinociception carried out in the rabbit and based on the return of
corneal anesthesia due to instillation of procaine (7). This test
has been previously used in our laboratory to investigate the effects of some natural and synthetic enkephalins injected into the
cerebral ventricles (1).

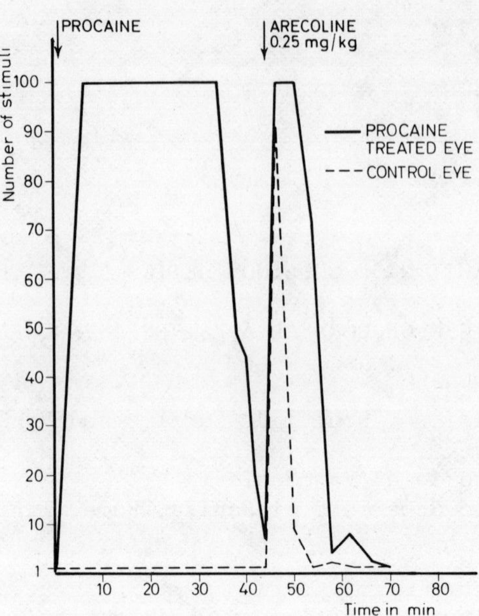

FIGURE 1: Return of procaine induced anesthesia after intravenous
administration of arecoline (0.25 mg/kg) to the rabbit.
Note the inhibition of the blinking response also present
in the untreated eye. Abscissa: time in min;
Ordinate: number of stimuli.

METHODS

Adult rabbits of both sexes were used. The animals were
placed in a restraining box, leaving the head free. The cornea
was brushed lightly with a calibrated horse hair, which induced
an immediate blinking response in control animals. After instillation in the conjunctival sac of 0.5 ml of 2% procaine, the number of touches necessary to provoke the closure of the eyelid was
assessed at intervals of 4 min. This number expressed the degree
of local anesthesia, which was considered complete when 100 stimuli failed to cause a blinking response (10). The other eye was
tested at the same time for control. Complete local anesthesia
was induced by procaine within 10 min and lasted 20-40 min; when
its effect had vanished, various doses of the drugs under study were
injected (i.v.), checking the eventual return of the local anesthesia. The ability of the drugs to restore the effect of the
local anesthetic agent was considered an expression of their antinociceptive action.

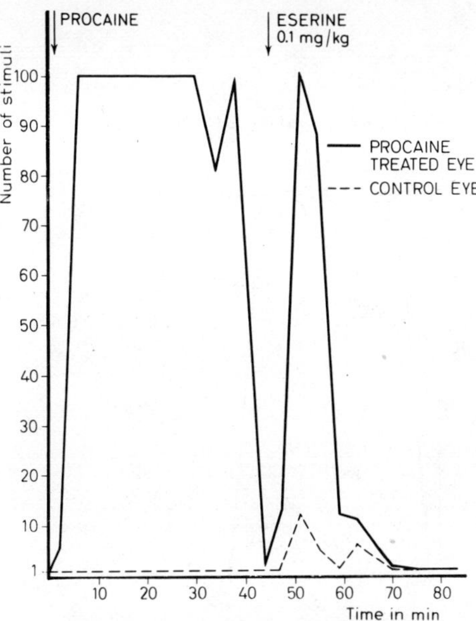

FIGURE 2: Return of procaine induced corneal anesthesia after
intravenous administration of eserine (0.1 mg/kg) to
the rabbit. Note the slight inhibition of the blinking
response in the untreated eye. For details see Fig. 1.

Compounds used were morphine sulphate, arecoline hydrobromide,
eserine salycilate, tremorine dihydrochloride, atropine sulphate,
pilocarpine hydrochloride and naloxone hydrochloride. All drugs
were dissolved in water and administered intravenously. The doses
in the text refer to the weight of the free base. To protect the
animals from the neurovegetative peripheral effects of the cholinomimetics, atropine methylbromide (0.5 mg/kg) was administered
subcutaneously 30 min before the experiments.

RESULTS

For each drug the minimal dose which restored the complete
anesthesia was assessed and compared with morphine.

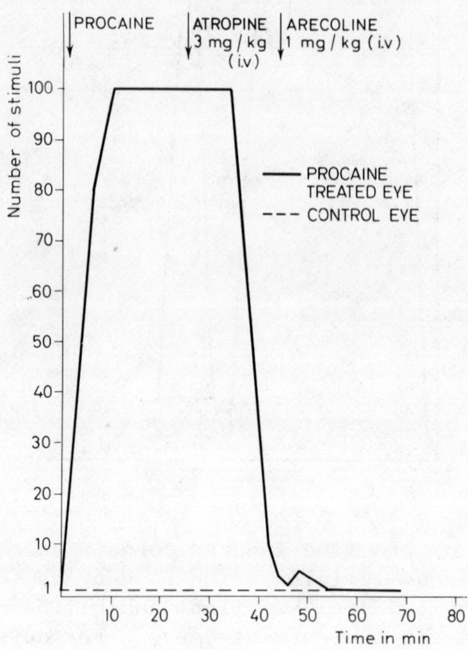

FIGURE 3: Antagonistic effect of atropine on the return of corneal
anesthesia induced by arecoline. Atropine (3 mg/kg i.v.)
administered 20 min before arecoline (1 mg/kg i.v.) inhibits the antinociceptive effect of the cholinomimetic
drug. In this experiment no influence on the blinking
response of the control eye is observed. For details
see Fig. 1.

Arecoline

Doses varying from 0.1 to 2 mg/kg were administered to a
total of 18 animals. The minimal dose restoring complete anesthesia was 0.25 mg/kg. The average duration of the effect in the
6 animals treated with such a dose was 8 min (Fig. 1). This dose
induced excitation and exophthalmus together with a partial inhibition of the blinking response in the untreated eye. This
phenomenon was more evident with higher doses of the drug (0.5 and
1 mg/kg), which induced complete anesthesia both in the treated
and untreated eye. With these doses the animals exhibited excitation and transient tremors. More intense tremors appeared after
the administration of 2 mg/kg, making it difficult to measure the
corneal reflex.

Eserine

This drug was administered to 8 animal in doses of 0.05, 0.1 and 0.15 mg/kg. The dose restoring complete anesthesia was 0.1 mg/kg; the duration of effect was 10 min (Fig. 2). A slight inhibition of the blinking response of the untreated eye was noticed with eserine. Despite pretreatment with the quaternary anticholinergic, atropine methylbromide, the highest dose of eserine induced marked signs of peripheral cholinergic stimulation, including salivation, defecation and paresis.

Tremorine

This drug was administered to 11 rabbits in doses varying from 0.1 to 0.4 mg/kg. The dose inducing the return of total anesthesia was 0.4 mg/kg, but the duration of the effect was very brief (5 min). At this dose, no influence on the responsiveness of the untreated eye was observed, nor was there any effect of the drug on motor function.

Pilocarpine

This drug was administered to 10 rabbits in doses of 5, 10 and 15 mg/kg i.v. The dose which restored complete anesthesia was 10 mg/kg and the duration óf its effect was 20-30 min. At the lower dose (5 mg/kg i.v.) a partial return of anesthesia was observed in 70% of the animals. Exophthalmus, excitation and chewing appeared after drug administration. At the highest dose (15 mg/kg) the very marked excitation made it difficult to measure the corneal reflex. No effect on the blinking response of the control eye was observed at the tested doses.

Morphine

The results obtained with morphine confirmed our previous findings (1). The return of complete anesthesia for 10 min was obtained with 3 mg/kg; doses of 5 mg/kg had a longer effect (20 min). At both doses lid closure of the other eye was not impaired; only after higher doses (40 mg/kg) was it possible to obtain a slight increase in the number of touches necessary to elicit the corneal reflex in the untreated eye.

Atropine and Naloxone

The inhibitory effects of atropine and naloxone were evaluated in a separate series of experiments. The antinociceptive effects

of the four cholinomimetics were abolished by atropine (3 mg/kg),
administered either before or at the peak of the effect (Fig. 3).
Atropine did not influence the return of anesthesia due to morphine,
which was instead antagonized by naloxone (0.5 mg/kg). On the other
hand, naloxone, administered in doses up to 10 mg/kg, was unable to
influence the antinociceptive effects of arecoline, pilocarpine and
eserine; it was instead effective against tremorine.

DISCUSSION

The antinociceptive action of cholinomimetics has been confirmed
in the present experiments since all drugs considered showed an influence on the corneal reflex similar to that exerted by the narcotic analgetics. Together with similarities, some differences
have been evidenced by the method employed. In contradistinction
to morphine, arecoline and eserine (in particular arecoline) inhibited the blinking reflex in the eye not treated with the local
anesthetic. The nature of this action has not been clarified. All
these drugs induce exophthalmus and this situation might mechanically hinder the close of the eyelids. It should be noted however that
in the untreated eye the presence of exophthalmus is not related to
the degree of inhibition of the blinking reflex observed after administration of the various drugs. The block of the corneal reflex
could be dependent on a general inhibition of reactive behavior
which has been attributed to cholinomimetics (cf. the alert, non-
motile behavior of Karczmar (6)), or it could be more specifically
related to a pain inhibiting effect. It should be noted in this
connection that Carrol and Lim (2) reported in the rat blockade of
the corneal reflex after 2.5 mg/kg (i.p.) of morphine, while in the
present experiments the alkaloid prevented lid closure only at doses
of 40 mg/kg. This effect never reached the criterium of complete
anesthesia (100 stimuli eliciting no response).

The similarities between the antinociceptive action of cholinomimetic agents and the morphine-like narcotics could indicate
that the substrate affected is not completely different and might
be related to the enkephalin receptor system. However, in spite of
these parallels there are other data which point to some basic differences. For example, in the present experiments the antinociceptive action of cholinomimetics is abolished by atropine and that
of morphine by naloxone, but not vice versa. From this point of
view, our observations confirm the data of Chen (3) and of Ireson
(5), but disagree somewhat with other published reports (4,9),
which described the inhibitory effects of several narcotic antagonists on the antinociceptive action of oxotremorine, eserine
and ACh.

ACKNOWLEDGEMENTS

The authors are grateful to Miss A. Pezzola for technical assistance.

REFERENCES

1. Aloisi, F. and Scotti de Carolis, A. (1980): Israel J. Med. Sci. (in press).
2. Carrol, M.N. Jr. and Lim, R.K.S. (1960): Arch. Int. Pharmacodyn. 125:383-403.
3. Chen, G. (1958): J. Pharmacol. Exp. Ther. 124:73-76.
4. Harris, L.S., Dewey, W.L., Howes, J.F., Kennedy, J.S. and Pars, H. (1969): J. Pharmacol. Exp. Ther. 169:17-22.
5. Ireson, J.D. (1970): Brit. J. Pharmacol. 40:92-101.
6. Karczmar, A.G. (1977): IN Cholinergic Mechanisms and Psychopharmacology (ed) D.J. Jenden, Plenum Press, New York, pp. 679-708.
7. Keil, W. and Pohls, F.H. (1936): Arch. Exp. Pathol. Pharmacol. 181:285-291.
8. Metys, J., Wayner, N., Metysova, J. and Herz, A. (1969): Int. J. Neuropharmacol. 8:413-425.
9. Pedigo, N.W., Dewey, W.L. and Harris, L.S. (1975): J. Pharmacol. Exp. Ther. 193:845-852.
10. Regnler, M.J. (1923): C.R. Soc. Biol. Paris 77:558-560.

DO GANGLIOSIDES AFFECT THE RECOVERY OF CHOLINERGIC ENZYMES

OF RATS AFTER SEPTAL LESIONS?

M. Wojcik and J. Ulas

Laboratory of Neurochemistry
Department of Biochemistry of Nervous System and Muscle
Nencki Institute of Experimental Biology
3 Pasteur Str., 02-093 Warsaw, Poland

INTRODUCTION

A partial progressive recovery of acetylcholinesterase (AChE) and choline acetyltransferase (CAT) activities has been shown after extensive medioventral septal lesions (7). It was suggested that the recovery of cholinergic enzymes activity could be due to sprouting of cholinergic nerve terminals from the intact part of the septum. Since gangliosides can stimulate the regeneration and re-innervation processes of cholinergic fibers in the peripheral nervous system (2), it seemed interesting to investigate the influence of gangliosides on the activity of cholinergic enzymes in the hippocampus of rats after medioventral septal lesions.

MATERIAL AND METHODS

Four groups of adult, male Wistar rats were used: 1) unoperated, untreated; 2) unoperated, injected i.m. with gangliosides; 3) operated, untreated; and 4) operated, injected i.m. with gangliosides. A mixture of gangliosides from cerebral cortex (GM_1 -32%; GD_{1a} -38%; GC_{1b} -17%; GT -13%) was obtained from Fidia Research Laboratories (Abano Terme, Italy).

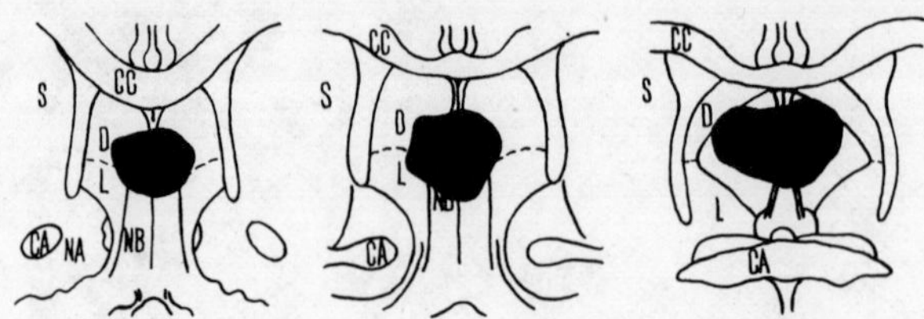

FIGURE 1: Example of a typical septal lesion. The extent of the
           damage is marked in black.   CA = anterior commissure,
           CC = corpus callosum; D = nucleus dorsalis, L = nucleus
           lateralis; NA = nucleus accumbens septum; NB = nucleus
           tractus diagonalis of Broca; NF = nucleus fimbriatus;
           S = striatum.

TABLE 1:   The effect of i.m. injection of gangliosides of AChE and
           CAT activities in the hippocampus of unoperated rats.

Group of Animals	AChE	CAT
1. Unoperated, untreated	5.67 ± 0.16(8)	4.23 ± 0.22(8)
2. Unoperated, injected i.m. with gangliosides	5.79 ± 0.38(6)	4.87 ± 0.40(6)

The data represent the results of a series of experiments performed
simultaneously on both groups of animals. Results are means ± S.E..
Number of cases in parentheses. Activity of AChE expressed in μmol
of AcTCh per mg protein per hr; activity of CAT expressed as μmol
ACh formed per 100 mg protein per hr. The animals were injected
with gangliosides eleven times during the period of 18 days.
Daily dose of gangliosides: 50 mg/kg. The differences between the
two groups are not statistically significant (Mann–Whitney U-test).

     Bilateral electrolytic lesions in the septum were made stereo-
taxically under sodium pentobarbital anesthesia. Animals of groups
[2] and [4] were injected with gangliosides eleven times during a
period of 18 days with doses of 50 mg/kg of gangliosides dissolved
in 0.1 ml of 0.01 M phosphate buffer. On the 18th day after surgery
the animals were decapitated under a light ether anesthesia and

TABLE 2:   The influence of ganglioside treatment on the activity of
           cholinergic enzymes in the hippocampus of septally lesioned
           rats.

	Condition	AChE	CAT
1.	Unoperated, untreated (control)	$5.03 \pm 0.26(29)$	$4.42 \pm 0.17(27)$
2.	Operated, untreated	$2.05 \pm 0.19(22)*$	$2.01 \pm 0.27(22)*$
3.	Operated, injected i.m. with gangliosides	$3.04 \pm 0.35(17)**$	$2.91 \pm 0.38(17)**$

See Table 1 for legend.   The control and the operated animals were
examined simultaneously in each experiment. The animals were injected with gangliosides eleven times between the first and the
18th day after surgery.   * = $p < 0.001$, with respect to control
group (Mann–Whitney U-test).   ** = $p < 0.05$, with respect to operated
uninjected group (Mann–Whitney U-test).

both hippocampi quickly removed.   The anterior part of each brain
was also removed and used for histological verification of the
lesions.   AChE activity was estimated by the method of Ellman et
al. (4), modified by Srebro et al. (10).   CAT activity was measured
according to Fonnum (5).   Protein content was estimated according
to Lowry et al. (6).   The Mann–Whitney U-test was used for the
statistical analysis.

RESULTS

The localization of a typical lesion is given in Fig. 1.   The
lesion generally destroyed most of the vertical limb of the diagonal
band, invaded the nucleus dorsalis and, slightly, the nucleus
lateralis.   Table 1 shows that repeated injection of gangliosides
had no effect on AChE and CAT activities in unoperated rats.   The
activities of AChE and CAT in the hippocampus of gangliosides injected rats were, respectively, 102% and 115% of the values of unoperated, uninjected rats.   The small differences between both
groups were not statistically significant.   Therefore, untreated
unoperated animals were used as a control group in further experiments on the effect of gangliosides in lesioned rats.

Table 2 shows the activity of AChE and CAT in three groups of
animals: control, operated uninjected and operated injected with
gangliosides. In the hippocampus of lesioned uninjected rats the
activity of AChE and CAT were 41% and 45%, respectively, of the control group. In the hippocampus of the operated animals injected
with gangliosides the decrease in activity of both enzymes was much
less pronounced. Activities of AChE and CAT were 60% and 66% of the
control values, respectively. The differences between operated
animals with and without ganglioside injection were statistically
significant for both CAT and AChE activities ($p < 0.05$, Mann-Whitney
U-test).

DISCUSSION

While not affecting the activity of both cholinergic enzymes
in control animals, gangliosides elevated the activity of these
enzymes in the hippocampus of the rats after medioventral septal
lesions.

Gangliosides, natural components of nerve terminals, are involved in a number of regulatory processes in the nervous system. It
was assumed that gangliosides which may be incorporated into the
neuronal membranes (11) play a role in a number of surface events,
like recognition phenomena, modulation of membranes, enzymes activities, ion binding (1,12) and neurotransmitter release (3).

If we accept the hypothesis that the recovery of the enzymic
activity was due to a neuronal sprouting mechanism, we can suggest
that gangliosides have a stimulating effect on the outgrowth of the
cholinergic nerve terminals, because they can markedly reduce the
effect of septal lesions. It was shown (8) that the ganglioside
GM_1, amounting to 32% of the brain cortex gangliosides mixture used
by us, crosses the blood brain barrier. It is probable that it may
be incorporated in vivo into neuronal membranes and can enhance
axonal sprouting. The suggestion that gangliosides exert a stimulating influence on outgrowth of the cholinergic nerve terminals
would agree with the results of Purpura et al. (9) and Ceccarelli
et al. (2). Purpura showed that in feline GM_1 gangliosidosis the
accumulation in the brain of gangliosides or of products of their
metabolism could induce neurite outgrowth from meganeurites. According to the results of Ceccarelli (2) gangliosides can stimulate
the regeneration and reinnervation processes of cholinergic nerve
fibers in the peripheral nervous system. The possibility that gangliosides might delay the degeneration process following the septal
lesion should also be taken into account.

ACKNOWLEDGEMENTS

We wish to acknowledge Professor Pepeu's suggestion to use the septo-hippocampal model to study the effect of gangliosides; Dr. Oderfeld-Nowak's involvement in elaboration of the data and in preparation of the manuscript; Dr. Korczynski's performance of septal lesions; and Mrs. Dabrowska's excellent technical help throughout the experiments. The investigation was supported by Project 10-4-1-01-8-pl of the Polish Academy of Sciences.

REFERENCES

1. Bonetti, A.C., Benvegnu, D., Leon, A., Facci, L., Toffano, C. and Tettamenti, G. (1979): IN Proc. Fifth Int. Symp. Glyco-conjugates, Kiel, pp. 576-579.

2. Ceccarelli, B., Aporti, F. and Finesso, M. (1978): IN Ganglio-side Function (eds) G. Porcellati, R. Ceccarelli and G. Tettamanti, Plenum Press, New York, pp. 275-293.

3. Cumar, F.A., Maggio, B. and Caputto, R. (1978): Biochem. Bio-phys. Res. Commun. 84:65-69.

4. Ellman, G.L., Courtney, K.D., Anders, V. and Featherstone, R. (1961): Biochem. Pharmacol. 7:88-95.

5. Fonnum, F. (1975): J. Neurochem. 24:407-409.

6. Lowry, O.H., Rosebrough, J.J., Farr, A.L. and Randall, R.J. (1951): J. Biol. Chem. 193:265-275.

7. Oderfeld-Nowak, B., Potempska, A. and Oderfeld, J. (1977): Neuroscience 2:641-648.

8. Orlando, P., Cocciante, G., Ippolito, G., Massari, P., Roberti, S. and Tettamanti, G. (1979): Pharmacol. Res. Commun. 11: 759-773.

9. Purpura, D.P. and Baker, H.J. (1978): Brain Res. 143:13-15.

10. Srebro, B., Oderfeld-Nowak, B., Klodos, I., Dobrowska, J. and Narkiewicz, O. (1973): Life Sci. 12:261-270.

11. Toffano, G., Benvegnu, D., Bonetti, A.C., Facci, L., Leon, A., Orlando, P. and Tettamanti, G. (1980): (in press).

12. Yamamawa, T. and Nagai, Y. (1978): TIBS 3:128.

COMPARATIVE STUDIES ON RAT BRAIN SOLUBLE ACETYLCHOLINESTERASE AND

ITS MOLECULAR FORMS DURING INTOXICATION BY DFP AND PARAOXON

H. Michalek, G.M. Bisso and A. Meneguz

Laboratory of Pharmacology
Istituto Superiore di Sanita
Rome 00161 Italy

INTRODUCTION

Previous reports from this laboratory (9) showed a differential
recovery of various forms of acetylcholinesterase (AChE) of the rat
brain after poisoning by DFP, the recovery being particularly rapid
for medium molecular weight forms, and the phenomenon markedly pronounced in the striatum (8). It is well known that DFP and paraoxon
cause substantially the same symptoms in the early phase of poisoning
although differing in their chemical structure, which determines the
type of phosphorylated AChE formed (diisopropyl- and diethylphos-
phoryl-, respectively) and the rate of subsequent recovery in vivo
and in vitro (for references see 1,4,7,12).

In order to shed some light on the mechanisms of such recovery
phenomena a comparative study of DFP and paraoxon at doses inducing
similar initial inhibition of brain AChE was carried out to compare
the percentage contribution of individual molecular forms during
12 days following an acute intoxication. Since the recovery in the
case of so-called irreversible inhibitors is presumed to depend
mainly on the biosynthesis of new enzyme molecules, the experiments
were performed using the soluble fraction of brain AChE where according to independent lines of evidence such process occurs (2,10,
13,14).

Furthermore, in order to ascertain whether there is a differential dephosphorylation of various molecular forms, some additional experiments were carried out on the in vitro reactivation

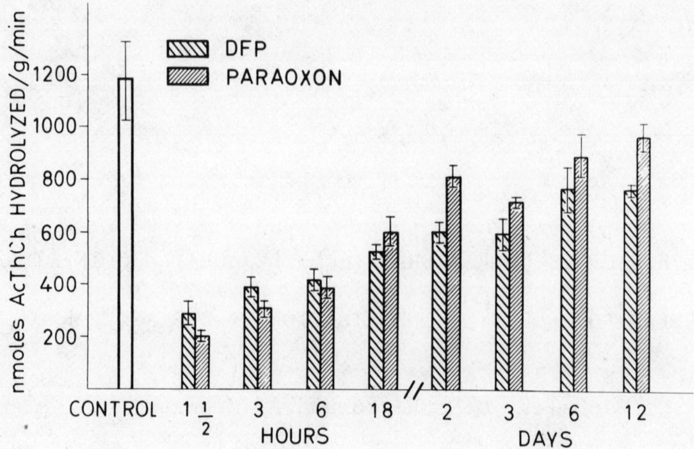

FIGURE 1: Recovery of rat brain soluble AChE following a single subcutaneous dose of DFP (1.1 mg/kg) or paraoxon (0.25 mg/kg).

(spontaneous and due to pralidoxime) of enzymatic preparations inhibited in vivo by DFP or paraoxon.

MATERIALS AND METHODS

Wistar male albino rats (180-220 g) were treated in balanced
replications subcutaneously with DFP (1.1 mg/kg in arachis oil) or
paraoxon (0.25 mg/kg in water), i.e. at doses resulting at 90 min
in a reduction of brain AChE to 24% and 17% of the control level,
respectively. Separate groups of six rats were sacrificed at 3, 6,
and 18 hr and 2, 3, 6, and 12 days after treatment. Appropriate
groups treated with the vehicle were run in parallel. Whole brain
including cerebellum was removed, frozen and thawed three times in
liquid air, homogenized with Tris-HCl buffer 0.038 M, pH 8.5 (tis-
sue:buffer ratio 1:6) and centrifuged for 1 hr at 100,000 g in a
Beckman L2 65B ultracentrifuge. The enzymatic activity determined
(6) in the supernatant (soluble AChE) corresponded to about 20% of
the activity present in crude homogenate. The separation of multiple molecular forms was carried out by disc polyacrylamide gel
electrophoresis. Samples (about 5-15 µl of supernatant) were run
in double layer rod gels (1 cm of 3% spacer gel and 9 cm of 7.5%
separating gel) using Tris-glycine buffer (0.05 M, pH 8.1) for
about 150 min at 1 mA/gel. The enzymatic reaction was performed

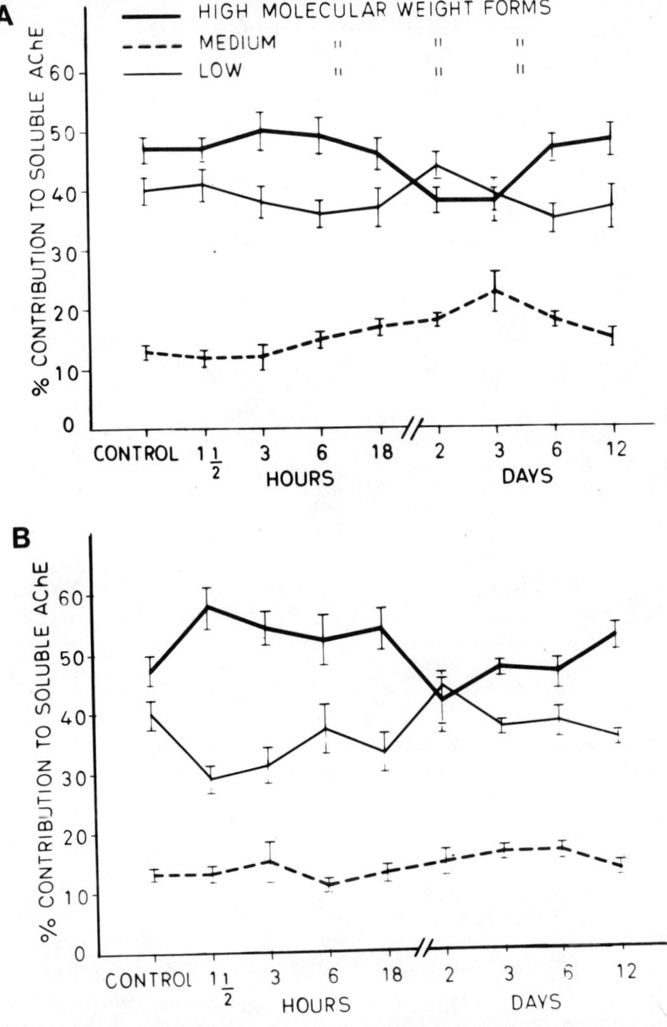

FIGURE 2: Percentage contribution of individual molecular forms to
soluble AChE during the intoxication by DFP (A) and
paraoxon (B).

at 30°C with shaking for 1 hr with acetylthiocholine iodide (0.0035 M)
in a solution containing 0.2 M maleic acid, 0.02 M $CuSO_4$ and 0.38 M
glycine at pH 6.5, followed by immersion of gels in 30% Na_2SO_4. The
gels were finally stained in a saturated solution of dithio-oxamide
and scanned at 600 nm in a Gilford 240 spectrometer. The peak areas

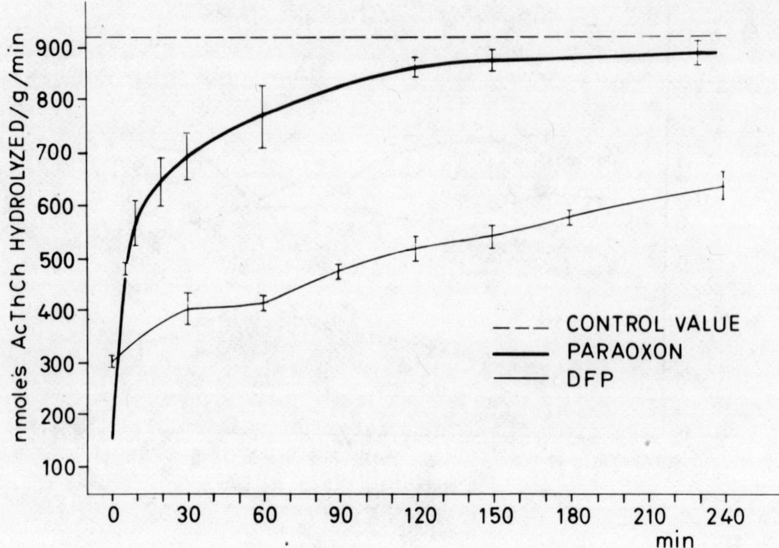

FIGURE 3: Reactivation _in vitro_ by pralidoxime (10^{-4} M) of soluble
AChE inhibited _in vivo_ by DFP or paraoxon.

were integrated using a Hewlett-Packard 9804 digitizer. The data
were elaborated statistically by analysis of variance (ANOVA).

RESULTS AND DISCUSSION

The data of Fig. 1 show some differences in the rate of recovery of soluble AChE, with somewhat higher enzyme levels after
DFP in the initial phase, followed by faster recovery after paraoxon
in subsequent days. In fact, a 2 x 8 ANOVA yielded a significant
interaction between type of organophosphate treatment and time of
measurement ($p < 0.02$).

The data of Fig. 2 indicate that during the initial phase of
intoxication the relative contribution to the enzymatic activity
of individual molecular forms does not differ from that found in
the controls in the case of DFP (A) but changes markedly in the
case of paraoxon (B): the low molecular weight forms, characteristic
of cytosol appear to be more depressed than high molecular weight
forms, probably in relation to the greater hydrosolubility of this
compound. As concerns the subsequent recovery phase, the DFP data
show a selective increase of medium molecular weight forms, reaching

a peak at 72 hr and less marked increase of low molecular weight
forms. These modifications of the pattern are partially reversed
by the sixth day and disappear by the 12th day. In the case of
paraoxon the low molecular weight forms, most depressed initially,
show a very fast recovery, while the increase of medium forms appears negligible and normalization of the pattern occurs by the
third day. These differences in trends were confirmed by 2 x 8
ANOVA's carried out separately for each molecular form showing
significant effects of treatment (T) and intervals (I), with no
interaction between the two, both in the case of high weight forms
(T: $p < 0.001$; I:$p < 0.001$) and in the case of the low weight forms
(T:$p < 0.02$; I:$p < 0.02$), while a significant T x I interaction appeared in the case of the medium weight forms ($p < 0.05$).

Additional experiments _in vitro_ using soluble AChE from animals
sacrificed 90 min after treatment indicated a negligible spontaneous
reactivation of DFP inhibited enzyme at 30°C for 24 hr while the
reactivation of paraoxon inhibited enzyme was considerable. Correspondingly, the reactivation by pralidoxime (Fig. 3) was markedly
faster in the case of paraoxon than in the case of DFP inhibited
enzyme. However, none of the individual molecular forms appeared
to be reactivated preferentially _in vitro_ after DFP or paraoxon
inhibition.

In conclusion, the transient increase in the activity of medium
and/or low molecular weight forms in the early phase of recovery
after DFP and paraoxon has to be considered as an initial step in
the biosynthesis of new molecules of AChE. Similar increases
(blocked by inhibitors of protein synthesis) have been described
for cultured chick embryo muscle cells treated by DFP (13) or
paraoxon (3), and for mouse neuroblastoma cells (11) or rat retina
(5) after DFP. An apparent difference between the recovery after
DFP and paraoxon emerging in the present study consists in a marked
increase of low molecular weight forms (but not medium weight forms)
only in the case of paraoxon. This may depend on the particularly
strong initial inhibition of such forms, with a feedback mechanism
controlling their recovery, or on different rates of interconversion
into high molecular weight forms. Moreover, the faster recovery of
paraoxon as compared to DFP inhibited enzyme, may partially depend
on differential rates of dephosphorylation as suggested by experiments on spontaneous and pralidoxime induced reactivation.

ACKNOWLEDGEMENT

This research was supported by the Commission of European
Communities Contract No. ENV/367/I-(S).

REFERENCES

1. Bignami, G., Rosic, N., Michalek, H., Milosevic, M. and Gatti, G.L. (1975): IN Behavioral Toxicology (eds) B. Weiss and V.G. Laties, Plenum Press, New York, pp. 155-215.

2. Bisso, G.M., Nemesio, R. and Michalek, H. (1980): IN A Multidisciplinary Approach to Brain Development (eds) C. Di Benedetta, R. Balazs, G. Gombos and G. Porcellati, Elsevier-North-Holland, Amsterdam, pp. 235-236.

3. Cisson, C.M. and Wilson, B.W. (1977): Biochem. Pharmacol. 26: 1955-1960.

4. Cohen, J.A. and Oosterbaan, R.A. (1963): IN Handbuch der Experimentellen Pharmakologie, Vol. 15 (ed) G.B. Koelle, Springer, Berlin, pp. 299-373.

5. Davis, G.A. and Agranoff, B.W. (1968): Nature 20:277-280.

6. Ellman, G.L., Courtney, K.D., Andres, V. Jr. and Featherstone, R.M. (1961): Biochem. Pharmacol. 7:88-95.

7. Main, A.R. (1979): Pharmacol. Ther. 6:579-628.

8. Meneguz, A., Bisso, G.M. and Michalek, H. (1981): Clin. Toxicol. (in press).

9. Michalek, H., Meneguz, A., Bisso, G.M., Carro-Ciampi, G., Gatti, G.L. and Bignami, G. (1979): IN Advances in Pharmacology and Therapeutics, Vol. 9 (ed) Y. Cohen, Pergamon Press, Oxford, pp. 187-201.

10. Niemierko, S. and Skangiel-Kramska, J. (1976): Bull. Acad. Pol. Sci. (Biol) 24:445-448.

11. Rieger, F., Faivre Bauman, A., Benda, P. and Vigny, M. (1976): J. Neurochem. 27:1059-1063.

12. Usdin, E. (1970): IN Anticholinesterase Agents (ed) A.G. Karczmar, Pergamon Press, Oxford, pp. 47-354.

13. Wilson, B.W. and Walker, C.R. (1974): Proc. Nat. Acad. Sci. 71:3194-3198.

14. Yaksh, T.L., Filbert, M.G., Harris, L.W. and Yamamura, H.I. (1975): J. Neurochem. 25:853-860.

BASIC PHENOMENA UNDERLYING NOVEL USE OF CHOLINERGIC AGENTS,

ANTICHOLINESTERASES AND PRECURSORS IN NEUROLOGICAL INCLUDING

PERIPHERAL AND PSYCHIATRIC DISEASE

A.G. Karczmar

Department of Pharmacology and
Institute for Neuropharmacology
Loyola University Medical Center
Maywood, Illinois   60153   USA

INTRODUCTION

This paper concerns the bases underlying clinical uses of
peripherally and centrally acting cholinergic drugs, described
elsewhere in this section. Certain additional uses of such drugs
that, while not described in this volume, deserve advocacy will
also be mentioned. Besides the clinical uses mention in the chapter there are some twenty classical indications for cholinergic and
anticholinergic drugs and anticholinesterases (antiChE's) that include glaucoma, gastric hypermotility, motion sickness and parkinsonian disease (49,64,65), as well as certain novel uses (as for
example in anxiety or fear); these cannot be discussed here for
lack of space.

Most of the clinical applications discussed in this section
conern the central nervous system (CNS). Several papers deal, however, with other than CNS aspects of the use of cholinergic drugs.
One of these papers concerns the somatic nervous system and the
data adduced will point out the continued clinical pertinence of
cholinergic drugs in the condition in question. Other papers (17,
29) deal with rbc as possible markers of choline (Ch) transport,
cholinergic function and/or antipsychotic therapy. These papers
serve to remind us that while the present volume concerns essentially neurotropic actions of cholinergic drugs, cholinergic processes
are involved also in membrane and trophic phenomena unrelated to
transmission (56).

853

Cholinergic Drugs and Myasthenia Gravis

For more than four decades antiChE's were drugs of choice in
the treatment of myasthenia gravis. Recently, it appeared that the
latter may be an autoimmune disease (19,31,87) that involves a
number of specific antibodies (113) and generates inflammatory
processes and pathology of both nerve terminal and of the endplate
(EP). Thus, the earlier treatment with ACTH and more recent em-
ployment of steroids such as prednisone appears to be both indicated
and rational. Yet, as side actions and a damage of body defenses
arises with such a treatment, in many cases of myasthenia antiChE's
may be preferred, at least for the initial and, sometimes, for prolonged treatment. However, this leads frequently to development
of refractoriness of the EP's to antiChE's, as pointed out early
(109; cf. also 42). It is very likely that what is involved in
this phenomenon is the desensitization (DS) of the EP to endogenous
acetylcholine (ACh)(86,108). Indeed, under experimental conditions,
repeated or prolonged pulses of ACh or repeated presynaptic stimulation causes DS (2,59,77,86), and antiChE's accelerate this process
by a direct EP effect that adds to the DS action which they induce
by causing accumulation of ACh released from the terminal (1,2).
In addition, myasthenic EP may be particularly vulnerable to DS
(Jacobs and Karczmar, unpublished data). It is of interest in this
context that of the many substances tested as possible antagonists
of DS (77) only one, which is probably not usable clinically, can
antagonize antiChE induced DS (1,2,89).

It is important then to employ antiChE's clinically so as to
minimize and delay development of refractoriness, i.e. to use minimal doses still capable of yielding effective concentrations in the
plasma and at the EP. There lies then the importance of the knowledge of the pharmacokinetics of antiChE's, and the study of
Eckernas et al. (cf. 22) fulfills this goal, particularly as it
shows that plasma concentrations of pyridostigmine and neostigmine
are variable, and that cholinergic crisis - a condition particularly prone to produce refractoriness and DS - may arise unexpectedly
with a dose that could be expected to be safe.

Cholinergic System and Transport Phenomena

It appears that the cholinergic system of the rbc may reflect
certain forms of mental disease (cf. 29), and, by extrapolation,
the cholinergic state of the CNS; there seems to be also an interaction at the level of both rbc's and the blood-brain barrier or
the CNS, between lithium and Ch uptake (17,38,88; cf. elsewhere in
this volume). The relevant basic facts concern the participation
via several mechanisms of the cholinergic system in non-transmittive

functions. These functions of the cholinergic system that may be
referred to as metabotropic (82) include the control of the permeability of blood-brain barrier, epithelium, exocrine glands, and
of the rbc, and structurization and development of the cells including neurons and their membranes (for references cf. 41,50,56,
63). A common mechanism may involve ACh or Ch activation of the
phosphaditic acid dependent sodium pump and phorphorylation processes (33). Particularly important in this context is the augmentation of blood-brain barrier permeability by antiChE's and
cholinergic agonists and the role of ChE's in the regulation of rbc
permeability (cf. 43). It may be then that alterations in metabotropic cholinergic system may lead to parallel changes on the one
hand, in brain function and, on the other, in influx-efflux regulation on the rbc Ch. Indeed, it is consistent with this admittedly far fetched speculation that abnormalities in transport of Ch
of the rbc's were reported in several neurological and mental conditions in which changes of central cholinergic function are suspected to occur (for references see 29). Furthermore, there may be
an interplay both at the blood-brain barrier and in the rbc's
between cholinergic system and Ch transport on the one hand, and
lithium on the other, which may underlie lithium action on the rbc
Ch (17) and on brain permeability to Ch (84).

Clinical Exploitation of Centrally Acting
Cholinergics and Anticholinergics

     Clinical studies of cholinergic agonists described in this
volume pertain to motor disorders, memory and primary affective
illness or depression. Basic facts underlying the relation of
those conditions to cholinergic system will be reviewed. Furthermore, two additional conditions that acquired recent pertinence in
the area of cholinergic therapy, pain and schizophrenic states will
be adduced.

     First, certain aspects of general relevance should be raised.
Distribution of cholinergic synapses is significant in the present
text, e.g. their radiation to basal ganglia and cerebellum is pertinent for the use of cholinergic drugs in motor dysfunction while
their distribution to limbic system, particularly hippocampus and
projection from nucleus basalis magnocellularis to neocortex and
frontal cortex (cf. 70) are pertinent for their uses in schizophrenia and other mental conditions, and in memory processes (see
also 46,83).

     Second, it is not easy to pinpoint mechanisms underlying par-
ticular clinical effects of cholinergic drugs, as they act via

several mechanisms thatconcern, moreover, various sites within cholinergic synapses. This is discussed more specifically elsewhere
in this volume (20). Furthermore, cholinergic drugs may exert their
effects indirectly via their action on other transmitter systems
(20,48). Altogether, these various effects and mechanisms interplay
in a complex manner with cholinergic circuitry and the circuitry of
other transmitter systems in inducing behavioral effects of interest
in the present context.

A particular problem in this context concerns the evaluation
of the effects of drugs presumed to be precursors of ACh and to
augment activity of cholinergic neurons (cf. 115). Presumably electrophysiological evaluation would constitute a direct means to establish their effectiveness; there are obvious difficulties in de-
vising an appropriate protocol. Evaluation of ACh turnover, which
also should reflect cholinergic activity, is similarly difficult.
Identification of ACh precursors and the knowledge of their kinetics
and uptake mechanisms are not complete (35,36), and there are several different methods of calculation of ACh turnover, all subject
to criticisms (cf. 37). Finally, measurement of ACh turnover and/
or appropriate electrophysiology cannot be carried out in man. Altogether, indirect indices are used by many investigators, whether
in man or in animals, to evaluate precursor effectiveness. For
instance, in animal studies an effective precursor may be expected
to prevent ACh depletion induced by atropine via a mechanism described by Szerb (106, see also 112) and, in man, metabolites resulting from increased catecholaminergic function and hypothalamic
and pituitary hormones may appear, following precursor treatment,
in blood, urine or csf (13). Two points of caution must be raised
in this context. First, there is controversy as to whether the use
of precursors can raise brain ACh levels or turnover (compare 21
and 115). Second, precursors, particularly Ch may exert their actions not be affecting ACh turnover but via their direct agonist
effect (43). In fact, Krnjevic and Reinhardt (43) demonstrated
such effects on cortical neurons following iontophoretic administration of Ch.

Cholinergic Drugs, REM Sleep and Motor Dysfunction

Elsewhere in this volume several investigators deal with centrally originating motor disorders. There is good evidence that
some of these conditions are concomitant with a differential loss
of cholinergic neurons or cholinergic function, as in the case of
Alzheimer's disease (for references cf. 4). Less compelling evidence is available for related conditions of tardive dyskinesia,
Huntington's chorea or aging (see 4,16).

An underlying condition may be faulty oxidative metabolism or
hypoxia to which ACh synthesis is very vulnerable (cf. 8). Besides

certain genetic conditions such as Friedreich's ataxia this may
arise in a number of encephalopathies (cf. 8). In addition, hypoxia
may produce an effect functionally similar to that arising from a
deficit in ACh synthesis, namely, the blockade of cholinergic, particularly muscarinic responses which are very vulnerable to low
oxygen tension and inhibitors of oxidations (cf. 55,67).

Of course, if the conditions in question are characterized by
a cholinergic deficit, they would be expected to benefit from the
use of cholinergic precursors of agonists which would attack directly their cause. However, cholinergic agonists may be expected to
ameliorate the conditions in question symptomatically, via their
specific behavioral effect termed "cholinergic alert non-mobile
behavior (CANMB)(47,53).

Depression of motor activity and/or exploratory activity by
cholinergic agonists and antiChE's may be readily demonstrated in
several paradigms, including their block of experimental hyper-
activity induced by amphetamine, septal lesions and/or lead poisoning
(98). It is of interest that, due to their lipophylicity and CNS
sequestration, organophosphorus drugs exert this activity at very
small doses, and that this effect can be demonstrated not only by
means of activity cages but also in vivarium conditions (Table 1).
That hypoactivity induced by cholinergic agonists is not coupled
with the depression of mental function can be demonstrated dramatically by employing inclined climbing screen avoidance paradigm
(58) in which both learning and motor (climbing activity) may be
evaluated simultaneously as under these conditions antiChE's attenuate climbing while improving learning (Table 1). It should be
added that cholinergic agonists also attenuate clonic behavior induced in guinea pigs by 5-hydroxytryptophan administration (Kindel
and Karczmar, unpublished data). This is pertinent with regard to
possible use of cholinergic agonists in Huntington's chorea (see 3).

In view of the component of mental awareness of the CANMB its
EEG concomitant is significant (51). Indeed, while cholinergic
agonists may also induce, under certain circumstances, slow EEG
and/or slow sleep (51,79,93), the characteristic finding with
these drugs is the EEG desynchronization (114) combined with hippocampal theta rhythm (73), which corresponds to mental arousal and/
or learning state (75). It must be emphasized that these behavioral
and EEG phenomena induced by cholinergics closely resemble spontaneous EEG status ("slow theta")(107) characteristic for an alert
but immobile animal; in fact, both phenomena are antagonizable by
atropine (see for e.g. 15,75).

Still another behavior may be considered as a component of
CANMB, as pointed out by Karczmar (45); this is Rapid Eye Movement
Sleep (REMS). Indeed, much evidence suggests the cholinergicity

TABLE 1: Effects of antiChE's and scopolamine on "mouse city" behavior, motor activity and conditioned avoidance response (CAR).

Drug & Dose	City Behaviors				Climbing Screen (CAR)	
(mg/kg)	Motor Activity	Exploration	Grooming	Steroetyp.	Learning	Climbing
Physostigmine (0.05)	−14	−26*	−15	−25**	+18**	−17*
DFP (0.25)	−7	−32**	−27**	−38**	+18**	−10
Scopolamine (1.0)	+20*	+18**			−25**	

Percent change from controls: * = $p < 0.05$; ** = $p < 0.01$.

of REMS, including the early demonstration by Jouvet (39) that REMS
may be blocked by atropine. This demonstration has since been
duplicated in man (25,103; for other evidence see 51).   It is of
interest, in fact, that while the role of catecholamines (CA) in
REMS is well known (40,95), cholinergic REMS may be induced in the
absence of CA's (54).

That REMS-like state is a part of CANMB emphasizes the specific
character of the latter. When Wikler (114) observed early cholinergic EEG dysynchronization and its conversion by atropine or
scopolamine into a slow rhythm, he stressed the "divorce" of either
pattern from the expected behavioral counterpart, namely behavioral
arousal and sleep, respectively. He could not have realized at
that time that what he was actually dealing with was two specific
behavioral patterns; CANMB on the one hand, and "confusion" and EEG
hypersynchrony (74) on the other. Altogether, CANMB and cholinergic
antagonism of clonic motion must be taken under consideration with
respect to cholinergic effects on hypermotility, mania and ataxic
or choreatic motion. Certain mental health and psychiatric aspects
of this matter will be referred to subsequently.

Cholinergic System, Memory and Learning

Several lines of evidence implicating cholinergicity of memory
were described by Karczmar (49,50).   Important pharmacological evidence is that provided by Drachman (18; see also 26) that senile
amnesia resembles, in a parametric analysis, that induced in younger
subjects by scopolamine, and that physostigmine facilitates memory
processes, particularly in aged population. Drachman (18) linked
this evidence with presumed differential loss of cholinergic neurons
in old age (see above). Animals studes are, generaly, corroborative.
In fact, not only muscarinic, but also nicotinic drugs may be effective (7,9). It must, however, be remembered, in keeping with the
concept of CANMB, that cholinergic drugs are particularly effective
in learning trials involving passive avoidance rather than in escape or active avoidance paradigms (cf. 46,49).   There is also
evidence based on ablation experiments implicating areas rich in
cholinergic synapses in memory function as well as functional and
neuropharmacological evidence associating hippocampal and limbic-
cortical interactions with memory processes (18,75,80,81).   Additional mechanisms include the dipole consisting of periventricular
grey and median forebrain bundle (90,105) which may consittúte the
anatomical substrates of the punishment-reward dipole. Within this
framework cholinergic agonists active the former, inhibiting and
preventing undue attention to trivial stimuli and thus reinforcing
responses to vital perceptions (11, cf. 45).   In view of this and
related evidence, particularly results demonstrating biphasic
action (57) of cholinergics on memory and other behavior, Deutsch
(14) proposed that memory is constituted by "cholinergic memory

synapses" that change in sensitivity during memory processes, thus
cholinergic agonists and antiChE's exerting facilitatory and blocking effects at different points in the course of memory phenomena.

It bears stressing that cholinergic synapses lend themselves
particularly to temporally extended modulations that may serve as
models of learning. For instance, in the ganglia long-lasting augmentation of the excitability of the muscarinic receptor may be
elicted by other ganglionic transmitters or modulators (20,72).

Pain Mechanisms

There are at this time no instances of clinical exploitation
of cholinergic analgesia although in normal (non-pained) subjects
pain threshold was augmented by physostigmine (101). An IND protocol (No.16,412) (52) was, however, approved by FDA and indubitively
some work in this area will be instituted soon.

Actually, the phenomenon is known for some 35 years (23,24,
104, cf. 50), when intense cholinergic analgesia, including potentiation of morphine and codeine analgesia were first reported. More
recently, it was shown that cholinergic agonists produce these effects given systemically as well as following their localized application to hypothalamus, limbic sites, medulla, and mesencephalon
(49). It is of particular interest that in animals analgesic action
of DFP is characterized by smallness of dose (amount to about 1/50
to 1/100 of LD_{50}) and long duration of the effect (61,62); this may
be a point of particular clinical interest.

There is a close topographical relation between 5-HT, substance
P, enkephalins and cholinergic neurons at a number of sites pertinent for nociception, including basal ganglia, limbic system and
spinal cord (s. gelatinosa for substance P, 5-HT and enkephalins;
dorsal cord for ACh)(see 32,66,110). Thus, these substances were
implicated in pain perception and analgesia. For example, raphe
lesions may prevent morphine from exerting analgesia (96), and inhibitors of serotonin synthesis may cause hyperalgesia (61).
Serotonergic mechanisms may be particularly significant in the context of DFP analgesia, as DFP was shown to increase very markedly
levels and turnover of serotonin in several brain parts (5,6,61).
Serotonin contribution to DFP analgesia is not likely, however.
When blockade of serotonin synthesis and of DFP augmentation of
serotonin levels was prevented by appropriate regimens of p-
chlorophenyl alanine, DFP analgesia was still present, and in fact,
augmented (61).

Yet cholinergic analgesia may be still indirectly elicited,
for instance, via interaction with endephalinergic system. Existing
evidence both supports and opposes this concept. Opiate antagonists

seem to exert some effect on cholinergic analgesia; while data are
not quite consistent (compare 92 with 92; see also 91), it appears
that there is some similarity of, but also some stereochemical
difference between, opiate and cholinergic receptors involved in
analgesia. It is of interest in this context that morphine tolerant
rats were rendered analgetic by DFP or pilocarpine (61); this fact
is of potential clinical interest.Conversely, Pedigo et al. (91)
reported that there was no cross tolerance to morphine in animals
rendered tolerant to analgetic effect of intraventricular administration of ACh. The questions thus raised will be answered in the
future in terms of studies of the opiate receptors, several types
of which were described, and of the effect of cholinergic agonists
on endogenous endorphins and enkephalins.

Affective and Depressive Illness and Schizophrnia

Elsewhere in this volume Sitaram and Gillin (102) suggest that
patients suffering from affective illness may exhibit cholinergic
supersensitivity, as arecoline induced in these patients the second
REMS period significantly faster than in normal subjects (see above).
This is consistent with the indication that the REMS latency of
these patients may be unduly short, and that they are helped particularly by those tricyclic antidepressants that exhibit potent
atropinic activity (69, cf. 49). It is of further interest that
these patients may be particularly helped by atropine, or scopolamine coma therapy ((27, cf. 49,76)(after a number of cycles in its
vogue, this treatment is used very seldom today.) Furthermore, in
normal subjects, cholinergic agonists and antiChE's induce depressive syndromes, hallucinations, and paranoid behavior (e.g. 10;
cf. 12,50,100). The concept that cholinergic excess is related to
depressive and defensive reactions may have as its basis animal
studies indicating that the periventricular grey, the "punishment"
substrate of the punishment reward dipole (see above), is sensitive
to cholinergic agonists and that the latter increase, while atropine
and scopolamine decrease, the defensive and fear-mediated responses
85,94); the concept of CANMB may also be pertinent in this context.

Should the primary depression and/or schizoaffective states
then represent, etiologically, the converse of schiphrenic states?
Indeed, cholinergic agonists were shown almost a quarter of a century ago (111) to alleviate momentarily hebephrenic schizophrenia,
and, more recently, Klawans et al. (60) pointed out that the incidence of schizophrenia may be unduly high among patients suffering
from Huntington's chorea. They proposed that the cholinergic
deficit and/or cholinergic-dopaminergic imbalance may extent in
these subjects from extrapyramidal system to structures involved
in schizophrenic processes. On pharmacological grounds, similar
proposals were made earlier (12,30,34). Let it be stressed, however, that an opposite viewpoint is also possible as in the late

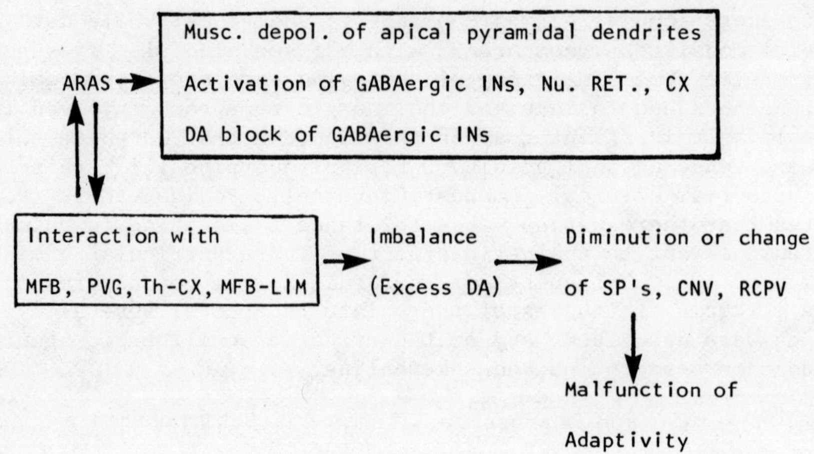

FIGURE 1: Diagrammatic representation of multitransmitter systems
engaged in adaptive function. Modified from Marczynski
(81,82). See text for further explanations.

50's Harold E. Himwich proposed that the anticholinergic component
of action of phenothiazines and butyrophenones was necessary for
their effectiveness. Furthermore, more recently, supportive clinical data were obtained by Neubauer et al. (88), and that antiChE's
may exacerbate schizophrenia (111).

Finally, Haase's (28) preliminary findings indicated that the
use of atropinics in conjunction with antipsychotic therapy for
prevention of extrapyramidal (parkinsonian) side effects of the
latter attenuated therapeutic effectiveness of the therapy, and
recently, Singh and his associates (99, see also 49). Data obtained with animal models of schizophrenia are pertinent. Compulsive and perseverant behavior characteristic for schizophrenia
may be speculatively considered as reflected in animal stereotypic
and self-stimulation phenomena, while avoidance paradigm may stimulate non-adaptive and paranoid characteristics of schizophrenia.
It is, therefore, of interest that while anti-psychotics block
self stimulation, avoidance and stereotypic behavior, cholinergic
agonists exert similar effects and/or synergize with anti-psychotics,
whereas atropinics either exacerbate the behaviors in question or
antagonize the effect of anti-psychotics (see 46, 100). Similar
interaction between anti-psychotics and cholinergic agonists and
antagonists may be demonstrated with regard to stereotypy induced
by amphetamine or apomorphine.

Theoretical bases for the animal data described above and for
the concept that cholinergic-dopaminergic imbalance underlies
schizophrenia were provided recently by Marczynski (80,81). Briefly,
Marczynski stresses the role in adaptive behavior of the interaction
between ascending reticular activating system (ARAS) and brainstem
reticular formation (BSRF) and relay systems such as median forebrain bundle (MFB) and periventricular grey (PVG: reward-punishment
axis), and ARAS and BSRF radiation to the limbic system and the loop
consisting of the thalamocortical pathway, perforant pathway of
Cajal, and limbic system. The functions of this organization and
contingent adaptive phenomena include evaluation of the status of
the organism resulting from selection of a particular response from
effector repertory as compared with goal perception. This is EEG-
wise, expressed in slow surface potentials (SP's), such as compound
EP's, contingent negative variation (CNV) or anticipatory wave
(Bereitschafft potential), and reward contingent positive variation
(RCPV), as well as synchronization-desynchronization dipole.
Marczynski (80,81) speculates that this dipole depends on the interaction between cholinergic ARAS which induces muscarinic depolarization of apical pyramidal dendrites as well as activation of inhibitory GABAergic interneurons in the nucleus reticularis thalami and
in the cortex on the one hand, and dopaminergic ARAS or BSRF which
inhibits the interneurons in question causing disinhibition (arousal)
on the other (Fig. 1). Slackening of the dopaminergic tonus increases inhibition, and relates to positive SP's and synchronization.
In schizophrenics, changes in SP's, slowing of their recovery upon
repeated stimulation, lower CNV amplitude, etc, result from imbalance between these two systems. For instance, muscarinic excess
or dopaminergic sensitization of muscarinic dendritic depolarization
may induce pathology of sensory input and its evaluation.

Obviously, the task of providing sound theoretical bases for
the understanding of either depressive or schizophrenic syndromes
is not even half over. Certain obvious and all-too-easy criticisms
may be raised at this time. The concept that dopaminergic-cholin-
ergic imbalance underlies schizophrenia is not much less simplistic
than that of this disease being due to dopaminergic excess, abnormal
indole metabolism, or related speculations proposed previously (see
69,97). Furthermore, the animal models used may have little relation to schizophrenia. Avoidance behavior represents a teleological
attempt of the animal to avoid or respond by emotional signs (CER)
to, noxious stimuli; why should the blockade of this behavior be
considered as analogous to the favorable effect of phenothiazines
on schizophrenic patients? In fact, the animal paradigm of "no-
goal," non-adpative compulsive behavior (78) may be a better model
of schizophrenia than avoidance or stereotypy, as in this case the
animal can be shown "to know better: and yet continues in his per-
servation" (71). It is of interest that many antipsychotics, anti-

depressants, sedatives and stimulants were tested, to no avail, in this paradigm (71).

SUMMARY

Two comments are appropriate. In keeping with the aims of this volume, the evidence presented in this chapter as well as clinical evaluations elsewhere in this volume primarily concern cholinergic systems; yet, as stressed in this paper, cholinergic systems cannot be considered independently of other neurotransmitters and of the circuitry involved, and the concepts presented here may be oversimplistic in nature. Second, when clinical situations result from neuronal loss, as in Alzheimer's disease, memory malfunction or tardive dyskinesia, no treatment can be helpful unless directed at neuronal growth - quite a challenge in the case of the CNS! In this context, it is possible that we deal in certain instances not only with cholinergic deficit but, indeed, with cholinergic supersensitivity, a point raised elsewhere in the volume by Sitaram and Gillin (102). Thus, cholinergic drugs may be too effective at one time and, at another time, when sufficient neuronal loss has occurred, not effective at all!

Thus, all is not well, as yet, with cholinergic therapy. However, to those of us who studied the cholinergic system for many years now and complained in the past that this system is clinically underexploited, the present status of the art constitutes a marked theoretical and empirical advance.

ACKNOWLEDGEMENTS

Published and unpublished investigations carried out in these laboratories and referred to in this paper were supported in part by USPHS NIH grant NS-06455, VA grant 4830 and a grant from the M.E. Ballweber Foundation.

REFERENCES

1.   Akasu, T. and Karczmar, A.G. (1979): Fed. Proc. 38:430.
2.   Akasu, T. and Karczmar, A.G. (1980): Neuropharmacology 19: 393-403.
3.   Aquilonius, S.M. and Eckernas, S.A. (1980): This volume.
4.   Barbeau, A., Growdon, J.H. and Wurtman, R.J., eds (1979): Choline and Lecithin in Brain Disorders, Nutrition and the Brain, Vol.5, Raven Press, New York.

5.   Barnes, L., Karczmar, A.G. and Ingerson, A. (1975): Pharma-
     cologist 17:180.
6.   Barnes, L., Karczmar, A.G. and Ingerson, A. (1976): Pharma-
     cologist 18:202.
7.   Battig, K. (1978): Behavioral Effects of Nicotine, Karger,
     Basel.
8.   Blass, J.P., Gibson, G.E., Duffy, T.E. and Plum, F. (1980):
     This volume.
9.   Bovet, D., Bovet-Nitti, F. and Oliverio, A. (1967): Ann. N.Y.
     Acad. Sci. 142:261-267.
10. Bowers, M.B. Jr., Goodman, E. and Sim, V.N. (1964): J. Nerv.
     Ment. Dis. 138:389.
11. Carlton, P.L. (1968): IN Psychopharmacology - A Review of
     Progress 1957-67 (ed) D.H. Efron, US Govt. Printing Office
     PHS Publ. No.836, Washington D.C., pp. 125-135.
12. Davis, K.L., Berger, P.A., Hollister, L.E. and Barchas, J.D.
     (1978): Life Sci. 22:1865-1872.
13. Davis, B.M. and Davis, K.L. (1979): IN Brain Acetylcholine and
     Neuropsychiatric Disease (eds) K.L. Davis and P.A. Berger,
     Plenum Press, New York, pp. 445-458.
14. Deutsch, J.A. (1973): The Psychological Basis of Memory, Acad-
     emic Press, New York.
15. DeRyck, M. and Teitelbaum, P. (1978): Physiol. Behav. 20:
     403-408.
16. Domino, E.F. (1978): IN Neuropsychopharmacology: Proc. 10th
     Int. Congr. CINP (eds) P. Deniker, C. Radouco-Thomas and
     A. Villeneuve, Pergamon Press, Oxford, pp. 1655-1658.
17. Domino, E.F., Mathews, B., Taib, S.K., Demetriou, S.K. and
     Fucek, F. (1980): This volume.
18. Drachman, D.A. (1978): IN Psychopharmacology: A Generation of
     Progress (eds) M.A. Liptin, A. Di Mascio and K.F. Killam,
     Raven Press, New York, 651-652.
19. Drachman, D.B. (1978): New Eng. J. Med. 298:136-186.
20. Dun, N.J. and Karczmar, A.G. (1970): This volume.
21. Eckernas, S.A. (1977): Acta Physiol. Scand. Suppl. 499:8-62.
22. Eckernas, S.A., Aquilonius, S.M., Hartvig, P., Lindstrom, B.
     and Osterman, P.O. (1980): This volume.
23. Flodmark, S. and Wrammer, T. (1945): Acta Physiol. Scand. 9:
     88-96.
24. Frommel, E., Beck, M., Beck, I., Melkonian, T.D., Wyss, R.,
     Vallette, F. and Ducommun, M. (1951): Actual. Pharmacol.
     3:157-187.
25. Gillin, J.C., Mendelson, W.B., Sitaram, N. and Wyatt, R.J.
     (1978): Ann. Rev. Pharmacol. Toxicol. 18:563-579.
26. Gillin, J.C., Sitaram, N. and Weingartner, H. (1980): This
     volume.
27. Goldner, R.D. (1967): Int. J. Neuropsychol. 3:234-238.
28. Haase, H.J. (1965): IN Actions of Neuroleptics. A Psychiatric,
     Neurologic and Pharmacological Investigation (eds) H.J. Haase
     and P.A.J. Jensen, North-Holland, Amsterdam.

29. Hanin, I., Spiker, D.G., Mallinger, A.G., Kopp, U., Himmelbach,
     J.M., Neil, J.F. and Kupfer, D.J. (1980): This volume.
30. Harris, C.M., Davis, J.M. and Janowsky, D.S. (1979): IN Choline
     and Lecithin in Brain Disorders (eds) A. Barbeau, J.H. Grow-
     don and R.J. Wurtman, Raven Press, New York, pp. 397-408.
31. Heilbronn, E. and Stalberg, E. (1978): J. Neurochem. 31:5-12.
32. Hokfelt, T., Elde, R., Johansson, O., Ljungdahl, A., Schultz-
     berg, M., Fuxe, K., Goldstein, M., Nilsson, G., Pernter, B.,
     Terenius, L., Ganten, D., Jeffeoate, S.L., Rehfeld, J. and
     Said, S. (1978): IN Psychopharmacology: A Generation of
     Progress (eds) M.A. Lipton, A. Di Mascio and K.F. Killam,
     Raven Press, New York, pp. 39-66.
33. Hokin-Neaverson, M. (1977): Adv. Exp. Med. Biol. 83:429-446.
34. Janowsky, D.S. and Davis, J.M. (1978): IN Psychiatric Disgnosis:
     Exploration of Biological Predictors (eds) H.S. Akiskal and
     W.L. Webb, Spectrum Publ., New York, pp. 263-278.
35. Jenden, D.J. (1977): Cholinergic Mechanisms and Psychopharma-
     cology (ed) D.J. Jenden, Plenum Press, New York, pp. 139-162.
36. Jenden, D.J. (1979): IN Brain Acetylcholine and Neuropsychia-
     tric Disease (eds) K.L. Davis and P.A. Berger, Plenum Press,
     New York, pp. 483-513.
37. Jenden, D.J. (1979): IN Choline and Lecithin in Brain Dis-
     orders: Nutrition and the Brain Vol. 5 (eds) A. Barbeau,
     J.H. Growdon and R.J. Wurtmen, Plenum Press, New York,
     pp. 13-24.
38. Jope, R.S., Jenden, D.J., Ehrlich, B.E. and Diamond, J.M.
     (1978): New Eng. J. Med. 299:833-834.
39. Jouvet, M. (1961): IN The Nature of Sleep (eds) G.E.W. Wolsten-
     holme and M. O'Connor, J&A Churchill, London, pp. 188-208.
40. Jouvet, M. (1972): IN Brain and Human Behavior (eds) A.G.
     Karczmar and J.C. Eccles, Springer-Verlag, Berlin, pp. 131-161.
41. Karczmar, A.G. (1963): IN Cholinesterases and Anticholinesterase
     Agents (ed) G.B. Koelle, Springer-Verlag, Berlin.
42. Karczmar, A.G. (1967): Ann. Rev. Pharmacol. 7:241-276.
43. Karczmar, A.G. (1967): IN Physiological Pharmacology (eds)
     W.S. Root and F.G. Hofman, Academic Press, New York, pp. 164-
     263.
44. Karczmar, A.G. (1970): Presentation at Regional Midwest EEG
     Meeting, April, 1971, Hines Vet. Adm. Hospital.
45. Karczmar, A.G. (1971): IN Advances in Neuropsychopharmacology
     (eds) O. Vinar, Z. Votava and P.B. Bradley, North-Holland,
     Amsterdam, pp. 655-680.
46. Karczmar, A.G. (1976): IN Biology of Cholinergic Function (eds)
     A.M. Goldberg and I. Hanin, Raven Press, New York, pp.345-449.
47. Karczmar, A.G. (1977): IN Cholinergic Mechanisms and Psycho-
     pharmacology (ed) D.J. Jenden, Plenum Press, New York,
     pp. 679-708.
48. Karczmar, A.G. (1978): IN Neuropsychopharmacology: Proc. 10th
     Congr. CINP (eds) P. Deniker, C. Radouco-Thomas and A. Ville-
     neuve, Pergamon Press, Oxford, pp. 581-608.

49. Karczmar, A.G. (1979): Dru Ther. 4:31-38.
50. Karczmar, A.G. (1979): IN Choline and Lecithin in Brain Dis-
        orders. Nutrition and the Brain, Vol. 5 (eds) A. Barbeau,
        J.H. Growdon and R.J. Wurtman, Raven Press, New York,
        pp. 161-175.
51. Karczmar, A.G. (1979): IN Brain Acetylcholine and Neuropsychia-
        tric Disease (eds) K.L. Davis and P.A. Berger, Plenum Press,
        New York, pp. 265-310.
52. Karczmar, A.G., Badrinath, K. and Koehn, G. (1979): HEW-IND
        No.16,412, Loyola University Medical Center.
53. Karczmar, A.G. and Koehn, G. (1980): IN Neurochemistry and
        Clinical Neurology (eds) L. Battistin, G. Hashim and A.
        Lajtha, Alan R. Liss, Inc., New York (in press).
54. Karczmar, A.G., Longo, V.G. and Scotti de Carolis, A. (1970):
        Physiol. Behav. 5:175-182.
55. Karczmar, A.G. and Nishi, S. (1971): IN Advances in Cytopharma-
        cology (eds) F. Clemente and B. Ceccarelli, pp. 301-317.
56. Karczmar, A.G., Nishi, S., Minota, S. and Kindel, G. (1980):
        Gen. Pharmacol. 11:127-134.
57. Karczmar, A.G. and Scudder, C.L. (1969): IN Aggressive Behavior
        (eds) S. Garattini and E.B. Sigg, Excerpta Medica, Amsterdam,
        pp. 223-229.
58. Karczmar, A.G., Scudder, C.L. and Richardson, D.L. (1973):
        Neurosci. Res. 159-244.
59. Kim, K.C. and Karczmar, A.G. (1967): Int. J. Neuropharmacol.
        6:51-61.
60. Klawans, H.A. Jr., Westheimer, R. and Goetz, C.G. (1976): Dis.
        Nerv. Syst. 36:267-275.
61. Koehn, G., Henderson, G. and Karczmar, A.G. (1980): Eur. J.
        Pharmacol. 61:167-174.
62. Koehn, G. and Karczmar, A.G. (1978): Prog. Neuropsychopharmacol.
        2:169-174.
63. Koelle, G.B. (1963): IN Cholinesterases and Anticholinesterase
        Agents (ed) Springer-Verlag, Berlin, pp. 187-298.
64. Koelle, G.B. (1975): IN The Pharmacological Basis of Therapeu-
        tics (eds) L.S. Goodman and A. Gilman, MacMillan Publ. Co.,
        New York, pp. 575-588.
65. Koelle, G.B. (1972): New Eng. J. Med. 286:1086-1088.
66. Koketsu, K., Karczmar, A.G. and Kitamura, R. (1969): Int. J.
        Neuropharmacol. 8:329-336.
67. Krnjevic, K., Puil, E. and Werman, P. (1976): Canad. J.
        Physiol. Pharmacol. 54:172-176.
68. Krnjevic, K. and Reinhardt, W. (1979): Science 206:1321-1322.
69. Kupfer, D.J. and Efwards, D.J. (1978): IN Interrelationship
        between Various Neurotransmitter Systems (eds) A.G. Karczmar
        and J. Glowinski, Pergamon Press, Oxford, pp. 609-622.
70. Lehmann, J. and Fibiger, H.C. (1980): This volume.
71. Liberson, W.T. and Karczmar, A.G. (1969): IN Neurophysiological
        and Behavioral Aspects of Psychotropic Drugs (eds) A.G.
        Karczmar and W.P. Koelle, C.C.Thomas, Springfield, pp.161-178.

72. Libet, B. (1976): IN SIF Cells: Structure and Function of the
     Small Intensely Fluorescent Sympathetic Cells (ed) Fogarty
     Int. Center, Proc. No.30, U.S. Govt. Printing Office,
     Washington, D.C., pp. 163-177.

73. Longo, V.G. (1962): Electroencephalographic Atlas for Pharm-
     acological Research, Elsevier, Amsterdam.

74. Longo, V.G. (1966): Pharmacol. Rev. 18:965-996.

75. Longo, V.G. and Loizzo, A. (1973): IN Brain, Nerves and Syn-
     apses (eds) F.E. Bloom and G.H. Acheson, Kargar, Basel,
     pp. 46-54.

76. Lynch, H.D. and Anderson, M.H. (1976): Dis. Nerv. Syst. 36:
     648-652.

77. Magazanik, L.G. and Vyskocil, F. (1976): IN Motor Innervation
     of Muscle (ed) S. Thesleff, Academic Press, New York,

78. Maier, N.R.F. (1949): The Study of Behavior without Goal,
     McGaw, New York.

79. Marczynski, T.J. (1967): Ergeb. Physiol. 59:86-97.

80. Marczynski, T.J. (1978): IN Multidisciplinary Perspectives in
     Event Related Brain Potential Research (ed) D. Otto,
     U.S. Environmental Protection Agency, Washington, D.C.

81. Marczynski, T.J. (1978): IN Multidisciplinary Perspectives in
     Event Related Brain Potential Research (ed) D. Otto,
     U.S. Environmental Protection Agency, Washington, D.C.

82. McGeer, P.L., Eccles, J.C. and McGeer, E.G. (1978): Molecular
     Neurobiology of the Mammalian Brain, Plenum Press, New York.

83. McGeer, P.L. and McGeer, E.G. (1979): IN Choline and Lecithin:
     Nutritrion and the Brain, Vol. 5 (eds) A. Barbeau, J.H.
     Growdon and R.J. Wurtman, Raven Press, New York, pp. 177-199.

84. Millington, W.B., McCall, A.L. and Wurtman, R.J. (1979): IN
     Choline and Lecithin in Brain Disorders: Nutrition and the
     Brain Vol. 5 (eds) A. Barbeau, J.H. Growdon and R.J. Wurt-
     man, Raven Press, New York, pp. 417-424.

85. Mollenauer, J., White, M., Plotnik, R. and Bradlee-Tiffany, P.
     (1979): Pharmacol. Biochem. Behav. 11:189.

86. Nastuk, W.L. (1967): Fed. Proc. 26:1639.

87. Nastuk, W.L., Plascia, O.J. and Osserman, K.E. (1960): Proc.
     Soc. Exp. Biol. Med. 105:177-183.

88. Neubauer, H., Adams, M. and Redfern, P. (1975): Med. Hypoth.
     1:32-34.

89. Ohta, Y. and Karczmar, A.G. (1980): Fed. Proc. 39:409.

90. Olds, J. (1958): Science 127:315-324.

91. Pedigo, N.W. and Dewey, W.L. (1980): This volume.

92. Pedigo, N.W., Dewey, W.L. and Harris, L.S. (1975): J. Pharma-
     col. Exp. Ther. 193:845-852.

93. Pfefferbaum, A., Davis, K.L., Coulter, C.L., Mohs, R.C. and
     Kopell, B.S. (1979): IN Brain Acetylcholine and Neuropsychia-
     tric Disease (eds) K.L. Davis and P.A. Berger, Plenum Press,
     New York, pp. 345-360.

94. Plotnik, R., Mollenauer, S.O. and Snyder, E.F. (1974): J. Comp.
     Physiol. Psychol. 86:1074-1078.

95. Pujol., J.F., Keane, P.E. and Jouvet, M. (1978): IN Interrela-
    tionships between Various Neurotransmitter Systems (eds)
    A.G. Karczmar and J. Glowinski, Pergamon Press, Oxford,
    pp. 559-568.

96. Samanin, R. and Valzelli, L. (1971): Eur. J. Pharmacol. 16:
    298-302.

97. Schildkraut, J.J. (1970): Neuropsychopharmacology and the
    Affective Disorders, Little, Brown & Co., Boston.

98. Silbergeld, E.K. and Goldberg, A.M. (1976): IN Biology of
    Cholinergic Function (eds) A.M. Goldberg and I. Hanin,
    Raven Press, New York, pp. 619-645.

99. Singh, M.M. and Kay, S.R. (1978): Neuropsychobiology 4:288-304.

100. Singh, M.M. and Lal, J. (1979): Proc. end World. Congr. Biol.
     Psych., Elsevier, Amersterdam.

101. Sitaram, N., Buchsbaum, M.S. and Christian, G.J. (1977): Eur.
     J. Pharmacol. 42:285-290.

102. Sitaram, N. and Gillin, J.C. (1980): This volume.

103. Sitaram, N., Wyatt, R.J., Dawson, S. and Gillin, J.C. (1976):
     Science 191:1281-1282.

104. Slaughter, D. and Gross, E.G. (1940): J. Pharmacol. Exp. Ther.
     68:96-103.

105. Stein, L. (1968): IN Psychopharmacology: A Review of Progress
     (ed) D. Efron, Govt. Printing Office, Publ. No. 836,
     Washington, D.C. 105-123.

106. Szerb, J.C. (1977): IN Cholinergic Mechanisms and Psycho-
     pharmacology (ed) D.J. Jenden, Plenum Press, New York,
     pp. 49-60.

107. Teitelbaum, H., Lee, J.F. and Johannessen, J.N. (1975):
     Science, 188:1114.

108. Thesleff, S. (1959): J. Physiol. 148:659-664.

109. Thesleff, S. and Quastel, D.M.J. (1965): Ann. Rev. Pharmacol.
     5:263-284.

110. Yang, H.Y., Hong, J.S. and Costa, E. (1977): Neuropharmacology
     16:303-308.

111. Van Andel, H. (1959): IN Neuropsychopharmacology (eds) P.B.
     Bradley, P. Deniker and C. Radouco-Thomas, Elsevier,
     Amsterdam, pp. 701-703.

112. Wecker, L. and Schmidt, D.E. (1980): Brain Res. 184:234-240.

113. Weinberg, C.B. and Hall, Z.W. (1979): Proc. Nat. Acad. Sci.
     76:504.

114. Wikler, A. (1952): Proc. Soc. Exp. Biol. Med. 79:261-265.

115. Wurtman, R.J. (1979): IN Choline and Lecithin in Brain Dis-
     orders: Nutrition and the Brain, Vol. 5 (eds) J.H. Growdon
     and R.J. Wurtman, Raven Press, New York, pp. 7-12.

CLINICAL TRIALS WITH CHOLINE AND 4-AMINOPYRIDINE

IN HUNTINGTON'S CHOREA

S.M. Aquilonius and S.A. Eckernas

Department of Neurology
University Hospital
S-750 14, Uppsala, Sweden

INTRODUCTION

The rare hereditary form of chorea described by George
Huntington in 1872 (23) has, especially during the past decade,
been the object of intense research by neuroscientists using the
principal approaches illustrated in Fig. 1. The figure also includes the different reasons to suggest a striatal cholinergic
hypofunction playing one pathophysiological role in the disease.
This field of research has recently been reviewed in detail (8).

Trials with Choline

Shortly following the proposal that increased plasma concentration of choline (Ch) in animal experiments was paralleled by
increased synthesis of acetylcholine (ACh) in the brain (12) we
could demonstrate (5) that plasma Ch in man can be dose dependently
increased following oral administration of large doses of Ch chloride. The need for a therapy in clinical neurology intended to increase cholinergic function within the brain is reflected in the
considerable number of trials of Ch therapy initiated in the last
few years. In addition to Huntington's chorea some other conditions, e.g. tardive dyskinesia, dementia disorders and different
forms of ataxia - which might involve central cholinergic hypofunction (see Ref. 16) - have been treated with oral Ch. However,
most studies of Ch therapy in movement disorders have been performed on small numbers of patients (Table 1). Furthermore, the
unpleasant "rotten fish oder" secondary to intestinal breakdown of
exogenous Ch has been one hindrance to placebo controlled investigations. We administered Ch chloride to five patients with

A. Neostriatal cholinergic interneurons degenerate in Huntington's chorea (6).
B. CSF Ch concentration is often low in Huntington's chorea (4).
C. Choreatic movements can be reduced following physostigmine (3,15,25).
D. Biochemical changes including reduced striatal choline acetyltransferase and increased locomotion can be produced in rats by striatal kainic acid lesions (14,32).

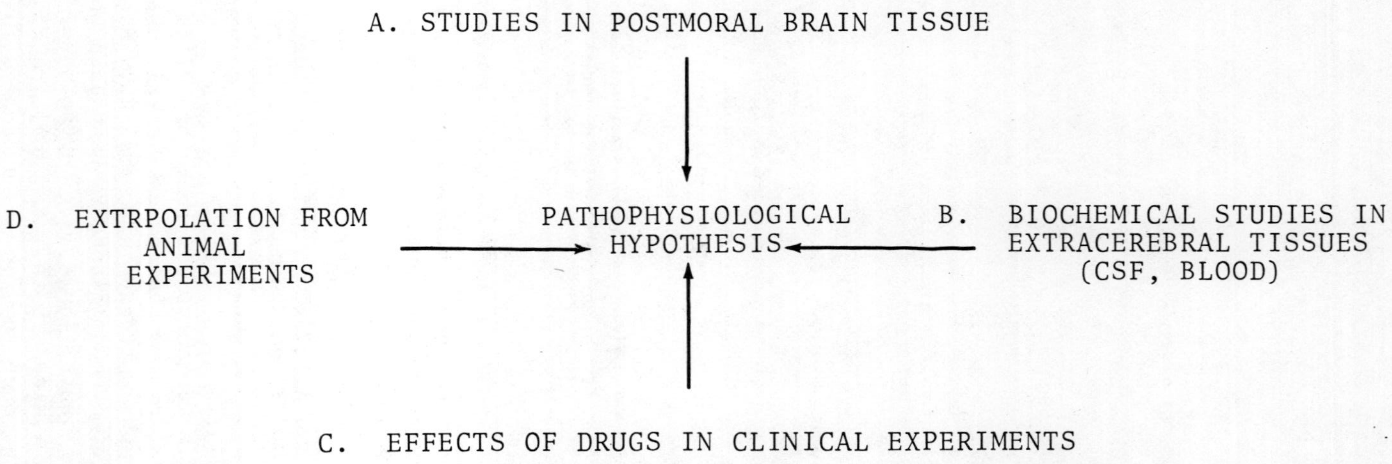

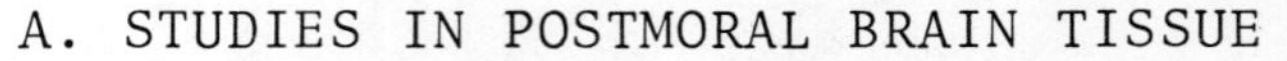

FIGURE 1: Principal approaches to cholinergic neuropharmacology and reasons to suggest a striatal cholinergic hypofunction in Huntington's chorea.

TABLE 1: Choline therapy in neurological disorders.

DISORDER	IMPROVED/TREATED	REFERENCE NUMBER
Huntington's chorea	0/5	5
	0/10	20
	3/6	16
	2/4	9
Tardive dyskinesia	3/4	15
	8/18	22
	1/2	9
Cerebral ataxia	1/1	27
	0/13	28
	6/12	29
Friedreich ataxia	6/6	9
	4/8	29
Dementia	0/7	11
	1/3	18
	0/8	37
	0/8	33
	0/18	36

The tardive dyskinesia group of Hirsch et al. (22) is heterogeneous and includes 13 patients with schizophrenia and 5 with Alzheimer's disease. In the study by Signoret et al. (37) a minor improvement might exist in patients with short disease duration. In the study by Mohs (33) memory impairment was mild to moderate.

Huntington's chorea (7). The drug was given three times a day and increased from 3 to 15 g every fourth day. Following this initial schedule long-term treatment was performed upon all patients for periods of 1 month to about 1 year. Hyperkinesia was recorded on videotape under standardized environmental conditions. In spite of an increase in plasma Ch concentration following 15 g daily doses, to the same level (25-30 μmol/liter) as that claimed to be paralleled by 45% increase in striatal ACh in the rat (13), hyperkinesia was not conclusively altered in the patients. In one patient receiving Ch therapy for more than one year, there was clear-cut progress of the disease.

Table 1 represents a summary of therapeutic effects in several Ch therapy trials reported in the literature (studies with lecithin or phosphorylcholine have not been included). It is obvious that such a non-graded condensation of results can lead to misinterpretation. However, the effects observed have seldom been marked but

slight or moderate. Following the evaluation of the clinical literature we feel doubtful that there is any conclusive therapeutic effect connected with Ch therapy in Huntington's chorea and in
dementia.

It is now far from settled whether the administration of large
doses of Ch increases cholinergic activity in the brain. In contrast to the increase in brain ACh concentration following Ch administration to laboratory animals reported by Cohen and Wurtman
(13) and Haubrich (21), in several recent studies similar treatments
have been made without any effect on brain ACh (19,35,39). Furthermore, a tenfold increase of plasma Ch in the rat has not produced
changes in brain regional turnover rates of ACh, while a slight
(11%) increase in striatal ACh concentration has been found (17).
The latter results could indicate a direct muscarinic agonistic
action as in recent experiments by Ladinsky et al. (26).

Trials with 4-Aminopyridine

4-Aminopyridine (4-AP) has been shown to enhance the stimulation evoked release of ACh in the peripheral nervous system (31).
In view of its lipid solubility, 4-AP should easily penetrate into
the central nervous system. Accordingly, we found it justified to
perform a pilot study of 4-AP therapy in Huntington's chorea.

Three patients with severe hyperkinesia took part in this
study. None of them was treated with neuroleptic drugs. 4-AP
(0.3 mg/kg), a dose known to ameliorate the neuromuscular function
in Eaton-Lambert syndrome (1,30) and restore neuromuscular block
following pancurone (10), was administered by slow intravenous injection. The patients were then observed under standardized conditions and videotape recordings were taken at regular intervals
over a 3 hr period following injection.

No reduction in hyperkinesia was observed in any of the
patients. In all three patients gait function deteriorated, apparently due to a general decrease in muscle tone. In two patients
periods of nausea and vomiting occurred during the first hour following 4-AP administration. No peripheral cholinomimetic symptoms
were observed. Two patients complained of "tingling" in lips and
fingers; this has been previously described following 4-AP (30).

The transmitter releasing potency of 4-AP seems not to be restricted to cholinergic neurons (24) and the drug has recently been
shown to increase noradrenaline turnover in the central nervous
system of the rat (2). It could well be that the central net effect
of 4-AP in Huntington's chorea is a functional overactivity in
monoamine systems.

DISCUSSION

So far, efforts to reduce the hyperkinetic movements in
Huntington's chorea by therapeutic approaches intended to increase
central cholinergic activity has turned out rather negative. Only
some patients get temporarily reduced chorea following physostigmine (15), while treatment with the direct cholinergic agonist
arecoline (34), rather than improving chorea, tended to exacerbate
the movements. It is known that the degeneration of cholinergic
neurons in the striatum is not uniform (6). Complete cholinergic
degeneration in one portion might cause a particular type of hyperkinesia resistent to therapy with a drug that premises some presynaptic activity (like physostigmine). A marked reduction of
muscarinic receptors which can take place in the disease (38) precludes the effectiveness of an agonistic drug like arecoline.

Our results and those reported in the literature do not favor
a clinical usefulness of high oral doses of Ch in the treatment of
Huntington's chorea. In view of the negative results in the pilot
trial with 4-AP an extended study of the pharmacologically interesting substance in Huntington's chorea does not seem to be
justified.

Our knowledge concerning the neuropharmacology of chorea has
increased considerably during the past few years (see Ref. 8) and
it seems probable that a striatal cholinergic hypofunction is only
one link in a complex disturbance of interaction between different
neuronal systems (i.e. glutamate, ACh, dopamine, neuropeptides).
Future progress depends on the possibilities to further utilize the
approaches presented in Fig. 1 and to add methods allowing a more
direct study of neuronal metabolism within the human brain. Ch
labelled with ^{11}C has recently been synthesized at the Department
of Chemistry, Uppsala, and we believe that positron emitting tomog-
raphy following administration of pertinent ^{11}C-labelled substances
could mark a new way in which to study the cholinergic neuropharmacology of neuropsychiatric diseases including Huntington's chorea.

REFERENCES

1. Agoston, S., van Weerden, T., Westra, P. and Broekert, A.
   (1978): Brit. J. Anesthesiol. 50:383-385.
2. Anden, N.E. and Leander, S. (1979): J. Neural Trans. 44:1-12.
3. Aquilonius, S.M. and Sjostrom, R. (1971): Life Sci. 10:405-414.
4. Aquilonius, S.M., Nystrom, B., Schuberth, J. and Sundwall, A.
   (1972): J. Neurol. Neurosurg. Psychiat. 35:720-725.
5. Aquilonius, S.M. and Eckernas, S.A. (1975): New Engl. J. Med.
   293:1105-1106.

6. Aquilonius, S.M., Eckernas, S.A. and Sundwall, A. (1975):
    J. Neurol. Neurosurg. Psychiat. 38:669–677.

7. Aquilonius, S.M. and Eckernas, S.A. (1977): Neurology 27:
    887–889.

8. Aquilonius, S.M. (1978): Acta Neurol. Scand. 57(Suppl. 67):
    115–122.

9. Barbeau, A. (1978): J. Canad. Sci. Neurol. 5:157–160.

10. Booij, L.H.D.J., Miller, R.D. and Crul, J.F. (1978):
    Anaesthesiol. Analg. 57:316–321.

11. Boyd, W.D., Graham-White, J., Blackwood, G., Glen, I. and
    McQueen, J. (1977): Lancet 2:711.

12. Cohen, E.L. and Wurtman, R.J. (1975): Life Sci. 16:1095–1102.

13. Cohen, E.L. and Wurtman, R.J. (1976): Science 191:561–562.

14. Coyle, J.T. and Schwarcz, R. (1976): Nature 263:244–246.

15. Davis, K.L., Hollister, L.E., Barchas, J.D. and Berger, P.A.
    (1976): Life Sci. 19:1507–1516.

16. Davis, K.L. and Berger, P.A. (1978): Biol. Psychiat. 13:23–49.

17. Eckernas, S.A., Sahlstrom, L. and Aquilonius, S.M. (1977):
    Acta Physiol. Scand. 101:404–410.

18. Etienne, P., Gauthier, S., Johnson, G., Collier, B., Mendis,
    T., Dastoor, D., Cole, M. and Muller, H.F. (1978): Lancet 1:
    508–509.

19. Flentge, F. and Van den Berg, C.J. (1979): J. Neurochem. 32:
    1331–1333.

20. Growden, J.H., Cohen, E.L. and Wurtman, R.J. (1977): Ann.
    Neurol. 1:418–422.

21. Haubrich, D.R., Wang, P.F., Clody, D.E. and Wedeking, P.W.
    (1975): Life Sci. 17:975–980.

22. Hirsch, M.J., Growden, J.H. and Wurtman, R.J. (1977):
    Neurology 27:391.

23. Huntington, G. (1872): Med. Surg. Rept., Philadelphia, 317–321.

24. Johns, A., Golko, D.S., Lauzon, P.A. and Paton, D.M. (1976):
    Eur. J. Pharmacol. 38:71–78.

25. Klawans, H.Z. and Rubovits, R. (1972): Neurology 22:107–116.

26. Ladinsky, H., Consolo, S. and Pugnetti, P. (1979): In Nutrition
    and the Brain, (eds) A. Barbeau, J.H. Growden and R.J.
    Wurtman, Raven Press, New York.

27. Legg, N.J. (1978): Brit. Med. J. 2:1403.

28. Legg, N.J. (1979): Brit. Med. J. 2:133.

29. Livingstone, I.R. and Mastaglia, F.L. (1979): Brit. Med. J.
    2:939.

30. Lundh, H., Nilsson, O. and Rosen, I. (1977): J. Neurol. Neuro-
    surg. Psychiat. 40:1109–1112.

31. Lundh, H. and Thesleff, S. (1977): Eur. J. Pharmacol. 42:
    411–412.

32. Mason, S.T. and Fibiger, H.C. (1979): Neuropharmacology 18:
    403–407.

33. Mohs, R.C., Davis, K.L., Tinklenberg, J.R., Hollister, L.E.,
     Yesavage, J.A. and Kopell, B.S. (1979): Amer. J. Psychiat.
     136:1275-1277.
34. Nutt, J.G., Rosin, A. and Chase, T.N. (1978): Neurology 28:
     1061-1064.
35. Pedata, F., Wieraszko, A. and Pepeu, G. (1977): Pharmacol.
     Res. Commun. 9:755-761.
36. Renvoize, E.B. and Jerram, T. (1979): New Engl. J. Med. 2:330.
37. Signoret, J.L., Whiteley, A. and L'Hermite, F. (1978): Lancet
     2:837.
38. Wastek, G.J. and Yamamura, H.I. (1978): Molec. Pharmacol.
     14:768-780.
39. Wecker, L. and Schmidt, D.E. (1979): Life Sci. 25:375-384.

PHARMACOKINETICS OF NEOSTIGMINE AND PYRIDOSTIGMINE IN MAN AND ITS

CORRELATION TO CLINICAL EFFECTS IN MYASTHENIA GRAVIS

S.-A. Eckernas, S.-M. Aquilonius, P. Hartvig[†] B. Lindstrom[§]
P.O. Osterman and E. Stalberg[*]

Departments of Neurology, Neurophysiology[*]& Pharmacy[+]
University Hospital, S-750 Uppsala, Sweden, and
Department of Drugs[§] National Board of Health & Welfare
Box 607, S-751, 25 Uppsala, Sweden

INTRODUCTION

Cholinesterase inhibitors, such as neostigmine and pyridostig-
ine, have been in clinical use for more than 40 years as the basic
medical treatment in myasthenia gravis. The first investigation of
the basic pharmacokinetics was performed by Kornfeld et al. (7)
with ^{14}C-labelled pyridostigmine. Later, two different gas chromatographic (GC) procedures for the assay of neostigmine and pyridostigmine were described by Chan et al. (3) and Cohen et al. (6).
None of these methods, however, seems to be sufficiently sensitive
or selective to permit accurate determination of neostigmine and
pyridostigmine in the plasma of myasthenic patients receiving oral
treatment.

In the present study the pharmacokinetics of pyridostigmine
and neostigmine was evaluated in man after intravenous and oral
administration. The assay was performed by the gas chromatographic
mass spectrometric (GCMS) method as recently described (1,2).

Since the disease myasthenia gravis offers the possibility to
measure the clinical effect by objective methods, e.g. decrement of
the evoked electrical muscle response of repetitive nerve stimulation we have also tried to correlate the plasma concentration to
the effect in a few patients.

MATERIAL AND METHODS

D_6-Neostigmine and D_6-pyridostigmine was used as internal standard. The assay was performed by a Finnigan 4000 GCMS with chemical ionization as earlier described (1,2). After collecting venous blood the sample was immediately cooled and the plasma was separated. To an aliquot of plasma the deuterated internal standard was added. The kinetic of neostigmine was studied in four patients who received 2.5-3 mg intravenously to antagonize a curare-like effect of pancurone used during anesthesia. In seven myasthenic patients the pharmacokinetics after oral administration was studied.

The pharmacokinetics of 2.5 mg of pyridostigmine following intravenous administration was studied in two healthy male volunteers. In five healthy male subjects the plasma concentration was analysed after the ingestion of 120 mg pyridostigmine during fasting and, in a separate experiment, together with food. The plasma concentration of pyridostigmine was determined every hour during an interval between doses in seven patients with myasthenia gravis on their ordinary dose schedule of cholinesterase inhibitors.

In a preliminary study two myasthenic patients were studied after a single dose of neostigmine (intravenously or orally). The decrement of the evoked muscle response from the deltoid muscle during stimulation of Erb's point was measured. Surface electrodes were used for stimulation and recording. By means of a computer (LSi-11) the amplitude and area of the first and fourth response was measured and the relative percent difference between them was calculated. A stimulation frequency of 2 Hz was used. The investigation was made in the morning before the patients had left the bed. The skin temperature was kept constant (35°C) by means of a thermostated heating lamp. The muscle temperature was equilibrated for 1 hr before the drug was given, during which time decrement measurements were also made every fifth or tenth minute. Tests were made every fifth minute after the intravenous administration of the drug and every tenth minute after peroral dose for 3 and 5 hr, respectively.

RESULTS

Pharmacokinetics of Neostigmine

Intravenous administration: The time course of the neostigmine concentration in plasma after intravenous injection is exemplified in Fig. 1. The calculated pharmacokinetic data are summarized in Table 1. The inter-individual differences in drug elimination were small and the mean plasma half-life was 0.89 $\pm$ 0.05 hr.

TABLE 1: Pharmacokinetic data for neostigmine following intravenous
          administration.

| | Patients | | | | Mean $\pm$ S.E. |
	1	2	3	4	
Dose, mg	3.0	2.5	3.0	3.0	–
Dose, μg x kg^{-1}	42	35	48	32	–
AUC, hr x ng x ml^{-1}	46.6	46.3	59.8	35.2	–
Volume of distribution 1 x kg^{-1}	1.31	0.89	0.92	1.21	1.08 $\pm$ 0.11
Plasma half-life, hr	1.01	0.82	0.79	0.92	0.89 $\pm$ 0.05
Plasma clearance, 1 x kg^{-1} x hr^{-1}	0.89	0.75	0.81	0.91	0.84 $\pm$ 0.04

AUC = area under curve

TABLE 2: Pharmacokinetic data for neostigmine following oral
          administration.

| | Patients | | | Mean $\pm$ S.E. |
	5	6	7	
Dose, mg	30	30	30	–
Dose, μg x kg^{-1}	492	357	476	–
AUC, hr x ng x ml^{-1}	13.7	9.8	7.2	–
AUC in % of expected mean i.v. value*	2.3	2.3	1.3	2.0 0.33
Volume of distribution 1 x kg^{-1}	1.09	0.86	1.24	1.06 $\pm$ 0.11
Plasma half-life, hr	0.91	0.71	1.0	0.87 $\pm$ 0.09
Plasma clearance, 1 x kg^{-1} x hr^{-1}	0.83	0.84	0.86	0.84 $\pm$ 0.01

* A linear relationship between the i.v. doses and the corresponding area under the curve (AUC) has been assumed. In that
case the AUC corresponding to an i.v. neostigmine dose of 1 μg/kg,
as calculated from the data in Table 1 is 1.20 $\pm$ 0.05 hr x ng x
ml^{-1}.

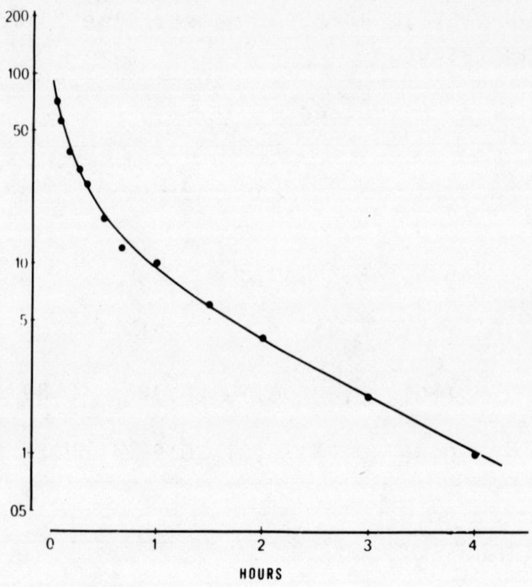

FIGURE 1: Plasma concentrations of neostigmine (ng/ml) following intravenous injection of 2.5 mg. Patient 4.

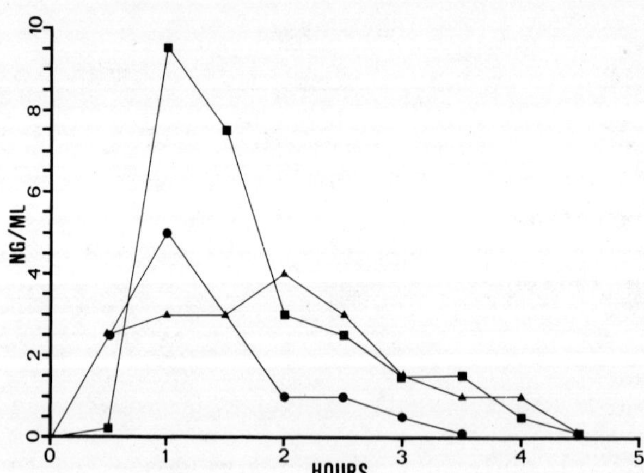

FIGURE 2: Plasma concentrations of neostigmine in three myasthenic patients following an oral dose of 30 mg in the morning. The patients had fasted and not taken any drugs overnight. Patients: 5) ■ — ■ ; 6) ▲ — ▲ ; 7) ● — ● .

TABLE 3: Pharmacokinetic constants for pyridostigmine following
          intravenous administration.

	Subjects	
	POO	SMA
Dose, mg	2.5	2.5
Dose, μg/kg	33	33
AUC, hr x ng/ml	51.7	50.3
Volume of distribution, liter/kg	1.62	1.23
Plasma half-life	1.74	1.29
Plasma clearance, liter/kg x hr	0.64	0.66

AUC = area under curve

Oral administration: The plasma concentration of neostigmine
following a single oral dose of 30 mg is shown in Fig. 2 and the
corresponding pharmacokinetic data shown in Table 2. Considerable
variation in absorption was found with the peak concentration occurring between 1 and 2 hr after intake of the drug. The biological
availability is calculated to be only 1-2% of the ingested dose.

Steady state plasma concentration: Although the variation in
daily dose was considerable in three of the patients taking repeated
doses of neostigmine and pyridostigmine, the plasma concentration
of neostigmine 4 hr after the first dose of the day was fairly
similar (about 3 ng/ml). In one patient, who was on repeated doses
of neostigmine 150 mg + pyridostigmine 60 mg, the plasma steady
state was about 10 times higher.

Pharmacokinetics of Pyridostigmine

Intravenous administration: In the two subjects to whom
pyridostigmine was given intravenously, the time course of the
plasma concentration was similar (Fig. 3). The pharmacokinetic
constants derived are given in Table 3. The elimination half-life
was estimated to be about 1.5 hr.

Oral administration: The peak plasma concentration appeared
1.7 hr on average after administration of 120 mg pyridostigmine
orally to 5 fasting subjects (Fig. 4 and Table 4). The elimination

TABLE 4: Pharmacokinetic constants for pyridostigmine after oral administration.

	Subjects					Mean ± S.E.
	BL	POO	PH	SMA	SAE	
Dose, mg	120	120	120	120	120	
Dose, μg/kg	1600	1600	1600	1600	1700	
AUC, hr x ng/ml	122.1	281.8	179.9	182.9	177.5	
AUC, percent of i.v. values	5.0	11.5	7.0	7.4	7.2	7.6 ± 2.4
Volume of distribution, liter/kg	1.35	1.72	1.96	1.32	1.85	1.65 ± 0.29
Plasma half-life, hr	1.43	1.83	2.09	1.70	1.84	1.78 ± 0.24
Plasma clearance, liter/kg x hr	0.655	0.653	0.652	0.644	0.699	0.66 ± 0.02

AUC = area under curve

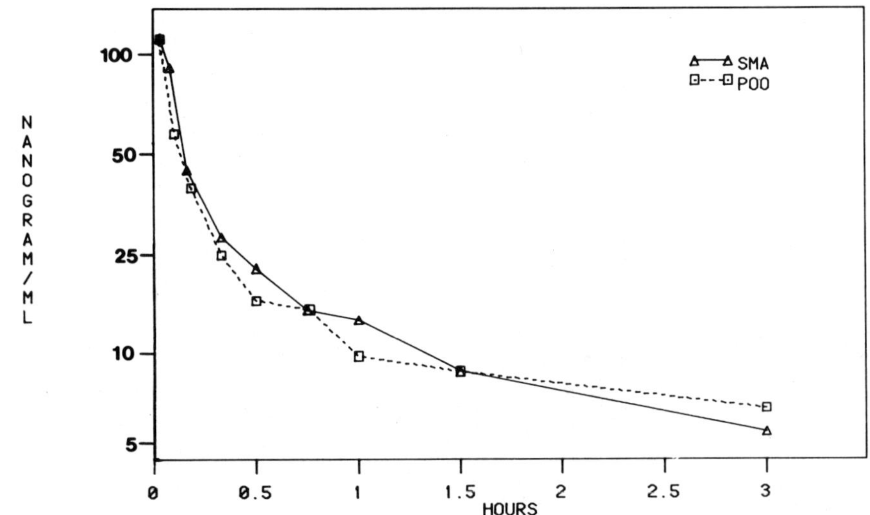

FIGURE 3: Plasma concentrations following intravenous administration of 2.5 mg of pyridostigmine.

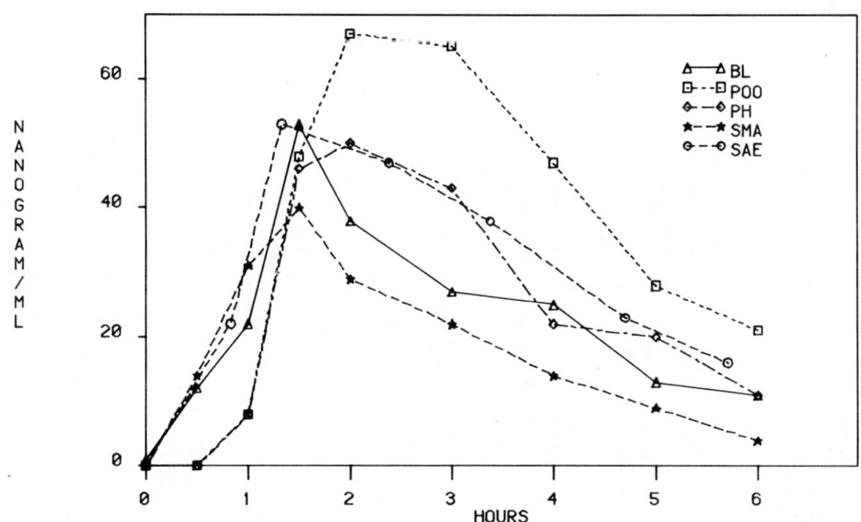

FIGURE 4: Plasma concentrations of pyridostigmine in 5 fasting subjects following an oral dose of 120 mg.

TABLE 5: Data for patients with myasthenia gravis treated with cholinesterase inhibitors.

Patient	Age (Years)	Sex	Body Weight (kg)	Serum Creatinine	Daily Dose of		Predose Steady-state Concentration (ng/ml)
					Pyridostigmine (mg)	Neostigmine (mg)	
A	71	M	85	226	60 x 8	---	340
B	83	M	74	124	60 x 12	15 x 3	220
C	23	F	76	56	60 x 6	---	57
D	72	F	58	85	60 x 12	---	116
E	58	M	76	93	60 x 6	---	58
F	69	F	54	60	60 x 12	---	242

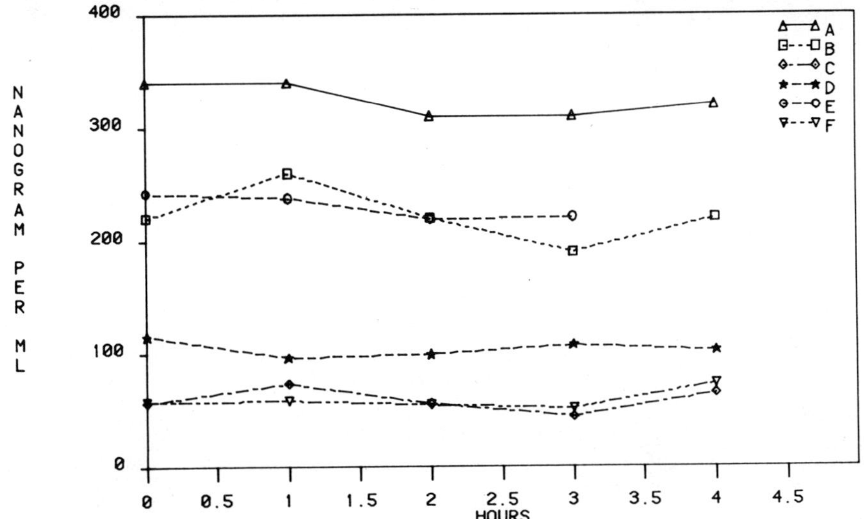

FIGURE 5: Plasma concentrations of pyridostigmine during an interval between doses in 6 myasthenic patients on their ordinary dose schedule. Data for patients and their daily
dosage are summarized in Table 5.

half-life was somewhat longer than after intravenous administration,
namely 1.8 hr. Together with food, the peak plasma concentration
was delayed and did appear 3.2 hr after the administration of the
drug.   After fasting, the mean oral bioavailability was 7.6% and
when pyridostigmine was ingested together with food, the bioavailability did not change significantly.

Steady state concentration: Plasma concentrations of pyridostigmine were measured during an interval between doses in myasthenic patients on their ordinary dose schedule of the drug. As
seen in Fig. 5, there was a sevenfold difference in the steady state
plasma concentration between patients. However, as can be seen in
Table 5, there is no correlation between the daily dose and the
plasma concentration.

Correlation to Decrement

As can be seen in Fig. 6, there seems to be a good correlation
between the plasma concentration and the effect on neuromuscular

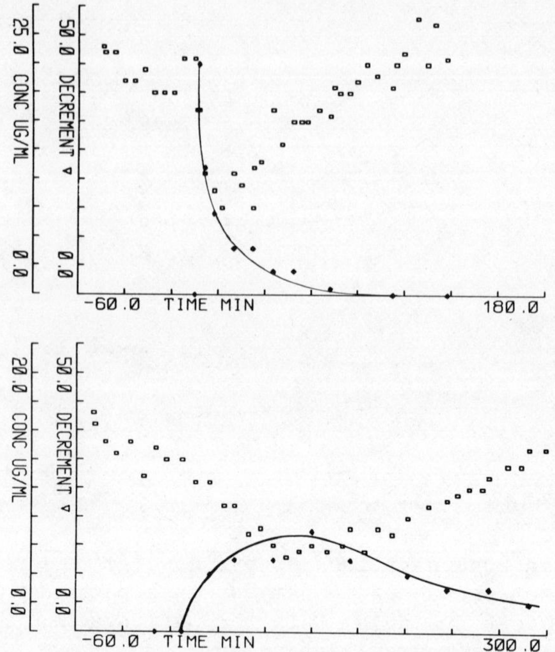

FIGURE 6: The relation between plasma concentration and decrement.
Upper: myasthenic patient receiving 1.5 mg neostigmine
i.v.   Lower: myasthenic patient receiving 30 mg neostigmine orally. ◆——◆ plasma concentration;
◻——◻ decrement in percent.

transmission. After the intravenous injection, the decrement decreased from about 40% to a minimal value of 16% after 14 min. The
decrement began to increase when the plasma was below 5 ng/ml. The
decrement is close to initial values after about 100 min when the
plasma concentration of neostigmine is zero. After the oral
dose the first effect is seen after 30 min. The maximal effect
with a change in the decrement from 33 to 14% is obtained after
about 120 min, coinciding with a maximal plasma concentration of
the drug.

DISCUSSION

In most instances pyridostigmine is used as a basic drug in the treatment of myasthenia gravis. It has been claimed to have a longer duration than neostigmine, although the pharmacokinetic evidence underlying this belief have been lacking. In Sweden, a combination therapy with pyridostigmine and neostigmine is used very frequently. From a pharmacokinetic point of view, however, pyridostigmine offers an advantage over neostigmine in long term treatment of myasthenia gravis. The lower bioavailability, chiefly the shorter plasma half-life of neostigmine, are drawbacks, as large doses must be taken at short intervals. Furthermore, the length of time up to the maximal peak concentration in fasting subjects is about the same for the two drugs. Thus, the clinical practice of giving preference to pryidostigmine over neostigmine is supported by this study.

Interindividual differences in the plasma concentration of pyridostigmine after a fixed dose have previously been explained by "malabsorption of the drug" (7). However, kidney function may also play a role in the overall disposition of pyridostigmine, since the compound is predominantly eliminated unchanged in the urine (4). One patient with impaired renal function showed the highest plasma concentration (Table 5).

Myasthenia gravis offers the opportunity to measure the effect of a drug on neuromuscular transmission. In the limited studies carried out hitherto, the pharmacological effect of pyridostigmine has been found to be closely correlated to its plasma concentration (5,6). In the present study, there is a good correlation between the plasma concentration of neostigmine and the effect on the decrement of the evoked muscle response. Further studies are needed to elucidate in detail the time-to-effect after intravenous administration of the drug, the dose-response curves in non-treated patients and in patients on long term therapy, and the interaction between different drugs. Reduced responsiveness due to morphologically abnormal moter endplates or due to desensitization may influence the relation between plasma concentration of the drug and the effect on neuromuscular transmission. It may well be that combination therapy with pyridostigmine and neostigmine may alter the pharmacokinetics of the individual drugs. Such effects are now under study.

REFERENCES

1.   Aquilonius, S.-M., Eckernas, S.-A., Hartvig, P., Lindstrom,
     B. and Osterman, P.O. (1979): Eur. J. Clin. Pharmacol.
     15:367-371.
2.   Aquilonius, S.-M., Eckernas, S.-A., Hartvig, P., Lindstrom,
     B. and Osterman, P.O. (1980): J. Clin. Pharmacol. 18:425-428.
3.   Chan, K., Williams, N.E., Baty, J.D. and Calvey, T.N. (1976):
     J. Chromatogr. 120:349-358.
4.   Chan, K. and Calvey, T.N. (1977): Eur. Neurol. 10:69-72.
5.   Chan, K. and Calvey, T.N. (1978): Clin. Pharmaco. Ther. 22:
     596-601.
6.   Cohan, S.L., Pholmann, L.W., Mikswewski, J. and O'Doherty,
     D.S. (1976): Neurology 26:536-539.
7.   Cohan, S.L., Dretchen, K.L. and Neal, A. (1977): Neurology
     27:299-301.
8.   Kornfeld, P., Samuels, A.J., Wolf, R.L. and Osserman, K.E.
     (1970): Neurology 20:634-641.

RED BLOOD CELL/PLASMA CHOLINE RATIO - A POSSIBLE BIOLOGICAL MARKER

OF LITHIUM THERAPY - CLINICAL CORRELATIONS AND LIMITATIONS

E.F. Domino, B. Mathews, S.K. Tait, S.K. Demetriou
and F. Fucek

Department of Pharmacology, Lafayette Clinic
Detroit, Michigan 48207 and,
University of Michigan, Ann Arbor, Michigan 48109 USA

INTRODUCTION

Martin (10) described the accumulation of choline (Ch) against
a concentration gradient by human red blood cells. Later he and
his associates (8,9,11) reported that lithium treatment strongly
inhibited Ch transport in human red cells. Furthermore, the Ch
transport system did not recover when red cell ghosts were prepared, a procedure which removed both intracellular as well as
extracellular lithium. After a patient was withdrawn from lithium
therapy, the Ch transport system in the red cells took about three
months to recover. Inasmuch as this time is about the life of red
cells in the blood compartment, they suggested that lithium therapy
produced an irreversible inhibition of Ch transport. The clinical
significance of this startling observation is unknown.

In the cholinergic mechanisms symposium of three years ago,
Hanin et al. (5) reported on the total levels of both Ch and acetylcholine (ACh) in human red cells as well as plasma. Although the
red cell/plasma Ch ratio varied widely, it usually was greater than
1 and less than 20. These findings were consistent with a concentrative Ch transport system into the red cell. Naively, we predicted that the red cell/plasma Ch ratio would be 1 in lithium
treated patients and were in the process of attempting to prove it.
Hence, it was most unexpected when Jope et al. (7) reported that
Ch accumulated in human red cells during lithium therapy. They
concluded that lithium inhibits Ch efflux more than its influx in
red cells. Martin and his associates showed that rat red blood

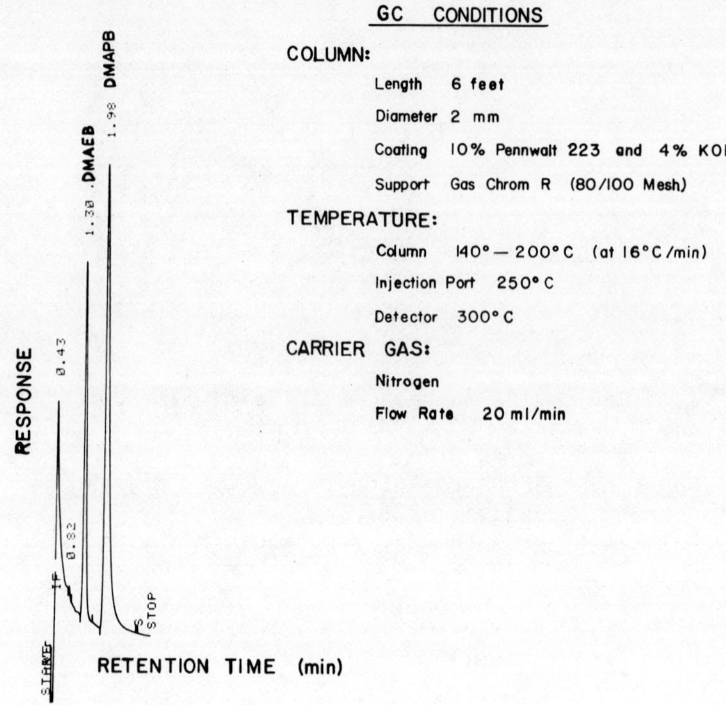

FIGURE 1: GC tracing of a typical human plasma Ch assay using
methylated preanol as the internal standard. The demethylated butyric acid esters of Ch (dimethylaminoethyl
butyrate) and methylated preanol (dimethylaminopropyl
butyrate) have the retention times (min) and the conditions shown.

was repeated for the red cells. The samples were then centrifuged
for 20 min at 16,000 x g and 4°C. The resulting plasma and red
cell supernatants were transferred to two more tubes and 255 and
180 µl of 7.5 N potassium acetate were added respectively. The
tubes were then capped, vortexed and frozen until the day they
were to be assayed.

cells do not respond to lithium like human red blood cells (8,9,
11). However, McCall et al. (12) were able to show that Ch influx
into rat brain was reduced by chronic lithium treatment. When Ch
plus lithium were given, rat brain Ch (and ACh) levels were significantly increased, more than after Ch treatment alone (13).
Millington et al. (13) postulated that lithium's effect on Ch efflux from the rat brain must be greater than its inhibitory effect
on Ch influx. Inasmuch as lithium seemed to inhibit Ch efflux from
both human red cells as well as rat brain, we thought that the
human red cell might be a useful model for the low affinity transport of Ch in the brain in patients treated with lithium. Perhaps
the red cell/plasma Ch ratio could be used as a biologic marker of
effectiveness of lithium therapy. This report describes some clinical correlations and limitations of this phenomenon in patients
treated with lithium carbonate.

MATERIALS AND METHODS

Subjects

The psychiatric patients used in this study were 8 males and
11 females, from 21 to 80 years old. All were on chronic lithium
carbonate therapy for at least one week. The majority of patients
had a major affective disorder and they were all actively monitored
for blood lithium levels. Psychiatric diagnosis and therapy were
independently assessed by their psychiatrist/therapist. Therapeutic
response to lithium carbonate was graded double-blind on an overall
scale of: no effect = 0, poor response = 1, fair = 2, good = 3,
and excellent = 4. Sixteen normal, adult volunteers who were either
workers or blood donors at University Hospital served as the control
subjects. The ages were from 20 to 47 years old and all were in
good physical and mental health. Three females (none on oral con-
traceptives) and 13 males were studied. A morning fasting blood
sample was obtained from all of the subjects. The patients were
instructed not to take lithium carbonate the morning of the blood
test. Venous samples were drawn into 10 ml Vacutainer green-stop-
pered, heparinized tubes and prepared within 6 hr for Ch and lithium
assays.

Preparation of Blood Samples for Ch Assay

Plasma and red cells were separated by centrifuging for 30 min
at 2500 x g and 4°C. A 1.5 ml aliquot of plasma was placed in a
polystyrene Falcon tube followed by 4.5 ml of ice-cold 0.4 N per-
chloric acid. The sample was thoroughly mixed. This same procedure

Ch Assay

The Ch samples were analyzed by the gas chromatographic (GC) assay of Jenden and Hanin (6), as modified by Zahniser et al. (14). Our procedure has been changed slightly in the choice of internal standard. As suggested by Buchanan et al. (1), we used methylated preanol (3-hydroxy-N,N,N-trimethyl propanaminium iodide) as the internal standard to match as closely as possible the chemical characteristics of Ch. The extraction procedure consisted of isolation of the Ch and internal standard from deproteinized plasma or red blood cells on Dowex 50W-X8 cation exchange resin, and elution with 1:1 2 M sodium chloride-methanol. This solution was evaporated to dryness, the compounds of interest extracted with acetonitrile and again evaporated. Ch and methylated preanol were first esterfied with butyryl chloride and then demethylated as described by Jenden and Hanin (6) to give stable, non-polar, volatile derivatives for GC.

GC analysis was with a Hewlett-Packard 5730 GC equipped with a nitrogen-phosphorus flame ionization detector. The GC column was a 6 ft x 2 mm (I.D.) glass column with 10% Pennwalt 223 and 4% KOH on GCR 80/100 packing (Applied Science Laboratories). Column conditions were: injection port 250°C, oven 150°C, detector 300°C, and column flow about 20 ml/min. A typical chromatogram is shown in Fig. 1.

Lithium Assay

The measurement of lithium ion was done with a Jarrell-Ash Dial Atom III Atomic Absorption unit. Standards included equivalent amounts of sodium, potassium and calcium as found in normal human plasma and red blood cells. The measurement of lithium in the red blood cells was done directly on erythrocyte hemolysates. Both plasma and red blood cells were diluted 1:50 with distilled H_2O for lithium analysis. Dade Monitrol I and II, as well as a pooled plasma lithium sample, were used as lithium controls.

Statistical Analysis

Data were analyzed on the Lafayette Clinic IBM 1800 computer using the Student t-test for independent means and the Pearson product-moment correlation. Linear regressions were run on an Olivetti Programma Desk Computer.

TABLE 1 – Relation of Psychiatric Diagnostic Dategories to Treatment Outcome and Lithium and Choline Plasma and Red Cell Levels

Psychiatric Illness Patient No.	Age	Sex	Therapeutic Response	Lithium mEq/1 Plasma	RBC	Ratio	Choline nmol/ml Plasma	RBC	Ratio	Current Li$^+$ Dosage	Duration of Li$^+$ Therapy
Manic Depressive											
Lithium Alone											
1 (M)	57	F	Good	0.61	0.40	0.66	17.5	402	23.0	900 mg	11 yrs
2 (M)	25	M	Excellent	1.30	0.50	0.40	5.54	596	108	1200 mg	9 yrs
3 (D)	30	M	Excellent	0.82	0.38	0.46	10.9	1240	114	1200 mg	1 yr
Manic Depressive											
Lithium + Tricyclic											
4 (D)	36	F	Fair	0.20	0.08	0.40	9.78	15.2	1.55	600 mg*	1 wk+
5 (D)	37	F	Good	0.88	0.25	0.28	13.0	381	29.3	1200 mg	3 wks
6 (D)	60	M	Excellent	0.57	0.28	0.49	10.4	738	71.0	1200 mg	4 yrs
7 (M)	44	M	Poor	0.53	0.20	0.38	23.4	520	22.2	1500 mg	1 yr+
			Good	0.60	0.20	0.33	20.8	948	45.6		
8 (D)	26	F	Fair	1.22	0.76	0.62	11.3	613	54.2	1500 mg	3 mos
Manic Depressive											
Lithium + Neuroleptic											
9 (M)	28	F	Poor	0.65	0.25	0.38	15.1	1300	85.8	600 mg	3 yrs
10 (D)	30	F	Poor	0.78	0.63	0.81	13.3	699	52.6	1200 mg	2 yrs
			Fair	1.22	0.64	0.52	12.2	355	29.1	900 mg	
11 (D)	22	M	Good	0.37	0.13	0.35	12.1	914	75.5	1800 mg	1 yr+
			Good	0.63	0.33	0.52	12.5	1360	109		
12 (D)	22	F	Fair	1.14	0.61	0.54	15.1	862	54.7	1200 mg	4 mos
13 (M)	35	F	Good	0.60	0.34	0.57	18.9	738	39.0	1200 mg	2 mos
14 (M)	27	M	Poor	1.66	0.68	0.41	12.2	719	58.9	300 mg	3 yrs
Atypical Affective											
Disorder											
Lithium + Neuroleptic											
15 (D)	22	F	No change	0.55	0.25	0.45	15.6	207	13.3	900 mg	2 yrs
Involutional Depression											
Lithium Alone											
16 (D)	63	F	Excellent	0.45	0.20	0.44	13.5	930	68.9	900 mg	4 yrs
Manic Depressive											
Organic Brain Syndrome											
Lithium + Tricyclic											
17 (D)	80	M	Poor	0.40	0.13	0.32	20.2	1040	51.3	300 mg	1 yr+
Schizoaffective											
Lithium Alone											
18 (D)	27	M	Excellent	0.23	0.11	0.48	12.7	1120	88.0	1200 mg	1 mo
Chronic Schizophrenic											
Lithium + Neuroleptic											
19 (D)	21	F	Good	0.80	0.38	0.48	22.8	496	21.8	600 mg	2 mos
			Excellent	0.73	0.27	0.37	7.95	472	59.4	900 mg	

*partially noncompliant

Patient's current diagnostic state: (M)=Manic (D)=Depressed

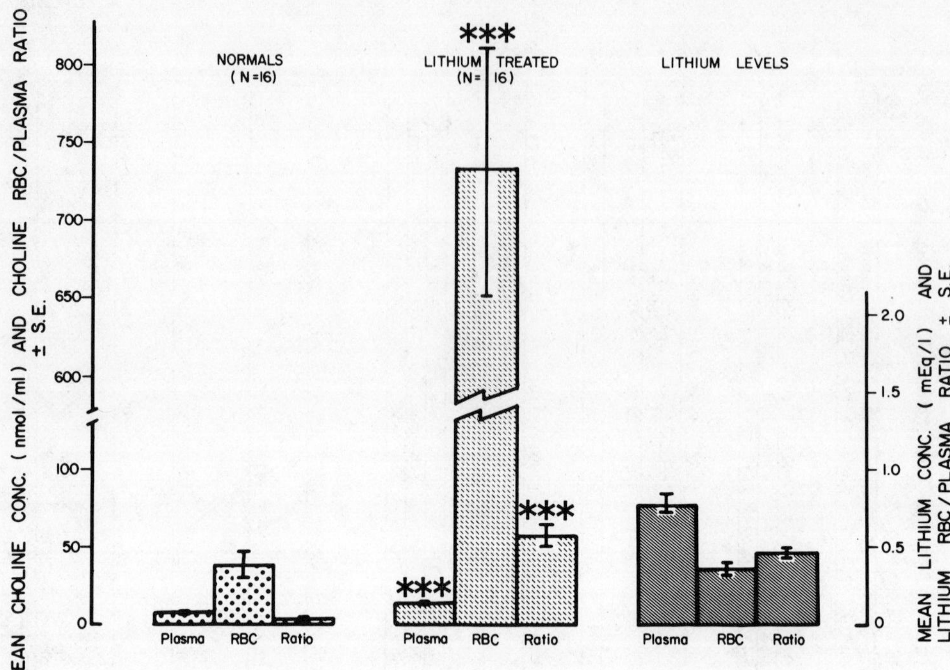

FIGURE 2: Mean plasma and red cell Ch concentrations in normal
volunteers and chronic lithium treated patients. All
were asked to fast prior to a morning venous blood
sample. The 16 normals were young adult hospital workers
and volunteers. The 16 lithium treated patients were on
chronic lithium therapy for at least 3 weeks. The
patients were asked to refrain from taking lithium carbonate the morning of the blood sample. A group comparison Student t-test was done comparing the normals
to the patients on plasma and red cell Ch and the red
cell/plasma Ch ratio. *** p < 0.001.

RESULTS

Relation of Psychiatric Diagnostic Categories,
Therapeutic Responses, and Lithium and Ch Plasma and Red Cell Levels

The data obtained are summarized in Table 1. Patients were
subdivided into various diagnostic categories. The medications
listed were given as lithium carbonate alone, plus tricyclic, plus
neuroleptic medications, etc. The therapeutic response listed was
determined on the same day as the blood draw.

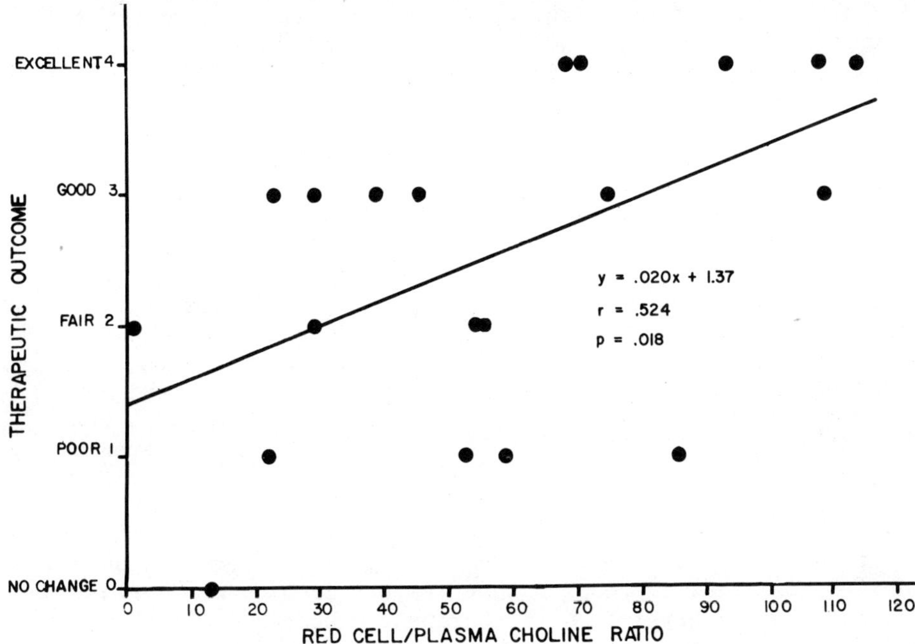

FIGURE 3: Red cell/plasma Ch ratio vs therapeutic outcome in
lithium treated patients with major affective disorders. The least squares regression line is y = 0.020X
+ 1.37 with r = .524. P = 0.018. Seventeen patients
are included. Two samples of at least a week apart are
used for three patients.

Plasma and Red Cell Ch Concentrations in Normal Volunteers and Chronic Lithium Treated Patients

A total of 16 normal subjects free of lithium, and 16 chronic
lithium carbonate treated patients from Table 1, who had a therapeutic plasma lithium level of at least 0.4 and below 1.7 mEq/1,
were studied. In patients from whom we had multiple samples, only
the first blood draw was included. Plasma, red cell and red cell/
plasma ratios were determined. The mean data are summarized in
Fig. 2. For the 16 normal subjects, the mean $\pm$ S.E. plasma Ch
was 8.54 $\pm$ 0.460 nmol/ml, and the red cell Ch was 39.5 $\pm$ 9.63 nmol/
ml with a red cell/plasma Ch ratio of 4.49 $\pm$ 0.931.

For the 16 chronic lithium treated patients, the mean $\pm$ S.E. plasma Ch was 15.5 $\pm$ 1.23 nmol/ml, the red cell Ch was 709 $\pm$ 75.6 nmol/ml, and the red cell/plasma Ch ratio was 52.4 $\pm$ 7.67. In each case, the increase above control was highly significant (P <0.001). In the patients the mean $\pm$ S.E. plasma lithium was 0.78 $\pm$ 0.07 mEq/l and the red cell lithium was 0.39 $\pm$ 0.05 mEq/l with a red cell/ plasma lithium ratio of 0.49 $\pm$ 0.03.

Of the 19 patients listed in Table 1, only 17 with major af- fective disorders were analyzed for inclusion in Fig. 3. This ex- cluded patients number 17 and 10 who were diagnosed as organic brain syndrome and chronic schizophrenic respectively. Figure 3 shows the correlation between red cell/plasma Ch ratio and therapeutic outcome of the patients, as analyzed by the Pearson product-moment correla- tion and linear regression analysis. This gave a least squares regression line of y = 0.020X + 1.37 with r = 0.524 and probability of P = 0.018.

When therapeutic outcome was analyzed vs each of the other Ch or lithium blood level parameters of Table 1, no correlation was found. Hence it is impressive that neither lithium or Ch plasma or red cell levels correlate with therapeutic response. Although the correlation of therapeutic response to the Ch red cell/plasma ratio was statistically significant, considerable individual variation was noted as seen in Fig. 3 where each dot represents a given patient. In some instances the same patient was used several times when the therapeutic response varied over time.

DISCUSSION

The search for a biologic marker that is predictive of the out- come of lithium therapy has been indeed frustrating (3,4). It is generally agreed that serum or plasma levels of lithium provide a useful index of whether most patients have an inadequate dose schedule of lithium salts or a possible dangerous excess. However, within the "therapeutic window" of approximately 0.4 to 1.6 mEq/l the serum or plasma lithium level does not correlate with therapeutic benefit. This certainly was the case with our small series of lithium treated patients.

It is a common experience of many physicians to observe some patients with very low "therapeutic" levels of lithium show toxicity and the more rare reverse situation of patients with very high levels of lithium show no toxicity. The hypothesis that the red blood cell/plasma ratio of lithium is a potential predictor of thera- peutic success has generated a great deal of research effort, but not much clinical benefit to the patient. It is for this reason

that it would be premature to suggest that the red cell/plasma Ch
ratio is the long sought answer for a biologic marker of treatment
outcome with lithium. In our series of lithium treated patients
with heterogenous affective disorders, we were pleasantly surprised
that a positive correlation of about 0.5 was found with therapeutic
outcome and the red cell/plasma Ch ratio. Obviously there was
great individual patient variability. The fact that Bunney et al.
(2) observed in a longitudinal study of two manic patients treated
with lithium carbonate and placebo a definite increase in mania
within 24 hr on withdrawal of lithium, makes us most cautious indeed.
After all, Martin and his associates (8-11) observed that the effects of lithium on Ch uptake lasts for as long as 3 months. That is
in agreement with our own unpublished observations of one patient
withdrawn from lithium therapy. It took about 3 months for the red
cell/plasma Ch ratio to return to normal. It may be that the red
cell/plasma Ch ratio may be useful in patients who show lithium
toxicity, even though their plasma or serum lithium levels are in
the very low therapeutic rang. Perhaps this preliminary report will
stimulate further research to provide a clinically relevant correlate to a most impressive biologic phenomenon. Whether the red
cell can be used as a model of brain transport of Ch remains un-
answered.

SUMMARY

    The purpose of this study was to correlate red cell/plasma Ch
ratios as a biologic marker of lithium treatment. Plasma and red
blood cell Ch and lithium levels were measured in psychiatric
patients during chronic lithium carbonate therapy. Ch was assayed
using chemical demethylation, GC and nitrogen detection. Lithium
levels were measured by atomic absorption. A total of 19 patients
were studies in whom lithium therapy was indicated for psychiatric
reasons. Most suffered from manic-depressive illness. The mean $\pm$
S.E. red cell/plasma Ch ratio was 4.49 $\pm$ 0.931 for the normal control subject and 54.5 $\pm$ 6.36 for those patients on chronic lithium
therapy. Although therapeutic benefit tended to correlate with
elevated ratios, this was not always the case. The correlation
coefficient of therapeutic outcome vs red cell/plasma Ch ratio was
0.524 with a P value = 0.018. It is concluded that the red cell/
plasma Ch ratio as a biologic marker of lithium therapy is of very
limited practical clinical value. Future studies are needed to
assess its role, if any, in clinical psychopharmacology.

ACKNOWLEDGEMENTS

The research protocol was approved by the Lafayette Clinic Institutional Review Board and by the Michigan Department of Mental Health's Technical Advisory Research Committee.

REFERENCES

1. Buchanan, D.N., Fucek, F.R. and Domino, E.F. (1980): J. Chromatog. 181:329–336.
2. Bunney, W.E., Jr., Goodwin, F.K., Davis, J.M. and Fawcett, J. A. (1968): Amer. J. Psychiat. 125:91–104.
3. Carroll, B.J. (1979): Arch. Gen. Psychiat. 36:870–878.
4. Frazer, A., Mendels, J., Brunswick, D., London, J., Pring, M., Ramsey, T.A. and Rybakowski, J. (1978): Amer. J. Psychiat. 135:1065–1069.
5. Hanin, I., Kopp, U., Zahniser, N.R., Shih, T.M., Spiker, D.G., Merikangas, J.R., Kupfer, D.J. and Foster, F.G. (1978): In Cholinergic Mechanisms and Psychopharmacology (ed) D.J. Jenden, Plenum Press, New York, pp. 181–195.
6. Jenden, D.J. and Hanin, I. (1974): In Choline and Acetylcholine: Handbook of Chemical Methods (ed) I. Hanin, Raven Press, New York, pp. 135–150.
7. Jope, R.S., Jenden, D.J., Ehrlich, B.E. and Diamond, J.M. (1978): New Eng. J. Med. 299:833–834.
8. Lee, G., Lingsch, C., Lyle, P.T. and Martin, K. (1974): Brit. J. Clin. Pharmacol. 1:365–370.
9. Lingsch, C. and Martin, K. (1976): Brit. J. Pharmacol. 57: 323–327.
10. Martin, K. (1968): J. Gen. Physiol. 51:497–517.
11. Martin, K. (1974): In Drugs and Transport Processes (ed). R.A. Callinghan, Macmillan, London.
12. McCall, A.L., Millington, W.R., Botticelli, L.J. and Wurtman, R.J. (1978): Proc. Soc. Neurosci. 4:1360.
13. Millington, W.R., McCall, A.L. and Wurtman, R.J. (1979): New Eng. J. Med. 300:196–197.
14. Zahniser, N.R., Chou, D. and Hanin, I. (1977): J. Pharmacol. Exp. Ther. 200:545–558.

BLOOD CHOLINE AND ITS MEANING IN PSYCHIATRIC

AND NEUROLOGIC DISEASE STATES

I. Hanin, D.G. Spiker, A.G. Mallinger, U. Kopp, J.M.
Himmelhoch, J.F. Neil and D.J. Kupfer

Western Psychiatric Institute and Clinic
Department of Psychiatry, School of Medicine
University of Pittsburgh
Pittsburgh, Pennsylvania   15261 USA

INTRODUCTION

The involvement of central cholinergic mechanisms has been
implicated in a variety of neurologic disease states. These include such syndromes as tardive dyskinesia (11,17,28,32,62), Huntington's disease (3-5,13,17,18,26,33,49,59,72,84,85,87,90),
Friedreich's ataxia (6,47), presenile dementia of Alzheimer's type
(10,14-16,25,63,64,69,77,89), and Gilles de laTourette's disease
(31,35,66). In all these syndromes a reduction in normal cholinergic activity has been implicated. As a result, attempts have
been initiated over the past five years, with variable success, to
alleviate some of the associated symptoms, using agents shown in
animal experiments to be effective cholinomimetic agents (8). To
date, however, use of direct cholinergic agonists and/or antagonists
in the treatment of various neurologic, and particularly psychiatric
disorders, is still in the experimental phase; such agents have not
yet superceded conventionally used psychotherapeutic agents.

Recently there has been a revival of interest in the role that
central cholinergic mechanisms may play in the etiology of a number
of mental and psychiatric disorders. Inferences regarding the involvement of acetylcholine (ACh) in these psychiatric disease states
have resulted from clinical observations indicating that central
drugs with known cholinergic agonistic properties will generally
induce a lowering in mood (representative papers fro a vast literature
include: 9,12,19,24,29,43,57,65,70,83,86, and others). These initial
clinical observations have been influenced by a wide resource of

animal-based experimental data, which have provided valuable information leading towards our further understanding of the interplay
between various psychotherapeutic agents and central cholinergic
mechanisms in animals.

The clinical inferences, mentioned above, for cholinergic
involvement in mental disease states, have also been bolstered by
studies showing, on a molecular level, that certain potent psychotropic agents commonly used in the psychiatric clinic, exert a
strong affinity for the muscarinic receptor in brain (67,68,73,78).
Thus these agents could conceivably interact in a competitive manner with central cholinergic mechanisms, and exert an effect on
mood and behavior, through some modification of endogenous cholinergic mechanisms. For example, it is possible that some psychiatric diseases may result from distortions in normal regulation of
central cholinergic terminals, and that the psychotropic agents
will restore such an imbalance to normal conditions.

In general, therefore, one can state, based on the available
information, that a wide variety of psychotherapeutic agents commonly used in clinical practice have distinct effects on central
cholinergic activity in vivo. However, whether this effect is of
primary relevance to the specific therapeutic function of these
different compounds has not yet been established. An investigation
of this question would necessarily involve well designed experiments
utilizing animals, that would parallel and supplement information
obtainable from human subjects.

It would also be ideal if one were to have access to a direct
and non-traumatic mechanism by which one could obtain, in humans,
an index of central cholinergic mechanisms in vivo. In order to
consider such a possibility it is first necessary to evaluate what
are pertinent biochemical measurements that one could conduct in
man, from which one might conceivably gain insight into central
cholinergic phenomena in vivo.

Potential Sources for Analysis
of Choline and Acetylcholine in vivo

Analysis of urinary excretion of choline (Ch), ACh, and/or
metabolites of Ch as an index of central nervous system (CNS) ACh
metabolic function is at present impractical. It is difficult to
ascertain from such analyses what proportion, if any, of the metabolites were contributed from CNS activity alone. It is conceivable
that at some later date a new metabolite of Ch may be discovered
which is formed primarily in the CNS and is excreted in the urine

(as has been purported to be the case with MHPG, the major CNS
metabolite of norepinephrine)(52,74,75). Unless such a metabolite
is discovered, however, one would have to utilize other approaches
in attempting to study normal and pathological _in vivo_ central
cholinergic functions in man.

There are a number of reports in the literature on studies of
levels of Ch, ACh, acetylcholinesterase (AChE), and choline acetyltransferase (CAT) in both cerebrospinal fluid (CSF) and plasma.
Essentially they all serve to emphasize the fact that, unfortunately, we still lack an effective and reliable biological marker of
in vivo central cholinergic function in man. This conclusion is
aptly summarized by Jenden (44), Domino (21) and by Davis and co-
investigators (19), who have effectively reviewed the available
data relevant to this question. They have provided convincing
evidence to indicate that the measurement of any one or even a combination of the above components in either CSF or plasma would have
to be approached with caution, since the available data are sometimes questionable and often subject to misinterpretation.

For example, in studying CSF CAT activity, Aquilonius and
Eckernas (2) have demonstrated that non-enzymatic synthesis of CSF
ACh may occur under the experimental conditions which they utilized
for analysis of CAT activity. This method conventionally calculates
the extent of CAT activity from the amount of ACh formed in a specified time period from an excess of the substrate, Ch. If non-
enzymatic synthesis of ACh also occurs; then one would obtain a
false value for the measured activity of this enzyme. In a comparable evaluation of CSF Ch levels, data obtained from the laboratories of Jenden and colleagues (76,88) and of Klaver et al. (48)
would suggest that CSF Ch is supplied from a number of sources,
including blood, apart from the hydrolysis of brain ACh (as implied
earlier: 1,45). Thus reports of CSF Ch levels as an index of brain
ACh release must also be viewed with a critical eye. Moreover, if
one were to review recent clinical reports in which plasma and/or
CSF Ch levels have been measured subsequent to treatment of patients
with Ch or lecithin, one would find that these levels provide an
excellent index of patient compliance in taking the prescribed
medication. However, there is at best only a tenuous relationship
between the measured levels of Ch, and clinical efficacy (for example: 7,20,27,32,42; plus chapters in reference 7 on the subject
of use of Ch and lecithin in movement disorders); i.e. in some cases
there was, but in others there was no improvement observed coincident with measured increases in plasma Ch levels. These observations
do not totally negate the utility of measuring Ch, ACh, CAT and/or
AChE in CSF or in plasma. They, nevertheless, point out the limitations of utilizing measured values of any of these parameters alone,
and reflect the need for a more effective approach to measure brain
cholinergic function _in vivo_, preferably utilizing a noninvasive
procedure.

Recent reports of studies with human subjects have, nevertheless, indicated that blood cholinergic parameters in certain categories of psychiatric and neurologic disease states may be different from such parameters measured in normal controls. Milstoc and co-investigators (56) have reported that patients with affective disorders exhibit low red blood cell (RBC) AChE activity.   Stavinoha and coinvestigators (81) have shown that Ch and ACh levels in whole blood are higher in depressive neurotics and in unipolar depressives, respectively, than they are in normal controls, schizophrenics or bipolar depressives.   In our own laboratories we have demonstrated that RBC Ch is significantly higher than normal in a subcategory of depressed patients (38). Finally, Jope et al. (46) have reported that treatment of bipolar patients with lithium results in marked elevations in RBC Ch levels. We have confirmed these latter findings, and moreover, will present in this report evidence that this effect would appear to be irreversible at the level of the RBC membrane. Similar observations have also been reported by Domino et al. (this volume; 22).

While still subject to confirmation in other laboratories, these reports collectively suggest that blood Ch may provide an index of altered cholinergic function *in vivo*, in individuals with certain mental and neurologic disorders.

Cholinergic Parameters in Plasma
and RBCs of Psychiatric Patients
and Normal Volunteers

We have measured baseline conditions of blood cholinergic parameters in large populations of both normal and mentally ill volunteers.   The goal of these studies was to determine definitively whether any correlation is to be found between plasma and RBC cholinergic parameters, and clinical state in a variety of psychiatric patients.

Categories of patients studied to date have included a) depressed inpatients undergoing a strict, specific protocol for antidepressant treatment with amitriptyline; b) outpatients with Gilles de la Tourette's syndrome; c) bipolar outpatients undergoing treatment with lithium carbonate and other drugs; and d) outpatients undergoing treatment for a variety of neurological disorders. Control subjects have included both volunteers and their families from among staff and faculty at WPIC, as well as from the local blood bank. Data obtained with these various groups are summarized in the following sections.

Choline Levels in Plasma
and RBCs of Depressed Patients

Patients were housed in the Clinical Research Unit (CRU) on
the 11th floor at WPIC, and were under the supervision of two of
the coauthors (DGS and JFN). These patients were concurrently
participating in an NIMH supported investigation of EEG sleep in
relation to affective illness (under the guidance of DJK), and
were part of a double-blind protocol comparing amitriptyline and
placebo. The patients had been carefully diagnosed as having major
depressive disorder according to the Research Diagnostic Criteria
(79). All patients had been drug free for a minimum of 14 days
prior to the study. Blood for Ch analysis was drawn from these
patients at the end of the drug free period and just prior to initiation of either amitriptyline or placebo treatment. To date we
have analyzed blood from 78 such patients. The spread of Ch levels
observed in RBC and plasma of these individuals is shown in Fig. 1.
For comparison see the data from 48 blood bank donors (Fig. 2), and
from 29 WPIC staff and faculty volunteers and their families (Fig.3).
It is evident from these observations that RBC Ch levels are much
more variable than plasma Ch levels in any given population; and
that a large concentration gradient can exist in some cases between
plasma and RBCs.

We have conducted a correlation analysis of some of the available patient and control data. This analysis includes 33 patients
to date, and involves only those individuals that have completed
the NIMH protocol mentioned above. Detailed clinical data, sleep
records, and blood amitriptyline and nortriptyline data have been
collected from all of these patients, and have, therefore, enabled
us to conduct in these patients a multivariate correlation between
plasma and RBC Ch levels and a wide variety of clinical data. Of
the 29 "known" controls (subjects personally known to use from
within the Institute and their families), 23 were included in this
analysis; individuals less than 17 years old were excluded, in order
to provide a population more reasonably matched in age to the other
two populations included in this analysis. At this point only data
obtained from patients when they were drug free have been analyzed.
Some of the results obtained are summarized in Table 1.

It is interesting to note that plasma Ch values, age and sex
profiles all appear to be comparable in the three populations
studied. RBC Ch levels, on the other hand, showed a marked difference in mean values calculated for the three groups.

Mean RBC Ch data obtained with blood bank donor bloods were
intermediate in value between those of controls and the depressed
patients. This is not surprising since one invariably has to deal
with an unknown population in terms of drug history and medical and

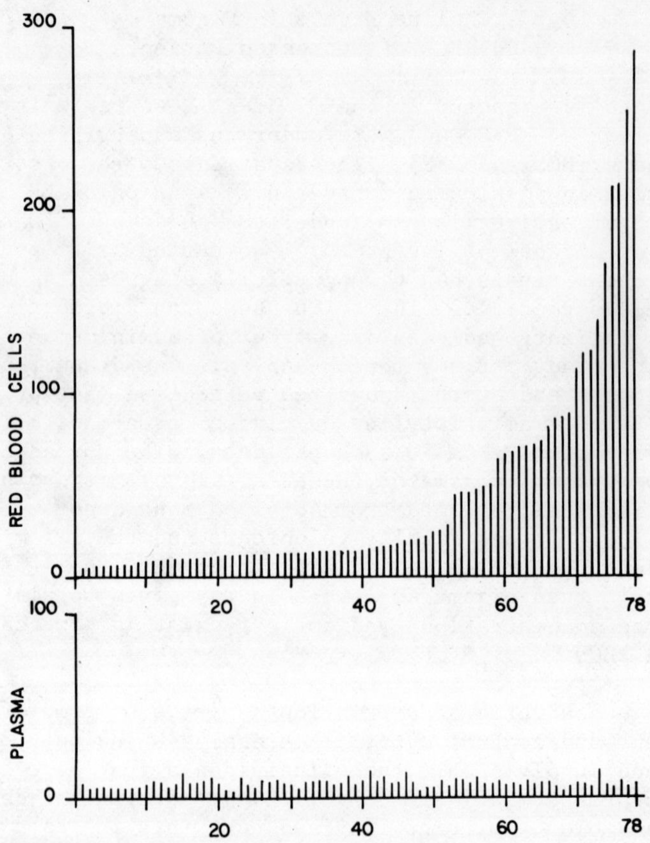

FIGURE 1: RBC and plasma Ch in a cross section of drug free primary
and secondary depressed inpatients. Ordinate represents
RBC and plasma Ch levels in nmol/ml. Abscissa represents
a cross section of individuals studied. Data have been
arranged in increasing order of measured RBC Ch values.
Below each RBC Ch value is the corresponding value of
plasma Ch measured in the same individual.

psychiatric background when blood bank donors are involved. Any
information available about these volunteers has to depend on the
volunteers' own self report which, at best, is unreliable. For this
reason, in subsequent analyses of the data obtained with our patients
only the group of "known" controls will be used for comparative purposes. The blood bank donor data have, nevertheless, been included
in this report to demonstrate, on the one hand, the existence of the

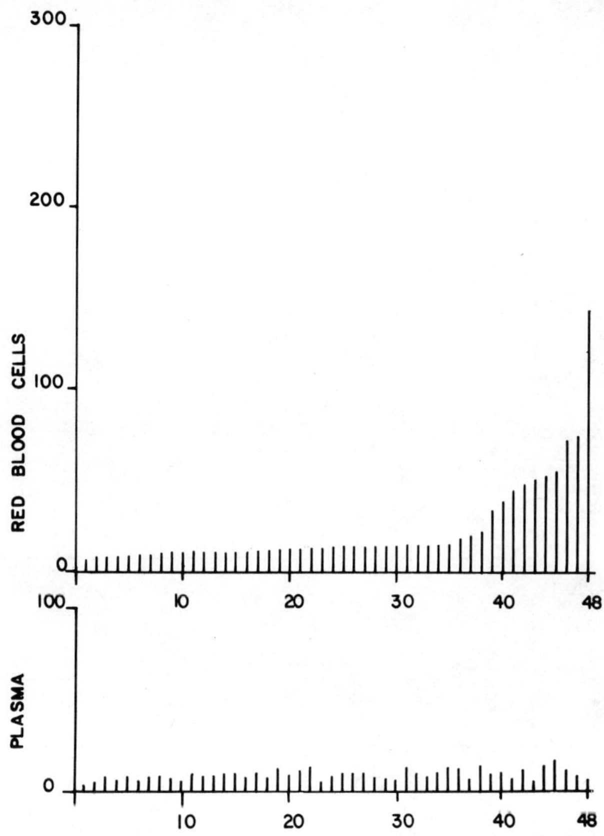

FIGURE 2: RBC and plasma Ch in a cross section of blood bank donors.
           Ordinate represents RBC and plasma Ch levels in nmol/ml.
           Abscissa represents a cross section of individuals studied.
           Data have been arranged in increasing order of measured
           RBC Ch values. Below each RBC Ch value is the corres-
           ponding value of plasma Ch measured in the same individual.

spread of RBC Ch data in this particular population as well (see
Fig. 2); and on the other hand, to accentuate quantitatively the
fact that blood bank donors are different from "known" controls in
terms of RBC Ch levels.

    The unusually high Ch levels measured in a segment of the de-
pressed population prompted us to investigate whether, within this
group of carefully diagnosed, drug free patients, one could distinguish clinically relevant differences between patients exhibiting

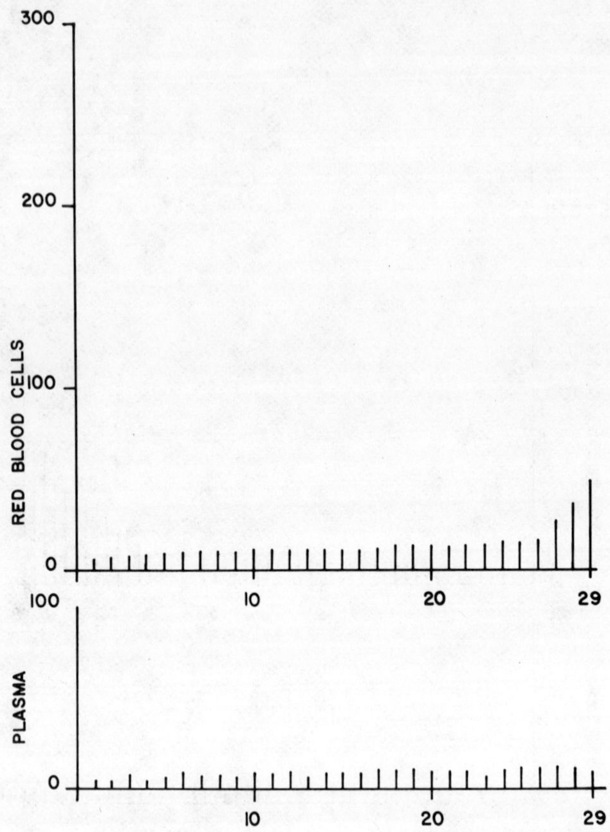

FIGURE 3: RBC and plasma Ch in a cross section of known controls.
Ordinate represents RBC and plasma Ch levels in nmol/ml.
Abscissa represents a cross section of individuals studied.
Data have been arranged in increasing order of measured
RBC Ch values. Below each RBC Ch value is the corres-
ponding value of plasma Ch measured in the same individual.

such different values of RBC Ch. Although preliminary at this
point, the data do reveal that the subgroup of patients with RBC
Ch levels exceeding 35 nmol/ml (arbitrarily selected as a natural
cutoff point, based on the spread of data obtained) generally appear clinically to be characterized by anergia, to have suffered a
more protracted illness, and to have exhibited fewer recurrent depressive episodes than patients with RBC Ch levels below 35 nmol/ml.
Moreover, there would appear to be a higher incidence of secondary

TABLE 1: Cross sectional analysis of RBC and plasma Ch, sex and age, in a clinically depressed and control population, and in a population of blood bank donors.

	RBC Ch (nmol/ml)	Plasma Ch (nmol/ml)	Age	Sex Male (%)	Female (%)	n
Depressed Patients						
Mean ± S.E.	30.69 ± 8.91	7.69 ± 0.36	35.36 ± 2.02			
Median	12.53	7.50	35.00	39	61	33
Range	6.10 – 285.50	4.20 – 12.70	18.00 – 58.0			
Blood Bank Donors						
Mean ± S.E.	21.80 ± 3.53	8.75 ± 0.36	36.86 ± 1.87			
Median	12.75	8.45	33.75	43	57	48
Range	6.70 – 144.20	4.30 – 17.20	21.00 – 68.0			
Known Controls						
Mean ± S.E.	14.27 ± 2.12	7.69 ± 0.38	33.61 ± 3.45			
Median	10.60	7.90	30.00	48	52	23
Range	5.80 – 49.00	4.60 – 12.00	17.00 – 85.0			

depressive patients with RBC Ch values exceeding 35 nmol/ml than
there is among primary depressive patients. This analysis would
however, require a considerably larger number of subjects before
any statistically significant evaluation could be made.

Choline Levels in Plasma and RBCs
of Patients with Gilles de la Tourette's Syndrome

Levels of plasma and RBC Ch in 20 drug free patients suffering
from Gilles de la Tourette's disease were compared with those in
age and sex matched controls. Although plasma Ch levels were identical, there was a significant difference in RBC Ch levels between
the two groups (39). The Tourette's patients exhibited a significantly higher mean value of RBC Ch levels than the controls (mean $\pm$
S.E.: patients - 38.96 $\pm$ 6.76 nmol/ml; controls - 12.24 $\pm$ 0.82 nmol/
ml). Consistent with these findings, a comparison of RBC/plasma Ch
ratios in the two groups showed a highly significant difference
between the groups ($p < 0.01$, Student's t-test with matched pairs),
since plasma levels were similar in both populations. If this
difference can be ascribed to differences in Ch transport, then
these data suggest that in Gilles de la Tourette's syndrome there
may be a defect in membrane characteristics for transport of Ch.

Effect of Lithium Treatment on Choline Levels
in Plasma and RBCs: _in vivo_ observations

Lithium exerts a significant trophic effect on Ch levels in
RBCs, but not in plasma. This effect occurs after administration
of even a single, 600 mg dose of lithium carbonate. Within hours
after administration of the drug, a definite increase in RBC Ch
levels is observed which, with continued chronic administration of
the compound, results in a significant and permanent elevation of
RBC Ch levels (to between 40- and 110-fold above normal). Table 2
summarizes the measured values of plasma and RBC Ch in 20 consecutive outpatients from the Affective Disorders Clinic at WPIC
(directed by JMH). It is evident that patients who had been chronically treated with lithium had significant elevations of their RBC
Ch levels. It is also interesting to note that in spite of these
tremendous elevations in RBC Ch levels, plasma Ch levels remained
within the normal range in the same individuals. The levels of RBC
Ch attained in a particular individual after lithium treatment would
probably be affected by the dose and duration of lithium administration, as well as by the Ch handling characteristics of the RBCs of

TABLE 2:   Effect of lithium treatment on plasma and RBC Ch levels
           (nmol/ml).

Patient	Plasma Ch	RBC Ch
No Lithium Treatment		
1	7.9	10.2
2	7.4	11.5
3	7.4	13.8
4	11.5	15.6
5	9.2	15.7
6	7.6	19.7
7	11.1	80.7
8	5.5	99.0
Mean ± S.E.	8.45 ±0.72	33.28 ±12.51
Lithium Treatment		
9	14.1	432.6
10	15.4	552.3
11	9.7	674.0
12	10.9	704.6
13	15.2	714.3
14	13.2	742.6
15	16.2	757.5
16	9.8	828.4
17	9.3	867.7
18	14.7	871.1
19	13.0	924.3
20	11.7	1080.6
Mean ± S.E.	12.77 ±0.70	762.5 ±49.35

the individual patients involved. While these studies were in
effect, Jope et al. (46) published a preliminary report showing the
same experimental observations. Domino et al. (22) have also obtained similar results with their patients. Thus, these findings
have now been replicated in at least three laboratories.

The effect of lithium on RBCs appears to be irreversible,
because 3-4 months (equivalent to the approximate life span of the
erythrocyte) were necessary for RBC Ch levels of an individual
treated with lithium to return to normal after discontinuation of

the drug. This finding is consistent with that of Lingsch and
Martin (51), who have demonstrated that Ch transport both into and
out of erythrocytes is inhibited by lithium administration to
patients, and that it requires a period of 3 months after discontinuation of lithium treatment for Ch transport into RBCs to recover to normal rates.

A likely explanation for these observations is that the effect
of lithium on RBC Ch levels occurs as a consequence of a direct
action on the RBC membrane Ch transport system. We have initiated
studies to test this premise in more detail by investigating RBCs
in vitro, utilizing gas chromatography mass spectrometry (GCMS) and
stable isotope methodology (36,37).

Effect of Lithium Carbonate on Cholinergic
Parameters in RBCs: in vitro observations

It has been demonstrated by Martin and his associates (50,51)
that Ch transport across the RBC membrane is energy dependent and
is inhibited by prior lithium treatment. In order to understand
more explicitly the striking effect of lithium administration on
RBC Ch levels in vivo, we decided to study, in further detail,
factors which affect Ch transport across RBC membranes in vitro,
and to evaluate the interaction of lithium with these factors. GCMS
has been used, in conjunction with stable isotope flux studies.

Transport of deuterium labelled Ch (D_4Ch; hydrogens in ethylene
chain replaced by 4 deuterium atoms) has been measured across the
RBC membrane in vitro, utilizing a modification of an already established system which we routinely use for the analysis of transport of lithium and other cations across the RBC membrane in vitro
(53).

Our data, although preliminary, have already demonstrated that:
a) it is possible under our experimental conditions to maintain RBCs
in vitro for at least 3 hr. No appreciable breakdown of phospholipids was observed, as reflected in minimal changes in free Ch
content in the medium and minimal hemolysis over the entire time
period of the experiment; b) an exchange of Ch between RBC and incubating medium occurs under physiological conditions; and c) the
rate of Ch exchange across the RBC membrane is inversely related to
the initial concentration of Ch in the RBC. Higher initial levels
of endogenous RBC Ch appear to be indicative of less exchange of
Ch between the RBC and its surrounding medium (Fig. 4).

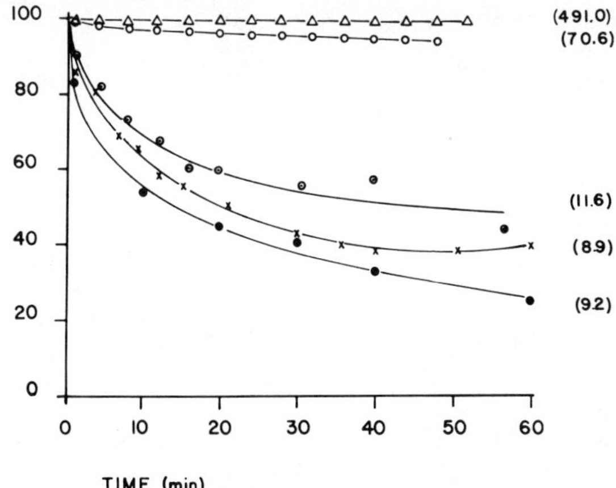

FIGURE 4: Relationship of initial Ch content of RBC to subsequent
release of Ch by RBC in vitro. Data are expressed as percent of initial RBC Ch content at different times after
incubation of RBCs with D4Ch. Each curve represents blood
drawn from a different individual. Numbers in brackets
represent corresponding initial RBC Ch levels in nmol/ml.
Ordinate: percent of initial RBC Ch level. Abscissa: time
of incubation of RBCs.

Effect of Various Therapeutic Agents
on RBC Choline levels in vivo

Because of the significant elevation of RBC Ch levels (as noted
above) in patients following treatment with lithium (see Table 2),
we sought to establish whether other therapeutic agents would also
have an effect on plasma and/or RBC Ch levels as well.

Over the past 5 years, we have analyzed additional plasma and
RBC Ch levels in a large number of individuals, which have not been
included in the data summaries listed above. These are cases where
individuals were not drug free at the time blood was drawn; or where
drug treatment was initiated after we had obtained a drug free
sample, and we subsequently elected to follow the patient's clinical
course in conjunction with analysis of his/her blood Ch levels.

Records were reviewed for individuals whose plasma and RBC Ch
levels did not exceed "normal" levels (according to Table 1). Drugs,
which they reported as having taken within 48 hr prior to drawing
of blood for Ch analysis, are listed in Table 3. By deductive

TABLE 3:   Medications which would appear not to affect RBC and
           plasma Ch levels.

Aspirin	Motrin
Aventyl (nortriptyline)	Mysoline
Caffeine	Nicotine
Clonazepam	Parnate
Cogentin	Prolixin
Dalmane	Stelazine
Dilantin	Tegretol
Elavil (amitriptyline)	Trilafon
Excedrin	Valium
Fiorinal	Valproic acid
Haldol	Vitamins
	Zarontin

reasoning we can state that the medications listed in Table 3 do
not affect RBC and plasma Ch levels - at least not to the extent we
have seen in the case of lithium and RBC Ch. However, until actually tested in a prospective analysis, this information is only suggestive and should be treated as such.

Potential Significance of Above Findings
in Psychiatric and Neurologic Disease States

Our findings, reviewed above, have demonstrated that RBC Ch is
significantly higher in a sub-category of depressed patients (35,38)
and in individuals with Gilles de la Tourette's disease (35). In
addition, Milstoc et al. (56) have demonstrated that patients with
affective disorders exhibit a low RBC AChE activity; and Stavinoha
and coinvestigators (81) have shown that Ch and ACh levels in whole
blood are higher in depressive neurotics and in unipolar depressives,
respectively, than they are in normal controls, schizophrenics or
bipolar depressives. Although still fragmentary, and requiring
confirmation from other laboratories, these combined reports do,
nevertheless, suggest the possibility that RBC cholinergic phenomena could provide an index of generalized cholinergic function _in
vivo_, in individuals with certain mental and neurological disorders.

In addition, lithium, an effective mood stabilizing agent, persistently elevates RBC Ch levels, even after a single dose. Since
the half life of lithium is short (hours), this delayed effect would

most probably not be due to the presence of lithium in the system.

Would such observations in RBCs provide an index of a similar
physiologic occurrence in brain? Alternatively stated: is it conceivable that certain specific neurologic and psychiatric illnesses
may be associated with a generalized defect in CNS nerve membrane
function which, conveniently, is also reflected in the RBC membrane? There actually is a precedent for such reasoning. For
example, it has been suggested by some investigators that lithium
transport across RBC membranes is reduced in certain bipolar patients
(60,61,71), as well as in those relatives of patients who have a
history of affective disorders (23). Furthermore, there is evidence
for a variety of RBC membrane function alterations in neuropsychiatric disorders (30,40,41,54,58,82).

Therefore all the facts provide, in combination with one another, information favoring the development of a more extensive and
systematic study of various RBC membrane related phenomena in humans,
if one is to further explain and/or understand the functional and
clinical significance of elevated RBC Ch levels found in certain
specific categories of psychiatric patients, and in individuals
treated with lithium carbonate. The RBC may, conceivably, be implicated both directly and indirectly. Specifically, the exchange of
Ch between plasma and RBC _in vivo_ may be an important factor determining the availability of free Ch for transport across the
blood-brain barrier, and thus its availability at cholinergic nerve
terminals in brain for subsequent synthesis of ACh. If this is the
case, the amount of endogenous RBC Ch, according to our above observations, could determine the amount of Ch that is accessible to
the blood-brain barrier. This would be a direct effect of the RBC.
On the other hand, the RBC may not be directly important at all in
regulating or "fine tuning" blood-to-brain transport of Ch. However, drug induced, or even endogenously related factors which
render the RBC membrane different from normal may provide indirect
clues for similar membrane effects at key nerve terminals in the
brain.

ACKNOWLEDGEMENTS

This work was supported by NIMH grants MH 26320 and MH 24652.
The authors would like to express their appreciation to the staffs
of the Clinical Research Unit and of the Affective Disorders Clinic,
without whose assistance this study would not have been completed.

REFERENCES

1. Aquilonius, S.M., Schuberth, J. and Sundwall, A. (1970): IN
   Drugs and Cholinergic Mechanisms in the CNS (eds) E. Heilbronn and A. Winter, Research Institute of National Defence,
   Stockholm, pp. 399-410.
2. Aquilonius, S.M. and Eckernas, S.A. (1976): J. Neurochem.
   27:317-318.
3. Aquilonius, S.M. and Eckernas, S.A. (1977): Neurology 27:887.
4. Barbeau, A. (1978): J. Canad. Sci. Neurol. (French) 5:157-160.
5. Barbeau, A. (1978): New Eng. J. Med. 299:200-201.
6. Barbeau, A. (1978): J. Canad. Sci. Neurol. (French) 5:161-165.
7. Barbeau, A., Growdon, J.G. and Wurtman, R.J. (eds)(1979): IN
   Nutrition and the Brain, Vol. 5, Choline and Lecithin in
   Brain Disorders, Raven Press, New York.
8. Berger, P.A., Davis, K.L. and Hollister, L.E. (1979): IN Brain
   Acetylcholine and Neuropsychiatric Disease (eds) K.L. Davis
   and P.A. Berger, Plenum Press, New York, pp. 15-32.
9. Bowers, M.B. Jr., Goodman, E. and Sim, O.M. (1964): J. Nerv.
   Ment. Dis. 138:383-389.
10. Boyd, W.D., Graham-White, J., Blackwood, G., Glen, I. and
    McQueen, J. (1977): Lancet 2:711.
11. Casey, D.E. (1977): Dis. Nerv. Syst. 38:7-15.
12. Collard, J., Lecoq, R. and Demaret, A. (1965): Acta Neurol.
    Psychiat. Belg. 65:122-126.
13. Coyle, J.T., Schwarcz, R., Bennett, J.P. and Campochiaro, P.
    (1977): Prog. Neuro-Psychopharmacol. 1:13-20.
14. Davies, P. (1979): Brain Res. 171:319-327.
15. Davies, P. and Verth, A.W. (1978): Brain Res. 13:385-392.
16. Davis, K.L. and Yesavage, J.A. (1978): IN Brain Acetylcholine
    and Neuropsychiatric Disease (eds) K.L. Davis and P.A. Berger,
    Plenum Press, New York, pp. 205-213.
17. Davis, K.L., Hollister, L.E., Barchas, J.D. and Berger, P.A.
    (1976): Life Sci. 19:1507-1516.
18. Davis, K.L., Berger, P.A., Hollister, L.E. and Barchas, J.D.
    (1978): Life Sci. 22:1865-1872.
19. Davis, K.L., Hollister, L.E. and Berger, P.A. (1979): Amer. J.
    Psychiat. 136:1581-1584.
20. Davis, K.L., Berger, P.A. and Hollister, L.E. (1979): IN
    Nutrition and the Brain, Vol. 5, Choline and Lecithin in
    Brain Disorders (eds) A. Barbeau, J.G. Growdon and R.J.
    Wurtman, Raven Press, New York, pp. 305-316.
21. Domino, E.F. (1979): IN Brain Acetylcholine and Neuropsychiatric
    Disease (eds) K.L. Davis and P.A. Berger, Plenum Press, New
    York, pp. 63-74.
22. Domino, E.F., Mathews, B., Tait, S.K., Demetriou, S.K. and
    Fucek, F. (1980): this volume.
23. Dorus, E., Pandey, G.N., Shaughnessy, R., Gaviria, M., Val, E.,
    Ericksen, S. and Davis, J.M. (1979): Science 205:932-933.

24. El-Yousef, M.K., Janowsky, D.S., Davis, J.M. and Rosenblatt,
     J.E. (1973): Brit. J. Addict. 68:321-325.
25. Etienne, P., Gauthier, S., Dastoor, D., Collier, B. and Ratner,
     J. (1979): IN Nutrition and the Brain, Vol. 5, Choline and
     Lecithin in Brain Disorders (eds) A. Barbeau, J.G. Growdon
     and R.J. Wurtman, Raven Press, New York, pp. 389-396.
26. Fahn, S., Mishkin, M.M. and Hofman, R.R. (1973): IN Advanced
     Neurology (eds) A. Barbeau, T.N. Chase and G.W. Paulson,
     Raven Press, New York, pp. 531-537.
27. Gelenberg, A.J., Doller-Wojcik, J.C. and Growdon, J.H. (1979):
     Amer. J. Psychiat. 136:772-776.
28. Gerlach, J., Reisby, N. and Randrup, A. (1974): Psychopharmaco-
     logia 34:21-35.
29. Gershon, S. and Shaw, F.H. (1961): Lancet 1:1371-1374.
30. Glen, A.I.M., Ongley, G.C. and Robinson, K. (1968): Lancet 2:
     241-243.
31. Gonge, M. and Barbeau, A. (1977): Union Med. Canad. 106:559-570.
32. Growdon, J.H., Cohen, E.L. and Wurtman, R.J. (1977): Ann.
     Neurol. 1:418-422.
33. Growdon, J.H., Hirsch, M.J., Wurtman, R.J. and Wiener, W.
     (1977): New Eng. J. Med. 297:524-527.
34. Growdon, J.H., Gelenberg, A.J., Doller, J., Hirsch, M.J. and
     Wurtman, R.J. (1978): New Eng. J. Med. 298:1029-1030.
35. Hanin, I. (1979): IN Brain Acetylcholine and Neuropsychiatric
     Disease (eds) K.L. Davis and P.A. Berger, Plenum Press, New
     York, pp. 75-86.
36. Hanin, I. and Schuberth, J. (1974): J. Neurochem. 23:819-824.
37. Hanin, I. and Skinner, R.F. (1975): Anal. Biochem. 66:568-583.
38. Hanin, I., Kopp, U., Spiker, D.G., Neil, J.F., Shaw, D.H. and
     Kupfer, D.J. (1980): Psychiat. Res. 3:345-355.
39. Hanin, I., Merikangas, J.R., Merikangas, K.R. and Kopp, U.
     (1979): New Eng. J. Med. 301:661-662.
40. Hesketh, J.E., Glen, A.I.M. and Reading, H.W. (1977): J. Neuro-
     chem. 28:1401-1402.
41. Hokin-Neaverson, M., Spiegel, D.A. and Lewis, W.C. (1974): Life
     Sci. 15:1739-1748.
42. Hollister, L.E., Jenden, D.J., Amaral, J.R.D., Barchas, J.D.,
     Davis, K.L. and Berger, P.A. (1978): Life Sci. 23:17-22.
43. Janowsky, D.S., El-Yousef, M.K. and Davis, J.M. (1974): Psycho-
     somat. Med. 36:248-257.
44. Jenden, D.J. (1979): IN Brain Acetylcholine and Neuropsychiatric
     Disease (eds) K.L. Davis and P.A. Berger, Plenum Press, New
     York, pp. 483-514.
45. Jonsson, L.E., Schuberth, J. and Sundwall, A. (1969): Life Sci.
     8:977-981.
46. Jope, R.S., Jenden, D.J., Ehrlich, B.E. and Diamond, J.M. (1978):
     New Eng. J. Med. 299:833-834.

47. Kark, R.A.P., Rodriguez-Budelli, M., Blass, J.P. and Spence, M.A. (1979): IN Brain Acetylcholine and Neuropsychiatric Disease (eds) K.L. Davis and P.A. Berger, Plenum Press, New York, pp. 143-153.

48. Klaver, M.M., Flentge, F., Nienhus-Kuiper, H.E. and Praag, H.M. (1979): Life Sci. 24:231-236.

49. Klawans, H.L. and Rubovits, R. (1972): Neurology 22:107-116.

50. Lee, G., Lingsch, C., Lyle, P.T. and Martin, K. (1974): Brit. J. Clin. Pharmacol. 1:365-370.

51. Lingsch, C. and Martin, K. (1976): Brit. J. Pharmacol. 57: 323-327.

52. Maas, J.W. and Landis, D.H. (1968): J. Pharmacol. Exp. Ther. 163:147-162.

53. Mallinger, A.G., Kupfer, D.J., Poust, R.I. and Hanin, I. (1975): Clin. Pharmacol. Ther. 18:467-474.

54. Mallinger, A.G., Mallinger, J.E., Himmelhoch, J.M., Neil, J.F. and Hanin, I. (1980): Psychopharmacology 68:249-255.

55. Martin, K. (1974): IN Drugs and Transport Processes (ed) B.H. Callingham, MacMillan, London, pp. 275-286.

56. Milstoc, M., Teodoru, C.V., Fieve, R.R. and Kumbaraci, T. (1975): Dis. Nerv. Syst. 36:197-199.

57. Modestin, J., Hunger, J. and Schwartz, R.B. (1973): Arch. Psychiat. Nervenkr. 218:67-77.

58. Naylor, G.J., Dick, D.A.T., Dick, E.G., LePoidevin, D. and Whyte, S.F. (1973): Psychol. Med. 3:502-508.

59. Nutt, J.G., Rosin, A. and Chase, T.N. (1978): Neurology 28: 1061-1064.

60. Ostrow, D.G., Pandey, G.N., Davis, J.M., Hurt, S.W. and Tosteson, D.C. (1978): Amer. J. Psychiat. 135:1070-1078.

61. Pandey, G.N., Dorus, E., Davis, J.M. and Tosteson, D.C. (1979): Arch. Gen. Psychiat. 36:902-908.

62. Penovich, P., Morgan, J.P., Kerzner, B., Karch, F. and Goldblatt, D. (1978): JAMA 239:1997-1998.

63. Perry, E.K., Gibson, P.H., Blessed, G., Perry, R.H. and Tomlinson, B.E. (1977): J. Neurol. Sci. 34:247-265.

64. Perry, E.K., Perry, R.H., Blessed, G. and Tomlinson, B.E. (1977): Lancet 1:189.

65. Pfeiffer, C.C. and Jenney, E.H. (1957): Ann. N.Y. Acad. Sci. 66:753-764.

66. Pinta, E.R. (1977): Dis. Nerv. Syst. 38:214-215.

67. Rehavi, M., Maayani, S. and Sokolovsky, M. (1977): Biochem Pharmacol. 26:1559-1567.

68. Richelson, E. (1979): Psychiat. Ann. 9:(4/April).

69. Reisine, T.D., Yamamura, H.I., Bird, E.D., Spokes, E. and Enna, S.J. (1978): Brain Res. 159:477-481.

70. Rowntree, D.W., Nevin, S. and Wilson, A. (1950): J. Neurol. Neurosurg. Psychiat. 13:47-62.

71. Rybakowski, J., Frazer, A., Mendels, J. and Ramsey, T.A. (1978): Commun. Psychopharmacol. 2:99-104.

72. Sadeh, M. and Brahman, J. (1979): Israel J. Med. Sci. 15:288.
73. Sayers, A.C. and Burki, H.R. (1976): J. Pharm. Pharmacol. 28:
     252-253.
74. Schanberg, S.M., Breese, G.R., Schildkraut, J.J., Gordon, E.K.
     and Kopin, I.J. (1968): Biochem. Pharmacol. 17:2006-2008.
75. Schanberg, S.M., Schildkraut, J.J., Breese, G.R. and Kopin,
     I.J. (1968): Biochem. Pharmacol. 17:247-254.
76. Schuberth, J. and Jenden, D.J. (1975): Brain Res. 84:245-256.
77. Smith, C.M. and Swase, M. (1979): Lancet 1:42.
78. Snyder, S.H. and Yamamura, H.I. (1977): Arch. Gen. Psychiat.
     34:236-239.
79. Spitzer, R.L., Endicott, J. and Robins, E. (1978): Arch. Gen.
     Psychiat. 35:773-782.
80. Stafford, R.J. and Fann, W.E. (1977): Dis. Nerv. Syst. 38:3-6.
81. Stavinoha, W.B., Modak, A.T. and Bowden, C.L. (1977): Soc.
     Neurosci. Abst. 3:1329.
82. Szentistvanyi, I. (1978): Acta Biol. Acad. Sci. Hung. 29:
     101-119.
83. Tamminga, C., Smith, R.C., Chang, S., Haraszti, J.S. and
     Davis, J.M. (1976): Lancet 2:905.
84. Tarsy, D. and Bralower, M. (1977): Arch. Neurol. 34:756-758.
85. Tarsy, D., Leopold, N. and Sax, D.S. (1974): Neurology 24:
     28-33.
86. Van Andel, H. (1959): IN Neuropsychopharmacology (eds) P.B.
     Bradley, P. Deniker and C. Raduoco-Thomas, Amsterdam,
     pp. 701-703.
87. Walker, J.E., Hochn, M., Sears, E. and Lewis, J. (1973):
     Lancet 1:1512.
88. Welch, M.J., Markham, C.H. and Jenden, D.J. (1976): J. Neurol.
     Neurosurg. Psychiat. 39:367-374.
89. White, P., Goodhardt, M.J., Keet, J.P., Hiley, C.P., Carasco,
     L.H., Williams, I.E.I. and Bowen, D.M. (1977): Lancet 1:
     668-671.
90. Yamamura, H.I., Wastek, G.J. and Hruska, R.E. (1979): IN Brain
     Acetylcholine and Neuropsychiatric Disease (eds) K.L. Davis
     and P.A. Berger, Plenum Press, New York, pp. 155-174.

CHOLINERGIC DYSFUNCTION: A COMMON DENOMINATOR IN METABOLIC

ENCEPHALOPATHIES

J.P. Blass, G.E. Gibson, T.E. Duffy and F. Plum

Department of Neurology, Cornell University Medical
College, New York, New York and
Dementia Research Service, Burke Rehabilitation Center
White Plains, New York, USA

The metabolic encephalopathies are a group of disorders of
brain function secondary to systemic metabolic derangements. Both
they and the disorders that cause them are common (Table 1). Clinically, the metabolic encephalopathies typically present as disorders of attention, cognition and arousal. The milder stages
usually lead to a loss of interest and attentiveness, and the associated inability to carry out a complex task. More severe stages
progress to obtundation, stupor and coma (25). Motor abnormalities
such as asterixis, motor resistance to passive stretch, and ataxia
also occur. Pupillary and oculomotor functions are commonly spared.
The key to identifying the specific cause of metabolic encephopathy
in an individual patient generally lies in the systemic manifestations or laboratory determinations since the neurological abnormalities tend to be stereotyped. This very stereotypy, in fact, suggests the hypothesis that the metabolic encephalopathies share at
least some common neurochemical and neuropharmacological derangements.

No single neurochemical change has so far been established that
links together the metabolic encephalopathies. Changes in overall
cerebral metabolism occur frequently, as measured by cerebral oxygen consumption (CMR_{O_2}), glucose utilization ($CMR_{glucose}$), or
cerebral blood flow (CBF). However, they are neither invariant nor
always in the same direction. Thus, Kety (20) found that hypoxia
induced by breathing 10% O_2 did not reduce CMR_{O_2} in healthy young
men, despite its marked impairment of higher brain functions.

921   .

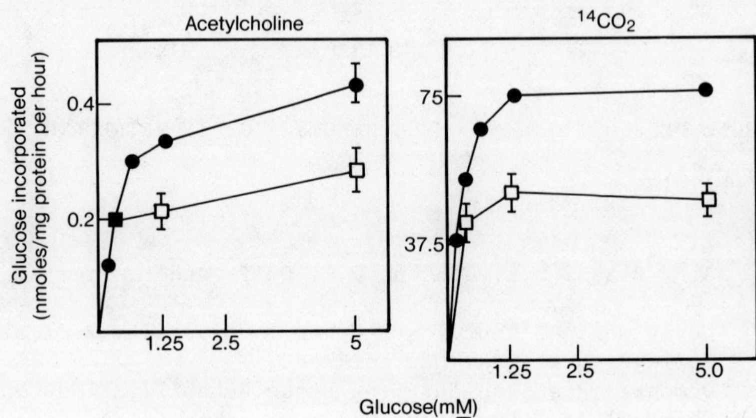

FIGURE 1: Acetylcholine synthesis and glucose oxidation. Rat brain
slices were incubated with varying concentrations of
[U-^{14}C]glucose in buffer containing either 31 mM K$^+$ (●)
or 5 mM K$^+$ (□), as described by Gibson and Blass (12).

TABLE 1:   Type of disorder causing metabolic encephalopathy.

1.	Inborn errors
2.	Nutritional
3.	Cardiovascular
4.	Pulmonary
5.	Hepatic
6.	Renal
7.	Endocrine
8.	Toxic (exogenous)

There is evidence for cholinergic deficiency in at least one example of each type of disorder. For a more detailed discussion,
see ref. 25.

Similar results have been observed in mildly hypoxic rats (14). Recently, Miyaoka et al. (22) reported that moderate hypoxia in rats increased 2-deoxyglucose utilization in subcortical structures and in white matter, but decreased it in cortex. There was no apparent net change in overall cerebral glucose consumption.

In contrast to the inconsistent changes observed in overall cerebral oxidative metabolism, several workers have identified alterations in acetylcholine (ACh) metabolism that suggest a possible common denominator in metabolic brain dysfunction. In 1940, Mann and Quastel (21) reported impaired synthesis of ACh in a nutritional encephalopathy. They studied minced brain from thiamine deficient and control rats, measuring ACh by bioassay with leech muscle. Thiamine deficient minces incubated without added K^+ accumulated about 70% of the ACh of controls, and the addition of thiamine had little effect. When 30 mM K^+ was added to stimulate ACh turnover, the thiamine deficient preparations produced only half as much of the transmitter as controls. The addition of thiamine cured the deficit. This brief but remarkable paper anticipated much future work on ACh metabolism. It suggested that moderate impairment of cerebral carbohydrate catabolism would impair ACh synthesis. It also indicated that a deficiency which appeared to be minimal under "basal" conditions could have profound effects in the face of increased metabolic or physiologic demands.

The rate of ACh synthesis is now known to be exquisitely sensitive to conditions which impair cerebral oxidative metabolism (3,5,15). In vitro, reducing the rate of oxidation in cerebral tissues produces a proportionate decline in ACh synthesis, even though the flux of carbon to the acetyl moiety of ACh is less than 1% that to CO_2. This dependence of ACh synthesis upon cellular oxidation has been demonstrated in brain slices and in synaptosomes; with high and low K^+ (Fig. 1); with labelled precursors of either the acetyl or the choline (Ch) moieties; with glucose, pyruvate, or ketone bodies as substrate; and with a wide variety of inhibitors of oxidative metabolism. In vivo, graded impairment of ACh synthesis accompanied hypoglycemia and graded anemic or histotoxic hypoxia (for review see 2). Subsequent experiments have demonstrated that ACh synthesis also decreases in brain during hypoxic hypoxia, induced by lowering the oxygen content in the inspired air (6,14). In the studies of Gibson and Duffy(14) a significant inhibition of ACh synthesis was observed in rat brain at an arterial oxygen tension (57 mm Hg) known to be associated with mild mental impairment in man.

The mechanisms linking cerebral metabolism to ACh synthesis are not yet fully understood. They appear to involve compartmentation of glucose metabolism with respect to ACh synthesis (13). They may involve alterations in redox potentials (4), although the latter

data are flawed by the lack of a good measure of NAD/NADH ratios
for brain mitochondria (2). As discussed by Tucek et al. (this
volume), ACh synthesis may relate more closely to the flux of pyruvate through the pyruvate dehydrogenase complex than to flux through
the Krebs tricarboxylic acid cycle. Jope et al. (19) have established that the linkage is not mediated by changes in Ch uptake,
except perhaps in such total, severe anoxia that tissue ATP levels
fall.

The physiological importance of reduced cholinergic activity
is becoming well established in metabolic encephalopathies. Studies
of superior cervical ganglia in several laboratories have shown
that transsynaptic cholinergic transmission is more sensitive to
metabolic insults than is axonal conduction (11,23,26). Perri et
al. (23) found that the transmission deficit could be brought out
in moderately thiamine deficient ganglia by stimulating them at a
relatively rapid rate (20 Hz), while the response to slower stimulation (2 Hz) remained normal. They postulated that the failure of
transmission was due to an inability of the thiamine deficient
ganglia to increase ACh synthesis in response to a metabolic demand. They subsequently confirmed this hypothesis (26) by direct
measurement of changes in ACh content. These experiments give
vigorous support to proposals put forward by Mann and Quastel (21)
some 40 years before.

Pharmacologic studies also support the significance of cholinergic deficiency in metabolic encephalopathies. Pretreatment
with physostigmine can reduce the number of animals dying from
hypoxic hypoxia, under carefully controlled conditions (28). Similar results have been reported previously for hypoglycemia (Table 2)
and chemical hypoxia (3,16). Recently, Barclay et al. (1) found
that treatment of rats with appropriate doses of physostigmine reversed certain behavioral effects of early thiamine deficiency
(Table 3). In these experiments, thiamine deficiency was induced
by applying a combination of the thiamine antagonist pyrithiamine
and a low thiamine diet. The measured behaviors included a motor
task (ability to move along a string) and general activity. Within
the limits of experimental error, physostigmine was as effective
in reversing these impairments as was thiamine itself, although
improvement was more consistently noted with thiamine than with
physostigmine.

These observations of impaired cholinergic function accompanying metabolic encephalopathy suggest several testable questions.
First, is cholinergic deficiency a general characteristic of metabolic encephalopathies? The apparent proportional relationship
between cerebral oxidative metabolism and ACh synthesis allows the
hypothesis that disorders which impair cerebral carbohydrate
metabolism will reduce ACh synthesis. Several references from the

TABLE 2:   Protection by physostigmine against hypoglycemia.

TREATMENT	NUMBER INJECTED	NUMBER DEAD
Insulin	40	14
Insulin + physostigmine	40	3

Mice were injected with a large dose of insulin (20 int. units/kg) and with either saline or physostigmine salicylate (0.5 mg/kg) as described in ref. 16. Survival was followed for 3 hr.

TABLE 3:   Effect of physostigmine in mild thiamine deficiency.

TREATMENT	MOTOR SCORE
None	7.3 ± 0.2
Pyrithiamine	1.1 ± 0.6
Pyrithiamine + saline	1.2 ± 0.6
Pyrithiamine + physostigmine	3.8 ± 0.3
Pyrithiamine + thiamine	4.1 ± 0.7

Animals were made thiamine deficient by treatment with pyrithiamine and a low thiamine diet. Motor performance was measured by a standardized "string test" (1). The dose of physostigmine was 0.05-0.5 mg/kg. Values are mean $\pm$ S.E.M. for 16-65 animals.

literature as well as our own studies support this hypothesis. In a number of other animal models of metabolic encephalopathy, brain ACh concentrations or turnover rates have been found abnormal. Takahashi et al. (29), for example, reported reduced levels of ACh in ammonia intoxicated rats. Decreased ACh turnover in the toxic encephalopathies induced by drugs have been amply demonstrated (18). Further studies of graded metabolic insults, analogous to those of graded hypoxia, now need to be done in which turnover is measured by modern isotopic techniques and the brain is rapidly fixed for chemical analysis by methods such as microwave inactivation or "brain-blowing." Studies of lipidoses will provide a rigorous test of whether or not ACh metabolism is consistently abnormal in metabolic encephalopathies. The gangliosidoses, for which there are now good animal models (8), are typified by a rapidly progressive and severe loss of mentation without any known lesion of carbohydrate oxidation.

What mechanisms link ACh synthesis to cerebral carbohydrate oxidation? Is the change cart or horse? As yet, no compelling evidence indicates whether the cholinergic decreases are a cause or a result of impaired brain activity.

Is the reduction in ACh synthesis only a reflection of generally deranged metabolism of neurotransmitters in metabolic encephalopathies? Relatively severe metabolic insults do impair the synthesis of glutamate and other putative amino acid neurotransmitters from glucose. Such insults include breathing 5% O_2 (30), fluoroacetate poisoning (9), or severe thiamine deficiency (7,24). It is not yet known whether milder insults which can impair ACh synthesis also affect these amino acids. The synthesis of catecholamines and serotonin is sensitive to even mild hypoxia, as expected since the K_m for O_2 of the synthetic hydroxylases is close to the physiological pO_2 of brain _in vivo_ (10,17). Whether or not the changes in catecholamine and serotonin synthesis accompanying hypoxia are important physiologically is less certain. They have not been reported in metabolic encephalopathies other than hypoxia. Since neurons containing different transmitters innervate each other in the brain, impairment of the synthesis of one neurotransmitter may be expected to alter the turnover of other transmitters. The crucial question which must be answered experimentally is whether neurons synthesizing ACh are selectively vulnerable to metabolic insults.

If cholinergic systems are selectively vulnerable, what is the physiological effect? One cannot presently be sure since cholinergic systems comprise much of the brainstem reticular activating system but also are diffusely distributed in the cerebral cortex. Either or both sets of neurons could affect the attentional mechanisms which are typically impaired in metabolic encephalopathies.

In order to obtain firm answers to these questions and speculations, three requirements must be met. First is the performance of extensive, basic mechanistic studies of ACh and other neurotransmitter metabolism in animal models of metabolic encephalopathies. Second is the development of acceptable techniques to detect cholinergic and other specific neurotransmitter deficiencies in man. Third is the introduction of a wider therapeutic armamentarium of cholinergic agonists and antagonists. At present, we are limited to a few cholinesterase inhibitors, only one of which acts centrally, to muscarinic blockers derived from belladonna, to one central muscarinic agonist (arecholine), and to precursor therapy (e.g. Ch or lecithin supplemented diets) of questionable clinical efficacy. In contrast, one has available a wide variety of medications for therapy of catecholaminergic or serotoninergic disorders. Effective and clinically tested central cholinergic drugs are likely to be widely used, especially if cholinergic deficiency proves to be an important part of the pathophysiology of the common metabolic encephalopathies.

ACKNOWLEDGEMENTS

Supported by the Winifred Masterson Burke Relief Foundation, the National Foundation (NO.6-215), and the National Institutes of Health (NS-15125, NS-16997, NS-15696, AA-03883, NS-03346) and the Will Rogers Institute.

REFERENCES

1.   Barclay, L.L., Gibson, G.E. and Blass, J.P. (1980): J. Clin. Invest. (in press).

2.   Berntman, L. and Siesjo, B.K. (1978): J. Neurochem. $\underline{31}$:1265-1276.

3.   Blass, J.P. and Gibson, G.E. (1978): IN Cholinergic Mechanisms and Psychopharmacology (ed) D.J. Jenden, Plenum Press, New York, pp. 791-803.

4.   Blass, J.P. and Gibson, G.E. (1979): Adv. Neurol. $\underline{26}$:229-250.

5.   Blass, J.P. and Gibson, G.E. (1979): IN Brain Acetylcholine and Neuropsychiatric Disease (eds) K.L. Davis and P.A. Berger, Plenum Press, New York, pp. 215-236.

6.   Blass, J.P., Gibson, G.E., Shimada, M., Kihara, T., Watanabe, M. and Kurinioto, K. (1980): IN Biochemistry of Dementia (ed) P.J. Roberts, John Wiley & Sons, London (in press).

7.   Butterworth, R.F., Hamel, E., Landreville, F. and Barbeau, A. (1979): Canad. J. Neurol. Sci. $\underline{6}$:217-222.

8.   Cork, L.C., Munnel, J.F., Lorenz, M.D., Murphy, J.V., Baker, H.J. and Rattazzi, M.C. (1977): Science $\underline{196}$:1014-1017.

9.   Clarke, D.D., Nicklas, W.J. and Berl, S. (1970): Biochem. J. $\underline{120}$:345-351.

10.   Davis, J.N. and Carlsson, A. (1973): J. Neurochem. $\underline{20}$:913-915.

11.   Dolivo, M. (1974): Fed. Proc. $\underline{33}$:1043-1048.

12.   Gibson, G.E. and Blass, J.P. (1976): J. Neurochem. $\underline{27}$:37-42.

13.   Gibson, G.E., Blass, J.P. and Jenden, D.J. (1977): J. Neurochem. $\underline{30}$:71-76.

14. Gibson, G.E. and Duffy, T.E. (1981): J. Neurochem. 37:28-33.
15. Gibson, G.E., Jope, R. and Blass, J.P. (1975): Biochem. J.
    148:17-23.
16. Gottstein, U. (1969): Verh. Dtsch. Ges. Inn. Med. 75:957-965.
17. Hedner, J. and Lundborg, P. (1979): Acta Physiol. Scand. 106:
    139-143.
18. Jenden, D.J., ed (1978): Cholinergic Mechanisms and Psycho-
    pharmacology, Plenum Press, New York.
19. Jope, R.S., Weiler, M.H. and Jenden, D.J. (1978): J. Neurochem.
    30:949-954.
20. Kety, S.S. (1962): IN Neurochemistry, 2nd Ed. (eds) K.A.C.
    Elliott, I.H. Page and J.H. Quastel, C.C. Thomas, Spring-
    field, pp. 113-127.
21. Mann, P.J.G. and Quastel, J.H. (1940): Nature 145:856-857.
22. Miyaoka, M., Shinohara, M., Kennedy, C. and Sokoloff, L. (1979):
    Ann. Neurol. 6:156-157.
23. Perri, V., Sacchi, O. and Casella, C. (1970): J. Exp. Physiol.
    55:25-35.
24. Pliatakis, A., Nicklas, W.G. and Desnick, R.J. (1980): Ann.
    Neurol. (in press).
25. Plum, F. and Posner, J.B. (1980): Diagnosis of Stupor and Coma,
    3rd Ed., F.A. Davis, Co., Philadelphia, pp. 177-303.
26. Sacchi, O., Ladinsky, H., Prigioni, I., Consolo, S., Perri, G.
    and Perri, V. (1978): Brain Res. 151:609-614.
27. Scheinberg, P. (1951): Blood 6:213-227.
28. Scremin, A.M.E. and Scremin, O.U. (1979): Stroke 10:142-143.
29. Takahashi, R., Nasu, T., Tamura, T. and Kariya, T. (1961):
    J. Neurochem. 7:103-112.
30. Yoshino, Y. and Elliott, K.A.C. (1970): Canad. J. Biochem.
    48:228-235.

CHOLINOMIMETIC AGENTS AND HUMAN MEMORY: PRELIMINARY OBSERVATIONS

IN ALZHEIMER'S DISEASE

K.L. Davis, R.C. Mohs, B.M. Davis, G.S. Rosenberg,
T.H. Horvath and Y. DeNigris

Mount Sinai School of Medicine, Departments of Psychiatry
and Pharmacology, New York, New York, and
Bronx Veterans Administration Medical Center, New York USA

INTRODUCTION

Two lines of evidence, one pharmacological and the other neurochemical, have come together to suggest an approach to the treatment of the Alzheimer's type of senile dementia. Neurochemical
investigation indicates that Alzheimer's disease is a primary degenerative nerve cell disorder affecting predominantly neocortical
cholinergic neurons (4,5,15,23,24,27,34).

This defect is manifested in a marked loss of choline acetyltransferase (CAT) activity as well as acetylcholinesterase (AChE)
activity (4,14,15,23,27,34), but relatively normal muscarinic receptor binding (13,23,25,34). However, there is some controversy
regarding the latter finding (27). CAT deficiency has been positively correlated with mental test scores, senile plaque formation,
and an actual in vitro diminution of acetylcholine (ACh) synthesis
from cortical biopsy samples (25,30). Pharmacological studies
have been equally convincing in implicating cholinergic neurotransmission in human memory, and an alteration in cholinergic
transmission underlying disordered memory.

The administration of scopolamine to young normal subjects
produces in those subjects a cognitive test profile that is very
similar to the test results obtained in Alzheimer's disease (17).
Furthermore, the specificity of scopolamine induced dementia is
indicated by the fact that it cannot be reversed by amphetamine (17).

Increasing central cholinergic activity by moderate doses of either
physostigmine or arecoline enhances just that aspect of memory dele-
teriously affected by scopolamine (12,31). This fact is particularly critical, suggesting that cholinergic agents might be able
to augment those memory functions most affected in patients with
Alzheimer's disease. It is because of the possibility that cholinomimetic agents might be of benefit in Alzheimer's disease that so
many investigators have attempted to treat this condition with
choline (Ch) or lecithin (6,9,26,29,32).

Unfortunately, results with these agents have not been very
promising. Our own data obtained in four double-blind studies of
various doses of Ch have been similarly discouraging (11,20-22).
Although occasionally a few subjects experienced a mild improvement
in memory, such improvement was not of a degree to be clinically
meaningful in patients with Alzheimer's disease. As a consequence
of these essentially negative findings with Ch chloride, a study
was designed to assess the effects upon memory of another cholinergic agent. Since physostigmine had previously been shown to enhance human memory in young normal subjects (12), a trial with this
agent was designed. This paper is a preliminary report on the
effect of physostigmine upon elderly subjects with and without
Alzheimer's type dementia. A portion of these data have been
previously reported (10).

METHODS

Eleven elderly patients between the ages of 55 and 68 parti-
cipated in this study. In seven patients the diagnosis of Alzheimer's disease was made with the aid of CAT scan, brain scan,
skull films, CSF analysis, serum analysis and a careful history
and physical examination. Particular care was attended to ruling
out cases of multi-infarct dementia. All patients with Alzheimer's
disease had a Memory and Information Test (MIT score < 10 or a
Dementia Rating Scale (DRS) score > 4. This criteria has been
shown to select patients with a high probability of Alzheimer's
disease as verified by histopathological examination upon autopsy
(18). Three subjects were non-demented, and one patient had a
dementia associated with Huntington's chorea.

Drug administration was divided into two phases. In the first,
or dose response phase, subjects received, under double-blind conditions 0 mg, 0.125 mg, 0.250 mg and 0.5 mg of physostigmine. The
drug was dissolved in 100 cc of saline and administered at a constant rate for 30 min. The order of infusions are randomized. In
the second, or replication phase of the study, the dose of physo-

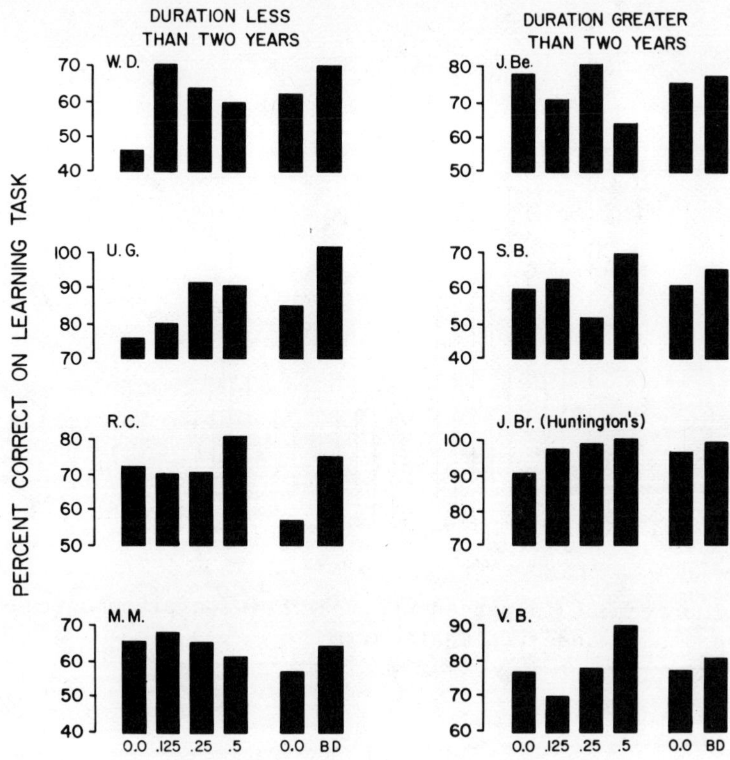

FIGURE 1: Alzheimer's patients.

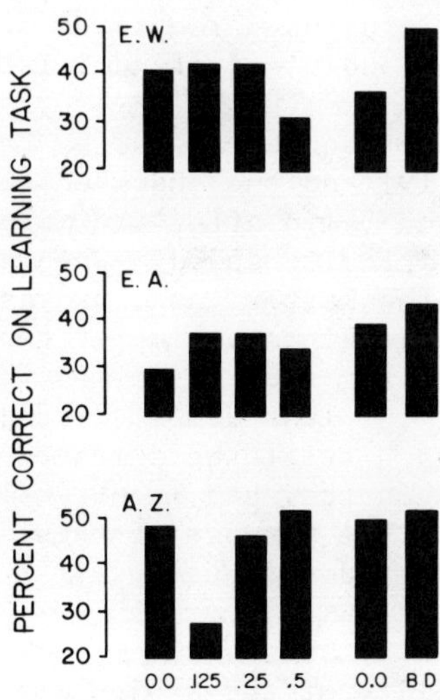

FIGURE 2: Non-demented patients.

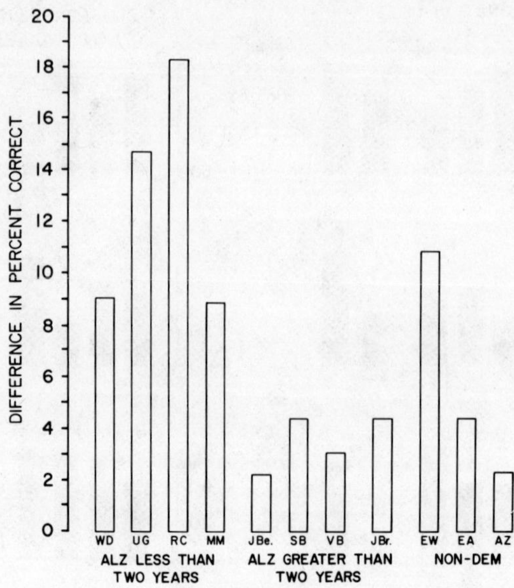

FIGURE 3: Improvement in percent correct between placebo regulation and best dose replication.

stigmine associated with the best performance on mental tests of storage into long term memory (LTM) was readministered (1,2), as was the placebo (0 mg) infusion. The order of these two infusions was randomized, and the conditions of administration double-blind.

Subjects memory function was assess by the digit span task, the famous faces test (19) and the Buschke selective reminding test (6,7). Testing was begun 5 min after the infusion began, and was concluded 15 min after infusion ended. When subjects could not complete or comprehend the latter task, as was true of all the demented subjects, they were given a simpler recognition memory task for either words or pictures (28). The most demented subjects recognized pictures and the less demented subjects recognized high imagery words. After each trial the demented subjects were reminded of just those items they had incorrectly recalled. Three trials were completed.   The purpose of these modifications of the Buschke paradigm was to produce a task with the following characteristics:

1. Measure the ability of the subject to encode information into LTM.

2. Be sensitive to an improvement or worsening in performance.
3. Be comprehensible to the subject, and be completed in the time period of physostigmine's biological activity.
4. Roughly equivalent or multiple forms of the test could be readily constructed.

RESULTS

Figure 1 describes the effects of physostigmine upon the ability of subjects to store information into LTM in the eight patients with dementia. Figure 2 illustrates the results of the three non-demented elderly subjects. The four Alzheimer's patients with a duration of dementia of less than two years are W.D., R.C., U.C. and M.M. All other patients had been demented for longer than two years. As determined during the dose response phase of the study, all subjects had their best performance in ability to encode information into LTM on some dose of physostigmine rather than the placebo saline infusion. However, the best dose of physostigmine was any of the doses. During the replcation phase of the study, all subjects again demonstrated a physostigmine related enhancement of LTM encoding (p < 0.001 by sign test). This effect was most pronounced in the patients suffering Alzheimer's disease for less than two years, was intermediate for the non-demented subjects, and was least for those patients with a duration of dementia for more than two years.

The percent improvement for each subject occurring during the replcation phase of the study is shown in Fig. 3. It should be noted that whenever the improvement was approximately 10% or greater, both the subject and the experimenter were aware of a rather marked, clinically meaningful, difference in the subjects qualitative performance.

The administration of physostigmine did not produce a consistent effect upon either the digit span task or the famous faces test. The effect was specifically limited to the encoding process of long term memory.

DISCUSSION

The ability of a low dose of physostigmine to transiently enhance the ability of patients with Alzheimer's disease and non-demented elderly subjects to encode information into long term memory is consistent with the similar effects of physostigmine and arecoline in young normal subjects (12,31).

The main difference between these effects is that elderly subjects have a greater variability in the dose of physostigmine that
enhances memory. However, a similar variability was found in the
response of elderly primates to physostigmine (3). The effect of
physostigmine to improve LTM is quite specific. The likelihood
that it reflects a general change in attention or performance is
decreased by the fact that in this study only encoding was affected
by cholinomimetics. Furthermore, attentional changes can be detected in the Sternberg paradigm (33), which is not influenced by
physostigmine (12). These results give added impetus to the suggestion that cholinergic neurons are critically linked to the encoding
of information into LTM ((16).

The present results also suggest that the cholinergic deficiency
underlying Alzheimer's disease can be partially corrected by cholinomimetic treatment. The fact that the duration of dementia might
predict treatment responsiveness could indicate that with continued
degeneration of cholinergic neurons anticholinesterase administration becomes a less viable strategy to increase cholinergic neurons.
Should muscarinic receptors remain intact in Alzheimer's disease,
treatment with muscarinic cholinergic agonist may prove more efficacious in patients with a prolonged period of disease.

Physostigmine administered intravenously had a very short duration of action. By this route the drug's half life could be as
short as 15 min (J.C. Gillin, personal communication). Thus a
therapeutic intervention based upon augmenting cholinergic activity
must await a safe, effective long acting oral cholinomimetic. The
oral administration of physostigmine or pilocarpine might partially
meet these requirements, although they are not ideal agents. However,
the possibility that tolerance could develop to chronic cholinomimetic therapy needs to be carefully considered. Should the eye
be a prototype, tolerance may decrease the effect of cholinergic
augmentation more than 50% (T. Mittag, this volume).

Despite these difficulties, investigators can be somewhat
encouraged by the fact that treatment of Alzheimer's disease can
follow a rational course.

REFERENCES

1.   Atkinson, R.C. and Shiffrin, R.M. (1968): IN The Psychology
     of Learning and Motivation: Advances in Research and Theory
     (eds) K.W. Spence and J.T. Spence, Academic Press, New York,
     pp. 89-195.

2.   Atkinson, R.C. and Shiffrin, R.M. (1971): Scientif. Amer.
      224:82-90.
3.   Bartus, R.T. (1979): J. Amer. Geriat. Soc. 23:7-12.
4.   Bowen, D.M., Smith, C.B., White, P. and Davison, A.N. (1976):
      Brain 99:459-496.
5.   Bowen, D.J., Spillane, J.A., Gurzon, G., Meier-Ruge, W., White,
      P., Goodhardt, M.J., Iwangoff, P. and Davison, A.N. (1979):
      Lancet 1:11-14.
6.   Boyd, W.D., Graham-White, J., Blackwood, G., Glen, I. and
      McQueen, J. (1977): Lancet 2:711.
7.   Buschke, H. (1973): J. Verb. Learn. Verb. Behav. 12:543-550.
8.   Buschke, H. and Fuld, P.A. (1974): Neurology 24:1019-1025.
9.   Christie, J.E., Blackburn, E.M., Glen, A.I.M., Zeisel, S.,
      Shering, A. and Yates, C.M. (1979): IN Nutrition and the
      Brain Vol. 5 (eds) A Barbeau, J.H. Growdon and R.J. Wurtman,
      Raven Press, New York, pp. 377-387.
10.   Davis, K.L., Mohs, R.C. and Tinklenberg,J.R. (1979): New Eng.
      J. Med. 301:946-947.
11.   Davis, K.L., Mohs, R.C., Tinklenberg, J.R., Hollister, L.E.
      Pfefferbaum, A. and Koppell, B.S. (1980): Arch. Neurol. 37:
      49-52.
12.   Davis, K.L., Mohs, R.C., Tinklenberg, J.R., Pfefferbaum, A.,
      Hollister, L.E. and Kopell, B.S. (1978): Science 201:272-274.
13.   Davies, P. and Verth, A.H. (1978): Brain Res. 138:385-392.
14.   Davies, P. (1979): Brain Res. 171:319-327.
15.   Davies, P. and Maloney, A.J.F. (1976): Lancet 2:1403.
16.   Drachman, D.A. (1977): Neurology 27:783-790.
17.   Drachman, D.A. and Leavitt, J. (1974): Arch. Neurol. 30:113-121.
18.   Kaye, D.W.K. (1977): IN Cognitive and Emotional Disturbance in
      the Elderly (eds) C. Eisdorfer and R.O. Friedel, Year Book
      Medical Publ., Chicago.
19.   Marslen-Wilson, W.D. and Teuber, H.L. (1975): Neuropsychologia
      13:353-354.
20.   Mohs, R.C. and Davis, K.L. (1980): Psychiat. Res. 2:149-152.
21.   Mohs, R.C., Davis, K.L., Tinklenberg, J.R., Hollister, L.E.,
      Yesavage, J.A. and Koppell, B.S. (1979): Amer. J. Psychiat.
      136:10-17.
22.   Mohs, R.C., Davis, K.L., Tinklenberg, J.R. and Hollister, L.E.
      (1980) Neurobiology of Aging (in press).
23.   Perry, E.K., Gibson, P.N., Blessed, G., Perry, R.H. and
      Tomlinson, B.E. (1977): J. Neurol. Sci. 34:247-265.
24.   Perry, E.K., Perry, R.H., Blessed, G. and Tomlinson, B.E.
      (1977): Lancet 1:189.
25.   Perry, E.K., Tomlinson, B.E., Blessed, G., Bergmann, K.,
      Gibson, P.H. and Perry, R.H. (1978): Brit. Med. J. 2:
      1457-1459.
26.   Peters, B.H. and Levin, H.S. (1979): Ann. Neurol. 16:219-222.
27.   Reisine, T.D., Yamamura, H.E., Bird, E.D., Spokes, E. and
      Enna, S.J. (1978): Brain Res. 159:477-482.

28. Shepard, R.N. (1967): J. Verb. Learn, Verb. Behav. 6:156-164.
29. Signoret, J.L., Whitely, A. and L'Hermite, F. (1978): Lancet
         2:837.
30. Sims, N.R., Smith, C.C.T., Davison, A.N., Bowen, D.M., Flack,
         R.H.A., Snowden, J.S. and Neary, D. (1980): Lancet 1:333-336.
31. Sitaram, N., Weingartner, H. and Gillin, J.C. (1978): Science
         261:274-276.
32. Spillane, J.A., Goodhardt, M.H., White, P., Bowen, D.M. and
         Davison, A.N. (1977): Lancet 2:826-827.
33. Sternberg, S. (1969): Amer. Sci. 57:421-432.
34. White, P., Hiley, C.R., Goodhardt, M.J., Carrasco, L.H., Keet,
         J.P., Williams, I.E.I. and Bowen, D.M. (1977): Lancet 1:
         668-670.

EFFECTS OF LECITHIN ON MEMORY AND PLASMA CHOLINE LEVELS:

A STUDY IN NORMAL VOLUNTEERS

J.C. Gillin[1,2], N. Sitaram[1], H. Weingartner[1], C. Crammer[3], and I. Hanin[3]

1.Unit on Sleep Studies, Biological Psychiatry Branch and
  Laborator of Psychology, NIMH, Bethesda, Maryland, and
2.Laboratory of Clinical Psychopharmacology, NIMH
  St. Elizabeths Hospital, Washington, D.C., and
3.Department of Psychiatry, School of Medicine,
  University of Pittsburgh, Pittsburgh, Pennsylvania

INTRODUCTION

Administration of choline (Ch) or lecithin (phosphatidylcholine), the biosynthetic precursors of acetylcholine (ACh) may increase functional cholinergic neurotransmission (1,14,15). Ch has been reported to increase plasma levels of Ch and brain levels of ACh (1, 14,15) and to be of some benefit in the treatment of tardive dyskinesia (7,11,12). Lecithin is similar but may be superior insofar as it does not cause one of the major side effects of Ch treatment, namely, an offensive "fishy" odor.

Since the available pharmacological agents for increasing central cholinergic activity, such as arecoline, physostigmine or diispropylflurophosphate (DFP), have a short duration of action or cause unacceptable side effects, Ch and lecithin could be extremely useful if it were shown that they are active and safe. One area of immediate application would be Alzheimer's disease, where there is strong evidence that loss of cholinergic function may play an important role in its pathophysiology (2,4-6,10,18,19,24).

Results of both animal and human studies indicate that ACh plays an important role in memory and learning (6,8,9,16,17). In earlier studies, we presented data suggesting that both arecoline, a muscarinic

agonist, and Ch increase the rate at which normal volunteers mastered
verbal learning tasks (20,21,25). In a complementary study, we found
that scopolamine had the opposite effect (20). Interestingly, in all
three studies, the major effects of these cholinergic drugs on learning and memory were seen in subjects who performed more poorly under
placebo conditions than those who did better. The so-called "poor
performers" showed the greatest improvement with arecoline and Ch
and the most impairment with scopolamine.

In the present study, we investigated the effect of lecithin
on human learning and upon plasm Ch levels in normal volunteers.
Since commercial lecithin may consist of widely varying amounts of
phosphatidylcholine and other impurities, we administered a relatively pure form, Phospholipon 100 (95% phosphatidylcholine: supplied
through the generosity of J. Eichberg, The American Lechin Co.,
Atlanta, Georgia).

METHODS

The subjects were normal volunteers (9 males, 6 females) aged
18 to 30. Each was screened by medical and psychiatric history,
physical examination, and mental status interview to exclude persons
with significant medical or psychiatric problems, concurrent use of
medications, or learning or occupational difficulties. All were
high school graduates. Five had completed college and were working
at jobs commensurate with thier education, or were furthering their
education, five were full time undergraduates, three were working
and in part time college studies, and the final two were working in
technical fields which required intensive training beyond the high
school level. All gave written informed consent. On each day of
testing subjects were instructed to refrain from eating foods with
high fat content. The experiments were begun about 7:00 PM following a light snack. During the experiments subjects also re-
frained from ingesting beverages high in caffeine and from smoking
cigarettes. The lecithin or a corn-oil placebo was ingested in a
milk shake (250 ml). Each subject was studied under four different
conditions: placebo, or lecithin (Phospholipon 100) 20 gm, 40 gm,
or 60 gm. The order of the conditions was randomized. At least
48 hr separated each condition.

Prior to the onset of the study, each subject was familiarized
with the procedures, including the two major cognitive tests, a
serial learning and the paired associates learning procedure followed
by retrieval testing. In the serial learning test, the experimenter
read each subject a list of 16 randomly (unrelated) words at a rate
of 1 word every 3 sec. The subject attempted to recall the entire

word list in the same order. This process was repeated for 5 learn-
ing-test trials or until the subject was able to recall the entire
list of words in correct order on two consecutive trials. The
average number of words recalled, in correct order, over 5 trials
was tallied for each subject.

In the second procedure, a modified paired associate learning
test, subjects were first trained to paired or associate words. Immediately afterwards, free recall was tested and a second free recall was tested 48 hr later. Because of the need to retest exactly
48 hr later, only 8 subjects were tested. The experimenter showed
20 index cards to the subject for 10 sec each. On each card were
two words, and the subject was instructed to learn the pair. Ten
of the word pairs came from a single category (i.e. furniture), and
the other 10 came from a second category (i.e. clothing), the two
categories were presented in alternate sequence from trial to trial.
Following the final presentations of the 20 word pairs, the experimenter gave the subject a set of 20 index cards, on each of which
was written one word from each of the 20 pairs. The subject was
asked to recall the other word. At 48 hr, subjects were tested for
free recall and cued recall (i.e. with the set of 20 index cards
with the cue word).

In order to measure the changes in plasma Ch induced by lecithin
we obtained blood samples in 8 subjects before and after the milk
shake. Blood samples (8 ml) were taken 30 min and immediately before ingestion of the milk shake, and then 30, 60, 120, 180 min,
and (in 4 subjects) 11.5 hr later. Plasma was assayed for Ch,
according to previously described methods (13).

Serial learning and the paired associates word tests were begun
2 hr after ingestion of the milk shake. In addition, mood and physical symptoms were assessed subjectively at that time and 3 hr
following ingestion (48 item check list: D. Murphy, unpublished
data).

RESULTS

A measure of side effects, based on the sum total of all positive side effects checked off by subjects, indicated that there was
a significant increase above "background" (placebo) levels only on
lecithin 60 gm (Fig. 1). The major problems were gastrointestinal,
principally nausea, vomiting, or diarrhea. For this reason, we have
eliminated the cognitive results for the 60 gm dose from further
statistical analysis. No specific effects on mood were noted at
any dose.

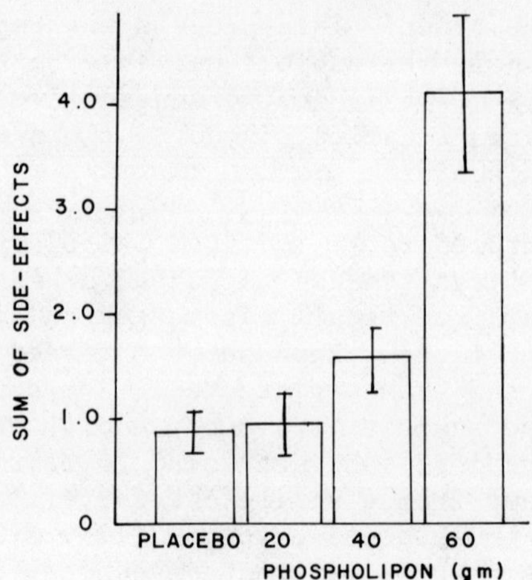

FIGURE 1: Side effects induced by phospholipon. Sum of side ef-
          fects (from 48 item check list, D.L. Murphy, unpublished
          data) with placebo and lecithin (phospholipon 100), 20,
          40 and 60 gm in 15 normal volunteers.

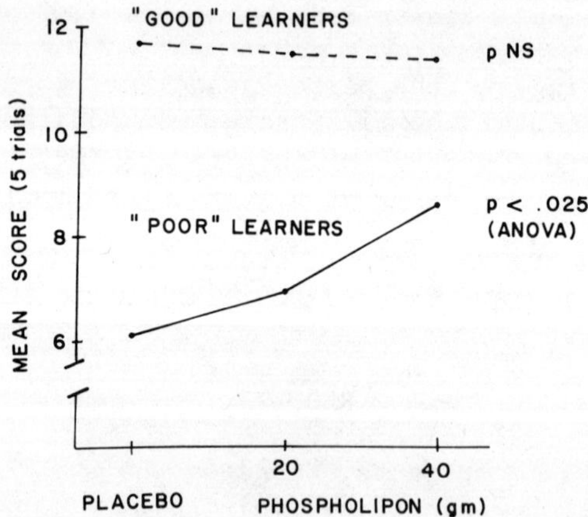

FIGURE 2: Phospholipon: serial learning improved in "poor" learners.
          Dose dependent improvement of performance on serial
          learning in "poor learners" (n = 7) but not in "good
          learners" (n = 9). The two groups were based on a median
          split of baseline performance under placebo condition.

Lecithin had no effect upon serial learning. The mean number
of words remembered on the 5 trials was 9.1 $\pm$ 3.2 (mean $\pm$ S.D.) on
placebo, 9.4 $\pm$ 3.1 on 20 gm, and 10.1 $\pm$ 2.3 on 40 gm (F = 1.41, df 2,
28, p < .3).

Since we had previously found greater effects of cholinergic
agents in "poor performers," we divided our sample of 15 subjects
into 7 "poor performers," and 8 "good performers" on the basis of
a median split of baseline performance. The average performance on
5 trials was 8.6 or less in the "poor performers" and 10.2 or greater
in the "good performers." As shown in Fig. 2, "poor performers"
tended to improve in a dose related way to lecithin (Phospholipon)
20 and 40 gm (p < 0.025, F 5.07, df 2, 14), whereas "good performers"
showed no change.

Since the major improvement in "poor performers" occurred at the
40 gm dose, we correlated baseline performance in all 15 subjects
with the change induced by lecithin 40 gm. The results were significant (p < 0.01, r = 0.7, y = -1.08 + 10.1), indicating the
greatest change was in the "poor performer" group.

Eight subjects were tested on the Associate Paired Words Test.
Lecithin had no effect on either immediate cued recall or on a variety
of measures 48 hr later (delayed recall)(Table 1). No significant
correlates emerged between baseline performance and changes associated with lecithin, perhaps because of the small number of subjects.

As shown in Fig. 3 lecithin did elevate plasma Ch in a dose
dependent way. Plasma Ch was essentially unchanged 30 min after
ingestion but began to increase at 60 min and was higher at 180 min
as compared with 120 min. In 4 subjects in whom we were about to
obtain blood samples 11.5 hr after ingestion, plasma Ch levels were
still rather high, being close to the 180 min values for the 40 gm
and 60 gm doses. These data indicate that lecithin produces a
rather prolonged elevation of plasma Ch. No significant correlations
were found between baseline Ch levels and performance or between
Ch levels at any dose of lecithin and performance at that dose.

DISCUSSION

These results suggest that lecithin has little or no overall
effects on serial learning in a group of young normal volunteers
with above average education and intelligence. Within this group,
however, lecithin benefited the so-called "poor performer." This
finding is consistent with our earlier observations that arecoline
and Ch improve, whereas scopolamine exacerbates, performances in
"poor performers." These drugs had no effect on good performers.

TABLE 1: Effect of lecithin (phospholipon) on paired associate word test (n = 8).

	Immediate Cued Recall (NW)	Long Term (48 hr)			
		Free Call (NW)	Number of Pairs Recalled	Number of Intrussions	(NW)
Placebo	12.3 ± 5.0	18.5 ± 5	4.4 ± 3.6	0.8 ± 1.0	8.3 ± 2.9
Lecithin (20 gm)	13.9 ± 4.3	18.3 ± 6.1	4.3 ± 2.1	1.1 ± 1.3	8.0 ± 5.2
Lecithin (40 gm)	13.1 ± 4.3	17.9 ± 8.3	4.3 ± 3.7	1 ± 1.4	8.1 ± 5.1

NW = number of words

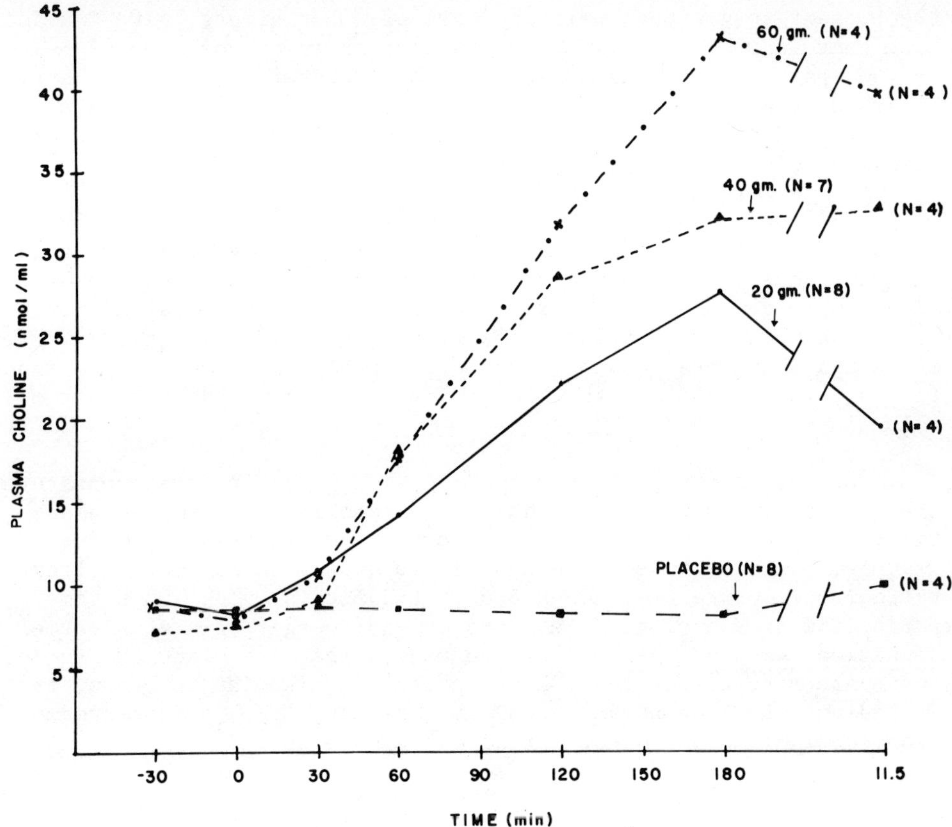

FIGURE 3: Phospholipon: dose dependent changes in plasma choline.
Dose dependent and time dependent elevation of plasma
choline in normal volunteers receiving placebo or
lecithin (phospholipon), 20, 40 or 60 gm.

Smith and Swash (23) have reported that Ch also effects poor performers more than good performers. In a subsample of subjects,
lecithin had no effect on the Paired Associates Word Test. Thus,
the effects of lecithin were certainly not robust in our normal
volunteers.

At the present time, a number of clinical trials are underway
investigating the effects of Ch or lecithin in patients with Alzheimer's disease (1,2,10,23,24). It is still too early to reach
any conclusions although it is possibly fair to say that no out-
standing dramatic improvements have yet been reported. Nevertheless,

the present results, as well as those of other groups, give cautious
hope that increased cholinergic neurotransmission could be of help
to some patients. It is noteworthy that Mohs et al. (16,17) have
suggested that there might be a "therapeutic window" for physostigmine, a finding which would be consistent with earlier results from
animal studies.

Cholinergic factors are obviously not the only ones involved
in normal cognition or in the pathophysiology of dementia. Biochemical investigations of brain autopsy samples obtained from patients with dementia have implicated dopamine, noradrenaline, serotonin, and monoamine oxidase as well as the cholinergic system. Even
if we had ideal pharmacological tools for altering cholinergic functioning, it is unlikely that major improvements would result from
this alone.

Whether or not precursor loading with Ch or lecithin consitutes
the best approach to increasing central cholinergic functional activity is controversial at this time. Casamenti et al. (3) for
example, found that lecithin had little effect on ACh output from
rat cerebral cortex, although phosphytidylserine did. When comparing the pharmacological effects of cholinergic agents on normal
cognition, we have the general impression that arecoline and scopolamine are more potent than Ch or lecithin. Likewise, we found that
Ch had no effect on human sleep, whereas arecoline, physostigmine
and scopolamine had potent, predictable effects (22).

ACKNOWLEDGEMENTS

This research was supported by USPHS, NIMH grant 26320. Ms. U.
Kopp provided invaluable assistance in conducting analysis of blood
Ch levels. Ms. A. Moore and C. Vanskiver contributed greatly to
the learning and memory studies. Ms. M. Adams expertly typed the
paper.

REFERENCES

1.   Barbeau, A., Growden, J.H. and Wurtman, R.J. (eds)(1979):
       Choline and Lecithin in Brain Disorders, Raven Press, New York.
2.   Boyd, W.D., Graham-White, J., Blackwood, G., Glen, I. and
       McQueen, J. (1977): Lancet 2:711.
3.   Casamenti, F., Mantovani, P., Amaducci, L. and Pepeu, E. (1979):
       J. Neurochem. 32:529-533.
4.   Davies, P. and Maloney, A.J.F. (1976): Lancet 2:1403.

5. Davies, P. and Verth, A.H. (1978): Brain Res. 138:385-392.
6. Davis, K. and Yamamura, H.I. (1978): Life Sci. 23:1729-1734.
7. Davis, K.L., Hollister, L.E., Barchas, J.D. and Berger, P.A.
      (1976): Life Sci. 19:1507-1516.
8. Deutsch, J.A. (1971): Science 174:788-794.
9. Drachman, D.A. (1977): Neurology 27:783-790.
10. Glen, A.I.M. and Whalley, L.J. (ed)(1979): Alzheimer's Disease:
      Recognition of Potentially Reversible Deficits, Churchill-
      Livingstone, Edinburgh.
11. Growdon, J.H., Gelenberg, A.J., Doller, J., Hirsh, M.J. and
      Wurtman, R.J. (1978): New Eng. J. Med. 298:1029-1030.
12. Growdon, J.H., Hirsch, M.J., Wurtman, R.J. and Weiner, W.
      (1977): New Eng. J. Med. 297:524-527.
13. Hanin, I. and Goldberg, A.M. (1976): IN Biology of Cholinergic
      Function, Appendix 1 (eds) A.M. Goldberg and I. Hanin, Raven
      Press, New York, pp. 647-654.
14. Haubrich, D.R. and Chippendale, T.J. (1977): Life Sci. 20:
      1465-1478.
15. Hirsch, M.J. and Wurtman, R.J. (1978): Science 202:223-225.
16. Mohs, R.C., Davis, K.C., Tinklenberg, J.R., Pfefferbaum, A.,
      Hollister, L.E. and Kopell, B.S. (1979): IN Brain Acetyl-
      choline and Neuropsychiatric Disease (eds) K.L. Davis and
      P.A. Berger, Plenum Press, New York, pp.
17. Mohs, R.C. and Davis, K.L. (1979): IN Alzheimer's Disease:
      Recognition of Potentially Reversible Deficits (eds) A.I.M.
      Glen and L.J. Whalley, Churchill-Livingstone, Edinburgh,
      pp. 154-158.
18. Perry, E.K., Perry, R.H., Blessed, G. and Tomlinson, B.E. (1978):
      Neuropathol. Appl. Neurobiol. 4:273-277.
19. Reisine, T.D., Bird, E.D., Spokes, E., Enna, J. and Yamamura,
      H.I. (1978): Trans. Amer. Soc. Neurochem. 9:203.
20. Sitaram, N., Weingartner, H. and Gillin, J.C. (1978): Science
      201:274-276.
21. Sitaram, N., Weingartner, H., Caine, E.D. and Gillin, J.C.
      (1978): Life Sci. 22:1555-1560.
22. Sitaram, N., Weingartner, H. and Gillin, J.C. (1979): IN Nutri-
      tion and the Brain (eds) A. Barbeau, J.H. Growdon, and R.J.
      Wurtman, Raven Press, New York, pp. 367-375.
23. Smith, C. and Swash, M. (1979): First Intl. Study Group in
      Pharmacol. of Memory Disorders Assoc. with Aging: Presentation.
24. Signoret, J.C., Whiteley, A. and Thermittee, C. (1978): Lancet
      2:837.
25. Weingartner, H., Sitaram, N. and Gillin, J.C. (1979): Bull.
      Psychonomic Soc. 13:9-11.

HYPERSENSITIVE CHOLINERGIC FUNCTIONING IN PRIMARY AFFECTIVE

ILLNESS

N. Sitaram, A.M. Moore, C. Vanskiver, J. Blendy, J.I.
Nurnberger Jr.*, E.S. Gershon*, and J.C. Gillin

Unit on Sleep Studies, and *Section on Psychogenetics
Biological Psychiatry Branch, National Institute of
Mental Health, National Institutes of Health
Bethesda, Maryland 20205 USA

INTRODUCTION

Primary affective illness is characterized by two distinct
mood states, depression and mania. While depressed, patients
typically show changes in mood (sadness, crying, feelings of worthlessness and hopelessness), physical activity (anergy, fatigue, psy-
chomotor retardation), and sleep changes (short fragmented sleep,
with decreased latency to the onset of the first REM period (8,18).
While manic, patients are euphoric, hyperactive and hyposomniac.

The role of acetylcholine (ACh) in the etiology of affective
illness is suggested by a number of pharmacological studies showing
that cholinomimetics have significant depressogenic and antimanic
activity (summarized in Table 1) in the depressed and manic state,
respectively. Further, as indicated in Table 1, cholinomimetics
appear to induce real mood changes (i.e. sadness, crying, feelings
of worthlessness and hopelessness, suicidal ideas) only in patients
with affective illness, but not in normal volunteers or non-af-
fective psychiatric control subjects. In remitted (euthymic)
patients, Oppenheim et al. (22) reported that 3 out of 5 patients
became quite depressed. None of the 5 normals tested experienced
the above depressive modd changes, although all of them became
anergic.   Recently, Janowski et al. (14) have reported that patients with primary affective illness have significantly greater
increases in serum prolactin levels, negative affect, behavioral

TABLE 1:   Effect of cholinomimetics on mood and behavior.

Subjects	Ref.	Drug	Effects Noted	
			Anergia	Depressed Mood
Normal Volunteers	10	DFP*	++	?("mild")
	23	DFP*	++	+
	5	DFP**	++	+
	1	EA01701***	+	−
	22	Physostigmine	+	−
	§	Physostigmine	+	−
Non-affective	23	DFP****	+	−
	13,14	Physostigmine*****	+	−
Affective Disorder	13,14	Physostigmine	++	++(depressogenic and antimanic effects)
	20,21			
	3			
	2			
	28			

§ = C. Risch et al. (unpublished data)

* = Uncontrolled studies, DFP given for 1 week (total dose from
    7-13 mg) induced "mild" depressive changes along with increased
    dreaming and insomnia (23) in contrast to severe depressive ef-
    fect in 4 hypomanic and 1 depresed patient. Two euthymic pat-
    ients were affected to the same degree as normals.

**= Exposure to insecticides for 1-1/2 years produced both depres-
    sive and schizophrenic changes.

***= EA01701 is an irreversible anticholinesterase.

****= Depressive and muscarinic side effects (vomiting, diarrhea)
    were much less in schizophrenic controls.

*****= Physostigmine induced significant anergic symptoms which
    correlated with increase in plasma prolactin. No clear-cut
    mood changes.

---

inhibition, anergy and nausea after physostigmine than patients without affective disorder and that the increase in prolactin and anergy
were positively correlated. These findings further reinforce their
previously formulated "cholinergic-adrenergic balance hypothesis"
according to which depression and mania are characterized by relative predominance of cholinergic and adrenergic activity, respectively (14).

The cholinergic-adrenergic hypothesis, although conceptually
attractive, presents certain problems: 1) it is mainly based on
the fact that cholinomimetics increase depression and decrease
mania in depressed and manic patients, respectively.   Whether the

psychological change induced by a drug like physostigmine is a
specific central cholinergic effect or the result of its rather
severe peripheral side effects such,as nausea and vomiting, is
still an open question. It is well known that mood changes could
be precipitated by a host of physical and psychological stimuli,
both conditioned and unconditioned. Even though mental change may
precede nausea and vomiting and may not be correlated with somatic
discomfort (20,21), the possibility remains that even prior to the
onset of nausea the subject may experience some unspecified symp-
tom or discomfort which serves as a "cue" to trigger a full-blown
depression. 2) The above hypothesis does not specify if the "increased cholinergic activity" in depression is due to pre- or postsynaptic factors or both.

     How does one circumvent the problem wherein the results of
any pharmacological manipulation associated with distressing side
effects becomes difficult to interpret in view of the extremely
subjective nature of any psychiatric illness? One method may be
the development of objective parameters of central cholinergic
activity which are not easily influenced by the patients subjective
state of mind or the outside environment. We chose to study the
following three psychophysiological response measures which may be
controlled at least partially by cholinergic mechanisms: 1) rapid
eye movement (REM) sleep, 2) pupillary diameter, and 3) temperature
regulation. Our purpose was to use the pharmacological challenge
strategy whereby changes in the above physiological measure(s) in
response to administration of a cholinomimetic are measured. These
changes may provide an index or reflect the status of in vivo central cholinergic functioning. After establishing that the responses were dose-dependent, we then used them as "probes" to compare normals and patients.

Sleep Studies

     The role of ACh in the modulation of REM sleep and cortical
arousal has previously been exclusively reviewed in this laboratory (6,34) and others (15,16). Our data on the sleep EEG effects
of cholinergic drugs may be summarized as follows: 1) Cholinomimetics such as physostigmine or arecoline, when infused intravenously during the first non-REM period of a sleeping subject,
significantly hastened the onset of the first REM period (7,29-32).
2) The experimentally induced REM period was accompanied by
dreaming (33). 3) The above effect was dose dependent with "higher"
doses (i.e. 1.0 mg physostigmine or 1.5 mg arecoline) producing
cortical arousal instead of REM sleep. 4) REM induction of arecoline was blocked by pretreatment with scopolamine but not methscopolamine (31). All the above studies were performed on normal

human volunteers who had given written informed consent.

Sleep EEG recordings were obtained using EEG, EOG (electro-oculogram) and EMG (electromyogram) leads and subjects "plugged in" at about 11:30 PM and were awakened at 7:00 AM. Prior to sleep a 21 gauge "butterfly" needle was inserted into a forearm vein and connected to a long (10 foot) polyethylene tubing extending outside the subject's room. A slow i.v. drip of normal saline was maintained to keep the vein open so that drug infusions could be given during discrete sleep cycles. All infusions of cholinomimetics were preceded by pretreatment with a peripheral anticholinergic drug (methscopolamine, 0.3–0.5 mg) given i.v. about 30 min before the injection of the cholinomimetic. In addition to demonstrating the direct pharmacological effects of cholinomimetics on sleep EEG, we also attempted to study the phenomenon of "disuse supersensitivity" (27). Prolonged interruption of nerve impulse flow to postsynaptic neurons or to an effector organ may result in the development of a supersensitive response to the neurotransmitter involved. For example, in mice treated for 5 days or more with scopolamine (a cholinergic muscarinic receptor antagonist), enhancement of the hypothermic effect of polocarpine (a cholinergic agonist) occurs.

Sagales et al. (25,26) have previously shown that intramuscular injection of scopolamine at bedtime produced a marked delay in the onset of REM sleep. Moreover, they noted that tolerance developed with 3 consecutive nights of scopolamine treatment. In order to develop a model of cholinergic muscarinic receptor supersensitivity we modified the experimental design employed by Sagales et al. (25, 26). We measured the sleep changes produced by daily administration of scopolamine on 3 consecutive mornings. The same intramuscular dose (6 μg/kg body weight) given by Sagales et al. was used but administered early in the morning (7:00 AM to 9:00 AM) rather than at bedtime. Based on prior pharmacological studies of normal volunteers (17,24), the central effects of the morning scopolamine injection were estimated to last 4 to 6 hr. Thus, by bedtime (about 11:00 PM), no central cholinergic receptor blockade should have been in effect. This expection was confirmed by the absence of cognitive effects of scopolamine at night. It was predicted, therefore, that central muscarinic supersensitivity would develop in the evening during the 3 day study.

As indicated in Fig. 1, the hypothesis was confirmed. REM latency (elapsed time from Stage II sleep onet to the first REM period)was significantly decrease on nights S_2 (50.1 $\pm$ 12 min, mean $\pm$ S.E.M.) and S_3 (44.7 $\pm$ 12.8 min) compared to baseline (B) night (96.4 $\pm$ 16.2 min, p < 0.005). There was a string but non-significant trend for REM latency to continue to be low on the first recovery (R_1) night; by the second recovery night (R_2) it had returned to near normal values.

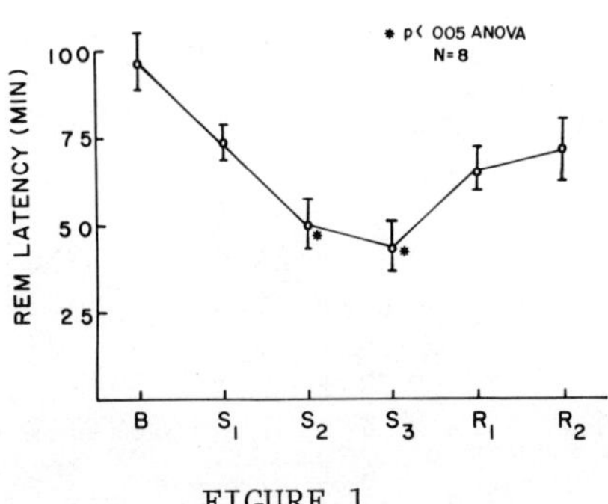

FIGURE 1

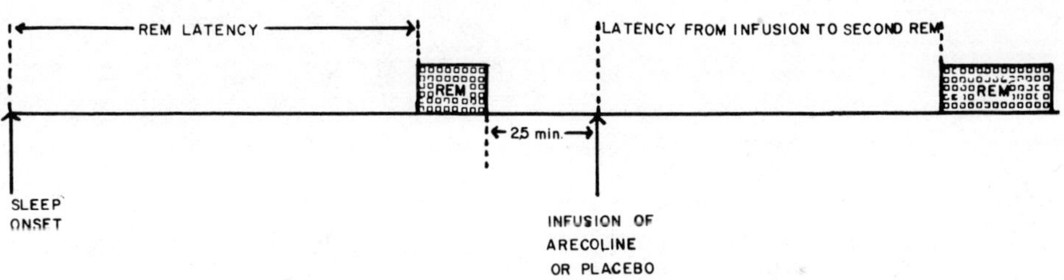

FIGURE 2: Cholinergic REM induction test.

In addition to the reduction in REM latency, sleep latency
increased, total sleep time fell, REM index and REM density (measures of number of eye movements during REM sleep) increased, and
sleep efficiency, non-REM sleep and Stage II fell (Table 2).

The above sleep changes were strikingly opposite to the direct
pharmacological effect of night-time scopolamine injection, namely,
increased REM latency, decreased eye movements, increased total
non-REM and Stage II sleep (25,26). In fact, the decrease in REM
latency seen on S_2 and S_3 nights was identical to the effect observed after intravenous administration of arecoline and physostigmine during the first non-REM period.

TABLE 2: Sleep before and after 3 days of scopolamine administration.

	Baseline	Scopolamine₃	Paired T-Test
Sleep latency	28.1 ± 2.6	46.4 ± 6.20	$p < 0.010$
Total sleep	403.5 ± 9.4	375.3 ± 7.60	$p < 0.025$
Early morning awake	2.9 ± 1.0	7.1 ± 2.50	NS
Awake	16.8 ± 4.2	20.4 ± 3.65	NS
Delta	59.1 ± 5.6	59.7 ± 7.20	NS
REM latency	94.0 ± 11.8	57.2 ± 12.40	$p < 0.025$
REM index	180.4 ± 24.9	230.7 ± 36.40	$p < 0.050$
REM time	95.4 ± 5.3	103.0 ± 7.40	NS
REM percent	23.6 ± 1.2	27.0 ± 1.40	$p < 0.050$
REM density	1.8 ± 0.2	2.2 ± 0.23	$p < 0.010$
Total recording period	451.4 ± 7.5	446.2 ± 8.00	NS
Sleep efficiency	89.4 ± 1.2	84.9 ± 1.40	$p < 0.050$
Non-REM time	311.8 ± 11.0	284.8 ± 11.20	$p < 0.020$
Stage II sleep	234.7 ± 13.7	196.2 ± 8.20	$p < 0.010$

Mean ± S.E.M.;   N = 11 subjects;   NS = not significant

TABLE 3: Classification of sleep records on baseline and scopol-
          amine night into either normal or depressed by dis-
          criminant analysis.

	BASELINE	S_1	S_2	S_3
Normal	11	9	5	4
Depressed	0	1	4	6
	(0%)	(11%)	(44%)	(60%)

Fishers exact probability $p < 0.05$. Data from Gillin et al. (9).

It was of interest that the sleep changes that were observed
after 3 days of morning scopolamine injections (i.e. decreased REM
latency, sleep efficiency, total sleep and increased sleep latency
and REM density) resembled the sleep abnormalities reported in
primary depressive illness (9,18) in many respects. In particular,

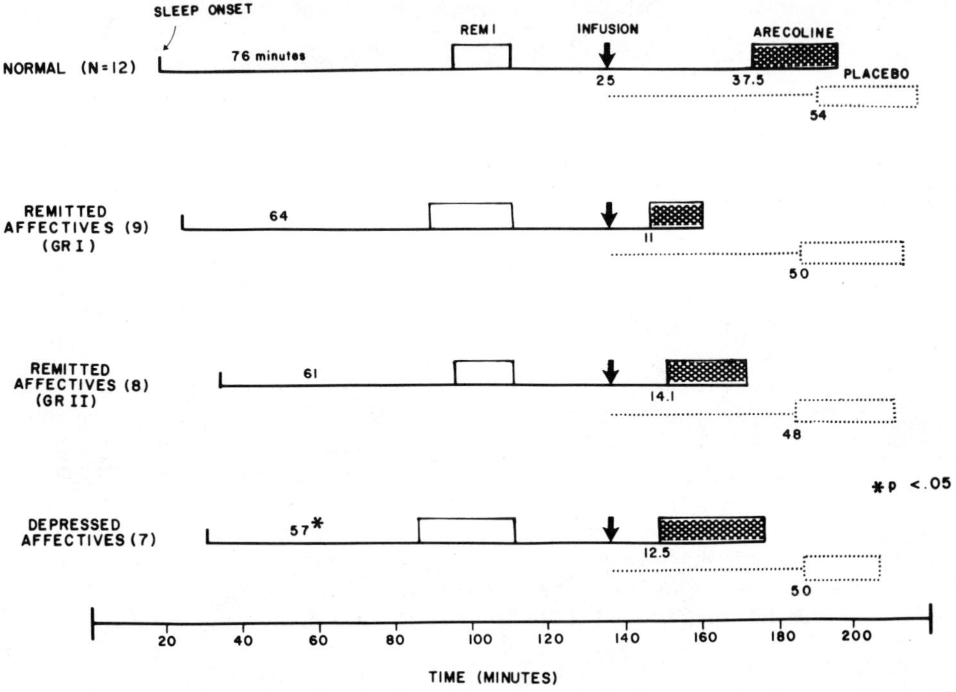

FIGURE 3: Faster REM induction by arecoline in primary affective illness.

depressed patients who are "hyposomniac" often have short, shallow and fragmented sleep. Using a multivariate discriminate analysis program (8), we analyzed the sleep data from the scopolamine study.

As shown in Table 3, compared to baseline there was a gradual increase in the percentage of records classified as "depressed" from the first to the third scopolamine night. Due to technical problems, not all the records from the subjects could be included in the above analysis. It is of greater interest that 60% of the records from the third scopolamine night were classified as depressed (9,36). This indicates that the sleep changes in normals induced by morning injections of scopolamine for 3 days may be consistent with cholinergic hypersensitivity and may provide a tentative pharmacological model for primary depressive illness (of the hyposomniac type). This model is necessarily incomplete since cholinergic supersensitive change in the sleep EEG was not accompanied by any change in mood or behavior in the subjects.

To further test the above hypothesis, normal control subjects were compared to remitted patients with primary affective illness using the "cholinergic REM induction test" (36-38). In this test we measured the speed with which REM sleep was induced by arecoline, a cholinergic muscarinic agonist administered i.v. during non-REM sleep. As illustrated in Fig. 2, arecoline was infused over a 3 min period and timed to end at about 25 min after the completion of the second REM period. Each subject received 1 infusion per night and was tested on several non-consecutive nights when he received either 1) placebo (isotonic saline), 2) arecoline 0.5 mg, or 3) arecoline 1.0 mg in random order. Methscopolamine was administered 30 min prior to the above infusions.

In the present experiments 12 normal subjects were compared with 1) two groups of remitted patients with primary affective illness; the first group (I) of 9 remitted patients were tested after all their regular psychotropic medications had been discontinued for at least 2 weeks and a second group (II) of 8 remitted patients specially recruited because they had either never received any somatic therapy whatsoever, or had been drug-free for more than 4 months; 2) a group of 7 depressed patients who had been taken off all their regular medications for 2 weeks or longer prior to testing.

The results for placebo and arecoline 0.5 mg dose are indicated in Fig. 3. The $Inf-REM_2$ latency (i.e. time elapsed from infusion to the onset of the second REM period) after placebo infusion was not defferent between the different groups (confirming the expectation that the second non-REM was not altered in depressive illness). Having shown that the placebo (baseline) $Inf-REM_2$ latencies were equal, it was now possible to directly compare the drug responses of the 4 groups. As seen in Fig. 3, the $Inf-REM_2$ latency after arecoline 0.5 mg in normals was 37.5 ± 21.9 min (mean $\pm$ S.D.) which was significantly shorter than the placebo $Inf-REM_2$ for normals (54 ± 14.1 min, paired t-test $p < 0.05$) showing that arecoline 0.5 mg doses produce a small but statistically significant REM inducing effect. The $Inf-REM_2$ latencies of all the various groups of patients, remitted affectives groups I and II and depressed affectives (Inf-REM latencies of 11 ± 6.3, 14.1 ± 7.9 and 12.9 ± 11.1 min, respectively), were all significantly different from that of normals ($p < 0.01$ ANOVA, 1-way repeated measures and Newman-Keuls comparison between normals and patients). Newman-Keuls comparison between the various groups of patients did not show any difference.

Data for the arecoline 1.0 mg dose (not indicated in Fig. 3) also showed the same effect. The 1.0 mg dose was administered only to the group I remitted patients (N = 9, Inf-REM latency = 7.1 ± 3.9 min) and normals (N = 10, $inf-REM_2$ latency = $19.3 \pm$

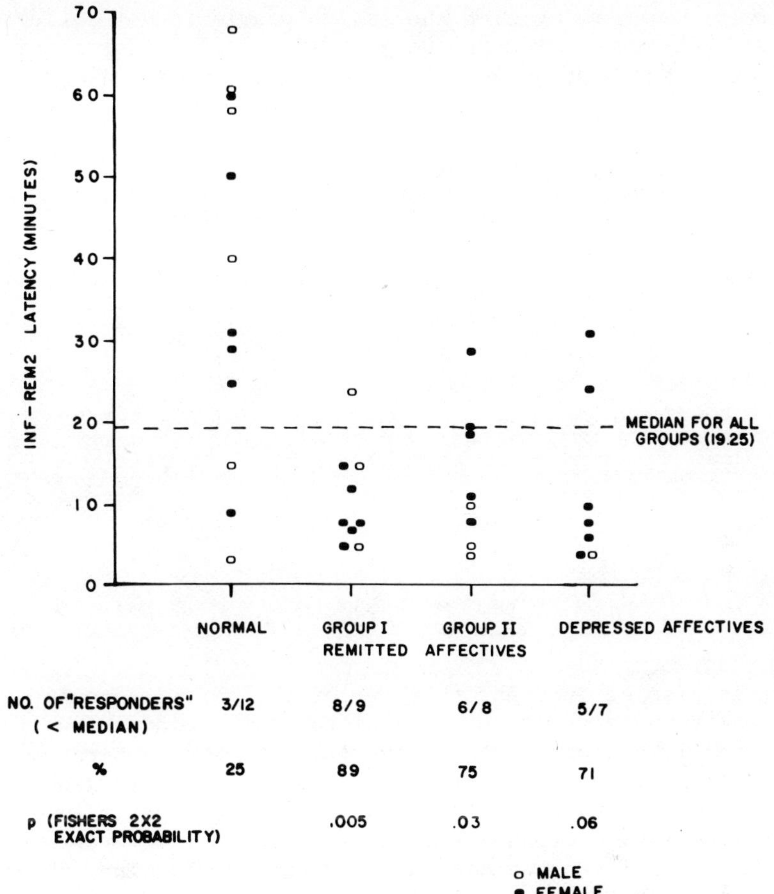

FIGURE 4: Discriminability of normal and affectives by the cholinergic REM induction test.

12 min) p < 0.01, 2-tailed t-test for independent groups. It can be seen that within normals the 1.0 mg had a markedly greater effect than the 0.5 mg dose (19 min vs 38 min).

Another method for analyzing the same data is illustrated in Fig. 4. Here we have plotted the Inf-REM$_2$ latency response to arecoline 0.5 mg of all the subjects and a median value for all the 36 data points was derived (19.25 min). All subjects whose Inf-REM latencies were below 19.25 min were considered "responders" and those above the median value "non-responders." As indicated in Fig. 3, 25% of the normals were responders as compared to 89% of remitted group I, 75% of remitted group II, and 71% of depressed

affectives. When we combine the entire population of patients
(N = 24), the number of responders is 19 (79.2%). Our findings
can be expressed also as follows:

$$
\begin{array}{llll}
1. & \text{True positives} & N = 19 & (79.2\%) \\
2. & \text{False positives} & N = 3 & (25.0\%) \\
3. & \text{True negatives} & N = 9 & (75.0\%) \\
4. & \text{False negatives} & N = 5 & (20.8\%)
\end{array}
$$

Calculating the predictability of a positive response as:

$$
= \frac{\text{True positives}}{\text{Total (true + false) positives}} \times 100
$$

$$
= \frac{19}{19 + 3} \times 100 = 86.4\%
$$

The predictability of a negative test is similarly:

$$
\frac{9}{14} \times 100 = 64.3\%
$$

It is also of interest that two of the three responders in
the normal volunteers group had a history of a minor affective illness or family history of affective illness. One subject, a 35
year old female, had a history of a six week postpartum depression
characterized by the loss of 10 lbs, decreased appetitie, difficulty sleeping, decreased self esteem, loss of interest in normal social activities and decreased energy level. The subject was
treated by psychotherapy by a psychologist and attained full remission. No medications were administered. This was the only
episode she reported. Another subject, a 31 year old male, had a
strong family history of major affective disorder. His father had
been hospitalized several times for depression and his mother was
described by the subject as a "cyclothymic personality" who would
experience periods of being very energetic, sleeping 3-4 hr, and
getting a lot accomplished, alternating with periods when she
would sleep for more than 10 hr and generally feel below par. She
had, however, never been treated by a psychiatrist. The subject's
maternal aunt was reported to be an alcoholic.

It must be emphasized that the two subjects described above
were included in the analysis since at the time they were recruited
we did not have any hypotheses or preconceptions regarding the
outcome of the cholinergic REM induction test. Thorough screening
of all the remaining "normal" subjects did not reveal any personal
or familial history of affective illness. Even so, the data with
those two subjects is presented here very tentatively and no firm

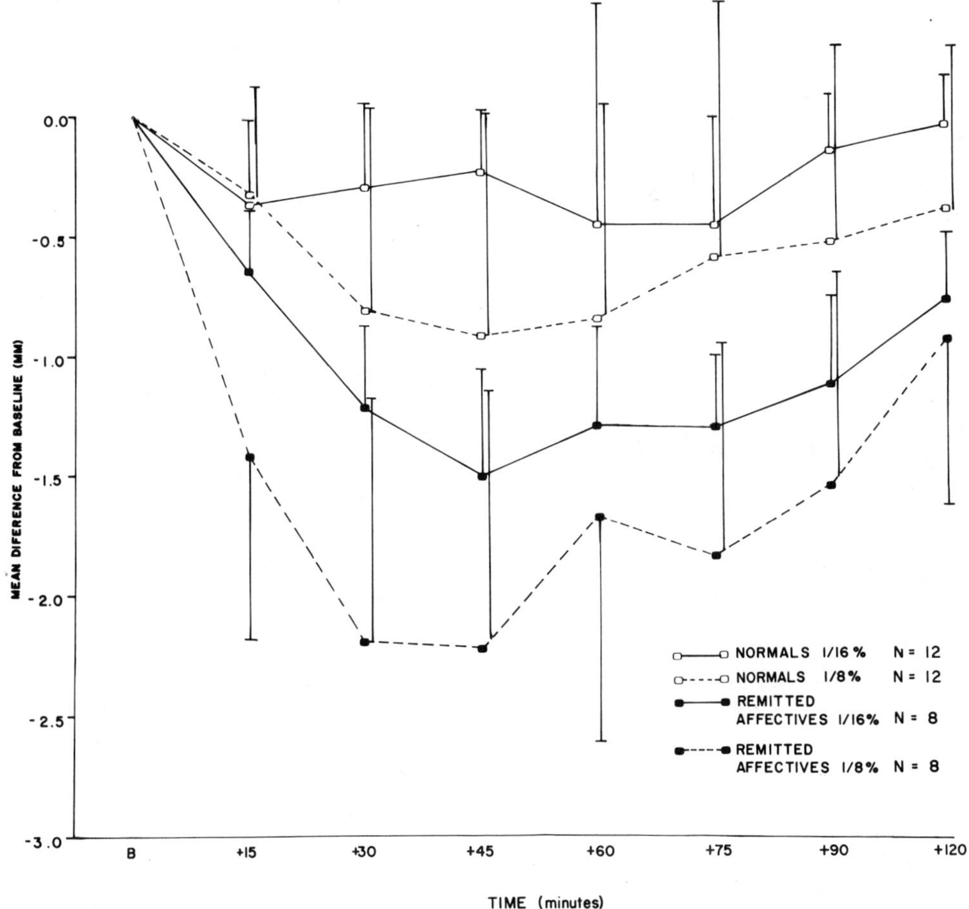

FIGURE 5: Greater miosisin remitted patients with affective illness after pilocarpine.

conclusions can be drawn at this time. A separate experiment
testing "normal" subjects with and without a history of affective
disorder is currently underway in our laboratory to determine if
the cholinergic REM induction test can successfully discriminate
subjects who have a history (personal or genetic) of affective
illness.

Pupillometry

The sphincter and dilator muscles of the iris are controlled by cholinergic and adrenergic mechanisms, respectively. There is a vast body of literature on ocular pharmacology, demonstrating dose-dependent effects of cholinergic and adrenergic drugs (12). We used the ocular miotic response to a cholinomimetic as the physiological measure of cholinergic muscarinic sensitivity.

Pilocarpine is a tertiary alkaloid and postsynaptic muscarinic agonist which readily penetrates the cornea on topical application and constricts the pupil (39). It acts postsynaptically on the iris sphincter as shown by the persisting miotic potency even after retrobulbar anesthesia (which blocks presynaptic release of ACh (11).

Twelve paid normal volunteers and 8 remitted patients with primary affective illness were studied. All patients were drug-free for at least 2 weeks prior to study. Subjects with a history of glaucoma or who wore contact lenses were excluded. Pupils were visualized on a split screen with two infrared television cameras, one for each eye, illuminated by infrared lighting on a split screen. Pupillary size was magnified 8 times by this method. The experiment was conducted in a 12 x 12 ft room in near total darkness. Subjects rested their chins on a stationary rest and their pupils were recorded in darkness for 90 sec, using a videotape recorder. Recordings were obtained at baseline and 15, 30, 45, 60, 75, 90 and 120 min after subconjunctival instillation of 1 drop (0.04 ml) of pilocarpine. Care was taken to ensure that the subject would close his eyes very slowly after instillation of the drop and keep his eyes closed for about 30 sec.

Measurements were made by playing back the videotape by an investigator who was blind to the dosage of pilocarpine administered. The first 30 sec of the playback was disregarded to allow the pupil to accommodate to the darkness. After this, each pupil was alternately measured in a fixed order 3 times, using stainless steel calipers with an accuracy of up to 0.001 in. The readings were then averaged and converted to millimeters. Repeat measurements rarely varied by more than 0.1 mm.

Mean baseline pupillary diameter (combined average of both eyes and different pretreatment baselines) was 6.5 ± 0.54 mm mean $\pm$ S.D. in normals and 6.36 ± 0.47 mm in patients (p = NS). The difference from baseline was calculated for each time point and plotted in Fig 5 for the 1/16 and 1/8 doses (12 normals and 8 patients). A 2-way ANOVA with repeated measures showed a highly significant difference between groups for both the 1/16% (p < 0.001) and 1/8% doses (p < 0.008). As seen in Fig. 5, pilocarpine

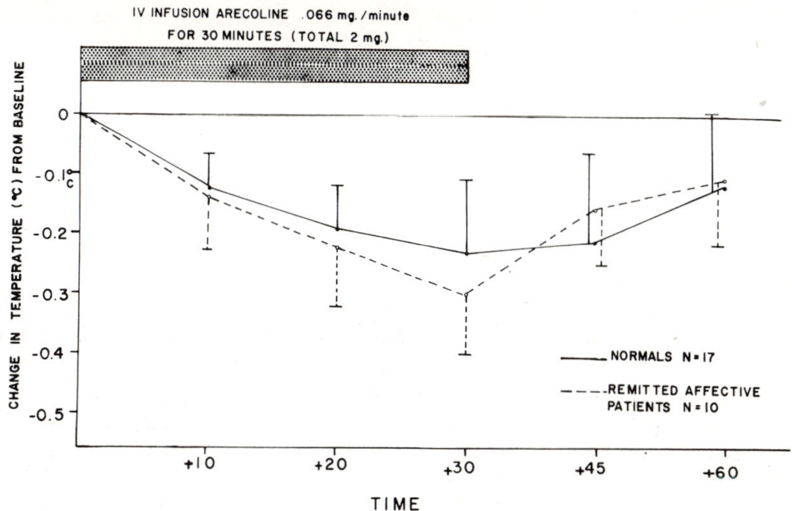

FIGURE 6: Hypothermic effect of arecoline in normal and remitted
affective patients.

---

produced a faster, more intense and longer-lasting miosis in
patients as compared to normals. Our data showing essentially
little or no response in normals to the 1/16% dose and about 0.91
mm constriction to the 1/8% dose is in excellent agreement with a
prior report (19). The response to the 1/16% dose in patients
exceeded the response of normals to 1/8% pilocarpine.

The results of the present cholinergic miosis test, like that
of the cholinergic REM induction test, are further evidence for a
state independent cholinergic involvement in affective illness.
Although pilocarpine exerts its effects predominantly as a muscarinic receptor agonist, we do not exclude the possibility that non-
cholinergic systems are also involved in either affective illness
or in the pupillary response.

Thermoregulation

The regulation of body temperature may be another CNS function
which may be under the partial control of ACh (4). We studied the
hypothermic response of arecoline in a group of 17 normals and
10 primary affective patients in complete remission and who were

drug-free for more than 2 weeks. Arecoline was administered as a
slow i.v. drip (0.066 mg per min, total dose = 2 mg over 30 min)
and oral temperature taken with a basal thermometer every 10 min.
As shown in Fig. 6, there was no difference between normals and
patients with respect to the hypothermic effect of arecoline.

DISCUSSION

Our data suggest that patients with primary affective illness
appear to be hyper-responsive to pharmacological challenge with
cholinergic agonists in 2 out of 3 measures tested; i.e. REM sleep
induction, and pupillary miosis. They did not differ from normals
with respect to the hypothermic effect of arecoline. The regulation of temperature like blood pressure is, of course, a very complex central and peripheral nervous system function. Thus, it may
not be an appropriate physiological system to measure specific
cholinergic effects.

As pointed out in the Introduction, our aim was to develop
pharmacological challenge strategies utilizing physiological end-
points that could be objectively measured and over which the
patient had no control. Thus, cholinergic activity could be assessed independent of the subject's motivation, expectations and
level of discomfort caused by side effects. Our data strongly
support the existence of a hyper-responsive cholinergic system in
affective illness, both during the depressed and euthymic (remitted)
states.

REFERENCES

1.   Bowers, M.B., Goodman, E. and Sim, V.M. (1964): J. Nerv.
         Ment. Dis. 138:383.
2.   Carroll, B.J., Frazer, A., Schess, A. and Mendels, A. (1973):
         Lancet 1:427-428.
3.   Davis, K.L., Berger, P.A., Hollister, L.E. and Defraites, E.
         (1978): Arch. Gen Psychiat. 35:119-122.
4.   Friedman, M.J., Jaffe, J.H. and Sharpless, S.K. (1969): J.
         Pharmacol. Exp. Ther. 167:45-55.
5.   Gershon, S. and Shaw, F.H. (1961): Lancet 1:1371-1374.
6.   Gillin, J.C., Mendelson, W.B. Sitaram, N. and Wyatt, R.J.
         (1978): Ann. Rev. Pharamcol. Toxicol. 18:563-579.
7.   Gillin, J.C., Sitaram, N., Mendelson, W.B. and Wyatt, R.J.
         (1978): Psychopharmacology 58:111-114.

8. Gillin, J.C., Duncan, W., Pettigrew, K.D., Frankel, B. and
     Synder, F. (1979): Arch. Gen. Psychiat. 36:85-90.
9. Gillin, J.C., Sitaram, N., Duncan, W.C. (1979): J. Psychiat.
     Res. 1:17-22.
10. Grob, D., Lilienthal, J.L., Harvey, A.M. and Jones, B.F. (1947):
     Bull. Johns Hopkins Hosp. 81:217-244.
11. Havener, W.H. (1974): IN Ocular Pharmacology (ed) W.H. Havener,
     C.V. Mosby Co., St. Louis, pp. 114-323.
12. Ishikawa, S., Dono, S. and Hikita, H. (1976): IN Drugs and
     Ocular Tissue (ed) S. Dikstein, Karger, Basel, pp. 288-382.
13. Janowsky, D.S. (1979): American College of Neuropharmacology
     Meeting, December.
14. Janowsky, D.S., El-Yousef, K., Davis, J.M. and Sekerke, H.T.
     (1972): Lancet 11:632-635.
15. Jouvet, M. (1975): IN Cholinergic Mechanisms and Sleep (ed)
     P.G. Waser, Raven Press, New York, pp. 265-310.
16. Karczmar, A.G. (1978): IN Brain Acetylcholine and Neuropsychia-
     tric Disease (eds) K.L. Davis and P.A. Berger, Plenum Press,
     New York, pp. 265-310.
17. Ketchum, J.S., Sidell, F.R., Crowell, E.R., Aghajanian, G.K.
     and Hayes, A.H. (1973): Psychopharmacology 28:121-145.
18. Kupfer, D.J. (1976): Biol. Psychiat. 11:159-174.
19. Lowenstein, I. and Lowenfeld, I.E. (1953): Arch. Opthal. 50:
     311-318.
20. Modestin, J.J., Hunger, R.B. and Schwartz, R.B. (1973): Arch.
     Psychiat. Nervenkr. 218:67-77.
21. Modestin, J.J., Schwartz, R.B. and Hunger, J. (1973):
     Pharmakopsychiatrie 9:300-304.
22. Oppenheim, G., Ebstein, R.P. and Belmaker, R.H. (1980): Psy-
     chiat. Res. (in press).
23. Rowntree, D.W., Neven, S. and Wilson, A. (1950): J. Neurol.
     Neurosurg. Psychiat. 13:47-62.
24. Safer, D.T. and Allen, R.P. (1971): Biol. Psychiat. 3:347-355.
25. Sagales, T., Erill, S. and Domino, E.F. (1967): Clin. Pharmacol.
     Ther. 10:522-529.
26. Sagales, T., Erill, S. and Domino, E.F. (1975): Clin. Pharmacol.
     Ther. 18:727-732.
27. Sharpless, S.K. (1964): Ann. Rev. Physiol. 26:357-388.
28. Shopsin, B., Janowsky, D.S., Davis, J.M. and Gershon, S. (1975):
     Neuropyschobiology 1:180-187.
29. Sitaram, N., Wyatt, R.J., Dawson, S. and Gillin, J.C. (1976):
     Science 191:1281-1283.
30. Sitaram, N., Mendelson, W.B., Wyatt, R.J. and Gillin, J.C.
     (1977): Brain Res. 122:562-567.
31. Sitaram, N., Moore, A.M. and Gillin, J.C. (1978): Sleep 1:83-90.
32. Sitaram, N., Moore, A.M. and Gillin, J.C. (1978): Nature 274:
     490-492.
33. Sitaram, N., Moore, A.M. and Gillin, J.C. (1978): Arch. Gen.
     Psychiat. 35:1239-1243.

34. Sitaram, N. and Gillin, J.C. (1979): IN Brain Acetylcholine
     and Neuropsychiatric Disease (eds) K.L. Davis and P.A. Berger,
     Plenum Press, New York, pp. 311-343.
35. Sitaram, N., Moore, A.M. and Gillin, J.C. (1979): J. Psychiat.
     Res. 1:9-16.
36. Sitaram, N. and Gillin, J.C. (1980): Biol. Psychiat. (in press).
37. Sitaram, N., Nurnberger, J.I., Gershon, E. and Gillin, J.C.
     (1980) Science 208:200-202.
38. Sitaram, N., Cohen, R.M., Rey, A., Murphy, D.L. and Gillin, J.
     C. (1980): Amer. J. Psychiat. (in press).
39. Van Rossum, J.M. (1960): Experientia 16:373-375.

ADDED IN PROOF:

Since submitting this chapter, we have studied additional
remitted depressed patients and normal control subjects with a
modified measure of pilocarpine-induced miosis, and no longer
find a statistically significant difference between the two groups.
We are conducting further studies at this time are are re-evaluating
the methodology.

PARTICIPANT LIST

Albanese, Alberto, Istituto di Neurologia, Universita Cattolica,
    Largo A. Gemelli, 8, 00168 Rome, Italy

Ansell, G. Brian, Department of Pharmacology, University of
    Birmingham, The Medical School, Birmingham B15 2TJ, UK

Aquilonius, Sten-Magnus, Department of Neurology, University
    Hospital, S-750 14 Uppsala, Sweden

Barker, Louis A., Department of Pharmacology, The Mount Sinai
    Medical Center, New York, New York, 10029 USA

Bartfai, Tamas, Department of Biochemistry, Arrhenius Laboratory,
    University of Stockholm, S-10691, Stockholm, Sweden

Bartolini, Alessandro, Department of Pharmacology, University of
    Florence, 50134, Florence, Italy

Beani, Lorenzo, Department of Pharmacology, School of Medicine,
    University of Ferrara, 44100 Ferrara, Italy

Bianchi, Beani, Clety, Department of Pharmacology, School of
    Pharmacy, University of Ferrara, 44100 Ferrara, Italy

Bigl, Volker, Abteilung fur Neurochemie, Karl Mark Universitat,
    Karl-Marx-Stadter Str. 50, DDR-7039 Leipzig, Germany

Birdsall, Nigel, J.M., National Institute for Medical Research,
    The Ridgeway, Mill Hill, London NW7 1AA, UK

Bisso, Guillermo Mario, Laboratorio di Farmacologia, Rep.
    Farmacologia Biochimica, Istituto Superiore di Sanita,
    Viale Regina Elena 299, Rome, Italy

Blass, John, Burke Rehabilitation Center, 385 Mammaroneck Avenue,
    White Plains, New York 10605, USA

Bonetti, Anna Chiara, Fidia Research Laboratories,
     35031 Abano Terme, Italy

Butcher, Larry L., Department of Psychology and The Brain Research
     Institute, University of California, Los Angeles, California
     90024, USA

Calderini, Gabriella, Fidia Research Laboratories, 35031 Abano
     Terme, Italy

Carmignoto, Giorgio, Fidia Research Laboratories, 35031 Abano
     Terme, Italy

Casamenti, Fiorella, Department of Pharmacology, School of
     Medicine, University of Florence, 50134 Florence, Italy

Ceccarelli, Bruno, Department of Pharmacology, University of
     Milan, 20129 Milan, Italy

Cheney, Darwin L., Laboratory of Preclinical Pharmacology, NIMH,
     St. Elizabeth's Hospital, Washington, D.C. 20032, USA

Collier, Brian, Department of Pharmacology, McIntyre Medical
     Sciences Bldg., McGill University, Quebec H3G 1Y6, Canada

Consolo Ladinksy, Silvana, Mario Negri Institute of Pharmacological
     Research, 20157 Milan, Italy

Cooper, Jack R., Department of Pharmacology, Yale University
     School of Medicine, 333 Cedar Street, New Haven, Connecticut
     06510 USA

Corradetti, Renato, Department of Pharmacology, School of Medicine,
     University of Florence, 50134 Florence, Italy

Corthay, Jacqueline, Department of Pharmacology, 20 rue de l'Ecole
     de Medicine, 1211 Geneva 4, Switzerland

Costa, Erminio, Laboratory of Preclinical Pharmacology, NIMH,
     St. Elizabeth's Hospital, Washington, D.C. 20032, USA

Dahlbom, Richard, Department of Pharmaceutical Chemistry,
     University of Uppsala, S-750 Uppsala, Sweden

Davis, Kenneth, Bronx Veteran's ADministration Hospital, Bronx,
     New York 10468 USA

De Biasi, Silvia, Fidia Research Laboratories, 35031 Abano Terme,
     Italy

DiGiamberardino, Luigi, CEA de Saclay-DB/LLPC, Boite Postale 2,
    91190 Gif-sur-Yvette, France

Dimpfel, W., Fidia Research Laboratories, 35031 Abano Terme,
    Italy

Domino, Edward F., Department of Pharmacology, University of
    Michigan, Medical School, Ann Arbor, Michigan 45105 USA

Dowdall, Michael J., Department of Biochemistry, The Medical
    School, University of Nottingham, Nottingham, UK

Dun, Nae J., Department of Pharmacology, Loyola University Stritch
    School of Medicine, 2160 S. First Avenue, Maywood, Illinois
    60153 USA

Dunant, Yves, Department of Pharmacology, 20 rue de l'Ecole de
    Medecine, 1211 Geneva 4, Switzerland

Eckernas, Sven-Ake, Department of Neurology, University Hospital,
    S-750 14 Uppsala, Sweden

Fesce, R., Department of Pharmacology, University of Milan,
    20129 Milan, Italy

Ferrari, Giovanna, Fidia Research Laboratories, 35031 Abano Terme,
    Italy

Fibiger, Hans C., University of British Columbia, Department of
    Psychiatry, Vancouver V6T 1W5, Canada

Filogamo, Guido, Istituto di Anatomia Umana Normale, University
    of Torino 10126 Torino, Italy

Flentge, F., Department of Biological Psychiatry, Academisch
    Ziekenhuis Groningen, 9700 RB Groningen, The Netherlands

Fonnum, Frode, Department of Toxicology, Norwegian Defence Research
    Establishment, N-2007, Kjeller, Norway

Froissart, Chantal, Certre de Neurochemie du CNRS, 11 rue Humann
    67085 Strasbourg cedex, France

Fulpius, Bernard W., University of Geneva, Department of Bio-
    chemistry, Sciences II, 1211 Geneva, Switzerland

Galli, Alessandro, Department of Pharmacology, School of Medicine,
    University of Florence 50134, Florence, Italy

Giacobini, Ezio, Department of Biobehavioral Sciences U-154,
    University of Connecticut, Storrs, Connecticut 06268 USA

Gibson, Gary, Department of Neurology, Cornell Medical College,
     Burke Rehabilitation Center, 785 Mamaroneck Avenue,
     White Plains, New York, 10605 USA

Gillin, J. Christian, National Institute of Mental Health, NIH,
     Bethesda, Maryland 20205 USA

Gorio, Alfredo, Fidia Research Laboratories, 35031 Abano Terme,
     Italy

Hanin, Israel, University of Pittsburgh, School of Medicine,
     Western Psychiatric Institute and Clinic, 3811 O'Hara Street,
     Pittsburgh, Pennsylvania 15261 USA

Haubrich, Dean R., Merck Institute for Therapeutic Research, West
     Point, Pennsylvania 19486 USA

Hedlund, Britta, Department of Biochemistry, Arrhenius Laboratory,
     University of Stockholm, S-10691 Stockholm, Sweden

Heilbronn, Edith, Unit of Neurochemistry and Neurotoxicology,
     University of Stockholm, c/o FOA Ursvik, S-104 50 Stockholm,
     Sweden

Hemsworth, Brain A., University of Aston, Department of Pharmacy,
     Gosta Green, Birmingham, B4 7ET, UK

Hersh, Louis B., Department of Biochemistry, The University of
     Texas, Health Science Center at Dallas, 5323 Harry Hines
     Blvd., Dallas, Texas 75235 USA

Hopff, Wolfgang, Department of Pharmacology, University of Zurich,
     Gloriastasse 32, CH-8005, Zurich, Switzerland

Howard, Bruce, Department of Biological Chemistry, School of
     Medicine, University of California, Los Angeles, California
     90024 USA

Israel, Maurice, CNRS, Laboratory of Cellular Neurobiology,
     Department of Neurochemistry, 91190 Gif-sur-Yvette, France

Jenden, Donald J., Department of Pharmacology, School of Medicine,
     University of California, Los Angeles, California 90024 USA

Jope, Richard S., Department of Pharmacology, School of Medicine
     University of California, Los Angeles, California 90024 USA

Karczmar, Alexander, G., Loyola University Stritch School of
     Medicine, Department of Pharmacology, Maywood, Illinois
     60153 USA

Kewitz, K., Freie Universitat Berlin, Institut fur Klinische
     Pharmakologie, Hindenburgdamm 30, D-1000 Berlin 45 Germany

Kharkevich, Dimitry A., Department of Pharmacology, First Medical
     Institute, 2/6 B. Pirogovskaja St., Moscow 119485 USSR

Kilbinger, Heinz, Department of Pharmacology, University of Mainz,
     D-6500 Mainz, Germany

Kimura, Hirosh, Kinsman Laboratory of Neurological Research,
     Department of Psychiatry, University of British Columbia,
     Vancouver, Canada

Koelle, George B., Department of Pharmacology, School of Medicine
     University of Pennsylvania, Philadelphis, Pennsylvania
     19174 USA

Koenig, Herbert, Universite P. et M. Curie, Laboratoire de Neuro-
     cytologie, 7 Quai St. Bernard, Bat A/3c, 75005 Paris, France

Lotas, A.M., Department of Pharmacology, Univesrity of London,
     Kings College, The Strand, London WC2R 2LS, UK

Kuhar, Michael J., Department of Pharmacology, John Hopkins
     University, School of Medicine, 725 N. Wolfe Street, Balti-
     more, Maryland 21205 USA

Ladinsky, Herbert, Mario Negri Institute of Pharmacological
     Research, 20157 Milan, Italy

Leon, Alberta, Fidia Research Laboratories, 35031 Abano Terme,
     Italy

Lo Conte, Giovanni, Department of Pharmacology, School of Medicine,
     University of Florence, 50134 Florence, Italy

Loffelholz, Konrad, Department of Pharmacology, University of
     Mainz, Obere Zahlbacher Strasse 67, 6500 Mainz, Germany

Longo, Vincenzo, Istituto Superiore di Sanita, Viale Regina
     Elena 299, 00100 Rome, Italy

Lowenadler, Bjorn, Department of Neurochemistry and Neurotoxicology,
     University of Stockholm, c.o FOA Ursvik, S-104 50 Stockholm,
     Sweden

Lqumani, Yunas A., Max-Planck-Institut, Biophysical Chemistry,
     Postfach 968, Gottingen, Germany

Malmberg, Aiello, Petra, Department of Pharmacology, University
     of Florence, 50134 Florence, Italy

Malthe-Sorenssen, Didrik, Division of Toxicology, Norwegian
     Defence Research Establishment, N-2007 Kjeller, Norway

Manaranche, M., CNRS, Laboratory of Cellular Neurobiology, Depart-
     ment of Neurochemistry, 91190 Gif-sur-Yvette, France

Mantovani, Piero, Department of Pharmacology, School of Medicine,
     University of Florence, 50134 Florence, Italy

Marchbanks, Roger M., Department of Biochemistry, Institute of
     Psychiatry, De Crespigny Park, London SE5 8AF, UK

Marchi, Mario, Department of Biobavioral Sciences U-154, University
     of Connecticut, Storrs, Connecticut 06268 USA

Massarelli, Raffaele, CNRS, Center of Neurochemistry, 11 rue Humann.
     67085 Strasbourg cedex, France

Mautner, Henry G., Department of Biochemistry and Pharmacology,
     School of Medicine, Tufts University, 136 Harrison Avenue,
     Boston, Massachusetts 02111 USA

Michalek, Moricca Hanna, Laboratorio di Farmacologia, Rep. Farma-
     cologia biochimica, Instituto Superiore di Sanita, Viale
     Regina Elena 299, 00100 Rome, Italy

Mittag, Tom, Pharmacology Department, Mt. Sinai School of Medicine,
     One Gustave Levy Place, New York, New York 10029 USA

Molenaar, Peter C., Department of Pharmacology, Sylvius Laboratories,
     University of Leiden, 72 Wassenaarseweg, Leiden, The Netherlands

Moroni, Flavio, Department of Pharmacology, School of Medicine,
     University of Florence, 50134 Florence, Italy

Neal, J., Department of Pharmacology, School of Pharmacy,
     29 Brunswick Square, London WC1, UK

Nordberg, Agneta, Department of Pharmacology, University of
     Uppsala, Box 578, 75123 Uppsala, Sweden

Oderfeld-Nowak, Barbara, Institute of Experimental Biology,
     Polish Academy of Sciences, Pasteur 8, 02093 Warsaw, Poland

Paggi, Paola, Institute of General Physiology, University of Rome,
     00100 Rome, Italy

Pauling, Peter, Chemistry Department, University College London,
     20 Gordon Street, London UC1, UK

Pedata, Felicita, Department of Pharmacology, School of Medicine,
    University of Florence, 50134 Florence, Italy

Pedigo, Norman W., Department of Pharmacology, University of
    Arizona Health Sciences Center, Tucson, Arizona 85724 USA

Pepeu, Giancarlo, Department of Pharmacology, School of Medicine,
    University of Florence, 50134 Florence, Italy

Perri, Virgilio, Institute of General Physiology, University of
    Ferrara, Via L. Borsari 46, 44100 Ferrara, Italy

Piazza Bartolini, Roselia, Department of Pharmacology, University
    of Florence, 50134 Florence, Italy

Polak, Robert, Medical Biology Laboratories, TNO, Ryswyk, The
    Netherlands

Procellati, Giuseepe, Institute of Chemical Biology, University
    of Perugia, 06100 Perugia, Italy

Prince, Alan Keith, Department of Pharmacology, University of
    London, Kings College, The Strand, London WC2R 2LS, UK

Racagni, Giorgio, Institute of Pharmacology and Pharmacognosia,
    University of Milan, 20129 Milan, Italy

Reynolds, R., Department of Pharmacology, Univesrity of London,
    Kings College, The Strand, London WC2R 2LS, UK

Renzi, Giovanni, Department of Pharmaceutical Chemistry, School
    of Pharmacy, University of Florence, 50121 Florence, Italy

Richardson, G. Peter, Max-Planck-Institut, Biophysical Chemistry,
    Postfach 968, 34 Gottingen, Germany

Riggio, G., Department of Pharmacology, University of Zurich,
    Gloriastrasse 32, CH-8006 Zurich, Switzerland

Robinson, Susan, Department of Pharmacology, College of Medicine,
    Texas A&M University, College Station, Texas 77843, USA

Scatton, Bernard, Synthelabo-LERS, 31 Avenue Paul-Vaillant
    Coutorier, 92220 Bagneux, France

Siniscalchi, Rossi, Anna, Department of Pharmacology, School of
    Medicine, University of Ferrara, 44100 Ferrara, Italy

Sitaram, Nastrai, National Institute of Mental Health, NIH,
    Bldg. 10, Room 3N224, Bethesda, Maryland 20014, USA

Sorbi, Sandro, Neurological Clinic, School of Medicine, University
of Florence, 50134 Florence, Italy

Spanner, Sheila, G., Department of Pharmacology, The Medical
School, University of Birmingham, Birmingham B15 2TJ, UK

Stadler, Herbert, Max-Planck-Institut, Biophysical Chemistry,
Postfach 968, 34 Gottingen, Germany

Stavinoha, William B., The University of Texas Medical School,
Department of Pharmacology, San Antonio, Texas, 78284, USA

Suszkiw, Janus B., The Biological Sciences Group, The University
of Connecticut, Storrs, Connecticut, 06268 USA

Szerb, John C., Department of Physiology and Biophysics, Dalhousie
University, Halifax, Nova Scotia, B3H 4H7 Canada

Tanganelli, Sergio, Department of Pharmacology, University of
Ferrara, 44100 Ferrara, Italy

Teolato, Sandra, Fidia Research Laboratories, 35031 Abano Terme,
Italy

Tirelli, Silvia, Mario Negro Institute of Pharmacological Research,
Via Eritrea 62, Milan, Italy

Toffano, Gino, Fidia Research Laboratories, 35031 Abano Terme,
Italy

Toschi, G., Institute of General Physiology, University of Rome,
00100 Rome, Italy

Tucek, Stanislav, Academy of Sciences, Insitute of Physiology,
Videnska 1083, 14220 Prague, Czechoslovakia

Ulas, Jolanta, Nencki Institute of Experimental Biology,
3 Pasteur Strasse, 02093 Warsaw, Poland

Vizi, Sylvester, Department of Pharmacology, Semmelweis University
of Medicine, H-1445 Budapest, Hungary

Wachtler, Klaus, Institut fur Zoologie, Der Tierartlishen Hochschule, 3 Hannover, Germany

Wahlstrom, Goran, University of Umea, Department of Pharmacology,
S-901,87 Umea, Sweden

Wecker, Lynn, Department of Pharmacology, Louisiana State University
Medical Center, New Orleans, Louisiana 70112 USA

Weiler, Molly, Laboratory for Neurochemistry, Max-Planck-Institut
    for Biophysical Chemistry, D-3400 Gottingen, Germany

Wenk, Hartmut, Abt. Neurochemie, Paul Flechsig Institut Hirnfor-
    schung, Karl Marx Stadtea Strasse 50, DDR-7039, Leipzig,
    Germany

Wessler, Ignaz, Department of Pharmacology, University of Mainz,
    D-6500, Mainz, Germany

Whittaker, Victor P., Laboratory for Neurochemistry, Max-Planck-
    Institut for Biophysical Chemistry, D-3400 Gottingen, Germany

Wong, Tuen Yee, CNRS, Center for Neurochemistry, rue Humann,
    67085 Strasbourg cedex, France

Wood, Paul L., Department of Pharmacology, Merck Frosst Laboratories,
    P.O. Box 1005, Pointe Claire-Dorval, Quebec H9R 4P8 Canada

Zimmerman, Herbert, Laboratory for Neurochemistry, Max-Planck-
    Institute for Biophysical Chemistry, D-3400 Gottingen,
    Germany

Zsilla, G., Department of Pharmacology, Semmelweis University of
    Medicine, H-1445 Budapest, Hungary